Algae Materials: Applications Benefitting Health

Developments in Applied Microbiology and Biotechnology

Algae Materials: Applications Benefitting Health

Edited by

K. Arunkumar
Department of Plant Sciences and Dean of Student Welfare, Central University of Kerala, Kasaragod, Kerala, India

A. Arun
Department of Microbiology, Alagappa University, Karaikudi, Tamilnadu, India

Rathinam Raja
Sree Balaji Medical College and Hospital (SBMCH)-BIHER, Chennai, Tamil Nadu, India

Ramasamy Palaniappan
Director of Research, Sree Balaji Medical College and Hospital, BIHER, Chennai, Tamil Nadu, India

ELSEVIER

ACADEMIC PRESS
An imprint of Elsevier

Academic Press is an imprint of Elsevier
125 London Wall, London EC2Y 5AS, United Kingdom
525 B Street, Suite 1650, San Diego, CA 92101, United States
50 Hampshire Street, 5th Floor, Cambridge, MA 02139, United States
The Boulevard, Langford Lane, Kidlington, Oxford OX5 1GB, United Kingdom

Notices

Knowledge and best practice in this field are constantly changing. As new research and experience broaden our understanding, changes in research methods, professional practices, or medical treatment may become necessary.

Practitioners and researchers must always rely on their own experience and knowledge in evaluating and using any information, methods, compounds, or experiments described herein. In using such information or methods they should be mindful of their own safety and the safety of others, including parties for whom they have a professional responsibility.

To the fullest extent of the law, neither the Publisher nor the authors, contributors, or editors, assume any liability for any injury and/or damage to persons or property as a matter of products liability, negligence or otherwise, or from any use or operation of any methods, products, instructions, or ideas contained in the material herein.

ISBN: 978-0-443-18816-9

For information on all Academic Press publications visit our website at https://www.elsevier.com/books-and-journals

Publisher: Stacy Masucci
Acquisitions Editor: Linda Versteeg-Buschman
Editorial Project Manager: Billie Jean Fernandez
Production Project Manager: Swapna Srinivasan
Cover Designer: Christian Bilbow

Typeset by TNQ Technologies

I wish to dedicate the book
To my father A. Rathinam

Contents

List of contributors

Quan (Sophia) He
Department of Engineering, Dalhousie University, Truro, NS, Canada

Sarah Amalina Adli
Department of Chemical Engineering and Sustainability, Kulliyyah of Engineering, International Islamic University Malaysia, Kuala Lumpur, Wilayah Persekutuan, Malaysia

Clélia Afonso
MARE—Marine and Environmental Sciences Centre, ESTM, Polytechnic of Leiria, Edifício CETEMARES Peniche, Portugal

Samuel Alejandro-Vega
Área de Toxicología, Universidad de La Laguna, Tenerife, Islas Canarias, Spain; Grupo Interuniversitario de Toxicología Alimentaria y Ambiental, Universidad de La Laguna, Tenerife, Islas Canarias, Spain

Fathilah Ali
Department of Chemical Engineering and Sustainability, Kulliyyah of Engineering, International Islamic University Malaysia, Kuala Lumpur, Wilayah Persekutuan, Malaysia

H.D. Ambika
Algae Ecologist and Taxonomist, Germany

K. Anjana
Phycoscience Laboratory, Department of Plant Science, School of Biological Sciences, Central University of Kerala, Kasaragod, Kerala, India

Hazleen Anuar
Department of Manufacturing and Materials Engineering, Kulliyyah of Engineering, International Islamic University Malaysia, Kuala Lumpur, Wilayah Persekutuan, Malaysia

Ajucarmelpricilla Arulprakasam
Department of Microbiology, Alagappa University, Karaikudi, Tamil Nadu, India

Kulanthaiyesu Arunkumar
Phycoscience Laboratory, Department of Plant Science, School of Biological Sciences, Central University of Kerala, Kasaragod, Kerala, India

Joana Assunção
CIIMAR — Interdisciplinary Centre of Marine and Environmental Research, University of Porto, Novo Edifício Do Terminal de Cruzeiros de Leixões, Matosinhos, Portugal; LEPABE — Laboratory of Engineering of Processes, Environment, Biotechnology and Energy, Faculty of Engineering of University of Porto, Porto, Portugal

Ishwarya Ayyanar
Department of Microbiology, Alagappa University, Karaikudi, Tamil Nadu, India

André Rolim Baby
Faculty of Pharmaceutical Sciences, University of São Paulo, São Paulo, Brazil

Golap Chandra Singha
Department of Environmental Sciences, Bharathiar University, Coimbatore, Tamil Nadu, India

Vikas Singh Chauhan
Plant Cell Biotechnology (PCBT) Department, CSIR-Central Food Technological Research Institute (CFTRI), Mysuru, Karnataka, India; Academy of Scientific and Innovative Research (AcSIR), Ghazibad, Uttar Pradesh, India

Mariany Costa Deprá
Bioprocess Intensification Group, Federal University of Santa Maria, UFSM, Santa Maria, Rio Grande do Sul, Brazil

João Cotas
University of Coimbra, MARE—Marine and Environmental Sciences Centre, ARNET - Aquatic Research Network, Department of Life Sciences, Coimbra, Portugal

João Carlos Monteiro de Carvalho
Faculty of Pharmaceutical Sciences, University of São Paulo, São Paulo, Brazil

Ramesh Desikan
Department of Renewable Energy Engineering, Agricultural Engineering College and Research Institute, Tamil Nadu Agricultural University, Coimbatore, Tamil Nadu, India

K. Nanthini Devi
Marine Planktonology & Aquaculture Laboratory, Department of Marine Science, School of Marine Sciences, Bharathidasan University, Tiruchirappalli, Tamil Nadu, India

G. Dharani
Marine Biotechnology Division, National Institute of Ocean Technology (MoES), Ministry of Earth Sciences, Pallikaranai, Chennai, Tamil Nadu, India

M. Divya
TNJFU-Fisheries Business School, Tamil Nadu Dr. J. Jayalalithaa Fisheries University, DIVA Campus, Chennai, Tamil Nadu, India

Bruna Bertoloni dos Santos
Faculty of Pharmaceutical Sciences, University of São Paulo, São Paulo, Brazil

Ahmed A. El-Refaey
Botany and Microbiology Department, Faculty of Science, Al-Azhar University, Nasr City, Cairo, Egypt

Aitazaz A. Farooque
Faculty of Sustainable Design Engineering, University of Prince Edward Island, Charlottetown, PE, Canada; School of Climate Change and Adaptation, University of Prince Edward Island, Charlottetown, PE, Canada

Duarte Fortunato
Lusalgae, Lda., Incubadora de Empresas da Figueira da Foz Figueira da Foz, Portugal

Sunandha Jeeva Bharathi G.
Department of Microbiology, Alagappa University, Karaikudi, Tamil Nadu, India

Louisa Gomes
University of Coimbra, MARE—Marine and Environmental Sciences Centre, ARNET - Aquatic Research Network, Department of Life Sciences, Coimbra, Portugal

Ana Marta Mendes Gonçalves
University of Coimbra, MARE—Marine and Environmental Sciences Centre, ARNET - Aquatic Research Network, Department of Life Sciences, Coimbra, Portugal; Department of Biology and CESAM, University of Aveiro, Aveiro, Portugal

A. Gowthami
Marine Planktonology & Aquaculture Laboratory, Department of Marine Science, School of Marine Sciences, Bharathidasan University, Tiruchirappalli, Tamil Nadu, India

S. Gowtham Kumar
Faculty of Allied Health Sciences, Chettinad Hospital & Research Institute, Chettinad Academy of Research and Education, Kelambakkam, Tamil Nadu, India

Kuljeet Grewal
Faculty of Sustainable Design Engineering, University of Prince Edward Island, Charlottetown, PE, Canada

A. Catarina Guedes
CIIMAR — Interdisciplinary Centre of Marine and Environmental Research, University of Porto, Novo Edifício Do Terminal de Cruzeiros de Leixões, Matosinhos, Portugal

Kelvii Guo
Department of Mechanical and Biomedical Engineering, City University of Hong Kong, Kowloon Tong, Hong Kong, China

Ángel Gutiérrez-Fernández
Área de Toxicología, Universidad de La Laguna, Tenerife, Islas Canarias, Spain; Grupo Interuniversitario de Toxicología Alimentaria y Ambiental, Universidad de La Laguna, Tenerife, Islas Canarias, Spain

Arturo Hardisson
Área de Toxicología, Universidad de La Laguna, Tenerife, Islas Canarias, Spain; Grupo Interuniversitario de Toxicología Alimentaria y Ambiental, Universidad de La Laguna, Tenerife, Islas Canarias, Spain

Rabeay Y.A. Hassan
Nanoscience Program, University of Science and Technology (UST), Zewail City of Science and Technology, Giza, Egypt

Zhongqi He
State Key Lab of Urban Water Resource and Environment, School of Civil and Environmental Engineering, Harbin Institute of Technology Shenzhen, Shenzhen, Guangdong, China; Shenzhen Engineering Laboratory of Microalgae Bioenergy, Harbin Institute of Technology Shenzhen, Shenzhen, Guangdong, China

Himamshu Hegde
Nitte (Deemed to be University), Nitte University Centre for Science Education and Research (NUCSER), Deralakatte, Mangaluru, Karnataka, India

Smitha Hegde
Nitte (Deemed to be University), Nitte University Centre for Science Education and Research (NUCSER), Deralakatte, Mangaluru, Karnataka, India

Yulin Hu
Faculty of Sustainable Design Engineering, University of Prince Edward Island, Charlottetown, PE, Canada

Eduardo Jacob-Lopes
Bioprocess Intensification Group, Federal University of Santa Maria, UFSM, Santa Maria, Rio Grande do Sul, Brazil

Jamarosliza Jamaluddin
Department of Chemical Engineering, School of Chemical and Energy Engineering, Universiti Teknologi Malaysia, Johor Bahru, Johor, Malaysia

Neetha Nirupam Jathanna
Nitte (Deemed to be University), Nitte University Centre for Science Education and Research (NUCSER), Deralakatte, Mangaluru, Karnataka, India

K.R. Jayappriyan
Centre for Advanced Studies in Botany, University of Madras, Chennai, Tamil Nadu, India

Wenbiao Jin
State Key Lab of Urban Water Resource and Environment, School of Civil and Environmental Engineering, Harbin Institute of Technology Shenzhen, Shenzhen, Guangdong, China; Shenzhen Engineering Laboratory of Microalgae Bioenergy, Harbin Institute of Technology Shenzhen, Shenzhen, Guangdong, China

Cordelia Mano John
Department of Biomedical Sciences, Sri Ramachandra Institute of Higher Education and Research, Chennai, Tamil Nadu, India

R. Kirubagaran
Deep Ocean Mission (DOM), Vertical VI, Marine Biotechnology Division, National Institute of Ocean Technology (MoES), Pallikaranai, Chennai, Tamil Nadu, India

N. Krishnaveni
Marine Planktonology & Aquaculture Laboratory, Department of Marine Science, School of Marine Sciences, Bharathidasan University, Tiruchirappalli, Tamil Nadu, India

Sharanya Kushalan
Nitte (Deemed to be University), Nitte University Centre for Science Education and Research (NUCSER), Deralakatte, Mangaluru, Karnataka, India

K. Langeswaran
Department of Biotechnology, Science Campus, Alagappa University, Karaikudi, Tamil Nadu, India

Fabiana Vieira Lima
Division of Faculty of Pharmacy, University of Espírito Santo, Espírito Santo, Brazil

Silvia Lomartire
University of Coimbra, MARE—Marine and Environmental Sciences Centre, ARNET - Aquatic Research Network, Department of Life Sciences, Coimbra, Portugal

Kalaivani M.K.
Sree Balaji Medical College and Hospital, Chennai, Tamil Nadu, India

Rehab H. Mahmoud
Water Pollution Research Department, National Research Centre (NRC), Giza, Egypt

João C. Marques
University of Coimbra, MARE—Marine and Environmental Sciences Centre, ARNET - Aquatic Research Network, Department of Life Sciences, Coimbra, Portugal

Rachel Carmelita Mathias
Nitte (Deemed to be University), Nitte University Centre for Science Education and Research (NUCSER), Deralakatte, Mangaluru, Karnataka, India

Raja Mohan
Chemical Industry Research Institute, University of Ulsan, Ulsan, South Korea

Pedro Monteiro
University of Coimbra, MARE—Marine and Environmental Sciences Centre, ARNET - Aquatic Research Network, Department of Life Sciences, Coimbra, Portugal

Tiago Morais
Lusalgae, Lda., Incubadora de Empresas da Figueira da Foz Figueira da Foz, Portugal; University of Coimbra, MARE—Marine and Environmental Sciences Centre, ARNET - Aquatic Research Network, Department of Life Sciences, Coimbra, Portugal

Ana Lucía Morocho-Jácome
Faculty of Pharmaceutical Sciences, University of São Paulo, São Paulo, Brazil

Mohammad Hossein Morowvat
Pharmaceutical Sciences Research Center, Shiraz University of Medical Sciences, Shiraz, Iran

T. Muralisankar
Aquatic Ecology Laboratory, Department of Zoology, Bharathiar University, Coimbatore, Tamil Nadu, India

K.T. Nachammai
Department of Biotechnology, Science Campus, Alagappa University, Karaikudi, Tamil Nadu, India

K. Nayana
Phycoscience Laboratory, Department of Plant Science, School of Biological Sciences, Central University of Kerala, Kasaragod, Kerala, India

Daniel Niebla-Canelo
Área de Toxicología, Universidad de La Laguna, Tenerife, Islas Canarias, Spain; Grupo Interuniversitario de Toxicología Alimentaria y Ambiental, Universidad de La Laguna, Tenerife, Islas Canarias, Spain

Fernando Pagels
CIIMAR — Interdisciplinary Centre of Marine and Environmental Research, University of Porto, Novo Edifício Do Terminal de Cruzeiros de Leixões, Matosinhos, Portugal; FCUP — Faculty of Science, University of Porto, Porto, Portugal

Soraya Paz-Montelongo
Área de Toxicología, Universidad de La Laguna, Tenerife, Islas Canarias, Spain; Grupo Interuniversitario de Toxicología Alimentaria y Ambiental, Universidad de La Laguna, Tenerife, Islas Canarias, Spain

Leonel Pereira
University of Coimbra, MARE—Marine and Environmental Sciences Centre, ARNET - Aquatic Research Network, Department of Life Sciences, Coimbra, Portugal

P. Perumal
Marine Planktonology & Aquaculture Laboratory, Department of Marine Science, School of Marine Sciences, Bharathidasan University, Tiruchirappalli, Tamil Nadu, India

Zulita Adriana Prieto
Faculty of Biological Sciences, National University of Trujillo, Trujillo, Peru

S. Pujithaa
Research and Development Wing, Sree Balaji Medical College and Hospital, Bharath Institute of Higher Education, Chennai, Tamil Nadu, India

Leila Queiroz Zepka
Bioprocess Intensification Group, Federal University of Santa Maria, UFSM, Santa Maria, Rio Grande do Sul, Brazil

Rathinam Raja
Research and Development Wing, Sree Balaji Medical College and Hospital (SBMCH), Bharath Institute of Higher Education and Research (BIHER), Chennai, Tamil Nadu, India

Rajapriya Rajendran
Department of Microbiology, Alagappa University, Karaikudi, Tamil Nadu, India

R. Rajkumar
Department of Environmental Sciences, Bharathiar University, Coimbatore, Tamil Nadu, India

Suchitra Rakesh
Department of Microbiology, School of Life Sciences, Central University of Tamil Nadu, Thiruvarur, Tamil Nadu, India

Bharathi Ramalingam
Suzhou Xishan Biotech Inc., VRL Asia, Suzhou, Jiangsu, China

Akila Ravindran
Department of Microbiology, Alagappa University, Karaikudi, Tamil Nadu, India

Mounika Munagala Srinivas Reddy
Plant Cell Biotechnology (PCBT) Department, CSIR-Central Food Technological Research Institute (CFTRI), Mysuru, Karnataka, India; Academy of Scientific and Innovative Research (AcSIR), Ghazibad, Uttar Pradesh, India

Carolina Pinheiro Rocha
University of Coimbra, MARE—Marine and Environmental Sciences Centre, ARNET - Aquatic Research Network, Department of Life Sciences, Coimbra, Portugal

Rosangela Rodrigues Dias
Bioprocess Intensification Group, Federal University of Santa Maria, UFSM, Santa Maria, Rio Grande do Sul, Brazil

J. Roopavathy
PG & Research Department of Zoology, Nirmala College for Women (Autonomous), Coimbatore, Tamil Nadu, India

Carmen Rubio-Armendáriz
Área de Toxicología, Universidad de La Laguna, Tenerife, Islas Canarias, Spain; Grupo Interuniversitario de Toxicología Alimentaria y Ambiental, Universidad de La Laguna, Tenerife, Islas Canarias, Spain

Salem S. Salem
Botany and Microbiology Department, Faculty of Science, Al-Azhar University, Nasr City, Cairo, Egypt

P. Sangavi
Department of Bioinformatics, Science Campus, Alagappa University, Karaikudi, Tamil Nadu, India

P. Santhanam
Marine Planktonology & Aquaculture Laboratory, Department of Marine Science, School of Marine Sciences, Bharathidasan University, Tiruchirappalli, Tamil Nadu, India

Raju Sekar
Department of Biological Sciences, School of Science, Xián Jiaotong-Liverpool University, Suzhou, Jiangsu, China

Sowmya Sharada P
Nitte (Deemed to be University), Nitte University Centre for Science Education and Research (NUCSER), Deralakatte, Mangaluru, Karnataka, India

M. Sreelakshmi
Phycoscience Laboratory, Department of Plant Science, School of Biological Sciences, Central University of Kerala, Kasaragod, Kerala, India

R. Srinithi
Department of Bioinformatics, Science Campus, Alagappa University, Karaikudi, Tamil Nadu, India

Karthikeyan Subburamu
Department of Renewable Energy Engineering, Agricultural Engineering College and Research Institute, Tamil Nadu Agricultural University, Coimbatore, Tamil Nadu, India

M.P. Sudhakar
Marine Biotechnology Division, National Institute of Ocean Technology (MoES), Ministry of Earth Sciences, Pallikaranai, Chennai, Tamil Nadu, India

Amina Tahreen
Department of Chemical Engineering and Sustainability, Kulliyyah of Engineering, International Islamic University Malaysia, Kuala Lumpur, Wilayah Persekutuan, Malaysia

João Rui Tanoeiro
MARE—Marine and Environmental Sciences Centre, ESTM, Polytechnic of Leiria, Edifício CETEMARES Peniche, Portugal

Adriane Terezinha Schneider
Bioprocess Intensification Group, Federal University of Santa Maria, UFSM, Santa Maria, Rio Grande do Sul, Brazil

Sonal Tiwari
Plant Cell Biotechnology (PCBT) Department, CSIR-Central Food Technological Research Institute (CFTRI), Mysuru, Karnataka, India

V.R. Umashree
Phycoscience Laboratory, Department of Plant Science, School of Biological Sciences, Central University of Kerala, Kasaragod, Kerala, India

Shailaja V.L.
Arham Bioresearch Technology Pvt. Ltd, Chennai, Tamil Nadu, India

Maria Valéria Robles Velasco
Faculty of Pharmaceutical Sciences, University of São Paulo, São Paulo, Brazil

Balasubramanian Vellaisamy
Department of Microbiology, Alagappa University, Karaikudi, Tamil Nadu, India

D. Vidya
Phycoscience Laboratory, Department of Plant Science, School of Biological Sciences, Central University of Kerala, Kasaragod, Kerala, India

M. Vijayakumar
Centre for Advanced Studies in Botany, University of Madras, Chennai, Tamil Nadu, India

B. Vinod
Phycoscience Laboratory, Department of Plant Science, School of Biological Sciences, Central University of Kerala, Kasaragod, Kerala, India

Shiyu Yin
State Key Lab of Urban Water Resource and Environment, School of Civil and Environmental Engineering, Harbin Institute of Technology Shenzhen, Shenzhen, Guangdong, China; Shenzhen Engineering Laboratory of Microalgae Bioenergy, Harbin Institute of Technology Shenzhen, Shenzhen, Guangdong, China

Xiaotian Zhao
Department of Biological Sciences, School of Science, Xián Jiaotong-Liverpool University, Suzhou, Jiangsu, China

Xu Zhou
State Key Lab of Urban Water Resource and Environment, School of Civil and Environmental Engineering, Harbin Institute of Technology Shenzhen, Shenzhen, Guangdong, China; Shenzhen Engineering Laboratory of Microalgae Bioenergy, Harbin Institute of Technology Shenzhen, Shenzhen, Guangdong, China

CHAPTER

An introduction to algae materials 1

V.R. Umashree[1], K. Anjana[1], D. Vidya[1], B. Vinod[1], K. Nayana[1], M. Sreelakshmi[1], Rathinam Raja[2] and Kulanthaiyesu Arunkumar[1]

[1]*Phycoscience Laboratory, Department of Plant Science, School of Biological Sciences, Central University of Kerala, Kasaragod, Kerala, India;* [2]*Research and Development Wing, Sree Balaji Medical College and Hospital (SBMCH), Bharath Institute of Higher Education and Research (BIHER), Chennai, Tamil Nadu, India*

1. Introduction

Due to enormous application potential in the biopharmaceutical, nutraceutical, and renewable energy industries, algae have recently gained significant interest on a global scale. Biofuels, bioactive medicines, and food ingredients can all be produced from algae in an environmental friendly, economically feasible manner (Sathasivam et al., 2019). The possibility for value-added products with exceptional pharmacological and biological properties has been explored for a number of algae species (Daroch et al., 2013). Algae-based materials can exist in the form of algal polymers, blends, and composites of algal biomass or biomolecules with other polymeric materials. Because of the characteristics such as biocompatibility, biodegradability, and therapeutic activities, algae-based materials have attracted great attention from the research community (Rehman et al., 2017). Bioethanol, biogas, biohydrogen, biodiesel, etc. are the several types of algae fuels. For a variety of purposes, including the synthesis of single cell proteins, pigments, bioactive compounds, medicines, and cosmetics, algae have been investigated as food sources (Becker, 2004). Algae have been extensively used in biosensing applications in addition to biomass generation and bioremediation (Antonacci & Scognamiglio, 2020). Algae naturally create polymers that have a high potential for bioplastic synthesis. Algae provide an environmentally benign substitute for oil-based plastics, which are seriously harming the environment (Carina et al., 2021). Algae-based biomaterials are promising candidates for the replacement of conventional, nonrenewable, fossil fuel-based polymeric materials. Algae-based polymers, blends, and composites have found many applications in numerous fields of human life, i.e., from food and medical to high-tech applications (Rehman et al., 2017). This chapter was written to shed light on the numerous biotechnological applications of algae as food, feed, fuel, and medicine as well as to give some details on various algae based materials.

Algae Materials: Applications Benefitting Health. https://doi.org/10.1016/B978-0-443-18816-9.00021-6

2. Algae materials for advanced biofuel production

The energy demands are increasing in daily life; presently, the price of fossil fuels is rising, and the burning of fossil fuels causes negative impacts on the environment. It is now important to find an alternative to fossil fuels for the world's energy supply. Biofuels find more attraction nowadays because it is a renewable energy source derived from biomass. Algae are considered a promising third-generation feedstock for biofuel production. This third-generation feedstock overcomes the drawbacks mainly found in first- and second-generation feedstocks because algal feedstocks do not compete with food supply and do not require any arable land and water resources (Kraan, 2012). Developing sophisticated algal biomass biofuel production technologies is critical for improving the cost-effectiveness of implementing a biomass-as-potential-renewable-energy approach (Alfarisi, 2020).

Biodiesel production by transesterification has also been widely investigated using heterogenous catalysts (base or acid), and heterogenous catalysts can be recycled, regenerated, and utilized again for a subsequent transesterification reaction, in contrast to homogenous catalysts (Klinthong et al., 2015). A catalyst made of an alkali or an acid is included in the homogenous catalyst. A solid acid, base, acid-base bifunctional, biomass waste-based, and nanocatalyst are all components of the heterogenous catalyst (Mandari & Devarai, 2021). It was discovered that the combination of ionic liquids with microwave irradiation had beneficial results, with only a minimal amount of ionic liquids needed to serve as a catalyst in the conversion of microalgae to biodiesel (Hasanudin et al., 2022). For in situ extractive transesterification of wet *Nannochloropsis* with methanol, 1-butyl-3-methylimidazolium hydrogen sulfate ([Bmim] [HSO_4]) is utilized as a solvent and an acid catalyst. It indicates that [Bmim] [HSO_4] catalyzed in situ transesterification is a cost-effective method of producing biodiesel (Sun et al., 2017). Other studies have reported different types of catalysts such as mordenite, meta-titanic acid, PTSA, and carbonized vegetable oil asphalt (Tabatabaei et al., 2015). According to Madison, ionic liquids with Brønsted acidity or basicity pose no environmental concern and produce a high biodiesel yield quickly and under mild reaction circumstances (Ong et al., 2021).

3. Valorizing algae biomass as materials for bioproducts and commercial application

Algal biomass is a viable feedstock for producing a variety of value-added compounds and products, such as food, feed, and fuel, in a sustainable manner. Algal extracts are excellent sources of biotechnologically viable chemicals that have antimicrobial, antioxidant, antiinflammatory, anticancer, and other medicinal and therapeutic activities. PUFA, essential oils, vitamins, antioxidants, carotenoids, and other metabolites derived from algal biomass have been reported to have exceptional nutritional and medicinal benefits (Sathasivam et al., 2019). According to studies, *Chlorococcum infusionum*, *Chlamydomonas reinhardtii* UTEX 90, *Chlorella vulgaris*, and *C. reinhardtii* are all viable options for large-scale bioethanol production (Daroch et al., 2013). *Spirulina* has achieved widespread appeal as a food supplement due to its high protein content and outstanding nutritious value (Sathasivam et al., 2019). The carotenoid produced by *Dunaliella* sp. and *Haematococcus* sp., as well as the lutein produced by *Muriellopsis* sp., is vital for fish larvae growth (Sathasivam et al., 2019).

The majority of carotenoids have medicinal properties, such as antiinflammatory and anticancer properties, which are mostly due to their powerful antioxidant impact, which protects organisms from oxidative stress. Essential fatty acids are produced by microalgal species such as *Crypthecodinium* and *Schizochytrium*. *Schizochytrium* strains that produce DHA and EPA are currently employed as an adult dietary supplement in cheeses, yoghurts, and breakfast cereals (Sathasivam et al., 2019). Algal extracts are utilized in various products, including antiaging creams, sunscreens, moisturizers, and emulsifiers (Aslam et al., 2021). Pimental et al. looked at existing commercial cosmetics derived from algae, including carrageenan, Gelcarin, alginate, and others, and found that these compounds are recognized as useful skin care products (Pimentel et al., 2018).

In general, algae include a variety of components that are thought to be extremely important and have a wide range of applications. Carbohydrates are utilized in the fermentation process; lipids are used in the creation of biodiesel, protein, and fatty acids; and pigments are employed in nutraceutical and pharmaceutical applications (Sudhakar et al., 2019). Because of their potential to lessen reliance on petroleum-based fuels and chemicals, interest in algae-based biofuels and chemicals has grown in recent years. Algae are widely regarded as the most ideal and long-term feedstock for producing green energy, as the entire process is carbon-neutral and may also be used for environmental cleanup (Trivedi et al., 2015). Algae have a lot of promise for usage as a biorefinery raw material because it can make a variety of products. As a result, rather than the traditional single product line strategy, a matrix method with many alternatives is preferable for successful algal biorefinery operation.

4. Algae materials for food and food packaging

Due to the modern lifestyle, people in most industrialized countries consume high-calorie foods, which contribute to obesity, diabetes, and heart problems. A healthy diet should include vitamins, minerals, polyunsaturated fatty acids (PUFAs), and other nutrients. Microalgae are a fantastic but understudied natural source of nutrients for a balanced diet. Several microalgae species have been found as being high in carbohydrates, proteins, lipids, and other nutritionally important components. Vitamins A, B1, B2, C, and E, as well as nicotinate, biotin, folic acid, pantothenic acid, niacin, iodine, potassium, iron, magnesium, and calcium, are rich in microalgae (Becker, 2004). *Nostoc* sp. was first taken as a food by the Chinese over 2000 years ago, and later, varieties of microalgae (*Chlorella* sp. and *Spirulina* sp.) were consumed as nutritious foods in Japan, Taiwan, and Mexico (Sathasivam et al., 2019). Most marketed microalgae products are now sold as healthy foods in the form of tablets, capsules, and liquids, and their products are mixed with snacks, sweets, beverages, and breakfast cereals. Some of the microalgae species extensively used as a human food source are *Aphanizomenon flos-aquae*, *Chlorella* sp., *Dunaliella salina*, *Dunaliella tertiolecta*, and *Spirulina platensis* (Sathasivam et al., 2019). Algae continue to be a novel food source (Mellor et al., 2022). However, understanding of the benefits of eating algae is needed to aware consumers to eat algal-based foods.

The modern food sector relies heavily on packaging. It is required for food preservation and ensures the food's safety and integrity. Packaging is an important aspect of our supply chain because of these benefits, but it also has drawbacks, such as contaminant residue migration, cost and energy efficiency, and sustainability. Glass, paper, various metals, and plastic are all regularly used materials for food packaging; nevertheless, plastics, after paper, are the most commonly utilized. Petroleum is used to make the majority of the plastics we use. But, even worse, it is tossed in the trash to be incinerated once it has been used. When it is burned, carbon dioxide is released into the atmosphere, which contributes to the greenhouse effect. Others simply discard plastic bottles and packaging into the environment, polluting ecosystem and affecting animal and human lives for ages because plastics take decades to disintegrate. Polymers that are biodegradable and biobased have been produced and are now available on the market to lessen these effects and establish a more sustainable approach to food packaging. Bioplastics manufactured from algae are nowadays proved to be a good food packaging material. In practice, it can be extracting a number of compounds from the biomass and converts them into plastic.

Seaweeds, which are high in polysaccharides, are thought to be a possible active agent or raw material. When mixed with any biodegradable polymer, they are better alternatives to traditional materials since they improve the product's sustainability, functionality, and sensory properties (Carina et al., 2021). Alginates are biopolymers made up of natural hydrophilic polysaccharide biopolymers derived primarily from marine brown algae. Alginates were first employed as a coating for perishable fresh fruits and vegetables to reduce respiration rate, but they can now be used to extend the shelf life of a wide range of foods (Kontominas, 2020). Alginates can work together to create a multifunction food packaging system, when employed as part of the green packaging technologies. *Cladophora* was chosen as a cellulose source because its cell walls contain a significant quantity of crystalline cellulose. When compared with other sources of cellulose, *Cladophora* algal cellulose has good mechanical characteristics. They have the potential to be developed as environmentally friendly, long-lasting packaging materials (Steven et al., 2020).

5. Superfood (algae)—in future

Microalgae have a wide range of potential for improving the nutritional content of traditional food preparations and acting as probiotic agents that benefit human and animal health (Becker, 2004). Microalgae are now commonly sold as a healthy food or supplement in the form of tablets, capsules, and liquids. Algae are also used in pasta, snacks, candy bars, and drinks, among other things, as a nutritional supplement or a natural food colorant. *Nannochloropsis*, *Tetraselmis*, *Isochrysis*, *Thalassiosira*, and *Chaetoceros* are microalgae that may synthesize long-chain fatty acids such as DHA and EPA, which are helpful as healthy food supplements (Chew et al., 2017; Priyadarshani & Rath, 2012). Carotenoids, chlorophylls, and phycobiliproteins are three primary groups of natural pigments in microalgae. These pigments have been used in food as vitamin precursors. In microalgae, phycobiliproteins are the most important photosynthetic accessory pigments. Natural colors in the food sector are one of the most common uses of phycobiliproteins. It possesses antioxidant, antiviral, anticancer, antiallergic, antiinflammatory, and neuroprotective characteristics, making it a promising substance for healthcare applications (Chew et al., 2017). Health-promoting and disease-suppressing metabolites abound in microalgae. Microalgae include a wide range of carotenoids, fatty acids, amino acids, antioxidants, and other secondary metabolites that boost the nutritional value of human and animal diets. Vitamins abound in microalgae. Microalgae such as *Dunaliella tertiolecta* and *Tetraselmis suecica* cultivated under nitrogen-deficient circumstances produce more vitamin E. Vitamin E is used in various ways. Cancer, heart disease, eye illness, Alzheimer's disease, Parkinson's disease, and other medical disorders are treated with it (Sathasivam et al., 2019).

Asiatic countries have employed green microalgae for hundreds of years as nutritional supplements or food sources. They are now consumed all around the world for their nutritional properties. The green algae *C. vulgaris*, *Haematococcus pluvialis*, *Dunaliella salina*, and the Cyanobacteria *Spirulina maxima* are among the most biotechnologically important microalgae (Priyadarshani & Rath, 2012). They are widely sold and used, primarily as nutritional supplements for humans. Millions of people eat seaweed worldwide, from Japan to Chile, South Africa to Belize, and even England. Seaweed has numerous advantages (Mellor et al., 2022). It is exceptionally nutritious since it contains numerous beneficial minerals and vitamins for the human body. Seaweed is also simple to grow, and because it is grown in water, it does not take up valuable land area. It develops swiftly and may thrive in dirty waters and in environments where typical food crops would perish. In the quest for more sustainable food systems for the future, microalgae are one of the most promising sustainable sources of food components.

6. Advanced instruments in algae materials: applications and characterization

Concerns about the environment and the finite supply of fossil fuels have driven research into biofuels and biomaterials. Biomaterials not only eliminate the problems associated with using traditional fossil fuel-based polymeric materials but also having characteristics that make them suited for a wider range of applications. Algae-based polymers, composites, and mixes are a significant class of biomaterials with applications in various sectors. Algal polymers, blends, and composites of algal biomass or biomolecules with other polymeric materials are all examples of algae-based materials (Kontominas, 2020).

The desire for a better alternative to present plastic products has given rise to algae-based bioplastic manufacture, which appears to be both biodegradable and cost-effective. A wide range of algal biopolymers attract a lot of attention as a way to combat the growing problem of plastic waste and environmental contamination. Polysaccharides derived from algae, including agar, alginate, and carrageenan, have been identified as interesting possibilities for delivery system design. The physicochemical characteristics of algal polysaccharides vary greatly. Furthermore, they have outstanding gelling qualities. These polysaccharides' physicochemical features allow them to be used in various targeted and controlled drug delivery methods, including particles, capsules, and gels (Rahmati et al., 2019).

For water quality analysis, a low-cost algae-based biosensor has been developed (Prudkin-Silva et al., 2021). Algal fuel cells (AFCs) are bioelectric devices that convert light and metabolic energy into electrical energy using photosynthetic organisms (Kannan & Donnellan, 2021). The potential of a fully biotic algal fuel cell is still being researched, which has led to a rethinking of the potential application of plant-based bioelectricity. Algae have been proved to be a viable fuel source for microbial fuel cells, either as an internally generated carbon source (due to their growth in an illuminated anodic compartment) or as an external carbon source (Walter et al., 2015). Because algae are good water quality indicators, studying them is an important part of water monitoring.

The measurement of chlorophyll-a, the primary photosynthetic pigment in algae, is one of several ways to estimate their quantity. Algal species have a variety of pigments, the detection of which can be used to determine algal composition based on the distinct light spectrum employed in the procedure. Fluorescence is a phenomenon in which a sample is activated by electromagnetic radiation at one wavelength and responds by emitting light at another slightly higher wavelength, which can be used to detect chlorophyll (Diána et al., n.d.). Today, equipment that uses fluorescence to estimate chlorophyll concentration and algal composition is accessible. In general, algal materials have a wide range of applications in research and industrial sectors.

7. Advancement in algae cultivation techniques

Autotrophs such as algae produce a variety of interesting compounds (Terrado et al., 2017). In Asia, algae have traditionally been viewed as a vital nutritional source, and many have eaten them in their diet, eventually gaining prominence in Europe, South America, North America, and Australia (Apurav et al., 2019). According to archaeological evidence found in Chile, this event took place 14,000 years ago (Dillehay et al., 2013). Nutraceuticals, pharmaceuticals, and cosmetics companies use algae for their bioactive compounds. Algae cultivation became more common as demand for raw materials increased (Bruna et al., 2015). More than 370 years have passed since algae have been cultivated for human consumption (Liang et al., 2015).

It is really challenging to cultivate algae on a commercial scale. Science and social acceptance are two challenges. They include the development of a cost-effective and eco-friendly farming system, which is adapted to optimum temperatures, healthy, and disease-resistant. In addition to cultivation onshore, algae were grown offshore as well. In the case of offshore cultivation, the method of cultivation differs from that of onshore cultivation. It is also an economical method to cultivate offshore. There has been a growing interest in offshore cultivation as a cost-effective method as well. Due to the lack of land use, regulating temperature, sunlight intensity, and nutrients is not required (Bak & Infante, 2020). During offline cultivation, environmental services are also provided since nitrogen and phosphorus are removed from algae, improving water quality for aquatic organisms. In addition, this nitrogen is also used to produce biofuels through combined algal processing. Economic and environmental sustainability of algal biorefineries are influenced by algal cultivation locations. In the pharmaceuticals, cosmetics, nutraceuticals, and biofuel industries, algal metabolites are important sources of lipids, polyunsaturated fatty acids, pigments, proteins, carbohydrates, vitamins, and antioxidants (Udayan et al., 2021).

Nutrient bioharvesting can be achieved with seaweed aquaculture. A suitable species for seaweed aquaculture should be selected depending on its purpose and efficiency. The limited supply of fossil fuels has created a high demand for clean, affordable, and sustainable energy sources. To realize the growing demand for energy and to achieve sustainable development, the algae industrial revolution has proved to be one of the most significant steps. Algal cultivation is currently used mostly to produce biodiesel from algal lipids. There are several benefits of using biodiesel as an alternative diesel fuel; it is biodegradable and nontoxic as well as highly lubricating (Das et al., 2020). A fast growth rate and high biomass productivity make algae an ideal feedstock for biofuel production. By cultivating algae in conjunction with wastewater and flue gas treatment, algae can simultaneously function as carbon sinks and purifiers. To determine whether it is worthwhile to scale up and commercialize algal cultivation for biofuel production, it becomes especially important to understand the overall supply chains for algal biofuels production (Lam & Lee, 2013).

8. Expanding algal polymers, proteins, and pigments as materials for industrial applications

There are a number of pigments present in algae such as chlorophyll, carotenoids, and phycobiliproteins, which exhibit colors ranging from green, yellow, brown, and red. Increasing awareness in society of the harmful effects of synthetic dyes as well as society's shift toward natural products has made algae a desirable resource. *Spirulina* and *Dunaliella* capsules are now commonly prescribed healthy foods to enhance human vitality and longevity. Because of their commercial value as natural colorants, algal pigments have great potential for use in nutraceutical, cosmetics, and pharmaceutical products (Chakdar & Pabbi, 2017).

Algae-based bioplastic production is a promising alternative to existing plastic products thanks to its excellent biodegradability and cost-effectiveness. As the surge of plastic waste and environmental pollution becomes more severe, algal biopolymers draw great attention. Polymers extracted from cyanobacteria include polyhydroxyalkanoates and poly-(hydroxybutyrate). Natural polymers include algal polysaccharides (alginate, laminarin, fucoidan, carrageenan, agar, ulvan) (Dove et al., 2020; Lee & Mooney, 2013; Santos & Grenha, 2015). A variety of microbes produce polyhydroxyalkanoates as esters, which constitute the structural foundation for a wide range of biomolecules (Hempel et al., 2011). Two types of cellulose acetate sheets were created by embedding two algal seaweeds (*Ulva fasciata* and *Sargassum dentifolium*) into cellulose acetate (CA) polymer; *Ulva* sheets (CA-U), and *Sargassum* sheets (CA-S) (Moghazy et al., 2020).

In the past hundreds of years, dried microalgae have been used as a food source. Recent efforts have focused on developing microalgae for use as food sources on a commercial scale because they accumulate large amounts of protein and oil. Biofuels seem to account for most of the oil production. FDA has designated algal protein as GRAS in the United States. Currently, its intensely green color limits its food applications (Zeece, n.d.). The food industry uses a very limited amount of algae proteins or purified protein fractions. As a blue chromoprotein in foods, phycobiliproteins serve as a family chemical used for food coloring in Japan. *Spirulina* sp. contains the pigment phycocyanin. The product is sold as "Lina Blue-A". In red algae, pigments of the same type, but in purple or red colors, phlorocyanins and phycoerythrins, are also present. These chromoproteins possess unique spectral properties, which make them ideal for use as food-coloring agents. In addition to possessing unique spectral properties, chromoproteins could be used to color foods (Sekar, 2008, pp. 113−136).

Commercially valuable pigments extracted from micro- and macroalgae available in the markets are beta carotene, astaxanthin, fucoxanthin, phycocyanin, phycoerythrin, lutein, and chlorophyll. To develop a sustainable and commercially viable microalgal bioindustry, it is essential to effectively recover these pigments. The increasing toxicity of textile effluents generated by synthetic dyes has prompted the search for alternatives without toxicity. These pigments have been found to possess antimicrobial and antioxidant properties, so dyeing fabric with algae can also be used as a food colorant. As meditextiles can be made from these pigments due to their antimicrobial properties, they have excellent potential for use in hospitals (Jamee & Siddique, 2019).

9. Algae for blue economy and blue carbon materials and their application in various fields

The term "blue " has recently been reframed in relation to human use of the ocean for economic purposes. In addition to being an alternative development paradigm, the blue economy promotes economic, social, and environmental benefits (Isurupremarathna, 2021). Blue economies have been defined in different ways. Here, they are understood as parts of the economic system and policies associated with sustainable use of oceanic resources. Oceanic sustainability includes sustainable fisheries, ecosystem health, and pollution, which is an important challenge of the blue economy. Taking into account this international policy context, the blue economy encompasses activities that provide access to the ocean's resources, explore, develop, and utilize the ocean's space, as well as conserve the ocean's ecosystems (Bari, 2017). A blue economy's other pillars, including improving human well-being and social equity as well as reducing environmental risks and ecological scarcities, are also well understood by science and technology (Hassanali, 2020).

Blue carbon is a part of the blue economy. Oceanic plants, such as seagrasses, mangroves, and tidal marshes, capture and store carbon in the form of blue carbon. These plants are highly productive and store high levels of carbon. Mangrove and saltmarsh carbon stocks, and not seagrass, may influence management actions and ecological processes depending on their position within the estuary (Cacho et al., 2021). Researchers conducted a study of salt marsh blue carbon stocks in Sri Lanka, and their findings expanded our understanding of global and regional saltmarsh carbon stocks (Kauffman et al., 2018).

10. Algae and farming applications—present and future scenario

As the demand for fossil fuels increases, the current reserves are insufficient to meet it and are very likely to be exhausted very soon. The quest for renewable energy sources has been sparked by global warming, pollution, and inflated oil prices. A viable and promising renewable resource for biofuels production is macroalgae (green, brown, and red marine seaweed). Macroalgae are believed to provide diverse bioproducts including biofuels and have been the subject of numerous research studies. Microalgae are an enviable source of biofuels because of their carbohydrate and lipid composition, along with their lack of lignin. In Europe, there are currently 447 production units for algae and *Spirulina* species. European countries have identified a variety of species, production methods, and commercial applications. The majority of macroalgae production units in Europe are harvested from wild stocks (68%). *Spirulina* sp. are primarily produced in open ponds (83%), while microalgae are mostly grown in photobioreactors (71%) (Araújo et al., 2021).

To boost the growth of the sector in Europe, the industry needs to ramp up its production volume and to progress in technological and market development. Currently in the precommercial phase, algae are one of the hottest emerging industries in the EU Blue Bioeconomy. In addition, using current methods and technologies to redefining primary energy usage, such as cooking and heating, is the key to the development of the future energy sector. In addition to stimulating global economic growth, algae-based energy has also helped advance the human civilization (Araújo et al., 2021).

It has been 25 years since the global demand for seaweed, and its products expanded exponentially. A great deal of potential exists for seaweed production on the continent of Africa as well as on its offshore islands. African seaweed production and consumption are lower than that of China and the rest of Asia. As far as red eucheumatoid seaweeds are concerned, Africa ranks third after Asia and Latin America. In the tropical, carrageenan-producing eucheumatoids of the rainforests to the temperate, carrageenan-producing kelp species of the southwest, seaweeds can be grown along the continent's diverse ecosystems. It has improved the livelihoods of coastal people in Africa through seaweed aquaculture production led mostly by women. A major constraint to the development of this industry includes disease outbreaks and pest outbreaks due to climate change. As the continent grows and expands its production and utilization of seaweeds, it can compete with China and Southeast Asia for global leadership (Msuya et al., 2022).

11. Algae materials for quorum sensing applications

Several physiological functions of microorganisms are controlled by quorum sensing (QS), including biofilm formation, bioluminescence, and antibiotic production (Rutherford & Bassler, 2012, pp. 1–25). It may be possible to alter the production of algicidal metabolites by the marine algicidal bacterium, *Ponticoccus* sp.PD-2, by inhibiting the QS system. In the process of disintegrating red tides, the QS-regulated algicidal system may be involved. To control red tides, QS might be a useful tool (Chi et al., 2017). Algae are ubiquitous in the marine environment, and the ways in which they interact with bacteria are of particular interest in marine ecology field. The interactions between primary producers and bacteria impact the physiology of both partners, alter the chemistry of their environment, and shape microbial diversity Although algal–bacterial interactions are well known and studied, information regarding the chemical–ecological role of this relationship remains limited, particularly with respect to QS, which is a system of stimuli and response correlated to population density (Lami, 2019). In the microbial biosphere, QS is pivotal in driving community structure and regulating behavioral ecology, including biofilm formation, virulence, antibiotic resistance, swarming motility, and secondary metabolite production. Many marine habitats, such as the phycosphere, harbor diverse populations of microorganisms and various signal languages (such as QS-based autoinducers) (Zhou et al., 2016).

QS-mediated interactions widely influence algal–bacterial symbiotic relationships, which in turn determine community organization, population structure, and ecosystem functioning. Understanding info chemicals-mediated ecological processes may shed light on the symbiotic interactions between algae host and associated microbes. In this chapter, we summarize current achievements about how QS modulates microbial behavior, affects symbiotic relationships, and regulates phytoplankton chemical ecological processes (Zhou et al., 2017) and role of microalgae-associated microorganisms with the QS system in algae–bacteria interactions and community succession of harmful algal blooms microalgae (Zhang et al., 2022). Algal biomass productivity increased by 2.25 times by QS molecules. QS molecule enhanced lipid content of *Chlorella sorokiniana* by 1.28. Bacterial QS molecules also improved the photosynthetic efficiency of *Chlorella sorokiniana*. Settling efficiency also increased by 2.5 times by these molecules. 74% higher CE for microbial fuel cell with QS induced lipid-extracted algae as anodic substrate (Das et al., 2019).

12. Algae-derived hard steel and building materials and application

Carbon fiber composite technology is an engineering tool for lowering the weight of vehicles and reducing their fuel consumption (Mathijsen, 2020). It requires the use of carbon fibers that can do exactly the same, or close enough to the automotive steel design manuals and the high-strength steel application design guidelines (Koumoulos et al., 2019). The production of carbon fiber requires fossil materials and a lot of energy. As a low-cost substitute, researchers identified that algae can be used for the production of carbon fibers. Then it would be a perfect green alternative to steel and aluminum (Mathijsen, 2020). Carbon fibers are, by far, the most widely used fiber in high-performance applications because of their lighter, stronger, and more durable solutions (Koumoulos et al., 2019). Researchers from the Technical University of Munich (TUM) developed a way to make carbon fibers, from halophilic algae, to remove CO_2 from the atmosphere and subsequently to make carbon fiber (Francis, 2019).

Algae oil consists of two basic components: a backbone of glycerin with fatty acids esterified onto the hydroxyl groups (Mathijsen, 2020). Algae convert atmospheric CO_2 into biomass and, subsequently, algae oil. The algae oil is produced by a nutrient depletion phase where nitrogen is limited in the culture medium, which triggers the accumulation of lipids. Algae oil is then hydrolyzed to get free fatty acids from the glycerol backbone (Francis, 2019). To convert glycerin into carbon fiber, the first step includes use of ammonia to turn the glycerin into acrylonitrile and then a polymerization reaction that produces polyacrylonitrile (PAN) (Mathijsen, 2020). PAN is the precursor for 90% of carbon fiber production (Francis, 2019). Polyacrylonitrile is turned into carbon fiber by heating up the fibers in a furnace to 2000 or 3000°C, depending on the desired properties (Mathijsen, 2020).

Applications include new crash-absorbing structures in automotive engineering, more efficient gas diffusion structures for fuel cells, adsorptive fibers for hydrogen high-pressure tanks, next-generation energy storage systems, cell-compatible fibers to replace nerve tracts in paraplegia, prosthetics with tailored properties to avoid stress shielding, reinforcement structures for carbon concrete, for wind turbine rotor blades, and many more (Koumoulos et al., 2019). Uwe Arnold, a world-renowned specialist for technology assessment and technoeconomical analyses, and Kolja Kuse came up with an innovative way of replacing cement by joining carbon fiber to natural stone. That is very relevant because cement accounts for up to 7% of all CO_2 emissions (Mathijsen, 2020).

13. Algae-derived materials for applications in automobile industries

The major application of algae derived materials on automobile industries relies on the production of biofuel. When compared with standard fuel Jet A, biofuels is better regarding the technical parameters such as freeze point, density, and viscosity of fuel. Moreover, biofuels are more efficient than petroleum fuels due to their high energy than standard fuels. Also, biofuels are environmentally friendly because during combustion of biodiesel the engine emits lesser amount of CO_2 and no sulfur oxides (Zabochnicka-Swiatek, 2010). Another interesting application is that algae-derived material can be used for designing biocatalytic converter that can eliminate the pollutants released by automobile engine exhaust (Ganji & Yenugula, 2017). It has been found that the addition of algal fillers to vinyl ester composites can improve its tensile strength and can be thus benefitted in the automobile industry (Bharathkumar et al., 2018). Microalgae *C. vulgaris* can be used as a potential candidate for the bioremediation of lead pollution caused by automobile industry (Chaudhuri & Sole Author-Chaudhuri, 2022).

14. Algae-derived biosensor materials and their applications

Biosensors are analytical devices that convert a biological response into an electrical signal. Biosensors have been used in different industries, including the food industry, the medical profession, and the marine sector, and they provide more stability and sensitivity than traditional approaches (Mehrotra, 2016). Algae have been extensively used in biosensing applications, in addition to biomass generation and bioremediation (Antonacci & Scognamiglio, 2020). Algae-based biosensors have shown promise in detecting analytes of agroenvironmental and security concerns in a sensitive and long-term. The availability of numerous algal bioreceptors, including whole cells and photosynthetic subcomponents, is one of its advantages. This can be incorporate into dual-transduction miniature devices, to monitor the environment continuously (Antonacci & Scognamiglio, 2020).

The key benefit of utilizing algae is that they are more adaptable to different settings than other bioreceptors since they are minor subjects to physicochemical changes. Researchers look into the prospects of using microalgae as sensor elements for biological pollution sensors. When microalgae-based biosensors are used for ecological monitoring of the aquatic environment, the condition of microalgal cells, as well as their uniformity and concentration, which is affected by the temperature, illumination, and chemical composition of the water, is found to influence the functioning of biological sensors (Voznesenskiy et al., 2016). In previous research, a system for water quality monitoring employing living green algae (*Chlorella kessleri*) and the intelligent mobile lab (IMOLA) was established (Umar et al., 2015).

Environmental applications of algal-based biosensors mainly include monitoring current issues such as water and air pollution, overfertilization, pesticide monitoring, etc. Even though many efficient biosensors for toxicity sensing using different aquatic organisms are used, the biosensors based on microalgae were identified as the most sensitive to waterborne chemicals compared with others. One of the main advantages of employing algae for environmental monitoring is the ability to develop enormous populations quickly. Algae respond to a wide range of pollutants quickly and consistently, providing potentially important early warning signs of deteriorating conditions.

In one study, *Scenedesmus acutus* and *Pseudokirchneriella subcapitata* were trapped in alginate beads and used as bioindicators (Prudkin-Silva et al., 2021). The device that was created was successfully tested. Without any scientific expertise, children, adults, and the elderly were able to construct the sensor and analyze the results. Despite challenges such as low stability and selectivity, algae-based biosensing is a viable option with some recent applications. The potential of algae-based sensors will be realized through the strategic use of cutting-edge technologies such as materials science, nanotechnology, etc.

15. Algae materials for bionanopesticides: nanoparticles and composites

The explosive increase in the global population leads to an increased demand for producing high-quality food. Limited agricultural land availability and loss of yield due to pest attacks were the major challenges for agricultural production. In this scenario, it is necessary to formulate a high efficient, eco-friendly pesticide with minimum investment. The application of nanotechnology for sustainable agricultural development by using bionanopesticides, bionanofertilizer, and bionanoherbicides will help increase agricultural productivity (Lade et al., 2019). For the safe and cost-effective production and use of pesticides, they should be effective against a wide variety of pests, nontoxic to farmers, and easy to prepare, must not get accumulated in the food chain, must not affect the quality of food, and should be safer to the environment.

The bionanopesticides attained special attention because of their eco-friendly behavior, easy delivery to the plants, and biodegradability. Due to the biocompatibility of nanoparticles, algae have now been exploited for synthesizing nanoparticles to produce bionanopesticides and herbicides. Nanopesticides such as Ag, Cu, SiO_2, and ZnO show better broad-spectrum pest protection efficiency and reduce water, soil, and environmental pollution than conventional pesticides (Lade et al., 2019). Most of the seaweed extracts exhibit insecticidal activity and can be used in integrated pest management. Halogenated monoterpenes extracted from red algae *Plocamium cartilaginous* shoed insecticidal property against tomato moth *Tuta absolute* and the cereal aphid, *Schizaphis graminum. Besides this, extracts from Spirulina platensis, Sargassum vulgar, Sargassum whitii, Padina* sp., *Dictyota* sp., *Caulerpa* sp., *Nostoc, Anabaena* contained phenols, tannins, and alkaloids. Further analysis of these extracts showed insecticidal activity against various insect pests (Asimakis et al., 2022).

The volatile oil from *Actinotrichia fragilis, Liagora ceranoides*, and *Colpomenia sinuosa* caused 90% mortality of pests in the stored products *Oryzaephilus mercator* and *Tribolium castaneum* (Pasdaranet al. 2016). Inhibition of photosynthesis, induction of plant defense responses, insecticidal activities via neurotoxicity, inhibition of quorum sensing, and virus entry are the pesticidal and herbicidal activities that the plants have demonstrated. The invention of antimetabolites for potential insecticidal activities is a recent advance in pesticide research. Algae represent an untapped resource for novel and safe biopesticide compounds.

16. Algae material for textile industries

It is identified that the textile industry contributes around half a million tons of plastic microfibers into the ocean annually, creating a threat to the environment (O'Connor, 2017). Most of the synthetic dyes used in the textile industries are toxic or cancerogenic and require harsh chemicals that are hazardous to human as well as environmental health (Mir et al., 2019). They cause allergy and toxicity to humans and generate hazardous waste, environmental pollution, and economical challenges (Venil et al., 2017). Thus, the algae that can be used for the extraction of biodegradable are gaining attention recently. These fibers can be used for the production of high-tech fabrics, which have the advantages of high stiffness and the ability to break without leaving rough and dangerous edges rather than low cost and ecofriendly. The textile applications for algae fibers already evolved from bandages and specialty wear to fashionable products (Gunter, 2010).

Together with this, environmentally friendly, perishable, and nontoxic natural dyes are gaining attraction (El-Khatib et al., 2016). Algae are a source of a wide range of photosynthetic pigments such as chlorophylls, carotenoids (carotenes and xanthophylls), and phycobilins. Macroalgae contain phycoerythrins and carotenoids, which represent valuable pigments for the textile finishing industry (Mona et al., 2019). Algae-based dyes for textile applications are explored recently (Mir et al., 2019). The usage of natural dyes in textile coloration as UV protectant and as antimicrobial agent has been explored by many researchers (Mona et al., 2019). Recently, natural dye obtained from green algae was used for dyeing wool fibers, which are used in handloom carpets. For enhancing the dye ability of wool fibers, pretreatment with chitosan and tannic acid was used, which are ecofriendly (El-Khatib et al., 2016). *Cladophora glomerata* L. (Chlorophyta) is an abundantly available fresh water alga. The relatively short time duration (45 min) was required for cotton fabric dyeing with alkaline extract of this alga to produce maximum color strength is suitable for industry (Mir et al., 2019).

3D printing is a powerful technology for fabrication of living functional materials. Researchers have used 3D printers and a novel bioprinting technique to develop a sustainable material made from algae. Bacterial cellulose, with unique mechanical properties, such as flexibility, toughness, strength, and ability to retain its shape, was used like the paper in a printer, whereas living microalgae were used as the ink and created material with tough and resilient while also eco-friendly, biodegradable, and simple and scalable to produce. The material could have a wide range of applications, from space exploration to the fashion industry, eventually. This would allow the fashion industry to become more circular and less wasteful (Koop, 2021).

17. Algae nanomaterials and nanostructures

The unique property of nanomaterials is recently attracted to targeted drug delivery, bio- and chemical sensing, catalysis, antimicrobial coatings, and optoelectronic devices. Unfortunately, the physicochemical synthesis of nanomaterials requires energy-intensive processes, high temperatures, toxic solvents, etc. Therefore, the necessity for the eco-friendly synthesis of nanomaterials becomes mandatory. So, the synthesis of nanomaterials from biological organisms became well established in the last two decades (Dahoumane et al., 2017). Algae are considered bionanofactories since both live and dead biomass can be utilized to synthesize the metallic nanoparticle. Depending on the species of algae and mode of operation, nanoparticles can be synthesized either intracellularly or extracellularly.

Properties such as chemical reactivity, surface chemistry, and availability of hydrophilic surface groups such as sulfate, carboxyl, and hydroxyl group on algae-mediated nanoparticles made the algae nanomaterials suitable and safe to use in medical treatment, imaging and sensing, biomedicine, catalyst, and cancer therapy. The algae of phyla Chlorophyta, Cyanophyta, Charophyta, Rhodophyta, Phaeophyta, Euglenophyta, Haptophyta, Ochrophyta, and Bacillariophyta were currently studied for the synthesis of nanomaterials (Dahoumane et al., 2017). Silver and gold nanoparticles, considered model elements, were produced from macroalgae and microalgae. Silver nanoparticles isolated from *Spirogyra hyaline* showed antibacterial, antifungal, insecticidal, antioxidant, and membrane damage activities (Al-Radadi et al., 2022). The brown alga *Bifurcate* has been used to synthesize copper nanoparticles. These nanoparticles showed antibacterial activity against *Enterobacter aerogens* and *Staphylococcus aureus* (Abboud et al., 2014).

Most of the algal nanoparticles were superior to the standard drugs in showing antibacterial, antifungal, and anticancer activities. Nanoparticle biosynthesis did not produce any hazardous chemicals compared with physicochemical synthesis, and only negligible waste materials were left after the synthesis. Apart from medical use, algal nanomaterials have applications in computing, optics, and cosmetics. Nanobiotechnology offers new disease identification and prevention methods by boosting human health and agriculture (Singh et al., 2018).

18. Algae material for drugs and drug delivery

Algae are a potential candidate for numerous bioactive compounds that provide several lifesaving properties. Bioactive compounds present in brown algae including phlorotannins, fucosterols, fucoidans, alginic acids, fucoxanthins, and phycocolloids are reported to have antidiabetic, antiinflammatory, cytotoxic, and antioxidant activities (Cikos et al., 2018). Phlorotannin extracted from marine brown algae has shown antidiabetic properties (Gunathilaka et al., 2020). Algae also serve as a source for various compounds such as polysaccharides and lectin that can fight against various viruses such as human immunodeficiency virus (HIV), dengue virus, and even SARS-CoV-2 (Sami et al., 2021). Compounds extracted from several marine seaweed have shown to mediate several inhibitory activities in cellular pathways namely apoptosis, angiogenesis, migration, and invasion suggesting their potential role in anticancer properties (Alves et al., 2018). Algal polysaccharides (alginate, fucoidan, carrageenans, and ulvan) derived from several marine seaweed can be considered as a potential candidate in modern medical due to their high biological activity, biocompatibility, biodegradability, nontoxicity, antioxidant, immunomodulatory, antimicrobial and anticoagulant, antitumor, antiviral, antibacterial, and angiogenic activities (Kuznetsova et al., 2020). Further carrageenan is considered for potential drug delivery applications because of their organized, ordered, and stable networks along with wide water pockets, which embeds, protects the active drug molecules, and releases them in a controlled and targeted manner (Qureshi et al., 2019). *Spirulina platensis* can be used as natural carriers to construct a drug-loaded system for targeted delivery and fluorescence imaging-guided chemotherapy on lung metastasis of breast cancer (Zhong, Zhang, et al., 2020).

19. Algae materials for clinical applications

Algae materials have tremendous clinical and biomedical applications, including the production of antibiotics, subunit vaccines, monoclonal antibodies, hepatotoxic and neurotoxic compounds, hormones, and enzymes (Zhong et al., 2021). Algae are now used in tissue engineering, bioanalysis, drug delivery, tumor therapy, wound healing, etc. Microalga *Thalassiosira weissflogii* is used for bioimaging and biosensing using fluorophores because of its complex surface chemistry with various micro- and nanostructures. The microalgae cell surface can selectively bind to antibodies, which helps use them as biosensors. Chemical composition analysis of sweat and blood samples is recently analyzed with the help of microalgae-based tablets or chips containing nanoparticles (Vona et al., 2017, pp. 3817–3823). In bone tissue engineering, algal-derived polysaccharides are widely used as polymer binding agents to help tissue growth and proliferation.

Microalgae species such as *C. reinhardtii* and *Spirulina* sp. have been studied for the in vitro delivery of chemotherapy drug to the tumor cells. *C. reinhardtii* cells modified with chitosan-coated iron oxide nanoparticles and chemotherapy drug showed high cancer cell uptake during cell culture experiments. Microalgae species *Chlorella* and *Spirulina* sp. have been recently studied for their utilization as oxygenator agent for treating hypoxic-associated cancer. This can be used for tumor therapy, including RT, PDT, or RT/PDT synergistic therapies (Zhong, Li, et al., 2020). Oxygen is significant in wound healing, cell metabolism, signaling, cell proliferation, reepithelialization, and antibacterial immune response. Increasing oxygen supply to the wound areas with the help of microalgae is a promising approach in regenerative therapies. The algal hydrogel containing *Spirulina platensis* and natural polymer carboxymethyl chitosan maintains the oxygen production capacity and promotes their binding to the wound cells (Li et al., 2021).

Microalgae are used as vaccine transporters as they are safe and contain a single chloroplast, which expresses proteins with high accumulation. E2 protein (vaccine against classical swine virus) D2-CTB fusion protein (oral vaccine against cholera toxin B subunit), and E7 protein are examples of these recombinant proteins (Khavari et al., 2021). The most common red algae, *S. filamentosa*, contain a wide variety of therapeutically essential phytochemicals such as alkaloids, phenols, flavonoids, and terpenoids. These phytochemicals possess potential antibacterial and antiviral activities against clinical pathogens (Valarmathi et al., 2020).

20. Algae material for cosmetics and cosmeceuticals

Cosmetics are chemical elements and their compounds in either natural or processed form, and mixtures intended to care for hygiene (clean, protect, reduce odors), enhance conditions, and beautify (perfume, alter appearance) external human body parts and the oral cavity, whereas cosmeceutical is a combination of cosmetics and pharmaceuticals and indicates that a specific product contains active ingredients retaining therapeutic value (Joshi et al., 2018; Michalak et al., 2020). Algae are naturally exposed to oxidative stress and develop several bioactive compounds that can act in cosmetics (Ariede et al., 2017). Traditionally algae have been used for cosmetic purposes in rituals from washing hair to thc nourishing of the body.

In recent years, the use of seaweed-derived ingredients in cosmetic products has increased because of the scientific studies that have proved the potential skincare properties of seaweed bioactives (Lopez-Hortas et al., 2021). These biologically active compounds include polysaccharides and sulfated polysaccharides, proteins and bioactive peptides, phenolic compounds, polyphenols and phlorotannins, pigments (e.g., fucoxanthin), and mycosporine-like amino acids—MAAs (Michalak et al., 2020). These algae-derived secondary metabolites benefit the skin through protection from UV radiations and the prevention of rough texture, wrinkles, and skin flaccidity. It also avoids skin aging due to the presence of antioxidant compounds (Ariede et al., 2017).

Products from diverse algal species have been used in the beauty industry as antioxidants, sunscreens, thickening agents, skin sensitizers, and moisturizing agents to strengthen skin abilities against abrasions, tanning, etc., and they are provided in various forms as simple toiletries such as body lotions, facial masks, and shampoos (El-Chaghaby & Rashad, 2021). Colorants for cosmetic formulations such as eye shadow, face makeup, and lipstick are currently obtained from red microalgae. Some types of brown and red macroalgae are used in cosmetics due to the presence of vitamins, minerals, amino acids, sugars, lipids, and other biologically active compounds (Ariede et al., 2017). Cosmeceuticals containing seaweeds are nonirritants and can perform different functions, such as improving psoriasis and preventing skin problems, especially atopic dermatitis, hyperpigmentation, acne, wrinkles, and hair loss (Lopez-Hortas et al., 2021). With the increase in demand for natural products for skin care and well-being treatments, algae may be a significant source of substances with beneficial effects for skin health. Its wealth of polysaccharides, carotenoids, unsaturated fatty acids, vitamins, and other antioxidants makes them potential raw materials for the development of cosmetics and cosmeceuticals (Mourelle et al., 2017).

21. Algal lipids, lipidomics, and biosurfactants

Algae produce an extensive array of commercially valuable unique bioactive, complex, exotic acyl lipids, and fatty acids in high quantity. Under stress conditions such as nitrogen limitation, salinity, and high temperature, algae produce an increased amount of storage lipids (TAGs) composed of saturated and monounsaturated fatty acids. Algal lipids are widely used in food formulations as an alternative polyunsaturated fatty acid (PUFA) source. Compared with terrestrial plants, algal biofuel feedstock has a significant advantage as it needs less land and produces more biomass (Da Costa et al., 2016). The yield of oil from algae is strain dependent, and the net productivity is much higher than that of oil crops (Morales et al., 2021).

Understanding lipid synthesis pathways and factors affecting lipid biosynthesis will help achieve lipid productivity in algae. Studies on algal lipids mainly focus on interpreting lipid and fatty acid composition, lipid signaling, the genes and enzymes involved, and their roles in stress response, innate immunity, and defense against pathogens. Lipid biosynthetic pathways and the quantity and quality of lipid can be intensely studied by lipidomics. A deep understanding of the algal lipid composition is necessary to produce biofuels and other related products (Anand et al., 2019). Systematic quantification of cell lipidome, lipid metabolism, and cell protein interactions by lipidomics helps to provide information regarding health and disease.

Biosurfactants are amphipathic molecules having both hydrophobic and hydrophilic regions. The ionic (cationic or anionic), nonionic, or amphoteric characteristics of the polar (hydrophilic) region enable them to possess a variety of beneficial properties such as low surface tension and effective hydrocarbon emulsion, and antimicrobial activity (Rahman et al., 2019). Biosurfactants display a variety of surface activities that allow the solubilization of hydrophobic substrates. Biosurfactants' unique properties and functionality at extremes of temperature, pH, and salinity confer lower toxicity and higher biodegradability (Paniagua-Michel et al., 2014). These characteristics allow biosurfactants to have several potential applications such as bioremediation, healthcare, and hygiene (Rahman et al., 2019). Low-molecular-weight compounds are biosurfactants, such as lipopeptides, glycolipids, and proteins. Polysaccharides in algae are essential producers of agar, alginate, carrageenan, dextran, gellan, pullulan, and xanthan gum. They have gelling, stabilizing, or thickening properties that are essential agents as food additives (Paniagua-Michel et al., 2014).

22. Positive and negative environmental impacts on algae

Algae are the primary precursor producer in aquatic environments and a critical component of the marine food web, whether in the ocean or on land and maintain ecological equilibrium (Wang et al.,. 2013). The algal culture depends on its growth and development factors, such as temperature, light, pH, nutrients, mixing, and pollution. Temperature: The influence of temperature on algal growth and reproduction is significant. The biological activity of algae and the effectiveness of nutrient use both respond to changes in water temperature, which impact algae development (Wang et al.,. 2013). Temperatures over the appropriate range cause growth inhibition and even death in some microalgae cells, and temperatures below the optimal range and above the frozen level do not kill microalgae. Still, they do make them inactive for growth (Gani et al., 2019). Light: For photosynthesis, microalgae require both light and dark regimes, and during the photoautotrophic development phase, light serves as the organisms' primary energy source as they transform CO_2 into organic molecules such as sugars (Chavan et al., 2014). However, excessive exposure to light intensity or oversaturation of the environment with light can result in photoinhabitation, which is caused by the production of reactive oxygen species, which are damaging to microalgae cells and indirectly reduce biomass productivity (Gani et al., 2019).

pH: In microalgal cultivation, hydrogen potential (pH) is significant since it controls how soluble minerals and carbon dioxide are in the medium and have an impact on the microalgae (Pedruzi et al., 2020, pp. 134–139). Although the pH of the initial culture an alga is suited to grow in is considered its optimum pH, most algal development happens in the vicinity of neutral pH (Chavan et al., 2014). Pollution: The susceptibility of different algae to a pollutant and various pollutants varies; however, most algae are sensitive to a variety of contaminants. Sewage discharges, hazardous trace metals, and oil pollution reduced sperm motility and fertilization and increased mortality in seaweed germlings (Agrawal, 2012).

Nutrients: The nutritional requirements of different microalgae species may differ, but the fundamental needs are the same for all. Microalgae's backbone is made up of macronutrients such as nitrogen, phosphorus, and carbon, which are all required for algal growth, and some marine microalgae species also require a macronutrient such as silicon (Khan et al., 2018). Mixing: mixing and aerating microalgae cultures ensure consistent distribution of nutrients, air, and CO_2, allowing light to penetrate and distribute uniformly throughout the culture and preventing biomass from settling and aggregating. If all of the other parameters are met but no mixing occurs, biomass productivity will be drastically reduced (Khan et al., 2018).

23. Genetic engineering of algae material: merits and demerits

Algae are being continuously explored for nutraceuticals, pharmaceuticals, and other industrially important bioactive compounds. To enhance the yield, strain improvement through genetic engineering could play an important role (Kumar et al., 2020). The aim of genetic modification (GM) is to increase and modify the accumulation of lipids, alcohols, hydrocarbons, polysaccharides, and other energy storage compounds (Radakovits et al., 2010). GM of green microalgae to improve CO_2 fixation will contribute to carbon sequestration (Barati et al., 2021). Successful genetic transformation has been reported for the green, red, and brown algae; diatoms; euglenids; and dinoflagellates. The green alga *C. reinhardtii* was the first alga to be genetically modified and has been the focus of most molecular and genetic phycological research (Radakovits et al., 2010). GM of microalgae is being explored to improve the commercial production of biofuels (Henley et al., 2013).

Genetic engineering of photosynthetic cells will considerably enhance the carbon uptake, leading to enhanced acclimation to rapid light changes and boost biomass productivity (Barati et al., 2021). In an algae called *Synechococcus elongates* PCC7942, one synthetic pathway called "malyl-CoA-glycerate (MCG)" was genetically engineered, resulted in two times higher CO_2 assimilation (Barati et al., 2021). Scientists have discovered new ways of using GE yeast for biofuel production, making yeasts more tolerant to the self-produced ethanol (Villarreal et al., 2020). GM of microalgae can increase biofuels production fivefold times (Flynn et al., 2013). So many improvements were achieved through GMs, which include increased lipid and carbohydrate production, improved H2 yields, and the diversion of central metabolic intermediates into fungible biofuels (Radakovits et al., 2010).

The genetic engineering of microalgae is limited mainly due to the unavailability of genetic information for robust and commercially suitable strains (Kumar et al., 2020). One important ecological and public health concern is how escaping mass-cultured GM algal strains may affect natural communities. There is a chance for accidently released GM algae to cause ecological harm and it cannot be ruled out, even with preventive safeguards (Henley et al., 2013). GM algae are generally less fit than their wild-type counterparts due to their special nutrient requirements and excess transgene-load. In addition, countermeasures can be incorporated into the design of GM algae to reduce their survival potential should they escape into the natural environment (Kumar, 2015).

24. Perspectives of algae materials 2.0

Due to the presence of numerous bioactive compounds such as lipids, carbohydrate, proteins, and pigments, algae are having enormous applications. Antioxidants, biological compounds produced by algal cells, can be used to prevent the oxidative damage caused by reactive oxygen species and can delay aging and several chronic conditions (de Souza et al., 2019). Microalgae are good source for natural antioxidants, but due to their enormous biodiversity, most of them remain largely unexplored and some are considered to be higher than some plants (Li et al., 2007). Algae are also good source for proteins. Proteins present in microalgae compete with that of conventional foods, such as soy, eggs, and fish in terms of quantity and quality (Ejike et al., 2016). Carbohydrates from algal biomass are generally used for the production of biofuels (de Souza et al., 2019). While some species are considered as a good source of biofuel as they can naturally accumulate a considerable amount of starch and lipid, some need to be genetically modified for improved biofuel production. Advancement in metabolic and genetic engineering could help in improving the algal strain for enhanced biofuel production (Khan & Fu, 2020). Pigments produced by algae mainly, carotenoids, chlorophylls, and phycobiliproteins are utilized as animal feed, additives, cosmetics, pharmaceutical, food colorants, and biomaterials (Chew et al., 2017). Pigment production is influenced by several environmental factors such as temperature, irradiation, wavelength, photoperiod, pH, nutrient limitation, nitrogen, salinity, pesticides, and heavy metals (Fatma, 2009). Traditionally, algae are used as fertilizer in agriculture. Nowadays, they are utilized as biostimulants that promote growth and tolerance to crops. Furthermore, they contain bioactive compounds that maintain the health of domestic animals (Fleurence, 2021).

Use of microalgae for the production of high-value products is generally safe; however, it must be grown in secure environments to avoid contamination (de Souza et al., 2019). Even though algal biomass has immense applications, biomass conversion to valuable products at industrial level is a bit challenging task (de Souza et al., 2019). It is important to minimize the culturing cost for making the biomass conversion process more feasible at industrial scale (Acién et al., 2012). Coupling algal biomass production with wastewater treatment is preferred as a sustainable choice for nutrient recycling and carbon sequestration (Renuka et al., 2021).

25. Algae biorefinery: strategies for a sustainable industry

Algae are a diverse group of autotrophic organisms from unicellular single-celled to large multicellular forms found in various environmental conditions. Algae biomass is rich in essential components such as carbohydrates, lipids, and protein, making it a viable feedstock for biofuels and commercially valuable by-products (Khoo et al., 2019); this process is known to be an algal biorefinery. Increased price and depletion of fossil fuels lead to the new sustainable renewable energy sources, which should be environmentally friendly and produced cost-effectively. Algae are known to be the third-generation feedstock and are a promising feedstock for biofuels production: algal biofuels, mainly bioethanol, biodiesel, biomethane, hydrogen, and hydrocarbons (Hessami et al., 2018). Microalgae contain large amounts of lipids; by transesterification, it converted into fatty acid methyl esters for biodiesel production (Singh & Dhar, 2019). Macroalgae or seaweeds contain 50%−70% carbohydrates on a dry weight basis, making them suitable feedstock for bioethanol production (Sudhakar et al., 2017). Compared with first- and second-generation feedstocks, algae do not compete with food crops and do not require any arable land and freshwater supplies (Kraan, 2012), which adds more value to algae as a promising feedstock for biofuel production. For the production of biofuels, different pretreatments and hydrolysis techniques were used to produce maximum yield.

Apart from biofuels from algae, it also produces many by-products, which found applications in food, cosmetics, and pharmaceutical/nutraceutical industries. Algae contain different pigments: chlorophyll in green, carotenoids in brown, and phycobiliproteins in red algae. Pigments, especially phycobiliproteins used as a food colorant, a natural dye that replaces artificial colorings, are reported to be toxic to the human body (Galetovi et al., n.d., pp. 1−13). Apart from food colorants reported various medicinal importance (Mysliwa-kurdziel & Solymosi, 2014). Like red algae, brown algae pigments also have many health benefits (Pangestuti & Kim, 2011). Algae contain proteins up to 60% in some algae; it found applications in animal/fish feeds, fertilizers, industrial enzymes, bioplastics, and surfactants. Algae have various types of polysaccharides, and it reported various economic and commercial importance in different fields (Kraan, 2012).

26. Algae material for the development of aquaculture feeds

Aquaculture is the fastest-growing segment of the food industry. Nearly 68% of commercial species rely on fish feed, and the demand for fish feed has increased by 300% in the past 10 years (Nagappan et al., 2021). Fish meal and fish oil are abundantly used in fish feed. The crop plant−derived protein sources commonly used in aquafeed have low digestibility and are deficient in certain essential amino acids such as lysine, methionine, threonine, and tryptophan (Norambuena et al., 2015). Feed for aquatic species from a single source will not contain all the nutrition such as carbohydrates, protein, fat, minerals, and vitamin levels so that it can fulfill the nutritional requirements of an aquatic species (Nagappan et al., 2021).

Algae are the natural foods of bivalves and the basis of almost all marine and freshwater aquatic food chains (Benemann, 1992). Algae produce the food resources that fish are adapted to consume. Its carbohydrate content, high-quality proteins, contains all the essential amino acids as well as lipids rich in "PUFAs," or polyunsaturated omega-3 and omega-6 fatty acids appropriate to protect the fish health (Henry, 2012). A readily digestible carbohydrate in microalgal species is content of starch. Species including *Tetraselmis subcordiformis*, *C. reinhardtii* and *C. vulgaris* comparatively have higher starch content (30%−49%) than other microalgae (Nagappan et al., 2021). Cultivated microalgae are used in hatchery production of many farmed finfish, shellfish, and other commercially important aquaculture species, while macroalgae are less widely used but provide an important source of nutrition for certain farmed invertebrates, such as sea urchins and abalone (Shields & Lupatsch, 2012). Species such as *Isochrysis* sp., *Pavlova* sp., and others are used in fish hatcheries as live feed for larvae and microalgae are being investigated as a potential bulk-feed ingredient for fingerlings and adult fishes.

Algae have been used fresh as a whole diet, and dried algal meal has been used as a partial or complete replacement of fishmeal protein in pelleted diets (Hasan & Chakrabarti, 2009, pp. 3−8). Inclusion of small amounts of algae in fish feed resulted in positive effects in growth performance and feed utilization efficiency. Algae meal has been also recommended as a good feed complement for counteracting intestinal inflammation produced by soybean meal (Norambuena et al., 2015). The choice of the algae is an important thing as an aquafeed. It must be low-cost and commercially available. Microalgae such as *Spirulina*, *Chlorella*, and *Dunaliella* can be produced by low-cost open pond technologies and are marketed as dry powders, and their nutritional profiles are well documented (Henry, 2012).

27. Algae and bioremediation

Algae are a wide variety of photosynthetic nonvascular plants, mostly aquatic organisms. Other than aquatics, they can grow in various other environments (Milledge et al., 2014). Based on the size, algae are classified as macro and micro and range in size from small, single-celled forms to large, complex ones. Apart from size, algae are categorized into green, brown, and red algae due to their pigments present. Algae are considered third-generation feedstocks significant in the food, cosmetic, and pharmaceutical/nutraceutical industries. Diverse natural resources, including water and soil, become contaminated due to the release of hazardous chemicals from human-made sources, becoming a major environmental threat (Singhal et al., 2021). Algae are promising for feedstocks for bioremediation of wastewater and other pollutants. Aeration is required for the traditional wastewater treatment methods, which consumes a lot of energy, and wastewater treatment using algae generates oxygen, which helps the aerobic bacteria (Shrestha et al., 2020). Algae have an effective way of nutrient uptake, and they also give aerobic bacteria oxygen through photosynthesis.

Algal bioremediation, the practice of removing excess dissolved nutrients/pollutants from aquaculture effluents using living algae, is commonly regarded as an effective and inexpensive wastewater treatment technique (Lawton et al., 2013) and does not form any secondary pollutants. Temperature, pH, redox potential, nutrient status, moisture, and the chemical makeup of heavy metals are some of the elements that affect and limit the effectiveness of bioremediation (Igiri et al., 2018). Choosing the suitable algae species and strains will enable bioremediation to reduce dissolved nutrient loads in aquaculture effluents as effectively as possible (Lawton et al., 2013). Algae should have a high growth rate and grow in a wide range of conditions, with large productivity and mass availability (Lawton et al., 2013).

Microalgae are capable of bioremediation of heavy metals, petroleum hydrocarbons, pesticides, toxins, polychlorinated biphenyls, polycyclic aromatic hydrocarbons, explosives, textile dyeing effluent, and radioactive compounds (Alaguprathana & Poonkothai, 2017; Ben Chekroun et al., 2014; Priyadarshani et al., 2012). Bioremediation is divided into two groups based on the transportation and removal of waste. With the use of procedures such as composting, bioreactors, etc., pollutants are treated outside of the contaminated site during ex situ bioremediation, and with the aid of techniques such as biostimulation, bioventing, bioaugmentation, etc., pollutants are treated inside the contaminant during in situ bioremediation (Singhal et al., 2021).

References

Abboud, Y., Saffaj, T., Chagraoui, A., El Bouari, A., Brouzi, K., Tanane, O., & Ihssane, B. (2014). Biosynthesis, characterization and antimicrobial activity of copper oxide nanoparticles (CONPs) produced using brown alga extract (*Bifurcaria bifurcata*). *Applied Nanoscience, 4*, 571–576. https://doi.org/10.1007/s13204-013-0233-x

Acién, F. G., Fernández, J. M., Magán, J. J., & Molina, E. (2012). Production cost of a real microalgae production plant and strategies to reduce it. *Biotechnology Advances, 30*, 1344–1353. https://doi.org/10.1016/j.biotechadv.2012.02.005

Agrawal, S. C. (2012). Factors controlling induction of reproduction in algae-review: The text. *Folia Microbiologica, 57*, 387–407. https://doi.org/10.1007/s12223-012-0147-0

Alaguprathana, M., & Poonkothai, M. (2017). Bioremediation of textile dyeing effluent using algae-A review. *Journal of Advances in Microbiology, 7*, 1–12. https://doi.org/10.9734/jamb/2017/37322

Alfarisi, M. S. (2020). Sustainable bioethanol production from microalgae through ionic liquid as a potential catalyst: Review. *AIP Conference Proceedings, 2223*. https://doi.org/10.1063/5.0000952

Al-Radadi, N. S., Hussain, T., Faisal, S., & Ali Raza Shah, S. (2022). Novel biosynthesis, characterization and biocatalytic potential of green algae (*Spirogyra hyalina*) mediated silver nanomaterials. *Saudi Journal of Biological Sciences, 29*, 411–419. https://doi.org/10.1016/j.sjbs.2021.09.013

Alves, C., Silva, J., Pinteus, S., Gaspar, H., Alpoim, M. C., Botana, L. M., & Pedrosa, R. (2018). From marine origin to therapeutics: The antitumor potential of marine algae-derived compounds. *Frontiers in Pharmacology, 9*, 1–24. https://doi.org/10.3389/fphar.2018.00777

Anand, V., Kashyap, M., Samadhiya, K., & Kiran, B. (2019). Strategies to unlock lipid production improvement in algae. *International Journal of Environmental Science and Technology, 16*, 1829–1838. https://doi.org/10.1007/s13762-018-2098-8

Antonacci, A., & Scognamiglio, V. (2020). Biotechnological advances in the design of algae-based biosensors. *Trends in Biotechnology, 38*, 334–347. https://doi.org/10.1016/j.tibtech.2019.10.005

Apurav, A., Koyande, K., & Chew, K. W. (2019). Microalgae: A potential alternative to health supplementation for humans. *Food Science and Human Wellness*. https://doi.org/10.1016/j.fshw.2019.03.001

Araújo, R., Calderón, F. V., López, J. S., Azevedo, I. C., Bruhn, A., Fluch, S., Tasende, M. G., Ghaderiardakani, F., Ilmjärv, T., Laurans, M., & Monagail, M. M. (2021). Current status of the algae production industry in Europe: An emerging sector of the blue. *Bioeconomy, 7*, 1–24. https://doi.org/10.3389/fmars.2020.626389

Ariede, M. B., Candido, T. M., Jacome, A. L. M., Velasco, M. V. R., de Carvalho, J. C. M., & Baby, A. R. (2017). Cosmetic attributes of algae-A review. *Algal Research, 25*, 483–487. https://doi.org/10.1016/j.algal.2017.05.019

Asimakis, E., Shehata, A. A., Eisenreich, W., Acheuk, F., Lasram, S., Basiouni, S., Emekci, M., Ntougias, S., Taner, G., May-Simera, H., Yilmaz, M., & Tsiamis, G. (2022). Algae and their metabolites as potential biopesticides. *Microorganisms, 10*. https://doi.org/10.3390/microorganisms10020307

Aslam, A., Bahadar, A., Liaquat, R., Saleem, M., Waqas, A., & Zwawi, M. (2021). Algae as an attractive source for cosmetics to counter environmental stress. *Science of the Total Environment, 772*, 144905. https://doi.org/10.1016/j.scitotenv.2020.144905

Bak, U. G., & Infante, J. (2020). Technical challenges for offshore cultivation of kelp species. *Lessons Learned and Future Directions, 63*, 341–353.

Barati, B., Zeng, K., Baeyens, J., Wang, S., Addy, M., Gan, S. Y., & El-Fatah Abomohra, A. (2021). Recent progress in genetically modified microalgae for enhanced carbon dioxide sequestration. *Biomass and Bioenergy, 145*, 105927. https://doi.org/10.1016/j.biombioe.2020.105927

Bari, A. (2017). Our Oceans and the blue economy: Opportunities and challenges. *Procedia Engineering, 194*, 5–11. https://doi.org/10.1016/j.proeng.2017.08.109

Becker, W. (2004). Microalgae in human and animal nutrition. In *Handbook of microalgal culture: Biotechnology and applied phycology* (pp. 312–351).

Ben Chekroun, K., Sánchez, E., & Baghour, M. (2014). The role of algae in bioremediation of organic pollutants. *International Research Journal of Public and Environmental Health, 1*, 19–32.

Benemann, J. R. (1992). Microalgae aquaculture feeds. *Journal of Applied Phycology, 4*, 233–245. https://doi.org/10.1007/BF02161209

Bharathkumar, B., Bharathiraja, G., & Vijayarangan, J. (2018). *Study of mechanical properties of algae filler vinylester composite*.

Bruna, A., Juliana, V., Moreira, B., Morais, G. De, Alberto, J., & Costa, V. (2015). Microalgae as a new source of bioactive compounds in food supplements. *Current Opinion in Food Science*. https://doi.org/10.1016/j.cofs.2015.12.006

Cacho, C. V., Conrad, S. R., Brown, D. R., Riggs, A., Gardner, K., Li, L., Laicher-edwards, D., Tischler, L., Hoffman, R., Brown, T., & Sanders, C. J. (2021). Local geomorphological gradients affect sedimentary organic carbon storage : A blue carbon case study from sub-tropical Australia. *Regional Studies in Marine Science, 45*, 101840. https://doi.org/10.1016/j.rsma.2021.101840

Carina, D., Sharma, S., Jaiswal, A. K., & Jaiswal, S. (2021). Seaweeds polysaccharides in active food packaging: A review of recent progress. *Trends in Food Science & Technology, 110*, 559–572. https://doi.org/10.1016/j.tifs.2021.02.022

Chakdar, H., & Pabbi, S. (2017). *Algal pigments for human health and cosmeceuticals, algal green chemistry.* Elsevier B.V. https://doi.org/10.1016/B978-0-444-63784-0.00009-6

Chavan, K. J., Chouhan, S., Jain, S., Singh, P., Yadav, M., & Tiwari, A. (2014). Environmental factors influencing algal biodiesel production. *Environmental Engineering Science, 31*, 602–611. https://doi.org/10.1089/ees.2014.0219

Chaudhuri, S., & Sole Author-Chaudhuri, S. (2022). The solution to pollution: effective lead removal via algal biosorption. *Youth and water Security in Africa*, 75.

Chew, K. W., Yap, J. Y., Show, P. L., Suan, N. H., Juan, J. C., Ling, T. C., Lee, D. J., & Chang, J. S. (2017). Microalgae biorefinery: High value products perspectives. *Bioresource Technology, 229*, 53–62. https://doi.org/10.1016/j.biortech.2017.01.006

Chi, W., Zheng, L., He, C., Han, B., Zheng, M., Gao, W., Sun, C., & Zhou, G. (2017). *Quorum sensing of microalgae associated marine Ponticoccus sp . PD _ 2 and its algicidal function regulation.* AMB Express. https://doi.org/10.1186/s13568-017-0357-6

Cikos, A. M., Jokić, S., Šubarić, D., & Jerković, I. (2018). Overview on the application of modern methods for the extraction of bioactive compounds from marine macroalgae. *Marine Drugs, 16*. https://doi.org/10.3390/md16100348

Da Costa, E., Silva, J., Mendonça, S. H., Abreu, M. H., & Domingues, M. R. (2016). Lipidomic approaches towards deciphering glycolipids from microalgae as a reservoir of bioactive lipids. *Marine Drugs, 14*. https://doi.org/10.3390/md14050101

Dahoumane, S. A., Mechouet, M., Wijesekera, K., Filipe, C. D. M., Sicard, C., Bazylinski, D. A., & Jeffryes, C. (2017). Algae-mediated biosynthesis of inorganic nanomaterials as a promising route in nanobiotechnology-a review. *Green Chemistry, 19*, 552–587. https://doi.org/10.1039/c6gc02346k

Daroch, M., Geng, S., & Wang, G. (2013). Recent advances in liquid biofuel production from algal feedstocks. *Applied Energy, 102*, 1371–1381. https://doi.org/10.1016/j.apenergy.2012.07.031

Das, S., Das, S., & Ghangrekar, M. M. (2019). Bioresource technology quorum-sensing mediated signals: A promising multi-functional modulators for separately enhancing algal yield and power generation in microbial fuel cell. *Bioresource Technology, 294*, 122138. https://doi.org/10.1016/j.biortech.2019.122138

Das, V., Tripathi, A. M., Borah, M. J., Dunford, N. T., & Deka, D. (2020). Cobalt-doped CaO catalyst synthesized and applied for algal biodiesel production. *Renewable Energy, 161*, 1110–1119. https://doi.org/10.1016/j.renene.2020.07.040

Diána, L., Szandra, K., Sándor, L., Attila, B., László, K., Nóra, A., Eszter, T., András, S., n.d. Chlorophyll fluorescence instrumentation for a rapid, in situ measurment of algal density, in: 26th international Symposium on analytical and environmental problems experimental. pp. 211–215.

Dillehay, T. D., Ramírez, C., Pino, M., Collins, M. B., Rossen, J., & Pino-Navarro, J. D. (2013). *Monte verde: Seaweed, food, medicine, and the peopling of South America 784*. https://doi.org/10.1126/science.1156533

Dove, A. P., Rodrigues, M. M., Pe, M. M., Mano, F., & Costa, A. M. S. (2020). *Modular functionalization of laminarin to create value-added naturally derived macromolecules*. https://doi.org/10.1021/jacs.0c09489

Ejike, C. E. C. C., Collins, S. A., Balasuriya, N., Swanson, A. K., Mason, B., & Udenigwe, C. C. (2016). Prospects of microalgae proteins in producing peptide-based functional foods for promoting cardiovascular health. *Trends in Food Science & Technology, 59*, 30–36. https://doi.org/10.1016/j.tifs.2016.10.026

El-Chaghaby, G., & Rashad, S. (2021). An overview of algae prospects in cosmeceuticals. *Journal of the Egyptian Women's Dermatologic Society, 18*, 161–166. https://doi.org/10.4103/jewd.jewd_22_21

El-Khatib, E. M., Ali, N. F., & El-Mohamedy, R. S. R. (2016). Enhancing dyeing of wool fibers with colorant pigment extracted from green algae. *Journal of Chemical and Pharmaceutical Research, 8*, 614–619.

Fatma, T. (2009). Screening of cyanobacteria for phycobiliproteins and effect of fferent environmental stress on its yield. *Bulletin of Environmental Contamination and Toxicology, 83*, 509–515. https://doi.org/10.1007/s00128-009-9837-y

Fleurence, J. (2021). Perspectives on the use of algae in agriculture and animal production. *Phycology, 1*, 79–82. https://doi.org/10.3390/phycology1020006

Flynn, K. J., Mitra, A., Greenwell, H. C., & Sui, J. (2013). Monster potential meets potential monster: Pros and cons of deploying genetically modified microalgae for biofuels production. *Interface Focus, 3*. https://doi.org/10.1098/rsfs.2012.0037

Francis, S. (2019). Technical University of Munich researchers explore using algae to make carbon fiber. *Composites World*, 2–3.

Galetovi, A., Seura, F., Gallardo, V., Cort, J., Benito, G., n.d. Use of phycobiliproteins from Atacama cyanobacteria as food colorants in a dairy beverage prototype 1–13.

Gani, P., Sunar, N. M., Matias-peralta, H. M., & Apandi, N. (2019). An overview of environmental Factor's effect on the growth of microalgae. *The Journal of the Applied Chemistry and Natural Resources, 1*, 1–5.

Ganji, R., & Yenugula, V. N. (2017). *Emission reduction in automobile engine exhaust using bio catalytic converter: An experimental study emission reduction in automobile engine exhaust using bio catalytic converter: An experimental study.* https://doi.org/10.9790/1684-1406022428

Gunathilaka, T. L., Samarakoon, K., Ranasinghe, P., & Peiris, L. D. C. (2020). Antidiabetic potential of marine Brown algae-a mini review. *Journal of Diabetes Research.* https://doi.org/10.1155/2020/1230218

Gunter, P. (2010). *Fibers from algae.*

Hasan, M. R., & Chakrabarti, R. (2009). *Alage FAO. Use algae aquat. macrophytes as feed small-scale aquac.*

Hasanudin, N., Abd Ghani, N., Ab Rahim, A. H., Azman, N. S., Rosdi, N. A., & Masri, A. N. (2022). Optimization and kinetic studies on biodiesel conversion from *Chlorella vulgaris* microalgae using pyrrolidinium-based ionic liquids as a catalyst. *Catalysts, 12*. https://doi.org/10.3390/catal12030277

Hassanali, K. (2020). CARICOM and the blue economy-Multiple understandings and their implications for global engagement. *Marine Policy, 120*, 104137. https://doi.org/10.1016/j.marpol.2020.104137

Hempel, F., Bozarth, A. S., Lindenkamp, N., Klingl, A., Zauner, S., & Linne, U. (2011). Microalgae as bioreactors for bioplastic production. *Microbial Cell Factories, 10*, 81. https://doi.org/10.1186/1475-2859-10-81

Henley, W. J., Litaker, R. W., Novoveská, L., Duke, C. S., Quemada, H. D., & Sayre, R. T. (2013). Initial risk assessment of genetically modified (GM) microalgae for commodity-scale biofuel cultivation. *Algal Research, 2*, 66–77. https://doi.org/10.1016/j.algal.2012.11.001

Henry, H. C. (2012). The use of algae in fish feeds as alternatives to fishmeal. *International Aquafeed, 1–27.*

Hessami, M. J., Phang, S.-M., Salleh, A., & Rabiei, R. (2018). Evaluation of tropical seaweeds as feedstock for bioethanol production. *International Journal of Environmental Science and Technology, 15*, 977–992. https://doi.org/10.1007/s13762-017-1455-3

Igiri, B. E., R., O. S. I., O., I. G., Akabuogu, E. P., Adeyi, A., & Ejiogu, I. K. (2018). Toxicity and bioremediation of heavy metals contaminated ecosystem from tannery wastewater: A review. *Journal of Toxicology, 7*, 609–616. https://doi.org/10.1007/978-3-319-00557-7_50

Isurupremarathna, P. K. B. (2021). *Blue economy strategic issues and opportunity in the Indian ocean: A study based on Sri Lanka blue.*

Jamee, R., & Siddique, R. (2019). *Biodegradation of synthetic dyes of textile effluent by microorganisms: An environmentally and economically sustainable approach.* https://doi.org/10.1556/1886.2019.00018

Joshi, S., Kumari, R., & Upasani, N. V. (2018). Engineering and technology (A high impact factor). *International Journal of Innovative Research Science, 7*, 1269–1278. https://doi.org/10.15680/IJIRSET.2018.0702038

Kannan, N., & Donnellan, P. (2021). Algae-assisted microbial fuel cells: A practical overview. *Bioresource Technology Reports, 15*, 100747. https://doi.org/10.1016/j.biteb.2021.100747

Kauffman, J. B., Bernardino, A. F., Ferreira, T. O., Giovannoni, L. R., Gomes, L. E. D. O., Romero, D. J., & Kauffman, J. B. (2018). *Carbon stocks of mangroves and salt marshes of the Amazon region, Brazil.*

Khan, S., & Fu, P. (2020). Biotechnological perspectives on algae: A viable option for next generation biofuels. *Current Opinion in Biotechnology, 62*, 146–152. https://doi.org/10.1016/j.copbio.2019.09.020

Khan, M. I., Shin, J. H., & Kim, J. D. (2018). The promising future of microalgae: Current status, challenges, and optimization of a sustainable and renewable industry for biofuels, feed, and other products. *Microbial Cell Factories, 17*, 1–21. https://doi.org/10.1186/s12934-018-0879-x

Khavari, F., Saidijam, M., Taheri, M., & Nouri, F. (2021). Microalgae: Therapeutic potentials and applications. *Molecular Biology Reports, 48*, 4757–4765. https://doi.org/10.1007/s11033-021-06422-w

Khoo, C. G., Dasan, Y. K., Lam, M. K., & Lee, K. T. (2019). Algae biorefinery: Review on a broad spectrum of downstream processes and products. *Bioresource Technology, 292*, 121964. https://doi.org/10.1016/j.biortech.2019.121964

Klinthong, W., Yang, Y. H., Huang, C. H., & Tan, C. S. (2015). A review: Microalgae and their applications in CO_2 capture and renewable energy. *Aerosol and Air Quality Research, 15*, 712–742. https://doi.org/10.4209/aaqr.2014.11.0299

Kontominas, M. G. (2020). Use of alginates as food packaging materials. *Foods, 9*. https://doi.org/10.3390/foods9101440

Koop, F. (2021). Algae-based material could revolutionize the fashion industry. *Environmental Issues*, 1–12.

Koumoulos, E. P., Trompeta, A. F., Santos, R. M., Martins, M., Dos Santos, C. M., Iglesias, V., Böhm, R., Gong, G., Chiminelli, A., Verpoest, I., Kiekens, P., & Charitidis, C. A. (2019). Research and development in carbon fibers and advanced high-performance composites supply chain in Europe: A roadmap for challenges and the industrial uptake. *Journal of Composites Science, 3*. https://doi.org/10.3390/jcs3030086

Kraan, S. (2012). *Algal polysaccharides, novel applications and Outlook.*

Kumar, S. (2015). GM algae for biofuel production: Biosafety and risk assessment. *Collection of Biosafety Review.*

Kumar, G., Shekh, A., Jakhu, S., Sharma, Y., Kapoor, R., & Sharma, T. R. (2020). Bioengineering of microalgae: Recent advances, perspectives, and regulatory challenges for industrial application. *Frontiers in Bioengineering and Biotechnology, 8*. https://doi.org/10.3389/fbioe.2020.00914

Kuznetsova, T. A., Andryukov, B. G., Besednova, N. N., Zaporozhets, T. S., & Kalinin, A. V. (2020). Marine algae polysaccharides as basis for wound dressings, drug delivery, and tissue engineering: A review. *Journal of Marine Science and Engineering, 8*. https://doi.org/10.3390/JMSE8070481

Lade, B. D., Gogle, D. P., Lade, D. B., Moon, G. M., Nandeshwar, S. B., & Kumbhare, S. D. (2019). Nano-biopesticide formulations: Application strategies today and future perspectives. *Nano-Biopesticides Today and Future Perspectives*, 179–206. https://doi.org/10.1016/B978-0-12-815829-6.00007-3

Lami, R. (2019). *Quorum sensing in marine biofilms and environments, quorum sensing: Molecular mechanism and biotechnological application.* https://doi.org/10.1016/B978-0-12-814905-8.00003-4

Lam, M. K., & Lee, K. T. (2013). *Scale-up and commercialization of algal cultivation and biofuel production, biofuels from algae.* Elsevier B.V. https://doi.org/10.1016/B978-0-444-59558-4.00012-7

Lawton, R. J., Mata, L., de Nys, R., & Paul, N. A. (2013). Algal bioremediation of waste waters from land-based aquaculture using *Ulva*: Selecting Target species and strains. *PLoS One, 8*. https://doi.org/10.1371/journal.pone.0077344

Lee, K. Y., & Mooney, D. J. (2013). *Alginate: Properties and biomedical applications* (Vol. 37, pp. 106–126). https://doi.org/10.1016/j.progpolymsci.2011.06.003.Alginate

Liang, Y., Kashdan, T., Sterner, C., Dombrowski, L., Kröger, M., & Höfer, R. (2015). *Algal biorefineries, industrial biorefineries and white biotechnology.* Elsevier B.V. https://doi.org/10.1016/B978-0-444-63453-5.00002-1

Li, H. B., Cheng, K. W., Wong, C. C., Fan, K. W., Chen, F., & Jiang, Y. (2007). Evaluation of antioxidant capacity and total phenolic content of different fractions of selected microalgae. *Food Chemistry, 102*, 771−776. https://doi.org/10.1016/j.foodchem.2006.06.022

Li, W., Wang, S., Zhong, D., Du, Z., & Zhou, M. (2021). A bioactive living hydrogel: Photosynthetic bacteria mediated Hypoxia elimination and bacteria-killing to promote infected wound healing. *Advances in Therapy, 4*, 1−9. https://doi.org/10.1002/adtp.202000107

Lopez-Hortas, L., Florez-Fernandez, N., Torres, M. D., Ferreira-Anta, T., Casas, M. P., Balboa, E. M., Falque, E., & Domínguez, H. (2021). Applying seaweed compounds in cosmetics, cosmeceuticals and nutricosmetics. *Marine Drugs, 19*, 1−30. https://doi.org/10.3390/md19100552

Mandari, V., & Devarai, S. K. (2021). Biodiesel production using homogeneous, heterogeneous, and enzyme catalysts *via* transesterification and esterification reactions: A critical review. *Bioenergy Resource*, 935−961. https://doi.org/10.1007/s12155-021-10333-w

Mathijsen, D. (2020). Algae to sustainably produce carbon fiber and simultaneously take CO_2 out of the atmosphere. *Reinforced Plastics, 64*, 50−53. https://doi.org/10.1016/j.repl.2019.04.070

Mehrotra, P. (2016). Biosensors and their applications - a review. *Journal of Oral Biology Craniofacial Research, 6*, 153−159. https://doi.org/10.1016/j.jobcr.2015.12.002

Mellor, C., Embling, R., Neilson, L., Rand all, T., Wakeham, C., Lee, M. D., & Wilkinson, L. L. (2022). Consumer knowledge and acceptance of algae as a protein alternative : A UK-based qualitative study. *Foods, 11*, 1−13.

Michalak, I., Dmytryk, A., & Chojnacka, K. (2020). Algae cosmetics. *Encyclopedia of Marine Biotechnology*, 65−85. https://doi.org/10.1002/9781119143802.ch3

Milledge, J. J., Smith, B., Dyer, P. W., & Harvey, P. (2014). Macroalgae-derived biofuel: A review of methods of energy extraction from seaweed biomass. *Energies, 7*, 7194−7222. https://doi.org/10.3390/en7117194

Mir, R. A., Adeel, S., Azeem, M., Batool, F., Khan, A. A., Gul, S., & Iqbal, N. (2019). Green algae, Cladophora glomerata L. Based natural colorants: Dyeing optimization and mordanting for textile processing. *Journal of Applied Phycology, 31*, 2541−2546. https://doi.org/10.1007/s10811-018-1717-6

Moghazy, R. M., Labena, A., Husien, S., Mansor, E. S., & Abdelhamid, A. E. (2020). International Journal of Biological Macromolecules Neoteric approach for efficient eco-friendly dye removal and recovery using algal-polymer biosorbent sheets: Characterization, factorial design, equilibrium and kinetics. *International Journal of Biological Macromolecules, 157*, 494−509. https://doi.org/10.1016/j.ijbiomac.2020.04.165

Mona, S., Yazhini, M., Fakhruddin Shaukat, P., & Chandra Sekarenthiran S, M. S. (2019). Extraction of algal pigments and their suitability as natural dyes. *Journal of Algal Biomass Utilization, 10*, 1−8.

Morales, M., Aflalo, C., & Bernard, O. (2021). Microalgal lipids: A review of lipids potential and quantification for 95 phytoplankton species. *Biomass and Bioenergy, 150*, 106108. https://doi.org/10.1016/j.biombioe.2021.106108

Mourelle, M. L., Gómez, C. P., & Legido, J. L. (2017). The potential use of marine microalgae and cyanobacteria in cosmetics and thalassotherapy. *Cosmetics, 4*. https://doi.org/10.3390/cosmetics4040046

Msuya, F. E., Bolton, J., Pascal, F., Narrain, K., Nyonje, B., & Cottier-Cook, E. J. (2022). Seaweed farming in Africa: Current status and future potential. *Journal of Applied Phycology, 34*, 985−1005. https://doi.org/10.1007/s10811-021-02676-w

Mysliwa-kurdziel, B., & Solymosi, K. (2014). *Phycobilins and phycobiliproteins used in food industry and medicine* (Vol. 1).

Nagappan, S., Das, P., AbdulQuadir, M., Thaher, M., Khan, S., Mahata, C., Al-Jabri, H., Vatland, A. K., & Kumar, G. (2021). Potential of microalgae as a sustainable feed ingredient for aquaculture. *Journal of Biotechnology, 341*, 1−20. https://doi.org/10.1016/j.jbiotec.2021.09.003

Norambuena, F., Hermon, K., Skrzypczyk, V., Emery, J. A., Sharon, Y., Beard, A., & Turchini, G. M. (2015). Algae in fish feed: Performances and fatty acid metabolism in juvenile Atlantic Salmon. *PLoS One, 10*, 1–17. https://doi.org/10.1371/journal.pone.0124042

O'Connor, M. C. (2017). Inside the lonely fight against the biggest environmental problem you've never heard of. *The Guardian*, 20–21.

Ong, H. C., Tiong, Y. W., Goh, B. H. H., Gan, Y. Y., Mofijur, M., Fattah, I. M. R., Chong, C. T., Alam, M. A., Lee, H. V., Silitonga, A. S., & Mahlia, T. M. I. (2021). Recent advances in biodiesel production from agricultural products and microalgae using ionic liquids: Opportunities and challenges. *Energy Conversion and Management, 228*, 113647. https://doi.org/10.1016/j.enconman.2020.113647

Pangestuti, R., & Kim, S. (2011). Biological activities and health benefit effects of natural pigments derived from marine algae. *Journal of Functional Foods, 3*, 255–266. https://doi.org/10.1016/j.jff.2011.07.001

Paniagua-Michel, J. de J., Olmos-Soto, J., & Morales-Guerrero, E. R. (2014). Algal and microbial exopolysaccharides: New insights as biosurfactants and bioemulsifiers. *Advances in Food & Nutrition Research*. https://doi.org/10.1016/B978-0-12-800268-1.00011-1

Pasdaran, A., Hamedi, A., & Mamedov, N. (2016). Antibacterial and insecticidal activity of volatile compounds of three algae species of Oman Sea. *International Journal of Secondary Metabolite*, 66–73. https://doi.org/10.21448/http-ijate-net-index-php-ijsm.243308

Pedruzi, G. O. L., Amorim, M. L., Santos, R. R., Martins, M. A., & Vaz, M. G. M. V. (2020). *Revista Brasileira de Engenharia Agrícola e Ambiental Biomass accumulation-influencing factors in microalgae farms*.

Pimentel, F. B., Alves, R. C., Rodrigues, F., & Oliveira, M. B. P. P. (2018). Macroalgae-derived ingredients for cosmetic industry-an update. *Cosmetics, 5*, 4–9. https://doi.org/10.3390/cosmetics5010002

Priyadarshani, I., & Rath, B. (2012). Commercial and industrial applications of micro algae-A review. *Journal of Algal Biomass Utilization, 3*, 89–100.

Priyadarshani, I., Sahu, D., & Rath, B. (2012). Microalgal bioremediation. *Current practices and perspectives, 3*, 299–304.

Prudkin-Silva, C., Lanzarotti, E., Álvarez, L., Vallerga, M. B., Factorovich, M., Morzan, U. N., Gómez, M. P., González, N. P., Acosta, Y. M., Carrizo, F., Carrizo, E., Galeano, S., Lagorio, M. G., Juárez, Á. B., Ithuralde, R. E., Romero, J. M., & Urdampilleta, C. M. (2021). A cost-effective algae-based biosensor for water quality analysis: Development and testing in collaboration with peasant communities. *Environmental Technology & Innovation, 22*, 101479. https://doi.org/10.1016/j.eti.2021.101479

Qureshi, D., Nayak, S. K., Maji, S., Kim, D., Banerjee, I., & Pal, K. (2019). Carrageenan: A wonder polymer from marine algae for potential drug delivery applications. *Current Pharmaceutical Design, 25*, 1172–1186. https://doi.org/10.2174/1381612825666190425190754

Radakovits, R., Jinkerson, R. E., Darzins, A., & Posewitz, M. C. (2010). Genetic engineering of algae for enhanced biofuel production. *Eukaryotic Cell, 9*, 486–501. https://doi.org/10.1128/EC.00364-09

Rahman, P. K. S. M., Mayat, A., Harvey, J. G. H., Randhawa, K. S., Relph, L. E., & Armstrong, M. C. (2019). Biosurfactants and bioemulsifiers from marine algae. *The role of microalgae wastewater treatment* (pp. 169–188). https://doi.org/10.1007/978-981-13-1586-2_13

Rahmati, M., Alipanahi, Z., & Mozafari, M. (2019). Emerging biomedical applications of algal polysaccharides. *Current Pharmaceutical Design, 25*, 1335–1344. https://doi.org/10.2174/1381612825666190423160357

Rehman, S., Ali, M., Zuber, M., Zia, K. M., & Iqbal, R. (2017). *Future prospects of algae-based materials, algae based polymers, blends, and composites: Chemistry, biotechnology and materials science*. Elsevier Inc. https://doi.org/10.1016/B978-0-12-812360-7.00020-3

Renuka, N., Ratha, S. K., Kader, F., Rawat, I., & Bux, F. (2021). Insights into the potential impact of algae-mediated wastewater beneficiation for the circular bioeconomy: A global perspective. *Journal of Environmental Management, 297*, 113257. https://doi.org/10.1016/j.jenvman.2021.113257

Rutherford, S. T., & Bassler, B. L. (2012). *Bacterial quorum sensing: Its role in virulence and possibilities for its control*.

Sami, N., Ahmad, R., & Fatma, T. (2021). *Exploring algae and cyanobacteria as a promising natural source of antiviral drug against SARS-CoV-2*.

Santos, M. A., & Grenha, A. (2015). Polysaccharide nanoparticles for protein and peptide delivery: Exploring less-known materials. In *Protein and peptide nanoparticles for drug delivery* (1st ed.). Elsevier Inc. https://doi.org/10.1016/bs.apcsb.2014.11.003

Sathasivam, R., Radhakrishnan, R., Hashem, A., & Abd Allah, E. F. (2019). Microalgae metabolites: A rich source for food and medicine. *Saudi Journal of Biological Sciences, 26*, 709−722. https://doi.org/10.1016/j.sjbs.2017.11.003

Sekar, S. (2008). *Phycobiliproteins as a commodity : Trends in applied research, patents and commercialization*. https://doi.org/10.1007/s10811-007-9188-1

Shields, R. J., & Lupatsch, I. (2012). Algae for aquaculture and animal feeds. *Technology of Theory and Practice, 21*.

Shrestha, S., Singh, R., & Chaurasiya, R. S. (2020). Bioremediation of yamuna water using algae. *Journal of Scientific and Industrial Research, 79*, 758−760.

Singhal, M., Jadhav, S., Sonone, S. S., & Sankhla, M. S. (2021). Microalgae based sustainable bioremediation of water contaminated by pesticides. *Biointerface Research in Applied Chemistry, 12*, 149−169. https://doi.org/10.33263/briac121.149169

Singh, J., & Dhar, D. W. (2019). Overview of carbon capture technology: Microalgal biorefinery concept and state-of-the-art. *Frontiers in Marine Science, 6*, 1−9. https://doi.org/10.3389/fmars.2019.00029

Singh, J., Dutta, T., Kim, K. H., Rawat, M., Samddar, P., & Kumar, P. (2018). Green synthesis of metals and their oxide nanoparticles: Applications for environmental remediation. *Journal of Nanobiotechnology, 16*, 1−24. https://doi.org/10.1186/s12951-018-0408-4

de Souza, M. P., Hoeltz, M., Gressler, P. D., Benitez, L. B., & Schneider, R. C. S. (2019). Potential of microalgal bioproducts: General perspectives and main challenges. *Waste and Biomass Valorization, 10*, 2139−2156. https://doi.org/10.1007/s12649-018-0253-6

Steven, S., Octiano, I., & Mardiyati, Y. (2020). Cladophora algae cellulose and starch based bio-composite as an alternative for environmentally friendly packaging material. *AIP Conference Proceedings, 2262*. https://doi.org/10.1063/5.0015845

Sudhakar, M. P., Jegatheesan, A., Poonam, C., Perumal, K., & Arunkumar, K. (2017). Biosacchari fi cation and ethanol production from spent seaweed biomass using marine bacteria and yeast. *Renewable Energy, 105*, 133−139.

Sudhakar, M. P., Kumar, B. R., Mathimani, T., & Arunkumar, K. (2019). A review on bioenergy and bioactive compounds from microalgae and macroalgae-sustainable energy perspective. *Journal of Cleaner Production, 228*, 1320−1333. https://doi.org/10.1016/j.jclepro.2019.04.287

Sun, Y., Cooke, P., Reddy, H. K., Muppaneni, T., Wang, J., Zeng, Z., & Deng, S. (2017). 1-Butyl-3-methylimidazolium hydrogen sulfate catalyzed in-situ transesterification of Nannochloropsis to fatty acid methyl esters. *Energy Conversion and Management, 132*, 213−220. https://doi.org/10.1016/j.enconman.2016.10.071

Tabatabaei, M., Karimi, K., Horváth, I. S., & Kumar, R. (2015). Recent trends in biodiesel production. *Biofuel Research Journal, 2*, 258−267. https://doi.org/10.18331/BRJ2015.2.3.4

Terrado, R., Pasulka, A. L., Lie, A. A., Orphan, V. J., Heidelberg, K. B., & Caron, D. A. (2017). *Autotrophic and heterotrophic acquisition of carbon and nitrogen by a mixotrophic chrysophyte established through stable isotope analysis* (Vol. 11, pp. 2022−2034). https://doi.org/10.1038/ismej.2017.68

Trivedi, J., Aila, M., Bangwal, D. P., Kaul, S., & Garg, M. O. (2015). *Algae based biorefinery-How to make sense?* (Vol. 47, pp. 295−307). https://doi.org/10.1016/j.rser.2015.03.052

Udayan, A., Pandey, A. K., Sharma, P., Sreekumar, N., & Kumar, S. (2021). Emerging industrial applications of microalgae: Challenges and future perspectives. *Systems Microbiology and Biomanufacturing, 1*, 411–431. https://doi.org/10.1007/s43393-021-00038-8

Umar, L., Alexander, F. A., & Wiest, J. (2015). Application of algae-biosensor for environmental monitoring. In *Proceedings of the annual international conference of the IEEE engineering in medicine and biology society* (pp. 7099–7102). EMBS. https://doi.org/10.1109/EMBC.2015.7320028

Valarmathi, N., Ameen, F., Almansob, A., Kumar, P., Arunprakash, S., & Govarthanan, M. (2020). Utilization of marine seaweed Spyridia filamentosa for silver nanoparticles synthesis and its clinical applications. *Materials Letters, 263*, 127244. https://doi.org/10.1016/j.matlet.2019.127244

Venil, C. K., Wahidin, M. A. B., Aruldass, C. A., & Ahmad, W. A. (2017). Production of bacterial pigments in low cost medium and formulation of biodegradable ink. *Indian Journal of Experimental Biology, 55*, 441–447.

Villarreal, J. V., Burgués, C., & Rösch, C. (2020). Acceptability of genetically engineered algae biofuels in Europe: Opinions of experts and stakeholders. *Biotechnology for Biofuels, 13*, 1–21. https://doi.org/10.1186/s13068-020-01730-y

Vona, D., Presti, M. L., Cicco, S. R., Palumbo, F., Ragni, R., & Farinola, G. M. (2017). *Light emitting silica nanostructures by surface functionalization of diatom algae shells with a triethoxysilane-functionalized π-conjugated fluorophore*.

Voznesenskiy, S. S., Popik, A. Y., Gamayunov, E. L., Orlova, T. Y., Markina, Z. V., & Kulchin, Y. N. (2016). Biosensors based on micro-algae for ecological monitoring of the aquatic environment. In *Algae-organisms for imminent biotechnology* (pp. 103–131). https://doi.org/10.5772/62801

Walter, X. A., Greenman, J., Taylor, B., & Ieropoulos, I. A. (2015). Microbial fuel cells continuously fuelled by untreated fresh algal biomass. *Algal Research, 11*, 103–107. https://doi.org/10.1016/j.algal.2015.06.003

Wang, J., Chen, J., & Meng, S. (2013). The effects of environmental factors on the growth and competition of algae. *Asian Agricultural Research, 5*, 86–89, 95.

Zabochnicka-Swiatek, M. (2010). Algae - feedstock of the future. *Archivum Combustionis, 30*, 225–236.

Zeece, M., n.d. Food systems and future directions. https://doi.org/10.1016/B978-0-12-809434-1.00009-8.

Zhang, Y., Zheng, L., Wang, S., Zhao, Y., Xu, X., & Han, B. (2022). *Quorum sensing bacteria in the phycosphere of HAB microalgae and their ecological functions related to cross-kingdom interactions*.

Zhong, D., Du, Z., & Zhou, M. (2021). Algae: A natural active material for biomedical applications. *View, 2*, 20200189. https://doi.org/10.1002/viw.20200189

Zhong, D., Li, W., Qi, Y., He, J., & Zhou, M. (2020). Photosynthetic biohybrid Nanoswimmers system to alleviate tumor Hypoxia for FL/PA/MR imaging-guided enhanced radio-photodynamic synergetic therapy. *Advances in Functional Materials, 30*, 1–11. https://doi.org/10.1002/adfm.201910395

Zhong, D., Zhang, D., Xie, T., & Zhou, M. (2020). Biodegradable microalgae-based carriers for targeted delivery and imaging-guided therapy toward lung metastasis of breast cancer. *Small, 16*, 1–10. https://doi.org/10.1002/smll.202000819

Zhou, J., Lyu, Y., Richlen, M. L., Anderson, D. M., & Cai, Z. (2016). Quorum sensing is a language of chemical signals and plays an ecological role in algal-bacterial interactions. *CRC Critical Reviews in Plant Sciences, 35*, 81–105. https://doi.org/10.1080/07352689.2016.1172461

Zhou, J., Lyu, Y., Richlen, M., Anderson, D. M., Cai, Z., Monitoring, E., Hole, W., & Hole, W. (2017). *Quorum sensing is a language of chemical signals and plays an ecological role in algal-bacterial interactions* (Vol. 35, pp. 81–105). https://doi.org/10.1080/07352689.2016.1172461.Quorum

CHAPTER

2

Algae materials for advanced biofuel production through the cost-effective process and integration of nanocatalysts

Golap Chandra Singha[1], M. Vijayakumar[2], R. Rajkumar[1], K.R. Jayappriyan[2] and S. Pujithaa[3]

[1]*Department of Environmental Sciences, Bharathiar University, Coimbatore, Tamil Nadu, India;* [2]*Centre for Advanced Studies in Botany, University of Madras, Chennai, Tamil Nadu, India;* [3]*Research and Development Wing, Sree Balaji Medical College and Hospital, Bharath Institute of Higher Education, Chennai, Tamil Nadu, India*

1. Introduction

A broad group of eukaryotic and prokaryotic organisms, from single-celled (*Chlorella*, diatoms, etc.) to multicelled, are called algae. Algae may grow to 50 m in length. A wide variety of aquatic ecosystems, including freshwater, seawater and even wastewater, support the growth of algae; these are photosynthetic creatures. They can withstand a wide range of pH, salinity, and temperature as well as different light intensities and conditions in growing environment and can grow alone or in association with other species (Khan et al., 2018). They require easily available sources such as sunlight, CO_2, and some micronutrients (nitrogen, potassium, and phosphorus) for growth. Moreover, algae are able to generate a variety of bioproducts such as lipids, protein, polysaccharides, vitamins, antioxidants, and biologically active compounds. Carbon compounds from algae may be used for biofuels as well as for health supplements, drugs, and cosmetics. The current focus is on microalgae as a promising raw material for producing bioenergy to meet the rising demand for renewable fuels, food, and other value-added products (VAPs) (Khan et al., 2018). Globally, algae are considered a suitable option for producing third-generation fuels, which can fulfill the demand for clean energy. Their viability as a bioresource for biobased products may be enhanced using growth-promoting strategies and genetic modification in the future (Fig. 2.1).

2. Biofuel

The world's need for fuel for energy is driven by its rapidly increasing population. The importance of renewable fuel sources has grown as a result of the unsustainability of nonrenewable energy sources and their constant depletion. Furthermore, excessive use of fossil fuels has resulted in environmental pollution, increased greenhouse gas emissions (GHG), global warming, receding glaciers, rising sea levels, and a loss of biodiversity. These factors have resulted in the demand for alternative, sustainable, cost-effective, and cleaner-burning energy sources. Biofuels are a developing possibility globally as an

Algae Materials: Applications Benefitting Health. https://doi.org/10.1016/B978-0-443-18816-9.00025-3

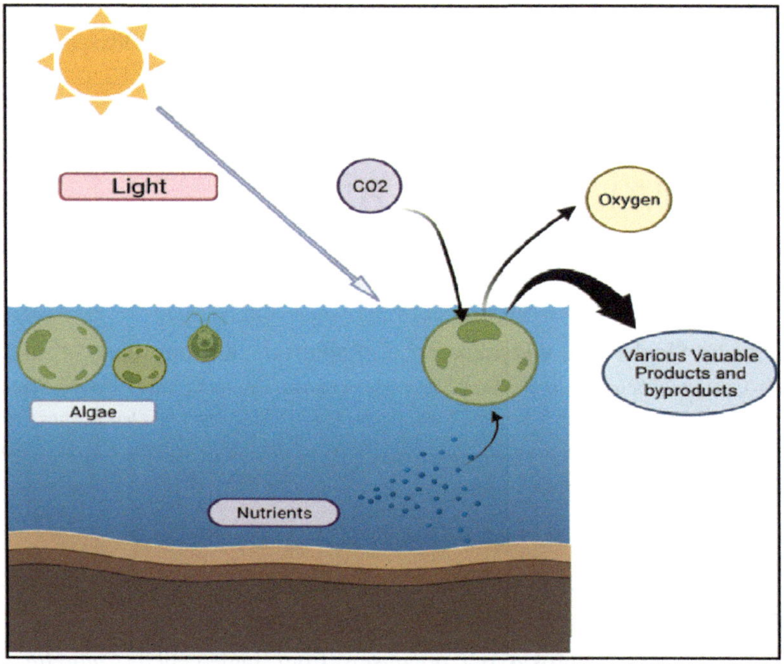

FIGURE 2.1

Microalgae Growth System. Created With BioRender.com.

important option to traditional fossil fuels because they can be produced in large quantities from abundant renewable biomass resources that can be used to generate heat and electricity. Biofuels are also an environmentally and economically sustainable source because they can replace petroleum based transportation fuels and reduce long-term CO_2 emissions. The primary source of increasing temperatures worldwide is the release of GHGs, largely CO_2, through the combustion of fossil fuels. Fossil fuels have released 35.3 billion tons of CO_2 into the atmosphere, or 29 gigatons of CO_2 every year. Biofuels such as fuels from algae have 10% – 45% of oxygen levels and extremely low sulfur emissions, whereas fossil fuels have oxygen levels of zero and substantial emissions of sulfur (Khan et al., 2018). Apart from these emissions, petroleum diesel combustion is a leading producer of additional air pollutants such as NOx, SOx, CO, particulate matter, and volatile organic compounds, all of which are harmful to the environment and cause air pollution. It is possible to eliminate these environmental problems by replacing petroleum diesel fuel with a renewable and sustainable biofuel (Bajhaiya et al., 2010).

The development of biofuel is divided into first, second, and third-generation biofuel, depending on the feedstocks used (Baldev et al., 2018). Food crops (oil seeds, soybeans, sugarcane, wheat, potato, coconut, etc.) are first-generation feedstocks that were used to manufacture biofuels between 2000 and 2008 to meet the world's energy needs. However, significant obstacles include the debate over food versus energy (fuel), the scarcity of water and land needed to produce food, and higher productive costs compared with petroleum. Hence, second-generation biofuel feedstocks have been identified, which cover inedible crops such as bagasse from sugarcane, cassava, miscanthus, leftovers from forest *Jatropha*, vegetational grasses, and municipal refuse garbage. The production of biofuel from this generation's feedstock is regarded as costly because a number of technical barriers need to overcome, as well as facilities for small-scale demonstrations.

Resources for biofuels ought to be less expensive than those for fossil fuels, able to cultivate a small area of land or water, environmentally sustainable, and require only a limited amount of water targeted for irrigation (Batool & Irfan, 2018). Third-generation energy crops, including microalgae and macroalgae, have attracted considerable attention as a novel biomass source for producing renewable energy such as biodiesel, biooil, biosyngas and biohydrogen, which can be produced along with reduced CO_2 emissions, wastewater treatment, and the production of high-value compounds. Advantages of algae over other biomass sources include their high biomass yield per unit of light area, their high starch or oil content, and their ability to grow in diverse habitats, in either marine or fresh waters. In addition, microalgal cultivation and downstream processing (such as harvesting, drying, and thermochemical processing) are expected to improve the cost-effectiveness of biofuel from microalgae (Li et al., 2008). Depending on the production technology and the part of the cells that is employed, algae can be transformed into a variety of fuels.

The lipid or oil content of the algal biomass can be extracted and processed into biodiesel in the same way as other vegetable oil or refined into "drop-in" replacements for petroleum-based fuels in a refinery. Alternatively, the carbohydrate content of algae can be fermented into bioethanol or butanol fuel after lipid extraction. Some types of biofuels include biodiesel, bioethanol, biobutanol, biogasoline, and biogas. Biofuel production also has a direct impact on GHG and economic sustainability. Biofuels have an integral part in developing low-carbon transport, which helps meet emission reduction targets and climate goals. Liquid biofuels, which require minimal changes to fuel distribution infrastructure or the transport fleet, can be rapidly deployed to reduce GHG (Stafford et al., 2017). Owing to the increase in oil demand and the depletion of oil reserves, the development of innovative techniques to produce biofuels from renewable biomass feedstock is gaining importance across the globe.

2.1 Potential of algae as source of biofuel production

The potential for algae in the production of biofuel is augmented by their rapid growth (up to 100 times faster than terrestrial crops). They reach maturity in as few as three days while producing more than half their weight in oil and the ability to thrive in diverse habitats without competing for arable land (Menetrez, 2012). Algae are able to produce a variety of biofuels and can withstand a wide range of environmental conditions, including temperature, dryness, salinity level, anaerobic, photooxidation, osmotic stress, and ultraviolet rays. Compared with land agricultural plants, microalgae are twice as capable of producing oil and they use sunlight more effectively. Because algae require phosphate, nitrogen, and CO_2, microalgae are also considered suitable for reducing pollution and toxic chemicals from waterways and alleviating CO_2 (Batool & Irfan, 2018). In addition, the short harvest cycle of algae is an advantage over conventional crops that are harvested once or twice a year. Algae are an alternative platform for generating renewable energy without adversely affecting agriculture. Algae use solar energy for photosynthesis, rapid growth, and biomass accumulation in the form of proteins, lipids, carbohydrates, hydrocarbons, secondary metabolites, and pigments.

Algae have the potential to convert 9%–10% of energy from sunlight (average solar radiation) theoretically into algal biomass with a yield of 77 g/biomass per m^2 per day (i.e., approximately 280 tons/hectare annually). Many microalga species are suitable for biofuel production because of their high lipid content, which ranges from 50% to 70%. Some, such as *Botryococcus braunii*, accumulate up to 80% of biomass as oil. A study by the Indian Institute of Technology found that algae can contain 15%–77% oil, which makes them an attractive candidate for biofuel production. Apart from lipid, a complete range of algal biomass constituents, such as carbohydrates and proteins, can be converted into biofuels using different manufacturing processes (Batool & Irfan, 2018). Algal biofuels are among the most advanced forms of renewable fuels derived from algal feedstock through biochemical, chemical, and thermochemical conversion methods. The high biomass productivity and high lipid content of algae make them an important biofuel source for the future. In addition to their oil mass, they have the ability to convert almost all feedstock energy into a range of valuable biofuels such as biodiesel, bioethanol, biohydrogen, biooil, and syngas via liquefaction and gasification (Adeniyi et al., 2018). Furthermore, microalgae help to clean the environment by converting carbon dioxide from the atmosphere into oxygen, reducing GHG and pollutants (nitrate and phosphate) and toxic chemicals in water bodies. According to reports, 1 kg of algal biomass is able to fix 1.83 kg of carbon dioxide. In addition, certain algae species use SOx and NOx as nutrients (Gendy & El-Temtamy, 2013). Because of their renewable and ecological features coupled with economic credibility in meeting the worldwide need for transportation fuels, algae have emerged as a promising contender in the quest to replace fossil fuels (Suganya et al., 2016).

2.2 Different techniques for producing biofuel

Several factors need to be considered regarding the cost-effectiveness of algae biofuel manufacturing. Biofuel manufacturing goes through numerous steps, including growing techniques and biomass collection. Several pretreatment activities must be performed after biomass collection to maximize yield. For example, fermentation is used to produce bioethanol. Transesterification and thermoprocesses are used to produce biodiesel and syngas, respectively (Ganesan et al., 2020). Thus, other biofuels are produced with their own respective processes (Fig. 2.2).

Some major types of biofuel production are discussed next.

2.2.1 Production of biodiesel

Biodiesel is an alternative fuel derived from renewable sources such as canola oil, soybean oil, sunflower oil, cottonseed oil, animal fats, and algal lipids. According to international agencies, the demand for energy may increase to 53% by 2030. The high heating values, high flashpoint, and kinematic viscidity of biodiesel make it a better choice over other fossil fuels. Biodiesel is a monoalkyl ester of a long-chain fatty acid derived from these oils used in proprietary diesel engines (Demirbas, 2009). Biodiesel is renewable, biodegradable, atoxic, noncombustible, and environmentally sustainable. It could be beneficial for reducing net CO_2 emissions by 78%. In addition to serving as a fuel on its own or combined with other fuels, its energy content and other characteristics are equal to those of traditional fossil diesel (Batool & Irfan, 2018). Most countries produce biodiesel from soybeans. Canola oil, animal fat, palm oil, corn oil, and waste cooking oil are other sources of commercial biodiesel. Research has shown that microalgae produce more oil than terrestrial plants such as palm, rapeseed, soybeans, and *Jatropha* (Bajhaiya et al., 2010). Studies to explore microalgae as a possible raw material for liquid fuels began in the mid-1980s. To solve the power crisis during 2nd World War, several German scientists attempted to extract lipids from diatoms (Cohen et al., 1995).

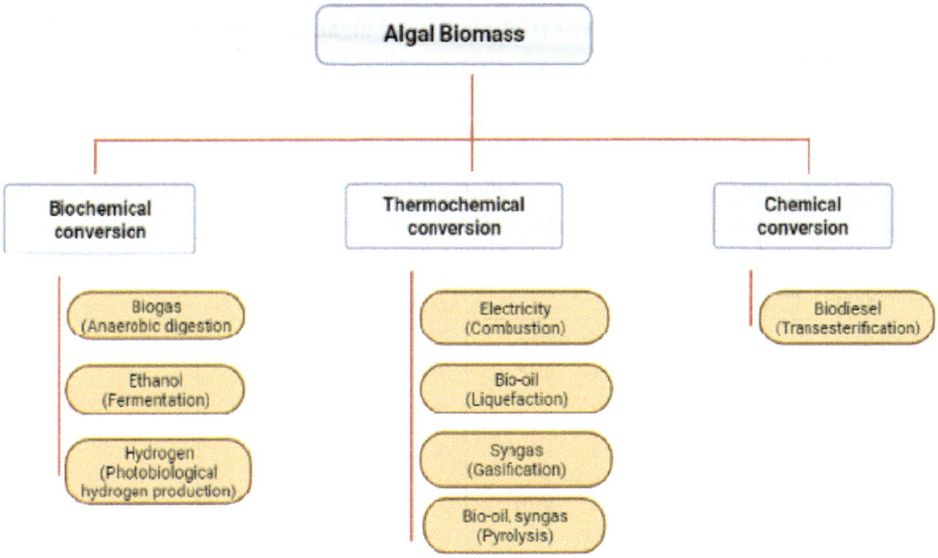

FIGURE 2.2

Different Approaches to Biofuel Production From Algal Biomass. Created With BioRender.com.

Most microalgae species are ideal for biodiesel production owing to their high lipid content of 50%–80%. For instance, the microalga *B. braunii*, accumulates a maximum of 80% oil in its biomass (Mata et al., 2010). Because of the high oil yield of algal biomass compared with oil seeds, it can be economically converted to biodiesel using a variety of techniques. Table 2.1 lists some of the comparative research on algae biomass, and terrestrial plants in generating biodiesel. Petroleum

Table 2.1 Comparative analysis of terrestrial plants and algal biomass for the generation of biodiesel.

Feedstock	Conditions for biodiesel production	Biodiesel (yield)	References
Algae			
Spirulina platensis	55°C at reaction time, concentration of catalysts: algal biomass to methanol ratio of 1:4, intensity of stirring: 450 rpm	60 g/kg of lipid	Nautiyal et al. (2014)
Scenedesmus sp.	Alkaline (NaOH) at 70°C	321.06 g/kg of lipid	Kim et al. (2014)
	Acidic catalysts at 70°C	282.23 g/kg of lipid	
Nannochloropsis salina	Freeze-dried biomass after extraction at 1:1 ratio of chloroform/methanol, and then alkali transesterification	180.78 g of biodiesel/kg of lipid	Muthukumar et al. (2012)

Continued

Table 2.1 Comparative analysis of terrestrial plants and algal biomass for the generation of biodiesel.—cont'd

Feedstock	Conditions for biodiesel production	Biodiesel (yield)	References
Nannochloropsis sp.	N-hexane used to extract oil, and then transesterification under acidic conditions	99 g of biodiesel/kg of lipid	Susilaningsih et al. (2009)
Terrestrial plants			
Pongamia pinnata	Transesterification with help of methanol, catalyst (NaOH), at 60°C	253 g of biodiesel/kg of lipid	Mamilla et al. (2011)
	Alkali-catalyzed esterification and acid-catalyzed esterification using H_2SO_4 (0.5% each)	193.2 g of biodiesel/kg of lipid	Naik et al. (2008)
Soybean	Use of basic catalyst: hydrotalcite, molar ratio of 20:1 (methanol/oil), 10 h reaction time.	189.6 g of biodiesel/kg of lipid	Martin et al. (2013)
Azadirachta indica	60-min reaction duration, 0.7% H_2SO_4 acid catalyst, 50°C reaction, and a 3:1 methanol/oil ratio	170 g of biodiesel/kg of lipid	Awolu and Layokun (2013)
Madhuca indica	Methanol into oil ratio of 0.30–0.35 (v/v), acid catalyst: 1% of v/v H_2SO_4, methanol 0.25 v/v, 0.7% w/v alkali catalyst (KOH)	186.2 g of biodiesel/kg of lipid	Ghadge and Raheman (2005)

productivity (the amount of petroleum generated from a specific unit of microalgae broth volume per day) relies on the algae species rate of growth and biomass level. To synthesize biodiesel, microalgae species with a high amount of polyunsaturated fatty acid methyl ester (FAME), such as *Chlamydocapsa bacillus, Kirchneriella lunaris, Ankistrodesmus falcates*, and *Ankistrodesmus fusiformis*, are usually employed (Nascimento et al., 2013). Approximately 5–15,000 gallons of biodiesel can be generated from algae biomass per acre annually, reflecting the potential (Chisti, 2007).

However, algae-based biodiesel has been approved in terms of physical and chemical properties as a substitute to fossil fuels (Brennan & Owende, 2010). The high content of polyunsaturated fatty acids in algae oil compared with vegetable oils makes it more susceptible to oxidation during storage, further limiting its use (Chisti, 2007). According to the reports of some researchers, differences in the benefits of algae biomass in the production of biodiesel are due to the larger growth rate of biomass and oil production compared with the maximum oil yield crops (Chisti, 2007; Hossain et al., 2008). The production of biodiesel from algae includes harvesting the biomass, drying, extracting oil, and transesterification, as will be discussed subsequently.

2.2.1.1 Algal biomass harvesting and the drying process

There are several methods for harvesting microalgal biomass for biodiesel production, including flocculation, filtration, gravity sedimentation, floatation, and centrifugation. Although the importance of harvesting is related to the economy, no specific standard method of harvesting has been introduced (Batool & Irfan, 2018). Single-celled microalgae build a cell wall that contains fatty acids and lipids, which distinguishes them from other plants and animals. It is important to harvest and dry the algal biomass before extraction using machinery and solvents to recover the bio-oil. Nets can be used to harvest macro-algae, which uses less energy, whereas the microalgae can be filtrated, aggregated, and centrifuged. They can be harvested in several traditional ways such as by foam separation, foam flotation, sedimentation, and separation by ultrasound (Bosma et al., 2003). The type of algae determines the harvesting technique.

To increase the shelf life of preharvest algal biomass and prevent postharvest rot, the process of drying the algal biomass is important (Munir et al., 2013). Drying methods have been applied to microalgal biomass include drum-drying, freeze-drying, sun-drying, and spray-drying (Williams & Laurens, 2010). As reported by Mata et al. (2010), because of the high water content of the biomass, sun-drying is not considered an effective method. *Spirulina* and *Scenedesmus*, which have a water content of roughly 90%, were dried using a conventional sun-drying apparatus. The drying temperature affects the triglyceride content and lipid production when extracting lipids from algal biomass.

2.2.1.2 Oil extraction from algal biomass

The cell wall of single-celled microalgae contains lipids and fatty acids, which differentiates them from other living organisms. Therefore, the process varies in its components. Oil is extracted from algae using mechanical and solvent extraction techniques (Li et al., 2014). However, this is an expensive and power-intensive technique for extracting lipids from microalgae.

2.2.1.2.1 Mechanical approaches to oil extraction. A press that is used to extract oil manually from seeds and nuts may also be used to extract microalgae. This technique requires the algal biomass to be dry. This approach does not require particular expertise and can recover around 75% of the oil (Munir et al., 2013). Using screw machinery, Khedkar et al. (2011) extracted oil from algae using a mechanical press (a piston) and osmotic pressure, resulting in a recovery of around 75% of oil. However, it requires more time for extraction than other approaches, which makes the operation less convenient and efficient.

2.2.1.2.2 Solvent-based oil extraction. Solvent extraction is another method of extracting lipid from microalgae. It includes a two-step solvent extraction system. Typically, solvent extraction extracts nearly all of the oil from the biomass, keeping just 0.5–0.7% of the oil. Thus, this process is considered a better technique for fat and oil extraction than mechanical approaches (Topare et al., 2011). The amount of lipids from extracted microalgal biomass mainly depends on the type of solvent used. Some organic solvents such as chloroform, hexane, cyclohexane, acetone, and benzene are used alone or in combination (Afify et al., 2010). Because organic solvents make oils more soluble compared with water, the solvent interacts with the algal cells to release oil from algal biomass. Hexane is thought to be the most efficient organic solvent owing to its lower cost and less toxicity. Similarly, according to the report of Lee et al. (1998), different solvent solutions were used to extract lipids from the green alga *B. braunii*, and a 2:1 chloroform/methanol ratio produced the highest lipid concentration. Hossain et al. (2008) mixed hexane and ether 1:1 (v/v) to extract oil, and the combination was allowed to stand for 24 h. Using hexane and ethanol in two rounds of extracting lipid and purification, the researchers obtained a lipid recovery rate of over 80% (Fajardo et al., 2007). Therefore, it is necessary to select the most appropriate solvent system for maximum oil extraction.

2.2.1.3 Transesterification

Numerous processes are involved in interactions between triglycerides and alcohol to generate biodiesel from algae oil. This reaction can be carried out with a variety of alcohols including methanol, butanol, propanol, amyl alcohol, and ethanol. Because of their low cost and physical and chemical advantages, ethanol and methanol are often employed in commercial development. The lipase enzymes or inorganic catalysts (alkali and acids) can be employed to accelerate the reaction. To make 3 mol of methyl esters (known as biodiesel) and 1 mol of glycerol (as a by-product), about 3 mol of alcohol is needed for every mole of triglyceride (Chisti, 2007). Because glycerol is thicker than biodiesel, it can be removed from the reactor continuously or regularly to promote equilibrium.

The conversion of triglycerides into methyl esters is exceedingly sluggish when employing acid catalysts, but there have been claims that the alkali-catalyzed transesterification is 44000 times faster than that of acid-catalyzed activity (Mazubert et al., 2013). Two commercially available alkali catalysts used are potassium and sodium hydroxides at a concentration of around 1% oil. Sodium methoxide, rather than sodium hydroxide, has proven to be a superior catalyst (Singh et al., 2006). *Scenedesmus* sp. was employed by Kim et al. (2014) to produce biodiesel using transesterification (both acid and alkali). They found 55.07% $\pm$ 2.18% lipid by weight of biodiesel conversion using NaOH as an alkaline catalyst, compared with 48.41% $\pm$ 0.21% biodiesel synthesis using H_2SO_4. Lipases as biocatalysts have distinct benefits over acid and alkali as catalysts owing to their adaptability, substrate selectivity, region selectivity, enantioselectivity, and strong catalytic performance at ambient pressure and temperature (Knezevic et al., 2004). Some lipases are incapable of hydrolyzing ester bonds at secondary locations, but some are able to hydrolyze both primary and secondary esters. Fatty acid selectivity in another type of lipase enables it to break ester bonds with particular fatty acids. After Luo et al. (2006) cloned the lipase gene lipB68, it was expressed in *Escherichia coli* BL21 and used as a biodiesel catalyst. After 12 h at 20°C, LipB68 could accelerate transesterification activity, yielding 92% biodiesel. The ability of the lipase enzyme to function at such low temperatures could result in significant energy savings. However, because of its exorbitant cost, it is rarely used (Sharma et al., 2001).

2.2.1.3.1 Extractive transesterification. Drying of the algal biomass, disruption of the algal cells, extraction of oil, transesterification, and biodiesel refinement are some procedures involved in this process for biodiesel production (Hidalgo et al., 2013). Primary problems are caused by the biomass over water content (greater than 80%), which drives up the overall expense of the operation.

2.2.1.3.2 In situ transesterification. The oil extraction stage is skipped in this method. Alcohol serves as an esterification reagent and solvent in the extraction process, increasing the cell membrane's porosity. The yields discovered are higher than those obtained using traditional methods, and waste is also decreased. The discharge of extraction solvent during industrial biodiesel synthesis adds to the formation of pollution and global warming. As a result, simplifying esterification processes can help to mitigate the downsides of appealing renewable fuel. Only one method may be tempting as a way to lessen the overall amount of chemicals and energy used to produce biodiesel (Patil et al., 2012). The differences between two transesterification processes are listed in Table 2.2.

Table 2.2 In situ and extraction transesterification processes.

Sl. No.	Extractive transesterification	In situ transesterification
1	Value of heat: low	High heat value
2	Low product yield	Product yield is high
3	Process is time-consuming and complex	Simple and quick process
4	Loss of lipid during operation	Lipid loss avoidable
5	Generates wastewater and pollutes environment	Minimal contaminating water pollutants
6	Expensive production	Cost lowers because of absence of harvesting and dewatering process

From Haas, M.J., & Wagner, K. (2011). Simplifying biodiesel production: the direct or in situ transesterification of algal biomass. European journal of lipid science and technology, 113*(10), 1219-1229..*

2.2.2 Bioethanol production

Bioethanol is the most common biofuel. However, technology is shifting toward the possibility of producing bioethanol from algal carbohydrates instead of sugars from corn and sugarcane, because they have less hemicellulose and lignin compared with plants with a higher lignocellulosic composition (Batool & Irfan, 2018). The high octane number of bioethanol, which reduces cylinder banging inside engines, is just one of its numerous benefits compared with petroleum-based fuels. Another benefit is that the use of bioethanol emits far fewer GHG, and it is the only biofuel that can be directly employed in modern automobiles with no modifications (Khan et al., 2018). Algae have a high concentration of carbohydrates, including glycogen, starch, agar, and cellulose, which can easily be converted into fermentable sugars for bioethanol production. The production of bioethanol from algae is an outstanding work in developing sustainable biofuels. However, achieving a huge amount of bioethanol production and its commercialization remains a big challenge.

Selecting the algal biomass is the main area in improving algae bioethanol technologies that may be enhanced for the industrialization of bioethanol, pretreatment, and an effective fermenting technique. In addition, biomass contains lignocellulose, an insoluble material that impedes pretreatment because it is difficult to convert. The amount of lignocellulose also increases the cost of pretreatment. Algal biomass is therefore acceptable when it contains little to no lignocellulosic content that could render it more affordable compared with other feedstocks such as terrestrial plants (Halim et al., 2012). As an alternative to conventional crops such as soybeans and corn, attempts have been made to employ algae as feedstock in a fermentation process to produce bioethanol (Nguyen & Vu, 2012). Table 2.3 compares terrestrial plants and algal biomass for the bioethanol production. *Gracilaria* sp., *Laminaria* sp., *Sargassum* sp., *Spirogyra* sp., *Chlorococcum* sp., *Prymnesium parvum*, and *Gelidium amansii* are some microalgae and macroalgae employed to produce bioethanol (Rajkumar et al., 2014). Cell lysis, which can be accomplished through enzymatic, acidic, and extraction methods, is necessary to remove

Table 2.3 Comparative analysis of terrestrial plants and algal biomass to generate bioethanol.

Feedstock	Conditions for bioethanol production	Bioethanol (yield)	References
Algae			
Chlorococcum infusionum	Pretreatment with alkaline condition at 120°C, introduce *Saccharomyces cerevisiae*	260 g/kg of ethanol	Harun et al. (2014)
Spirogyra	*Aspergillus niger* biomass saccharification, synthesized media growth, alkali conditions pretreatment, and fermentation (by *S. cerevisiae*)	80 g/kg of ethanol	Eshaq et al. (2010)
Chlorococcum humicola	Acid pretreatment, 160°C, *S. cerevisiae*	520 g/kg of ethanol	Harun and Danquah (2011a)
Terrestrial plants			
Manihot esculenta	Termamyl and amyloglucosidase enzyme, HCl of 1 N, *S. cerevisiae*, Ca-alginate immobilization	189 ± 301 g ethanol/kg of flour cassava	Behera et al. (2011)
Rice straw	Cellulase, β-glucosidase, solid-state fermentation, *Trichoderma reesei* RUT C30, and *Aspergillus niger* MTCC 7956 strains	93 g/kg of ethanol (rice straw—pretreated)	Sukumaran et al. (2010)
Sugarcane bagasse	Acid hydrolysis by H_2SO_4, *Kluyveromyces* sp. IIPE453, and fermentation	165 g ethanol/kg	Kumar et al. (2014)
Madhuca latifolia	*Zymomonas mobilis* MTCC 92 (strain), immobilize in luffa cylindrical sponge disks, maintained temperature at 30°C	251.1 ± 0.012 g ethanol/kg	Behera et al. (2011)

carbohydrates from algal cells. To convert the obtained sugar compound into monomer forms for use in fermentation to generate bioethanol and be separated by distillation, the sugars must first undergo pretreatment (Fig. 2.3).

Several studies show that microalgae can be used to produce bioethanol. For example, the cell walls of several green algae species such as *Spirogyra* and *Chlorococcum* possess a high quantity of

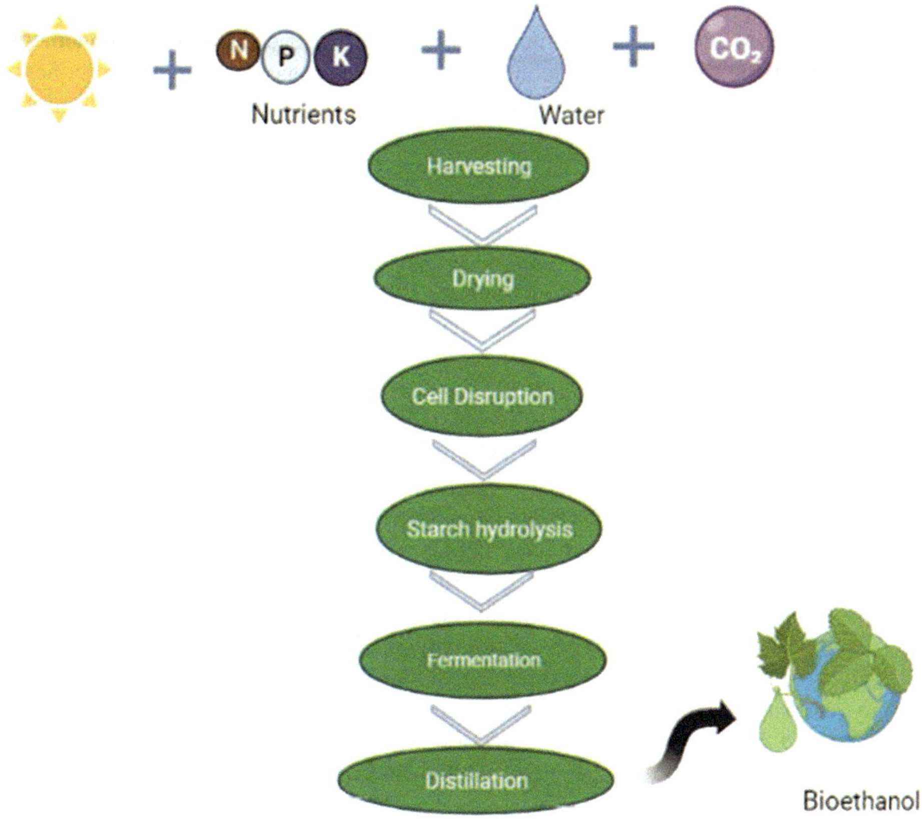

FIGURE 2.3

Microalgae Bioethanol Manufacturing Method. Created With BioRender.com.

polysaccharides. This starch can be used to produce bioethanol (Batool & Irfan, 2018). On a dry weight basis, microalgae such as *C. vulgaris* contain roughly 37% starch, which makes them the greatest source of bioethanol, with 65% conversion efficiency (Eshaq et al., 2010). However, before fermentation, the polysaccharide-based algal biomass needs to go through extra processing such as pretreatment and hydrolysis (Harun et al., 2010). Strains that produce amylase enzyme are used for the simultaneous hydrolysis of cellulose and fermentation to generate bioethanol in a single step. The method that can produce bioethanol from microalgae is the same as the first-generation methods that generate bioethanol from feedstocks based on corn.

Pretreatment is a crucial step involving degrading or disrupting the algal biomass in which the lipid and carbohydrate are converted, accumulated, and processed. Pretreatment uses little energy and is efficient (Harun & Danquah, 2011a, b). According to Yazdani et al. (2011), pretreating brown macroalga *Nizimuddinia zanardini* with 7% (w/w) H_2SO_4 is the most effective way to maximize sugar production while limiting the development of an inhibitor. The generation of bioethanol from *Eucheuma cottonii*, red seaweed, through acid pretreatment was studied by Candra et al. (2011). About

15.8 g/L of sugars was produced in this experiment using a 5% H_2SO_4 concentration for 120 min at 100°C. To turn sugar into ethanol, the accumulated sugars must be saccharified into monomeric units, because only fermentable sugars such as monosaccharides, disaccharides, and generally hexose and pentose can be digested by fermentation microbes. Saccharification is the process of breaking glycosidic linkages between monomers, such as α–($1 \rightarrow 4$), α–($1 \rightarrow 6$), α–($1 \rightarrow 3$), and α–($1 \rightarrow 6$) (Rajarapu & Scharf, 2017).

After 0.5 N HNO_3 and H_2SO_4 pretreatment, Markou et al. (2013) saccharified *Spirulina* (*Arthrospira platensis*) biomass and then fermented the hydrolysate to produce ethanol with a higher production of 16.32% and 16.27% (g ethanol/g biomass), respectively. Chigaiso (*Alaria crassifolia*), sea lettuce (*Ulva pertusa*), and agar weed (*Gelidium elegans*) were three species of seaweeds that Yanagisawa et al. (2011) investigated for polysaccharide content. These seaweeds are free of lignin, which indicates that polysaccharides can be hydrolyzed with no pretreatment. Singh and Trivedi (2013) used *Spirogyra* biomass to make bioethanol with *Zymomonas mobilis* with *Saccharomyces cerevisiae*. They used an acidic pretreatment of algal-derived biomass and then saccharified it with *Aspergillus niger*, which produces amylase. In addition, the biomass was instantly saccharified with no sample pretreatment. Compared with pretreatment and the saccharified algal-derived biomass, direct saccharification was able to produce 2% (w/w) more alcohol. This study found that *Spirogyra* does not require pretreatment with various chemicals, demonstrating its economic benefit for producing ethanol. Cellulase enzyme has also been used to saccharify algal biomass that contains cellulose. However, unlike amylases and glucoamylases, this enzyme system is costly. Furthermore, high doses are often needed for effective cellulose saccharification. Trivedi et al. (2013) investigated the saccharification of *Ulva* green algae using different cellulases. The maximum conversion efficiency of biomass into reducing sugars was achieved by using cellulase 22119 rather than Viscozyme L, cellulase 22086, and 22128. Based on the results of this experiment, they found a maximum yield of sugar 206.82 $\pm$ 14.96 mg/g with 2% (v/v) enzyme loading for 36 h at a temperature of 45°C.

2.2.2.1 Fermentation

Fermentation is a biological process that uses microorganisms that ferment sugars to produce bioethanol. In fermentation, microorganisms use sugars as substrates, including glucose, fructose, maltose, and rhamnose, which convert them into ethanol and carbon dioxide (Deesuth et al., 2012). Microorganisms such as yeast, bacteria, and fungi can be employed for the anaerobic fermentation of pretreated and saccharified algal biomass to produce bioethanol (Nguyen, 2012). The pH, temperature, level of oxygen, substrate concentration, and type of fermenting organisms employed all have an impact on the fermentation, significantly influencing the production and quality of the bioethanol. *S. cerevisiae* and *Z. mobilis* are well-known microorganisms used to ferment bioethanol. However, these microorganisms cannot anaerobically ferment mannitol, a polymer present in some algae. The anaerobic fermentation of mannitol requires oxygen throughout fermentation, which is attainable only with *Zymobacter palmae* (Horn et al., 2000).

The galactose and galactopyranose-based polymer known as agar found in certain coastal red algae can be used to produce bioethanol (Yoon et al., 2012). The biomass of red algae can easily be broken down into monomeric carbohydrates such as galactose and glucose. Along with mannitol and glucose, brown seaweed also contains 14% carbohydrates in the form of alginate (Wargacki et al., 2012). Cellulose, alginate, laminaran, mannitol, and fucoidan were found in several brown seaweeds, according to

Horn et al. (2000), which are good sources of sugar. The bacteria *Z. palmae* was used to ferment brown seaweed extract containing mannitol and produced 0.38 g ethanol/g mannitol ethanol yield.

Numerous academic publications describe microalgae as a suitable substrate for producing bioethanol. Hon-Nami (2006) obtained several by-products from *Chlamydomonas perigranulata* algae mass cultivation, including butanediol and ethanol. In addition to obtaining glucose and galactose from saccharified agar weed (red seaweed), which contains galactan and glucan, Yanagisawa et al. (2011) acquired 5.5% ethanol using *S. cerevisiae* fermentation. According to Harun et al. (2010), *Chlorococcum* sp. microalgal biomass produced 60% more ethanol when it was lipid extracted. This highlights the importance of algal biomass in manufacturing biodiesel and bioethanol.

2.2.3 Production of biogas

There has been a significant increase in the production of biogas from algae by anaerobic digestion owing to the existence of rich polysaccharides (carrageenan, mannitol, laminarin, agar, and alginate) with no lignin and a poor cellulose content. Several researchers have fermented various algae species including *Spirulina*, *Euglena*, *Scenedesmus*, and *Ulva* to generate biogas (Zhong et al., 2012). The yield of biogas produced from algal biomass and various terrestrial plants is listed in Table 2.4. Altogether, algae are thought to be an ideal feedstock for biogas production.

Biogas is generated when organic material in biocompostable feedstock such as marine algae undergoes anaerobic transformation, releasing gas such as methane, carbon dioxide, and traces of hydrogen sulfide. There are four main phases in anaerobic conversion. In the first stage, obligate anaerobes such as *Clostridia* and *Streptococci* use an enzyme to hydrolyze insoluble organic materials and biomolecules with a larger molecular mass, such as proteins, lipids, and carbohydrates, into soluble organic compounds. Acidogenesis is the second phase; it involves the conversion of soluble organic molecules to volatile fatty acids (VFAs) and alcohols with the help of enzymes released by acidogenic bacteria. These VFAs and alcohols also undergo acetogenesis, in which acetogenic bacteria convert them into acetic acid and hydrogen, which are then broken down by methanogens into methane and carbon dioxide (Brennan & Owende, 2010).

Green alga *Chaetomorpha litorea* was digested anaerobically with the generation of 80.5 L of biogas/kg of dry biomass under 299 psi, according to Sangeetha et al. (2011). In a two-phase anaerobic digestion process, Vergara-Fernandez et al. (2008) examined the digestion of marine algae *Macrocystis pyrifera* and *Durvillaea antarctica* and found comparable biogas generation rates of 180.4 (1.5) mL/g dry algae/day with an estimated 65% methane concentration. However, the same methane level was found for an algae blend, along with poor biogas generation. Mussgnug et al. obtained 587 and 287 mL biogas/g volatile solids in 2010 from a few green algae species such as *Chlamydomonas reinhardtii* and *Scenedesmus obliquus*, respectively. Furthermore, a few studies with microalgae reveal the influence of several pretreatments such as heat, microwave, and ultrasound for high-generation biogas (Passos et al., 2013).

However, some limitations, such as the requirement for more land space, infrastructure, and heat for the digesters, restrict the production of biogas (Jones & Mayfield, 2012). Algal cell proteins increase ammonium synthesis, resulting in a low carbon to nitrogen ratio, which affects biogas production by preventing the proliferation of anaerobic microbes. In addition, sodium ions suppress anaerobic microbes. Therefore, it is recommended to select microorganisms that can withstand salt for the anaerobic digestion of the algae biomass (Brennan & Owende, 2010; Jones & Mayfield, 2012).

Table 2.4 Comparative analysis of terrestrial plants and algal biomass to produce biogas.

Feedstock	Conditions for biogas production	Biogas (yield)	References
Algae			
Blue algae	pH 6.8, microcystin biological degradation	189.89 mL/g volatile solid (VS)	Yuan et al. (2011)
Chlamydomonas reinhardtii	Pretreatment-based dried, batch fermentation, 38°C maintained temperature	587 mL/g of VS	Mussgnug et al. (2010)
Ulva sp.	In batch reactor, codigestion with help of bovine slurry, maintained temperature of 35°C	191 mL/g of VS	Vanegas and Bartlett (2013)
Terrestrial plants			
Date palm tree waste	Pretreatment: alkaline, particle size of 2–5 mm, maintained temperature of 40°C	342.2 mL/g of VS	Al-Juhaimi et al. (2014)
Banana stem	Pretreatment: 54 h with 6% of NaOH at maintained temperature of 55°C; 37 ± 1°C, batch for 40 days	357.9 mL/g of VS	Zhang et al. (2013)
Rice straw	Pretreatment with ammonia concentration of 4% and moisture content of 70%, maintained temperature of 35 ± 2°C, 120 rpm, batch for 65 days	341.35 mL/g of VS	Yuan et al. (2014)
Saline creeping wild ryegrass	33 days at maintained temperature of 35°C	251 mL/g of VS	Zheng (2009)

2.2.4 Production of biohydrogen

The manufacture of liquid biofuels such ethanol and diesel had previously received greater attention, but now algae are also used to make biohydrogen and biogas (biomethane), which are then employed in a variety of applications, including the production of electricity. Because of their faster growth rates, ability to grow in saltwater (marine), and lack of cross-linking lignin in structure, macroalgae are the primary source for generating electricity-producing biogas. Through anaerobic fermentation followed by ferredoxin oxidation in the presence of hydrogenase enzymes, autotrophic microalgae and cyanobacteria are able to make biohydrogen directly. It was determined that generating algal biohydrogen is a common resource that may be used to provide energy or gaseous fuels. There are several ways to produce biohydrogen, including photosynthesis and biophotolysis (Shaishav et al., 2013). Algal biomass can produce biohydrogen with an efficiency similar to that of terrestrial plants (Table 2.5). Park et al. (2011) identified red algae (*Gelidium amansii*) as a potential biomass source for anaerobic

Table 2.5 Comparative analysis of terrestrial plants and algal biomass to produce biohydrogen.

Feedstock	Conditions for biohydrogen production	Biohydrogen (yield)	References
Algae			
Gelidium amansii	Hydrolysis at maintained temperature of 150°C	53.5 mL H_2/g of algae (dry)	Park et al. (2011)
Laminaria japonica	Mesophilic condition at temperature of 35 ± 1°C, maintained at pH 7.5, batch reactor (anaerobic sequencing), 6-day hydraulic retention time	712.04 mL H_2/g of algae (dry)	Shi et al. (2011)
Terrestrial plants			
Wheat straw	Pretreatment (acid), Simultaneous saccharification and fermentation (SSF) process	141 mL/g of VS	Nasirian et al. (2011)
Pretreated wheat straw	Strain *Caldicellulosiruptor saccharolyticus*, maintained temperature and pH of 70°C and 7.2, respectively	44.7 mL/g of wheat straw (dry)	Ivanova et al. (2009)
Bagasse	Strain *Klebsiella oxytoca* HP1, maintained temperature and pH of 37.5°C and 7, respectively	107.8 ± 7.5 mL H_2/g of bagasse	Wu et al. (2010)
Cornstalk	Maintained temperature of 55°C and pH of 7.4	61.4 mL of H_2/g of cornstalk	Cheng and Liu (2011)

fermentation to generate biohydrogen. They were still able to extract 53.5 mL of hydrogen from 1 g of dried algae, with a daily hydrogen production capacity of 0.518 L H_2/g VSS per day. The scientists found an inhibitor called 5-hydroxymethylfurfural, which is formed when the acid hydrolyzes *G. amansii* and lowers hydrogen production by about 50%.

To reduce the lag time for hydrogen production using the microalgae *Chlamydomonas reinhardtii*, Saleem et al. (2012) used an optical fiber as an artificial source of light. That study found that exogenic glucose and optical fiber increased hydrogen generation at a maximum rate of 6 mL/L culture/h, which is much greater than previously reported values.

Glycogen replaces starch in the cells of several microalgae, such as blue-green algae. This is an exception because it requires the hydrogenase enzyme to oxidize ferredoxin under anaerobic conditions to produce hydrogen. The detachment of electrons is another function of this enzyme. Thus, numerous researchers have focused on determining which enzyme activities interact with ferredoxin and other metabolic processes to produce micro-algal photobiohydrogen. They are also attempting to modify these interactions genetically to increase the production of bio-hydrogen.

2.2.4.1 Production of biogas and syngas

Algal biomass is converted into liquid bio-oil under anaerobic conditions at high temperatures. The composition of biooil changes depending on the feedstock and the processing conditions used in pyrolysis (Li et al., 2008). Water and the ash content, biomass composition, pyrolysis temperature, and vapor residence time are factors that influence biooil productivity (Fahmi et al., 2008). Crude biooil, on the other hand, cannot be used as a fuel because of the existence of phenolic, unsaturated oxygen, and water molecules. Therefore, certain treatments are necessary to improve its quality. External combustion, organic Rankine cycles, and the Stirling engine can all be used to convert biooils for power generation. Internal combustion engines such as diesel and gas turbine engines can be used to generate power (Chiaramonti et al., 2007). Compared with lignocellulosic biomass, there are only a few studies on algal pyrolysis in the literature. Although fluidized-bed fast pyrolysis techniques produce significant yields of biooil, a number of pyrolysis modes have been created to combat the drawbacks of high carrier gas flow and energy inputs (Oyedun et al., 2012). According to Demirbas and Balat (2006), who assessed the suitability of microalgal biomass for manufacturing biooil, it was superior in terms of quality compared with wood. After lipid extraction, Porphy and Farid (2012) produced biooil from the pyrolysis of algae (*Nannochloropsis* sp.) at 300°C, which featured 19 wt% aromatics such pyrazine and pyrrole and 50 wt% acetone. *Saccharina japonica*, a species of brown algae, was subjected to a pyrolysis experiment by Choia et al. (2014) at 450°C; the researchers found that roughly 47% of biooil was produced.

Gasification is a partial oxidation process that transforms biomass into a combustible gas mixture, known as syngas or producer gas, at higher temperatures of 800−1000°C. A mixture of gases including CH_4, H_2, CO, CO_2, CH_4, and N_2 is known as syngas. It can be made from woody biomass through conventional gasification. Here, biomass combines with oxygen and water (steam) to produce syngas. It is a low-calorific gas that can be used in gas turbines or directly as a source of energy. Gasification can be used to produce energy from a wide range of biomass feedstocks, which has a significant benefits (Lv et al., 2007).

2.2.5 Role of nanocatalysts in biofuel production

Catalysts have an essential role in the kinetics of reactions. Nonnanocatalysts, such as homogeneous catalysts including acid, alkali, transition-metal catalysts, and heterogeneous catalysts, have a number of benefits but also some drawbacks. Nanocatalysts can be developed depending on the superior properties of both homogeneous and heterogeneous nonnanocatalysts. Nanocatalysts are unique because they are extremely selective, highly active, and simple to recover (Jiang et al., 2014). Chemical contact between substrate (biomass) catalysts is improved by the nanosized catalyst owing to its high surface area to volume ratio. Some nanocatalysts used to produce biofuel, biodiesel, and bioethanol are discussed subsequently.

2.2.5.1 Application of nanocatalysts to produce biodiesel

The cost of biodiesel production is influenced by the properties of nanocatalysts, including their general stability, specificity, and catalytic activity. Physical, mechanical, and thermal deterioration impair catalytic activity. Powdered acidic and basic catalysts are unsuitable for industrial use owing to their small particle size, which has a significant impact on catalytic activity and methyl ester yield. In heterogeneous catalysis, a catalyst's structural and morphologic properties are crucial. Catalysts are spherical and more stable and simple to extract from the reaction mixture. Another way to affect reaction conditions is to design nanocatalysts to be more stable at high temperatures, because this enhances the likelihood of recycling and reusing the nanocatalyst (Rasouli & Esmaeili, 2019). The fabrication of nanocatalysts for biodiesel has attracted a lot of interest. Producing catalysts on a large scale has a few obstacles posed by the use of conventional catalysts, which must be overcome (Deepalakshmi et al., 2015). Different types of catalysts have been developed in accordance with inorganic compounds, metals, and so on (Bano et al., 2020). There are several difficulties with solid-based catalysts, such as mass transfer resistance, time requirements, quick deactivation, and the ineffectiveness of catalysts for prolonged use (Rizwanul Fattah et al., 2020). By providing composites with Lewis acidic or basic characteristics, enhanced acidity, and active sites, nanocatalysts can be used to address the drawbacks of solid-based catalysts and increase their effectiveness in catalyzing the transesterification reaction. The use of various materials to modify nanocatalysts is gaining popularity because it increases both porosity and surface area, enhances acidic and thermal stability, and decreases metal ion leaching.

Many researchers planned to develop nanocatalysts composed of polymer, zeolite, and carbon-based nanomaterials, which have good catalytic activity and require little reaction time and heat. Consequently, considerable amounts of biofuel could be produced by creating and manipulating nanocatalysts (Marwaha et al., 2018). Several processes, including chemical vapor deposition (CVD), impregnation, precipitation, and electrochemical deposition, can be used to produce the new nanocatalyst. Although impregnation and precipitation are efficient and affordable, they have restrictions on nanocatalysts in their overall shape and regular size control. Electrochemical and CVD techniques are expensive and need high temperatures, but they generate nanoparticles with a well-controlled shape and size. Several types of nanocatalysts can overcome these obstacles. Their significance in increasing the production of biofuel is addressed next (Saoud, 2018):

2.2.5.1.1 Metal oxide-based catalysts. Metal oxide catalysts have emerged as the most viable of available heterogenous catalysts. As a result, they have been extensively researched for the manufacture of biodiesel from a variety of feedstocks (Ingle et al., 2020). Solid catalysts based on metal oxides are often used to produce biodiesel. These catalysts are effectively used in large-scale conversion of processed material and are essential in the homogeneous catalytic process. This corresponds to the significant generation of biodiesel. After their use, these catalysts cannot be separated owing to the longer reaction times; hence, their activity is hampered. To catalyze the reaction and increase production, it is essential to create new, improved nanocatalysts, depending on the oxides of different metals, such as Ca, Zn, and Mg. Table 2.4 lists metal oxide-based nanocatalysts to enhance biodiesel synthesis.

Bharti and Singh (2019) reported that calcium oxide can be used as a nanocatalyst to convert soybean oil to biodiesel. This nanocatalyst has more efficient and improved active sites, and high area to volume ratio, and a small particle size of 8 nm, in compliance with the nanometric scale. Compared with commercially available CaO catalysts, the diameter of the Brunauer–Emmett–Teller (BET) specific surface area was higher and significant at 67.78 m^2/g and 3.302 nm, respectively, and was readily accessible on the market. The smaller particle size contributes to the increased surface area, which enhances the rate of reaction. A biodiesel conversion efficiency of 97.61% was attained after a number of optimization parameters were considered, including catalyst loading, the methanol/oil ratio, and the temperature of the reaction. The CaO nanocatalyst can convert a variety of lipids and oils; however, because of leaching and catalytically active obstruction, it is not recommended for prolonged use. Doping of metals and metal oxides can greatly reduce its effectiveness.

Madhuvilakku and Piraman (2013) reported that a synthesized TiO_2–ZnO nanocatalyst actively involved in processing palm oil was transesterified to generate FAME. In that study, 98% of oil conversion was achieved. The following reaction factors resulted in a yield of biodiesel of about 92%: a 200-mg catalyst, a methanol/oil molar ratio of 6:1, a temperature of 60°C, and a 5-hour period. Similar studies were performed to demonstrate the synthesis of a sulfate-doped TiO_2 solid acid nanocatalyst that can be used in petroleum refinery.

Borah et al. (2019) studied the effect of an eggshell CaO nanocatalyst with Zn doping for the transesterification of waste cooking oil. This nanocatalyst was prepared by changing the Zn^{2+} content (0.5–2wt%) with CaO using a wet impregnation process. The nanocatalyst with 1% Zn/CaO had a round morphology and irregular particle size of 30–42 nm, leading to more active sites and improved performance. The nanocatalyst activity declined as the Zn^{2+} dosage increased and the spherical morphology, resulting in a reduction in the active site number. In addition, to boost biodiesel synthesis, the catalyst concentration was raised to 5 wt%. The nanocatalyst was able to create a maximum of 96.74% biodiesel under the prescribed operating conditions, which included a 55 wt% catalyst quantity, a 20:1 of methanol/oil molar ratio at a 65°C drying reaction, and a 4-h reaction. It was used again for five more runs, however as a result of the hazardous catalyst and metal ion leaching, the yield gradually decreased.

Todorovic et al. (2019) used a CaO nanocatalyst and raw biodiesel as cosolvents. The methanol/oil ratio (7.1:1), catalytic concentration (0.7 M), and temperature of the reaction (52°C) were among the reaction variables that were tuned for biodiesel synthesis. A 99.8% success rate for biodiesel generation was attained in that study. The catalyst was reportedly employed four times under various reaction circumstances, in which the second cycle produced 97.7% biodiesel conversion in 5 hours. Owing to the high surface area to volumetric ratio and strong basic nature, many studies have claimed that the manufacture of CaO and modified CaO nanoparticles might be employed as nanocatalysts that enhance the conversion of lipids to fatty acid methyl esters.

According to Rahman et al. (2019), ideal reaction conditions such as the methanol/oil ratio (6:1), the amount of catalyst (5 wt%), temperature (in the reaction, of 65°C), and response time (2.5 h) were significant for optimizing biodiesel production when doped CaO was used as a nanocatalyst. It was reported that the Zn/CaO nanocatalyst had the maximum yield of biodiesel and was able to be reused in up to seven continuous cycles. However, it had a negative impact on conversion efficiency owing to leaching. Hydrolysis of the catalyst caused leaching and led to saponification, which reduced the total catalytic activity. Wen et al. (2010) stated that 93.9% of biodiesel could be produced using an Li-doped MgO nanocatalyst. Because of the addition of Li, these catalysts have higher basic site generation, which enhances the production of biofuel. Other nanocatalysts, such as Na-doping ZnO and Li-doping ZnO, were developed by Xie, Yang, et al. (2007) for use in transesterification and showed excellent catalytic activity depending on the amount of lithium loading.

Mguni et al. (2012) used sunflower oil as the feedstock, and the nanocatalyst MgO on TiO_2 served as a binary metal oxide. By varying the catalyst's quantity and temperature, the created nanocatalyst was used to produce biodiesel. About 10, 20, and 30 wt% concentrations of MgO were placed on top of TiO_2 nanoparticles. The temperature had an impact because the nanocatalysts were used to produce biodiesel with yields of 84%, 91%, and 95% at 250°C and 15%, 35%, and 42% at 150°C, respectively. With less catalytic activity and leaching loss, MgO supplementation boosts catalytic performance.

An $MgO/MgAl_2O_4$ nanocatalyst was synthesized by Vahid and Haghighi (2016) in accordance with the magnesium spinel structure, which was predominantly created by combustion followed by impregnation. The catalyst had a high surface area and good morphologic characteristics, and it was used to generate biodiesel from sunflower oil. While the catalyst was produced, the impact of the oil ratio was examined. It was clear that the size, width, and specific surface area of the pores, as well as their shape, are all affected by the fuel ratio. The ideal oil ratio of 1.5 had an average pore diameter of 6.3 nm and a specific surface area of 60.6 m^2/g. As the fuel ratio rises, the catalyst's crystallinity and surface area increase, indicating that the precursors have been entirely transformed into end products. With an energy conversion potential of 95.7%, the improved catalyst was tested to produce biodiesel under reaction operating parameters (110°C during a reaction time of 3 h at a catalyst quantity 3 wt% and methanol/oil ratio of 12:1). Reuse testing of the catalyst revealed a 5%−9% drop in catalytic performance after the third run, which was caused by leaching of the active species MgO. After the third attempt, however, there was no discernible decline in catalytic activity. The study found that MgO doping significantly increased the biodiesel output.

Nanocatalyst-based on TiO_2 was also extensively studied, with a wide range of variations. In the production of biodiesel, Gardy et al. (2017) modified TiO_2-based nanocatalysts. With the use of a grafting technique, a mesoporous $TiO_2/PrSO_3H$ acid nanocatalyst was developed. Propyl sulfonic acid was also used for esterification using waste cooking oil as a feedstock. It had a 98.3% FAME conversion rate. The hydrophilic sulfonic groups of the nanostructured material can increase the acidic strength and active site of the nanocatalyst, enabling enhanced interactions between the substrate oil and methanol, leading to greater conversion. It is possible to recycle and reuse the nanocatalysts four times without losing catalytic activity.

By functionalizing TiO_2 nanoparticles, new solid and acid magnetized-based TiO_2 nanoparticles were synthesized. Hematite provides the magnetic property whereas alumina serves as a layer of buffer in the $SO_4/Fe\text{-}Al\text{-}TiO_2$ catalyst. Because the magnetic material loses activity in acidic environments, a buffer layer is often used to prevent pushback. The stability of the catalyst as a whole can be improved by interspersing oxides of metal with polymers and stable components. Then, Brønsted acidity was supplied by sulfation with chlorosulfonic acid. With a conversion rate of 96%, waste cooking oil was transesterified using the new catalyst. Sulfate was added to the catalyst to increase its catalytic performance, and magnetization helps in separation with minimum loss during recovery. Before losing effectiveness, the catalysts were employed for 10 cycles. The catalyst was efficient for feedstocks with a high proportion of free fatty acids and simultaneous transesterification and esterification processes (Gardy et al., 2018). Ambat et al. (2018) employed an impregnation procedure to create TiO_2 nanoparticles doped with potassium tartrate and subsequently used it to generate biodiesel with linseed oil. With a peak output of 98.9%, the potassium tartrate loading greatly raised the nanocatalyst's basicity. The catalyst operated effectively for five cycles with no decline in activity.

Various magnetite-based nanocatalysts have been designed and evaluated for triglyceride transesterification in biodiesel synthesis (Shi et al., 2017). Using the nanocatalyst $Cs/Al/Fe_3O_4$, Feyzi et al. (2013) studied the transesterification of sunflower oil; the result was 94.8% biodiesel synthesis and excellent catalytic activity. The effect of various molar ratios of Cs-Al and Cs-Fe on the nanocatalyst, as well as temperatures of calcination and durations, were investigated to increase the overall effectiveness of the manufactured nanocatalyst. The magnetic nanocatalysts have shown outstanding catalytic activity at an ideal calcination temperature of 500°C and a duration of 7 hrs. Beyond the optimal conditions, the nanocatalyst exhibits morphological and textural changes due to agglomeration caused by sintering, and the surface area has also decreased, as confirmed by BET surface area analysis. Optimizing the nanocatalyst, followed by adjusting the reaction parameters, results in a significant increase in biodiesel production (Feyzi et al., 2013).

Future studies might focus on examining the overall phase structure of TiO_2 nanoparticles as well as other metallic oxides while doping them with polymeric and carbon-based nanomaterials to boost the lifetime and catalytic performance of the nanocatalyst dramatically, which would lead to higher biodiesel generation.

2.2.5.1.2 Carbon-based nanocatalysts. Graphene, nanotubes, and reduced graphene oxides are examples of carbon materials used to make nanocatalysts. The excellent chemical and physical properties of the materials can fit the shapes and diameters formed as nanocomposites (Shu et al., 2009). To analyze the transesterification reaction sequentially, Poonjarernsilp et al. (2014) used multiwalled porous nanohorn carbon nanotubes (CNTs) dispersed in Fe/Fe_2O_3 mixed nanoparticles. CNTs are allotropes of carbon formed by rolling graphene sheets into a cylinder. These nanotubes are mainly used to manufacture biosensors and microbial fuel cells because they have the potential to transport redox reactions and because of the electron transfer rate theory. There are two types of CNTs: multiwalled CNTs (MWCNTs) that contain multiple layers of graphene and single-walled CNTs that are composed of a single atomic layer of carbon atoms. CNTs have characteristic properties such as stability, a high surface area, and low toxicity, and are used as catalysts for biofuel production (Liu et al., 2012). Guan et al. (2017) employed a nanocatalyst made of sulfonated MWCNTs to transesterify glycerides. At 150°C, 97.8% biodiesel was produced in 1 hour with 3.7 wt% catalysts. Nanocatalysts with an acidic nature, significant porosity, and a high surface area had improved catalytic activity. The use of carbon-based nanocatalysts in improved biodiesel synthesis is shown in Table 2.6.

Because stainless steel is simple to separate and recycle, Zhang et al. (2014) reported carbon-based nanostructured materials synthesis using it as a substrate. Multiple techniques were used to multi-functionalize the CNT. For 95 h, the nanocatalyst converted 85% of the biodiesel without losing its structural integrity or shape. Another report (Asri et al., 2020) evaluated how to make MWCNTs with varying ZnO nanoparticles. With only 20% ZnO-loaded MWCNT, the generated heterogeneous solid acid nanocatalysts were employed to transesterify kesambi oil, yielding 13.82% methyl esters. CNTs have a huge specific surface area; however, when ZnO is added, the surface area is substantially reduced. The active sites mostly increase, resulting in an increase in the nanocatalyst's reaction rate and overall activity, as well as a significant decrease in mass transfer.

Shuit et al. (2015) used sulfonated MWCNTs (s-MWCNTs) to esterify palm fatty acid distillate. Compared with metal oxide-treated CNT, the s-MWCNT displayed greater catalytic performance, yielding 93.5% methyl esters with a minimal loading of 3 wt%, a reaction time of 2 h, and a temperature of 170°C. A 75% yield was achieved by using the nanocatalysts five times in succession. With sulfuric acid, it may also be reused, yielding 86.2%. Maintaining the loss of a sulfonic group throughout the leaching phase depends on the regeneration stage. Ibrahim et al. (2020) described a method for transesterifying used cooking oil by impregnating CNTs with sodium oxide. The addition of sodium and oxide ions to the nanocatalysts resulted in exceptional catalytic activity. Cations add Lewis acid and anions act as Brønsted bases, increasing catalytic activity and increasing biodiesel yields by 97%. It becomes unstable after three cycles owing to exudation, leading to inactivation of the active site.

CNTs were used as a catalyst for uniform alkaline catalysts, according to Shankar et al. (2017). A transesterification technique employing a nanocatalyst (homogeneous alkaline-based CNTs) was used to generate biodiesel from cottonseed oil, giving a 95% yield in 60–120 min. Immobilized lipase

Table 2.6 How carbon-based nanocatalysts have contributed to increasing biodiesel production.

Nanocatalyst	Synthesis mode	Feedstock	Amount of catalyst (wt%)	Methanol/ oil ratio	Retention time (h) and reaction temperature	Biodiesel yield (%)	Recyclability	Reference
Sulfonated biochar and activated carbon	Pyrolysis	Vegetable oil	2–8	6:1	6/55–60	97	6	Kastner et al. (2012)
KOH-loaded multiwalled carbon nanotubes	Impregnation method	Canola oil	3.02	13.56:1	3.34/60	94.23	3	Omraei n et al. (2013)
Sulfonated multiwalled carbon nanotubes		Oleic acid	0.20	5.8:1	1.5/135	95.46		Shu et al. (2009)
Silicon carbide/ sodium hydroxide-graphene oxide	In situ impregnation method	Rapeseed oil oleic acid	5	48:1	0.1/65	96	3	Loy et al. (2019)
Na_2O/carbon nanotubes	Impregnation method	Used cooking oil	3	20:1	3/65	97	3	Ibrahim et al. (2020)

Source: Bano et al. (2020).

enzymes are also used to produce biodiesel instead of MWCNTs owing to their high-temperature stability and recyclability (Fan et al., 2017). In addition, developed MWCNTs were bound to polyamide amine dendrimers before loading iron oxide and immobilizing onto *Burkholderia cepacia* lipase. The catalyst produced has a production potential of 92.8%, an exceptional recycled content of approximately 20 runs, and a catalytic activity of over 90%. This activity is boosted by the existence of massive dendrimers coupled with the exterior of the nanotubes. Because magnetic iron oxide is present, the nanocatalyst may easily be separated. Properly handled, this biotechnological application can be applied on a large scale, lowering the use of nanocatalysts.

Graphene oxide showed higher fatty acid conversion because of its larger surface area and various functional groups, Borah et al. (2018) reported the heterogeneous catalysis of waste cooking oil via transesterification to produce in situ TiO_2/reduced graphene oxide nanocomposites. The best conditions were determined to be a 12:1 methanol/oil molar ratio at 65°C for 3 h with a catalyst loading of 1.5 wt%, resulting in 98% conversion efficiency. The catalyst was dried at 100°C for 2 h before being reused three times with 78.86% efficiency.

Nanocatalysts made from biochar are essential for catalysis in the production of biodiesel. It is produced through the hydrothermal carbonization and pyrolysis of various biomass sources, and then functionalized with sulfonated groups, nitrogen-containing groups, and metal-metal oxide nanoparticles to increase catalytic activity, the effective surface area, porosity, and active sites. Fly ash and nanomaterials derived from sugars can be used to carbonize other catalysts that are carbon-based, which can then be evaluated for transesterification. Although biodiesel production has grown, the catalyst is thermally unstable and requires a lengthy reaction period (Gardy et al., 2019).

Rashid et al. (2019) produced biodiesel using palm kernel shell as a feedstock material and a sulfonated zirconium catalyst that is biochar-based. They achieved 94.3% fatty acid methyl ester production with just 3 wt% catalyst for 3 h at 75°C. In addition, the catalyst was effectively reusable for five consecutive cycles with 80% conversion efficiency.

2.2.5.1.3 Zeolite-based nanocatalyst. Different types of catalysts, such as zeolites, are employed in the commercial synthesis of biodiesel. Zeolite materials are being used more frequently these days as their benefits become apparent. They are widely investigated owing to their many specific characteristics, such as high thermal stability and surface area, as well as the existence of active acidic and basic sites, high catalytic activity, and easy modification using different functionalities (Zhang et al., 2019). Commercially, synthetic-based materials are more widely employed than natural zeolite products. The proportions of silicate and aluminate materials, as well as the exchange of different metal and metal oxide nanoparticles, can be altered to produce clean crystalline, homogeneous pores and particle diameter materials. They are used as a carrier or support system when impregnated with various nanoparticles of metal oxides, such as CaO, MgO, and Fe_2O_3. Beta zeolites and ZSM-5,X,Y are commercially available synthetic zeolites primarily used in biodiesel synthesis (Abukhadra et al., 2019). Few scientists have created nanozeolite–enzyme compounds, which are the most effective catalysts for enhancing biodiesel generation. Using nanozeolites as a stable support to immobilize catalytic enzymes such as lipase increases the immobilized enzymes' catalytic activity compared with free enzymes. The increased catalytic activity of immobilized enzymes is believed to be primarily due to their increased stability to high temperatures and various environmental conditions. In addition, their capacity to be reused is the main benefit of using these nanozeolite–enzyme combinations. In a continuous cycle of transesterification of triacylglycerides, catalytic enzymes can be recovered and reused because they are immobilized on solid nanocarriers (de Vasconcellos et al., 2018).

The transesterification of soybean oil as a feedstock was successful using KOH-loaded NaX zeolite (Xie, Huang, et al., 2007). The optimized reaction conditions yielded 3% catalyst with a 10:1 methanol/oil ratio and a reaction time of 8 h and 65°C, respectively. With a composition of x and y, it was discovered that 85.6% of biodiesel was produced. To produce the most biodiesel possible, this zeolite was produced and examined. Zeolites are acidic because of an imbalance of charges on the alumina and silica units, which principally contribute to the aluminosilicate framework (Fattahi et al., 2019).

Another group discovered that the desired outcomes can be achieved by employing KOH/NaY and NaY zeolites as catalysts in the transesterification of *Madhuca indica* oil. The conversion efficiency of these zeolites is 98.9%. These catalysts are easily activated and can be used repeatedly by soaking them in methanol and calcining them in a muffle furnace for a predetermined time. Their catalytic activity is greatly increased as a result. To achieve maximal conversion, recyclability, and reusability of the catalyst used in transesterification, nanocatalysts made with zeolite are synthesized (Pratap et al., 2015). With zeolite-supported CaO as a strong base catalyst, soybean oil is transesterified into biodiesel (Wu et al., 2013).

Castor oil was employed as a feedstock for fatty acid methyl biosynthesis, as described by Li et al. (2019), who also reported the hydrothermal and microemulsion production of Li/Na-based zeolites. High catalytic activity was achieved by calcining the produced particles at various temperatures, which is perfect for boosting biodiesel conversion. The biodiesel production was 98.6% when an ideal catalyst amount of 3 wt% and an alcohol/oil molar ratio of 18:1 were used, with reaction temperatures and durations of 75°C and 2 h, respectively. The catalyst's potential for reuse and recycling was also proven.

Martinez et al. (2011) successfully transesterified sunflower oil using CaO nanoparticles and NaX zeolite under ideal circumstances, providing 93.5% biodiesel. CaO/NaY zeolites were shown by Wu et al. (2013) to have 95% conversion efficiency. On the other hand, leaching and instability at high temperatures can cause CaO to lose activity. Firouzjaee and Taghizadeh (2017) developed a CaO/NaY$-Fe_3O_4$ nanomagnetic zeolite catalyst that is stable at high temperatures and easily separated, to address these difficulties. Canola oil, which is used as a feedstock, has 95.37% conversion efficiency, according to several characteristics being studied.

ZSM-based materials have been synthesized and altered using desilication accompanied by impregnation to increase the output of biofuel production. Alaba et al. (2016) synthesized mesoporous zeolites designated 0.3 and 0.4 mesoZBio using a modified version of ZSM-5 (ZBio). The kinetics of biodiesel generation from shea butter employing these catalysts were investigated using pyrolysis. Because of the existence of acid active sites, ZBio has greater activation energy, resulting in substantial biodiesel production. In addition, the 87% biodiesel yield is closely correlated with the straight functional groups' acidity and porosity (Alaba et al., 2016; Vieira et al., 2013). Moreover, zeolites were treated with MoO_3 nanoparticles by impregnating them after hydrothermally modification. Molybdenum impregnation mostly reduces acid site strength while increasing acidic active site concentration. The spherical shape, high crystallinity, and mesoporous structure of the optimized 25% MoO_3 considerably increase conversion, which has 98% efficiency. Because of the high crystallinity and elevated acidity, catalytic activity is enhanced. Functionalizing zeolites with nanocatalysts provides fresh insights into enhancing biodiesel production on a wide scale (Mohebbi et al., 2020).

2.2.5.2 Role of nanocatalyst in producing bioethanol

The environment is experiencing serious challenges, including increasing fossil fuel consumption and its devastating consequences, such as global warming and climate change. Biofuels are a viable and sustainable alternative to fossil fuels. A lot of research has been done on their production in past several decades, and there are many benefits to using them. Nanotechnology has a lot of potential to make biofuel production methods more cost-effective and efficient, with still more potential for significant advances in the future. Nanotechnology is accelerating the production of biofuels and increasing the amount of biofuels produced from nonedible sources (Arya et al., 2021). Reaction temperatures during the reaction, the weight of the catalyst, and the ratio of alcohol to oil all decrease when a nanocatalyst is used (Bano et al., 2020). Nanomaterials (nanocatalysts) have outstanding qualities such as a high surface area, catalytic efficiency, crystallinity, durability, and energy storage capabilities, which make them ideal for enhancing biofuel production systems. Nanocatalysts could potentially be used for recovery, reuse, and recycling.

Studies have focused on enhancing nanocatalytic production and development for biofuel process optimization. Increasing biodiesel production with nanocatalysts includes producing biodiesel from several sources using multiple technologies such as pyrolysis, microemulsions, direct mixing, and transesterification. Several nanocatalysts can be used to speed biofuel production. In the creation of biodiesel, nanocatalysts are crucial. They can increase the response rate by providing more active sites, because their increased surface area to volume ratio makes them more likely to capture and activate the receptor. They are important in the biodiesel synthesis of transesterification for different oils. Immobilizing lipase enzyme for biodiesel production is an example of how nanoparticles can be used. These catalysts were shown to be affordable and environmentally friendly in many investigations. Nanoparticles have an important role in how radiation therapy works (Cherian et al., 2022).

Since its inception, biodiesel production has used direct blending, pyrolysis, microemulsification, and transesterification. All methods are feasible to some extent, as detailed subsequently, but they have certain advantages and disadvantages. Direct blending, pyrolysis, and microemulsifying biodiesel production technologies are inadequate for production on a huge scale. The best of these processes is transesterification, which has various advantages such as enhanced productivity, simple separation, and mass transfer resistance restriction. It progresses by a catalytic reaction between fatty acid and methanol throughout the presence of a catalyst. Catalysis is further classified as homogeneous or heterogeneous. Acid- and base-catalyzed reactions are used in homogeneous catalysis. A reaction catalyzed by a base is much faster than one catalyzed by an acid, with the highest yield. NaOH, KOH, $NaOCH_3$, and $KOCH_3$ are the catalysts most often employed (Gardy et al., 2017), which require a low temperature and generate high production (Gardy et al., 2019). Alkali, acid, and enzymatic catalysts are most commonly used in biodiesel production (Ingle et al., 2020). Metal oxides (Fe_3O_4, CaO, MgO, $Ca/Al/Fe_3O_4$, KF/Al_2O_3, SnO_2, etc.) and others such as sulfated oxides, hydrotalcites, zeolites, zirconia, and other solid nanocatalysts have all been used to synthesize biodiesel (Zuliani et al., 2018).

Ionic liquids (ILs) are among the alternatives with the most potential for traditional homogeneous catalysts. They are a form of organic salts with melting points below 100°C (Liu et al., 2012). They are mostly nonvolatile, soluble in organic and inorganic materials, and much less toxic than standard organic solvents. They also have a wide temperature range of liquids and high thermal and chemical stability (Fauzi & Amin, 2012). More crucially, by modifying the type of cation or anion in their structure, the properties of ILs can be adjusted and tailored for a given purpose (Luo et al., 2013). Introducing ILs as potential catalysts for transesterification into biodiesel could result in higher yields with a simpler manufacturing method, as well as reduced equipment corrosion, saponification, and waste formation.

2.3 Need for cost-effectiveness in biofuel production

Biofuels made from microalgae can have a role in addressing global energy insecurity and climate change. The commercialization of global algae production faces significant financial challenges. Studies have ignored cost-cutting measures, especially for microalgae farming. This part provides important information on the technology and economy of large-scale cost-cutting strategies. About half of biofuels have higher environmental costs than petroleum fuels. Therefore, to meet our energy needs, we need cheaper and more environmentally friendly solutions. There have been three generations of biofuels. Algae-based fuels are a third-generation technology to replace traditional fossil fuels. The production of biofuels from algae can be seen as a promising form of energy. Advances in technology and energy issues have sparked interest in agriculture to produce different types of biofuels from wasted agricultural resources. Algal fuels have appealing properties among all biofuels, such as the ability to be grown with low impact on freshwater supplies and the lack of a specialized growing environment. Numerous species of microalgae are used around the world to make green biodiesel that can be used in diesel-operated engines (Singh et al., 2018). A more affordable substitute for fossil diesel is biodiesel.

A pilot economic analysis of the potential for microalgae-derived biodiesel in Hubei province found that it cost \$2.29/kg to produce compared with \$1.08/kg for petroleum diesel (Sun et al., 2019). The higher costs of producing and operating are significant barriers to commercial exploitation based on the productive capacity, the procedures employed, the biomass, and the lipid content. Production of microalgae-derived biodiesel may be more expensive compared with producing fuel from traditional sources.

Microalgae cultivation in aquaculture or wastewater, or biogas upgrading, as well as coproduction of VAPs, including protein, astaxanthin, and exopolysaccharides, can significantly lower biofuel production costs (Rafa et al., 2021).

2.3.1 Cost-effective approaches to biofuel production

Biofuels are often advertised as a cheap and environmentally friendly alternative to petroleum and other fossil fuels, especially given rising oil prices and growing concerns about the impact of fossil fuels on global warming. The cultivation of microalgae in aquaculture is becoming increasingly popular in industrial applications owing to its various advantages. Antioxidants from marine microalgae have been demonstrated to increase the immunity of fish, reduce morbidity, and avoid the misuse of antibiotics (Li et al., 2021). Other VAPs from microalgal cells may synthesize vitamins, terpenoids, astaxanthin, phytosterols, antioxidants, polyhydroxy butyrate, hormones, protein, polyunsaturated fatty acids, polysaccharides, colors, alginate, and agar, in addition to biofuel. Microalgae-derived VAPs are used to manufacture medicine and personal care products as well as cholesterol-lowering, anticancer, antiinflammatory, antimicrobial, and anticoagulant agents (Kothari et al., 2017). The cost of microalgae biorefineries can be greatly reduced by producing various co-VAPs along with biodiesel. In fact, a substantial number of carbohydrates remain in the cells after lipid extraction. These can be used to produce other products such as ethanol and biogas (Zhang et al., 2018), improving the potential for microalgal products to generate energy. The biogas industry is expected to grow as the relevance of CO_2 emission reduction is acknowledged, with uses from cooking to heating, the generation of power, and transportation. To enhance the methane content of biogas (CH_4), autotrophic microalgae can capture CO_2 and transform it into lipids. The purchase and supply of additional CO_2 to increase biomass levels is a key factors contributing to the high production costs of biodiesel (Chia et al., 2018).

2.4 Conclusion

The mainstream commercialization of microalgae-based biofuel has been limited by the costs associated with their technology. The cost-cutting strategies of producing biofuel from algal biomass, such as aqua farming, growing in wastewater, and biogas upgrading, as well as VAPs as coproducts, show tremendous potential that make the practices workable and generating revenue from other by-products. These offer good economic opportunities for practical biorefinery schemes and revenue growth. If wastewater is used for algal growth or VAPs are coproduced, the cost of generating microalgal biofuel can be reduced. Furthermore, open raceway pond—based cultivation, flocculation-based harvesting, and extraction with mixed solvents can help achieve the maximum extraction of biofuel and minimize the cost. The production of biofuel using several advanced nanocatalysts has good potential to generate the maximum yield of biofuel from algae and other feedstocks. These can help to reduce the cost so that it may be possible to commercialize biofuel.

References

Abukhadra, M. R., Ibrahim, S. M., Yakout, S. M., El-Zaidy, M. E., & Abdeltawab, A. A. (2019). Synthesis of Na+ trapped bentonite/zeolite-P composite as a novel catalyst for effective production of biodiesel from palm oil; Effect of ultrasonic irradiation and mechanism. *Energy Conversion and Management, 196*, 739—750.

Adeniyi, O. M., Azimov, U., & Burluka, A. (2018). Algae biofuel: Current status and future applications. *Renewable and Sustainable Energy Reviews, 90*, 316—335.

Afify, A. E. M. M., Shalaby, E. A., & Shanab, S. M. (2010). Enhancement of biodiesel production from different species of algae. *Grasas Y Aceites, 61*(4), 416—422.

Alaba, A. P., Sani, Y. M., Mohd, W., & Daud, W. M. W. A. (2016). A comparative study on thermal decomposition behavior of biodiesel samples produced from shea butter over micro- and mesoporous ZSM-5 zeolites using different kinetic models. *Journal of Thermal Analysis and Calorimetry, 126*(2), 943—948. https://doi.org/10.1007/s10973-016-5505-8

Al-Juhaimi, F.Y., Hamad, S.H., Al-Ahaideb, I.S., Al-Otaibi, M.M., Ghafoor, K., Abbasi, T., & Abbasi, S.A. (2014). Biogas production through the anaerobic digestion of date palm tree wastes-process optimization. *BioResources, 9*(2), 3323—3333.

Ambat, I., Srivastava, V., Haapaniemi, E., & Sillanpaa, M. (2018). Application of potassium ion impregnated titanium dioxide as nanocatalyst for transesterification of linseed oil. *Energy & Fuels, 32*(11), 11645—11655.

Arya, S., Mahajan, P., Mahajan, S., Khosla, A., Datt, R., Gupta, V., Young, S. J., & Oruganti, S. K. (2021). Influence of processing parameters to control morphology and optical properties of Sol-Gel synthesized ZnO nanoparticles. *ECS Journal of Solid State Science and Technology, 10*(2), 023002.

Asri, N. P., Yuniati, Y., Hindarso, H., & Yogaswara, R. R. (2020). Biodiesel production from Kesambi (*Schleicheraoleosa*) oil using multi-walled carbon nanotubes supported zinc oxide as a solid acid catalyst. In *IOP conference series: Earth and environmental science, 456* IOP Publishing, 1, 012003.

Awolu, O. O., & Layokun, S. K. (2013). Optimization of two-step transesterification production of biodiesel from neem (Azadirachta indica) oil. *International Journal of Energy and Environmental Engineering, 4*(1), 1—9.

Bajhaiya, A. K., Mandotra, S. K., Suseela, M. R., Toppo, K., & Ranade, S. (2010). Algal biodiesel: The next generation biofuel for India. *Asian Journal of Experimental Biological Sciences, 4*, 728—739.

Baldev, E., Mubarakali, D., Saravanakumar, K., Arutselvan, C., Alharbi, N. S., Alharbi, S. A., Sivasubramanian, V., & Thajuddin, N. (2018). Unveiling algal cultivation using raceway ponds for biodiesel production and its quality assessment. *Renewable Energy, 123*, 486—498.

Bano, S., Ganie, A. S., Sultana, S., Sabir, S., & Khan, M. Z. (2020). Fabrication and optimization of nanocatalyst for biodiesel production: An overview. *Frontiers in Energy Research, 350.*

Batool, T., & Irfan, M. (2018). Algae as a potential candidate for biofuel production: A review: Biofuel production from algae. *Proceedings of the Pakistan Academy of Sciences: B. Life and Environmental Sciences, 55*(1), 1–14.

Behera, S., Mohanty, R. C., & Ray, R. C. (2011). Ethanol production from mahula (*Madhuca latifolia* L.) flowers with immobilized cells of Saccharomyces cerevisiae in Luffa cylindrica L. sponge discs. *Applied Energy, 88*(1), 212–215.

Bharti, P., Singh, B., & Dey, R. K. (2019). Process optimization of biodiesel production catalyzed by CaOnanocatalyst using response surface methodology. *Journal of Nanostructure in Chemistry, 9*(4), 269–280.

Borah, M. J., Das, A., Das, V., Bhuyan, N., & Deka, D. (2019). Transesterification of waste cooking oil for biodiesel production catalyzed by Zn substituted waste egg shell derived CaOnanocatalyst. *Fuel, 242*, 345–354.

Borah, M. J., Devi, A., Saikia, R. A., & Deka, D. (2018). Biodiesel production from waste cooking oil catalyzed by in-situ decorated TiO2 on reduced graphene oxide nanocomposite. *Energy, 158*, 881–889.

Bosma, R., Van Spronsen, W. A., Tramper, J., & Wijffels, R. H. (2003). Ultrasound, a new separation technique to harvest microalgae. *Journal of Applied Phycology, 15*(2), 143–153.

Brennan, L., & Owende, P. (2010). Biofuels from microalgae- a review of technologies for production, processing, and extractions of biofuels and co-products. *Renewable and Sustainable Energy Reviews, 14*(2), 557–577.

Candra, K. P., Sarwono, S., & Sarinah, S. (2011). *Study on bioethanol production using red seaweed Eucheuma cottonii from Bontang sea water.*

Cheng, X. Y., & Liu, C. Z. (2011). Hydrogen production via thermophilic fermentation of cornstalk by *Clostridium thermocellum. Energy & Fuels, 25*(4), 1714–1720.

Cherian, E., Kalavathy, G., Joshi, T. J., Phoebe, M. G. L., & Gurunathan, B. (2022). Importance of nanocatalyst and its role in biofuel production. In *Biofuels and bioenergy* (pp. 171–182). Elsevier.

Chia, S. R., Chew, K. W., Show, P. L., Yap, Y. J., Ong, H. C., Ling, T. C., & Chang, J. S. (2018). Analysis of economic and environmental aspects of microalgae biorefinery for biofuels production: A review. *Biotechnology Journal, 13*(6), 1700618.

Chiaramonti, D., Oasmaa, A., & Solantausta, Y. (2007). Power generation using fast pyrolysis liquids from biomass. *Renewable and Sustainable Energy Reviews, 11*(6), 1056–1086.

Chisti, Y. (2007). Biodiesel from microalgae. *Biotechnology Advances, 25*(3), 294–306.

Choia, J. H., Woob, H. C., & Suha, D. J. (2014). Pyrolysis of seaweeds for bio-oil and bio-char production. *Chemical Engineering, 37*, 121–126.

Cohen, Z., Norman, H. A., & Heimer, Y. M. (1995). Microalgae as a source of ω3 fatty Acids1. *Plants in Human Nutrition, 77*, 1–31.

Deepalakshmi, S., Sivalingam, A., Thirumarimurugan, M., Sivakumar, P., & Ashokkumar, V. (2015). Optimization of biodiesel synthesis from *Calophyllum inophyllum. Energy Sources, Part A: Recovery, Utilization, and Environmental Effects, 37*(23), 2601–2608.

Deesuth, O., Laopaiboon, P., Jaisil, P., & Laopaiboon, L. (2012). Optimization of nitrogen and metal ions supplementation for very high gravity bioethanol fermentation from sweet sorghum juice using an orthogonal array design. *Energies, 5*(9), 3178–3197.

Demirbas, A. (2009). Progress and recent trends in biodiesel fuels. *Energy Conversion and Management, 50*(1), 14–34.

Demirbas, M. F., & Balat, M. (2006). Recent advances on the production and utilization trends of bio-fuels: A global perspective. *Energy Conversion and Management, 47*(15–16), 2371–2381.

Eshaq, F. S., Ali, M. N., & Mohd, M. K. (2010). Spirogyra biomass a renewable source for biofuel (bioethanol) production. *International Journal of Engineering Science and Technology, 2*(12), 7045–7054.

Fahmi, R., Bridgwater, A. V., Donnison, I., Yates, N., & Jones, J. M. (2008). The effect of lignin and inorganic species in biomass on pyrolysis oil yields, quality and stability. *Fuel, 87*(7), 1230–1240.

Fajardo, A. R., Cerdan, L. E., Medina, A. R., Fernández, F. G. A., Moreno, P. A. G., & Grima, E. M. (2007). Lipid extraction from the microalga Phaeodactylumtricornutum. *European Journal of Lipid Science and Technology, 109*(2), 120–126.

Fan, Y., Su, F., Li, K., Ke, C., & Yan, Y. (2017). Carbon nanotube filled with magnetic iron oxide and modified with polyamidoamine dendrimers for immobilizing lipase toward application in biodiesel production. *Scientific Reports, 7*(1), 1–13.

Fattahi, N., Triantafyllidis, K., Luque, R., & Ramazani, A. (2019). Zeolite-based catalysts: A valuable approach toward ester bond formation. *Catalysts, 9*(9), 758.

Fauzi, A. H. M., & Amin, N. A. S. (2012). An overview of ionic liquids as solvents in biodiesel synthesis. *Renewable and Sustainable Energy Reviews, 16*(8), 5770–5786.

Feyzi, M., Hassankhani, A., & Rafiee, H. R. (2013). Preparation and characterization of Cs/Al/Fe_3O_4 nanocatalysts for biodiesel production. *Energy Conversion and Management, 71*, 62–68.

Firouzjaee, M. H., & Taghizadeh, M. (2017). Optimization of process variables for biodiesel production using the nanomagnetic catalyst CaO/NaY-Fe3O4. *Chemical Engineering & Technology, 40*(6), 1140–1148.

Ganesan, R., Manigandan, S., Samuel, M. S., Shanmuganathan, R., Brindhadevi, K., Chi, N. T. L., Duc, P. A., & Pugazhendhi, A. (2020). A review on prospective production of biofuel from microalgae. *Biotechnology Reports, 27*, e00509.

Gardy, J., Hassanpour, A., Lai, X., Ahmed, M. H., & Rehan, M. (2017). Biodiesel production from used cooking oil using a novel surface functionalised TiO2 nano-catalyst. *Applied Catalysis B: Environmental, 207*, 297–310.

Gardy, J., Osatiashtiani, A., Cespedes, O., Hassanpour, A., Lai, X., Lee, A. F., Wilson, K., & Rehan, M. (2018). A magnetically separable SO_4/Fe-Al-TiO_2 solid acid catalyst for biodiesel production from waste cooking oil. *Applied Catalysis B: Environmental, 234*, 268–278.

Gardy, J., Rehan, M., Hassanpour, A., Lai, X., & Nizami, A. S. (2019). Advances in nano-catalysts based biodiesel production from non-food feedstocks. *Journal of Environmental Management, 249*, 109316.

Gendy, T. S., & El-Temtamy, S. A. (2013). Commercialization potential aspects of microalgae for biofuel production: An overview. *Egyptian Journal of Petroleum, 22*(1), 43–51.

Ghadge, S. V., & Raheman, H. (2005). Biodiesel production from mahua (*Madhuca indica*) oil having high free fatty acids. *Biomass and Bioenergy, 28*(6), 601–605.

Guan, Q., Li, Y., Chen, Y., Shi, Y., Gu, J., Li, B., Miao, R., Chen, Q., & Ning, P. (2017). Sulfonated multi-walled carbon nanotubes for biodiesel production through triglycerides transesterification. *RSC Advances, 7*(12), 7250–7258.

Haas, M.J., & Wagner, K. (2011). Simplifying biodiesel production: the direct or in situ transesterification of algal biomass. *European journal of lipid science and technology, 113*(10), 1219–1229.

Halim, R., Danquah, M. K., & Webley, P. A. (2012). Extraction of oil from microalgae for biodiesel production: A review. *Biotechnology Advances, 30*(3), 709–732.

Harun, R., & Danquah, M. K. (2011a). Influence of acid pre-treatment on microalgal biomass for bioethanol production. *Process Biochemistry (Oxford, United Kingdom), 46*, 304–309.

Harun, R., & Danquah, M. K. (2011b). Enzymatic hydrolysis of microalgae biomass for bioethanol production. *Chemical Engineering Journal, 168*, 1079–1084.

Harun, R., Danquah, M. K., & Forde, G. M. (2010). Microalgal biomass as a fermentation feedstock for bioethanol production. *Journal of Chemical Technology & Biotechnology, 85*(2), 199–203.

Harun, R., Yip, J. W., Thiruvenkadam, S., Ghani, W. A., Cherrington, T., & Danquah, M. K. (2014). Algal biomass conversion to bioethanol—a step-by-step assessment. *Biotechnology Journal, 9*(1), 73–86.

Hidalgo, P., Toro, C., & Navia, R. (2013). Advances in direct transesterification of microalgal biomass for biodiesel production. *Reviews in Environmental Science and Biotechnology, 12*(2), 179–199.

Hon-Nami, K. (2006). A unique feature of hydrogen recovery in endogenous starch-to-alcohol fermentation of the marine microalga, *Chlamydomonas perigranulata*. In *Twenty-seventh symposium on biotechnology for fuels and chemicals* (pp. 808–828). Humana Press.

Horn, S. J., Aasen, I. M., & Østgaard, K. (2000). Production of ethanol from mannitol by *Zymobacter palmae*. *Journal of Industrial Microbiology and Biotechnology, 24*(1), 51–57.

Hossain, A. S., Salleh, A., Boyce, A. N., Chowdhury, P., & Naqiuddin, M. (2008). Biodiesel fuel production from algae as renewable energy. *American Journal of Biochemistry and Biotechnology, 4*(3), 250–254.

Ibrahim, M. L., Khalil, N. N. A. N. A., Islam, A., Rashid, U., Ibrahim, S. F., Mashuri, S. I. S., & Taufiq-Yap, Y. H. (2020). Preparation of Na_2O supported CNTs nanocatalyst for efficient biodiesel production from waste-oil. *Energy Conversion and Management, 205*, 112445.

Ingle, A. P., Chandel, A. K., Philippini, R., Martiniano, S. E., & da Silva, S. S. (2020). Advances in nanocatalysts mediated biodiesel production: A critical appraisal. *Symmetry, 12*(2), 256.

Ivanova, G., Rákhely, G., & Kovács, K. L. (2009). Thermophilic biohydrogen production from energy plants by Caldicellulosiruptor saccharolyticus and comparison with related studies. *International Journal of Hydrogen Energy, 34*(9), 3659–3670.

Jiang, C., Jia, J., & Zhai, S. (2014). Mechanistic understanding of toxicity from nanocatalysts. *International Journal of Molecular Sciences, 15*(8), 13967–13992.

Jones, C. S., & Mayfield, S. P. (2012). Algae biofuels: Versatility for the future of bioenergy. *Current Opinion in Biotechnology, 23*(3), 346–351.

Kastner, J. R., Miller, J., Geller, D. P., Locklin, J., Keith, L. H., & Johnson, T. (2012). Catalytic esterification of fatty acids using solid acid catalysts generated from biochar and activated carbon. *Catalysis Today, 190*(1), 122–132.

Khan, M. I., Shin, J. H., & Kim, J. D. (2018). The promising future of microalgae: Current status, challenges, and optimization of a sustainable and renewable industry for biofuels, feed, and other products. *Microbial Cell Factories, 17*(1), 1–21.

Khedkar, S. V., Chavan, Y. P., & Bhagat, S. L. (2011). Extraction of oil from algae by solvent extraction and oil expeller method. *International Journal of Chemical Science, 9*(4), 1746–1750.

Kim, G. V., Choi, W., Kang, D., Lee, S., & Lee, H. (2014). Enhancement of biodiesel production from marine alga, *Scenedesmus* sp. through in situ transesterification process associated with acidic catalyst. *BioMed Research International, 2014*.

Knezevic, Z. (2004). Immobilized lipases as practical catalysts.[Doctoral dissertation]. *BibliotekaDissertatio-ZaduzbinaAndrejevic (Serbia and Montenegro)*.

Kothari, R., Pandey, A., Ahmad, S., Kumar, A., Pathak, V. V., & Tyagi, V. V. (2017). Microalgal cultivation for value-added products: A critical enviro-economical assessment. *3 Biotech, 7*(4), 1–15.

Kumar, A., Negi, Y. S., Choudhary, V., & Bhardwaj, N. K. (2014). Sugarcane bagasse: A promising source for the production of nanocellulose. *Journal Polymer Composites, 2*(3), 23–27.

Lee, S. J., Yoon, B. D., & Oh, H. M. (1998). Rapid method for the determination of lipid from the green alga *Botryococcus braunii*. *Biotechnology Techniques, 12*(7), 553–556.

Li, H., Chen, S., Liao, K., Lu, Q., & Zhou, W. (2021). Microalgae biotechnology as a promising pathway to ecofriendly aquaculture: A state-of-the-art review. *Journal of Chemical Technology & Biotechnology, 96*(4), 837–852.

Li, Z., Ding, S., Chen, C., Qu, S., Du, L., Lu, J., & Ding, J. (2019). Recyclable Li/NaY zeolite as a heterogeneous alkaline catalyst for biodiesel production: Process optimization and kinetics study. *Energy Conversion and Management, 192*, 335–345.

Li, Y., Ghasemi Naghdi, F., Garg, S., Adarme-Vega, T. C., Thurecht, K. J., Ghafor, W. A., Tannock, S., & Schenk, P. M. (2014). A comparative study: The impact of different lipid extraction methods on current microalgal lipid research. *Microbial Cell Factories, 13*(1), 1–9.

Li, Y., Horsman, M., Wu, N., Lan, C. Q., & Dubois-Calero, N. (2008). Biofuels from microalgae. *Biotechnology Progress, 24*(4), 815–820.

Liu, H., Su, L., Shao, Y., & Zou, L. (2012). Biodiesel production catalyzed by cinder supported CaO/KF particle catalyst. *Fuel, 97*, 651–657.

Loy, A. C. M., Quitain, A. T., Lam, M. K., Yusup, S., Sasaki, M., & Kida, T. (2019). Development of high microwave-absorptive bifunctional graphene oxide-based catalyst for biodiesel production. *Energy Conversion and Management, 180*, 1013–1025.

Luo, H., Fan, W., Li, Y., & Nan, G. (2013). Biodiesel production using alkaline ionic liquid and adopted as lubricity additive for low-sulfur diesel fuel. *Bioresource Technology, 140*, 337–341.

Luo, Y., Zheng, Y., Jiang, Z., Ma, Y., & Wei, D. (2006). A novel psychrophilic lipase from Pseudomonas fluorescens with unique property in chiral resolution and biodiesel production via transesterification. *Applied Microbiology and Biotechnology, 73*(2), 349–355.

Lv, P., Yuan, Z., Wu, C., Ma, L., Chen, Y., & Tsubaki, N. (2007). Biosyngas production from biomass catalytic gasification. *Energy Conversion and Management, 48*, 1132–1139.

Madhuvilakku, R., & Piraman, S. (2013). Biodiesel synthesis by TiO2–ZnO mixed oxide nanocatalyst catalyzed palm oil transesterification process. *Bioresource Technology, 150*, 55–59.

Mamilla, V. R., Mallikarjun, M. V., & Rao, G. L. N. (2011). Preparation of biodiesel from Karanja oil. *International Journal of Energy Engineering, 1*(2), 94–100.

Markou, G., Angelidaki, I., Nerantzis, E., & Georgakakis, D. (2013). Bioethanol production by carbohydrate-enriched biomass of *Arthrospira (Spirulina) platensis*. *Energies, 6*(8), 3937–3950.

Martínez, S. L., Romero, R., Romero, A., Víctor, S., & Natividad, R. (2011). Preparation and characterization of CaO nanoparticles/NaX zeolite catalysts for the transesterification of sunflower oil. *Industrial & Engineering Chemistry Research, 50*, 2665–2670. https://doi.org/10.1021/ie1006867

Martin, M., & Grossmann, I. E. (2013). ASI: Toward the optimal integrated production of biodiesel with internal recycling of methanol produced from glycerol. *Environmental Progress & Sustainable Energy, 32*(4), 891–901.

Marwaha, A., Dhir, A., Mahla, S. K., & Mohapatra, S. K. (2018). An overview of solid base heterogeneous catalysts for biodiesel production. *Catalysis Reviews, 60*(4), 594–628.

Mata, T. M., Martins, A. A., & Caetano, N. S. (2010). Microalgae for biodiesel production and other applications: A review. *Renewable and Sustainable Energy Reviews, 14*(1), 217–232.

Mazubert, A., Poux, M., & Aubin, J. (2013). Intensified processes for FAME production from waste cooking oil: A technological review. *Chemical Engineering Journal, 233*, 201–223.

Menetrez, M. Y. (2012). An overview of algae biofuel production and potential environmental impact. *Environmental Science & Technology, 46*(13), 7073–7085.

Mguni, L. L., Meijboom, R., & Jalama, K. (2012). Biodiesel production over nano-MgO supported on titania. *International Journal of Materials and Metallurgical Engineering, 6*(4), 380–384.

Mohebbi, S., Rostamizadeh, M., & Kahforoushan, D. (2020). Effect of molybdenum promoter on performance of high silica MoO_3/B-ZSM-5 nanocatalyst in biodiesel production. *Fuel, 266*, 117063.

Munir, M., Sharif, N.A.D.I.A., Naz, S.H.A.G.U.F.T.A., Saleem, F.A.I.Z.A., & Manzoor, F.A.R.K.H.A.N.D.A. (2013). Harvesting and processing of microalgae biomass fractions for biodiesel production. *Science, Technology and Development, 32*(3), 235–243.

Mussgnug, J. H., Klassen, V., Schlüter, A., & Kruse, O. (2010). Microalgae as substrates for fermentative biogas production in a combined biorefinery concept. *Journal of Biotechnology, 150*(1), 51–56.

Muthukumar, A., Elayaraja, S., Ajithkumar, T. T., Kumaresan, S., & Balasubramanian, T. (2012). Biodiesel production from marine microalgae *Chlorella marina* and *Nannochloropsis salina*. *Journal of Petroleum Technology and Alternative Fuels, 3*(5), 58–62.

Naik, M., Meher, L. C., Naik, S. N., & Das, L. M. (2008). Production of biodiesel from high free fatty acid Karanja (*Pongamia pinnata*) oil. *Biomass and Bioenergy, 32*(4), 354–357.

Nascimento, I. A., Marques, S. S. I., Cabanelas, I. T. D., Pereira, S. A., Druzian, J. I., de Souza, C. O., Vich, D. V., Carvalho, G. C.de., & Nascimento, M. A. (2013). Screening microalgae strains for biodiesel production: Lipid productivity and estimation of fuel quality based on fatty acids profiles as selective criteria. *Bioenergy Research, 6*(1), 1–13.

Nasirian, N., Almassi, M., Minaei, S., & Widmann, R. (2011). Development of a method for biohydrogen production from wheat straw by dark fermentation. *International Journal of Hydrogen Energy, 36*(1), 411–420.

Nautiyal, P., Subramanian, K. A., & Dastidar, M. G. (2014). Production and characterization of biodiesel from algae. *Fuel Processing Technology, 120*, 79–88.

Nguyen, T. H. M. (2012). Bioethanol production from marine algae biomass: Prospect and troubles. *Journal of Vietnamese Environment, 3*(1), 25–29.

Omraei, M., Sheibani, S., Sadrameli, S. M., & Tow, J. (2013). Preparation of biodiesel using KOH-MWCNT catalysts: An optimization study. *Industrial & Engineering Chemistry Research, 52*, 1829–1835. https://doi.org/10.1021/ie301418y

Oyedun, A. O., Lam, K. L., Gebreegziabher, T., Lee, H. K. M., & Hui, C. W. (2012). Optimisation of operating parameters in multi-stage pyrolysis. *Chemical Engineering Transactions, 29*, 655–660.

Park, J. H., Yoon, J. J., Park, H. D., Kim, Y. J., Lim, D. J., & Kim, S. H. (2011). Feasibility of biohydrogen production from *Gelidium amansii*. *International Journal of Hydrogen Energy, 36*(21), 13997–14003.

Passos, F., García, J., & Ferrer, I. (2013). Impact of low temperature pretreatment on the anaerobic digestion of microalgal biomass. *Bioresource Technology, 138*, 79–86.

Patil, P. D., Gude, V. G., Mannarswamy, A., Cooke, P., Nirmalakhandan, N., Lammers, P., & Deng, S. (2012). Comparison of direct transesterification of algal biomass under supercritical methanol and microwave irradiation conditions. *Fuel, 97*, 822–831.

Poonjarernsilp, C., Sano, N., & Tamon, H. (2014). Hydrothermally sulfonated single-walled carbon nanohorns for use as solid catalysts in biodiesel production by esterification of palmitic acid. *Applied Catalysis B: Environmental, 147*, 726–732.

Porphy, S. J., & Farid, M. M. (2012). Feasibility study for production of biofuel and chemicals from marine microalgae *Nannochloropsis* sp. based on basic mass and energy analysis. *International Scholarly Research Notices, 2012*.

Pratap, S. R., Shamshuddin, S. Z. M., Thimmaraju, N., Shyamsundar, M., & Reena, S. S. (2015). Kinetics of transesterification of *Madhuca indica* oil over modified zeolites: Biodiesel synthesis. *Bangladesh Journal of Scientific & Industrial Research, 50*(4), 271–278.

Rafa, N., Ahmed, S. F., Badruddin, I. A., Mofijur, M., & Kamangar, S. (2021). Strategies to produce cost-effective third-generation biofuel from microalgae. *Frontiers in Energy Research*. https://doi.org/10.3389/fenrg.2021.749968

Rahman, W. U., Fatima, A., Anwer, A. H., Athar, M., Khan, M. Z., Khan, N. A., & Halder, G. (2019). Biodiesel synthesis from eucalyptus oil by utilizing waste egg shell derived calcium based metal oxide catalyst. *Process Safety and Environmental Protection, 122*, 313–319.

Rajarapu, S. P., & Scharf, M. E. (2017). Saccharification of agricultural lignocellulose feedstocks and protein-level responses by a termite gut-microbe bioreactor. *Frontiers in Energy Research, 5*, 5.

Rajkumar, R., Yaakob, Z., & Takriff, M. S. (2014). Potential of micro and macro algae for biofuel production: A brief review. *Bioresources, 9*(1), 1606–1633.

Rashid, U., Soltani, S., Yaw Choong, T. S., Nehdi, I. A., Ahmad, J., & Ngamcharussrivichai, C. (2019). Palm biochar-based sulphated zirconium (Zr-AC-HSO3) catalyst for methyl ester production from palm fatty acid distillate. *Catalysts, 9*(12), 1029.

Rasouli, H., & Esmaeili, H. (2019). Characterization of MgO nanocatalyst to produce biodiesel from goat fat using transesterification process. *3 Biotech, 9*(11), 1–11.

Rizwanul Fattah, I. M., Ong, H. C., Mahlia, T. M. I., Mofijur, M., Silitonga, A. S., Rahman, S. A., & Ahmad, A. (2020). State of the art of catalysts for biodiesel production. *Frontiers in Energy Research, 8*, 101.

Saleem, M., Chakrabarti, M. H., Raman, A. A. A., Daud, W. M. A. W., & Mustafa, A. (2012). Hydrogen production by *Chlamydomonas reinhardtii* in a two-stage process with and without illumination at alkaline pH. *International Journal of Hydrogen Energy, 37*(6), 4930–4934.

Sangeetha, P., Babu, S., & Rangasamy, R. (2011). Potential of green alga *Chaetomorpha litorea* (Harvey) for biogas production. *International Journal of Current Science, 1*, 24–29.

Saoud, K., Rola Al Soubaihi, S. S., Bensalah, N., Al-Fandi, M., & Singh, T. (2018). Heterogeneous Ag and ZnOBased photocatalytic for waste water treatment under different irradiation conditions. *Journal of Materials and Environmental Science, 9*, 400–413.

Shaishav S., Singh R.N., & Satyendra T. (2013). Biohydrogen from algae: fuel of the future. *International Research Journal of Environmental Sciences, 2*, 44–47.

Shankar, A. A., Pentapati, P. R., & Prasad, R. K. (2017). Biodiesel synthesis from cottonseed oil using homogeneous alkali catalyst and using heterogeneous multi walled carbon nanotubes: Characterization and blending studies. *Egyptian Journal of Petroleum, 26*(1), 125–133.

Sharma, R., Chisti, Y., & Banerjee, U. C. (2001). Production, purification, characterization, and applications of lipases. *Biotechnology Advances, 19*(8), 627–662.

Shi, X., Jung, K. W., Kim, D. H., Ahn, Y. T., & Shin, H. S. (2011). Direct fermentation of Laminaria japonica for biohydrogen production by anaerobic mixed cultures. *International Journal of Hydrogen Energy, 36*(10), 5857–5864.

Shi, M., Zhang, P., Fan, M., Jiang, P., & Dong, Y. (2017). Influence of crystal of Fe_2O_3 in magnetism and activity of nanoparticle CaO@ Fe_2O_3 for biodiesel production. *Fuel, 197*, 343–347.

Shuit, S. H., Ng, E. P., & Tan, S. H. (2015). A facile and acid-free approach towards the preparation of sulphonated multi-walled carbon nanotubes as a strong protonic acid catalyst for biodiesel production. *Journal of the Taiwan Institute of Chemical Engineers, 52*, 100–108.

Shu, Q., Zhang, Q., Xu, G., & Wang, J. (2009). Preparation of biodiesel using s-MWCNT catalysts and the coupling of reaction and separation. *Food and Bioproducts Processing, 87*(3), 164–170.

Singh, A., He, B., Thompson, J., & Van Gerpen, J. (2006). Process optimization of biodiesel production using alkaline catalysts. *Applied Engineering in Agriculture, 22*(4), 597–600.

Singh, D., Subramanian, K. A., & Garg, M. O. (2018). Comprehensive review of combustion, performance and emissions characteristics of a compression ignition engine fueled with hydroprocessed renewable diesel. *Renewable and Sustainable Energy Reviews, 81*, 2947–2954.

Singh, D. P., & Trivedi, R. K. (2013). Production of biofuel from algae: An economic and eco-friendly resource. *International Journal of Science and Research, 2*, 352–357.

Stafford, W., Lotter, A., Brent, A., & von Maltitz, G. (2017). Biofuels technology: A look forward (No. 2017/87). WIDER Working Paper.

Suganya, T., Varman, M., Masjuki, H. H., & Renganathan, S. (2016). Macroalgae and microalgae as a potential source for commercial applications along with biofuels production: A biorefinery approach. *Renewable and Sustainable Energy Reviews, 55*, 909–941.

Sukumaran, R. K., Surender, V. J., Sindhu, R., Binod, P., Janu, K. U., Sajna, K. V., Rajasree, K. P., & Pandey, A. (2010). Lignocellulosic ethanol in India: Prospects, challenges and feedstock availability. *Bioresource Technology, 101*(13), 4826–4833.

Sun, J., Xiong, X., Wang, M., Du, H., Li, J., Zhou, D., & Zuo, J. (2019). Microalgae biodiesel production in China: A preliminary economic analysis. *Renewable and Sustainable Energy Reviews, 104*, 296–306.

Susilaningsih, D., Djohan, A. C., Widyaningrum, D. N., & Anam, K. (2009). Biodiesel from indigenous Indonesian marine microalgae *Nanochloropsis* sp. *Journal of Biotechnology Research Tropical Region, 2*, 1–4.

Todorovic, Z. B., Troter, D. Z., Dokic-Stojanovic, D. R., Velickovic, A. V., Avramovic, J. M., Stamenkovic, O. S., Veselinovic, L., & Veljkovic, V. B. (2019). Optimization of CaO-catalyzed sunflower oil methanolysis with crude biodiesel as a cosolvent. *Fuel, 237*, 903–910.

Topare, N. S., Renge, V. C., Khedkar, S. V., Chavan, Y. P., & Bhagat, S. L. (2011). Biodiesel from algae oil as an alternative fuel for diesel engine. *International Journal of Chemical Environmental and Pharmaceutical Research, 2*, 116–120.

Trivedi, N., Gupta, V., Reddy, C. R. K., & Jha, B. (2013). Enzymatic hydrolysis and production of bioethanol from common macrophytic green alga *Ulva fasciata* Delile. *Bioresource Technology, 150*, 106–112.

Vahid, B. R., & Haghighi, M. (2016). Urea-nitrate combustion synthesis of $MgO/MgAl_2O_4$ nanocatalyst used in biodiesel production from sunflower oil: Influence of fuel ratio on catalytic properties and performance. *Energy Conversion and Management, 126*, 362–372.

de Vasconcellos, A., Miller, A. H., Aranda, D. A., & Nery, J. G. (2018). Biocatalysts based on nanozeolite-enzyme complexes: Effects of alkoxysilane surface functionalization and biofuel production using microalgae lipids feedstock. *Colloids and Surfaces B: Biointerfaces, 165*, 150–157.

Vanegas, C. H., & Bartlett, J. (2013). Green energy from marine algae: biogas production and composition from the anaerobic digestion of Irish seaweed species. *Environmental technology, 34*(15), 2277–2283.

Vergara-Fernández, A., Vargas, G., Alarcón, N., & Velasco, A. (2008). Evaluation of marine algae as a source of biogas in a two-stage anaerobic reactor system. *Biomass and Bioenergy, 32*(4), 338–344.

Vieira, S. S., Magriotis, Z. M., Santos, N. A., Saczk, A. A., Hori, C. E., & Arroyo, P. A. (2013). Biodiesel production by free fatty acid esterification using lanthanum (La3+) and HZSM-5 based catalysts. *Bioresource Technology, 133*, 248–255.

Wargacki, A. J., Leonard, E., Win, M. N., Regitsky, D. D., Santos, C. N. S., Kim, P. B., Cooper, S. R., Raisner, R. M., Herman, A., Sivitz, A. B., Lakshmanaswamy, A., Kashiyama, Y., Baker, D., & Yoshikuni, Y. (2012). An engineered microbial platform for direct biofuel production from brown macroalgae. *Science, 335*(6066), 308–313.

Wen, Z., Yu, X., Tu, S. T., Yan, J., & Dahlquist, E. (2010). Synthesis of biodiesel from vegetable oil with methanol catalyzed by Li-doped magnesium oxide catalysts. *Applied Energy, 87*(3), 743–748.

Williams, P. J. L. B., & Laurens, L. M. (2010). Microalgae as biodiesel & biomass feedstocks: Review & analysis of the biochemistry, energetics & economics. *Energy & Environmental Science, 3*(5), 554–590.

Wu, H., Zhang, J., Wei, Q., Zheng, J., & Zhang, J. (2013). Transesterification of soybean oil to biodiesel using zeolite supported CaO as strong base catalysts. *Fuel Processing Technology, 109*, 13–18. https://doi.org/10.1016/j.fuproc.2012.09.032

Wu, X., Li, Q., Dieudonne, M., Cong, Y., Zhou, J., & Long, M. (2010). Enhanced H2 gas production from bagasse using adhE inactivated *Klebsiella oxytoca* HP1 by sequential dark-photo fermentations. *Bioresource Technology, 101*(24), 9605–9611.

Wu, H., Zhang, J., Wei, Q., Zheng, J., & Zhang, J. (2013). Transesterification of soybean oil to biodiesel using zeolite supported CaO as strong base catalysts. *Fuel Processing Technology, 109*, 13–18.

Xie, W., Huang, X., & Li, H. (2007). Soybean oil methyl esters preparation using NaX zeolites loaded with KOH as a heterogeneous catalyst. *Bioresource Technology, 98*(4), 936–939.

Xie, W., Yang, Z., & Chun, H. (2007). Catalytic properties of lithium-doped ZnO catalysts used for biodiesel preparations. *Industrial & Engineering Chemistry Research, 46*(24), 7942–7949.

Yanagisawa, M., Nakamura, K., Ariga, O., & Nakasaki, K. (2011). Production of high concentrations of bioethanol from seaweeds that contain easily hydrolyzable polysaccharides. *Process Biochemistry, 46*(11), 2111–2116.

Yazdani, P., Karimi, K., & Taherzadeh, M. J. (2011). Improvement of enzymatic hydrolysis of a marine macroalga by dilute acid hydrolysis pretreatment. In *WREC11 world renewable energy conference 2011* (pp. 1–6). Linköping, Sweden: Linköping University.

Yoon, M., Choi, J. I., Lee, J. W., & Park, D. H. (2012). Improvement of saccharification process for bioethanol production from *Undaria* sp. by gamma irradiation. *Radiation Physics and Chemistry, 81*(8), 999–1002.

Yuan, H., Zhang, Y., Li, X., Meng, Y., Liu, C., Zou, D., & Liu, Y. (2014). Effects of ammoniation pretreatment at low moisture content on anaerobic digestion performance of rice straw. *Bioresources, 9*(4), 6707–6718.

Yuan, X., Shi, X., Zhang, D., Qiu, Y., Guo, R., & Wang, L. (2011). Biogas production and microcystin biodegradation in anaerobic digestion of blue algae. *Energy & Environmental Science, 4*(4), 1511–1515.

Zhang, C., Li, J., Liu, C., Liu, X., Wang, J., Li, S., & Zhang, L. (2013). Alkaline pretreatment for enhancement of biogas production from banana stem and swine manure by anaerobic codigestion. *Bioresource technology, 149*, 353–358.

Zhang, L., Cheng, J., Pei, H., Pan, J., Jiang, L., Hou, Q., & Han, F. (2018). Cultivation of microalgae using anaerobically digested effluent from kitchen waste as a nutrient source for biodiesel production. *Renewable Energy, 115*, 276–287.

Zhang, D., Wei, D., Li, Q., Ge, X., Guo, X., Xie, Z., & Ding, W. (2014). High performance catalytic distillation using CNTs-based holistic catalyst for production of high quality biodiesel. *Scientific Reports, 4*(1), 1–5.

Zhang, Z., Yuan, X., Miao, S., Li, H., Shan, W., & Jia, M. (2019). Effect of Fe additives on the catalytic performance of Ion-exchanged CsX zeolites for side-chain Alkylation of Toluene with methanol. *Catalysts, 9*, 829. https://doi.org/10.3390/catal9100829

Zheng, Y., Pan, Z., Zhang, R., El-Mashad, H. M., Pan, J., & Jenkins, B. M. (2009). Anaerobic digestion of saline creeping wild ryegrass for biogas production and pretreatment of particleboard material. *Bioresource Technology, 100*(4), 1582–1588.

Zhong, W., Zhang, Z., Luo, Y., Qiao, W., Xiao, M., & Zhang, M. (2012). Biogas productivity by co-digesting Taihu blue algae with corn straw as an external carbon source. *Bioresource Technology, 114*, 281–286.

Zuliani, A., Ivars, F., & Luque, R. (2018). Advances in nanocatalyst design for biofuel production. *ChemCatChem, 10*(9), 1968–1981.

CHAPTER

3

Valorizing algae biomass as materials for bioproducts and commercial applications

Yulin Hu[1], Kuljeet Grewal[1], Aitazaz A. Farooque[1,2] and Quan (Sophia) He[3]

[1]*Faculty of Sustainable Design Engineering, University of Prince Edward Island, Charlottetown, PE, Canada;*
[2]*School of Climate Change and Adaptation, University of Prince Edward Island, Charlottetown, PE, Canada;*
[3]*Department of Engineering, Dalhousie University, Truro, NS, Canada*

1. Introduction

Concerns regarding the depletion of fossil fuels and a massive amount of greenhouse gas (GHG) emissions are caused by the use of fossil fuels; the need for replacing them with renewable and sustainable biomass energy and bioproducts has become increased. As suggested by Prabakaran et al. (Prabakaran et al., 2022), the current use of fossil fuels accounts for around 80% of global energy consumption, i.e., coal (20%), natural gas (27%), and oil (33%); however, the major contributor for GHG and CO_2 emissions is found to be fossil fuels, and thus the extreme weather conditions nowadays are linked to the consumption of the fossil fuels. Annual global CO_2 emissions caused by using fossil fuels have continuously increased over the years, as shown in Fig. 3.1. Thus, it is urgent to lower the

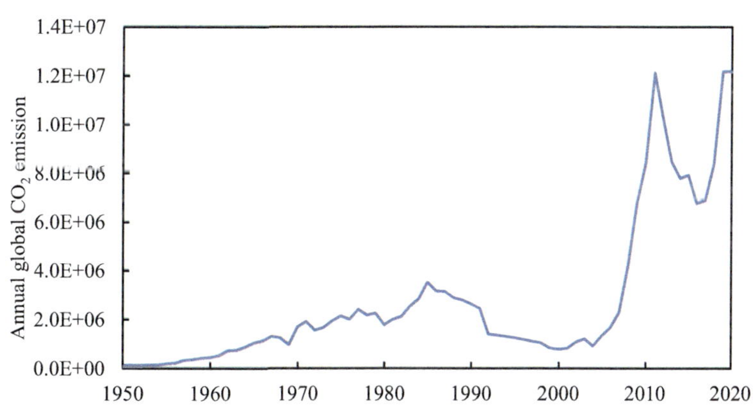

FIGURE 3.1

Annual Global CO_2 Emission.

Source: Our World in Data.

Algae Materials: Applications Benefitting Health. https://doi.org/10.1016/B978-0-443-18816-9.00011-3

use of fossil fuels and start using renewable and environmentally friendly biomass-derived biofuels and biochemicals like obtained from algal biomass.

To compare with other first- and second-generation biomass, algae have been considered as more suitable resources for producing valuable bioproducts owing to their inherent benefits such as high biomass yield, fast growth rate, the presence of value-added biomolecules (e.g., lipid, protein, polysaccharides, omega-3-fatty acid, and pigments), the ability to be grown in nonarable land, as well as carbon sequestration (Hu et al., 2019). This has led to the applications of algae in a range of sectors, as displayed in Fig. 3.2.

In general, algae can be categorized into two main groups including microalgae and macroalgae based on their cell structure. Macroalgae or called as seaweed are primarily composed of brown macroalgae (e.g., *Laminaria digitate* and *Ascophyllum nodosum*), red macroalgae (e.g., *Gracilaria Salicornia* and *Gelidium elegans*), and green macroalgae (e.g., *Ulva lactuca* and *Enteromorpha intestinalis*), which is based on their color from chlorophyll and natural pigments. While there are up to 50,000 microalgae, species have been found so far and they are composed of unicellular structure (Zhang et al., 2021). Because of the renewable and sustainable properties of algal biomas, a number of studies have been carried out by converting algae into value-added biofuels including solid, liquid, and gaseous forms, biochemicals such as lutein, astaxanthin, and β-carotene with high nutritious and medical value, and other high-value biomaterials. Therefore, this chapter is outlined as follows: (1) algae as resources to prepare a highly invaluable and versatile material——biochar; (2) converting algae into liquid and gaseous forms of biofuels such as biodiesel and H_2; and (3) the synthesis and extraction of carotenoids (i.e., astaxanthin and lutein) from algae.

2. Biochar from algae

Biochar is a carbonaceous material that can be produced from biomass via slow pyrolysis, which can be potentially applied in a range of areas such as catalytic reactions, heating source, carbon sequestration, wastewater purification, soil fertilizer, and fuel cell electrode (Fig. 3.3). Slow pyrolysis is one type of pyrolysis processes operating in oxygen-free condition at a low heating rate of 0.1−1°C/s and residence time ranging from several minutes to hours, during which dehydration, devolatilization, and

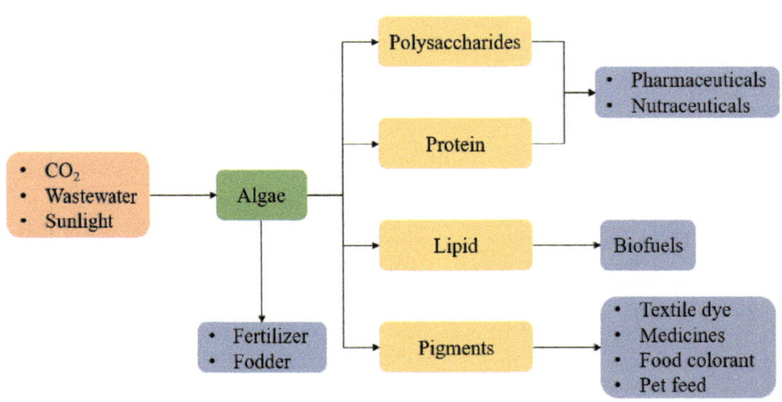

FIGURE 3.2

A Summary of the Applications of Algae (Nayana et al., 2022).

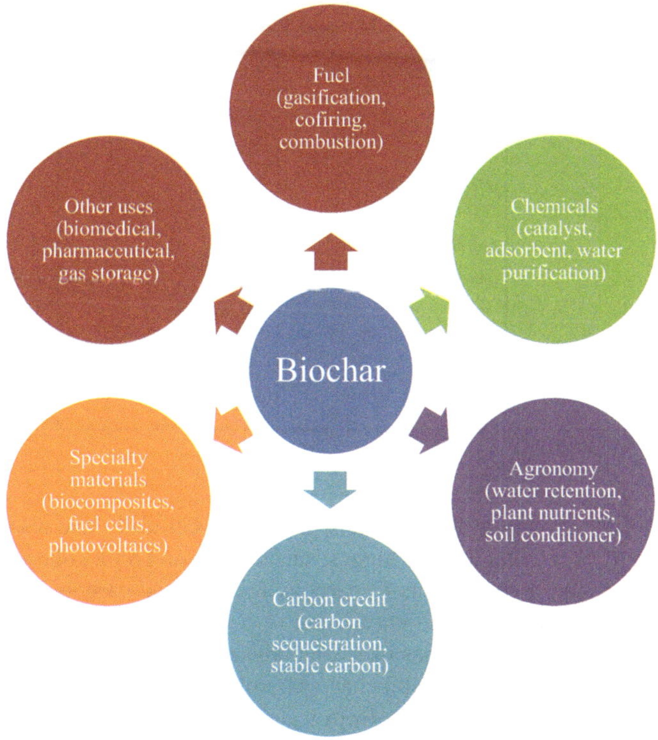

FIGURE 3.3

A Summary of the Potential Applications of Biochar (Varma, 2019).

disintegration occur (Sun et al., 2022). As previously suggested by Lee et al. (2020), the determination of the chemical composition (i.e., protein, lipid, and carbohydrates) as well as the content of moisture, volatiles, fixed carbon, and ash is capable for providing insights of algae as a potential pyrolysis feedstock to produce biochar. For instance, volatile matter (VM) is defined as the combustible gases or vapors released from the biomass during the heating, and the fixed carbon (FC) is defined as the solid residue remaining pyrolysis at high temperature minus the amount of ash content. Table 3.1 summarizes the VM and FC content of different algae strains. As shown in Table 3.1, it is clearly stated that microalgae contain a higher VM content than that of macroalgae, and an opposite trend is found in the FC content of microalgae and macroalgae. In typical, VM/FC ratio is an indicator for biochar yield, and a higher VM/FC ratio tends to lead to a lower yield of biochar (Sun et al., 2022). The following paragraphs will describe the recent studies on the production of algae-derived biochar by slow pyrolysis and how the biomass characteristics and reaction conditions will affect the properties of biochar and its application in the respective fields.

Babatabar et al. (2022) compared pyrolysis of six types of feedstocks (i.e., rice husk, coconut shell, walnut shell, *Cladophora glomerata*, *Gracilaria gracilis*, and *Azolla filiculoides*) in a fixed-bed reactor at 500°C, and the influences of chemical composition on the product yields and properties were investigated. In general, when compared with biochar obtained from lignocellulose, it was found that the biochar derived from algal bloom showed more nitrogen-containing compounds because of the

Table 3.1 A summary of volatile matter (VM) and fixed carbon (FC) content of algae.

Species	VM (wt.%)	FC (wt.%)	Reference
Chlorella	84.3	10.4	Jabeen et al. (2020)
Spirulina	82.3	10.4	Jabeen et al. (2020)
Kappaphycus alvarezii	61.3	61.5	Nallasivam et al. (2022)
Eucheuma denticulatum	0.4	23.9	Nallasivam et al. (2022)
Sargassum natans	48.9	11.6	Tobio-perez et al. (2022)
Sargassum pallidum	44.6	8.3	Li et al. (2010)
Laminaria japonica	41.6	13.3	Li et al. (2010)
Gracilaria gracilis	53.1	10.9	Parsa et al. (2018)
Dunaliella tertiolecta	54.5	27.0	Shuping et al. (2010)
Chlamydomonas reinhardtii	70.8	11.6	Andrade et al. (2018)

thermal degradation of the protein in the pyrolysis process. Specifically, the results showed that the biochar obtained from *A. filiculoides* had the highest surface area and pore volume. The *C. glomerata*-derived biochar demonstrated high contents of carboxylic acids and carbonyl groups, indicating its potential to be used for removing lead and chromium.

Nguyen et al. (2022) applied three types of marine algae (i.e., *Ulva ohnoi*, *Agardhiella subulata*, and *Sargassum hemiphyllum*) to synthesize biochar by pyrolysis and chemical activation using $ZnCl_2$, and their abilities for adsorbing ciprofloxacin from water were determined. It was observed that the activated biochar obtained from *S. hemiphyllum* contained mesopores and excellent physiochemical characteristics such as surface area, pore volume, and pore diameter, which ensures this biochar sample the best ciprofloxacin adsorption ability. Besides, the electrostatic interaction, $\pi-\pi$ EDA interaction, pore filling, and hydrogen bonding were reported to be the adsorption mechanism of the algae-derived biochar for removing ciprofloxacin.

Farobie et al. (2022) prepared biochar from *Sargassum* sp. by slow pyrolysis at 400–600°C for 10–30 min in a batch reactor, and the higher energy density (23.12–25.89 MJ/kg) of the biochar sample than low-ranked coals indicated its ability to be employed as the solid fuel. To compare microalgae (i.e., *Chlorella vulgaris*), biochar obtained from *Sargassum* sp. (one type of macroalgae or called seaweed) porous than that obtained from *C. vulgaris*, which could be due to the higher content of carbohydrates in the macroalgae than microalgae. In addition, as revealed by the BET surface area (4.28–5.68 m^2/g) and the mean pore diameter (10.83–12.20 nm), the biochar obtained from *Sargassum* sp. as mesoporous materials.

Furthermore, apart from the use of biochar as sorbent and solid fuel, it can be utilized as the environmentally friendly anode. Previously, Kiminaitė et al. (2022) fabricated and characterized the anode that is prepared from *Rhodophyta*-derived activated biochar combining with a flaxseeds mucilage binder at varying ratios (i.e., 5 wt.%, 10 wt.%, 20 wt.%, and 30 wt.%), as displayed in Fig. 3.4. The incorporation of activated biochar and 10 wt.% binder produced the best anode material to be used in the microbial fuel cell application. The authors observed that the effect of binder addition on the thermal stability of anode was insignificant at temperatures up to 200°C. On the other hand, the

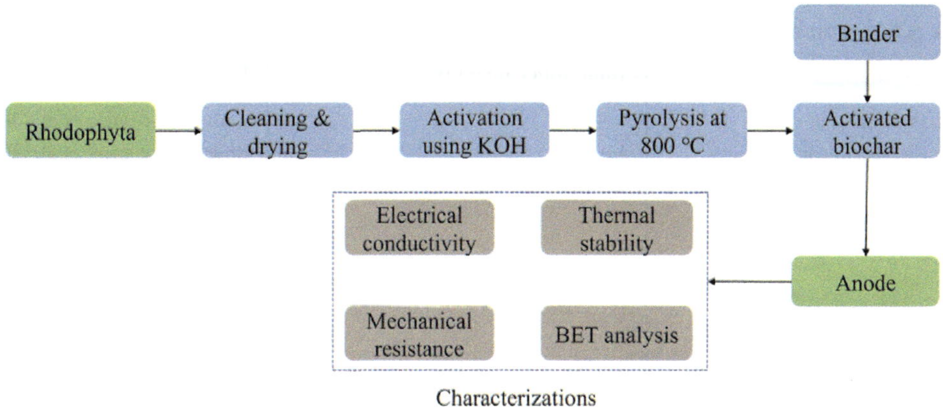

FIGURE 3.4

The Flowchart for Preparing Anode from Algae-derived Activated Biochar (Kiminaitė et al., 2022).

binder concentration demonstrated a great impact on the mechanical resistance and electrical conductivity of the algal biochar-based anode, i.e., the highest compressive strength of anode of 104 bar was found when adding 30 wt.% of binder, while the lowest binder concentration (i.e., 5 wt.%) resulted in the highest electrical conductivity of anode of 58 S/m.

In summary, the preparation of biochar from algae has a variety of uses, and this synthesis process often consists of slow pyrolysis and postprocessing (e.g., physical activation, chemical activation, heteroatom doping, and biochar—nanostructure composites). Although these postmodification/functionalization steps can significantly enhance the property of algae-derived biochar, the cost of current methods is extremely high, and thus, it is utmost of importance to develop cheap and environmentally friendly technique (Sun et al., 2022).

3. Biofuels from algae

3.1 Biodiesel

Biodiesel can be produced from vegetable oil or animal fat via transesterification, during which methanol or ethanol and a homogenous/heterogenous catalyst should be added. The produced biodiesel can be applied to substitute petroleum-based diesel fuel or mix with it (Kale & Patle, 2022). The biodiesel can be used in the diesel engines with some precautions in different sectors such as on-road vehicles (e.g., fleet, heavy-duty track, school bus, and urban transit bus), off-road vehicles (e.g., agricultural, construction, mining, and forestry equipment, trains, and marine vessels), and stationary equipment (e.g., electricity generator and furnace). Except for the feedstocks as earlier mentioned, algae are alternative and excellent resources for biodiesel production due to the high lipid content, high biomass productivity, and tailorable lipid accumulation in the cell. The lipid content of various algae strains is summarized in Table 3.2. Despite these advantages, the use of algae as the feedstocks to produce biodiesel is still technology-driven instead of market-driven (Shokravi et al., 2022). The flowchart for producing biodiesel from algae is illustrated in Fig. 3.5.

Table 3.2 A summary of total lipid content of different algae species.

Algae specie	Total lipid content (wt.%)	Reference
Coelastrella sp.	44–46	Rodriguez-palacio et al. (2022)
Verrucodesmus verrucosus	43–44	Rodriguez-palacio et al. (2022)
Neochloris oleoabundans	35–37	Rodriguez-palacio et al. (2022)
Azolla filiculoides	11.7	Golzary et al. (2021)
Nannochloropsis sp.	33	Islam et al. (2022)
Tetraselmis sp.	19	Islam et al. (2022)
Chaetoceros sp.	18	Islam et al. (2022)
Leptolyngbya	32.1	Cao et al. (2022)
Geitlerinema	16.5	Cao et al. (2022)
Chlorella vulgaris	13	Hu et al. (2017)

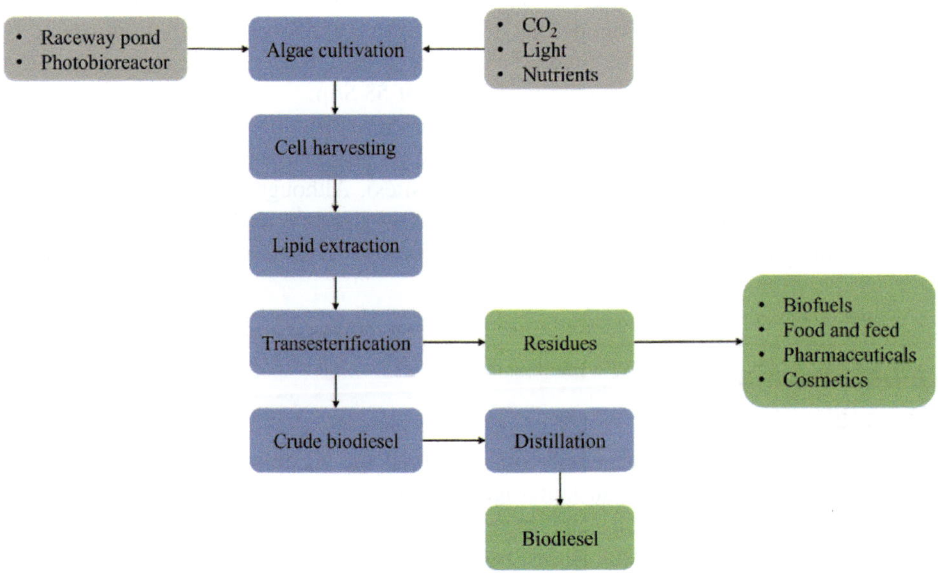

FIGURE 3.5

The Flowchart for Biodiesel Production from Algae (Liu, 2011, p. Ch. 7).

Recently, novel lipid extraction methods have been investigated for algae, especially when considering the high cost and time-consuming of conventional approach (Folch/Bligh and Dryer extraction method of dried algae) (Enamala et al., 2018). Supercritical CO_2 was adopted by Giri and

Majidian (2022) to extract lipid from dried *Dunaliella tertiolecta* at 40–80°C for 200–370 bar and CO_2 flow rate of 100–200 g/min, followed by centrifugation, removal of pigments and polar fats, and water removal to produce neutral fats. The highest fatty acid content/composition was obtained at 40°C and 370 bar, and the lowest value was achieved at 80°C and 200 bar. Additionally, palmitic acid (C16:0) and stearic acid (C18:0) contained the highest concentration in the extracted lipid from *D. tertiolecta* by supercritical CO_2. In another study, 24% of lipid and 90% of extraction efficiency were attained from supercritical CO_2 extraction of *Nannochloropsis oculata* (Crampon et al., 2013).

Aron et al. (2022) investigated the use of liquid biphasic flotation method to recover lipid from *Chlorella sorokiniana*, and mechanism of this extraction process is depicted in Fig. 3.6. This new extraction approach is to integrate a salt (ammonium sulfate) and an alcohol with mass transfer mode of solvent sublation that makes the transfer of biomolecules from the lower layer to upper layer easy. In this study, the tested parameters included types of alcohol (ethanol, methanol, 1-propanol, and 2-propanol), concentration of alcohol (20%–100%), algae concentration (12.5%–100%), and flotation time (10–30 min). The results indicated that an approximately 74% of lipid was recovered from *C. sorokiniana* using 100 vol.% of ethanol and ammonium sulfate for 20 min. The most common fatty acid composition identified in the extracted lipid was short-chain fatty acid like C5:0, medium chain fatty acid like C12:0, and long chain fatty acid like C20:0. Subsequently, the capacity of this liquid biphasic flotation system was scaled up from 50 to 500 mL, and the lipid extraction obtained from 50 to 500 mL was similar, implying that the liquid biphasic flotation for lipid extraction can be conducted at an industrial scale with ease of operation.

Despite dried algal biomass to be used as the feedstock for producing biodiesel, wet algae have been investigated and fed directly into the reactor to produce biodiesel via in situ transesterification. This method is an integration of extraction and transesterification occurring in a combined reactive

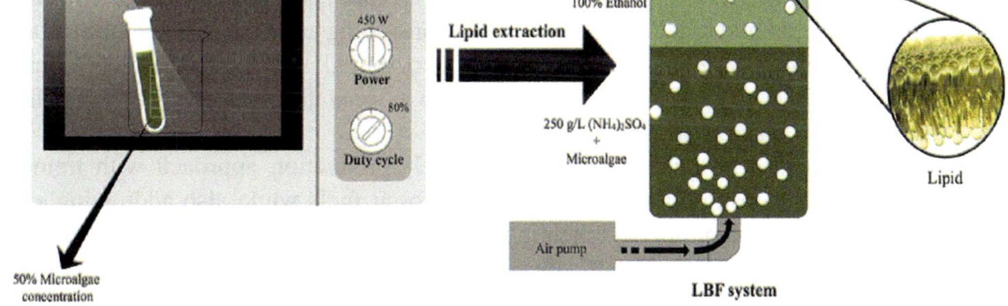

FIGURE 3.6

The Mechanisms Used for Lipid Extraction from algae by Liquid Biphasic Flotation Method (Mat Aron et al., 2022) (permission received from Elsevier).

extraction environment, which allows to eliminate the energy-intensive predrying step (Go et al., 2016). Moreover, the oil recovery and refining stages before actual biodiesel conversion stage can be eliminated by performing in situ transesterification (Park et al., 2015). Felix et al. (2019) produced biodiesel from *C. vulgaris* with 80% of moisture via noncatalytic subcritical methanol at 200–250°C, 1–3 h, and solvent to feedstock ratio of 2–8 $ml_{methanol}/g_{wet\ feedstock}$. It was observed that the optimal conditions were temperature of 220°C, residence time of 2 h, and solvent to feedstock ratio of 8 $ml_{methanol}/g_{wet\ feedstock}$, leading to a maximum yield of fatty acid methyl ester of 74.6% with respect to the highest lipid concentration of *C. vulgaris*.

Except for the need for improving current lipid extraction and biodiesel production methods for algae, the harvesting and cultivation of algae are another important factors affecting the quality of algae-derived biodiesel. For example, the fatty acid profile and lipid yield are prone to be affected by variability during the harvesting, and the algal cultivation largely determines the feasibility of the commercial algal biofuel production. Undoubtedly, it is of utmost importance to integrate cultivation, and harvesting is a key toward the commercialization of algal biodiesel production; however, most previous studies only limit their investigation on the algal harvesting and cultivation on the lab scale, and future research work regarding the upscaling is a necessity (Shokravi et al., 2022).

3.2 H_2

H_2 is a clean energy carrier since only water is generated as the by-product upon combustion. Aside from its wide application in the oil refinery (representing >33% of the H_2 consumption), H_2 is also utilized in the production of ammonia, methanol, and steel. Market forecasts underscore that the global H_2 production will increase dramatically from 60 million metric tons in 2018 to 300 million metric tons in 2030. Presently, the majority (>98%) of H_2 is produced from either steam methane reforming (SMR) (represents 76% of the global H_2 production) or coal gasification (represents 22% of the global H_2 production) (Office of Fossil Energy USD of E, 2020), whereas these technologies lead to a massive amount of CO_2 emission, i.e., SMR releases around 7 kg of CO_2 per kg of H_2 produced. Instead of using nonrenewable fossil fuels as the feed and reducing the associated CO_2 emissions, algae could be promising resources for H_2 production. The recent studies have focused on H_2 production from algae by thermochemical (e.g., gasification and supercritical water gasification) and biological approaches (e.g., dark fermentation and photofermentation). An overview of the production routes is presented in Fig. 3.7. In comparison, biological approaches are energy-saving and greener than thermochemical methods. Nevertheless, biological methods are very time-consuming and thus might not be suitable for scaling up when considering the relatively lower H_2 yield compared with thermochemical conversion routes (Farobie et al., 2021). As a result, this subsection is only focused on the thermochemical conversion for producing H_2 from algae.

Biomass gasification is a promising and alternative H_2 production approach with tremendous possibilities to reduce the consumption of nonrenewable fossil fuels while also addressing environmental issues in long-term planning and helping to achieve a sustainable development manner (Narnaware & Panwar, 2022). Gasification is a thermochemical conversion method to produce gaseous products at elevated temperatures using gasifying agents (e.g., air, O_2, steam, or their combinations). Babatabar & Saidi (2021) developed a process consisting of steam gasification and water gas shift reaction to produce H_2 from algae. Initially, algae were gasified in a fluidized bed reactor using steam as the gasifying agent, followed by a cyclone to separate solid and a reforming reactor to crack tar.

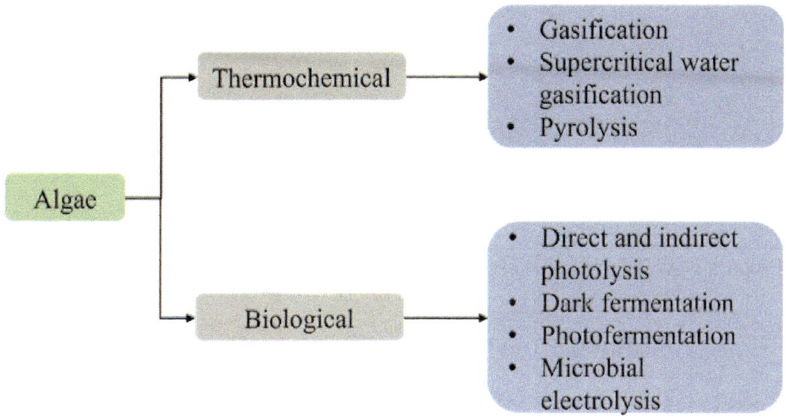

FIGURE 3.7

A Summary of the Thermochemical and Biological Conversion Routes for H_2 Production from Algae (Kumar Sharma et al., 2021).

After this, the gas contaminants in the product gas such as HCl, H_2S, and NH_3 were removed by a gas cleanup unit. To enhance the H_2/CO ratio, the product gas was charged into a water shift reactor. The simulation results showed that the highest H_2 content was achieved at the highest steam to biomass ratio and water gas shift reactor and lowest temperature in water gas shift reaction. By calculating the cold gas efficiency, the tested feedstocks such as woody biomass, *Spirulina*, corn cob, and rice husk increased with increasing temperature and steam to biomass ratio in the gasification treatment. To compare, the optimal gasification conversion was obtained when using woody biomass (83.6%), followed by *Spirulina* (83.5%), corn cob (74.9%), and rice husk (69.9%).

Aside from the conventional biomass gasification methods such as using steam, O_2, air, or their mixture over a catalyst or not, supercritical water gasification (SCWG) is an alternative and more promising approach for converting algae into H_2 when considering its ability to gasify feedstocks in supercritical water and the elimination of predrying treatment. At supercritical conditions of water (i.e., temperature $\geq$374°C and pressure $\geq$22.1 MPa), water behaves like a solvent with a high diffusivity, mass transfer, and dissolving power, and its physiochemical properties such as density, dielectric constant, ionic product, and thermal conductivity drastically differ from them at ambient conditions. In addition, since water can behave like both a liquid and a gas at supercritical state, the reaction for the gasification of liquid and gas systems can achieve a homogenous condition without any phase boundaries (Farobie et al., 2021). A simplified process flow diagram of H_2 production from algae via SCWG is depicted in Fig. 3.8.

In SCWG, steam reforming, water–gas shift, and methanation reaction are the most common reactions, as shown in Eq. (3.1) (Hu et al., 2019; Zhang et al., 2021). In supercritical water conditions, the organic fraction of feedstock can be decomposed to produce CO and H_2 at high temperatures due to the thermodynamically endothermic nature of CO and H_2 formation. The produced CO can be further converted into H_2 and CO_2 by water–gas shift reaction (Eq. 3.2), or CO can react with H_2 to form CH_4 (Eq. 3.3). On the contrary, thermodynamically speaking, CO_2 and CH_4 are more favorable to form at lower temperatures and higher biomass loading (Yan et al., 2006).

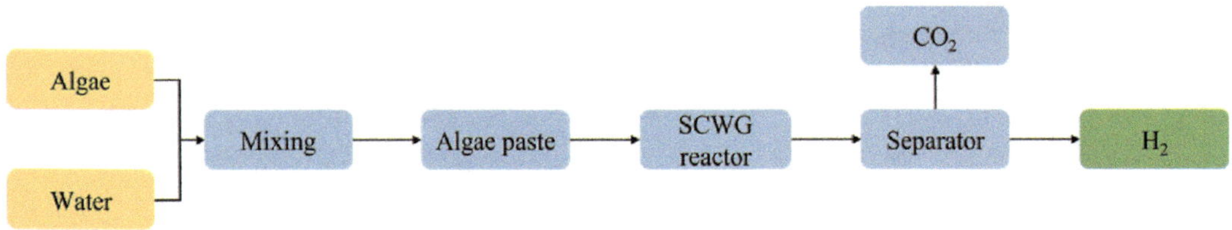

FIGURE 3.8

The Simplified Process Flow Diagram of H_2 Production from Algae via SCWG.

$$C_nH_mO_y + (n\text{-}y)H_2O \rightarrow nCO + (n - y + m/2)H_2 \tag{3.1}$$

$$CO + H_2O \rightleftharpoons CO_2 + H_2 \tag{3.2}$$

$$CO + 3H_2 \rightleftharpoons CH_4 + H_2O \tag{3.3}$$

To compare with conventional gasification, SCWG offers several advantages: (1) no requirement for drying as the pretreatment; (2) no mass transfer constrain due to the special properties of supercritical water; (3) better heat transfer at supercritical water conditions than that in the liquid and gas; (4) higher reaction rate; and (5) better flowability to lower the coke yield and thus extend the lifetime of the catalyst (Farobie et al., 2017; Guo et al., 2010; Hu et al., 2020). To date, SCWG of biomass has been performed in either a batch reactor mode (Duan et al., 2018) (Fig. 3.9) or a continuous reactor mode (Samanmulya et al., 2017) (Fig. 3.10). The most recent studies regarding the SCGW of algae for H_2 production are summarized in Table 3.3.

Even though SCWG is an environmentally benign and promising approach that can be employed to gasify algae in supercritical water to produce H_2, there are several technical challenges that must be addressed at small- and large-scale operations:

(i) Low energy efficiency: A high energy demand is required to maintain high temperature and pressure throughout SCWG treatment and also the endothermic nature of biomass degradation. To lower the energy consumption, an effective heat exchanger might be needed to recover the heat caused by the water heating and the application of nonconventional energy source such as solar energy (Onigbajumo et al., 2022).

(ii) High operating and capital cost: Due to the high temperature and pressure applied in the SCWG process, advanced construction materials are needed like alloy-type materials, and thus, the total operating and capital cost will be dramatically increased.

(iii) Reactor plugging and char formation: Reactor plugging and char formation are two most critical challenges faced by SCWG process. The reactor plugging is mainly caused by the decreased solubility of inorganics present in the biomass at higher temperature (Hu et al., 2020). Char formation could be resulted from the incomplete biomass decomposition, and it can be reduced by using higher heating rates or organic acid as a radical scavenger.

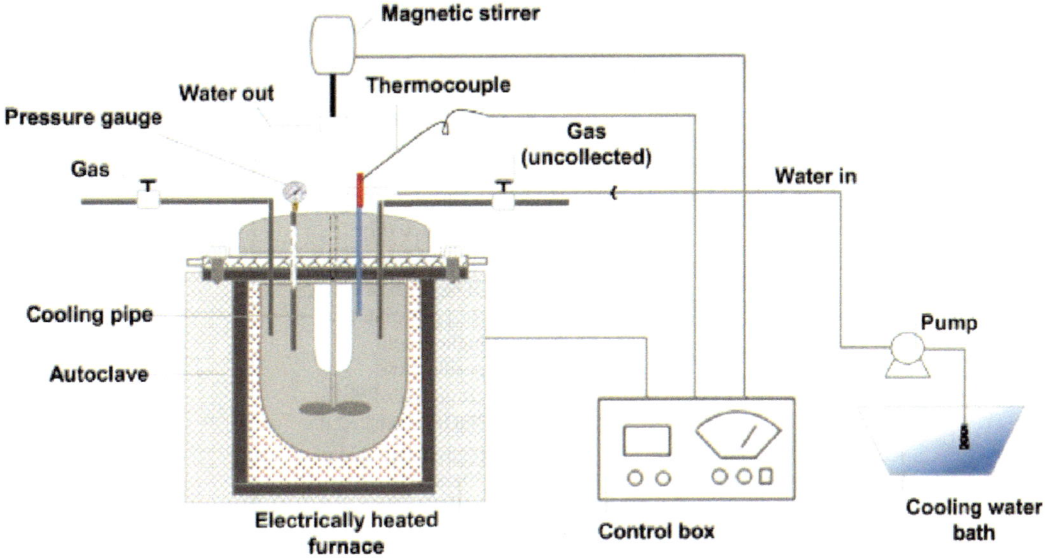

FIGURE 3.9

The batch reactor used for SCWG for H_2 production (Farobie et al., 2021) (permission received from elsevier).

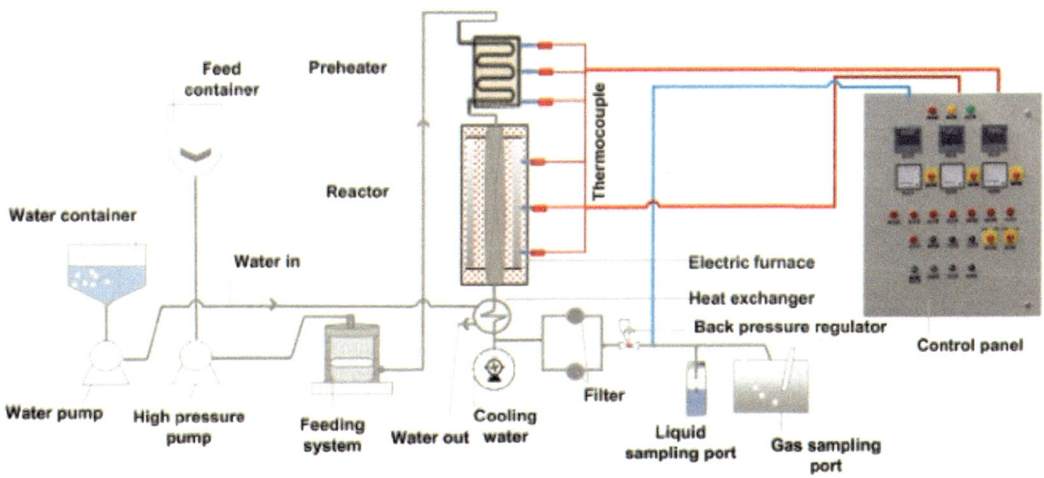

FIGURE 3.10

A continuous reactor for H_2 production by SCWG (Farobie et al., 2021) (permission received from elsevier).

(iv) Corrosion of the reactor: Ni-based materials are commonly used as the construction materials in SCWG of biomass; however, the occurrence of unwanted side reactions between Ni and feedstock and the harsh reaction conditions could cause corrosion to the reactor material (Yan et al., 2021).

Table 3.3 A summary of the recent studies on SCWG of algae for H_2 production.

Algae specie	Conditions	Main conclusions	References
Laminaria hyperborea	500°C; 35 min; 6.6 wt.% feedstock concentration; catalyst type of Ru/Al_2O_3, Ni/Al_2O_3, and NaOH.	• Product gas yield was proportional to temperature, catalyst loading, and residence time. • Ru/Al_2O_3 was found to be the best catalyst for producing the highest cold gasification efficiency of 98%. • NaOH produced the highest H_2 yield of 30 mol/kg.	Cherad et al. (2014)
Posidonia oceanica	300–600°C; 60 min; 0.04–0.12 g/mL feedstock loading	• in the product gas, H_2, CH_{4s}, and CO_2 were the main gaseous products. • The highest H_2 yield of 10.37 mol/kg was obtained at 600°C and 0.08 g/mL feedstock loading.	Deniz et al. (2015)
Ulva	400–550°C; 230–270 bar; 7–120 min; 7–16.4 wt.% feedstock loading	• H_2 formation was favorable at higher temperatures but unfavorable at higher feedstock loading. • overall, the gasification efficiency reached around 100%, which was supported by a 0.4 wt.% of dissolved organic carbon in the liquid phase.	Graz et al. (2016)
Enteromorpha intestinalis	440°C; 250 bar; 1 wt.% feedstock loading; $Fe\text{-}Ni\text{-}Ru/\gamma\text{-}al_2O_3$	• H_2 production was enhanced by adding $Fe\text{-}Ni\text{-}Ru/\gamma\text{-}al_2O_3$ as the catalyst. • The highest H2 yield of 12.28 mmol/g was obtained using	Norouzi et al. (2017)

Table 3.3 A summary of the recent studies on SCWG of algae for H_2 production.—cont'd

Algae specie	Conditions	Main conclusions	References
		2 wt.% Ru promoted catalyst, along with a H_2 selectivity of 0.74.	
Chlorella vulgaris; *Spirulina platensis*; *Saccharina latissimi*	500°C; 36 MPa; 30 min; NaOH and Ni/Al_2O_3 as the catalyst	• *Saccharina* led to the highest H_2 yield of 15.1 mol/kg. • in terms the composition of tar, protein-rich algae such as *Chlorella* and *Spirulina* mainly contained pyridines, pyrroles, indoles, etc. • While the tar obtained from saccharina primarily contained phenols and benzenes.	Onwudili et al. (2013)
Cladophora glomerata	380–460°C; 5–30 min; 23.2–27 MPa; algal hydrochar as the catalyst	• The addition of algal hydrochar as the catalyst promoted the H_2 formation. • The highest H_2 yield of 9.63 mmol/g was achieved when adding hydrochar as the catalyst with its weight ratio to feedstock of 0.4.	Safari et al. (2016)
Fucus serratus; *Laminaria digitata*; *Alaria esculenta*; *Bifurcaria bifurcata*	500°C; 60 min; 5 wt.% feedstock loading	• The feedstocks mainly containing lignocellulose and protein led to a lower coke yield. • The main components identified in the product gas were H_2, CH_4, and CO_2. • H2 yield was in the range of 11.8–16 $gH_2/kg_{seaweed}$, and CH_4 yield was between 39 and 104 g $CH_4/kg_{seedweed}$.	Schumacher et al. (2011)

4. Carotenoids from algae

4.1 Astaxanthin

Algae have been explored as cell factories for producing value-added products such as carotenoids that can be used in pharmaceuticals, foods, and cosmetics (Gong & Bassi, 2016). Among them, astaxanthin (Fig. 3.11) has demonstrated the strongest antioxidant capacity, i.e., 1000 times of vitamin E, 65 times of vitamin C, 17 times of grape seed, and 10 times of other carotenoids (e.g., β-carotene, lutein, and canthaxanthin). In addition to its strong antioxidant property, astaxanthin offers several desirable properties such as antiinflammatory, anticancer, antihypertensive, and antiaging (Liu et al., 2021). Although natural astaxanthin has been approved by the Food and Drug Administration (FDA) for direct human consumption, chemically synthesized astaxanthin still occupies the majority of the market. It is undoubtedly that the algal biomass productivity and astaxanthin yield must be improved to industrialize the production of algae-derived astaxanthin. Several previous studies have performed to synthesize astaxanthin using algae such as Pitacco et al. (2022), Cheirslip et al. (2022), Lv et al. (2022), Parkes et al. (2022), Yu et al. (2022), Indrayani et al. (2022), Sirotiya et al. (2022), Wei et al. (2022), Sung and Sim (2022), Zhu et al. (2022), and Viazau et al. (2021).

Astaxanthin (3,3′-dihydroxy-β,β-carotene-4,4′-dione) is a secondary carotenoid and can be biosynthesized by algae such as *Haematococcus pluvialis* and yeast such as *Phaffia rhodozyma*, whereas the extraction and purification of astaxanthin from *H. pluvialis* is conventionally to be time-consuming and labor-intensive due to the following reasons: (i) the high chemical instability of astaxanthin over long period of extraction or purification and under harsh conditions such as pH, O_2, pH, and temperature, which can damage the medical and nutritious properties of astaxanthin; (ii) the rigid cell wall structure of algae thus makes the penetration of solvent to extract astaxanthin very challenging; (iii) astaxanthin is surrounded by lipid droplet, and these two molecules have a similar solubility and thus deter the extraction; (iv) the need of pretreatment before extraction such as dewatering. Another challenge is the use of hazardous solvents in the traditional extraction method. This leads to the switch

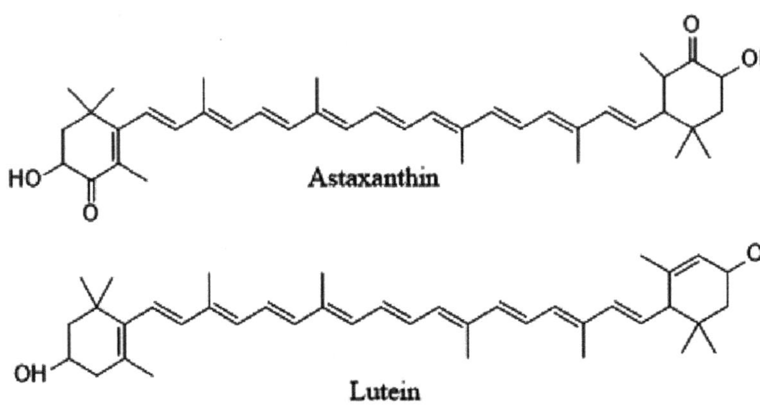

FIGURE 3.11

Chemical structure of astaxanthin and lutein (Arteni et al., 2015; Bachar et al., 2020).

toward the application of greener and safer solvents such as ethanol and ethyl acetone and unconventional solvents such as supercritical CO_2, ionic liquid, deep eutectic solvent, and vegetable oil.

Till now, one challenge faced by the algal astaxanthin production is related to the fact that the accumulation of the secondary carotenoids such as astaxanthin is often induced by the stressed conditions to mediate the cellular redox imbalance, which contradicts the cultivation conditions required for optimizing the algal biomass productivity (Nisar et al., 2015). Therefore, commercially, the astaxanthin production is performed in a two-stage cultivation system that consists of a "greening" phase for optimizing biomass productivity and a "reddening" phase under stressed conditions (Sun et al., 2018). However, this cultivation approach could cause an increase in capital and operating costs (Liyanaarachchi et al., 2021). Thus, previous studies have adopted genetic engineering tools to boost the astaxanthin accumulation in algae (Mehariya et al., 2021), whereas the key genes and enzymes of astaxanthin biosynthesis present in various types of algae have not been well understood. Although phytoene synthase is well accepted to be the rate-limiting enzyme in the upstream carotenoids synthesis, downstream carotenoids biosynthetic pathways diverge across algal species (Narang et al., 2021).

4.2 Lutein

Unlike astaxanthin as one of the most recognized carotenoids in the market, lutein (Fig. 3.11) has attracted a great deal of attention in recent years, which is resulted from its beneficial use to prevent cataract and age-related macular degradation, anticancer, and anticardiovascular properties (Gong & Bassi, 2016). Lutein demonstrates a strong antioxidant property and is able to prevent approximately 60 chronic diseases. In terms of the lutein synthesis, lutein can be not naturally synthesized by human and is commended by the health experts that an amount of 1.5 mg/day is required for human health. Several algal species have been tested as the feedstock to synthesize and extract lutein such as *Scenedesmus almeriensis*, *Chlorella zofingiensis*, and *Muriellopsis* sp. (Mary Leema et al., 2022).

For example, Li et al. (2022) simultaneously triggered the accumulation of lutein and lipid in *C. vulgaris* by controlling light and nitrogen supply to the culture, such as high light, moderate light, or low light integrating with nitrogen depletion, nitrogen limitation, and nitrogen repletion. The results indicated that the highest lutein yield of 5.39 mg/g was achieved when applying high light coupled with nitrogen repletion tactic. Besides, the use of high light coupled with nitrogen depletion tactic promoted the genetic expression of the key genes for both lutein (e.g., *psy*, *lcye*, and *pds*) and lipid (e.g., *me*, *acc*, and *kas*) formation in the algae, and this cultivation approach maximized the coproduction of lutein and lipid, when compared with other methods. Gao et al. (2022) isolated a new algae strain, which belongs to *Barrancaceae*. The effects of various cultivation parameters (e.g., light density and nitrogen levels) on the accumulation of lutein were evaluated, and the highest yield of lutein was found to be 3.33 mg/g DW.

Similar to the problems faced by the astaxanthin, there is still a need to develop green and alternative extraction methods for algae-derived lutein because the organic solvents are usually used in the lutein extraction from algae. This is typically due to the rigid cell wall/membrane structure, especially for unicellular structure of microalgae. Another reason is that lutein is firmly bounded by other biomolecules (e.g., fatty acids and proteins), thereby making the mass transfer of extraction solvent very difficult (Saini & Keum, 2018). Consequently, new alternative extraction methods based on the ultrasonication, microwave irradiation, and supercritical fluid have been developed for maximizing the lutein extraction from algae. For instance, Mary Leema et al. (2022) exposed the *C. sorokiniana* to the microwave irradiation to facilitate lutein extraction, Gayathri et al. (2018) investigated the ultrasonication-assisted lutein extraction from *Chlorella salina*, Nobre et al. (2006) utilized supercritical CO_2 for lutein recovery from *H. pluvialis*, as well as the combination of two extraction methods was explored such as ultrasonication-enhanced subcritical CO_2 extraction by Fan et al. (2015).

In short, undoubtedly, algae are excellent sources for synthesizing value-added carotenoids (e.g., astaxanthin and lutein) and then gradually phasing out the chemically synthesized carotenoids. However, the cost of production of most algae-based carotenoids is extremely high; particularly the challenges relating to the downstream treatment such as harvesting and extraction process remain as major concerns (Gong & Bassi, 2016).

5. Conclusions

Under the circumstance of pressing GHG emission, climate change, and other environmental problems and the depletion of nonrenewable fossil fuels (oil, natural gas, and coal), it is an urgent need to search for alternative and sustainable resources. Unlike first- and second-generation biomass, algae including microalgae and macroalgae are promising feedstocks for producing biofuels and biochemicals, which can gradually lead to the phase out the fossil fuel–derived fuels and chemicals. In this chapter, the conversion of algae into biomaterials such as biochar was discussed, and the applications of algae-derived biochar as electrode, solid fuel, and sorbents were covered. Following this, the recent studies conducting on the production of biofuels from algae were focused on the biodiesel and H_2. Owing to the high lipid content of certain types of algae and the tailorable nature of lipid content, algae have been widely considered as excellent sources for biodiesel production. Except for liquid biofuels, gaseous form of biofuels (i.e., H_2) was also presented in this chapter, with a focus on the thermochemical conversion of algae into H_2 via conventional gasification and SCWG. Furthermore, the synthesis of carotenoids including astaxanthin and lutein in the algae was covered.

Acknowledgment

The authors would like to thank the Natura Science and Engineering Research Council of Canada (NSERC) and the Startup Fund from the University of Prince Edward Island (UPEI) for funding this research.

References

Andrade, L. A., Batista, F. R. X., Lira, T. S., Barrozo, M. A. S., & Vieira, L. G. M. (2018). Characterization and product formation during the catalytic and non-catalytic pyrolysis of the green microalgae Chlamydomonas reinhardtii. *Renew Energy, 119*(x), 731–740. https://doi.org/10.1016/j.renene.2017.12.056. Available from:.

Arteni, A.-A., Fradot, M., Galzerano, D., Mendes-Pinto, M. M., Sahel, J.-A., Picaud, S., et al. (2015). Structure and conformation of the carotenoids in human retinal macular pigment. *PLoS One, 10*(8), e0135779. https://doi.org/10.1371/journal.pone.0135779. Available from:.

Babatabar, M. A., & Saidi, M. (2021). Hydrogen production via integrated configuration of steam gasification process of biomass and water-gas shift reaction: Process simulation and optimization. *International Journal of Energy Research, 45*(13), 19378–19394. https://doi.org/10.1002/er.7087. Available from:.

Babatabar, M. A., Yousefian, F., Mousavi, M. V., Hosseini, M., & Tavasoli, A. (2022). Pyrolysis of lignocellulosic and algal biomasses in a fixed-bed reactor: A comparative study on the composition and application potential of bioproducts. *International Journal of Energy Research, 46*(7), 9836–9850. https://doi.org/10.1002/er.7855. Available from:.

Bachar, S. C., Bachar, R., Jannat, K., Jahan, R., & Rahmatullah, M. (2020). Chapter seven - hepatoprotective natural products. In S. D. Sarker, & M. C. Nahar LBT-AR in (Eds.), *Medicinal natural products: A disease-focused approach* (pp. 207–249). Academic Press. Available from: https://www.sciencedirect.com/science/article/pii/S0065774320300294.

Cao, J., Chen, F., Fang, Z., Gu, Y., Wang, H., Lu, J., et al. (2022). Effect of filamentous algae in a microalgal-bacterial granular sludge system treating saline wastewater: Assessing stability, lipid production and nutrients removal. *Bioresource Technology, 354*, 127182. Available from: https://www.sciencedirect.com/science/article/pii/S0960852422005119.

Cheirsilp, B., Wantip, K., Chai-issarapap, N., Maneechote, W., Pekkoh, J., Duangjan, K., et al. (2022). Enhanced production of astaxanthin and co-bioproducts from microalga Haematococcus sp. integrated with valorization of industrial wastewater under two-stage LED light illumination strategy. *Environment Technology and Innovation, 28*, 102620. Available from: https://www.sciencedirect.com/science/article/pii/S2352186422001821.

Cherad, R., Onwudili, J. A., Williams, P. T., & Ross, A. B. (2014). A parametric study on supercritical water gasification of Laminaria hyperborea: A carbohydrate-rich macroalga. *Bioresource Technology, 169*, 573–580. Available from: https://www.sciencedirect.com/science/article/pii/S0960852414010335.

Crampon, C., Mouahid, A., Toudji, S.-A. A., Lépine, O., & Badens, E. (2013). Influence of pretreatment on supercritical CO2 extraction from Nannochloropsis oculata. *Journal of Supercritical Fluids, 79*, 337–344. Available from: https://www.sciencedirect.com/science/article/pii/S0896844612003932.

Deniz, I., Vardar-Sukan, F., Yüksel, M., Saglam, M., Ballice, L., & Yesil-Celiktas, O. (2015). Hydrogen production from marine biomass by hydrothermal gasification. *Energy Conversion and Management, 96*, 124–130. Available from: https://www.sciencedirect.com/science/article/pii/S0196890415001612.

Duan, P. G., Yang, S. K., Xu, Y. P., Wang, F., Zhao, D., Weng, Y. J., et al. (2018). Integration of hydrothermal liquefaction and supercritical water gasification for improvement of energy recovery from algal biomass. *Energy, 155*, 734–745. https://doi.org/10.1016/j.energy.2018.05.044. Available from:.

Enamala, M. K., Enamala, S., Chavali, M., Donepudi, J., Yadavalli, R., Kolapalli, B., et al. (2018). Production of biofuels from microalgae - a review on cultivation, harvesting, lipid extraction, and numerous applications of microalgae. *Renewable and Sustainable Energy Reviews, 94*(June), 49–68. https://doi.org/10.1016/j.rser.2018.05.012. Available from:.

Fan, X.-D., Hou, Y., Huang, X.-X., Qiu, T.-Q., & Jiang, J.-G. (2015). Ultrasound-enhanced subcritical CO2 extraction of lutein from Chlorella pyrenoidosa. *Journal of Agriculture Food Chemistry, 63*(18), 4597–4605. https://doi.org/10.1021/acs.jafc.5b00461. Available from:.

Farobie, O., Amrullah, A., Bayu, A., Syaftika, N., Anis, L. A., & Hartulistiyoso, E. (2022). In-depth study of bio-oil and biochar production from macroalgae Sargassum sp. via slow pyrolysis. *RSC Advances, 12*(16), 9567–9578. Available from: https://pubmed.ncbi.nlm.nih.gov/35424963.

Farobie, O., Changkiendee, P., Inoue, S., Inoue, T., Kawai, Y., Noguchi, T., et al. (2017). Effect of the heating rate on the supercritical water gasification of a glucose/guaiacol mixture. *Industrial and Engineering Chemistry Research, 56*, 6401–6407.

Farobie, O., Matsumura, Y., Syaftika, N., Amrullah, A., Hartulistiyoso, E., Bayu, A., et al. (2021). Recent advancement on hydrogen production from macroalgae via supercritical water gasification. *Bioresource Technology Reports, 16*, 100844. Available from: https://www.sciencedirect.com/science/article/pii/S2589014X2100222X.

Felix, C., Ubando, A., Madrazo, C., Gue, H. I., Sutanto, S., Tran-Nguyen, P. L., et al. (2019). Non-catalytic in-situ (trans) esterification of lipids in wet microalgae Chlorella vulgaris under subcritical conditions for the synthesis of fatty acid methyl esters. *Applied Energy, 248*, 526–537. https://doi.org/10.1016/j.apenergy.2019.04.149. Available from:.

Gao, B., Dai, C., Zhang, H., & Zhang, C. (2022). Evaluation of a novel oleaginous filamentous green alga, Barranca yajiagengensis (Chlorophyta, Chaetophorales) for biomass, lipids and pigments production. *Algal Research, 64*, 102681. Available from: https://www.sciencedirect.com/science/article/pii/S2211926422000522.

Gayathri, S., Rajasree Radhika, S. R., Suman, T. Y., & Aranganathan, L. (2018). Ultrasound-assisted micro-extraction of β, ε-carotene-3, 3′-diol (lutein) from marine microalgae Chlorella salina: Effect of different extraction parameters. *Biomass Conversion and Biorefinery, 8*(4), 791–797. https://doi.org/10.1007/s13399-018-0331-9. Available from:.

Giri, M. S., & Majidian, N. (2022). Design and optimization of lipids extraction process based on supercritical CO2 using Dunaliella Tertiolecta microalga for biodiesel production. *Arab Journal of Chemistry, 15*(7), 103920. Available from: https://www.sciencedirect.com/science/article/pii/S1878535222002362.

Golzary, A., Hosseini, A., & Saber, M. (2021). *Azolla filiculoides* as a feedstock for biofuel production: Cultivation condition optimization. *International Journal Energy Water Resource, 5*(1), 85–94. https://doi.org/10.1007/s42108-020-00092-3. Available from:.

Gong, M., & Bassi, A. (2016). Carotenoids from microalgae: A review of recent developments. *Biotechnology Advances, 34*(8), 1396–1412. https://doi.org/10.1016/j.biotechadv.2016.10.005. Available from:.

Go, A. W., Sutanto, S., Ong, L. K., Tran-Nguyen, P. L., Ismadji, S., & Ju, Y.-H. (2016). Developments in in-situ (trans) esterification for biodiesel production: A critical review. *Renewable and Sustainable Energy Reviews, 60*, 284–305. Available from: https://www.sciencedirect.com/science/article/pii/S1364032116001003.

Graz, Y., Bostyn, S., Richard, T., Bocanegra, P. E., de Bilbao, E., Poirier, J., et al. (2016). Hydrothermal conversion of ulva macro algae in supercritical water. *Journal of Supercritical Fluids, 107*, 182–188. Available from: https://www.sciencedirect.com/science/article/pii/S0896844615300917.

Guo, Y., Wang, S. Z., Xu, D. H., Gong, Y. M., Ma, H. H., & Tang, X. Y. (2010). Review of catalytic supercritical water gasification for hydrogen production from biomass. *Renewable and Sustainable Energy Reviews, 14*(1), 334–343.

Hu, Y., Feng, S., Yuan, Z., Xu, C. C., & Bassi, A. (2017). Investigation of aqueous phase recycling for improving bio-crude oil yield in hydrothermal liquefaction of algae. *Bioresource Technology, 239*.

Hu, Y., Gong, M., Feng, S., Xu, C. C., & Bassi, A. (2019). A review of recent developments of pre-treatment technologies and hydrothermal liquefaction of microalgae for bio-crude oil production. *Renewable and Sustainable Energy Reviews, 101*, 476–492.

Hu, Y., Gong, M., Xing, X., Wang, H., Zeng, Y., & Xu, C. (2020). Supercritical water gasification of biomass model compounds: A review. *Renewable and Sustainable Energy Reviews, 118*, 109529. https://doi.org/10.1016/j.rser.2019.109529. Available from:.

Indrayani, I., Egeland, E. S., Moheimani, N. R., & Borowitzka, M. A. (2022). Carotenoid production of Botryococcus braunii CCAP 807/2 under different growth conditions. *Journal Applied Phycology, 34*(3), 1177–1188. https://doi.org/10.1007/s10811-022-02682-6. Available from:.

Islam, Z., Khatoon, H., Rahman, M. R., Barman, S. K., Hossain, S., Usha, S. Z., et al. (2022). Growth, productivity and proximate profiling of indigenous marine microalgae from southern coast of Bangladesh as potential feedstuff for animal feed. *Bioresource Technology Reports, 18*, 101025. Available from: https://www.sciencedirect.com/science/article/pii/S2589014X22000822.

Jabeen, S., Gao, X., Altarawneh, M., Hayashi, J., Zhang, M., & Dlugogorski, B. Z. (2020). Analytical procedure for proximate analysis of algal biomass: Case study for Spirulina platensis and Chlorella vulgaris. *Energy & Fuels, 34*(1), 474–482. https://doi.org/10.1021/acs.energyfuels.9b03156. Available from:.

Kale, B. N., & Patle, S. D. (2022). State of art review of algal biodiesel and its blends influence on performance and emission characteristics of compression ignition engine. *Cleaner Engineering Technology, 7*, 100431. Available from: https://www.sciencedirect.com/science/article/pii/S2666790822000362.

Kiminaitė, I., Lisauskas, A., Striūgas, N., & Kryževičius, Ž. (2022). Fabrication and characterization of environmentally friendly biochar anode. *Energies, 15*.

Kumar Sharma, A., Kumar Ghodke, P., Manna, S., & Chen, W.-H. (2021). Emerging technologies for sustainable production of biohydrogen production from microalgae: A state-of-the-art review of upstream and downstream processes. *Bioresource Technology, 342*, 126057. Available from: https://www.sciencedirect.com/science/article/pii/S0960852421013997.

Lee, X. J., Ong, H. C., Gan, Y. Y., Chen, W.-H., & Mahlia, T. M. I. (2020). State of art review on conventional and advanced pyrolysis of macroalgae and microalgae for biochar, bio-oil and bio-syngas production. *Energy Conversion and Management, 210*, 112707. Available from: https://www.sciencedirect.com/science/article/pii/S0196890420302454.

Li, D., Chen, L., Yi, X., Zhang, X., & Ye, N. (2010). Pyrolytic characteristics and kinetics of two brown algae and sodium alginate. *Bioresource Technology, 101*(18), 7131–7136. Available from: https://www.sciencedirect.com/science/article/pii/S0960852410006681.

Li, Y., Qi, Z., Fan, Y., Tang, Y., & Zhou, R. (2022). The concurrent production of lipid and lutein in Chlorella vulgaris triggered by light coupling nitrogen tactics. *Biochemical Engineering Journal, 182*, 108435. Available from: https://www.sciencedirect.com/science/article/pii/S1369703X22001048.

Liu, J. (2011). Microalgae as feedstocks for biodiesel production. In J. Huang (Ed.), *Rijeka: IntechOpen* (p. Ch. 7). https://doi.org/10.5772/25600. Available from:.

Liu, C., Hu, B., Cheng, Y., Guo, Y., Yao, W., & Qian, H. (2021). Carotenoids from fungi and microalgae: A review on their recent production, extraction, and developments. *Bioresource Technology, 337*, 125398. https://doi.org/10.1016/j.biortech.2021.125398. Available from:.

Liyanaarachchi, V. C., Premaratne, M., Ariyadasa, T. U., Nimarshana, P. H. V., & Malik, A. (2021). Two-stage cultivation of microalgae for production of high-value compounds and biofuels: A review. *Algal Research, 57*(May), 102353. https://doi.org/10.1016/j.algal.2021.102353. Available from:.

Lv, R., Liu, K., Chen, F., Xing, H., Xu, N., Sun, X., et al. (2022). Buffering culture solution significantly improves astaxanthin production efficiency of mixotrophic Haematococcus pluvialis. *Bioresource Technology, 354*, 127175. Available from: https://www.sciencedirect.com/science/article/pii/S0960852422005041.

Mary Leema, J. T., Persia Jothy, T., & Dharani, G. (2022). Rapid green microwave assisted extraction of lutein from Chlorella sorokiniana (NIOT-2) – process optimization. *Food Chemistry, 372*, 131151. Available from: https://www.sciencedirect.com/science/article/pii/S0308814621021579.

Mat Aron, N. S., Chew, K. W., Ang, W. L., Ratchahat, S., Rinklebe, J., & Show, P. L. (2022). Recovery of microalgae biodiesel using liquid biphasic flotation system. *Fuel, 317*, 123368. Available from: https://www.sciencedirect.com/science/article/pii/S0016236122002368.

Mehariya, S., Goswami, R. K., Karthikeysan, O. P., & Verma, P. (2021). Microalgae for high-value products: A way towards green nutraceutical and pharmaceutical compounds. *Chemosphere, 280*, 130553. https://doi.org/10.1016/j.chemosphere.2021.130553. Available from:.

Nallasivam, J., Francis Prashanth, P., Harisankar, S., Nori, S., Suryanarayan, S., Chakravarthy, S. R., et al. (2022). Valorization of red macroalgae biomass via hydrothermal liquefaction using homogeneous catalysts. *Bioresource Technology, 346*, 126515. Available from: https://www.sciencedirect.com/science/article/pii/S0960852421018575.

Narang, P. K., Dey, J., Mahapatra, S. R., Ghosh, M., Misra, N., Suar, M., et al. (2021). Metabolic engineering for carotenoid production using eukaryotic microalgae and prokaryotic cyanobacteria. In *Carotenoids: Biosynthetic and biofunctional approaches* (pp. 121–135). Elsevier B.V. https://doi.org/10.1016/j.sajb.2021.04.014. Available from:.

Narnaware, S. L., & Panwar, N. L. (2022). Biomass gasification for climate change mitigation and policy framework in India: A review. *Bioresource Technology Reports, 17*, 100892. Available from: https://www.sciencedirect.com/science/article/pii/S2589014X2100270X.

Nayana, K., Sudhakar, M. P., & Arunkumar, K. (2022). Biorefinery potential of Coelastrella biomass for fuel and bioproducts—a review. *Biomass Conversion and Biorefinery*. https://doi.org/10.1007/s13399-022-02519-9. Available from:.

Nguyen, T.-B., Truong, Q.-M., Chen, C.-W., Chen, W.-H., & Dong, C.-D. (2022). Pyrolysis of marine algae for biochar production for adsorption of Ciprofloxacin from aqueous solutions. *Bioresource Technology, 351*, 127043. Available from: https://www.sciencedirect.com/science/article/pii/S0960852422003728.

Nisar, N., Li, L., Lu, S., Khin, N. C., & Pogson, B. J. (2015). Carotenoid metabolism in plants. *Molecular Plant, 8*(1), 68−82.

Nobre, B., Marcelo, F., Passos, R., Beirão, L., Palavra, A., Gouveia, L., et al. (2006). Supercritical carbon dioxide extraction of astaxanthin and other carotenoids from the microalga Haematococcus pluvialis. *Eur Food Res Technol = Zeitschrift für Leb und -Forschung A, 223*(6), 787−790. Available from: https://www.lib.uwo.ca/cgi-bin/ezpauthn.cgi?url=http://search.proquest.com/scholarly-journals/supercritical-carbon-dioxide-extraction/docview/632565919/se-2?accountid=15115.

Norouzi, O., Safari, F., Jafarian, S., Tavasoli, A., & Karimi, A. (2017). Hydrothermal gasification performance of Enteromorpha intestinalis as an algal biomass for hydrogen-rich gas production using Ru promoted Fe−Ni/Γ-Al2O3 nanocatalysts. *Energy Conversion and Management, 141*, 63−71. https://doi.org/10.1016/j.enconman.2016.04.083. Available from:.

Office of Fossil Energy USD of E. (2020). Enabling a low-carbon economy (Vol 24). Available from: https://www.energy.gov/sites/prod/files/2020/07/f76/USDOE_FE_Hydrogen_Strategy_July2020.pdf.

Onigbajumo, A., Taghipour, A., Will, G., Chu Van, T., Couperthwaite, S., Steinberg, T., et al. (2022). Effects of process-thermal configuration on energy, exergy, and thermo-economic performance of solar driven supercritical water gasification. *Energy Conversion and Management, 251*, 115002. Available from: https://www.sciencedirect.com/science/article/pii/S019689042101178X.

Onwudili, J. A., Lea-Langton, A. R., Ross, A. B., & Williams, P. T. (2013). Catalytic hydrothermal gasification of algae for hydrogen production: Ccomposition of reaction products and potential for nutrient recycling. *Bioresource Technology, 127*, 72−80. Available from: https://www.sciencedirect.com/science/article/pii/S0960852412015222.

Parkes, R., Barone, M. E., Herbert, H., Gillespie, E., & Touzet, N. (2022). Antioxidant activity and carotenoid content responses of three Haematococcus sp. (chlorophyta) strains exposed to multiple stressors. *Applied Biochemistry and Biotechnology*, 1−19. https://doi.org/10.1007/s12010-022-03926-4. Available from:.

Park, J.-Y., Park, M. S., Lee, Y.-C., & Yang, J.-W. (2015). Advances in direct transesterification of algal oils from wet biomass. *Bioresource Technology, 184*, 267−275. Available from: https://www.sciencedirect.com/science/article/pii/S0960852414015211.

Parsa, M., Jalilzadeh, H., Pazoki, M., Ghasemzadeh, R., & Abduli, M. A. (2018). Hydrothermal liquefaction of Gracilaria gracilis and Cladophora glomerata macro-algae for biocrude production. *Bioresource Technology, 250*(August 2017), 26−34.

Pitacco, W., Samorì, C., Pezzolesi, L., Gori, V., Grillo, A., Tiecco, M., et al. (2022). Extraction of astaxanthin from Haematococcus pluvialis with hydrophobic deep eutectic solvents based on oleic acid. *Food Chemistry, 379*, 132156. Available from: https://www.sciencedirect.com/science/article/pii/S0308814622001170.

Prabakaran, S., Mohanraj, T., Arumugam, A., & Sudalai, S. (2022). A state-of-the-art review on the environmental benefits and prospects of Azolla in biofuel, bioremediation and biofertilizer applications. *Industrial Crops and Products, 183*, 114942. Available from: https://www.sciencedirect.com/science/article/pii/S0926669022004253.

Rodríguez-Palacio, M. C., Cabrera-Cruz, R. B. E., Rolón-Aguilar, J. C., Tobías-Jaramillo, R., Martínez-Hernández, M., & Lozano-Ramírez, C. (2022). The cultivation of five microalgae species and their potential for biodiesel production. *Energy Sustainability and Society, 12*(1), 10. https://doi.org/10.1186/s13705-022-00337-5. Available from:.

Safari, F., Norouzi, O., & Tavasoli, A. (2016). Hydrothermal gasification of Cladophora glomerata macroalgae over its hydrochar as a catalyst for hydrogen-rich gas production. *Bioresource Technology, 222*, 232–241. https://doi.org/10.1016/j.biortech.2016.09.082. Available from:.

Saini, R. K., & Keum, Y. S. (2018). Carotenoid extraction methods: A review of recent developments. *Food Chemistry, 240*, 90–103. https://doi.org/10.1016/j.foodchem.2017.07.099. Available from:.

Samanmulya, T., Farobie, O., & Matsumura, Y. (2017). Gasification characteristics of histidine and 4-methylimidazole under supercritical water conditions. *Biomass Conversion and Biorefinery, 7*, 487–494.

Schumacher, M., Yanık, J., Sınağ, A., & Kruse, A. (2011). Hydrothermal conversion of seaweeds in a batch autoclave. *Journal of Supercritical Fluids, 58*(1), 131–135. Available from: https://www.sciencedirect.com/science/article/pii/S0896844611001549.

Shokravi, Z., Shokravi, H., Atabani, A. E., Lau, W. J., Chyuan, O. H., & Ismail, A. F. (2022). Impacts of the harvesting process on microalgae fatty acid profiles and lipid yields: Implications for biodiesel production. *Renewable and Sustainable Energy Reviews, 161*, 112410. Available from: https://www.sciencedirect.com/science/article/pii/S1364032122003185.

Shuping, Z., Yulong, W., Mingde, Y., Kaleem, I., Chun, L., & Tong, J. (2010). Production and characterization of bio-oil from hydrothermal liquefaction of microalgae Dunaliella tertiolecta cake. *Energy, 35*(12), 5406–5411. https://doi.org/10.1016/j.energy.2010.07.013. Available from:.

Sirotiya, V., Ahirwar, A., Mourya, M., Khan, M. J., Rai, A., Kwatra, R., et al. (2022). Astaxanthin bioaccumulation in microalgae under environmental stress simulated in industrial effluents highlighting prospects of Haematococcus pluvialis: Knowledge gaps and prospective approaches. *Phytochemistry Reviews*. https://doi.org/10.1007/s11101-022-09807-2. Available from:.

Sung, Y. J., & Sim, S. J. (2022). Multifaceted strategies for economic production of microalgae Haematococcus pluvialis-derived astaxanthin via direct conversion of CO2. *Bioresource Technology, 344*, 126255. Available from: https://www.sciencedirect.com/science/article/pii/S0960852421015972.

Sun, J., Norouzi, O., & Mašek, O. (2022). A state-of-the-art review on algae pyrolysis for bioenergy and biochar production. *Bioresource Technology, 346*, 126258. Available from: https://www.sciencedirect.com/science/article/pii/S096085242101600X.

Sun, X. M., Ren, L. J., Zhao, Q. Y., Ji, X. J., & Huang, H. (2018). Microalgae for the production of lipid and carotenoids: A review with focus on stress regulation and adaptation. *Biotechnology Biofuels, 11*, 1–16. https://doi.org/10.1186/s13068-018-1275-9. Available from:.

Tobío-Pérez, I., Alfonso-Cardero, A., Díaz-Domínguez, Y., Pohl, S., Piloto-Rodríguez, R., & Lapuerta, M. (2022). Thermochemical conversion of Sargassum for energy production: A comprehensive review. *BioEnergy Research*. https://doi.org/10.1007/s12155-021-10382-1. Available from:.

Varma, R. S. (2019). Biomass-derived renewable carbonaceous materials for sustainable chemical and environmental applications. *ACS Sustainable Chemistry Engineering, 7*(7), 6458–6470. https://doi.org/10.1021/acssuschemeng.8b06550. Available from:.

Viazau, Y. V., Goncharik, R. G., Kulikov, E. A., & Selishcheva, A. A. (2021). Pigment composition of Haematococcus pluvialis green alga under the action of several inducers of astaxanthin accumulation. *Applied Biochemistry Microbiology, 57*(8), 886–889. https://doi.org/10.1134/S0003683821080081. Available from:.

Wei, J., Zhang, X., Ai, S., Huang, Y., Yang, X., Mei, Y., et al. (2022). The effective astaxanthin productivities of immobilized Haematococcus pluvialis with bacterial cellulose. *Bioresource Technology, 344*, 126317. Available from: https://www.sciencedirect.com/science/article/pii/S096085242101659X.

Yan, Q., Guo, L., & Lu, Y. (2006). Thermodynamic analysis of hydrogen production from biomass gasification in supercritical water. *Energy Conversion and Management, 47*(11), 1515–1528. Available from: https://www.sciencedirect.com/science/article/pii/S0196890405001950.

Yan, M., Liu, J., Yu, C., Zhou, X., Hantoko, D., & Kanchanatip, E. (2021). Experimental investigation on gasification of food waste in supercritical water: Effects of NaCl on syngas production and corrosion of reactor. *Environmental Technology and Innovation, 23*, 101538. https://doi.org/10.1016/j.eti.2021.101538. Available from:.

Yu, W., Zhang, L., Zhao, J., & Liu, J. (2022). Enhancement of astaxanthin accumulation in Haematococcus pluvialis by exogenous oxaloacetate combined with nitrogen deficiency. *Bioresource Technology, 345*, 126484. Available from: https://www.sciencedirect.com/science/article/pii/S0960852421018265.

Zhang, K., Zhang, F., & Wu, Y.-R. (2021). Emerging technologies for conversion of sustainable algal biomass into value-added products: A state-of-the-art review. *Science of the Total Environmental, 784*, 147024. Available from: https://www.sciencedirect.com/science/article/pii/S0048969721020945.

Zhu, X., Meng, C., Du, H., Chen, L., Sun, F., Chen, W., et al. (2022). Enhancement of astaxanthin production in Schizochytrium limacinum B4D1 under ethanol induction. *Algal Research, 61*, 102537. Available from: https://www.sciencedirect.com/science/article/pii/S2211926421003568.

CHAPTER

Algae materials for food and food packaging

4

Amina Tahreen[1], Sarah Amalina Adli[1], Hazleen Anuar[2], Jamarosliza Jamaluddin[3], Raja Mohan[4] and Fathilah Ali[1]

[1]*Department of Chemical Engineering and Sustainability, Kulliyyah of Engineering, International Islamic University Malaysia, Kuala Lumpur, Wilayah Persekutuan, Malaysia;* [2]*Department of Manufacturing and Materials Engineering, Kulliyyah of Engineering, International Islamic University Malaysia, Kuala Lumpur, Wilayah Persekutuan, Malaysia;* [3]*Department of Chemical Engineering, School of Chemical and Energy Engineering, Universiti Teknologi Malaysia, Johor Bahru, Johor, Malaysia;* [4]*Chemical Industry Research Institute, University of Ulsan, Ulsan, South Korea*

1. Algae for food

With growing interests in plant-based protein diet with considerable nutritional value, algae and algae-based food and supplements are gaining global popularity. With its prokaryotic and eukaryotic forms, algae emerged as a rich bioresource for bioactive compounds for both therapeutic and nutraceutical applications (Brown et al., 2014; Cicero et al., 2017; Mogany et al., 2019). With comprehensive nutritional and toxicological analyses, algal proteins have demonstrated high nutritional value while being harmless to the human body and environment (Becker, 2007). Although decades ago algae-based food and supplements were limited for consumption due to high production costs, the advancements in technology and algae-based health supplements supply are more widely increasing along with the rising interests on its potential as plant-based protein-rich alternative to meat (Michel et al., 2021). Moreover, algae-derived biocompounds have depicted considerable antimicrobial, antiviral, antiinflammatory, and immune boosting abilities (Pagels et al., 2019), which they were considered for research to develop a drug against severe acute respiratory syndrome (SARS) coronavirus (CoV) 2, especially due to the presence of angiotensin-converting enzyme (ACE) inhibitor peptides in algae (Ratha et al., 2021). Hence, being a rich bioresource with nutritional values, algae potentially play a significant role in the food industry. Moreover, algae are a potential alternative for protein, bioactive compound, or nutrient extraction from vegetarian source for consumers who prefer a strict vegetarian diet. Hence, this section provides an overview of the application of algae and its resources in the food and consumption sector.

1.1 Seaweed

In Southeast Asia, marine algae are typically harvested largely as seaweed for consumption (Jensen, 1993). It is a prime component of many prominent local dishes in Southeast Asia. Many popular dishes

Algae Materials: Applications Benefitting Health. https://doi.org/10.1016/B978-0-443-18816-9.00006-X

such as different types of sushi (sticky rice wrapped with seaweed) require the addition of seaweed. The seaweed is even fermented and seasoned to serve as crispy snacks. It is not only enjoyed as a delicacy but also eaten for nutritional values present as bioactive compounds. It can be harvested as sea vegetable for direct consumption, or used for the extraction of food additives (such as phycocolloids) (Fleurence, 2016). Currently, other countries throughout the world including Asia, Europe, North and South America, Africa and Pacific Island nations, besides Southeast Asian countries, have also been known to produce their own seaweed-based food, food additives, and nutritional supplements (Fleurence, 2016).

1.2 Supplements

A widely consumed form of algae is in terms of dietary supplements, where the algae species used are typically *Spirulina* and *Chlorella* (Ścieszka & Klewicka, 2019). *Spirulina* powder is a commercially available form of nutritious blue-green algae, typically mixed with water or juice to boost the immune system. However, health precautions and physician consultation are required before starting any supplement, as people who may be allergic to sea food or iodine may be affected with the algae-based supplement. Moreover, *Spirulina* powder can be also added to coffee to serve as a delightful blue nutritious drink. The increased popularity of *Spirulina* resulted from proven studies that support both innate and adaptive immunity boost (Chei et al., 2020; Hirahashi et al., 2002).

On the other hand, a strong demand for eicosapentaenoic acid (EPA)–enriched and docosahexaenoic (DHA)-enriched oil (typically derived from fish) as a health supplement imposes a pressure on fish population and the marine ecosystem. However, algae being the primary producers in the water are the source of omega 3 fatty acids, which are passed onto fish on consumption. Hence, harvesting algae oil–based omega 3 fatty acid is a vegetarian alternative to the fish oil supplements, which is a necessary health supplement with high demand for regular maintenance of a healthy immune system (Lenihan-Geels et al., 2013).

1.3 Others

Beside seaweed and essential health supplements, algae-based protein is an alluring topic in the vegan and vegetarian food industries. Thorough examinations depicted that algal protein was comparable with vegetable protein (Becker, 2007). Although plant-based protein burgers constantly require producers to compete with the taste of meat (which is favorable to consumers), it is an emerging protein source with rising interest for its nutritional favors and eco-friendliness compared with meat (Michel et al., 2021).

Research on algae as sustainable nutritious food source in space missions is considered to be able to create a bioregenerative life support system (Vinayak, 2022). Moreover, algae-based additives or fermented algae have depicted improved food quality in terms of resistance to spoilage and improved food nutritional value (Ścieszka & Klewicka, 2019). Hence, algae-derived polysaccharides such as agar, alginate, and carrageenan have major applications in the food industry in terms of stabilizers, preservatives, and edible flavored gel-based food such as fruit jams, jellies, and marmalades.

2. Algae for food packaging

2.1 Polymers as packaging materials

Plastics have been the main source of packaging materials in the industry nowadays. Some of the common examples include polyethylene (PET), polyvinyl chloride (PVC), and polystyrene (PS). However, all of the conventional plastics used in the packaging industry are derived from petroleum, which is a diminishing natural resource. The reason why these conventional plastics are favorable is because of their durability, which is the main functionality of creating a packaging. Other reasons include stability to biotic and abiotic stress and easy processing. Although these plastics have a high durability, they are all nonbiodegradable. These result in accumulation of waste on landfill as they took hundreds or thousands of years to decomposed.

The best solution to counteract this problem is to come up with a biodegradable polymer, which means that they will decompose into the soil without affecting the environment. Biodegradable polymer is opted more nowadays based on their attractiveness in terms of biodegradability, cost reduction, and also safety. Biodegradable polymer is defined as a polymeric material that converts to carbon dioxide, nitrogen, and biomass when degrades. There are three groups of biopolymers, namely agropolymer from biomass (e.g., polysaccharide), biopolyester synthesized by microorganism (e.g., polyhydroxyalkanoates and polyhydroxybutyrate), and lastly polymer synthesized from naturally occurring monomer (polylactic acid, PLA) (Mensitieri et al., 2011).

PLA was first discovered by Wallace Carothers in 1932. They were mainly produced from basic materials such as sugarcane, corn, whey, or cellulose biomass. The basic materials are first processed into lactic acid via bacterial fermentation. Lactic acid is the basic monomer of PLA. PLA can be prepared through three methods. The first is by ring-opening polymerization (ROP) of the dehydrated ring-formed dimer or dilactide, second by polycondensation and manipulation of the equilibrium between lactic acid and polylactide by removal of reaction water using drying agents, or third by polycondensation and linking of lactic acid monomers. PLA is preferred due to its processability and biocompatibility; however, PLA also possesses certain weakness per say brittleness, poor melt strength, low thermal stability, slow degradation rate, and high cost.

Therefore, some steps are required to modify PLA to meet its function. There are steps available in improvizing the PLA such as polymer—polymer blending, polymer—fillers blending, and polymer—plasticizer blending. Polymer is usually blended with other polymers to improve mechanical properties, like what has been done by Yen et al. (Wu et al., 2016) in blending PLA with polymethyl methacrylate (PMMA) to improve the toughness of the brittle PLA, while the addition of fillers contributes to the reduction of cost and increases flexibility of the composites. One example of filler used for this purpose was starch where the elongation at break increases up to 400% (Hassan et al., 2013).

2.1.1 Modification of polymers

PLA is one of the plastics that shows a major potential in replacing conventional plastic as a packaging material. However, due to its slow degradation rate and brittleness, PLA lags behind conventional ones (Burgos et al., 2014). Many research studies had been done to improve PLA properties, and the most preferable one is by making PLA composites. PLA composites can be made by three different blending methods; they are by blending PLA with polymers, PLA with fillers, and PLA with plasticizers.

Extensive research had been done on all these three different methods and not only focused on PLA but also other biopolymers such as PHA or PHB.

Some approaches for PLA–polymer blending were on PLA/polyhydroxybutyrate blend (Musioł et al., 2016) and PLA/polycaprolactone blend (Matta et al., 2014). For polymer–filler blends, some of the studies were on using cellulose (Cichorek et al., 2017) and algae (Wu, 2015). Lastly, for the addition of plasticizers, adipate and citrate esters (Shirai et al., 2015) and oligomers of lactic acid (OLAs) (Burgos et al., 2014). The blends are usually prepared by either melt blending or solvent casting method. The selection of method used is generally based on the properties of material used and the desired characteristic at the end of the process.

2.2 Algae as filler in polymers for packaging materials

Other than the examples mentioned earlier, algae can be used as a filler. It is available abundantly in nature, or it can be obtained in the form of biomass. In addition, algae can be prepared in many sizes, up to nanosize particles. However, not much studies had been done on algae as fillers. Nonetheless, Bulota and Budtova (2016) had prepared PLA/algae composites, and Wu (2015) had prepared polycaprolactone (PCL)/marine algae. Even though both studies showed promising results as the objectives were achieved, which was to improve the flexibility and mechanical properties of the polymers, plasticizers were not used in both research studies. As there is past research that showed promising result in blending PLA with plasticizer (EPO) (Ali et al., 2016), this research was conducted to study the properties of PLA blended with EPO as plasticizer and algae as a filler.

Algae are a filler that consist of a stiff and strong cellulose-based fibrils. They exist in many groups, but the common ones are red, green, and brown algae. Many researchers nowadays have diverted their attention toward discovering the alternatives of algae as a filler because they can reduce the cost and utilize the algae waste. Two approaches are approached in using algae in composites: first as fillers (not treated, mild, and pulverized) and second as a reinforcing fibers (treated, bleached to keep the cellulosic fibers) (Bulota & Budtova, 2015).

A study by Bulota and Budtova (2016) on PLA/algae composites showed many insights. The most important notes obtained by this study were that algae will undergo thermal degradation if it is heated above the limited temperature. For the green algae such as spirulina, the degradation of chlorophyll starts at 60°C followed by hemicellulose and cellulose degradation at 220 and 315°C, respectively. Upon heat treatment, particle breakage happens because of weakening of the algae flake-like structure, caused by degradation of chlorophyll. Based on this study, it should be noted that blending of algae polymer cannot be done by melt mixing method, as thermal degradation occurred altered the chemical composition.

Putting the disadvantages aside, Young's modulus of composites with 40 wt% large green algae particles exceeded that of the neat PLA. The results obtained show that depending on the application, algae can be used as filler in thermoplastic composites up to at least 40 wt% loading, since the stiffness of materials can be maintained. Table 4.1 summarizes some of the fillers used in polymer blends.

Table 4.1 Fillers used in polymer blends.

Fillers	Content (%)	Strength	Flexibility	Reference
Cellulose	10	↑	↓	Fortunati et al. (2012)
Algae	40	↓	↑	Bulota and Budtova (2016)
Starch	10	↑	↑	Kiangkitiwan and Srikulkit (2013)
Kenaf fiber	6	↑	↑	Zainuddin et al. (2013)

↑, increased, ↓, decreased.

3. Conclusion

Certainly, algae are a rich renewable bioresource with enormous potential in many sectors, with food being one of the significant ones. Although algae consumption is already in existence in the food industry, numerous studies are emerging on further exploring and expanding its potential to a larger and more efficient scale, from exploiting its nutritional benefits to reducing the stress on fish and meat to meet the regular ongoing protein or supplement demands. Hence, a huge role can be played by algae in regulating a sustainable nutrient supply as a renewable and environmentally friendly protein-rich vegetarian source. Algae can also further be researched to be used as a filler in packaging materials.

References

Ali, F. B., Awale, R. J., Fakhruldin, H., & Anuar, H. (2016). Plasticizing poly(lactic acid) using epoxidized palm oil for environmental friendly packaging material. *Malaysian Journal of Analytical Sciences, 20*, 1153–1158. https://doi.org/10.17576/mjas-2016-2005-22

Becker, E. W. (2007). Micro-algae as a source of protein. *Biotechnology Advances, 25*, 207–210. https://doi.org/10.1016/j.biotechadv.2006.11.002

Brown, E. M., Allsopp, P. J., Magee, P. J., Gill, C. I., Nitecki, S., Strain, C. R., & Mcsorley, E. M. (2014). Seaweed and human health. *Nutrition Reviews, 72*, 205–216. https://doi.org/10.1111/nure.12091

Bulota, M., & Budtova, T. (2015). Composites: Part A PLA/algae composites : Morphology and mechanical properties. *Composites Part A, 73*, 109–115. https://doi.org/10.1016/j.compositesa.2015.03.001

Bulota, M., & Budtova, T. (2016). Valorisation of macroalgae industrial by-product as filler in thermoplastic polymer composites. *Composites Part A: Applied Science and Manufacturing, 90*, 271–277. https://doi.org/10.1016/j.compositesa.2016.07.010

Burgos, N., Tolaguera, D., Fiori, S., & Jiménez, A. (2014). Synthesis and characterization of lactic acid oligomers: Evaluation of performance as poly(lactic acid) plasticizers. *Journal of Polymers and the Environment, 22*, 227–235. https://doi.org/10.1007/s10924-013-0628-5

Chei, S., Oh, H. J., Song, J. H., Seo, Y. J., Lee, K., Kim, K. J., & Lee, B. Y. (2020). Spirulina maxima extract prevents activation of the NLRP3 inflammasome by inhibiting ERK signaling. *Scientific Reports, 10*, 1–10. https://doi.org/10.1038/s41598-020-58896-6

Cicero, A. F. G., Colletti, A., Bajraktari, G., Descamps, O., Djuric, D. M., Ezhov, M., Fras, Z., Katsiki, N., Langlois, M., Latkovskis, G., Panagiotakos, D. B., Paragh, G., Mikhailidis, D. P., Mitchenko, O.,

Paulweber, B., Pella, D., Pitsavos, C., Reiner, Ž., Ray, K. K., ... Banach, M. (2017). Lipid-lowering nutraceuticals in clinical practice: Position paper from an international lipid expert panel. *Nutrition Reviews, 75*, 731–767. https://doi.org/10.1093/nutrit/nux047

Cichorek, M., Piorkowska, E., & Krasnikova, N. (2017). Stiff biodegradable polylactide composites with ultrafine cellulose filler. *Journal of Polymers and the Environment, 25*, 74–80. https://doi.org/10.1007/s10924-016-0788-1

Fleurence, J. (2016). Seaweeds as food. In *Seaweed Health Disease and Prevention* (pp. 149–167). Elsevier Inc.. https://doi.org/10.1016/B978-0-12-802772-1.00005-1

Fortunati, E., Armentano, I., Zhou, Q., Puglia, D., Terenzi, A., Berglund, L. A., & Kenny, J. M. (2012). Microstructure and nonisothermal cold crystallization of PLA composites based on silver nanoparticles and nanocrystalline cellulose. *Polymer Degradation and Stability, 97*, 2027–2036. https://doi.org/10.1016/j.polymdegradstab.2012.03.027

Hassan, A., Balakrishnan, H., & Akbari, A. (2013). In S. Thomas, P. M. Visakh, & A. P. Mathew (Eds.), *Polylactic acid based blends, composites and nanocomposites BT - Advances in natural polymers: Composites and nanocomposites* (pp. 361–396). Springer Berlin Heidelberg. https://doi.org/10.1007/978-3-642-20940-6_11

Hirahashi, T., Matsumoto, M., Hazeki, K., Saeki, Y., Ui, M., & Seya, T. (2002). Activation of the human innate immune system by Spirulina: Augmentation of interferon production and NK cytotoxicity by oral administration of hot water extract of Spirulina platensis. *International Immunopharmacology, 2*, 423–434. https://doi.org/10.1016/S1567-5769(01)00166-7

Jensen, A. (1993). Present and future needs for algae and algal products. *Hydrobiologia, 260–261*, 15–23. https://doi.org/10.1007/BF00048998

Kiangkitiwan, N., & Srikulkit, K. (2013). Poly(Lactic acid) filled with cassava starch-g-soybean oil maleate. *Science World Journal, 2013*, 860487. https://doi.org/10.1155/2013/860487

Lenihan-Geels, G., Bishop, K. S., & Ferguson, L. R. (2013). Alternative sources of omega-3 fats: Can we find a sustainable substitute for fish? *Nutrients, 5*, 1301–1315. https://doi.org/10.3390/nu5041301

Matta, A. K., Rao, R. U., Suman, K. N. S., & Rambabu, V. (2014). Preparation and characterization of biodegradable PLA/PCL polymeric blends. *Procedia Material Science, 6*, 1266–1270. https://doi.org/10.1016/j.mspro.2014.07.201

Mensitieri, G., Di Maio, E., Buonocore, G. G., Nedi, I., Oliviero, M., Sansone, L., & Iannace, S. (2011). Processing and shelf life issues of selected food packaging materials and structures from renewable resources. *Trends in Food Science & Technology, 22*, 72–80. https://doi.org/10.1016/J.TIFS.2010.10.001

Michel, F., Knaapila, A., Hartmann, C., & Siegrist, M. (2021). A multi-national comparison of meat eaters' attitudes and expectations for burgers containing beef, pea or algae protein. *Food Quality and Preference, 91*, 104195. https://doi.org/10.1016/j.foodqual.2021.104195

Mogany, T., Kumari, S., Swalaha, F. M., & Bux, F. (2019). Extraction and characterisation of analytical grade C-phycocyanin from *Euhalothece* sp. *Journal of Applied Phycology, 31*, 1661–1674. https://doi.org/10.1007/s10811-018-1661-5

Musioł, M., Sikorska, W., Adamus, G., Janeczek, H., Kowalczuk, M., & Rydz, J. (2016). (Bio)degradable polymers as a potential material for food packaging: Studies on the (bio)degradation process of PLA/(R,S)-PHB rigid foils under industrial composting conditions. *European Food Research and Technology, 242*, 815–823. https://doi.org/10.1007/s00217-015-2611-y

Pagels, F., Guedes, A. C., Amaro, H. M., Kijjoa, A., & Vasconcelos, V. (2019). Phycobiliproteins from cyanobacteria: Chemistry and biotechnological applications. *Biotechnology Advances, 37*, 422–443. https://doi.org/10.1016/j.biotechadv.2019.02.010

Ratha, S. K., Renuka, N., Rawat, I., & Bux, F. (2021). Prospective options of algae-derived nutraceuticals as supplements to combat COVID-19 and human coronavirus diseases. *Nutrition, 83*, 111089. https://doi.org/10.1016/j.nut.2020.111089

Ścieszka, S., & Klewicka, E. (2019). Algae in food: A general review. *Critical Reviews in Food Science and Nutrition, 59*, 3538–3547. https://doi.org/10.1080/10408398.2018.1496319

Shirai, M. A., Müller, C. M. O., Grossmann, M. V. E., & Yamashita, F. (2015). Adipate and citrate esters as plasticizers for poly(lactic acid)/thermoplastic starch sheets. *Journal of Polymers and the Environment, 23*, 54–61. https://doi.org/10.1007/s10924-014-0680-9

Vinayak, V. (2022). Algae as sustainable food in space missions. *Biomass, Biofuels, Biochemical: Circular Bioeconomy*, 517–540. https://doi.org/10.1016/B978-0-323-89855-3.00018-2

Wu, C.-S. (2015). Preparation, characterisation, and controlled-release of biodegradable polyester and marine-algae composite. *Journal of Polymers and the Environment, 23*, 356–366. https://doi.org/10.1007/s10924-015-0712-0

Wu, J.-H., Yen, M.-S., Chen, C.-W., & Kuo, M. C. (2016). Mechanical properties and isothermal crystallization behaviour of poly(lactide)/poly(methyl methacrylate)/α-Cellulose composites. *Journal of Polymers and the Environment, 24*, 318–327. https://doi.org/10.1007/s10924-016-0775-6

Zainuddin, S. Y. Z., Ahmad, I., Kargarzadeh, H., Abdullah, I., & Dufresne, A. (2013). Potential of using multiscale kenaf fibers as reinforcing filler in cassava starch-kenaf biocomposites. *Carbohydrate Polymers, 92*, 2299–2305. https://doi.org/10.1016/j.carbpol.2012.11.106

CHAPTER

Microalgae as superfood

5

Adriane Terezinha Schneider, Mariany Costa Deprá, Rosangela Rodrigues Dias, Leila Queiroz Zepka and Eduardo Jacob-Lopes

Bioprocess Intensification Group, Federal University of Santa Maria, UFSM, Santa Maria, Rio Grande do Sul, Brazil

1. Introduction

A good diet is important to ensure nutrition and the proper functioning of the body, being a crucial act for survival (Ferreira et al., 2021). The rapid growth of the human population in the coming years tends to drastically increase the demand for food, which could lead to a greater number of people in a situation of food and nutrition insecurity (UNICEF, 2021). Achieving global food and nutrition security requires many efforts. More and more research is aimed at finding new sources of healthier foods to solve food shortages. Food safety and quality require a global agreement, as they are complex choices and of paramount importance to the world (Bené et al., 2019). Food and nutrition insecurity causes several health problems, including mental illness, as well as is directly linked to a lower life expectancy (Elgar et al., 2021).

Consumers are increasingly concerned about their food, demanding increasingly healthy and sustainable diets. Current food production has several problems throughout the system. Greenhouse gas emissions, scarcity of freshwater and climate change, and a decrease in arable land are some environmental problems that make it difficult to produce large-scale and quality food (Hosseinkhani et al., 2022). Generally, foods that have health claims, with the use of terms such as superfruit, supergrain, and superfoods, have a greater preference for consumer choice (Liu et al., 2021). The so-called superfoods are classified this way because they have high levels of nutrients capable of helping and transforming the body, promoting a better quality of life (Proestos, 2018).

In this context, the diversity of food benefits from microalgae has been reported in the literature. Microalgae are a very promising superfood, as they are natural and sustainable, with high nutritional value, and are gaining space in the diet of a portion of the population (Vrenna et al., 2021). They have a wealth of nutrients and bioactive compounds such as amino acids, vitamins, proteins, essential phytochemicals, and polyunsaturated fatty acids (PUFAs). The high levels of protein found in microalgae make them great sources of this nutrient, exceeding the concentrations found in common sources (Huy et al., 2021). It is noteworthy that microalgae produce biomass through the use of carbon dioxide (CO_2), light, and nutrients (nitrogen, potassium, and phosphorus), and some species also use organic carbon sources (Hosseinkhani et al., 2022).

Algae Materials: Applications Benefitting Health. https://doi.org/10.1016/B978-0-443-18816-9.00023-X

That said, the objective of this chapter is to present the properties of microalgae, with a focus on *Arthrospira* and *Chlorella*, as future superfoods and the difficulties of their large-scale production.

2. Superfoods

In recent years, several terminologies have emerged to define the term superfood; however, there is still no official definition that meets this set of foods. According to Fernández-Ríos et al. (2021) and Proestos (2018), the term superfood is used to define foods that have a greater amount of significant properties, with greater concentrations of important nutrients, such as vitamins, proteins, minerals, fibers, and antioxidants, contributing to a healthy body. However, most foods known as superfoods have a long history of use by a large number of indigenous communities for the purpose of disease prevention. Currently, there are a wide variety of foods called superfoods, and the search for them is intensifying, as consumers are increasingly demanding in terms of health (Magrach & Sanz, 2020).

The consumption of superfoods, even in small amounts, results in benefits for the body, since the content of beneficial substances contained in these foods is high (Proestos, 2018). According to Magrach and Sanz (2020), the superfoods with the highest consumption worldwide are avocado, açaí, almonds, cocoa, coconuts, and quinoa. In addition to these, you can also find in the literature the name of superfoods kefir, corn, blueberries, royal jelly, ginger, donkey milk, pomegranate, blackberry, Brazil nut, hemp seed, watercress, maca, amla, camel milk, jackfruit, and goji berry, among others (Jagdale et al., 2021; Proestos, 2018).

As can be seen, a wide variety of foods exist called superfoods. Blueberries have a wide composition of compounds, such as B vitamins, vitamin C and E, minerals and trace elements, omega-3 and omega-6 fatty acids, antioxidants such as flavonoids, and carotenoids. Corn is distinguished by its gluten quality and high lysine content (Proestos, 2018). Quinoa has a high nutritional value, being a source of dietary fiber, vitamins, and minerals, as well as providing all the essential amino acids (Melini & Melini, 2021).

Another superfood that is gaining prominence is microalgae. In turn, microalgae are a great source of vitamins, pigments, high-quality proteins, carbohydrates, lipids, essential amino acids, and minerals (Hannan et al., 2020; Vrenna et al., 2021). Because they are rich in micronutrients such as vitamins A, B1, B2, B3, B6, B12, C, E, K, and nicotinic, they are important actors in the immune system due to their antioxidant activities. In relation to PUFAs, these comprise the ω-3 fatty acids, being linolenic acid, eicosapentaenoic acid (EPA), and docosahexaenoic acid (DHA), and the ω-6 fatty acids, being them linoleic acid, arachidonic acid (ARA), and gamma-linolenic acid (GLA) (Jacob-Lopes et al., 2019). Of note, essential fatty acids are extremely important because they play crucial roles in the functioning and development of the body, the nervous and immune system, cardiovascular functions, and hormone synthesis, among other functions. As for the growth rate, compared with oilseeds such as corn and soybeans, microalgae have a biomass production of around 7 to 20 times higher per land unit (Hosseinkhani et al., 2022). In short, microalgae have been considered one of the best foods for the future due to their wide nutritional variety, quality, and quantity of compounds, in addition to their great appeal in terms of sustainability.

3. Microalgae strains with potential

3.1 Arthrospira

The *Arthrospira* is a multicellular filamentous cyanobacterium, with spontaneous growth, mainly in equatorial, tropical, and subtropical areas (Dias et al., 2021; Vrenna et al., 2021). There are reports of *Arthrospira* being harvested from Lake Texcoco for later consumption in the Tenochtitlan (modern-day Mexico City) trade during the 16th century. That gives a great history of consumption of this microalgae by humans, being generally marketed as cakes called "tecuitlatl" and dry bread called "dihe," the latter still today (Jung et al., 2019; Lafarga et al., 2020).

In 1967, the International Association for Applied Microbiology recognized *Arthrospira* as a future food source, since the benefits arising from its consumption make *Arthrospira* a potential ingredient for the development of foods with functional characteristics (Lafarga et al., 2020). In the 1970s, commercial cultivation of *Arthrospira* was started with the main objective of providing single-cell protein (whole dried biomass), with applications directed to food and prophylactic use (Jacob-Lopes et al., 2019). In 2003, the FDA (US Food and Drug Administration) classified *Arthrospira* as "generally recognized as " (GRAS), and the FAO (Food and Agriculture Organization of the United Nations) holds these microalgae as a great promise for malnutrition due to high concentration of good quality nutrients in their composition (Branyikova & Lucakova, 2021). The United Nations Organization (UNO), in turn, elected the *Arthrospira* as the best food for the future of humanity. In parallel, after rigorous studies and reviews of toxicological data in vitro and in vivo, clinical cases, and adverse events, the Dietary Supplements Information Expert Committee (DSI-EC) of the United States Pharmacopeial Convention (USP) classified *Arthrospira* in "Class A," certifying its safety for human consumption when under controlled cultivation conditions (Trotta et al., 2022).

However, its use as a food ingredient still has hindrances. One of the biggest challenges in the use of microalgae by the food industry is the intense odor, color, and flavor, which are not being well accepted by consumers. In that regard, technological efforts are being directed toward obtaining a product with improved characteristics. BYAS-F601 is a powdered product obtained from *Arthrospira* that guarantees the nutritional benefits of its use without adding unpleasant taste and odor to foods when added (Hosseinkhani et al., 2022). Of note, the different concentrations of chlorophyll resulting from the selection and breeding of microalgae can obtain an ingredient with perfected organoleptic characteristics (Algenuity, 2020).

Furthermore, the *Arthrospira* microalgae have in its composition natural antioxidants such as carotenoids and polyphenols resulting in a great antioxidant capacity and can be used for therapeutic applications (Abdel-Moneim et al., 2022). The use of *Arthrospira* encapsulation techniques for use as an ingredient in yogurts was studied by da Silva et al. (2019), who obtained a positive result, with greater antioxidant and antiinflammatory activity, as well as the greater thermal stability of microalgae when compared with the use without encapsulation techniques.

Table 5.1 presents the nutritional profile of the *Arthrospira* microalgae, which contains proteins, essential amino acids, vitamins, mainly vitamin B12, carotenoids, minerals such as iron, zinc, magnesium, calcium, potassium, and manganese, essential fatty acids such as gamma-linolenic acid, polysaccharides, and glycolipids (Sorrenti et al., 2021).

However, in the data presented in Table 5.1, it should be taken into account that the variation between phyla, environment, and cultivation conditions is responsible for the variation in the values of

Table 5.1 Nutritional profile of the *Arthrospira* microalgae dry biomass.

Macronutrients (g/100 g)		Total saturated fatty acids (g/100 g)	
Protein	57.5	SFA 16:0	2.5
Carbohydrate	23.9	SFA 18:0	0.077
Total lipids	7.72	SFA 14:0	0.075
Ashes	6.23	**Total monounsaturated fatty acids (g/100 g)**	
Water	4.68	MUFA18:1	0.347
Fiber	3.6	MUFA 16:1	0.328
Minerals (mg/100 g)		**Total polyunsaturated fatty acids (g/100 g)**	
Potassium	1360.0	PUFA 18:2	1.25
Sodium	1050.0	PUFA 18:3	0.823
Magnesium	195.0	**Amino acids (g/100 g)**	
Calcium	120.0	Glutamic acid	8.39
Phosphor	118.0	Aspartic acid	5.79
Iron	28.5	Leucine	4.95
Copper	6.1	Alanine	4.52
Zinc	2.0	Arginine	4.15
Manganese	1.9	Valine	3.51
Selenium	0.007	Isoleucine	3.21
Vitamins (mg/100 g)		Glycine	3.1
Vitamin A, IU	570.0 IU	Lysine	3.02
Vitamin A, RAE	0.029	Serine	3.0
Vitamin B8	66.0	Threonine	2.97
Vitamin B3	12.8	Phenylalanine	2.78
Vitamin C	10.1	Tyrosine	2.58
Vitamin E	5.0	Proline	2.38
Vitamin B2	3.67	Methionine	1.15
Vitamin B5	3.48	Histidine	1.08
Vitamin B1	2.38	Tryptophan	0.929
Vitamin B9	0.094	Cystine	0.662
Vitamin K	0.025		
Vitamin B6	0.364		

Source: US Department of Agriculture, FoodData Central.

the composition of its biomass (Lafarga et al., 2020). Finally, it is noteworthy that the use of *Arthrospira* is currently more directed toward a self-care approach in several countries (Koli et al., 2022). But, apart from that, they are also a viable source of protein similar to plant and animal protein (Bleakley & Hayes, 2017).

3.2 Chlorella

Chlorella microalgae are spherical, eukaryotic, and unicellular, with an annual biomass production of approximately 5000 tons per year (Vieira et al., 2020). In the 1960s, in Japan and Taiwan, *Chlorella* microalgae were already cultivated on a commercial scale to produce healthier foods (Dragone, 2022). To date, few microalgae species have GRAS certification carried out by the US Food and Drug Administration, but *Chlorella* is one of these few species. This microalga has a high content of proteins, carotenoids such as astaxanthin, lutein, β-carotene, violaxanthin, and zeaxanthin, vitamins, polysaccharides, and fatty acids, giving it great use in the food and pharmaceutical industry (Junior et al., 2020; Khavari et al., 2021).

Of note, the protein composition of *Chlorella* biomass has essential amino acids, which are similar to conventional sources of vegetable and animal protein. When it comes to vitamins, *Chlorella* stands out for its vitamin B12 content. According to studies reported by Fu et al. (2021), people who adhere to veganism and who obtained supplementation with *Chlorella* had twice the serum levels of vitamin B12 compared with those who did not receive this supplementation. Evidently, the food industry can come to insert microalgae as a promising functional ingredient for vitamin B12 supplementation.

Again, it stands out that the values contained in Table 5.2 may vary according to external aspects, such as cultivation method and microalgae phylum, and should be compared with caution (Lafarga et al., 2020). In sum, just like for *Arthrospira*, the use of *Chlorella* microalgae in food poses some

Table 5.2 Nutritional profile of *Chlorella* microalgae dry biomass.

Macronutrients (mg/100 g)		Amino acids (g/100 g)	
Proteins	90.46	Glutamic acid	11.71
Carbohydrates	46.86	Aspartic acid	8.75
Total lipids	36.24	Leucine	8.25
Magnesium	37.3	Alanine	7.98
Calcium	33.04	Arginine	4.56
Iron	31.88	Isoleucine	2.79
Copper	1.86	Glycine	5.40
Zinc	5.44	Lysine	5.06
Manganese	11.32	Serine	4.45
Vitamins (mg/100 g)		Threonine	4.22
Vitamin A	2.08	Phenylalanine	4.87
Vitamin B12	0.21	Tyrosine	3.17
Vitamin B3	13.3	Proline	4.26
Vitamin B2	4.1	Methionine	1.98
Vitamin B1	0.62	Histidine	2.41
Vitamin K	11.21	Tryptophan	2.03
Vitamin B6	1.4		

Adapted from Alagawany, M., Taha, A.E., Noreldin, A., El-Tarabily, K.A., & Abd El-Hack, M.E. (2021). Nutritional applications of species of spirulina and Chlorella in farmed fish: A review. Aquaculture, 542, *736841. https://doi.org/10.1016/j.aquaculture.2021.736841.*

challenges in terms of consumer acceptance. Studies have reported that its compounds have peculiar aromas such as ink, fresh shrimp, cooked cereal, and sulfur. However, new research has emerged with the aim of reducing or even eliminating these attributes not approved by buyers. *Chlorella Colors*, for example, is a product based on stable varieties of *Chlorella vulgaris* created through plant breeding, thus obtaining neutral biomass in terms of taste, odor, and color, facilitating its use in a range of food products (Hosseinkhani et al., 2022).

Finally, the nutritional characteristics presented before make *Arthrospira* and *Chlorella* microalgae a promising solution for the vast food and nutritional needs of humanity. Their great content of nutrients and bioactive compounds make them increasingly interesting, making studies in the area important sources of knowledge for greater use with increasingly better returns.

4. Microalgae superfoods: issues to consider

As seen in the previous sections, food futurologists are convinced that microalgae as superfoods will be an option that is easily consumed and accessible on market shelves (Vrenna et al., 2021). However, the bottlenecks that haunt microalgal biotechnology still persist. And, therefore, there is still much to be done for this perspective to become a reality.

For this purpose, significant efforts have been made by researchers around the world to circumvent the technical problems arising from the processes of microalgal cultures. Among them, the main barrier still found is in relation to limited biomass production. Currently, laboratory-scale tests have evolved a lot in terms of productivity. Through research associated with genetic and metabolic engineering, microalgal yields are reaching values in the range of 3–5 g/L (Muthukrishnan, 2022). However, so far, these values are much lower than commercial cultures, which reach values of up to 100 g/L, as is the case with some bacteria of commercial interest. However, the scenario worsens when the work volumes of microalgae cultures are expanded to an industrial scale. Generally, these culture systems generate very dilute cultures (0.08–3.6 g/L); once the process control is complex on an industrial scale, parameters such as the climatic variations, as well as factors such as light availability and temperature control, end up strongly influencing cell performance (Gifuni et al., 2019).

Given this scenario, to try to solve the productivity gaps, several photobioreactor configurations were developed to achieve scale ensure characteristics proportional to industrial demands (Deprá et al., 2019). Among them, closed systems such as tubular bioreactors, bubble columns, and hybrid models have gained prominence due to scientific advances. However, most of these new projects bring together operational complexities and excessive demands on energy, which make the high workload difficult, making large-scale activity unfeasible. Furthermore, the dynamics behind the engineering and modeling of photobioreactors are still not well understood, requiring further investigation of the processes of optimizing and upgrading cultivation systems (Tham et al., 2022).

Under this same bias, problems associated with energy demand emerge. Both the upstream and downstream process steps are energy-intensive. Furthermore, more specifically, the harvesting, drying, and extraction are responsible for up to 60% of the total demands of the process, making it clear that this is a critical point attributed to microalgal processes and products (Deprá et al., 2020). Hence, since the largest source of energy is derived from coal, the environmental burdens put to the test, the real environmental viability of these products. Therefore, some transition strategies for renewable energies, such as the integration of solar panels and wind energy, have been discovered to reduce the emission of pollutants, and also generate products with low carbon seals and sustainable (Dias et al., 2022).

Additionally, beyond productivity, scalability, and energy demand issues, the variation in total production costs is also restrictive factor for microalgal biomass to be widely consumed by the population. This is because, in practice, the proposal to be a superfood is also associated with the tangible access to this food product by the neediest populations. However, operational maintenance factors, the high requirements for nutrients, materials, and energy, in addition to aspects related to logistics and location, as well as certifications such as organic and sustainable products, lead to price inflation of these microalgae superfoods (Verdelho Vieira et al., 2022). Proof of this is that market estimates indicate that the most recent prices for bulk products of *Chlorella* and *Arthrospira* range between 25 and 70 €/kg (Araújo et al., 2021). By way of comparison, the price of beef, a product that is considered equivalent in terms of protein content, is currently quoted between 3.44 and 46.65 €/kg (European Comission, 2022).

Also, despite these technological and economic obstacles, there are still complex and abstract discussions about the sensory attributes of several microalgal species, referring to fish flavor and odor characteristics and unpleasant color intensity for the majority of the population. These individual impasses raised by consumers test whether there is, in fact, space and inclination of consumers to purchase these future microalgae superproducts (Hosseinkhani et al., 2022).

Therefore, when considering all these barriers associated with microalgal biotechnology, we can conclude that—currently—microalgae as a superfood of the future face several bottlenecks (Cheng et al., 2022). However, the development of integrated technologies, the consecration of more efficient culture systems, and strategies for the production of multiple products—as biorefinery models—can be robust and effective measures to circumvent these hurdles hitherto imposed and lead to the consolidation of the market of these superproducts.

5. Conclusion

Microalgae have a great ability to become a superfood of great importance for future human nutrition. From a nutritional and quality point of view, microalgae products can drastically reduce the lack of nutrients in the diet. However, by the economic bias, due to their relatively high production cost, considerably low biomass productivity, and energy-intensive processes, these products are far from being available to the poorest part of the population, which is the most affected by the lack of safe and adequate food and nutrients. In this sense, it is first necessary to untie the technological knots for these products to reach everyone's table. It is hoped that with the rapid development of technology, this can be achieved soon.

References

Abdel-Moneim, A. M. E., El-Saadony, M. T., Shehata, A. M., Saad, A. M., Aldhumri, S. A., Ouda, S. M., & Mesalam, N. M. (2022). Antioxidant and antimicrobial activities of *Spirulina platensis* extracts and biogenic selenium nanoparticles against selected pathogenic bacteria and fungi. *Saudi Journal of Biological Sciences, 29*(2), 1197–1209. https://doi.org/10.1016/j.sjbs.2021.09.046

Alagawany, M., Taha, A. E., Noreldin, A., El-Tarabily, K. A., & Abd El-Hack, M. E. (2021). Nutritional applications of species of *spirulina* and *Chlorella* in farmed fish: A review. *Aquaculture, 542*, 736841. https://doi.org/10.1016/j.aquaculture.2021.736841

Algenuity. (2020). *Food and beverage*. Available in: https://www.algenuity.com/food-and-beverage.

Araújo, R., Vázquez Calderón, F., Sánchez López, J., Azevedo, I. C., Bruhn, A., Fluch, S., … Ullmann, J. (2021). Current status of the algae production industry in Europe: An emerging sector of the blue bioeconomy. *Frontiers in Marine Science, 7*, 1247. https://doi.org/10.3389/fmars.2020.626389

Béné, C., Prager, S. D., Achicanoy, H. A., Toro, P. A., Lamotte, L., Bonilla, C., & Mapes, B. R. (2019). Global map and indicators of food system sustainability. *Scientific Data, 6*(1), 1–15. https://doi.org/10.1038/s41597-019-0301-5

Bleakley, S., & Hayes, M. (2017). Algal proteins: Extraction, application, and challenges concerning production. *Foods, 6*(5), 33. https://doi.org/10.3390/foods6050033

Branyikova, I., & Lucakova, S. (2021). Technical and physiological aspects of microalgae cultivation and productivity—*spirulina* as a promising and feasible choice. *Organic Agriculture, 11*(2), 269–276. https://doi.org/10.1007/s13165-020-00323-1

Cheng, P., Li, Y., Wang, C., Guo, J., Zhou, C., Zhang, R., … Ruan, R. (2022). Integrated marine microalgae biorefineries for improved bioactive compounds: A review. *Science of the Total Environment, 817*, 152895. https://doi.org/10.1016/j.scitotenv.2021.152895

Deprá, M. C., Mérida, L. G., de Menezes, C. R., Zepka, L. Q., & Jacob-Lopes, E. (2019). A new hybrid photobioreactor design for microalgae culture. *Chemical Engineering Research and Design, 144*, 1–10. https://doi.org/10.1016/j.cherd.2019.01.023

Deprá, M. C., Severo, I. A., dos Santos, A. M., Zepka, L. Q., & Jacob-Lopes, E. (2020). Environmental impacts on commercial microalgae-based products: Sustainability metrics and indicators. *Algal Research, 51*. https://doi.org/10.1016/j.algal.2020.102056

Dias, R. R., Deprá, M. C., Severo, I. A., Zepka, L. Q., & Jacob-Lopes, E. (2022). Smart override of the energy matrix in commercial microalgae facilities: A transition path to a low-carbon bioeconomy. *Sustainable Energy Technologies and Assessments, 52*, 102073. https://doi.org/10.1016/j.seta.2022.102073

Dias, R. R., Lasta, P., Vendruscolo, R. G., Wagner, R., Zepka, L. Q., & Jacob-Lopes, E. (2021). Mapping the performance of photobioreactors for microalgae cultivation. Part II: Equatorial and tropical climate zone. *Journal of Chemical Technology & Biotechnology, 96*(3), 613–621. https://doi.org/10.1002/jctb.6574

Dragone, G. (2022). Challenges and opportunities to increase economic feasibility and sustainability of mixotrophic cultivation of green microalgae of the genus *Chlorella*. *Renewable and Sustainable Energy Reviews, 160*, 112284. https://doi.org/10.1016/j.rser.2022.112284

Elgar, F. J., Pickett, W., Pförtner, T. K., Gariépy, G., Gordon, D., Georgiades, K., … Melgar-Quiñonez, H. R. (2021). Relative food insecurity, mental health and wellbeing in 160 countries. *Social Science & Medicine, 268*, 113556. https://doi.org/10.1016/j.socscimed.2020.113556

European Comission. (2022). Meat market observatory – beef and veal. https://ec.europa.eu/info/sites/default/files/food-farming-fisheries/farming/documents/beef-weekly-prices_en.pdf.

Fernández-Ríos, A., Laso, J., Ruiz-Salmón, I., Hoehn, D., Campos, C., Cristóbal, J., … Aldaco, R. (2021). The environmental impact of 'superfoods': A space for debate and joint reflection. https://doi.org/10.3390/Foods2021-11022

Ferreira, A., Guerra, I., Costa, M., Silva, J., & Gouveia, L. (2021). Future perspectives of microalgae in the food industry. In *Cultured microalgae for the food industry* (pp. 387–433). Academic Press. https://doi.org/10.1016/B978-0-12-821080-2.00008-3

Fu, Y., Chen, T., Chen, S. H. Y., Liu, B., Sun, P., Sun, H., & Chen, F. (2021). The potentials and challenges of using microalgae as an ingredient to produce meat analogues. *Trends in Food Science & Technology, 112*, 188–200. https://doi.org/10.1016/j.tifs.2021.03.050Get

Gifuni, I., Pollio, A., Safi, C., Marzocchella, A., & Olivieri, G. (2019). Current bottlenecks and challenges of the microalgal biorefinery. *Trends in Biotechnology, 37*(3), 242–252. https://doi.org/10.1016/j.tibtech.2018.09.006

Hannan, M., Dash, R., Haque, M., Mohibbullah, M., Sohag, A. A. M., Rahman, M., … Moon, I. S. (2020). Neuroprotective potentials of marine algae and their bioactive metabolites: Pharmacological insights and therapeutic advances. *Marine Drugs, 18*(7), 347. https://doi.org/10.3390/md18070347

Hosseinkhani, N., McCauley, J. I., & Ralph, P. J. (2022). Key challenges for the commercial expansion of ingredients from algae into human food products. *Algal Research, 64*, 102696. https://doi.org/10.1016/j.algal.2022.102696

Huy, M., Vatland, A. K., & Kumar, G. (2021). Nutraceutical productions from microalgal derived compounds via circular bioeconomy perspective. *Bioresource Technology, 347*, 126575. https://doi.org/10.1016/j.biortech.2021.126575

Jacob-Lopes, E., Maroneze, M. M., Deprá, M. C., Sartori, R. B., Dias, R. R., & Zepka, L. Q. (2019). Bioactive food compounds from microalgae: An innovative framework on industrial biorefineries. *Current Opinion in Food Science, 25*, 1–7. https://doi.org/10.1016/j.cofs.2018.12.003

Jagdale, Y. D., Mahale, S. V., Zohra, B., Nayik, G. A., Dar, A. H., Khan, K. A., … Karabagias, I. K. (2021). Nutritional profile and potential health benefits of super foods: A review. *Sustainability, 13*(16), 9240. https://doi.org/10.3390/su13169240

Jung, F., Krüger-Genge, A., Waldeck, P., & Küpper, J. H. (2019). *Spirulina platensis*, a super food? *Journal of Cellular Biotechnology, 5*(1), 43–54. https://doi.org/10.3233/JCB-189012

Junior, W. G. M., Gorgich, M., Corrêa, P. S., Martins, A. A., Mata, T. M., & Caetano, N. S. (2020). Microalgae for biotechnological applications: Cultivation, harvesting and biomass processing. *Aquaculture, 528*, 735562. https://doi.org/10.1016/j.aquaculture.2020.735562

Khavari, F., Saidijam, M., Taheri, M., & Nouri, F. (2021). Microalgae: Therapeutic potentials and applications. *Molecular Biology Reports, 48*(5), 4757–4765. https://doi.org/10.1007/s11033-021-06422-w

Koli, D. K., Rudra, S. G., Bhowmik, A., & Pabbi, S. (2022). Nutritional, functional, textural and sensory evaluation of *spirulina* enriched green pasta: A potential dietary and health supplement. *Foods, 11*(7), 979. https://doi.org/10.3390/foods11070979

Lafarga, T., Fernández-Sevilla, J. M., González-López, C., & Acién-Fernández, F. G. (2020). *Spirulina* for the food and functional food industries. *Food Research International, 137*, 109356. https://doi.org/10.1016/j.foodres.2020.109356

Liu, H., Meng-Lewis, Y., Ibrahim, F., & Zhu, X. (2021). Superfoods, super healthy: Myth or reality? Examining consumers' repurchase and wom intention regarding superfoods: A theory of consumption values perspective. *Journal of Business Research, 137*, 69–88. https://doi.org/10.1016/j.jbusres.2021.08.018

Magrach, A., & Sanz, M. J. (2020). Environmental and social consequences of the increase in the demand for 'superfoods' world-wide. *People and Nature, 2*(2), 267–278. https://doi.org/10.1002/pan3.10085

Melini, V., & Melini, F. (2021). Functional components and anti-nutritional factors in gluten-free grains: A focus on quinoa seeds. *Foods, 10*(2), 351. https://doi.org/10.3390/foods10020351

Muthukrishnan, L. (2022). Bio-engineering of microalgae: Challenges and future prospects toward industrial and environmental applications. *Journal of Basic Microbiology, 62*. https://doi.org/10.1002/jobm.202100417

Proestos, C. (2018). Superfoods: Recent data on their role in the prevention of diseases. *Current Research in Nutrition and Food Science Journal, 6*(3), 576–593. https://doi.org/10.12944/CRNFSJ.6.3.02

da Silva, S. C., Fernandes, I. P., Barros, L., Fernandes, Â., Alves, M. J., Calhelha, R. C., … Barreiro, M. F. (2019). Spray-dried *Spirulina platensis* as an effective ingredient to improve yogurt formulations: Testing different encapsulating solutions. *Journal of Functional Foods, 60*, 103427. https://doi.org/10.1016/j.jff.2019.103427

Sorrenti, V., Castagna, D. A., Fortinguerra, S., Buriani, A., Scapagnini, G., & Willcox, D. C. (2021). *Spirulina* microalgae and brain health: A scoping review of experimental and clinical evidence. *Marine Drugs, 19*(6), 293. https://doi.org/10.3390/md19060293

Tham, P. E., Ng, Y. J., Vadivelu, N., Lim, H. R., Khoo, K. S., Chew, K. W., & Show, P. L. (2022). Sustainable smart photobioreactor for continuous cultivation of microalgae embedded with Internet of Things. *Bioresource Technology, 346*, 126558. https://doi.org/10.1016/j.biortech.2021.126558

Trotta, T., Porro, C., Cianciulli, A., & Panaro, M. A. (2022). Beneficial effects of *spirulina* consumption on brain health. *Nutrients, 14*(3), 676. https://doi.org/10.3390/nu14030676

UNICEF. (2021). The state of food security and nutrition in the world 2021.

Verdelho Vieira, V., Cadoret, J. P., Acien, F. G., & Benemann, J. (2022). Clarification of most relevant concepts related to the microalgae production sector. *Processes, 10*(1), 175. https://doi.org/10.3390/pr10010175

Vieira, R. C., Medeiros, J. A., do Nascimento, M. A. A., de Souza Abud, A. K., Raymundo, A., & de Farias Silva, C. E. (2020). Microalgae as sustainable food: Incorporation as strategy in the formulation of functional food. In *New and future developments in microbial biotechnology and bioengineering* (pp. 19–30). Elsevier. https://doi.org/10.1016/B978-0-12-820528-0.00003-X

Vrenna, M., Peruccio, P. P., Liu, X., Zhong, F., & Sun, Y. (2021). Microalgae as future superfoods: Fostering adoption through practice-based design research. *Sustainability, 13*(5), 2848. https://doi.org/10.3390/su13052848

Further reading

Koyande, A. K., Chew, K. W., Rambabu, K., Tao, Y., Chu, D. T., & Show, P. L. (2019). Microalgae: A potential alternative to health supplementation for humans. *Food Science and Human Wellness, 8*(1), 16–24. https://doi.org/10.1016/j.fshw.2019.03.001

US Department of Agriculture. (2019). FoodData central. https://fdc.nal.usda.gov/fdc-app.html#/food-details/170495/nutrients.

Yadav, M., Kumar, V., Sandal, N., & Chauhan, M. K. (2022). Quantitative evaluation of *Chlorella vulgaris* for removal of toxic metals from body. *Journal of Applied Phycology*, 1–12. https://doi.org/10.1007/s10811-021-02640-8

CHAPTER

6 Advanced instruments in algae materials: applications and characterization of algae-based nanoparticles for cancer therapies

Pedro Monteiro[1], João Cotas[1], Silvia Lomartire[1], Carolina Pinheiro Rocha[1], Leonel Pereira[1] and Ana Marta Mendes Gonçalves[1,2]

[1]*University of Coimbra, MARE–Marine and Environmental Sciences Centre, ARNET - Aquatic Research Network, Department of Life Sciences, Coimbra, Portugal;* [2]*Department of Biology and CESAM, University of Aveiro, Aveiro, Portugal*

1. Introduction

Cancer is currently one of the most common causes of death worldwide, second only to heart disease. Exposure to carcinogens such as UV radiation, chemical compounds, stress, hereditary conditions, and unhealthy diets are among the causes for the occurrence of the disease (Gavas et al., 2021). Overall, the most common type of tumors are lung, colon, liver, stomach, and breast tumors, and usually the only effective treatment options are chemotherapy and radiotherapy (Siegel et al., 2022). Although these therapies are successful and have developed over the years to increase efficiency, drawbacks of therapy such as low specificity, low drug bioavailability, drug resistance, and disease recurrence remain as a challenge for the effective treatment of the pathology. In that scenario, nanotechnology can improve disease treatment. Nanotechnology in cancer therapy has the potential to improve treatment, improving drug solubility, bioavailability, and targeting, efficiently transporting drugs to the cancer sites, reducing side metabolizations, and overall improving the pharmacokinetic and pharmacodynamic profile of chemotherapeutics (Khan et al., 2022). Nanotechnology is the synthesis of nanometer-scale materials (NMs), with the direct control of morphology and dimension during formation (Fawcett et al., 2017). NMs have at least one dimension less than 100 nm and can have a myriad of geometrical shapes such as plates, sheets, tubes, wires, and particles. Several reports highlight how nanomaterials exhibit unique chemical, physical, electronic, optic, thermal, mechanical, and biological properties that significantly differ from their bulk counterpart. NMs can be broadly defined into two types: organic and inorganic. Organic NMs are carbon based, while inorganic NMs include noble metals (gold, platinum and silver), magnetic materials (iron oxide Fe_3O_4), and semiconductors such as titanium dioxide and zinc oxide. The larger surface area-to-volume ratio is the major contributing factor that influences the biochemical properties. At the nanometer scale, there is a significant change in both surface chemistry and chemical reactivity. Nanoparticle applications can include biosensors,

Algae Materials: Applications Benefitting Health. https://doi.org/10.1016/B978-0-443-18816-9.00002-2

catalysts, environmental remediation, immunoassays, hyperthermia treatment of tumors, antibacterial drugs, and vector delivering of therapeutic drugs for cancer treatment (Jeevanandam et al., 2016). Reports have shown that the properties of manufactured nanoparticles are heavily influenced by the specific parameters such as particle size, morphology, and size distribution, which emphasize the importance of controlling particle shape and size during manufacture.

Nevertheless, the production of nanoparticles is far from optimized. The traditional manufacture of nanoparticles can be done in two approaches: the top-down approach, which involves a size reduction through physical or chemical processes and during this size reduction, the resulting particle size, shape and structure is heavily dependent of the technique used. Size reduction also tends to introduce surface imperfections, which can impact the overall physiochemical properties of the fabricated nanoparticle. The second approach is through a bottom-up approach via the assembly of atoms, molecules, and smaller particles or monomers. Both approaches have chemical and physical drawbacks such as low material conversion rates, are technically complex, have high energy requirements, and are costly to perform (Fawcett et al., 2017). Furthermore, many of these processes use hazardous chemical reagents as reducing agents, organic solvents, and nonbiodegradable stabilizing agents. Processes such as chemical precipitation and pyrolysis often result in toxic chemical species being deposited in the surface of newly formed particles. Conventional chemical and physical manufacturing processes use chemical hazardous such as surfactants, which act as sculpturing agents that direct particle growth. Capping agents are generally used to stabilize and prevent nanoparticle aggregation (Fawcett et al., 2017).

Due to these disadvantages and with focus on sustainability, there is a large interest in the formulation of these interesting nanomaterials through utilization of natural resources. Biosynthesis through biological entities such as microbes, plants, and marine algae has gained considerable interest, as they do not require chemical solvents and several molecules and active compounds act as reducing and stabilizing agents for the synthesis of these nanoparticles (Fawcett et al., 2017).

There is a growing body of literature focusing on the nanoparticles formulated through macroalgae and their biological activities as medical applications. Notably, there is a strong basis for the valorization of these marine nanoparticles as potential cancer therapeutics (Baghban et al., 2022; Gavas et al., 2021; Khan et al., 2022). However, a clear notion of the drawbacks of these nanoparticles in biomedical applications needs to be clarified for the development of efficient therapies that do not endanger or compromise patient's safety.

This chapter focuses on the advantages and disadvantages of nanoparticle formulation through macroalgae reducing power, applications in cancer therapy of conventional and macroalgal-based nanoparticles, focusing on silver, gold, and zinc nanoparticles. This chapter also addresses nanotoxicity, a hot research topic that focuses on clarifying the true potential and limitations of nanotechnology medical applicability.

2. Macroalgae as nanoparticle factories

Macroalgae are broadly classified into three groups, based on algae body and thallus pigmentation. Green (phylum Chlorophyta), red (phylum Rhodophyta), and brown (phylum Ochrophyta, class Phaeophyceae) macroalgae possess several biomolecules and bioproperties that assist in the formation of nanoparticles. There are several reports highlighting the properties of macroalgae-based nanoparticles, with sizes, structures, and overall characteristics of these nanomaterials according to the macroalgae used (Fawcett et al., 2017) According to the algae used, several conformations of the nanoparticles can be found.

A wide variety of macroalgae have been used in the formulation of nanoparticles, and reports highlight how the extracts of seaweeds act as reductant and capping agents, crucial factors in the formulation of nanoparticles. It is believed that phytochemicals such as carboxylic acids, carbohydrates, and proteins, act as encapsulating and reducing agents in the formulation of the nanoparticles (Ghosh, 2020). Biosynthesis of nanoparticles via biological entities has the potential to deliver clean and ecofriendly approaches; in particular, biomolecules present in macroalgae and seagrasses have the ability to synthesize a variety of nanoparticles with low to no toxicity. Biosynthesis of nanoparticles via marine algae has the potential to be a clean, nontoxic, and ecofriendly chemical approach with the potential to deliver a wide variety of morphological distinct nanoparticles (Ghosh, 2020).

So far, there is no clear understanding of the mechanisms by which macroalgae form nanoparticles. Studies have shown that many of the biomolecules present in the cell walls act as catalysts that assist in the reduction of precursor metal salts to nucleate metals and metal oxide nanoparticles, while other amphiphilic molecules act as surfactants, directing and controlling nanoparticle growth (Fawcett et al., 2017). The biogenic formation of nanoparticles is a straightforward process, as it is obtained by simply mixing the metal solution with the algae extract and observed a colorimetric reaction, which indicates the formation of the nanoparticles (Fawcett et al., 2017).

It is believed that the growth of nanoparticles in solution begins with the metal ion being converted from their mono- or divalent oxidation states to the zero-valent states. This is followed by metal ion reduction and subsequent nucleation. Following initial nucleation, a kinetically controlled process takes place in which smaller neighboring particles attach to low energy faces of the forming crystals to create larger thermodynamically stable nanoparticles. As growth progresses, biomolecules contained within the macroalgae extract act as natural surfactants on specific facets of the nanoparticle. The adsorption of the capping agents and subsequent interactions on the crystal facets reduce interfacial energy and lower surface tension (Fawcett et al., 2017). Typical morphologies produced through biosynthesis include cubes, hexagons, pentagons, rods, spheres, triangles, and wires. Biomolecules present in macroalgae can influence particle morphology, and there is a need for a considerable amount of research to identify and determine the role of specific biomolecules involved in the formation process and influence of individual biomolecules in the formation of nanoparticles (Fawcett et al., 2017; Ghosh, 2020; Jeevanandam et al., 2016).

Overall, the focus has been on applications of noble metal and oxide metal nanoparticles, such as silver, gold, and zinc nanoparticles. These demonstrate several interesting properties as potential cancer therapeutics, which are described in the following.

3. Macroalgae-based nanoparticles as cancer therapies

3.1 Zinc nanoparticles

Zinc is an essential trace element in living organisms and essential for human life activity. Zinc has been proven in studies to have a key role in DNA synthesis, gene expression, and apoptosis induction (Liu et al., 2013). Under alkaline circumstances, polysaccharides may react with zinc chloride to produce polysaccharide–zinc complexes, and the hydroxyl on polysaccharide chains is replaced by zinc (Liao et al., 2015). Zinc in polysaccharide–zinc complexes has anticancer properties by lowering

ROS production and preventing oxidative damage (Harmaza & Slobozhanina, 2014). Several investigations have been conducted to develop polysaccharide–zinc complexes based on the combined anticancer effect of polysaccharide and zinc.

Nanoparticles based on *Gracilaria edulis* (Rhodophyta) aqueous extract and zinc oxide were assayed against cervical cancer cell lines (SiHa cells). The particles had a size between 20 and 50 nm. The stability of the nanoparticles in the dispersion medium was demonstrated by a zeta potential value of 28.2 mV. The anticancer potential of this nanoparticle is in a dosage-dependent manner, with an IC_{50} value of 35 μg/mL. This effect is based on both apoptotic- and necrotic-mediated death in the cancer cell line tested (Asik et al., 2019).

The effect of these zinc oxide–*G. edulis* extract nanoparticles is similar to that of the medically applied cisplatin; moreover, these nanoparticles demonstrate nontoxic effect up to the dosage of 200 μg/mL in in vitro systems, the human PBMC cells, or hemolytic assays. These types of nanoparticles based in zinc oxide and *G. edulis* aqueous extract demonstrated the same potential in prostate cancer cell line (PC-3). The zinc oxide NPs demonstrated a higher potential against the cancer cell line than the silver NPs (IC_{50}: 28.55 and 39.60 μg/mL, respectively), and zinc oxide NPs demonstrated lower cytotoxicity against VERO cell line (IC_{50}: 88.05 and 68.49 μg/mL, respectively). Also, the proportion of apoptotic bodies was found to be 62% (silver NPs) and 70% (zinc oxide NPs). In this study, the most noticeable morphological alterations in silver NPs and zinc oxide NPs-treated cells were the cytoplasmic condensation, cell shrinkage, and the development of multiple cell surface swellings at the plasma membrane, demonstrating that these nanoparticles have potential for reduction of tumor growth (Priyadharshini et al., 2014).

The cytotoxicity and apoptotic characteristics of zinc oxide nanoparticles (ZnO NPs) produced by *Sargassum muticum* (Phaeophyceae) extraction were tested on a human liver cancer cell line (HepG2), showing a concentration-dependent reduction in cell viability (IC_{50} near 175 μg/mL), and demonstrated enhanced growth inhibition in a time- and dose-dependent manner (IC_{50} near 150 μg/mL). This nanoparticle demonstrated inhibition of angiogenesis and promoted apoptosis in the human liver cancer cell line (Sanaeimehr et al., 2018).

In another study, Rajathi et al. (2021) demonstrated the fucoidan zinc oxide nanoparticle potential against the prostate cancer cell line (PC-3). The biosynthesized fucoidan zinc oxide nanoparticle has an IC_{50} value of 234.21 μg/mL, where the major effect was the inhibition of cancer cellular growth and inducing cancer cells apoptosis (Rajathi et al., 2021).

3.2 Gold nanoparticles

Production of gold nanoparticles (AuNPs) has been perfected since the middle of the 19th century, with Michael Faraday publishing the first known research paper on the chemically controlled production of gold nanoparticles (Faraday, 1857). From the beginning of the 21st century, research on the effectiveness of gold nanoparticles for bioengineering, molecular biology procedures, medical diagnosis, and treatment has gained impetus (Ansari et al., 2021; Chen et al., 2008; Park et al., 2008; Shukla et al., 2005). Gold nanoparticles have particular potentialities, as ability to achieve ultrasmall sizes (as of 5 nm) (Bajak et al., 2015), high reactivity to other components, enabling the successful production of nanoparticle–drug complexes and surface plasmon resonance bands. The most common and widely investigated medical application of gold particles is their utility in cancer diagnosis and therapy, through the transportation and local delivery of drugs. Gold nanoparticles have been reported to enter tumoral tissues, accumulating in the extracellular matrix and then entering tumoral cells (Sun et al., 2018). The potentialities of using fucoidan/chitosan complexes for the production of gold nanoparticles have been reported by numerous authors (Manivasagan et al., 2016; Rasin et al., 2021).

Venkatpurwar and Pokharkar (2011) described a one-pot size-controlled green synthesis of AuNPs utilizing porphyran. AuNPs containing porphyran were stable throughout a wide pH range and electrolyte concentrations, suggesting that they might be effective drug carriers. Furthermore, these AuNPs could be efficiently loaded with anticancer medications, and the cytotoxicity of these porphyran-capped AuNPs loaded with anticancer pharmaceuticals increased in LN-229 cell lines.

Manivasagan et al. (2016) established the utility of chitosan/fucoidan dual polymer–coated gold nanorods in cancer photothermal treatment. Their findings showed that mice with MDA-MB-231 tumor cells recovered entirely by the 20th day of therapy after being treated with chitosan/fucoidan multilayer–coated gold nanorods (CS/F-GNRs) and laser irradiation, implying that this treatment might be employed for anticancer therapy in the future.

AuNPs produced with resort to fucoidan extracted from *Padina tetrastromatica* (Rajeshkumar et al., 2021) were tested in human alveolar (A549) and liver (hepG2) cancer cell lines, with effective toxicity (IC_{50} of 100 μM).

Carrageenan extracted from marine red algae was used by Chen et al. (2018) to produce AuNPs aimed at testing antitumoral efficacy in human breast cancer cells (MDS-MB-231) and human colon cancer cells (HCT-116). IC_{50} values for gold particles–carrageenan complexes reported for the different cell lines were of 34.4 ± 1.7 μg/mL for HCT-116 and of 129.2 ± 1.7 μg/mL for MDA-MB-231. The authors tested, as well, for cytotoxicity induced by exposure to the carrageenan extracts alone, reporting higher IC_{50} values compared with the complexes. These findings supported, as in other already mentioned cases, the higher effectiveness of nanoparticles in locally induced therapies: The complex of nanoparticle–carrageenan fabricated by the authors presented both higher cytotoxicity to cancerous cells and higher effectiveness in HCT-116 cells compared with human breast cell lines due to the higher accumulation in targeted cells and induced-receptor mediated endocytosis of galactose (Chen et al., 2018).

González-Ballesteros et al. (2019) used phenols extracted from *Ulva lactuca* to produce AuNPs that were then used as vehicle for colorectal cancer therapy research, using colorectal cancer Caco-2 and HT-29 cell lines. Cytotoxicity of gold nanoparticles–macroalgae carbohydrate extracts was verified at IC_{50} of 23 μM for HT-29 cells; in Caco-2 cell lines, however, viability was close to 80%, even at the highest dosage used (of 200 μM) (González-Ballesteros et al., 2019). Other studies, such as the one by Bajak et al. (2015), have also reported Caco-2 viability over 80% at even higher doses, of 300 μM. González-Ballesteros et al. (2017), on other hand, have shown cytotoxicity of Caco-2 cells at IC_{50} of 79.02 μM using carbohydrate extracts from *Gongolaria baccata* (formerly *Cystoseira baccata*), suggesting that the inefficiency in reducing proliferation of Caco-2 cells is likely due to the characteristics of the extracts from *U. lactuca*.

Hasan et al. (2022) showed the effectiveness of producing AuNPs from *Turbinaria decurrens* extracts, including phenolic components, flavonoids, and soluble sugars, against four human cancer cell lines. Cytotoxicity was observed in all cell lines, and the IC_{50} values reported were of 23.8 ± 0.04 μg/mL for MCF-7 (human breast cancer), 60.8 ± 0.007 μg/mL for hepG2, 98 ± 0.057 μg/mL for HCT-116, and 195.7 ± 0.07 μg/mL for A549.

AuNPs fabricated from macroalgae extracts of various compositions have proven to be quite effective in therapeutic methodologies of various cancer cell lines and enable higher locally defined action of the complexes, with promising results. This "green" methodology of producing cancer therapies may not only reduce hazardous waste in the production of the nanoparticles themselves (as has been for the long path) but also allows developing therapeutics form natural products, potentially diminishing side effects and toxicity associated with other forms of chemotherapy for cancer control and treatment.

3.3 Silver nanoparticles

Several investigations have found that silver nanoparticles demonstrate anticancer properties (Srinithya et al., 2016; Ullah Khan et al., 2018). Devi (2012) and Srinithya (2016) reported that there is a connection between silver nanoparticles activity and the aqueous *U. lactuca* (Chlorophyta) extracts. These authors achieve an HT-29 IC_{50} value of 49 μg/mL (Devi, 2012; Srinithya et al., 2016). Additionally, research using silver compounds created from various algal and plant extracts has demonstrated a greater IC_{50} against HT-29 and other cancer cell lines (Palanisamy et al., 2017). *U. lactuca* was utilized for the synthesis of gold and silver nanoparticles to evaluate their antitumoral activity. Recent studies discovered the potential use of *U. lactuca* in different application such as tissues engineering (Manivasagan & Oh, 2016; Venkatesan et al., 2015), drug delivery (Venkatesan et al., 2015), and antitumoral activity has been also studied (Abd-elrahman et al., 2017; Ahmed, 2014). Silver nanoparticles with *U. lactuca* were prepared at an extract concentration of 0.5 g/mL and final silver concentration of 0.17 mM. At higher concentration, it was observed that the nanoparticles aggregated and precipitated. The extract was heated at reflux, and then $AgNO_3$ 0.005 M was added. The cytotoxic activity of *U. lactuca* extracts, silver and gold nanoparticles, was tested in the human colon cancer cell lines Caco-2 and HT-29. The normal cell lines PCS-201−010 and CCD-112CoN were used as control of the effect in healthy cells. This study shows that *U. lactuca* extract did not induce cytotoxicity in healthy cell lines, but it was induced in the colorectal cancer cell lines at a dosage of 1 μg/mL with viabilities of $49.8 \pm 1.1\%$ and $37.7 \pm 2.3\%$ for Caco-2 and HT-29, respectively. In terms of the colorectal cancer cell lines, the analyses revealed that silver nanoparticles presented the higher effect in HT-29 cells at 13 μM, compared with gold nanoparticles. Therefore, the cytotoxic effect of gold nanoparticles and silver nanoparticles was specific to the colorectal cancer cell lines and did not produce significant toxicities at IC_{50} doses on the healthy cells, increasing the therapeutic interest of the new synthetized compounds. In HT-29 compared with Caco-2, a 50-fold higher proportion of cells were found to be in late apoptosis or necrosis. Comparing the effects of the two substances on the induction of early apoptosis, silver nanoparticles influenced both colorectal cancer cell lines, which were almost nine times greater than that of gold nanoparticles (González-Ballesteros et al., 2019).

Khalifa et al. (2016) evaluate the anticancer activity on Ehrlich ascites carcinoma (EAC) *in vitro* of silver nanoparticles biosynthesized by the reduction of silver nitrate in the algal aqueous extracts of the following seaweeds: *U. lactuca* (as *Ulva fasciata*) (Chlorophyta), *Ellisolandia elongata* (as *Corallina elongata*), *Gelidium crinale, Laurencia obtusa* (Rhodophyta), *Polycladia myrica* (as *Cystoseira myrica*), and *Turbinaria turbinata* (Phaeophyceae). Results showed that the maximum inhibition percentage of EAC cells were (94% and 99%) with 98 μg/mL AgNPs biosynthesized by *U. lactuca* and *T. turbinata*, respectively (Khalifa et al., 2016).

The anticancer effectiveness of silver nanoparticles has been investigated *in vitro* from nanoparticles synthesized from marine algae *Chaetomorpha linum* (Chlorophyta) against colon cancer cell HCT-116 treated with seaweed-based silver nanoparticles in different concentrations at 24 and 48 h.

Using MTT assay, the authors demonstrated the dose-dependent cytotoxic effect of silver nanoparticles in human colorectal carcinoma cells (HCT-116). Using the PCR technique, the immunoblotting analysis of caspases and the expression of pro- and antiapoptotic proteins were examined to comprehend the underlying molecular mechanism of silver nanoparticles on cancer cells. The apoptotic studies revealed an increase in the expression of proapoptotic proteins such Bcl-2 and Bcl-xl as well as proapoptotic caspase 3, caspase 9, BH3-interacting domain death agonist (Bid), and Bax. Results reveal that HCT-116 colon cancer cells are inhibited by treatment with silver nanoparticles at concentrations of 48.84 μg/mL and 32.75 μg/mL for the treatment periods of 24 and 48 h, respectively. This suggests that biosynthesized silver nanoparticles are effective anticancer agents that can trigger apoptosis in the HCT-116 colon cells (Acharya et al., 2021).

Azeem et al. (2022) evaluated the free radical scavenging activity DPPH, ABTS, and anticancer activity in human liver carcinoma (HepG2) cell line from macroalgae-based silver nanoparticles. Results showed that, dose of 200 μg/mL, the DPPH radical scavenging capacity was maximized for nanoparticles from *Turbinaria ornata* (67.26%); however, nanoparticles from *Ulva flexuosa* (as *Enteromorpha flexuosa*) (Chlorophyta) were the most potent as ABTs scavenger (97.74%). Additionally, nanoparticles from *Galaxaura elongata* (Rhodophyta) had the maximum proteinase inhibitory action with 59.78%. The 1000 μg/mL of *G. elongata* nanoparticles, *T. ornata* nanoparticles, and *Ulva lactuca* nanoparticles revealed significant inhibitions of cell growth of HepG2, resulting in cell viabilities at 5.92%, 4.44%, and 11.33%, respectively. These findings suggest that macroalgae bio-capped silver nanoparticles have magnificent biological potentials for safe biomedical applications (Azeem et al., 2022).

Silver nanoparticles synthesized using extracts of different marine algae species, including *Ulva rigida* (Chlorophyta), *P. myrica* (Phaeophyceae), and *Gracilaria foliifera* (Rhodophyta), have been tested as reducing and capping agents. The physiochemical properties, cytotoxicity, anticancer and antimicrobial activities of the biosynthesized silver nanoparticles were assessed using nauplii of *Artemia salina*, normal skin cell lines (HFb-4), and breast cancer cell lines (MCF-7 cell line). The lethality was found to be directly proportional to the silver nanoparticles concentration. The IC_{50} values of *P. myrica* and *G. foliifera* silver nanoparticles against nauplii of *Artemia salina* were 5 and 10 μg/mL after 4 and 16 h, respectively, whereas *U. rigida* silver nanoparticles did not exhibit cytotoxic effects. Anticancer activity of the biosynthesized silver nanoparticles was dose dependent. The IC_{50} values of the biosynthesized silver nanoparticles were 13, 13, and 43 μg/mL for *U. rigida, P. myrica,* and *G. foliifera*, respectively. *U. rigida* silver nanoparticles particularly exhibited potent anticancer activity (92.62%) against a human breast adenocarcinoma cell line MCF-7 with high selectivity compared with the normal cells ($IC_{50} = 13$ μg/mL), followed by *P. myrica* silver nanoparticles ($IC_{50} = 13$ μg/mL). This study suggests that extracts of *U. rigida* could be effectively used as reducing agents for the green synthesis of silver nanoparticles with interesting anticancer properties (Algotiml et al., 2022).

The *Sargassum wightii* (Phaeophyceae) aqueous extract and green−synthesized silver nanoparticles were evaluated by MTT method against HeLa cancer cell lines for their effect on cell viability at different concentrations of 10, 20, 30, 40, and 50 μg/mL (Suganya et al., 2020). The tested aqueous extract and biosynthesized silver nanoparticles showed a prompt decrease in cell viability at the end of 24h. According to the results, the extract of *S. wightii*-synthesized nanoparticles showed the best anticancer activity. At the same time, increasing the incubation time up to 48 h showed the loss of cell viability. In the case of aqueous extract, the inhibition of cell viability to

50% was observed at 107.88 μg/mL with an exposure time of 24 h, and silver nanoparticles showed 50% inhibition of cell viability at 47.48 μg/mL in 24 h. On the other hand, inhibition of cell viability to 50% was observed at 39.59 μg/mL of 48 h in aqueous extract and at 6.84 μg/mL of silver nanoparticles at 48 h. From MTT assay and direct cell count, it was observed that for 24 and 48 h exposure to seaweed extracts, a clear inhibition of the HeLa cell was recorded. The morphological changes in cells indicated that cells underwent apoptosis at 24 h after incubation with seaweed. *S. wightii* were characterized by a very small size and uniform shape. The authors reported that these properties contributed to the cytotoxicity of the silver nanoparticles. Thus, it is suggested that the macroalga *S. wightii* could be a potent source of natural anticancer drug, which is of great importance as a therapeutic agent in inhibiting the growth of oncogenic cells (Suganya et al., 2020).

Most of the research demonstrating the efficacy of these novel macroalgae nanoparticles are conducted in vitro or in vivo on animal models, which poses a limitation for their clinical application. Therefore, concerns exist regarding potential negative effects on people from their use in medical settings. Addressing these productions to clinical studies might make it easier to use these drugs clinically in the future.

4. Associated toxicity

Due to the growing interest in nanomedicine, safeguarding their toxicity is paramount (Machado et al., 2021). Overall, larger surface area-to-volume ratios mean that more parameters must be considered when compared with conventional bulk scale materials. Parameters such as particle size, morphology, composition, concentration, chemical reactivity, dispersion, and aggregation can all directly influence the behavior and interaction between nanoparticles and specific environments. Studies have shown that particles around 10 nm can induce greater human cell death rates when compared with larger particles ranging from 50 to 100 nm. Additionally, particles of the same composition can display behavioral differences when interacting with different environments (Gavas et al., 2021; Machado et al., 2021; Tripathi & Goshisht, 2022).

Naked nanoparticles do not exist for very long in biological or environmental surroundings. Their high surface energies attract atoms and molecules, which rapidly coat and change surface properties. In biological settings, biomolecules such as enzymes and proteins rapidly attach to nanoparticles to form surface layering or "corona." The difficulty in determining toxicity arises from how nanoparticles bind and interact with biological materials and living organisms. How these binding and interaction mechanisms change nanoparticles surface properties and their subsequent influence in living organisms are not fully understood (Machado et al., 2021).

For example, silver has been used as an antimicrobial agent for centuries, and the use of silver nanoparticles has increased the effectiveness of silver as a medicinal preparation. However, using high concentration of Ag nanoparticles can induce toxicity and consequently produce a variety of health problems. Furthermore, release of Ag nanoparticles to the environment can produce significant ecological problems. Toxicity results not only from particle size, morphology, composition, and surface reactivity but also from the presence of surface contaminants (Zhao et al., 2022). This added complication makes it necessary to understand the interaction of formation chemicals during synthesis and ultimately the interaction occurring in environment to determine potential toxicity issues. As such, to reduce the toxicity concerns, there is a need to pay attention to find alternative, clean nontoxic, and green chemistry approaches for manufacturing nanoparticles. Importantly, the interactions between a nanoparticle and its surrounding environment are a surface phenomenon. Surface and surface contaminants are actively involved in the nanoparticle/environment rather than the nanoparticle material core. Medical use of nanoparticle-based pharmaceuticals will be evaluated by their therapeutic benefit against the toxicity risks (Machado et al., 2021; Zhao et al., 2022).

5. Challenges for the successful implementation of nanoparticles as therapeutical tools

There are advantages of using seaweed extracts in the formation of nanoparticles for human therapy, such as biocompatibility, lower toxicity when compared with chemical formulated nanoparticles, low extraction and purification costs, and manipulation for drug timed control release (Khare et al., 2020; Zhang et al., 2016). Furthermore, certain properties such as pH responsiveness, gastroresistance, degradability by microbiome, and mucoadhesiveness can be improved by utilization of macroalgal biomass and its compositional phytochemicals, further providing significant benefits to micro/nanoparticulate systems in overcoming oral delivery barriers and achieving specific-site distribution.

Despite their significant promise in the biological area, certain negative health consequences have been connected to their use (Wang et al., 2013). For example, agglomeration is one of the most significant challenges in turning this treatment into pharmaceuticals due to associated toxicity in organ systems. Even when not agglomerated, nanoparticles can provoke cellular damage (Fischer & Chan, 2007). The toxicity of NPs is mainly ascribed to their shape and surface reactivity. Nevertheless, the toxicity of NPs can be reduced due to their manufacturing flexibility, by incorporating free groups on their surfaces, such as $-COOH$ groups, which are less toxic than $-OH$ and $-NH_2$ groups (Wahab et al., 2013). Toxicity can also be reduced by altering the size (30–100 nm) of metal NPs (Andleeb et al., 2021). Understanding the various interactions between biological systems is critical for the specialized and targeted usage of nanomaterials (Sharma et al., 2018). Understanding the probable interactions between biological systems and nanoparticles is critical for specialized and focused usage of nanomaterials, since aggressive responses may be prevented. Because of the presence of biocompatible phytoconstituents, biological production of NPs is favored to decrease toxicity (Handy et al., 2008; Nel et al., 2009).

Acknowledgments

This work was financed by national funds through the FCT—Foundation for Science and Technology, I.P., within the scope of the projects LA/P/0069/2020 granted to the Associate Laboratory ARNET, UIDB/04292/2020 granted to MARE—Marine and Environmental Sciences Centre, and UIDP/50017/2020 + UIDB/50017/2020 (by FCT/MTCES) granted to CESAM—Centre for Environmental and Marine Studies. This research was cofinanced by the project AlgaMar4antivirus—Marine MacroalgaE in the prevention and treatment of ANTIviral diseases (Coronavirus SARSCoV-2) (FA_07_2018_009), funded by the Blue Fund under Public Notice No. 7/2018—Scientific Research and Marine Technology. Pedro Monteiro thanks the project MENU—Marine Macroalgae: Alternative recipes for a daily nutritional diet (FA_05_2017_011), funded by the Blue Fund under Public Notice No. 5—Blue Biotechnology which cofinanced this research. Silvia Lomartire thanks Foundation for Science and Technology (FCT) for the financial support granted through the doctoral grant 2021.05005.BD. C. P. Rocha thanks FCT for the financial support provided through doctoral grant SFRH/BD/140922/2018 cofunded by the Human Potential Operational Programme (National Strategic Reference Framework (2007–13), European Social Fund

(EU), and the program POPH/FSE. João Cotas thanks the European Regional Development Fund through the Interreg Atlantic Area Program under the project NASPA (EAPA_451/2016). Ana M. M. Gonçalves acknowledges the University of Coimbra for the contract IT057-18−7253.

References

Abd-elrahman, G. E., Ahmed, O., Abdel-Reheim, E., & Abdel-Hamid, A.-H. (2017). *Ulva lactuca* polysaccharides prevent Wistar rat breast carcinogenesis through the augmentation of apoptosis, enhancement of antioxidant defense system, and suppression of inflammation. *Breast Cancer: Targets and Therapy, 9*, 67−83. https://doi.org/10.2147/BCTT.S125165

Acharya, D., Satapathy, S., Somu, P., Parida, U. K., & Mishra, G. (2021). Apoptotic effect and anticancer activity of biosynthesized silver nanoparticles from marine algae *Chaetomorpha linum* extract against human colon cancer cell HCT-116. *Biological Trace Element Research, 199*(5), 1812−1822. https://doi.org/10.1007/s12011-020-02304-7

Algotiml, R., Gab-Alla, A., Seoudi, R., Abulreesh, H. H., El-Readi, M. Z., & Elbanna, K. (2022). Anticancer and antimicrobial activity of biosynthesized Red Sea marine algal silver nanoparticles. *Scientific Reports, 12*(1), 1−18. https://doi.org/10.1038/s41598-022-06412-3

Andleeb, A., Andleeb, A., Asghar, S., Zaman, G., Tariq, M., Mehmood, A., Nadeem, M., Hano, C., Lorenzo, J. M., & Abbasi, B. H. (2021). A systematic review of biosynthesized metallic nanoparticles as a promising anti-cancer-strategy. *Cancers, 13*(11), 2818. https://doi.org/10.3390/cancers13112818

Asik, R. M., Gowdhami, B., Jaabir, M. S., Archunan, G., & Suganthy, N. (2019). Anticancer potential of zinc oxide nanoparticles against cervical carcinoma cells synthesized via biogenic route using aqueous extract of *Gracilaria edulis*. *Materials Science and Engineering: C, 103*, 109840. https://doi.org/10.1016/j.msec.2019.109840

Azeem, M. N. A., Ahmed, O. M., Shaban, M., & Elsayed, K. N. M. (2022). In vitro antioxidant, anticancer, anti-inflammatory, anti-diabetic and anti-Alzheimer potentials of innovative macroalgae bio-capped silver nanoparticles. *Environmental Science and Pollution Research*. https://doi.org/10.1007/s11356-022-20039-x, 0123456789.

Baghban, N., Khoradmehr, A., Nabipour, I., Tamadon, A., & Ullah, M. (2022). The potential of marine-based gold nanomaterials in cancer therapy: A mini-review. *Gold Bulletin, 55*(1), 53−63. https://doi.org/10.1007/s13404-021-00304-6

Bajak, E., Fabbri, M., Ponti, J., Gioria, S., Ojea-Jiménez, I., Collotta, A., Mariani, V., Gilliland, D., Rossi, F., & Gribaldo, L. (2015). Changes in Caco-2 cells transcriptome profiles upon exposure to gold nanoparticles. *Toxicology Letters, 233*(2), 187−199. https://doi.org/10.1016/j.toxlet.2014.12.008

Chen, H., Roco, M. C., Li, X., & Lin, Y. (2008). Trends in nanotechnology patents. *Nature Nanotechnology, 3*(3), 123−125. https://doi.org/10.1038/nnano.2008.51

Chen, X., Zhao, X., Gao, Y., Yin, J., Bai, M., & Wang, F. (2018). Green synthesis of gold nanoparticles using carrageenan oligosaccharide and their in vitro antitumor activity. *Marine Drugs, 16*(8), 277. https://doi.org/10.3390/md16080277

Devi, J. S. (2012). Anticancer activity of silver nanoparticles synthesized by the seaweed *Ulva lactuca* in vitro. *Journal of Nanomedicine & Biotherapeutic Discovery, 02*(03). https://doi.org/10.4172/scientificreports.242

Faraday, M. (1857). The bakerian lecture: Experimental relations of gold (and other metals) to light. *Philosophical Transactions of the Royal Society of London, 147*, 145−181. http://www.jstor.org/stable/108616.

Fawcett, D., Verduin, J. J., Shah, M., Sharma, S. B., & Poinern, G. E. J. (2017). A review of current research into the biogenic synthesis of metal and metal oxide nanoparticles via marine algae and seagrasses. *Journal of Nanoscience*, 1−15. https://doi.org/10.1155/2017/8013850

Fischer, H. C., & Chan, W. C. (2007). Nanotoxicity: The growing need for in vivo study. *Current Opinion in Biotechnology, 18*(6), 565–571. https://doi.org/10.1016/j.copbio.2007.11.008

Gavas, S., Quazi, S., & Karpiński, T. M. (2021). Nanoparticles for cancer therapy: Current progress and challenges. *Nanoscale Research Letters, 16*(1). https://doi.org/10.1186/s11671-021-03628-6

Ghosh, V. (2020). Marine bioresources as potential source for synthesis of nanoparticles. *Encyclopedia of Marine Biotechnology*, 1521–1534. https://doi.org/10.1002/9781119143802.ch64

González-Ballesteros, N., Prado-López, S., Rodríguez-González, J. B., Lastra, M., & Rodríguez-Argüelles, M. C. (2017). Green synthesis of gold nanoparticles using brown algae *Cystoseira baccata* : Its activity in colon cancer cells. *Colloids and Surfaces B: Biointerfaces, 153*, 190–198. https://doi.org/10.1016/j.colsurfb.2017.02.020

González-Ballesteros, N., Rodríguez-Argüelles, M. C., Prado-López, S., Lastra, M., Grimaldi, M., Cavazza, A., Nasi, L., Salviati, G., & Bigi, F. (2019). Macroalgae to nanoparticles: Study of *Ulva lactuca* L. role in biosynthesis of gold and silver nanoparticles and of their cytotoxicity on colon cancer cell lines. *Materials Science and Engineering: C, 97*, 498–509. https://doi.org/10.1016/j.msec.2018.12.066

Handy, R. D., von der Kammer, F., Lead, J. R., Hassellöv, M., Owen, R., & Crane, M. (2008). The ecotoxicology and chemistry of manufactured nanoparticles. *Ecotoxicology, 17*(4), 287–314. https://doi.org/10.1007/s10646-008-0199-8

Harmaza, Y. M., & Slobozhanina, E. I. (2014). Zinc essentiality and toxicity. Biophysical aspects. *Biophysics, 59*(2), 264–275. https://doi.org/10.1134/S0006350914020092

Hasan, E. A., El-Hashash, M. A., Zahran, M. K., & El-Rafie, H. M. (2022). Comparative study of chemical composition, antioxidant and anticancer activities of both *Turbinaria decurrens* Bory methanol extract and its biosynthesized gold nanoparticles. *Journal of Drug Delivery Science and Technology, 67*, 103005. https://doi.org/10.1016/j.jddst.2021.103005

Jeevanandam, J., Chan, Y. S., & Danquah, M. K. (2016). Nano-formulations of drugs: Recent developments, impact and challenges. *Biochimie, 128–129*, 99–112. https://doi.org/10.1016/j.biochi.2016.07.008

Khalifa, K. S., Hamouda, R. A., Hanafy, D., & Hamza, A. (2016). In vitro antitumor activity of silver nanoparticles biosynthesized by marine algae. *Digest Journal of Nanomaterials and Biostructures, 11*(1), 213–221.

Khan, M. I., Hossain, M. I., Hossain, M. K., Rubel, M. H. K., Hossain, K. M., Mahfuz, A. M. U. B., & Anik, M. I. (2022). Recent progress in nanostructured smart drug delivery systems for cancer therapy: A review. *ACS Applied Bio Materials, 5*(3), 971–1012. https://doi.org/10.1021/acsabm.2c00002

Khare, T., Palakurthi, S. S., Shah, B. M., Palakurthi, S., & Khare, S. (2020). Natural product-based nanomedicine in treatment of inflammatory bowel disease. *International Journal of Molecular Sciences, 21*(11), 3956. https://doi.org/10.3390/ijms21113956

Liao, W., Lu, Y., Fu, J., Ning, Z., Yang, J., & Ren, J. (2015). Preparation and characterization of d*ictyophora indusiata* polysaccharide–zinc complex and its augmented antiproliferative activity on human cancer cells. *Journal of Agricultural and Food Chemistry, 63*(29), 6525–6534. https://doi.org/10.1021/acs.jafc.5b00614

Liu, S., Cao, W., Yu, L., Zheng, W., Li, L., Fan, C., & Chen, T. (2013). Zinc(ii) complexes containing bis-benzimidazole derivatives as a new class of apoptosis inducers that trigger DNA damage-mediated p53 phosphorylation in cancer cells. *Dalton Transactions, 42*(16), 5932. https://doi.org/10.1039/c3dt33077j

M Ahmed, O. (2014). Anti-proliferative and apoptotic efficacies of ulvan polysaccharides against different types of carcinoma cells in vitro and in vivo. *Journal of Cancer Science & Therapy, 06*(06). https://doi.org/10.4172/1948-5956.1000272

Machado, S., González-Ballesteros, N., Gonçalves, A., Magalhães, L., de Passos, M. S. P., Rodríguez-Argüelles, M. C., & Gomes, A. C. (2021). Toxicity in vitro and in zebrafish embryonic development of gold nanoparticles biosynthesized using cystoseira macroalgae extracts. *International Journal of Nanomedicine, 16*(July), 5017–5036. https://doi.org/10.2147/IJN.S300674

Manivasagan, P., Bharathiraja, S., Bui, N. Q., Jang, B., Oh, Y.-O., Lim, I. G., & Oh, J. (2016). Doxorubicin-loaded fucoidan capped gold nanoparticles for drug delivery and photoacoustic imaging. *International Journal of Biological Macromolecules, 91*, 578–588. https://doi.org/10.1016/j.ijbiomac.2016.06.007

Manivasagan, P., & Oh, J. (2016). Marine polysaccharide-based nanomaterials as a novel source of nanobiotechnological applications. *International Journal of Biological Macromolecules, 82*, 315–327. https://doi.org/10.1016/j.ijbiomac.2015.10.081

Nel, A. E., Mädler, L., Velegol, D., Xia, T., Hoek, E. M. V., Somasundaran, P., Klaessig, F., Castranova, V., & Thompson, M. (2009). Understanding biophysicochemical interactions at the nano–bio interface. *Nature Materials, 8*(7), 543–557. https://doi.org/10.1038/nmat2442

Palanisamy, S., Rajasekar, P., Vijayaprasath, G., Ravi, G., Manikandan, R., & Marimuthu Prabhu, N. (2017). A green route to synthesis silver nanoparticles using Sargassum polycystum and its antioxidant and cytotoxic effects: An in vitro analysis. *Materials Letters, 189*, 196–200. https://doi.org/10.1016/j.matlet.2016.12.005

Park, J.-H., von Maltzahn, G., Ruoslahti, E., Bhatia, S. N., & Sailor, M. J. (2008). Micellar hybrid nanoparticles for simultaneous magnetofluorescent imaging and drug delivery. *Angewandte Chemie International Edition, 47*(38), 7284–7288. https://doi.org/10.1002/anie.200801810

Priyadharshini, R. I., Prasannaraj, G., Geetha, N., & Venkatachalam, P. (2014). Microwave-mediated extracellular synthesis of metallic silver and zinc oxide nanoparticles using macro-algae (*Gracilaria edulis*) extracts and its anticancer activity against human PC3 cell lines. *Applied Biochemistry and Biotechnology, 174*(8), 2777–2790. https://doi.org/10.1007/s12010-014-1225-3

Rajathi, K., Muthubharathi, M., & Leneeygreen, K. B. (2021). *Effect of fucoidan mediated zinc oxide nanoparticle on the TRAIL induced apoptosis in prostate cancer Pc-3 cell line* (Vol 8, pp. 391–398), 4.

Rajeshkumar, S., Sherif, M. H., Malarkodi, C., Ponnanikajamideen, M., Arasu, M. V., Al-Dhabi, N. A., & Roopan, S. M. (2021). Cytotoxicity behaviour of response surface model optimized gold nanoparticles by utilizing fucoidan extracted from *Padina tetrastromatica*. *Journal of Molecular Structure, 1228*, 129440. https://doi.org/10.1016/j.molstruc.2020.129440

Rasin, A. B., Shevchenko, N. M., Silchenko, A. S., Kusaykin, M. I., Likhatskaya, G. N., Zvyagintseva, T. N., & Ermakova, S. P. (2021). Relationship between the structure of a highly regular fucoidan from *Fucus evanescens* and its ability to form nanoparticles. *International Journal of Biological Macromolecules, 185*, 679–687. https://doi.org/10.1016/j.ijbiomac.2021.06.180

Sanaeimehr, Z., Javadi, I., & Namvar, F. (2018). Antiangiogenic and antiapoptotic effects of green-synthesized zinc oxide nanoparticles using *Sargassum muticum* algae extraction. *Cancer Nanotechnology, 9*(1), 3. https://doi.org/10.1186/s12645-018-0037-5

Sharma, A., Goyal, A. K., & Rath, G. (2018). Recent advances in metal nanoparticles in cancer therapy. *Journal of Drug Targeting, 26*(8), 617–632. https://doi.org/10.1080/1061186X.2017.1400553

Shukla, R., Bansal, V., Chaudhary, M., Basu, A., Bhonde, R. R., & Sastry, M. (2005). Biocompatibility of gold nanoparticles and their endocytotic fate inside the cellular compartment: A microscopic overview. *Langmuir, 21*(23), 10644–10654. https://doi.org/10.1021/la0513712

Siegel, R. L., Miller, K. D., Fuchs, H. E., & Jemal, A. (2022). Cancer statistics, 2022. *CA: A Cancer Journal for Clinicians, 72*(1), 7–33. https://doi.org/10.3322/caac.21708

Srinithya, B., Kumar, V. V., Vadivel, V., Pemaiah, B., Anthony, S. P., & Muthuraman, M. S. (2016). Synthesis of biofunctionalized AgNPs using medicinally important *Sida cordifolia* leaf extract for enhanced antioxidant and anticancer activities. *Materials Letters, 170*, 101–104. https://doi.org/10.1016/j.matlet.2016.02.019

Suganya, S., Dhanalakshmi, B., Dinesh Kumar, S., & Santhanam, P. (2020). Cytotoxic effect of silver nanoparticles synthesized from *Sargassum wightii* on cervical cancer cell line. *Proceedings of the National Academy of Sciences, India – Section B: Biological Sciences, 90*(4), 811–818. https://doi.org/10.1007/s40011-019-01152-3

Sun, H., Jia, J., Jiang, C., & Zhai, S. (2018). Gold nanoparticle-induced cell death and potential applications in nanomedicine. *International Journal of Molecular Sciences, 19*(3), 754. https://doi.org/10.3390/ijms19030754

Tripathi, N., & Goshisht, M. K. (2022). Recent advances and mechanistic insights into antibacterial activity, antibiofilm activity, and cytotoxicity of silver nanoparticles. *ACS Applied Bio Materials, 5*(4), 1391−1463. https://doi.org/10.1021/acsabm.2c00014

Ullah Khan, S., Saleh, T. A., Wahab, A., Ullah Khan, M. H., Khan, D., Ullah Khan, W., Rahim, A., Kamal, S., Ullah Khan, F., & Fahad, S. (2018). Nanosilver: New ageless and versatile biomedical therapeutic scaffold. *International Journal of Nanomedicine, 13*, 733−762. https://doi.org/10.2147/IJN.S153167

Venkatesan, J., Lowe, B., Anil, S., Manivasagan, P., Kheraif, A. A., Kang, K.-H., & Kim, S.-K. (2015). Seaweed polysaccharides and their potential biomedical applications. *Starch - Stärke, 67*(5−6), 381−390. https://doi.org/10.1002/star.201400127

Venkatpurwar, V., & Pokharkar, V. (2011). Green synthesis of silver nanoparticles using marine polysaccharide: Study of in-vitro antibacterial activity. *Materials Letters, 65*(6), 999−1002. https://doi.org/10.1016/j.matlet.2010.12.057

Wahab, R., Kaushik, N. K., Kaushik, N., Choi, E. H., Umar, A., Dwivedi, S., Musarrat, J., & Al-Khedhairy, A. A. (2013). ZnO nanoparticles induces cell death in malignant human T98G gliomas, KB and non-malignant HEK cells. *Journal of Biomedical Nanotechnology, 9*(7), 1181−1189. https://doi.org/10.1166/jbn.2013.1652

Wang, X., Reece, S. P., & Brown, J. M. (2013). Immunotoxicological impact of engineered nanomaterial exposure: Mechanisms of immune cell modulation. *Toxicology Mechanisms and Methods, 23*(3), 168−177. https://doi.org/10.3109/15376516.2012.757686

Zhang, L., Sang, Y., Feng, J., Li, Z., & Zhao, A. (2016). Polysaccharide-based micro/nanocarriers for oral colon-targeted drug delivery. *Journal of Drug Targeting, 24*(7), 579−589. https://doi.org/10.3109/1061186X.2015.1128941

Zhao, R., Xiang, J., Wang, B., Chen, L., & Tan, S. (2022). Recent advances in the development of noble metal NPs for cancer therapy. *Bioinorganic Chemistry and Applications*. https://doi.org/10.1155/2022/2444516

Further reading

Mahmood Ansari, S., Saquib, Q., De Matteis, V., Awad Alwathnani, H., Ali Alharbi, S., & Ali Al-Khedhairy, A. (2021). Marine macroalgae display bioreductant efficacy for fabricating metallic nanoparticles: Intra/extracellular mechanism and potential biomedical applications. *Bioinorganic Chemistry and Applications*, 1−26. https://doi.org/10.1155/2021/5985377

CHAPTER

Advanced techniques for cultivating algae with regard to the industrialization

7

João Rui Tanoeiro[1], Duarte Fortunato[2], Tiago Morais[2,3], João Cotas[3], Ana Marta Mendes Gonçalves[3,4], Clélia Afonso[1] and Leonel Pereira[3]

[1]*MARE—Marine and Environmental Sciences Centre, ESTM, Polytechnic of Leiria, Edifício CETEMARES Peniche, Portugal;* [2]*Lusalgae, Lda., Incubadora de Empresas da Figueira da Foz Figueira da Foz, Portugal;* [3]*University of Coimbra, MARE—Marine and Environmental Sciences Centre, ARNET - Aquatic Research Network, Department of Life Sciences, Coimbra, Portugal;* [4]*Department of Biology and CESAM, University of Aveiro, Aveiro, Portugal*

1. Introduction

Seaweed has long been a part of the human diet in the western hemisphere. There have been even discoveries of partially consumed seaweeds at a 14,000-year-old site in southern Chile (Dillehay et al., 2008). Despite this, a large portion of the world's population is still unaware of the benefits of seaweed and holds a variety of misunderstandings about it. Thinking on it as a weed, as the name implies, yet the term "seaweed" is a misnomer in itself. A weed is a plant that grows rapidly and harms the environment in which it develops. Seaweeds, on the other hand, are one of the most nutrient-dense food sources on the planet, serving as both food and habitat for a variety of marine life forms (Nehal, 2014).

Microalgae are attributed to a diverse group of microorganisms that can be found in water, air (Sahu & Tangutur, 2015), soil (Subashchandrabose et al., 2011), tree bark microhabitats (Wicker & Bhatnagar, 2020), and in some cases, even on animals (Pauli et al., 2014). Microalgae are classified as eukaryotic microorganisms containing chlorophyll *a* and can be observed in nature as individual cells or small colonies (Phwan et al., 2018).

Compared with other microorganisms and even to terrestrial plants, microalgae exhibit several advantages. They are photosynthesis-capable primary producers which absorb sunlight (photons) and assimilate carbon dioxide (CO_2) from the atmosphere to generate biomass. In comparison, most bacteria and fungi that lack photosynthetic apparatus rely on organic materials to thrive. Unlike plants, microalgae growth is not limited to freshwater or arable lands (Sajjadi et al., 2018). They can even be grown on unproductive land, such as infertile, dry regions, or polluted soils that are unsuitable for conventional farming (Zhu et al., 2013). These microbes can also grow in nutrient-rich effluent and saline water. Furthermore, microalgae cultivation is not seasonal; it can be done all year and harvested daily (Gouveia & Oliveira, 2009).

Algae Materials: Applications Benefitting Health. https://doi.org/10.1016/B978-0-443-18816-9.00007-1

Macroalgae or seaweeds are multicellular, macroscopic, eukaryotic, autotrophic organisms and are organized taxonomically into three large and distinct groups, based on the color of their thallus. Green algae are part of phylum Chlorophyta, red algae are part of phylum Rhodophyta, and brown algae are part of phylum Ochrophyta, class Phaeophyceae (Leandro et al., 2019).

As energy storage, all algae accumulate starch and other large molecular chain polysaccharides in their cells. Green algae produce ulvan, which is pigmented with carotene, xanthophylls, and chlorophylls *a* and *b*, supporting the theory that they are ancestral to plants. Chlorophylls *a* and *d*, as well as carotenoids, are found in red algae, and their coloring is caused by the presence of phycoerythrin (red pigment) in their cells. The pigments chlorophyll *a* and *c*, and carotenoids, as well as lipids and polysaccharides (such as laminarin or fucoidan), are found in brown algae (Barsanti & Gualtieri, 2005; Vidotti & Rollemberg, 2004).

Seaweeds, such as plants on land, provide similar ecological roles in aquatic environments. Some species of macroalgae can act as bioindicators of water quality, and others can perform bioremediation through biosorption and bioaccumulation (Henriques et al., 2017; Neveux et al., 2018; Yu et al., 2016). These capabilities are suitable to become a service. In 2012, a team led by de Groot estimated the ecoservices provided by macroalgae as being valued at 28,000 $/ha/year, indicating, at the time, a developing trend in the global market (de Groot et al., 2012).

Like other plants, algae are primary producers and the base of the marine food chain, supporting a variety of benthic animal groups. In addition to the need for carbon dioxide and water to thrive, they compete for light, nutrients, and space. Algae and plants, for example, create the same storage chemicals and employ similar methods of defense against predators and parasites (Barsanti & Gualtieri, 2005).

To survive biotic hazards such as bacteria, viruses, and fungal diseases, over millions of years of development, seaweeds have devised strategies. Algae have evolved to live under changeable, harsh, and hostile abiotic environmental and stress circumstances, such as temperature and salt variations, environmental toxins, or exposure to UV radiation, from sessile vegetables. As a result, these organisms can produce a wide spectrum of "secondary metabolites," including pigments, vitamins, phenolic compounds, sterols, and other bioactive substances. Aside from that, they create amino acids and proteins, saturated and unsaturated fatty acids, and polysaccharides that are directly involved in the development, growth, and reproduction of organisms to perform physiological activities. Thus, in addition to their ecological relevance, marine macroalgae have a significant commercial value based on the production of these compounds. That is why, a few years ago, there was a surge interest in the production and exploitation of macroalgae in all its forms. Many countries now use seaweeds for a variety of applications, including industrial extraction of phycocolloids and the extraction of antiviral, antibacterial, and anticancer compounds (Shannon & Abu-Ghannam, 2016). They can also be used for food and feed (livestock) or agricultural uses (biofertilizers), either directly or indirectly (Pereira, 2016).

Several of the compounds generated by algae have significant potential to be exploited in areas such as the pharmaceutical, cosmetic, and food industries, even though there is still much to learn and discover about these living organisms. As their popularity, cultivation, and applications grow, their market value also increases. This value is expected to be more than double by 2024 from what it was in 2017 (Leandro et al., 2019).

By weight, total seaweed aquaculture production occupies about 20% of global world marine aquaculture production, with an annual value of US $6.7 billion in 2013 (Cottier-Cook et al., 2016; World Bank Group, 2016).

Seaweed farming has been an alternative to depleting natural resources. It is currently economically significant in Asia and is becoming increasingly important in Europe. The global need for vast amounts of seaweed will increase in the coming years. Yet, continuous optimization of the cultivation system is still ongoing to meet this expanding demand, ensuring sustainable algal production (Buschmann et al., 2017; Hafting et al., 2015; Knowler et al., 2020).

Through research and development centers (R&D), collaboration between Academia and the aquaculture industry has led to the development of research initiatives to find possibilities and technologies to improve the efficiency and productivity of algal aquaculture systems, seeking to make them more eco-sustainable and fit for the blue economy (García-Poza et al., 2020; Hafting et al., 2011).

That said, there is a clear interest in new advanced algae cultivation techniques and their industrialization. Therefore, this chapter focuses on several of these innovative techniques.

2. Cultivation classic techniques

Seaweed farming can be done offshore, onshore, or in integrated aquaculture systems. The species, location of the farm, and production facilities all influence seaweed culture. In other situations, the procedures are the same, but the proportions differ owing to space constraints. These techniques are already applied in their full capability, as demonstrated in the following for the most commercial seaweeds nowadays, such as some kelps (Phaeophyceae), *Kappaphycus*, *Porphyra*, and *Gracilaria* (Rhodophyta).

2.1 Porphyra

Porphyra sp. is an edible red seaweed, commonly known as Nori. Cultivation of these species is widespread across Asia, where the classical method is mastered with minor optimizations by Japanese, Korean, and Chinese populations. Those modifications are linked to the cultivation substrate (Kim et al., 2017). The activity of those three countries, in 2017, accounted for more than 90% of the cultivation of *Porphyra* species, being also the most commercially success seaweed (almost a billion US dollars' worth of business volume) cultivated by these countries (Kim et al., 2017).

As for control methods, most Chinese farms, as well as some Korean and Japanese farms, use desiccation control methods to kill epiphytes and competing organisms (e.g., *Ulva* spp.) by exposing *Porphyra* nets to air, whereas most Korean and Japanese farmers prefer to use a pH control method to do so, by applying organic acids to the nets, although this method is costly (Kim et al., 2014; Pereira & Yarish, 2010).

The desiccation process has no (or very little) additional expense and can even improve the protein content of the tissues (Kim et al., 2013). On the other hand, this is not as effective as the pH control method.

On the other hand, in western countries, since the 1980s, there is interest in the cultivation of *Porphyra* sp. but with limited success (Kim et al., 2017). In 1994, the unsuccessful attempt to establish *Porphyra* culture in the State of Maine, United States, started a vast number of initiatives aiming to successfully establish and optimize this species' culture, which are now used as guidelines in aquacultures worldwide (Kim et al., 2017).

2.2 Gracilaria

Another widely cultivated edible red seaweed is *Gracilaria* sp., which records about 3.8 million tons produced annually (Araújo et al., 2022). This is valued at over one billion US dollars in net value. This seaweed is mostly cultivated in Asia, with China and Indonesia leading the worldwide production (70% and 28%, respectively). In the western world, Chile is the top producer, producing around 12,8 tons yearly, evaluated in 29 million US dollars (Araújo et al., 2022). *Gracilaria* production worldwide is used as source for food grade agar (Pereira & Yarish, 2008) and as animal feed (Johnson et al., 2014; Qi et al., 2010). Easy cultivating methods and relatively high growth rates have made *Gracilaria* sp. in attractive seaweed for aquacultures.

Open-water rope culture, near-shore bottom cultivation, pond culture, and tank culture are the most common methods to cultivate *Gracilaria* (Oliveira et al., 2008; Pereira & Yarish, 2008; Sahoo & Yarish, 2005). Providing sustainable seedstock is crucial in any of these approaches. Most *Gracilaria* seedstock is now sourced from nature (either by collecting healthy *Gracilaria* branches from natural stock or by selecting breeding plants to harvest spores (carpospores or tetraspores) for seeding (Buschmann et al., 2008).

Biological variances (e.g., growth, agar content, etc.) in seedstock can be a major concern if relying on natural stocks. Nursery (tank culture) systems are another way that could be employed to provide sufficient seedstock through vegetative propagation (Abreu et al., 2011; Kim & Yarish, 2014). Tank cultivation has several advantages, one of which is the ease with which the culture system may be controlled (Pereira et al., 2013).

This ensures that production meets stringent quality and biosafety requirements for human consumption, as well as other high-value applications in the cosmeceutical and pharmaceutical industries. However, significant management expenses are a constraint for tank culture (Caines et al., 2014).

Cost-effective options, such as CO_2 injection and the use of commercial fertilizers, have been suggested (Kim & Yarish, 2014). The adoption of LED lighting could help to further lower tank cultivation costs (Kim et al., 2015).

Gracilaria aquacultures usually suffer a lot due to fouling. Fouling organisms, such as epiphytes, tunicates, and hydroids, are observed in cultivated organisms. The development of methods and technology to solve this problem is a path that should be crossed in the near future (Kim & Yarish, 2014; Kimet al., 2017).

2.3 *Kappaphycus* and *Eucheuma*

These two red macroalgae are major sources of carrageenan (80% of world production), which makes them commercially very attractive. Worldwide, 10,75 million tons are produced per year, for a total of 1,9 billion US dollars in 2014 (FAO, 2017). Indonesia and the Philippines are the world's leading producers (83% and 13%, respectively). Other producers include Malaysia, Cambodia/Vietnam, China, Tanzania/Madagascar, Belize, and Brazil (Kim et al., 2017; Valderrama et al., 2015; World Bank Group, 2016).

Kappaphycus alvarezii and *Eucheuma denticulatum* are the most farmed of the six taxonomically recognized *Kappaphycus,* and 30 *Eucheuma* species. The fixed, off-bottom line method, the floating raft method, and the basket method are all used in the cultivation of these species (Ask & Azanza, 2002; Hayashi et al., 2010; Pereira & Yarish, 2008).

2.4 Kelps (*Saccharina and Undaria*)

Kelp could refer to the genera *Saccharina* sp. and *Undaria* sp. Worldwide, over 8 million tons of Kelp are produced, totaling 1.4 billion US dollars. Most of this production occurs in Asia, particularly in China and Korea (88.3% and 11% of global production, respectively) (Food and Agriculture Organization of the United Nations, 2017). It is mainly used as human food. However, lately there is a growing interest in using Kelp species as animal feed, due to low production costs (Kim et al., 2017). Some species of Kelp are also farmed in Western countries, in cold temperate waters. This occurs mainly in the North Atlantic Ocean, in countries such as the United States, Canada, Iceland, Norway, Scotland, Ireland, Sweden, and Germany (Kim et al., 2017), making it one of the fastest growing industries in the world (Kim et al., 2017).

The cultivation method of *Saccharina* and *Undaria* begins with sowing zoospores (meiospores). Due to the nursery capacities and scale of operations of open water farms, seeding processes differ slightly between Asia (use of seed frames) and the West (use of seeds pools) (Pereira & Yarish, 2008; Redmond et al., 2013).

Open water farming techniques using longlines, on the other hand, are extremely comparable. Kelp thalli can grow up to 2–5 m in length after being outplanted at open water farms, but they can also grow up to 10 m (Redmond et al., 2013).

The traditional aquaculture of Kelp species is particular sensible to climatic changes. In one hand, selective breeding and intense selection done in Asia (to increase growth, high temperature, and disease tolerance led to reduced genetic variety and narrow germplasm breeding (Kim et al., 2017)). On the other hand, increasing water temperature will be a challenge for aquaculture farming of cold Kelp species (Kim et al., 2017). The solution for this kind of aquaculture, located mainly in the North Atlantic, is to develop the farming of strains with tolerance to higher temperatures, by maintaining a seed bank of the same and not depending on natural seed banks (Kim et al., 2017).

2.5 Sargassum

Sargassum sp. is one of the most found brown seaweeds in the world, as it has adapted to various conditions with multiple reproductive strategies (Kim et al., 2017). It is mainly used for feeding purposes in Asian countries, due to its nutritional value and nutty flavor (Kim et al., 2017).

Traditionally, wild seedlings obtained from natural beds were used in *Sargassum* growing methods. At intervals of 5–10 cm, groups of 3–4 seedlings with a length of 5–10 cm would be inserted into seeding rope. From November to May, this smaller seeding line was connected to a primary longline planted at depths of 2–3 m and cultivated (Hwang et al., 2006; Redmond et al., 2013).

Due to this reliance on wild seedlings, natural beds have been overharvested, necessitating the development of new growing methods. The use of holdfast-derived seeding was the first step in developing *Sargassum* culture procedures. This culture takes advantage of the holdfast's perennial nature, allowing farmers to reuse holdfasts from previous year's crops (Hwang et al., 1999).

While it is still possible to obtain plants from wild beds, the associated holdfasts can be maintained for the following season's crop by oversummering in the sea until the next growing season. While this allows for the reuse of existing cultured plants, the harvestable biomass produced each year tends to reduce.

Juvenile plants obtained from reproductive adults are now used to seed *Sargassum* lines. Seedlings obtained through sexual reproduction enable a large production of new plants for seeding, resulting in higher biomass yields (Peng et al., 2013; Redmond et al., 2014).

Fertilized eggs are collected from mature fronds and "seeded" onto seedstring, letting juveniles with freshly formed rhizoids to connect to seed lines. Seedlings are attached and nurtured in a nursery until they are ready to be outplanted at sea, where they are transported to submerged long lines until harvested. The attached holdfasts can also be reused for several years without the need for additional culture rope initiation. This is a cost-effective cultivation method, although fouling organisms pose an issue. Technological development to prevent fouling is critical for the long-term viability and growth of the *Sargassum* aquaculture industry (Kim et al., 2017).

Sargassum sp. is mainly wild-harvested and farmed in Japan, China, and Korea, which has led to a low production status. However, the commercial value of this seaweed is increasing (Kim et al., 2017).

Fouling is a problematic for a whatsoever economically reasonable aquaculture method. Thus, more research on antifouling solutions for this aquaculture systems will be needed in the short term (Kim et al., 2017).

3. Cultivation innovative techniques

The global need for significant volumes of seaweed will increase in the coming years; yet, there is still ongoing cultivation system optimization required to meet this expanding demand for sustainable seaweed production and chemicals. However, collaboration between Academia and the aquaculture industry via research and development centers (R&D) has resulted in the development of joint research initiatives to identify new opportunities and new technologies to improve the efficiency and productivity of seaweed aquaculture systems, making them more eco-sustainable and fit for the blue economy (García-Poza et al., 2020).

3.1 Cultivation of seaweeds in wastewater to boost their growth and nitrogen content

It has been observed that cultivating seaweeds in wastewater considerably boosts both growth and nitrogen content. Studies show that wastewater can boost growth and nitrogen content in all types of seaweeds, either seaweeds grown on land or in the ocean, and that wastewater-based seaweed farming should not be limited to fish aquaculture wastewater (Stedt et al., 2022).

These findings clearly demonstrate the ability to generate protein-rich biomass from seaweed cultivation in a variety of wastewaters, resulting in sustainable and healthful food products for the growing world population (Stedt et al., 2022).

Green seaweeds specially benefit from wastewater culture in terms of growth and nitrogen content. Many green seaweeds are opportunistic "green tide" species, which grow quickly and can create mats that cover enormous areas when nutrients are abundant (Back et al., 2000; Ye et al., 2011). They are often hardy and resilient, with the ability to withstand considerable changes in environmental parameters such as salinity, temperature, light, and nutrient concentrations (Morand & Merceron, 2005; Xu & Lin, 2008).

This makes them perfect candidates for wastewater cultivations. Brown and red seaweeds also grow better in wastewaters, albeit the effect is not as strong as for green seaweeds. Additionally, red seaweeds demonstrate a rise in nitrogen content when grown in wastewaters. A nonsignificant favorable influence was observed for brown seaweeds (Stedt et al., 2022).

3.2 Development of offshore seaweed cultivation: food safety, cultivation, ecology, and economy

North Sea Farm (NSF) has developed and implemented many offshore seaweeds farming pilot projects since 2014. In December 2017, they started a new pilot scale trial. To cultivate *Saccharina latissima* for this pilot-scale trial, two scalable macroalgae cultivation (SMAC) modules with a total of 2000 m of seaweed production lines were deployed. The purpose of this pilot-scale trial was to gain a better understanding of the environmental and food safety impacts, as well as to optimize cultivation methods and generate scenarios for upscaling and performing further pilots within offshore wind parks.

The project's key issues were as follows: What is the impact of offshore seaweed farming (*S. latissima*) on ecology and food safety? What are the prospects for developing a sustainable large-scale offshore seaweed farming in the Dutch North Sea (Jansen et al., 2019)?

Sensors were placed at the culture module during the project to monitor abiotic factors (temperature and light). NSF has been monitoring seaweed growth monthly, using a procedure developed and verified by Wageningen Plant Research.

The use of sensors can be noted as using modern technology to our advantage in the world of aquaculture.

3.3 Seaweed and mussel cocultivation

Seaweed and shellfish cocultivation (also known as integrated culture) is cited as a multipurpose technique to optimize space in offshore wind parks. The possibility for seaweed to benefit from additional nutrients (particularly ammonium) excreted by mussels is one of the many interactions between these two organisms. As a result, growth and nitrogen assimilation procedures for the green seaweed *Ulva* spp. in monoculture and coculture with blue mussels *Mytilus edulis* were explored (Jansen, 2019). Cocultivation is projected to directly increase growth rates of *Ulva* sp. under nitrogen-limited environments; however, in light-limited situations, growth or nitrogen uptake may still be enhanced because NH_4 excreted by mussels requires less energy (ATP) to absorb than NO_3, which is naturally more abundant in marine waters.

Although the findings of this study do not clearly illustrate a benefit of growing seaweed with mussels, they do not rule out the possibility of future benefits from combined seaweed−shellfish production. Furthermore, these findings highlight the complexity of the connections between seaweed and shellfish ecosystems, as well as the importance of environmental factors in joint production.

3.4 Marine macroalgal nursery: a model for sustainable seedling production for large-scale farming

Vegetative fragments or reproductive cells (spores or gametes) are generally used as a source of seedlings (propagules) for propagation in seaweed farming (Carl et al., 2014; Ganesan et al., 2011; Gupta et al., 2018; Kraan, 2013; Ohno et al., 2009; Radulovich et al., 2015).

Vegetative propagation has surely been successful for seaweed species with higher potentials for proliferation of vegetative fragment, such as *Kappaphycus, Gracilaria, Gelidiella, Gelidium*, and others (Mantri et al., 2017; Santelices & Doty, 1989; Tseng, 2019). However, a significant portion of the crop production (one-fourth of total harvest) is used as seed material for future farming (Radulovich et al., 2015). Furthermore, the use of seedlings from the same genotype repeatedly causes loss of vigor, decline in productivity, and susceptibility to a range of diseases and pests (Buschmann et al., 2017; Gachon et al., 2010; Hurtado et al., 2013; Liu et al., 2012).

Spores, on the other hand, have been exploited as a source of seeding material for farming of seaweed species with higher reproductive potentials (*Porphyra, Pyropia, Ulva, Monostroma, Saccharina, Undaria*, and so on) (Blouin et al., 2011; Carl et al., 2016; Pereira & Yarish, 2008; Peteiro et al., 2016). The spores are sown on artificial substrates and grown under controlled conditions in a land-based seedling rearing facility until they reach a suitable size for transplanting (Camus et al., 2016; Conitz et al., 2013; Hafting et al., 2015). Ropes with growing plantlets were outplanted in the open sea for field cultivation after the seeding procedures (Camus et al., 2016; Kraan, 2013).

This type of approach could provide continuous feedstock for cultivation, but success is mostly driven by the reproductive potentials of the species, which is regarded as a major limiting factor. The development of a combination of several abiotic elements that might induce reproductive maturity and subsequent sporulation, independent of the normal life cycle, resulted from extensive knowledge on the reproductive biology of seaweeds (Carl et al., 2014, 2016). To maximize the conditions that promote maturation and sporulation in each species; however, extensive experimentation over long periods of time is required. Furthermore, the set of optimal sporulation conditions is extremely species-specific, fluctuates with plant age and physiological background, and is therefore not reproducible (Carl et al., 2014; Mantri et al., 2011).

As a result, it is extremely desirable to develop alternative seedling sources and methods that are sustainable, scalable, yield high-quality plantlets quickly, and are completely free of spores and vegetative fragments.

Protoplasts are a type of homogenous cell population that lacks cell walls and can regenerate into new plants (Reddy et al., 2008). Isolation and regeneration of protoplasts from various seaweeds have made tremendous progress. The use of protoplast for seedling formation in *Monostroma* and *Porphyra* has been proven in previous research (Dipakkore et al., 2005; Reddy et al., 2006). The advantages of using protoplasts as seed material include the ability to produce multiple plantlets from a small amount of donor biomass, a homogenous population, bypassing the optimization of stressor combinations, and the ability to produce many plantlets quickly and without fouling contaminants. However, its potential for the construction of a land-based nursery for plantlet rearing has been underutilized.

This study confirms and illustrates an integrated model for producing a large number of protoplast-based seedlings from the green alga *Ulva* as a source of seed stock. This research also indicates the scope and feasibility of using protoplasts for bulk plantlet generation at will, eliminating the need for traditional spore manufacturing for seedling production. Depending on need and preference, the plantlets can be further grown in the open sea, in a land-based tank, or in a photobioreactor system. For the first time, the enzyme reusability for five cycles was demonstrated without jeopardizing the yields and regeneration potentials of the protoplasts produced. This would considerably improve the long-term viability and appeal of the proposed protoplast-based macroalgal nursery.

The novelty of this model is the ability to produce a significant number of plantlets, up to 10^6–10^7, from biomass as small as 100 mg fresh weight and a culture space as small as 0.15 m^2. The model depicts a seedling bank-like nursery that can continually feed seedlings for open-water cultivation, land-based cultivation, or photobioreactor cultivation. When there is a full loss of harvest due to natural occurrences, this model becomes a great opportunity.

3.5 Opportunities, challenges, and future directions of open-water seaweed aquaculture in the United States—seaweed farming in Alaska

The private firm Premium Oceanic (Blue Evolution of San Francisco, California, United States) funded applied seaweed mariculture research at the University of Alaska in 2015. Its attention was directed to kelp species, which were outplanted on longlines using seeded strings.

Briefly, in late summer, mother plants are harvested, and spore release is induced onto seedstring-wrapped PVC pipes. These are incubated until the sporophyte blades are ready for outplanting under specified temperature, photoperiod, light, and nutrition conditions.

Late fall was discovered to be the best time for outplanting, after upwelling (which increases the ambient seawater nutrient concentration). Longline outplants grew best around 2–3 m below the surface, attaining their peak size and quality in April–May.

Healthy, fouling-free fronds were among the quality criteria.

Blue Evolution secured a commercial "hatchery" permit in partnership with the University of Alaska to produce seedstring of *S. latissima, Alaria*, and other kelp species to sell to farmers.

The state approved three seaweed farms in 2016: two in Kodiak and one in Ketchikan. The *Saccharina* and *Alaria* outplants were both successful, with over 5000 kg of wet weight collected and sold by Blue Evolution.

3.6 Automation concepts for industrial-scale seaweed production

SES is based in Trondheim and on Frya Island off the Mid-Norway coast. Their major production here is on horizontal carrier ropes, at shallow depths. They produced around 30 metric tons per hectare in 2017, with an average of 1 m carrier rope per square meter. SES has also developed a two-dimensional vertical substrate, but it is not currently in use, although it may be an option for future aquaculture attempts. Seaweed from Norway (SFN) is a company that farms *S. latissima* and *Alaria esculenta* for human consumption and is located in Bulandet on the west coast of Norway's. The farming system Buland 10 was created by SFN in partnership with Vaerlandet fiskeredskap. This system features a cultivation area of 1 ha and 2500 m of carrier rope. A horizontal carrier rope system is tensioned by an external rope rig or frame system. Multiple modules can be linked together to expand the system. In a collaboration between several industrial partners, the Norwegian project Seaweed vessel 2020 on the development of a large-scale, specialized seaweed culture vessel has been completed.

The vessel, which is currently in the design stage, will be used in all stages of cultivation, from offsetting up farm facilities to harvesting and transport in biomass. This type of vessel allows remote sites in offshore regions to connect.

AT-sea/SIOEN Algae sheet is a spin-off from the AT-Sea EU project and is manufactured by SIOEN in Belgium. The system consists of 10 3.2 m sheets that are joined to make a 100 m long unit. In winter crops around the Irish coast, the technique has been reported to produce up to 14 kg/m^2 of *S. latissima*. At-Sea, currently known as AtSeaNova, is a Belgian firm that provides comprehensive seaweed farm solutions. They released a product called the SeaHarvester I in late January 2020, which automates the seeding, harvesting, and cleaning procedures for both 2D substrata and long lines. AtSeaNova also provides direct seeding solutions for brown and green seaweeds, utilizing an algae binder to adhere juveniles to the growth substrate.

4. Conclusion

Concluding this analysis, the seaweed aquaculture industry has a lot of room for optimization. Depending on the type of cultivation, different problems may arise. We mainly identify three problems, on which aquaculture biotechnology should focus its work:

1. Climatic changes—To cope with climatic changes as we are watching nowadays, knowing that the trend is to get worst, the solutions spam from depending on seeding banks, rather than natural ones, to genetically developing strains with higher tolerance to heat.
2. Fouling agents and diseases—This is a problem where there are some solutions practiced nowadays, although limited. Urgent research in this area is required.
3. Bioabsorption and bioaccumulation—Seaweeds are capable of accumulating some toxic materials. This can be a major problem for open-water cultivation, which needs to be addressed when dealing with seaweed as a product (in the food chain, for example).

But these are not the only challenges for the industry. Nowadays, we can already see several solutions to increase the yield and production of seaweed. As this is an increasingly researched and explored industry, it is expected that these innovative methods will be refined and increasingly used by 2030. So, the next challenge will be to generate a business model that allows aquaculture to produce seaweed at low production costs. The solution for this may lie in the use of IMTA systems, which can boost the production, and remotely manage the aquaculture through a 4.0 system. In this way, we can effectively reduce the cost of seaweed production.

Seaweeds can also help us through the practice of bioremediation of ecosystems. This is a rising trend in the industry since it was estimated, 10 years ago, that this kind of service is valued at 28,000 $/ton/year.

As we already know, seaweed is a rising industry, capable of generating products to multiple market segments. This can help us face many of the challenges facing us, be it climatic change or the hunger crisis, just to name some.

Acknowledgments

This work is financed by national funds through FCT—Foundation for Science and Technology, I.P., within the scope of the projects LA/P/0069/2020 granted to the Associate Laboratory ARNET, UIDB/04292/2020—granted to MARE—Marine and Environmental Sciences Centre and UIDP/50017/2020+UIDB/50017/2020 (by FCT/MTCES) granted to CESAM—Centre for Environmental and Marine Studies. This research was cofinanced by the project MENU—Marine Macroalgae: Alternative recipes for a daily nutritional diet (FA_05_2017_011), funded by the Blue Fund under Public Notice No. 5-Blue Biotechnology and by the project AlgaMar4antivirus—Marine MacroalgaE in the prevention and treatment of ANTIviral diseases (Coronavirus SARS-CoV-2) (FA_07_2018_009), funded by the Blue Fund under Public Notice No. 7/2018—Scientific Research and Marine Technology. João Cotas thanks to the European Regional Development Fund through the Interreg Atlantic Area Program, under the project NASPA (EAPA_451/2016). Ana Marta Mendes Gonçalves acknowledges the University of Coimbra for the contract IT057-18-7253.

References

Abreu, M. H., Pereira, R., Buschmann, A. H., Sousa-Pinto, I., & Yarish, C. (2011). Nitrogen uptake responses of Gracilaria vermiculophylla (Ohmi) Papenfuss under combined and single addition of nitrate and ammonium. *Journal of Experimental Marine Biology and Ecology, 407*(2), 190–199. https://doi.org/10.1016/J.JEMBE.2011.06.034

Araújo, G. S., Morais, T., Cotas, J., García-Poza, S., Silva, J. W. A., Gonçalves, A. M. M., & Pereira, L. (2022). In A. Ranga Rao, & G. A. Ravishankar (Eds.), *Sustainable Global Resources Of Seaweeds, 1* pp. 63–73). Cham: Springer International Publishing. https://doi.org/10.1007/978-3-030-91955-9_4

Ask, E. I., & Azanza, R.v. (2002). Advances in cultivation technology of commercial eucheumatoid species: A review with suggestions for future research. *Aquaculture, 206*(3–4), 257–277. https://doi.org/10.1016/S0044-8486(01)00724-4

Back, S., Lehvo, A., & Blomster, J. (2000). *Mass occurrence of unattached* Enteromorpha *intestinalis on the Finnish Baltic sea coast.* https://www.jstor.org/stable/23726896.

Barsanti, L., & Gualtieri, P. (2005). *Algae : Anatomy, biochemistry, and biotechnology.* https://doi.org/10.1201/9780203492598

Blouin, N. A., Brodie, J. A., Grossman, A. C., Xu, P., & Brawley, S. H. (2011). Porphyra: A marine crop shaped by stress. *Trends in Plant Science, 16*(1), 29–37. https://doi.org/10.1016/J.TPLANTS.2010.10.004

Buschmann, A. H., Camus, C., Infante, J., Neori, A., Israel, Á., Hernández-González, M. C., Pereda, S.v., Gomez-Pinchetti, J. L., Golberg, A., Tadmor-Shalev, N., & Critchley, A. T. (2017). Seaweed production: Overview of the global state of exploitation, farming and emerging research activity. *European Journal of Phycology, 52*(4), 391–406. https://doi.org/10.1080/09670262.2017.1365175

Buschmann, A. H., Hernández-González, M. D. C., & Varela, D. (2008). Seaweed future cultivation in Chile: Perspectives and challenges. *International Journal of Environment and Pollution, 33*(4), 432–456. https://doi.org/10.1504/IJEP.2008.020571

Caines, S., Manríquez-Hernández, J. A., Duston, J., Corey, P., & Garbary, D. J. (2014). Intermittent aeration affects the bioremediation potential of two red algae cultured in finfish effluent. *Journal of Applied Phycology, 26*(5), 2173–2181. https://doi.org/10.1007/S10811-014-0247-0/TABLES/5

Camus, C., Ballerino, P., Delgado, R., Olivera-Nappa, Á., Leyton, C., & Buschmann, A. H. (2016). Scaling up bioethanol production from the farmed brown macroalga Macrocystis pyrifera in Chile. *Biofuels, Bioproducts and Biorefining, 10*(6), 673–685. https://doi.org/10.1002/BBB.1708

Carl, C., de Nys, R., & Paul, N. A. (2014). The seeding and cultivation of a tropical species of filamentous Ulva for algal biomass production. *PLoS One, 9*(6), e98700. https://doi.org/10.1371/JOURNAL.PONE.0098700

Carl, C., Lawton, R. J., Paul, N. A., & de Nys, R. (2016). Reproductive output and productivity of filamentous tropical Ulva over time. *Journal of Applied Phycology, 28*(1), 429–438. https://doi.org/10.1007/S10811-015-0589-2/FIGURES/5

Conitz, J. M., Fagen, R., & Stekoll, M. S. (2013). Effects of density and substrate type on recruitment and growth of Pyropia torta (Rhodophyta) gametophytes. *Botanica Marina, 56*(5–6), 525–533. https://doi.org/10.1515/BOT-2013-0067/MACHINEREADABLECITATION/RIS

Cottier-Cook, E. J., Nagabhatla, N., Badis, Y., Campbell, M. L., Chopin, T., Dai, W., Fang, J., He, P., Hewitt, C. L., Kim, G. H., Huo, Y., Jiang, Z., Kema, G., Li, X., Liu, F., Liu, H., Liu, Y., Lu, Q., Luo, Q., … Gachon, C. M. M. (2016). *Safeguarding the future of the global seaweed aquaculture industry.* www.sams.ac.uk/globalseaweed.

Dillehay, T. D., Ramírez, C., Pino, M., Collins, M. B., Rossen, J., & Pino-Navarro, J. D. (2008). Monte verde: Seaweed, food, medicine, and the peopling of south America. *Science, 320*(5877), 784–786. https://doi.org/10.1126/SCIENCE.1156533/SUPPL_FILE/DILLEHAY.SOM.PDF

Dipakkore, S., Reddy, C. R. K., & Jha, B. (2005). Production and seeding of protoplasts of Porphyra okhaensis (Bangiales, Rhodophyta) in laboratory culture. *Journal of Applied Phycology, 17*, 331–337. https://doi.org/10.1007/s10811-005-7291-8

FAO. (2017). *Fishery and Aquaculture Statistics*. Rome, Italy: FAO.

Gachon, C. M. M., Sime-Ngando, T., Strittmatter, M., Chambouvet, A., & Kim, G. H. (2010). Algal diseases: Spotlight on a black box. *Trends in Plant Science, 15*(11), 633–640. https://doi.org/10.1016/J.TPLANTS.2010.08.005

Ganesan, M., Thiruppathi, S., Eswaran, K., Reddy, C. R. K., & Jha, B. (2011). Development of an improved method of cultivation to obtain high biomass of the red alga Gelidiella acerosa (Gelidiales, Rhodophyta) in the open sea. *Biomass and Bioenergy, 35*(7), 2729–2736. https://doi.org/10.1016/J.BIOMBIOE.2011.03.014

García-Poza, S., Leandro, A., Cotas, C., Cotas, J., Marques, J. C., Pereira, L., & Gonçalves, A. M. M. (2020). The evolution road of seaweed aquaculture: Cultivation technologies and the industry 4.0. *International Journal of Environmental Research and Public Health, 17*(18), 6528. https://doi.org/10.3390/IJERPH17186528

Gouveia, L., & Oliveira, A. C. (2009). Microalgae as a raw material for biofuels production. *Journal of Industrial Microbiology and Biotechnology, 36*(2), 269–274. https://doi.org/10.1007/S10295-008-0495-6

de Groot, R., Brander, L., van der Ploeg, S., Costanza, R., Bernard, F., Braat, L., Christie, M., Crossman, N., Ghermandi, A., Hein, L., Hussain, S., Kumar, P., McVittie, A., Portela, R., Rodriguez, L.C., ten Brink, P., van Beukering, P. (2012). Global estimates of the value of ecosystems and their services in monetary units. *Ecosyst. Serv.,* 1, 50–61. 10.1016/j.ecoser.2012.07.005

Gupta, V., Trivedi, N., Simoni, S., & Reddy, C. R. K. (2018). Marine macroalgal nursery: A model for sustainable production of seedlings for large scale farming. *Algal Research, 31*, 463–468. https://doi.org/10.1016/J.ALGAL.2018.02.032

Hafting, J. T., Craigie, J. S., Stengel, D. B., Loureiro, R. R., Buschmann, A. H., Yarish, C., Edwards, M. D., & Critchley, A. T. (2015). Prospects and challenges for industrial production of seaweed bioactives. *Journal of Phycology, 51*(5), 821–837. https://doi.org/10.1111/JPY.12326

Hafting, J. T., Critchley, A. T., Cornish, M. L., Hubley, S. A., & Archibald, A. F. (2011). On-land cultivation of functional seaweed products for human usage. *Journal of Applied Phycology, 24*(3), 385–392. https://doi.org/10.1007/S10811-011-9720-1

Hayashi, L., Hurtado, A. Q., Msuya, F. E., Bleicher-Lhonneur, G., & Critchley, A. T. (2010). A review of Kappaphycus farming: Prospects and constraints. *Seaweeds and their role in globally changing environments* (pp. 251–283). https://doi.org/10.1007/978-90-481-8569-6_15

Henriques, B., Lopes, C. B., Figueira, P., Rocha, L. S., Duarte, A. C., Vale, C., Pardal, M. A., & Pereira, E. (2017). Bioaccumulation of Hg, Cd and Pb by Fucus vesiculosus in single and multi-metal contamination scenarios and its effect on growth rate. *Chemosphere, 171*, 208–222. https://doi.org/10.1016/J.CHEMOSPHERE.2016.12.086

Hurtado, A. Q., Montaño, M. N. E., & Martinez-Goss, M. R. (2013). Commercial production of carrageenophytes in the Philippines: Ensuring long-term sustainability for the industry. *Journal of Applied Phycology, 25*(3), 733–742. https://doi.org/10.1007/S10811-012-9945-7/FIGURES/6

Hwang, E. K., Cho, Y. C., & Sohn, chul H. (1999). *Reuse of holdfasts in hizikia cultivation.* https://www.semanticscholar.org/paper/Reuse-of-Holdfasts-in-Hizikia-Cultivation-Kyoung-Chul/bb725ed73cb3748bf519f954448f077fd981b450.

Hwang, E.-K., Ha, D.-S., Baek, J.-M., Wee, M.-Y., & Park, C.-S. (2006). Effects of pH and salinity on the cultivated brown alga sargassumfulvellum and associated animals. *Algae, 21*(3), 317–321. https://doi.org/10.4490/ALGAE.2006.21.3.317

Jansen, H. (2019). *Co-cultivation of the seaweed Ulva sp. and* Mytilus edulis. https://doi.org/10.18174/470705

Jansen, H. M., Tonk, L., van der Werf, A., van der Meer, I., van Tuinen, S., van der Burg, S., Veen, J., Bronswijk, L., & Brouwers, E. (2019). *Development of offshore seaweed cultivation: Food safety, cultivation, ecology and economy : Synthesis report 2018*. https://doi.org/10.18174/470706

Johnson, R. B., Kim, J. K., Armbruster, L. C., & Yarish, C. (2014). Nitrogen allocation of Gracilaria tikvahiae grown in urbanized estuaries of Long Island Sound and New York City, USA: a preliminary evaluation of ocean farmed Gracilaria for alternative fish feeds. *ALGAE, 29*, 227–235. https://doi.org/10.4490/algae.2014.29.3.227

Kim, J. K., Kraemer, G. P., & Yarish, C. (2013). Emersion induces nitrogen release and alteration of nitrogen metabolism in the intertidal genus Porphyra. *PLoS One, 8*(7), e69961. https://doi.org/10.1371/JOURNAL.PONE.0069961

Kim, J. K., Mao, Y., Kraemer, G., & Yarish, C. (2015). Growth and pigment content of Gracilaria tikvahiae McLachlan under fluorescent and LED lighting. *Aquaculture, 436*, 52–57. https://doi.org/10.1016/J.AQUACULTURE.2014.10.037

Kim, G. H., Moon, K. H., Kim, J. Y., Shim, J., & Klochkova, T. A. (2014). A revaluation of algal diseases in Korean Pyropia (Porphyra) sea farms and their economic impact. *Algae, 29*(4), 249–265. https://doi.org/10.4490/ALGAE.2014.29.4.249

Kim, J. K., & Yarish, C. (2014). Development of a sustainable land-based Gracilaria cultivation system. *Algae, 29*(3), 217–225. https://doi.org/10.4490/ALGAE.2014.29.3.217

Kim, J. K., Yarish, C., Hwang, E. K., Park, M., Kim, Y., Kim, J. K., Yarish, C., Hwang, E. K., Park, M., & Kim, Y. (2017). Seaweed aquaculture: Cultivation technologies, challenges and its ecosystem services. *Algae, 32*(1), 1–13. https://doi.org/10.4490/ALGAE.2017.32.3.3

Knowler, D., Chopin, T., Martínez-Espiñeira, R., Neori, A., Nobre, A., Noce, A., & Reid, G. (2020). The economics of integrated multi-trophic aquaculture: Where are we now and where do we need to go? *Reviews in Aquaculture, 12*(3), 1579–1594. https://doi.org/10.1111/RAQ.12399

Kraan, S. (2013). Mass-cultivation of carbohydrate rich macroalgae, a possible solution for sustainable biofuel production. *Mitigation and Adaptation Strategies for Global Change, 18*(1), 27–46. https://doi.org/10.1007/S11027-010-9275-5/FIGURES/2

Leandro, A., Pereira, L., & Gonçalves, A. M. M. (2019). Diverse applications of marine macroalgae. *Marine Drugs, 18*(1), 17. https://doi.org/10.3390/MD18010017

Liu, F., Yao, J., Wang, X., Repnikova, A., Galanin, D. A., & Duan, D. (2012). Genetic diversity and structure within and between wild and cultivated Saccharina japonica (Laminariales, Phaeophyta) revealed by SSR markers. *Aquaculture, 358–359*, 139–145. https://doi.org/10.1016/J.AQUACULTURE.2012.06.022

Mantri, V. A., Eswaran, K., Shanmugam, M., Ganesan, M., Veeragurunathan, V., Thiruppathi, S., Reddy, C. R. K., & Seth, A. (2017). An appraisal on commercial farming of Kappaphycus alvarezii in India: Success in diversification of livelihood and prospects. *Journal of Applied Phycology, 29*(1), 335–357. https://doi.org/10.1007/S10811-016-0948-7/FIGURES/6

Mantri, V. A., Singh, R. P., Bijo, A. J., Kumari, P., Reddy, C. R. K., & Jha, B. (2011). Differential response of varying salinity and temperature on zoospore induction, regeneration and daily growth rate in Ulva fasciata (Chlorophyta, Ulvales). *Journal of Applied Phycology, 23*(2), 243–250. https://doi.org/10.1007/S10811-010-9544-4/FIGURES/4

Morand, P., & Merceron, M. (2005). Macroalgal population and sustainability. *Journal of Coastal Research, 2005*(215), 1009–1020. https://doi.org/10.2112/04-700A.1

Nehal, N. (2014). *Seaweed: A potential "superfood" unexplored and untapped*. https://www.academia.edu/8795700/Seaweed_A_potential_Superfood_unexplored_and_untapped.

Neveux, N., Bolton, J. J., Bruhn, A., Roberts, D. A., & Ras, M. (2018). The bioremediation potential of seaweeds: Recycling nitrogen, phosphorus, and other waste products. *Blue Biotechnology*, 217–239. https://doi.org/10.1002/9783527801718.CH7

Ohno, T., Uramoto, H., Masuda, Y., Kubota, H., & Ishihara, T. (2009). Influence of aging on stress ulcer formation in rats. *Journal of Gastroenterology, 24*(S162), 95–99. https://doi.org/10.3109/00365528909091134

Oliveira, E. C., Alveal, K., & Anderson, R. J. (2008). Mariculture of the agar-producing gracilarioid red algae. *Reviews in Fisheries Science, 8*(4), 345–377. https://doi.org/10.1080/10408340308951116

Pauli, J. N., Mendoza, J. E., Steffan, S. A., Carey, C. C., Weimer, P. J., & Peery, M. Z. (2014). A syndrome of mutualism reinforces the lifestyle of a sloth. *Proceedings of the Royal Society B: Biological Sciences, 281*(1778). https://doi.org/10.1098/RSPB.2013.3006

Peng, Y., Xie, E., Zheng, K., Fredimoses, M., Yang, X., Zhou, X., Wang, Y., Yang, B., Lin, X., Liu, J., & Liu, Y. (2013). Nutritional and chemical composition and antiviral activity of cultivated seaweed Sargassum naozhouense Tseng et Lu. *Marine Drugs, 11*(1), 20. https://doi.org/10.3390/MD11010020

Pereira, L. (2016). Edible seaweeds of the world. *Edible Seaweeds of the World.* https://doi.org/10.1201/B19970

Pereira, R., & Yarish, C. (2008). Mass production of marine macroalgae. *Encyclopedia of Ecology, Five-Volume Set*, 2236–2247. https://doi.org/10.1016/B978-008045405-4.00066-5

Pereira, R., & Yarish, C. (2010). *The role of in sustainable culture systems: physiology and applications*, 339–354. https://doi.org/10.1007/978-90-481-8569-6_19

Pereira, R., Yarish, C., & Critchley, A. T. (2013). Seaweed aquaculture for human foods in land-based and IMTA systems. *Sustainable Food Production*, 1405–1424. https://doi.org/10.1007/978-1-4614-5797-8_189

Peteiro, C., Sánchez, N., & Martínez, B. (2016). Mariculture of the Asian kelp Undaria pinnatifida and the native kelp Saccharina latissima along the Atlantic coast of Southern Europe: An overview. *Algal Research, 15*, 9–23. https://doi.org/10.1016/J.ALGAL.2016.01.012

Phwan, C. K., Ong, H. C., Chen, W. H., Ling, T. C., Ng, E. P., & Show, P. L. (2018). Overview: Comparison of pretreatment technologies and fermentation processes of bioethanol from microalgae. *Energy Conversion and Management, 173*, 81–94. https://doi.org/10.1016/J.ENCONMAN.2018.07.054

Qi, Z., Liu, H., Li, B., Mao, Y., Jiang, Z., Zhang, J., & Fang, J. (2010). Suitability of two seaweeds, *Gracilaria lemaneiformis* and *Sargassum pallidum*, as feed for the abalone Haliotis discus hannai Ino. *Aquaculture, 300*, 189–193. https://doi.org/10.1016/j.aquaculture.2010.01.019

Radulovich, R., Neori, A., Valderrama, D., Reddy, C. R. K., Cronin, H., & Forster, J. (2015). Farming of seaweeds. *Seaweed Sustainability: Food and Non-Food Applications* (Vols. 27–59). https://doi.org/10.1016/B978-0-12-418697-2.00003-9

Reddy, C. R. K., Dipakkore, S., Kumar, G. R., Jha, B., Cheney, D. P., & Fujita, Y. (2006). An improved enzyme preparation for rapid mass production of protoplasts as seed stock for aquaculture of macrophytic marine green algae. *Aquaculture, 260*(1–4), 290–297. https://doi.org/10.1016/J.AQUACULTURE.2006.06.034

Reddy, C. R. K., Gupta, M. K., Mantri, V. A., & Jha, B. (2008). Seaweed protoplasts: Status, biotechnological perspectives and needs. *Journal of Applied Phycology, 20*(5), 619–632. https://doi.org/10.1007/S10811-007-9237-9/TABLES/5

Redmond, S., Green-Gavrielidis, L., & Yarish, C. (2013). *New England seaweed culture handbook: nursery systems*. https://www.researchgate.net/publication/262798428_New_England_Seaweed_Culture_Handbook_Nursery_Systems.

Redmond, S., Kim, J. K., Yarish, C., Pietrak, M., Bricknell, I., Redmond, S., Kim, J. K., Yarish, C., & Pietrak, M. (2014). *Culture of Sargassum in Korea techniques and potential for culture in the U.S.*

Sahoo, D., & Yarish, C. (2005). Mariculture of seaweeds. *Algal Culturing Techniques*, 219–237. https://doi.org/10.1016/B978-012088426-1/50016-0

Sahu, N., & Tangutur, A. D. (2015). Airborne algae: Overview of the current status and its implications on the environment. *Aerobiologia, 31*(1), 89–97. https://doi.org/10.1007/S10453-014-9349-Z

Sajjadi, B., Chen, W. Y., Raman, A. A. A., & Ibrahim, S. (2018). Microalgae lipid and biomass for biofuel production: A comprehensive review on lipid enhancement strategies and their effects on fatty acid composition. *Renewable and Sustainable Energy Reviews, 97*, 200–232. https://doi.org/10.1016/J.RSER.2018.07.050

Santelices, B., & Doty, M. S. (1989). A review of Gracilaria farming. *Aquaculture, 78*(2), 95–133. https://doi.org/10.1016/0044-8486(89)90026-4

Shannon, E., & Abu-Ghannam, N. (2016). Antibacterial derivatives of marine algae: An overview of pharmacological mechanisms and applications. *Marine Drugs, 14*(4), 81. https://doi.org/10.3390/MD14040081

Stedt, K., Pavia, H., & Toth, G. B. (2022). Cultivation in wastewater increases growth and nitrogen content of seaweeds: A meta-analysis. *Algal Research, 61*, 102573. https://doi.org/10.1016/J.ALGAL.2021.102573

Subashchandrabose, S. R., Ramakrishnan, B., Megharaj, M., Venkateswarlu, K., & Naidu, R. (2011). Consortia of cyanobacteria/microalgae and bacteria: Biotechnological potential. *Biotechnology Advances, 29*(6), 896–907. https://doi.org/10.1016/J.BIOTECHADV.2011.07.009

Tseng, C. K. (2019). Marine Phycoculturein China. In *International seaweed symposium (xth)* (pp. 123–152). https://doi.org/10.1515/9783110865271-010/HTML

Vidotti, E. C., & Rollemberg, M. D. C. E. (2004). Algas: Da economia nos ambientes aquáticos à bioremediação e à química analítica. *Química Nova, 27*(1), 139–145. https://doi.org/10.1590/S0100-40422004000100024

Wicker, R., & Bhatnagar, A. (2020). Application of Nordic microalgal-bacterial consortia for nutrient removal from wastewater. *Chemical Engineering Journal, 398*, 125567. https://doi.org/10.1016/J.CEJ.2020.125567

World Bank Group. (2016). Seaweed aquaculture for food security, income generation and environmental health in tropical developing countries. *Seaweed Aquaculture for Food Security, Income Generation and Environmental Health in Tropical Developing Countries*. https://doi.org/10.1596/24919

Xu, Y., & Lin, J. (2008). Effect of temperature, salinity, and light intensity on the growth of the green macroalga, chaetomorpha linum. *Journal of the World Aquaculture Society, 39*(6), 847–851. https://doi.org/10.1111/J.1749-7345.2008.00223.X

Ye, N., Zhang, hao, wen, X., Mao, ze, Y., Liang, C., wei, Xu, D., Zou, J., Zhuang, Z. meng, & Wang, Q. yin (2011). 'Green tides' are overwhelming the coastline of our blue planet: Taking the world's largest example. *Ecological Research, 26*(3), 477–485. https://doi.org/10.1007/S11284-011-0821-8

Yu, Z., Robinson, S. M. C., Xia, J., Sun, H., & Hu, C. (2016). Growth, bioaccumulation and fodder potentials of the seaweed Sargassum hemiphyllum grown in oyster and fish farms of South China. *Aquaculture, 464*, 459–468. https://doi.org/10.1016/J.AQUACULTURE.2016.07.031

Zhu, J., Rong, J., & Zong, B. (2013). Factors in mass cultivation of microalgae for biodiesel. *Chinese Journal of Catalysis, 34*(1), 80–100. https://doi.org/10.1016/S1872-2067(11)60497-X

Further reading

Kim, J. K., Stekoll, M., & Yarish, C. (2019). Opportunities, challenges and future directions of open-water seaweed aquaculture in the United States. *Phycologia, 58*(5), 446–461. https://doi.org/10.1080/00318884.2019.1625611

Solvang, T., Bale, E. S., Broch, O. J., Handå, A., & Alver, M. O. (2021). Automation concepts for industrial-scale production of seaweed. *Frontiers in Marine Science, 8*, 411. https://doi.org/10.3389/FMARS.2021.613093/BIBTEX

CHAPTER

8 Algae-based bioplastics: Expanding algal polymers as materials for industrial applications

Silvia Lomartire[1], Louisa Gomes[1], João Cotas[1], João C. Marques[1], Leonel Pereira[1] and Ana Marta Mendes Gonçalves[1,2]

[1]*University of Coimbra, MARE—Marine and Environmental Sciences Centre, ARNET - Aquatic Research Network, Department of Life Sciences, Coimbra, Portugal;* [2]*Department of Biology and CESAM, University of Aveiro, Aveiro, Portugal*

1. Introduction

Plastics are widely used because of its outstanding features in durability, permeability to water, lightness, stability, low cost widely employed in industry, and daily life since their invention (Schmaltz et al., 2020; Shen et al., 2020; Turner, 2018). Every year, more than 300 million tons of plastic are produced, with half of it being single use (Allouzi et al., 2021). Plastic waste is mistreated in most areas of the world because of its corrosion resilience, chemical stability, and difficulty of decomposing, which takes over 1000 years to break down. Plastics are essentially polymers, which are chains of repeating units joined together. The chains can be branched or cross-linked, and the addition of other chemical monomers can change their physicochemical characteristics (Price-Whelan et al., 2014; Waring et al., 2018).

Many types of plastic now available may be customized to meet specific needs. Next, the types of plastic that can be frequently encountered in daily life and some examples are described. Polyethylene (PE) in low density (LDPE) is used, for example, in bin bags, nursery bags, small squeeze bags, while PE in high density (HDPE) can be found in containers, bottle covers, injection-molded products such as mugs. Polypropylene (PP) is used in straws, sheets medical containers, hair dryer; polyvinyl chloride (PVC) is found in cables, wires, building and construction industries, window frames; polystyrene rigid (PS) is used in food pots, toys, circuit boards, and switch plates, whereas polystyrene expanded (EPS) is used in packaging material (Shen et al., 2020; Shent et al., 1999).

Mechanical, chemical, and biological activities, such as mechanical degradation, biodegradation, heat action, UV degradation, photodegradation, friction, turbulence, and other processes, cause large polymers to degrade when they interact with the natural environment. Macroplastics, microplastics (MPs), or nanoplastics (NPs) get into the environment directly through personal care items (e.g., shampoo, detergents, cosmetics, paints) or are degraded from bigger plastic particles present in ecosystems (Kumar, Chen, et al., 2021; Van Cauwenberghe et al., 2015). Macroplastics, MPs, and NPs can be found in soil, groundwater, atmosphere, rivers, and oceans; they can be also found in high-

Algae Materials: Applications Benefitting Health. https://doi.org/10.1016/B978-0-443-18816-9.00024-1

altitude ecosystems (e.g., mountains, glaciers) (Li et al., 2021). The presence of these plastics in the environment has been causing pollution problems, with social and health effects. While pollution refers to the environmental contamination caused by discarded plastic, this pollutant accumulates in all sorts of environmental media, including aquatic and terrestrial ecosystems and in the atmosphere (Adam et al., 2021; Chen et al., 2021; Kumar, Chen, et al., 2021; Price-Whelan et al., 2014; Shen et al., 2020). The presence of MPs and NPs in terrestrial ecosystems diminishes the capacity of the soil to store carbon, alterating one of the primary ecosystem services, which provides to keep the carbon cycle running. They also affect nutrient cycles such as nitrogen and phosphorus. These nutrient cycles are imperative to the productivity of the soil; if this is diminished, there will be a negative impact on the agricultural productivity (Hu et al., 2019; Kumar, Sharma, et al., 2021; Wang et al., 2021). Aquatic environment is highly affected by macroplastics, MPs, and NPs.

Macroplastics affect large animals, such as turtles that consume the plastic thinking their eating food. This ingestion can lead to the death of several species of animals; others entrap themselves in plastic materials being unable to free themselves (Revel et al., 2018; Waring et al., 2018; Xu et al., 2020). MPs and NPs can be consumed by zooplanktonic organisms and travel in the food chain in that way (Revel et al., 2018). The presence of plastics in the marine environment can bring consequences such as death, helping the reduction of biodiversity, or can bring problems in several organs of the animals. There are studies correlating the lower egg-laying capacity with the ingestion of MPs and NPs (Oliveira et al., 2019). There are also reports of the presence of MPs and NPs in the atmosphere, becoming other alternative way of entry in the human organism (Waring et al., 2018). The inhalation of MPs by humans can originate serious health concerns in humans, such as lung cancer, ulcers, and diverse nasal and olfactory diseases. The presence of these pollutants in food chain is also a growing problem because its ingestion can bring other kind of health problems (Revel et al., 2018; Wang et al., 2020). Microplastic contamination in the worldwide ecosystem illustrates the widespread use of fossil-based polymers.

The increasing concerns about the plastic pollution and the consequences in the ecosystems and human health have been motivating the need to find new biobased plastics, materials, and products that are being actively developed as alternatives to fossil-based plastics.

The European Union (EU) (García-Depraect et al., 2021) has identified bioplastics as a viable alternative to conventional plastics in concern of scarce fossil resources, a healthy environment, and climate change (Samantaray et al., 2020). It is possible to categorize bioplastics into four groups based on their chemical composition, source, and manufacturing method: (1) polymers derived directly from biomass (e.g., protein, starch, cellulose); (2) polymers made by chemical synthesis using bioderived monomers (e.g., bio-PE, polyhydroxybutyrate (PHB), polylactide (PLA)); (3) polymers made by microbial fermentation (e.g., polyhydroxyalkanoates); (4) polymers made out of chemical synthesis using both petroleum-based and bioderived monomers (e.g., polytrimethylene terephthalate, polybutylene succinate). Bioplastics can be created from different sources such as plants (e.g., starch-based source: rice, corn, barley, and cellulose derivates), bacteria, and algae (Srinivasan & Kulshreshtha, 2020; Thiruchelvi et al., 2020).

This chapter focuses on the bioplastic originated from seaweeds. Because of their rapid growth rate and broad environmental tolerance, seaweeds have been explored as a viable source to produce bioplastic, reducing competition for food and water resources. Furthermore, seaweeds may effectively offset carbon dioxide emissions from flue gas, as well as accumulate substantial amounts of polysaccharides, proteins, and lipids that can be utilized as potential biobased plastics

(Zhang et al., 2019b). Polysaccharides (Castilho et al., 2009) such as agar, alginate, and carrageenan are found in the most regularly utilized seaweed varieties in industry (Alaswad et al., 2015; Castilho et al., 2009).

2. Overview of seaweeds as renewable sources of biopolymers for industrial applications

Seaweed polysaccharide content, such as agar, alginate, and carrageenan, is useful industrially (Khalil et al., 2017; Masarin et al., 2016). These polysaccharides can be used to make bioplastics, biomaterials, and bioscaffolds. Biodegradable and recycled polymers, such as seaweed polysaccharides, are alternatives to petrochemical-based plastics (Nur Hanani & Aelma Husna, 2018). Seaweed polysaccharides are extremely colloidal in nature, have a cheap manufacturing cost, and may be employed in a range of industrial applications (Sudhakar et al., 2020).

Seaweeds have the potential to produce high biomass output while allowing fertile land to be utilized for other purposes (Jang et al., 2013). Algal polymers are a natural source with a variety of advantages, including good heat insulation and heat power, complete biodegradability, and heavy carbon dioxide fixation. Fibrils offer energy-absorbing qualities, resulting in good insulation, sound absorption, and flame retardancy (Hassan et al., 2008).

2.1 Alginate

Alginate (Fig. 8.1) is a polysaccharide found in brown seaweeds (Ochrophyta, Phaeophyceae) found in the ocean, such as *Macrocystis pyrifera* and *Saccharina japonica* (formerly known as *Laminaria japonica*), accounting for 22%–44% of their dry cell weight (Li, Wang, et al., 2019; Li, Zhu, et al., 2019; Lin et al., 2018; Venkatesan et al., 2016). Alginate is a linear copolymer made up of 1, 4-glycosidic linkages connecting mannuronic acid (M) and guluronic acid (G) (Pereira & Cotas, 2020). The ratios of alternating GG, MM, and MG blocks represent various physicochemical qualities. The greater the G content, the more effective the moisture barriers and the lower the water vapor permeability (Oms-Oliu et al., 2008). Alginate has been widely employed in biomedical applications

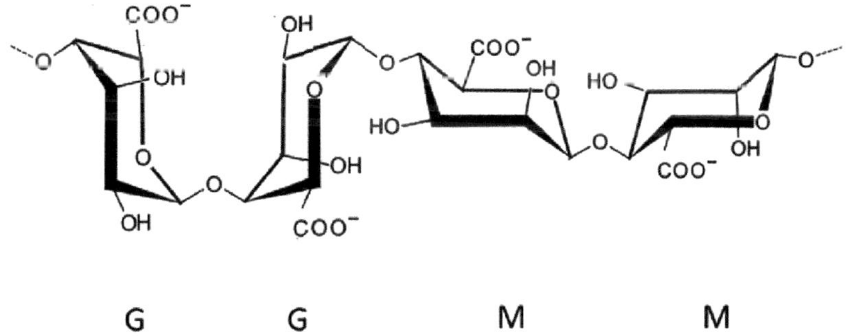

FIGURE 8.1

Idealized Structure of the Chemical Units of Alginate.

such as drug delivery systems, wound dressings, and tissue regeneration due to its great biocompatibility. Additionally, the calcium alginate–polyacrylamide combination can increase stiffness, recoverability, and durability, making it a potential cartilage alternative (Sun et al., 2012).

For example, alginates can cross-link with cations to make hydrogels or packing films, with Ca^{2+} having a stronger binding power, mechanical strength, and water barrier than Na^{+} (Cazón et al., 2017). Following cation exchange with Ca^{2+}, the strength and breakage rate of straw–alginate reached 216 g and 4%, respectively (Xue et al., 2019). With the addition of Ca^{2+} to the cellulose–alginate complex, swelling inhibition was found to be greater than 95%, and the Young's modulus and tensile strength were determined to be 135 and 17 MPa, respectively (Benselfelt et al., 2018). As a result, cations linked to alginate have the potential to be utilized in the production of a variety of ecologically acceptable packaging materials.

2.2 Carrageenan

Carrageenan is a polysaccharide composed of D-galactose and 3,6-anhydro-D-galactose that is extracted from red seaweeds and can account for up to 30%–50% of their dry weight (DW) (Rupérez & Saura-Calixto, 2001) in species, such as *Chondrus crispus* (Azevedo et al., 2015) and *Mastocarpus stellatus* (Rhodophyta) (Torres et al., 2016). Three main types of carrageenan fractions have been discovered based on the amount and placement of sulfate ester groups: kappa (κ), iota (ι), and lambda (λ) (Pereira et al., 2009) (Fig. 8.2). Carrageenan's double helix shape, like alginate, provides anion-hosting sites for improved gel structure. Monovalent ions (such as K^{+}) are often tolerated by carrageenan, but divalent ions (such as Ca^{2+}) are preferred. Many carrageenan bioplastics aim to make edible packaging sheets (Cazón et al., 2017). In this regard, ι-carrageenan, for example, might be an excellent food-coating ingredient for ham preservation (Carocho et al., 2019).

PVA has also been utilized as a synthetic material to improve the material's weak mechanical qualities. Carrageenan–PVA films were discovered to have 439.31% increase in elongation (Meng et al., 2018), while the addition of 5% SiO_2 nanoparticles enhanced the tensile strength of carrageenan–gelatine composites to 10.32 MPa (Hashemi Tabatabaei et al., 2018). The worldwide carrageenan market now accounts for the bulk of the global hydrocolloids market, which is predicted to reach $1 billion by 2024 (Zhang et al., 2019a). However, more than 80% of the products are used in food applications such as processed meals, dairy, sweets, and jellies (Campbell & Hotchkiss, 2017).

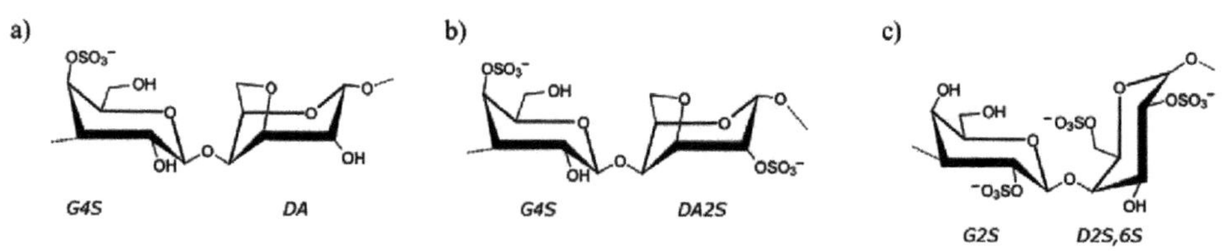

FIGURE 8.2

Idealized Structure of the Chemical Units of (A) Kappa-Carrageenan; (B) Iota-Carrageenan; (C) Lambda-Carrageenan.

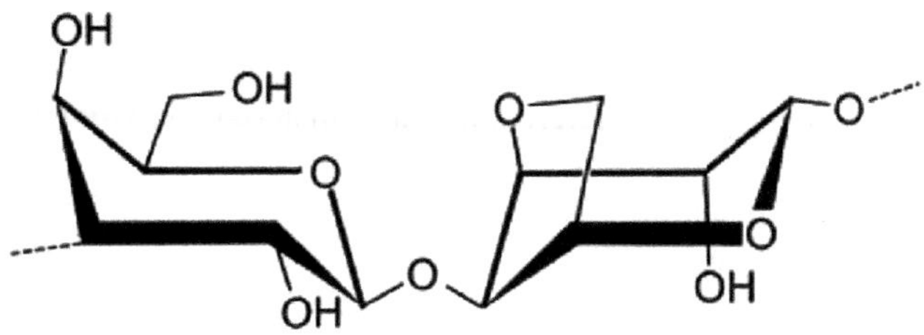

FIGURE 8.3

Idealized Structure of the Chemical Units of Agar.

2.3 Agar

Agar is a polysaccharide found in the red seaweed family Gracilariaceae and Gelidiaceae (Rhodophyta). The two most essential components are agarose and agaropectin (Fig. 8.3) (Arvizu-Higuera et al., 2007). This is a hydrophilic colloid that, when cooled from a hot aqueous solution, may produce reversible gels (Guerrero et al., 2014). Agar was originally employed in culinary, biotechnology, and pharmaceutical applications before being exploited as a raw material for the creation of bioplastic film production. Agar-based films, according to Phan et al. (2009), are clear, strong, and adaptable. It is also heat-sealable, making it excellent for use in food packaging.

2.4 Seaweed polymer composites

To improve the consistency of these polymers, many algal polysaccharides can be utilized (i.e., mechanical, thermal, and antimicrobial properties). They not only increase hydrophilicity and mechanical qualities such as tensile strength and elongation but also allow to be used as an active packaging material (Sharma et al., 2021). This is due to the naturally occurring antioxidant capabilities of seaweed, which can reduce free radicals that can be carcinogenic, mutagenic, or cytotoxic while also increasing the shelf life and nutritional value of food (Dima et al., 2014; Vital et al., 2016).

Seaweeds are viable alternative for bioplastics for the growth conditions shown before because the utilization of cultivable marine resources eliminates the need to use land and resources intended for crops, as well as the possibility for deforestation, as seaweed growing is done outside or in a laboratory (Asrofi et al., 2020; Hafting et al., 2015; Somleva et al., 2013). Furthermore, large-scale seaweed cultivation lowers prices and boosts compound availability. Alginate, carrageenan, and agar are the best seaweed polysaccharides for making biofilms or bioplastics (Kanmani & Rhim, 2014b; Tavassoli-Kafrani et al., 2016).

3. Effect of seaweed biopolymers incorporation in biofilm

3.1 Physicochemical characteristics of seaweed phycocolloids

Mechanical properties such as tensile strength, elongation at break, heat resistance, and water vapor permeability must be addressed defining the quality of an optimum bioplastic. The maximum stress that a material can endure while being stretched or pushed before breaking is known as tensile strength. The tensile strength and Young's modulus of plant fibers generally rise as the cellulose content of the fibers increases. The ratio between the modified length and the starting length of the tested material after breakage is known as elongation at break. It expresses the capacity to withstand shape changes without crack formation in a matrix comprising natural polymers. Thermal resistance is a thermal attribute that represents the temperature differential that a material can withstand (Djafari Petroudy, 2017). The ability of a material to enable a vapor (such as water vapor or, indeed, any gas) to pass through it is known as vapor permeability. Water vapor permeability is defined as "a feature of a textile material or composite that is dependent on water vapor resistance," according to ISO 11092:1993 (European Committee for Standardization, 2005). The faster water and vapor can move through a material with a higher permeability value, the best.

Phycocolloids extracted from seaweeds possess gelling properties different from each other, as these properties are dependent on their structure, concentration in a solution, temperature, pH, and syneresis potential (Lee et al., 2017; Liu et al., 2016; Ramdhan et al., 2019). Species, environmental conditions, method of extraction, and treatment processes are all factors that heavily influence the physicochemical characteristics of phycocolloids (Lee et al., 2017; Porse & Rudolph, 2017). Therefore, it is important to have a deep knowledge of these factors that will allow us to discover the most efficient techniques to obtain high-quality phycocolloids that can be used to develop bioplastics that maintain functional mechanical properties.

Despite the requirement of hot water for optimal extraction, agar and carrageenan are industrially extracted with alkali extraction, since hot water extraction weakens their rheological properties, decreasing the quality of these phycocolloids for biofilms incorporation. In the work of Khalil et al. (2018) is reported the percentages of gel strength obtained with treated and untreated extraction of agar and carrageenan, and their results show that, with alkali extraction, the gel strength is optimized, but the yield of production decreases.

Carrageenans possess high gel strength in form of salts complex and obtained gels with κ-carrageenan and ι-carrageenan stable at room temperature (Ilias et al., 2017). Pure carrageenans enhance gel formation and strength if enriched with cations, as Robal et al. (2017) demonstrated. Paula et al. (2015) studied the physical properties of glycerol-plasticized edible films made with κ-carrageenan, ι-carrageenan, and alginate; κ-carrageenan exhibited higher tensile strength and elasticity, higher moisture permeability, and lower opacity compared with ι-carrageenan, while alginate films revealed higher transparency (Paula et al., 2015). On the other hand, λ-carrageenan type has typical nongelling property and is the only cold-water-soluble carrageenan in its native form.

Similar to carrageenan, alginate also forms rigid and stable gel matrix in the presence of cations, especially Ca^{2+} (Yoon et al., 2014). Films obtained from sodium alginate with 1%–3% (*w/v*) calcium chloride solution exhibited an increase in tensile strength and elongation properties, and reduced opacity (Kok & Wong, 2018; Liling et al., 2016).

Agar films, compared with carrageenan and alginate films, have a lower tensile strength and water vapor permeability. Agar exhibits higher elongation and elasticity value of κ-carrageenan film (Kanmani & Rhim, 2014a; Rhim, 2012). Due to their gelling properties and viscosity, seaweed phycocolloids are widely used as stabilizers and/or thickening and gelling agents in the manufacturing of food, pharmaceuticals, and cosmetics.

Due to these characteristics, seaweed phycocolloids are potential compounds to be incorporated into biofilms and develop new biodegradable packaging (Khalil et al., 2017; Rao et al., 2018; Song et al., 2009; Stevens, 2001). Methods and processes carried out to produce biofilm have significant effects on the physical properties and microstructures of the film. Li, Wang, et al. (2019), Li, Zhu, et al. (2019) produced chitosan–alginate films through different processes; their results showed that biofilm prepared through layer-by-layer assembly combined with ferulic acid cross-linking has enhanced mechanical properties, opacity, and hydrophobicity compared with films prepared by direct mixing, cross-linking alone, and layer-by-layer assembly alone. The incorporation of seaweeds into other polymers changes the mechanical, thermal, optical, and chemical properties of the materials; therefore, further tests need to be performed before developing new bioplastic materials.

3.2 Seaweed-based bioplastics

Hydrocolloids are polymers with the ability to absorb water and be soluble in water; therefore, this ability allows them to increase their volume or form a gel in solution that can form film sheets when dried (Zhang et al., 2018). When compared with plastics made from other biopolymers, the development of carrageenan-based bioplastics is still in its early stages. The only method of production mentioned is pouring or casting the solvent with carrageenan, and then drying it to remove the water and leave the bioplastic solids. Clear, elastic, and soft gels are produced by carrageenan bioplastics containing calcium salts, whereas strong, rigid, and brittle films are produced by other carrageenans (Farhan & Hani, 2020; Sedayu et al., 2019). Similarly, the bioplastic properties or films produced from carrageenan exhibit a compact film and high gelling ability (Wu & Imai, 2012), which enhance the mechanical properties and water barrier properties (Paula et al., 2015). To improve the physical and mechanical properties of bioplastics, the addition of plasticizers is recommended, as the literature review research conducted by Budiman (Budiman et al., 2022) suggested.

Seaweed alginate, agar, carrageenan, and cellulose have great film-forming capabilities and are relatively easy to process; hence, seaweed studied for new biomaterials is primarily focused on these polymers (Rinaudo, 2008). Alginate extracted from the brown alga *Sargassum siliquosum* has been studied as a raw material for bioplastic film fabrication (Lim et al., 2018). Alginate was combined with sago starch, sorbitol, and calcium chloride during the treatment procedure ($CaCl_2$). The biofilm created using 2 g of alginate powder from *S. siliquosum* and 15% w/w sorbitol treated with 75% w/w $CaCl_2$ appears to have appropriate physical and mechanical qualities (tensile strength, elongation at break, water vapor permeability, and water solubility). Even though it is important to further analyze alginate behavior in the presence of inorganic compounds, this study revealed that alginate from *S. siliquosum* is a potential candidate for the manufacture of bioplastic films (Lim et al., 2018).

Doh et al. (2020) used crude extracts and cellulose nanocrystals from *S. japonica* (formerly *Laminaria japonica*) and *Sargassum natans* to create biofilms. It was discovered that the presence of cellulose nanocrystals improved the biofilm's physicochemical, mechanical, and thermal properties, indicating a promising future for the use of cellulose in biopackaging.

Biofilms constructed from pure k-carrageenan and whole *Kappaphycus* sp. seaweed were studied by Hanry and Surugau (2020). The purpose of this study was to look at the characteristics of both biofilms and determine if carrageenan extraction might be avoided. Biofilms constructed from the entire algae were less brittle, but they were weaker due to a lack of carrageenan, which resulted in reduced binding intermolecular forces. On the other hand, the biofilms that were generated could be used to replace petroleum-based plastic and nonbiodegradable containers (Hanry & Surugau, 2020). The most common application for this type of bioplastic is single-use packaging for powders, quick foods, candies, or to carry daily-use drugs such as tablets or pills, which do not require high mechanical properties and are easy to open.

To evaluate carrageenans made from the red seaweed *Kappaphycus alvarezii* (formerly *Eucheuma cottonii*) as bioplastic raw materials, researchers looked at their physical and biological features. Semirefined carrageenan flour was provided by *K. alvarezii*, whereas refined carrageenan flour was purchased. Bioplastic produced from removed carrageenans demonstrated higher antibacterial activity than refined carrageenans, but it was also weaker. Additionally, bioplastic made from refined carrageenans has a higher heat resistance. On the other hand, both bioplastics met market requirements (Wullandari et al., 2021).

To investigate its physicochemical properties as a bioplastic raw material, agar from *Gracilaria salicornia* was extracted. Two different agar extraction procedures were tested, and many biofilms were produced to see how well they worked. The biofilm obtained by photobleaching extraction (PB) had higher tensile strength and percent elongation than biofilm obtained by alkali extraction (AE), but the AE agar film had superior heat stability. Furthermore, after 30 days, the AE agar film disintegrated completely in the soil burial test (Ling et al., 2016). As a result, agar generated from *G. salicornia* has the potential to be used in commercial bioplastic films in the future.

Biofilm created from *Gracilaria vermiculophylla* agar possessed the same transparency and clarity as the commercial counterpart (Phan et al., 2005), according to Sousa et al. (2010). Agar-based films gain flexibility and mechanical strength when plasticizer glycerin is added, making them suitable for packing foods or covering pharmaceuticals. Agar-based film, which is heat resistant and antibacterial, is another intriguing alternative; nevertheless, more precise research studies are needed to determine the ideal formulation for resistant biopackaging (Kadar et al., 2021).

Darni et al. (2019) developed bioplastics with a sorghum stalk filler, *Eucheuma denticulatum* (formerly *Eucheuma spinosum*) polysaccharides, and glycerol as a plasticizer. The addition of filler and plasticizer to bioplastic synthesis increased its physical and mechanical properties, suggesting that *E. denticulatum* could be a source for developing alternative formulations for plastic packaging.

Green seaweed extracts have film-forming capabilities and could be used as a filler or reinforcement in medical and cosmetic applications (Lakshmi et al., 2020). Ulvans are a form of polysaccharide with a high viscosity and gelling ability, extracted from some species of green algae (Chlorophyta), which makes them a potential biofilm generator. Despite the fact that no research into the synthesis of bioplastic from ulvan has been done yet, these extracts have excellent thermal resistance and mechanical strength, which are both characteristics of perfect bioplastics (Usman et al., 2017).

Among all the polysaccharides found in green algae, cellulose was found to be the best for making biodegradable plastic; its properties enable it to form hydrocolloids in a suitable solvent solution, allowing it to have good mechanical performance (Baghel et al., 2021). It is also inexpensive, biodegradable, and renewable. *Cladophora* sp. cellulose is chemically inert and resistant, for example. The material's tenacity, as well as its mechanical, thermal, and morphological properties, makes it a suitable choice for food or pharmaceutical biopackaging (Mihranyan, 2011). Green seaweed compounds should be studied further since they offer exciting properties that make them potential biodegradable plastic candidates.

In the study of Oyekanmi et al. (2021), seaweed-based bioplastics were used as control to investigate the tensile, thermal, and the antimicrobial properties of seaweed-based bioplastics enriched with 5% w/w of *Azadirachta indica* (neem) extracts. Biofilms were irradiated with different dosages of gamma radiation (0.5, 1, 1.5, and 2 kGy). The tensile strength of the control seaweed film was significantly lower compared with seaweed–neem films. The increase in the tensile strength observed in the seaweed–neem film indicated good dispersibility as a result of the interaction of the hydroxyl groups in the matrix (Resano-Goizueta et al., 2018; Uthaya Kumar et al., 2019). Similarly, seaweed film elongation at break (EB) was found to be 17.6%. The addition of 5% w/w neem leaves extract in the seaweed film, on the other hand, boosted the EB to 19.4%. The addition of neem leaf extract to the matrix could potentially improve the interfacial bonding interaction (Din et al., 2021). Furthermore, the antibacterial properties of the seaweed-based films evaluated against *Staphylococcus aureus* and *Bacillus subtilis* revealed no inhibitory zones. The addition of 5% w/w neem leaves extract to the control film, on the other hand, showed inhibitory effects against *Staphylococcus aureus* and *Bacillus subtilis*. The inclusion of effective components such as carotenoids, phenolic compounds, flavonoids, triterpenoids, ketones, steroids, and tetra-triterpenoids azadirachtin in neem leaf extract-based seaweed films may explain their antibacterial activity (Jafarzadeh et al., 2020). The results showed that increasing the dose of irradiation of the seaweed–neem films improved the inhibitory zones against *S. aureus* and *B. subtilis*. Furthermore, when compared with *B. subtilis*, increasing the irradiation dosage greatly improved the antibacterial properties of the films against *S. aureus*. The irradiation of the films and increased bonding interactions between neem extracts in the seaweed matrix resulted in a release of free radicals in the film from both seaweed and neem, improving the antibacterial activity of the film (Oyekanmi et al., 2021). Therefore, this study successfully demonstrates the implementation of physical and biological properties of the developed seaweed-based packaging with the incorporation of neem extracts.

3.3 Active packaging

Biological properties exhibited by seaweeds bioactive compounds are multiple, and they can be exploited to produce bioactive packaging that assures the safety of the packaged product (Zhang et al., 2019a). Seaweed biopolymers are natural and nontoxic, they possess antimicrobial antioxidant properties, and therefore, they are considered as excellent candidates to develop biofilms for food and pharmaceuticals purposes (Ibrahim et al., 2017). The idea of "active packaging" is taking foot rapidly into society. The demand for packaging that interacts with the product, improving its safety, shelf life, and sensory properties, is increasing in society (Carina et al., 2021).

Active packaging systems are divided into two categories: active scavenging systems (absorbers) and active releasing systems (emitters). While the absorber eliminates undesired molecules from food or its environment, such as moisture, carbon dioxide, oxygen, ethylene, or odor, the emitters provide antimicrobial chemicals, flavors, ethylene, carbon dioxide, antioxidants, or ethanol to packed food (Yildirim et al., 2018). The use of active packaging instead of adding active compounds such as antimicrobials and antioxidants directly to the food may reduce the amount of these substances required. Active packaging possesses biological metabolites that express interesting biological properties released into the biofilm formulation; therefore, the packaging developed can provide a defense for the product against bacteria, oxidation, and UV rays (Perera et al., 2021).

Antimicrobial packaging attempts to limit, block, or postpone the growth of germs by including or coating antimicrobial compounds in/on food packaging, thus lengthening the lag phase and reducing microorganism growth. Several studies demonstrated interesting biological properties expressed by seaweed polysaccharides. Mixed polysaccharides extracted from the brown seaweed *Nizamuddinia zanardinii* inhibited the growth of *Escherichia coli* and *Pseudomonas aeruginosa*, while crude fucoidans extract from the *Sargassum polycystum* exhibited inhibitory effects against *Vibrio harveyi, S. aureus*, and *E. coli* (Alboofetileh et al., 2019; Chotigeat et al., 2004). Crude polysaccharides extracted from the red alga *Gracilaria ornata* showed antimicrobial activity against *E. coli* (dos Santos Amorim et al., 2012). Kanatt et al. (2015) examined the antimicrobial activity of pure κ-carrageenan from *K. alvarezii* against gram-positive bacteria *S. aureus* and *Bacillus cereus,* and the gram-negative bacteria *E. coli* and *Pseudomonas fluorescens.* In vitro results show antimicrobial activity only against gram-positive bacteria, and no growth inhibition of gram-negative bacteria. κ-carrageenan was included in polyvinyl acetate (PVA) film, and results showed an effective zone of inhibition against *S. aureus* and *B. cereus*, proving that the antimicrobial activity of the pure extract could be retained and affect the film, protecting the wrapped products from bacteria.

Crude extracts of *Amphiroa anceps*, *Corallina officinalis* (Rhodophyta), and *Sargassum filipendula* (Phaeophyceae) have been investigated for their antibacterial activity against *E. coli* ATCC 8739, *Salmonella typhimurium* ATCC 14028, and *S. aureus* ATCC 6538 (Mofeed et al., 2022). Among all the seaweeds examined, a different antimicrobial activity has been detected: *S. filipendula* extract had the highest antibacterial activity against *E. coli*, with 91.4% inhibition, followed by *C. officinalis* with 85.7% inhibition, while *A. anceps* extract had only 57.1% inhibition against *E. coli* cells. At a concentration of 100 g/mL, the extract of *C. officinalis* inhibited 100% of *S. aureus* cells, followed by *S. filipendula* (82.5%) and *A. anceps* (75%). The extract of *C. officinalis* inhibits *S. typhimurium* the most with a percentage of 80%, followed by *S. filipendula* (65%) and *A. anceps* (60%) (Mofeed et al., 2022).

Han and Wang (2017) used sodium alginate and carboxymethyl cellulose, glycerine, $CaCl_2$, and the natural antibacterial component pyrogallic acid to create an antimicrobial film (PA). Their findings indicate that the UV barrier property, moisture content, water vapor permeability, and oxygen permeability have all increased. Furthermore, films containing a high concentration of PA were more effective against *E. coli* and *S. aureus*.

Food deterioration and pathogenic microbe contamination pose a serious threat to human health, food security, and sustainability. As a result, there has been a surge interest in investigating alternative antibacterial and antioxidant agents such as essential oils.

Essential oils derived from *Undaria pinnatifida*, an edible seaweed, could be used as a natural source of antioxidant and antimicrobial chemicals, with applications for food preservation. They have been tested against two foodborne pathogenic bacteria, gram-positive *S. aureus* ATCC 12600 and gram-negative *Typhimurium* ATCC 43174 (Patra et al., 2017). Antioxidant potential was determined in vitro assays by ABTS radical scavenging, DPPH radical scavenging, hydroxyl radical scavenging, nitric oxide scavenging, superoxide radical scavenging, inhibition of lipid peroxidation, and reducing power assay. The findings of this study indicated that the essential oil derived from this seaweed had antioxidant properties as well as antibacterial action against foodborne pathogenic bacteria; therefore, the use of *U. pinnatifida* for development of food packaging can preserve the food from deterioration (Patra et al., 2017).

Essential oils are effective antimicrobials and antioxidants, but they are chemically and physiologically labile, and they have strong scents, limiting their use as antibacterial additions in food. Although encapsulated essential oils have primarily been studied as direct food preservatives, their utility in active packaging might be a novel idea and should be further investigated. The development of packaging involves a lot of heat, which can lead to the loss of heat-labile essential oils. As a result, it is critical to optimize manufacturing procedures and use more appropriate processing conditions to guarantee a minimum loss of essential oils. Therefore, new techniques should be investigated to exploit the antimicrobial and antioxidant properties of essential oils (Bustos et al., 2016; Mukurumbira et al., 2022).

Seaweed metabolites that express antioxidant activity are considered good candidates to constitute active packaging. Lipid oxidation has a significant impact on food quality, as it can reduce shelf life and nutritional value (Ahmed et al., 2016). The formulation of antioxidant packaging leads to a decrease in the amount of lipid oxidation and protein degradation within the packaging, avoiding degradation of the product coated on the film.

K. alvarezii aqueous extract has been investigated for its antioxidant activity. Extracts have been incorporated into PVA film, compared with pure PVA film, and results showed an antioxidant activity for the developed seaweed-based film (Kanatt et al., 2015). The antioxidant activity of *Ulva lactuca* (Chlorophyta), *Gracilariopsis lemaneiformis* (formerly *Gracilaria lemaneiformis*), *Sarcodia ceylonensis* (Rhodophyta), and *Durvillaea antarctica* (Phaeophyceae) has been investigated by He et al. (2016). Their results showed that crude polysaccharide from the green seaweed showed the highest activity, followed by *D. antarctica* and *S. ceylonensis*. Thus, these algae might be considered potential candidates for the incorporation of antioxidant extracts in active packaging.

Andrade et al. (2021) used a whey protein concentrate combined with *Fucus vesiculosus* extract to prepare an active biofilm to prevent the oxidation of lipids in chicken breasts. The thickness, tensile strength, and elastic modulus of biofilm increased. For 25 days of storage, the active film also reduced lipid oxidation in chicken breasts.

Seaweed-derived biopolymers are reported to be more resistant to microwave radiation due to the photoprotection properties of seaweed compounds and thus can conserve the quality of packaging for food or pharmaceuticals (Rajendran et al., 2012). Therefore, incorporation of seaweed components in biofilm can lead to a photoprotective effect on packaging; some species possess photoprotective compounds capable of absorbing UV rays. Methanol extracts of *Sargassum* sp. and *K. alvarezii* (as *Eucheuma cottonii*) showed photoprotective activity against UV radiation, probably due to the presence of flavonoids, phenols, and triterpenoids (Nurjanah et al., 2017); they can potentially be used as a raw material for sunscreen products or be included in biofilm formulation, enhancing the properties of active packaging.

Although the biological activities of pure seaweed compounds/extracts incorporated into biofilms should be further investigated, several studies have shown that most seaweed polysaccharides exhibit antioxidant, antimicrobial, and photoprotection activities, suggesting that their inclusion in biofilms can lead to the manufacture of safe and active packaging.

4. Bioplastics for a sustainable and circular economy

Plastic contamination in the environment has become a top problem for major global organizations such as the United Nations, the World Economic Forum (WEF), the World Health Organization (WHO), and the European Union (EU) (Rosenboom et al., 2022). Packaging is considered the most significant source of garbage globally, with 146 million tons created in 2015, 141 million tons of which remained unrecycled (96.6%). Packaging also has the shortest average working life of any industrial plastic industry (Geyer et al., 2017). The operating life of single-use plastics, from use to disposal, can be as brief as a few minutes. Because of inadequate waste management, 1%–5% of all plastic ends up as garbage in terrestrial and maritime habitats. Approximately 80% of marine plastic waste originates on land, primarily from poorly managed landfills and kerbsides that are ravaged by sea tides and wind (Bucknall, 2020).

As part of a circular economy in which virgin polymers are generated from renewable or recycled raw materials, bioplastics—mainly plastics made from biobased polymers—have the potential to contribute to more sustainable commercial plastic life cycles. Carbon-neutral energy is utilized in production, and items are reused or recycled when they reach the end of their useful life (EOL). Biobased plastics may have a lower carbon footprint and advantageous material properties when compared with fossil-based plastics; additionally, they may be compatible with existing recycling streams, and some may offer biodegradation as an EOL scenario if performed in controlled or predictable environments. However, there may be trade-offs, such as unfavorable agricultural consequences, rivalry with food production, ambiguous EOL management, and greater expenses. Emerging chemical and biological technologies have the potential to "upcycle" growing amounts of heterogenous plastic and bioplastic waste into higher-quality products. Existing (bio)plastic identification standards and life cycle assessment criteria must be revised and homogenized to advise converters and consumers in their purchase decisions. Furthermore, clear regulation and financial incentives are required to go from specialty polymers to large-scale bioplastic market uses that are really sustainable impact (Rosenboom et al., 2022).

Bioplastics that are 100% biobased are currently produced at a scale of 2 million tons per year and are thought to be a part of future circular economies to help achieve some of the United Nations' (UN) Sustainable Development Goals, such as diverting from fossil resources, introducing new recycling or degradation pathways, and using less toxic reagents and solvents in manufacturing processes (Gandini et al., 2016; Karan et al., 2019; Zhu et al., 2016).

Bioplastics, depending on their kind, can provide enhanced circularity by renewable (nonfossil) resources, a smaller carbon footprint, biodegradation as an alternative end-of-life (EOL) option, and improved material qualities. These advantages, however, are extremely reliant on several elements, including the chemical structure, manufacturing method, and most likely EOL situation. All of these issues must be examined throughout the life cycle, utilizing metrics such as climate effect, ecotoxicity, and recyclability, and methodologies such as a life cycle assessment (LCA) to determine the environmental advantage over alternatives (Hottle et al., 2013; Miller, 2020; Walker & Rothman, 2020). Bioplastics, like standard plastics, present issues about the leaching of monomers, oligomers, and additives and hence demand the same level of attention in product design and formulation (Zimmermann et al., 2020).

Biodegradation is a viable EOL alternative for readily hydrolysable polymers such as seaweed polymers, but it should be carried out only in regulated industrial conditions to assure full digestion without uncontrolled side effects such as contaminant leakage or microplastic production, guaranteeing the circular and eco-friendly approach (Mülhaupt, 2013; Organization for Economic Co-operation and Development, 2018).

4.1 Seaweed as next step toward of the blue economy in the bioplastic technology

Recent developments in marine culture techniques (such as the creation of low-cost sustainable techniques or the use of technologies 4.0 in aquaculture) have resulted in a rise in the production of seaweeds as a real "marine culture" (Amin Alamsjah, 2018). As a result, efforts have been undertaken to investigate strategies for sustainable harvesting (hand collection of the seaweed blade to permit the seaweed regeneration). This is done, for example, for *Ascophyllum nodosum* (Phaeophyceae) and

Chondrus crispus (Rhodophyta) in Ireland of natural seaweed resources and the development of efficient farming techniques to satisfy rising worldwide demand (Kim et al., 2017; Mac Monagail et al., 2017).

Ocean farms appear to be more sustainable than terrestrial agriculture since seaweed farming does not require freshwater, chemical fertilizers, or land, which are some of the major drawbacks of terrestrial agriculture (Tiwari & Troy, 2015).

Novel marine natural compound applications may reduce reliance on terrestrial regions, allowing for more dynamic use of soil and terrestrial ecosystems (FAO, 2018; García-Poza et al., 2020).

Seaweed farming is an industry that may significantly benefit the economies of the producing countries by generating foreign cash and improving the socioeconomic position of the coastal population (Rebours et al., 2014).

Algae-based bioplastic is regarded as a long-term answer to the circular economy. Following the use of bioplastic bags, they may be composted and used as a peat substitute to end the life cycle of these bioplastics (Khoo & Tan, 2010). As a result, algae-based bioplastics are a viable path to a circular economy because they improve cultivation flexibility on nonarable land, conserve chemical-based medium using wastewater resources, sequester CO_2, and have a positive impact on both land and marine ecosystems at the end of the life cycle.

Because of the natural-based feedstock features, algae-based bioplastics are biodegradable in nature; nevertheless, it must be noted that these bioplastics will have an environmental effect owing to the additives added to improve the mechanical properties of the material (Hahladakis et al., 2018).

In the absence of a sustainable recycling path toward a sustainable circular economy, petroleum-based polymers may face a similar problem (e.g., ecotoxicity and accumulation in ecosystem) (Thushari & Senevirathna, 2020; Zimmermann et al., 2020).

The primary goal of recycling bioplastics waste is to utilize it as recycled feedstock in the future biopolymer manufacturing process. This lowers the cost of following biopolymer manufacturing, increases biomaterial reuse efficiency, efficiently minimizes environmental load, and supports the cyclic process of bioeconomy (Tsang et al., 2019; Vu et al., 2020).

The main purpose of recycling bioplastics waste is that it can be reused as recycled feedstock for the subsequent biopolymer production process (Wojnowska-Baryła et al., 2020). This reduces the production cost of the subsequent biopolymer, improves the reuse efficiency of biomaterial, effectively reduces the environmental burden, and simultaneously promotes the cyclic process of bioeconomy (Tsang et al., 2019; Vu et al., 2020).

Mechanical recycling, chemical recycling, and biological recycling are the three recycling routes (Wojnowska-Baryła et al., 2020). Mechanical recycling involves the elimination of contaminants by screw extrusion, injection molding, and compression molding. The primary advantage of mechanical recycling over chemical recycling is the use of nonrenewable energy, which can help to mitigate the effects of global warming (Lamberti et al., 2020; Schyns & Shaver, 2021). The disadvantage of this technique is that it converts bioplastics into less demanding products since recycled biopolymer material does not preserve the same quality as virgin biopolymer material (Alaerts et al., 2018). As a result, recycled biopolymers through mechanical recycling are often combined with virgin biopolymer material, which may then be utilized in noncritical applications (for example, garbage bags and nonpressure pipes) (Grigore, 2017; Hopewell et al., 2009).

Chemical recycling (e.g., chemical depolymerization, pyrolysis, thermal cracking, gasification, chemolysis, and enzymatic hydrolysis) is the process by which polymers are decomposed into low-molecular-weight monomers (Grigore, 2017; Vu et al., 2020). Chemical recycling, according to researchers, offers the benefit of recovering monomers that can be replicated to biopolymers with comparable material qualities, potentially reducing the demand for virgin biopolymer materials for further biopolymer manufacture (Ariffin et al., 2010; Lamberti et al., 2020). However, one of the drawbacks of employing chemical recycling is that it requires high temperatures and pressure to decompose polymers into monomer units (e.g., hydrolysis depolymerization takes 150–200°C temperature with 5–1000 kPa) (Niaounakis, 2019).

Biological recycling is a recycling method for algae-based bioplastics that uses both aerobic (e.g., composting) and anaerobic processes (e.g., fermentation) to convert the carbon in algae-based bioplastics into gas. It takes roughly three months for algae-based bioplastics to convert 62.9% of their carbon into gas under anaerobic room temperature conditions. Similarly, in seawater aerobic room temperature conditions, it takes around 6.5 months to lose 83% of its carbon (Lamberti et al., 2020). However, as compared with mechanical and chemical recycling, biological recycling requires a longer recycling period to acquire the recovered product. Biological recycling of algae-based bioplastics is regarded the best recycling route among mechanical and chemical recycling techniques since the biological approach preserves the recovered biopolymer characteristics for further manufacturing. Thus, the seaweed-based plastic can be closed cycle without residues and be environmental neutral, thus being a next step for circular and blue economy, although more research needs to be developed toward this goal.

5. Conclusions

The use of seaweed for biodegradable, safe, and hygienic packaging has caught the interest of the marine pharmacology and food industries, which may be able to not only exploit the potential of seaweed-derived bioactive compounds for the development of novel functional food or natural drugs, as well as provide sustainable packaging that preserves and does not alter their properties in a sustainable and environmentally friendly manner. Unfortunately, traditional phycocolloid extraction has some drawbacks that can be overcome by using more eco-friendly extraction methods that have already been proven; however, they still require ongoing research to lower implementation costs and persuade industrial investment, as well as additional studies to develop these sustainable technologies. The English start-up company based in London, called "Nupla," has developed a natural, edible, plastic-like substance that degrades in 4–6 weeks. The membrane is manufactured from northern French seaweed that is first dried and processed into powder and then turned into a thick, gloopy fluid that dries to form a plastic-like substance. This natural plastic has the potential to reduce global plastic pollution, which now stands at 300 million tons per year. Another experiment incorporating sustainable plastic is Mak-Pak, which was launched by Alfred Wegener Institute researchers, Hochschule Bremerhaven University, and the fast food chain Nordsee. This project developed a sustainable macroalgae-based fast food packaging that has already been tested by real clients.

However, because to their weak mechanical and water barrier qualities, packaging manufactured from pure phycocolloids does not meet the criteria for commercial use bioplastic. The addition of biopolymers or plasticizers to the biomaterial will improve its performance, while the antioxidant and antibacterial characteristics of bioactive substances obtained from seaweed will improve the preservation and shelf life of food and medications, developing an active packaging that will save the environment.

Acknowledgments

This work is financed by national funds through FCT—Foundation for Science and Technology, I.P., within the scope of the projects LA/P/0069/2020 granted to the Associate Laboratory ARNET, UIDB/04292/2020—granted to MARE—Marine and Environmental Sciences Centre and UIDP/50017/2020+UIDB/50017/2020 (by FCT/MTCES) granted to CESAM—Centre for Environmental and Marine Studies. This research was cofinanced by the project MENU—Marine Macroalgae: Alternative recipes for a daily nutritional diet (FA_05_2017_011), funded by the Blue Fund under Public Notice No. 5-Blue Biotechnology and by the project AlgaMar4antivirus—Marine MacroalgaE in the prevention and treatment of ANTIviral diseases (Coronavirus SARS-CoV-2) (FA_07_2018_009), funded by the Blue Fund under Public Notice No. 7/2018—Scientific Research and Marine Technology. Silvia Lomartire thanks Foundation for Science and Technology (FCT) for the financial support granted through the doctoral grant 2021.05005.BD. João Cotas thanks to the European Regional Development Fund through the Interreg Atlantic Area Program, under the project NASPA (EAPA_451/2016). Ana Marta Mendes Gonçalves acknowledges University of Coimbra for the contract IT057-18-7253.

References

Abdul Khalil, H. P. S., Lai, T. K., Tye, Y. Y., Rizal, S., Chong, E. W. N., Yap, S. W., Hamzah, A. A., Nurul Fazita, M. R., & Paridah, M. T. (2018). A review of extractions of seaweed hydrocolloids: Properties and applications. *Express Polymer Letters, 12*(4), 296–317. https://doi.org/10.3144/expresspolymlett.2018.27

Abdul Khalil, H. P. S., Tye, Y. Y., Saurabh, C. K., Leh, C. P., Lai, T. K., Chong, E. W. N., Nurul Fazita, M. R., Mohd Hafiidz, J., Banerjee, A., & Syakir, M. I. (2017). Biodegradable polymer films from seaweed polysaccharides: A review on cellulose as a reinforcement material. *Express Polymer Letters, 11*(4), 244–265. https://doi.org/10.3144/expresspolymlett.2017.26

Adam, M. G., Tran, P. T. M., Bolan, N., & Balasubramanian, R. (2021). Biomass burning-derived airborne particulate matter in southeast asia: A critical review. *Journal of Hazardous Materials, 407*, 124760. https://doi.org/10.1016/j.jhazmat.2020.124760

Ahmed, M., Pickova, J., Ahmad, T., Liaquat, M., Farid, A., & Jahangir, M. (2016). Oxidation of lipids in foods. *Sarhad Journal of Agriculture, 32*(3), 230–238. https://doi.org/10.17582/journal.sja/2016.32.3.230.238

Alaerts, L., Augustinus, M., & Van Acker, K. (2018). Impact of bio-based plastics on current recycling of plastics. *Sustainability, 10*(5). https://doi.org/10.3390/su10051487

Alaswad, A., Dassisti, M., Prescott, T., & Olabi, A. G. (2015). Technologies and developments of third generation biofuel production. *Renewable and Sustainable Energy Reviews, 51*, 1446–1460. https://doi.org/10.1016/j.rser.2015.07.058

Alboofetileh, M., Rezaei, M., Tabarsa, M., Rittà, M., Donalisio, M., Mariatti, F., You, S. G., Lembo, D., & Cravotto, G. (2019). Effect of different non-conventional extraction methods on the antibacterial and antiviral activity of fucoidans extracted from Nizamuddinia zanardinii. *International Journal of Biological Macromolecules, 124*, 131–137. https://doi.org/10.1016/j.ijbiomac.2018.11.201

Allouzi, M. M. A., Tang, D. Y. Y., Chew, K. W., Rinklebe, J., Bolan, N., Allouzi, S. M. A., & Show, P. L. (2021). Micro (nano) plastic pollution: The ecological influence on soil-plant system and human health. *Science of the Total Environment, 788*, 147815. https://doi.org/10.1016/j.scitotenv.2021.147815

Amin Alamsjah, M. (2018). *An overview of the seaweed cultivation in several countries: Technology and challenge (Icpsuas 2017)* (Vol 98, pp. 6–12). https://doi.org/10.2991/icpsuas-17.2018.2

Amorim, R. D. N. D. S., Rodrigues, J. A. G., Holanda, M. L., Quinderé, A. L. G., de Paula, R. C. M., Melo, V. M. M., & Benevides, N. M. B. (2012). Antimicrobial effect of a crude sulfated polysaccharide from the red seaweed *Gracilaria ornata*. *Brazilian Archives of Biology and Technology, 55*(2), 171–181. https://doi.org/10.1590/S1516-89132012000200001

Andrade, M. A., Barbosa, C. H., Souza, V. G. L., Coelhoso, I. M., Reboleira, J., Bernardino, S., Ganhão, R., Mendes, S., Fernando, A. L., Vilarinho, F., Silva, A. S., & Ramos, F. (2021). Novel active food packaging films based on whey protein incorporated with seaweed extract: Development, characterization, and application in fresh poultry meat. *Coatings, 11*(2), 1–12. https://doi.org/10.3390/coatings11020229

Ariffin, H., Nishida, H., Hassan, M. A., & Shirai, Y. (2010). Chemical recycling of polyhydroxyalkanoates as a method towards sustainable development. *Biotechnology Journal, 5*(5), 484–492. https://doi.org/10.1002/biot.200900293

Arvizu-Higuera, D. L., Rodríguez-Montesinos, Y. E., Murillo-Álvarez, J. I., Muñoz-Ochoa, M., & Hernández-Carmona, G. (2007). Effect of alkali treatment time and extraction time on agar from Gracilaria vermiculophylla. In *Nineteenth international seaweed symposium* (pp. 65–69). Springer Netherlands. https://doi.org/10.1007/978-1-4020-9619-8_9

Asrofi, M., Sapuan, S. M., Ilyas, R. A., & Ramesh, M. (2020). Characteristic of composite bioplastics from tapioca starch and sugarcane bagasse fiber: Effect of time duration of ultrasonication (Bath-Type). *Materials Today Proceedings, 46*(xxxx), 1626–1630. https://doi.org/10.1016/j.matpr.2020.07.254

Azevedo, G., Torres, M. D., Sousa-Pinto, I., & Hilliou, L. (2015). Effect of pre-extraction alkali treatment on the chemical structure and gelling properties of extracted hybrid carrageenan from *Chondrus crispus* and *Ahnfeltiopsis devoniensis*. *Food Hydrocolloids, 50*, 150–158. https://doi.org/10.1016/j.foodhyd.2015.03.029

Baghel, R. S., Reddy, C. R. K., & Singh, R. P. (2021). Seaweed-based cellulose: Applications, and future perspectives. *Carbohydrate Polymers, 267*(May), 118241. https://doi.org/10.1016/j.carbpol.2021.118241

Benselfelt, T., Engström, J., & Wågberg, L. (2018). Supramolecular double networks of cellulose nanofibrils and algal polysaccharides with excellent wet mechanical properties. *Green Chemistry, 20*(11), 2558–2570. https://doi.org/10.1039/C8GC00590G

Bucknall, D. G. (2020). Plastics as a materials system in a circular economy: Plastics in the circular economy. *Philosophical Transactions of the Royal Society A: Mathematical, Physical & Engineering Sciences, 378*(2176). https://doi.org/10.1098/rsta.2019.0268

Budiman, M. A., Uju, & Tarman, K. (2022). A Review on the difference of physical and mechanical properties of bioplastic from seaweed hydrocolloids with various plasticizers. *IOP Conference Series: Earth and Environmental Science, 967*(1). https://doi.org/10.1088/1755-1315/967/1/012012

Bustos C, R. O., Alberti R, F. V., & Matiacevich, S. B. (2016). Edible antimicrobial films based on microencapsulated lemongrass oil. *Journal of Food Science & Technology, 53*(1), 832–839. https://doi.org/10.1007/s13197-015-2027-5

Campbell, R., & Hotchkiss, S. (2017). Carrageenan industry market overview. *Tropical seaweed farming trends, problems and opportunities* (pp. 193–205). Springer International Publishing. https://doi.org/10.1007/978-3-319-63498-2_13

Carina, D., Sharma, S., Jaiswal, A. K., & Jaiswal, S. (2021). Seaweeds polysaccharides in active food packaging: A review of recent progress. *Trends in Food Science & Technology, 110*, 559–572. https://doi.org/10.1016/j.tifs.2021.02.022

Carocho, M., Heleno, S., Rodrigues, P., Barreiro, M. F., Barros, L., & Ferreira, I. C. F. R. (2019). A novel natural coating for food preservation: Effectiveness on microbial growth and physicochemical parameters. *LWT, 104*, 76–83. https://doi.org/10.1016/j.lwt.2019.01.031

Castilho, L. R., Mitchell, D. A., & Freire, D. M. G. (2009). Production of polyhydroxyalkanoates (PHAs) from waste materials and by-products by submerged and solid-state fermentation. *Bioresource Technology, 100*(23), 5996–6009. https://doi.org/10.1016/j.biortech.2009.03.088

Cazón, P., Velazquez, G., Ramírez, J. A., & Vázquez, M. (2017). Polysaccharide-based films and coatings for food packaging: A review. *Food Hydrocolloids, 68*, 136–148. https://doi.org/10.1016/j.foodhyd.2016.09.009

Chen, Y., Awasthi, A. K., Wei, F., Tan, Q., & Li, J. (2021). Single-use plastics: Production, usage, disposal, and adverse impacts. *Science of the Total Environment, 752*, 141772. https://doi.org/10.1016/j.scitotenv.2020.141772

Chotigeat, W., Tongsupa, S., Supamataya, K., & Phongdara, A. (2004). Effect of fucoidan on disease resistance of black tiger shrimp. *Aquaculture, 233*(1−4), 23−30. https://doi.org/10.1016/j.aquaculture.2003.09.025

Darni, Y., Sumartini, S., Lismeri, L., Hanif, M., & Lesmana, D. (2019). Bioplastics synthesis based on sorghum-*Eucheuma spinosum* modified with sorghum stalk powder. *Journal of Physics: Conference Series, 1376*(1). https://doi.org/10.1088/1742-6596/1376/1/012042

Dima, C., Cotârlet, M., Alexe, P., & Dima, S. (2014). Microencapsulation of essential oil of pimento [*Pimenta dioica* (L.) Merr.] by chitosan/k-carrageenan complex coacervation method. *Innovative Food Science & Emerging Technologies, 22*, 203−211. https://doi.org/10.1016/j.ifset.2013.12.020

Din, M. I., Sehar, R., Hussain, Z., Khalid, R., & Shah, A. T. (2021). Synthesis of biodegradable semolina starch plastic films reinforced with biogenically synthesized ZnO nanoparticles. *Inorganic and Nano-Metal Chemistry, 51*(7), 985−994. https://doi.org/10.1080/24701556.2020.1813768

Djafari Petroudy, S. R. (2017). Physical and mechanical properties of natural fibers. *Advanced high strength natural fibre composites in construction*. Elsevier Ltd. https://doi.org/10.1016/B978-0-08-100411-1.00003-0

Doh, H., Dunno, K. D., & Whiteside, W. S. (2020). Preparation of novel seaweed nanocomposite film from brown seaweeds *Laminaria japonica* and *Sargassum natans*. *Food Hydrocolloids, 105*(February), 105744. https://doi.org/10.1016/j.foodhyd.2020.105744

European Committee for Standardization. (2005). *Textiles - determination of physiological properties - measurement of thermal and water - vapour resistance under steady-state conditions (sweating guarded - hotplate test) (ISO 11092:1993).*

Farhan, A., & Hani, N. M. (2020). Active edible films based on semi-refined κ-carrageenan: Antioxidant and color properties and application in chicken breast packaging. *Food Packaging and Shelf Life, 24*(January), 100476. https://doi.org/10.1016/j.fpsl.2020.100476

Food and Agriculture Organization of the United Nations (FAO). (2018). The global status of seaweed production, trade and utilization. *FAO Globefish Research Programme, 124*, 120.

Gandini, A., Lacerda, T. M., Carvalho, A. J. F., & Trovatti, E. (2016). Progress of polymers from renewable resources: Furans, vegetable oils, and polysaccharides. *Chemical Reviews, 116*(3), 1637−1669. https://doi.org/10.1021/acs.chemrev.5b00264

García-Depraect, O., Bordel, S., Lebrero, R., Santos-Beneit, F., Börner, R. A., Börner, T., & Muñoz, R. (2021). Inspired by nature: Microbial production, degradation and valorization of biodegradable bioplastics for life-cycle-engineered products. *Biotechnology Advances, 53*(May). https://doi.org/10.1016/j.biotechadv.2021.107772

García-Poza, S., Leandro, A., Cotas, C., Cotas, J., Marques, J. C., Pereira, L., & Gonçalves, A. M. M. (2020). The evolution road of seaweed aquaculture: Cultivation technologies and the industry 4.0. *International Journal of Environmental Research and Public Health, 17*(18), 6528. https://doi.org/10.3390/ijerph17186528

Geyer, R., Jambeck, J. R., & Law, K. L. (2017). Production, use, and fate of all plastics ever made. *Science Advances, 3*(7), 25−29. https://doi.org/10.1126/sciadv.1700782

Grigore, M. E. (2017). Methods of recycling, properties and applications of recycled thermoplastic polymers. *Recycling, 2*(4), 1−11. https://doi.org/10.3390/recycling2040024

Guerrero, P., Etxabide, A., Leceta, I., Peñalba, M., & de la Caba, K. (2014). Extraction of agar from *Gelidium sesquipedale* (Rodhopyta) and surface characterization of agar based films. *Carbohydrate Polymers, 99*, 491−498. https://doi.org/10.1016/j.carbpol.2013.08.049

Hafting, J. T., Craigie, J. S., Stengel, D. B., Loureiro, R. R., Buschmann, A. H., Yarish, C., Edwards, M. D., & Critchley, A. T. (2015). Prospects and challenges for industrial production of seaweed bioactives. *Journal of Phycology, 51*(5), 821−837. https://doi.org/10.1111/jpy.12326

Hahladakis, J. N., Velis, C. A., Weber, R., Iacovidou, E., & Purnell, P. (2018). An overview of chemical additives present in plastics: Migration, release, fate and environmental impact during their use, disposal and recycling. *Journal of Hazardous Materials, 344*, 179–199. https://doi.org/10.1016/j.jhazmat.2017.10.014

Hanry, E. L., & Surugau, N. (2020). Characteristics and properties of biofilms made from pure carrageenan powder and whole seaweed (*Kappaphycus* sp.). *Journal of Advanced Research in Fluid Mechanics and Thermal Sciences, 76*(2), 99–110. https://doi.org/10.37934/arfmts.76.2.99110

Han, Y., & Wang, L. (2017). Sodium alginate/carboxymethyl cellulose films containing pyrogallic acid: Physical and antibacterial properties. *Journal of the Science of Food and Agriculture, 97*(4), 1295–1301. https://doi.org/10.1002/jsfa.7863

Hashemi Tabatabaei, R., Jafari, S. M., Mirzaei, H., Mohammadi Nafchi, A., & Dehnad, D. (2018). Preparation and characterization of nano-SiO2 reinforced gelatin-k-carrageenan biocomposites. *International Journal of Biological Macromolecules, 111*, 1091–1099. https://doi.org/10.1016/j.ijbiomac.2018.01.116

Hassan, M. M., Mueller, M., & Wagners, M. H. (2008). Exploratory study on seaweed as novel filler in polypropylene composite. *Journal of Applied Polymer Science, 109*(2), 1242–1247. https://doi.org/10.1002/app.28287

He, J., Xu, Y., Chen, H., & Sun, P. (2016). Extraction, structural characterization, and potential antioxidant activity of the polysaccharides from four seaweeds. *International Journal of Molecular Sciences, 17*(12). https://doi.org/10.3390/ijms17121988

Hopewell, J., Dvorak, R., & Kosior, E. (2009). Plastics recycling: Challenges and opportunities. *Philosophical Transactions of the Royal Society B: Biological Sciences, 364*(1526), 2115–2126. https://doi.org/10.1098/rstb.2008.0311

Hottle, T. A., Bilec, M. M., & Landis, A. E. (2013). Sustainability assessments of bio-based polymers. *Polymer Degradation and Stability, 98*(9), 1898–1907. https://doi.org/10.1016/j.polymdegradstab.2013.06.016

Hu, D., Shen, M., Zhang, Y., Li, H., & Zeng, G. (2019). Microplastics and nanoplastics: Would they affect global biodiversity change? *Environmental Science and Pollution Research, 26*(19), 19997–20002. https://doi.org/10.1007/s11356-019-05414-5

Ibrahim, M., Salman, M., Kamal, S., Rehman, S., Razzaq, A., & Akash, S. H. (2017). Chapter 6: Algae-based biologically active compounds. *Algae based polymers, blends, and composites: Chemistry, biotechnology and materials science*. Elsevier Inc. https://doi.org/10.1016/B978-0-12-812360-7.00006-9

Ilias, M. A., Ismail, A., & Othman, R. (2017). Analysis of carrageenan yield and gel strength of Kappaphycus species in Semporna Sabah. *Journal of Tropical Plant Physiology, 9*(1), 14–23.

Jafarzadeh, S., Jafari, S. M., Salehabadi, A., Nafchi, A, M., Uthaya Kumar, U. S., & Khalil, H. P. S. A. (2020). Biodegradable green packaging with antimicrobial functions based on the bioactive compounds from tropical plants and their by-products. *Trends in Food Science & Technology, 100*(April), 262–277. https://doi.org/10.1016/j.tifs.2020.04.017

Jang, Y. H., Han, S. O., Sim, I. N., & Kim, H.-I. (2013). Pretreatment effects of seaweed on the thermal and mechanical properties of seaweed/polypropylene biocomposites. *Composites Part A: Applied Science and Manufacturing, 47*, 83–90. https://doi.org/10.1016/j.compositesa.2012.11.016

Kadar, N. A. H. A., Rahim2, N. S., Yusof, R., Nasir, N. A. H. A., & Hamid, A. H. (2021). A review on potential of algae in producing biodegradable plastic. *International Journal of Engineering Advanced Research, 3*(1), 13–26.

Kanatt, S. R., Lahare, P., Chawla, S. P., & Sharma, A. (2015). Kappaphycus alvarezii: Its antioxidant potential and use in bioactive packaging films. *Journal of Microbiology, Biotechnology and Food Sciences, 05*(01), 1–6. https://doi.org/10.15414/jmbfs.2015.5.1.1-6

Kanmani, P., & Rhim, J. W. (2014a). Antimicrobial and physical-mechanical properties of agar-based films incorporated with grapefruit seed extract. *Carbohydrate Polymers, 102*(1), 708–716. https://doi.org/10.1016/j.carbpol.2013.10.099

Kanmani, P., & Rhim, J. W. (2014b). Development and characterization of carrageenan/grapefruit seed extract composite films for active packaging. *International Journal of Biological Macromolecules, 68*, 258–266. https://doi.org/10.1016/j.ijbiomac.2014.05.011

Karan, H., Funk, C., Grabert, M., Oey, M., & Hankamer, B. (2019). Green bioplastics as part of a circular bioeconomy. *Trends in Plant Science, 24*(3), 237–249. https://doi.org/10.1016/j.tplants.2018.11.010

Khoo, H. H., & Tan, R. B. H. (2010). Environmental impacts of conventional plastic and bio-based carrier bags: Part 2: End-of-life options. *International Journal of Life Cycle Assessment, 15*(4), 338–345. https://doi.org/10.1007/s11367-010-0163-8

Kim, J. K., Yarish, C., Hwang, E. K., Park, M., & Kim, Y. (2017). Seaweed aquaculture: Cultivation technologies, challenges and its ecosystem services. *Algae, 32*(1), 1–13. https://doi.org/10.4490/algae.2017.32.3.3

Kok, J. M. L., & Wong, C. L. (2018). Physicochemical properties of edible alginate film from Malaysian *Sargassum polycystum* C. Agardh. *Sustainable Chemistry and Pharmacy, 9*(July), 87–94. https://doi.org/10.1016/j.scp.2018.07.001

Kumar, M., Chen, H., Sarsaiya, S., Qin, S., Liu, H., Awasthi, M. K., Kumar, S., Singh, L., Zhang, Z., Bolan, N. S., Pandey, A., Varjani, S., & Taherzadeh, M. J. (2021). Current research trends on micro- and nano-plastics as an emerging threat to global environment: A review. *Journal of Hazardous Materials, 409*(November 2020), 124967. https://doi.org/10.1016/j.jhazmat.2020.124967

Kumar, R., Sharma, P., Manna, C., & Jain, M. (2021). Abundance, interaction, ingestion, ecological concerns, and mitigation policies of microplastic pollution in riverine ecosystem: A review. *Science of the Total Environment, 782*, 146695. https://doi.org/10.1016/j.scitotenv.2021.146695

Lakshmi, D. S., Sankaranarayanan, S., Gajaria, T. K., Li, G., Kujawski, W., Kujawa, J., & Navia, R. (2020). A short review on the valorization of green seaweeds and ulvan: Feedstock for chemicals and biomaterials. *Biomolecules, 10*(7), 1–20. https://doi.org/10.3390/biom10070991

Lamberti, F. M., Román-Ramírez, L. A., & Wood, J. (2020). Recycling of bioplastics: Routes and benefits. *Journal of Polymers and the Environment, 28*(10), 2551–2571. https://doi.org/10.1007/s10924-020-01795-8

Lee, W. K., Lim, Y. Y., Leow, A. T. C., Namasivayam, P., Abdullah, J. O., & Ho, C. L. (2017). Factors affecting yield and gelling properties of agar. *Journal of Applied Phycology, 29*(3), 1527–1540. https://doi.org/10.1007/s10811-016-1009-y

Liling, G., Di, Z., Jiachao, X., Xin, G., Xiaoting, F., & Qing, Z. (2016). Effects of ionic crosslinking on physical and mechanical properties of alginate mulching films. *Carbohydrate Polymers, 136*, 259–265. https://doi.org/10.1016/j.carbpol.2015.09.034

Lim, J. Y., Hii, S. L., Chee, S. Y., & Wong, C. L. (2018). *Sargassum siliquosum* J. Agardh extract as potential material for synthesis of bioplastic film. *Journal of Applied Phycology, 30*(6), 3285–3297. https://doi.org/10.1007/s10811-018-1603-2

Lin, J. D., Lemay, M. A., & Parfrey, L. W. (2018). Diverse bacteria utilize alginate within the microbiome of the giant kelp Macrocystis pyrifera. *Frontiers in Microbiology, 9*. https://doi.org/10.3389/fmicb.2018.01914

Liu, S., Li, H., Tang, B., Bi, S., & Li, L. (2016). Scaling law and microstructure of alginate hydrogel. *Carbohydrate Polymers, 135*, 101–109. https://doi.org/10.1016/j.carbpol.2015.08.086

Li, P., Wang, X., Su, M., Zou, X., Duan, L., & Zhang, H. (2021). Characteristics of plastic pollution in the environment: A review. *Bulletin of Environmental Contamination and Toxicology, 107*(4), 577–584. https://doi.org/10.1007/s00128-020-02820-1

Li, S.-Y., Wang, Z.-P., Wang, L.-N., Peng, J.-X., Wang, Y.-N., Han, Y.-T., & Zhao, S.-F. (2019). Combined enzymatic hydrolysis and selective fermentation for green production of alginate oligosaccharides from *Laminaria japonica*. *Bioresource Technology, 281*, 84–89. https://doi.org/10.1016/j.biortech.2019.02.056

Li, K., Zhu, J., Guan, G., & Wu, H. (2019). Preparation of chitosan-sodium alginate films through layer-by-layer assembly and ferulic acid crosslinking: Film properties, characterization, and formation mechanism. *International Journal of Biological Macromolecules, 122*, 485–492. https://doi.org/10.1016/j.ijbiomac.2018.10.188

Mac Monagail, M., Cornish, L., Morrison, L., Araújo, R., & Critchley, A. T. (2017). Sustainable harvesting of wild seaweed resources. *European Journal of Phycology, 52*(4), 371–390. https://doi.org/10.1080/09670262.2017.1365273

Masarin, F., Cedeno, F. R. P., Chavez, E. G. S., de Oliveira, L. E., Gelli, V. C., & Monti, R. (2016). Chemical analysis and biorefinery of red algae *Kappaphycus alvarezii* for efficient production of glucose from residue of carrageenan extraction process. *Biotechnology for Biofuels, 9*(1), 122. https://doi.org/10.1186/s13068-016-0535-9

Meng, F., Zhang, Y., Xiong, Z., Wang, G., Li, F., & Zhang, L. (2018). Mechanical, hydrophobic and thermal properties of an organic-inorganic hybrid carrageenan-polyvinyl alcohol composite film. *Composites Part B: Engineering, 143*, 1–8. https://doi.org/10.1016/j.compositesb.2017.12.009

Mihranyan, A. (2011). Cellulose from cladophorales green algae: From environmental problem to high-tech composite materials. *Journal of Applied Polymer Science, 119*(4), 2449–2460. https://doi.org/10.1002/app.32959

Miller, S. A. (2020). Five misperceptions surrounding the environmental impacts of single-use plastic. *Environmental Science and Technology, 54*(22), 14143–14151. https://doi.org/10.1021/acs.est.0c05295

Mofeed, J., Deyab, M., Mohamed, A., Moustafa, M., Negm, S., & El-Bilawy, E. (2022). Antimicrobial activities of three seaweeds extract against some human viral and bacterial pathogens. *Biocell, 46*(1), 247–261. https://doi.org/10.32604/biocell.2022.015966

Mukurumbira, A. R., Shellie, R. A., Keast, R., Palombo, E. A., & Jadhav, S. R. (2022). Encapsulation of essential oils and their application in antimicrobial active packaging. *Food Control, 136*(October 2021), 108883. https://doi.org/10.1016/j.foodcont.2022.108883

Mülhaupt, R. (2013). Green polymer chemistry and bio-based plastics: Dreams and reality. *Macromolecular Chemistry and Physics, 214*(2), 159–174. https://doi.org/10.1002/macp.201200439

Niaounakis, M. (2019). Recycling of biopolymers – the patent perspective. *European Polymer Journal, 114*(February), 464–475. https://doi.org/10.1016/j.eurpolymj.2019.02.027

Nur Hanani, Z. A., & Aelma Husna, A. B. (2018). Effect of different types and concentrations of emulsifier on the characteristics of kappa-carrageenan films. *International Journal of Biological Macromolecules, 114*, 710–716. https://doi.org/10.1016/j.ijbiomac.2018.03.163

Nurjanah, Nurilmala, M., Anwar, E., Luthfiyana, N., & Hidayat, T. (2017). Identification of bioactive compounds of seaweed *Sargassum* sp. and *Eucheuma cottonii* doty as a raw sunscreen cream. *Proceedings of the Pakistan Academy of Sciences Part B, 54*(4), 311–318.

Oliveira, M., Ameixa, O. M. C. C., & Soares, A. M. V. M. (2019). Are ecosystem services provided by insects "bugged" by micro (nano)plastics? *TrAC, Trends in Analytical Chemistry, 113*, 317–320. https://doi.org/10.1016/j.trac.2019.02.018

Oms-Oliu, G., Soliva-Fortuny, R., & Martín-Belloso, O. (2008). Using polysaccharide-based edible coatings to enhance quality and antioxidant properties of fresh-cut melon. *LWT - Food Science and Technology, 41*(10), 1862–1870. https://doi.org/10.1016/j.lwt.2008.01.007

Organisation for Economic Co-operation and Development. (2018). Improving plastics management: Trends, policy responses, and the role of international co-operation and trade. *Environmental Policy Paper No. 12, 12*, 20.

Oyekanmi, A. A., Kumar, U. S. U., Abdul Khalil, H. P. S., Olaiya, N. G., Amirul, A. A., Rahman, A. A., Nuryawan, A., Abdullah, C. K., & Rizal, S. (2021). Functional properties of antimicrobial neem leaves extract based macroalgae biofilms for potential use as active dry packaging applications. *Polymers, 13*(10), 1–22. https://doi.org/10.3390/polym13101664

Patra, J. K., Lee, S. W., Park, J. G., & Baek, K. H. (2017). Antioxidant and antibacterial properties of essential oil extracted from an edible seaweed *Undaria pinnatifida*. *Journal of Food Biochemistry, 41*(1). https://doi.org/10.1111/jfbc.12278

Paula, G. A., Benevides, N. M. B., Cunha, A. P., de Oliveira, A. V., Pinto, A. M. B., Morais, J. P. S., & Azeredo, H. M. C. (2015). Development and characterization of edible films from mixtures ofκ-carrageenan, I-carrageenan, and alginate. *Food Hydrocolloids, 47*, 140–145. https://doi.org/10.1016/j.foodhyd.2015.01.004

Pereira, L., Amado, A. M., Critchley, A. T., van de Velde, F., & Ribeiro-Claro, P. J. A.a. (2009). Identification of selected seaweed polysaccharides (phycocolloids) by vibrational spectroscopy (FTIR-ATR and FT-Raman). *Food Hydrocolloids, 23*(7), 1903–1909. https://doi.org/10.1016/j.foodhyd.2008.11.014

Pereira, L., & Cotas, J. (2020). Introductory chapter: Alginates - a general overview. *Alginates - recent uses of this natural polymer*. IntechOpen. https://doi.org/10.5772/intechopen.88381

Perera, K. Y., Sharma, S., Pradhan, D., Jaiswal, A. K., & Jaiswal, S. (2021). Seaweed polysaccharide in food contact materials (Active packaging, intelligent packaging, edible films, and coatings). *Foods, 10*(9), 1–22. https://doi.org/10.3390/foods10092088

Phan The, D., Debeaufort, F., Voilley, A., & Luu, D. (2009). Biopolymer interactions affect the functional properties of edible films based on agar, cassava starch and arabinoxylan blends. *Journal of Food Engineering, 90*(4), 548–558. https://doi.org/10.1016/j.jfoodeng.2008.07.023

Phan, T. D., Debeaufort, F., Luu, D., & Voilley, A. (2005). Functional properties of edible agar-based and starch-based films for food quality preservation. *Journal of Agricultural and Food Chemistry, 53*(4), 973–981. https://doi.org/10.1021/jf040309s

Porse, H., & Rudolph, B. (2017). The seaweed hydrocolloid industry: 2016 updates, requirements, and outlook. *Journal of Applied Phycology, 29*(5), 2187–2200. https://doi.org/10.1007/s10811-017-1144-0

Price-Whelan, A. M., Agüeros, M. A., Fournier, A. P., Street, R., Ofek, E. O., Covey, K. R., Levitan, D., Laher, R. R., Sesar, B., & Surace, J. (2014). Statistical searches for microlensing events in large, non-uniformly sampled time-domain surveys: A test using palomar transient factory data. *The Astrophysical Journal, 781*(1). https://doi.org/10.1088/0004-637X/781/1/35

Rajendran, N., Puppala, S., Raj, M.,S., Angeeleena, B.,R., & Rajam, C. (2012). Seaweeds can be a new source for bioplastics. *Journal of Pharmacy Research, 5*(3), 1476–1479. http://jprsolutions.info.

Ramdhan, T., Ching, S. H., Prakash, S., & Bhandari, B. (2019). Time dependent gelling properties of cuboid alginate gels made by external gelation method: Effects of alginate-CaCl2 solution ratios and pH. *Food Hydrocolloids, 90*, 232–240. https://doi.org/10.1016/j.foodhyd.2018.12.022 (August 2018).

Rao, S. P. V., Peryasamy, C., Kumar, K. S., Rao, A. S., & Anantharaman, P. (2018). Chapter 6. Seaweeds: Distribution, production and uses. In M. N. Noor, S. K. Bhatnagar, & Shashi K. Sinha (Eds.), *Bioprospecting of algae* (pp. 59–78).

Rebours, C., Marinho-Soriano, E., Zertuche-González, J. A., Hayashi, L., Vásquez, J. A., Kradolfer, P., Soriano, G., Ugarte, R., Abreu, M. H., Bay-Larsen, I., Hovelsrud, G., Rødven, R., & Robledo, D. (2014). Seaweeds: An opportunity for wealth and sustainable livelihood for coastal communities. *Journal of Applied Phycology, 26*(5), 1939–1951. https://doi.org/10.1007/s10811-014-0304-8

Resano-Goizueta, I., Ashokan, B. K., Trezza, T. A., & Padua, G. W. (2018). Effect of nano-fillers on tensile properties of biopolymer films. *Journal of Polymers and the Environment, 26*(9), 3817–3823. https://doi.org/10.1007/s10924-018-1260-1

Revel, M., Châtel, A., & Mouneyrac, C. (2018). Micro(nano)plastics: A threat to human health? *Current Opinion in Environmental Science and Health, 1*, 17–23. https://doi.org/10.1016/j.coesh.2017.10.003

Rhim, J. W. (2012). Physical-mechanical properties of agar/κ-carrageenan blend film and derived clay nanocomposite film. *Journal of Food Science, 77*(12). https://doi.org/10.1111/j.1750-3841.2012.02988.x

Rinaudo, M. (2008). Main properties and current applications of some polysaccharides as biomaterials. *Polymer International, 57*(7), 397–430. https://doi.org/10.1002/pi

Robal, M., Brenner, T., Matsukawa, S., Ogawa, H., Truus, K., Rudolph, B., & Tuvikene, R. (2017). Monocationic salts of carrageenans: Preparation and physico-chemical properties. *Food Hydrocolloids, 63*, 656–667. https://doi.org/10.1016/j.foodhyd.2016.09.032

Rosenboom, J. G., Langer, R., & Traverso, G. (2022). Bioplastics for a circular economy. *Nature Reviews Materials, 7*(2), 117–137. https://doi.org/10.1038/s41578-021-00407-8

Rupérez, P., & Saura-Calixto, F. (2001). Dietary fibre and physicochemical properties of edible Spanish seaweeds. *European Food Research and Technology, 212*(3), 349–354. https://doi.org/10.1007/s002170000264

Samantaray, P. K., Little, A., Haddleton, D. M., McNally, T., Tan, B., Sun, Z., Huang, W., Ji, Y., & Wan, C. (2020). Poly(glycolic acid) (PGA): A versatile building block expanding high performance and sustainable bioplastic applications. *Green Chemistry, 22*(13), 4055–4081. https://doi.org/10.1039/d0gc01394c

Schmaltz, E., Melvin, E. C., Diana, Z., Gunady, E. F., Rittschof, D., Somarelli, J. A., Virdin, J., & Dunphy-Daly, M. M. (2020). Plastic pollution solutions: Emerging technologies to prevent and collect marine plastic pollution. *Environment International, 144*(August). https://doi.org/10.1016/j.envint.2020.106067

Schyns, Z. O. G., & Shaver, M. P. (2021). Mechanical recycling of packaging plastics: A review. *Macromolecular Rapid Communications, 42*(3), 1–27. https://doi.org/10.1002/marc.202000415

Sedayu, B. B., Cran, M. J., & Bigger, S. W. (2019). A review of property enhancement techniques for carrageenan-based films and coatings. *Carbohydrate Polymers, 216*(March), 287–302. https://doi.org/10.1016/j.carbpol.2019.04.021

Sharma, S., Barkauskaite, S., Jaiswal, A. K., & Jaiswal, S. (2021). Essential oils as additives in active food packaging. *Food Chemistry, 343*, 128403. https://doi.org/10.1016/j.foodchem.2020.128403

Shen, M., Song, B., Zeng, G., Zhang, Y., Huang, W., Wen, X., & Tang, W. (2020). Are biodegradable plastics a promising solution to solve the global plastic pollution? *Environmental Pollution, 263*, 114469. https://doi.org/10.1016/j.envpol.2020.114469

Shent, H., Pugh, R. J., & Forssberg, E. (1999). A review of plastics waste recycling and the flotation of plastics. *Resources, Conservation and Recycling, 25*(2), 85–109. https://doi.org/10.1016/S0921-3449(98)00017-2

Siew Ling, H., Lim, J.-Y., Ong, W.-T., & Wong, C.-L. (2016). Agar from Malaysian red seaweed as potential material for synthesis of bioplastic film. *Journal of Engineering Science & Technology, 11*, 1–15.

Somleva, M. N., Peoples, O. P., & Snell, K. D. (2013). PHA bioplastics, biochemicals, and energy from crops. *Plant Biotechnology Journal, 11*(2), 233–252. https://doi.org/10.1111/pbi.12039

Song, J. H., Murphy, R. J., Narayan, R., & Davies, G. B. H. (2009). Biodegradable and compostable alternatives to conventional plastics. *Philosophical Transactions of the Royal Society B: Biological Sciences, 364*(1526), 2127–2139. https://doi.org/10.1098/rstb.2008.0289

Sousa, A. M. M., Sereno, A. M., Hilliou, L., & Gonçalves, M. P. (2010). Biodegradable agar extracted from *Gracilaria vermiculophylla*: Film properties and application to edible coating. *Materials Science Forum, 636–637*, 739–744. https://doi.org/10.4028/www.scientific.net/MSF.636-637.739

Srinivasan, B., & Kulshreshtha, G. (2020). Recent developments in food-based bioplastics production. *Handbook of Environmental Chemistry, 104*, 107–127. https://doi.org/10.1007/698_2020_578

Stevens, E. S. (2001). Green plastics: An introduction to the new science of biodegradable plastics. *Journal of Chemical Education, 79*(9), 1072. https://doi.org/10.1300/j354v16n02_12

Sudhakar, M. P., Magesh Peter, D., & Dharani, G. (2020). Studies on the development and characterization of bioplastic film from the red seaweed (*Kappaphycus alvarezii*). *Environmental Science and Pollution Research*. https://doi.org/10.1007/s11356-020-10010-z

Sun, J.-Y., Zhao, X., Illeperuma, W. R. K., Chaudhuri, O., Oh, K. H., Mooney, D. J., Vlassak, J. J., & Suo, Z. (2012). Highly stretchable and tough hydrogels. *Nature, 489*(7414), 133–136. https://doi.org/10.1038/nature11409

Tavassoli-Kafrani, E., Shekarchizadeh, H., & Masoudpour-Behabadi, M. (2016). Development of edible films and coatings from alginates and carrageenans. *Carbohydrate Polymers, 137*, 360–374. https://doi.org/10.1016/j.carbpol.2015.10.074

Thiruchelvi, R., Das, A., & Sikdar, E. (2020). Bioplastics as better alternative to petro plastic. *Materials Today Proceedings, 37*(Part 2), 1634–1639. https://doi.org/10.1016/j.matpr.2020.07.176

Thushari, G. G. N., & Senevirathna, J. D. M. (2020). Plastic pollution in the marine environment. *Heliyon, 6*(8), e04709. https://doi.org/10.1016/j.heliyon.2020.e04709

Tiwari, B. K., & Troy, D. J. (2015). Seaweed sustainability: Food and non-food applications. In *Seaweed sustainability: Food and non-food applications*. Elsevier Inc. https://doi.org/10.1016/C2013-0-12836-X

Torres, M. D., Chenlo, F., & Moreira, R. (2016). Rheology of κ/ι-hybrid carrageenan from *Mastocarpus stellatus*: Critical parameters for the gel formation. *International Journal of Biological Macromolecules, 86*, 418–424. https://doi.org/10.1016/j.ijbiomac.2016.01.096

Tsang, Y. F., Kumar, V., Samadar, P., Yang, Y., Lee, J., Ok, Y. S., Song, H., Kim, K. H., Kwon, E. E., & Jeon, Y. J. (2019). Production of bioplastic through food waste valorization. *Environment International, 127*(March), 625–644. https://doi.org/10.1016/j.envint.2019.03.076

Turner, A. (2018). Mobilisation kinetics of hazardous elements in marine plastics subject to an avian physiologically-based extraction test. *Environmental Pollution, 236*, 1020–1026. https://doi.org/10.1016/j.envpol.2018.01.023

Usman, A., Khalid, S., Usman, A., Hussain, Z., & Wang, Y. (2017). Algal polysaccharides, novel application, and outlook. In *Algae based polymers, blends, and composites: Chemistry, biotechnology and materials science*. Elsevier Inc. https://doi.org/10.1016/B978-0-12-812360-7.00005-7

Uthaya Kumar, U. S., Paridah, M. T., Owolabi, F. A. T., Gopakumar, D. A., Rizal, S., Amirul, A. A., Rahman, A. A., Alfatah, T., Mistar, E. M., Aprilia, N. A. S., & Khalil, H. P. S. A. (2019). Neem leaves extract based seaweed bio-degradable composite films with excellent antimicrobial activity for sustainable packaging material. *Bioresources, 14*(1), 700–713. https://doi.org/10.15376/biores.14.1.700-713

Van Cauwenberghe, L., Devriese, L., Galgani, F., Robbens, J., & Janssen, C. R. (2015). Microplastics in sediments: A review of techniques, occurrence and effects. *Marine Environmental Research, 111*, 5–17. https://doi.org/10.1016/j.marenvres.2015.06.007

Venkatesan, J., Anil, S., Kim, S. K., & Shim, M. S. (2016). Seaweed polysaccharide-based nanoparticles: Preparation and applications for drug delivery. *Polymers, 8*(2), 1–25. https://doi.org/10.3390/polym8020030

Vital, A. C. P., Guerrero, A., Monteschio, J. de O., Valero, M. V., Carvalho, C. B., de Abreu Filho, B. A., Madrona, G. S., & do Prado, I. N. (2016). Effect of edible and active coating (with rosemary and oregano essential oils) on beef characteristics and consumer acceptability. *PLoS One, 11*(8), e0160535. https://doi.org/10.1371/journal.pone.0160535

Vu, D. H., Åkesson, D., Taherzadeh, M. J., & Ferreira, J. A. (2020). Recycling strategies for polyhydroxyalkanoate-based waste materials: An overview, *Vol 298. Bioresource technology*. Elsevier Ltd. https://doi.org/10.1016/j.biortech.2019.122393

Walker, S., & Rothman, R. (2020). Life cycle assessment of bio-based and fossil-based plastic: A review. *Journal of Cleaner Production, 261*, 121158. https://doi.org/10.1016/j.jclepro.2020.121158

Wang, Y. L., Lee, Y. H., Chiu, I. J., Lin, Y. F., & Chiu, H. W. (2020). Potent impact of plastic nanomaterials and micromaterials on the food chain and human health. *International Journal of Molecular Sciences, 21*(5), 1–14. https://doi.org/10.3390/ijms21051727

Wang, J., Peng, C., Li, H., Zhang, P., & Liu, X. (2021). The impact of microplastic-microbe interactions on animal health and biogeochemical cycles: A mini-review. *Science of the Total Environment, 773*, 145697. https://doi.org/10.1016/j.scitotenv.2021.145697

Waring, R. H., Harris, R. M., & Mitchell, S. C. (2018). Plastic contamination of the food chain: A threat to human health? *Maturitas, 115*, 64–68. https://doi.org/10.1016/j.maturitas.2018.06.010

Wojnowska-Baryła, I., Kulikowska, D., & Bernat, K. (2020). Effect of bio-based products on waste management. *Sustainability, 12*(5), 1–12. https://doi.org/10.3390/su12052088

Wu, P., & Imai, M. (2012). Outstanding molecular size recognition and regulation of water permeability on κ-carrageenan-pullulan membrane involved in synergistic design of composite polysaccharides-structure. *Procedia Engineering, 42*(August), 1313–1325. https://doi.org/10.1016/j.proeng.2012.07.523

Wullandari, P., Sedayu, B. B., Novianto, T. D., & Prasetyo, A. W. (2021). Characteristic of semi refined and refined carrageenan flours used in the making of biofilm (bioplastic). *IOP Conference Series: Earth and Environmental Science, 733*(1). https://doi.org/10.1088/1755-1315/733/1/012112

Xue, J., Wu, Y., Shi, K., Xiao, X., Gao, Y., Li, L., & Qiao, Y. (2019). Study on the degradation performance and kinetics of immobilized cells in straw-alginate beads in marine environment. *Bioresource Technology, 280*, 88–94. https://doi.org/10.1016/j.biortech.2019.02.019

Xu, S., Ma, J., Ji, R., Pan, K., & Miao, A. J. (2020). Microplastics in aquatic environments: Occurrence, accumulation, and biological effects. *Science of the Total Environment, 703*, 134699. https://doi.org/10.1016/j.scitotenv.2019.134699

Yildirim, S., Röcker, B., Pettersen, M. K., Nilsen-Nygaard, J., Ayhan, Z., Rutkaite, R., Radusin, T., Suminska, P., Marcos, B., & Coma, V. (2018). Active packaging applications for food. *Comprehensive Reviews in Food Science and Food Safety, 17*(1), 165–199. https://doi.org/10.1111/1541-4337.12322

Yoon, J., Oh, D. X., Jo, C., Lee, J., & Hwang, D. S. (2014). Improvement of desolvation and resilience of alginate binders for Si-based anodes in a lithium ion battery by calcium-mediated cross-linking. *Physical Chemistry Chemical Physics, 16*(46), 25628–25635. https://doi.org/10.1039/c4cp03499f

Zhang, C., Show, P.-L., & Ho, S.-H. (2019). Progress and perspective on algal plastics – a critical review. *Bioresource Technology, 289*, 121700. https://doi.org/10.1016/j.biortech.2019.121700

Zhang, D., Zhang, M., & Gu, X. (2018). Seaweed-derived hydrocolloids as food coating and encapsulation agents. In *Bioactive seaweeds for food applications: Natural ingredients for healthy diets*. Elsevier Inc. https://doi.org/10.1016/B978-0-12-813312-5.00008-X

Zhu, Y., Romain, C., & Williams, C. K. (2016). Sustainable polymers from renewable resources. *Nature, 540*(7633), 354–362. https://doi.org/10.1038/nature21001

Zimmermann, L., Dombrowski, A., Völker, C., & Wagner, M. (2020). Are bioplastics and plant-based materials safer than conventional plastics? In vitro toxicity and chemical composition. *Environment International, 145*(June), 106066. https://doi.org/10.1016/j.envint.2020.106066

CHAPTER

9 Blue economy and blue carbon materials applications in various fields with respect to marine algae

M.P. Sudhakar[1], Kulanthaiyesu Arunkumar[2], G. Dharani[1] and R. Kirubagaran[3]

[1]*Marine Biotechnology Division, National Institute of Ocean Technology (MoES), Ministry of Earth Sciences, Pallikaranai, Chennai, Tamil Nadu, India;* [2]*Phycoscience Laboratory, Department of Plant Science, School of Biological Sciences, Central University of Kerala, Kasaragod, Kerala, India;* [3]*Deep Ocean Mission (DOM), Vertical VI, Marine Biotechnology Division, National Institute of Ocean Technology (MoES), Pallikaranai, Chennai, Tamil Nadu, India*

1. Introduction

The United Nations Framework Convention on Climate Change developed policies on climate change to restore carbon on earth by mitigating carbon dioxide sequestration using renewable biomass, especially salt marshes, seagrasses, and mangroves, which together act as potential carbon sinks (Wylie et al., 2016). "Blue carbon" is a term coined in 2009 to restore the marine ecosystem (Nellemann & GRID–Arendal., 2009). Blue carbon is carbon generated from marine ecosystems through CO_2 sequestration, which ultimately boosts the surrounding economy (https://www.iucn.org/sites/dev/files/blue_carbon_issues_brief.pdf). Blue carbon is part of the blue economy. The blue economy and its development in the communal sector are emerging. Economic development through the ocean's resources and its benefits to social welfare are called the blue economy. Different definitions have been suggested by many international agencies on blue economy (Lovelock & Duarte, 2019; Macreadie et al., 2019; Steven et al., 2019). All definitions focus on oceanic bioresources for the welfare of humans. Blue carbon is simply carbon-rich materials derived from the marine environment for different industrial applications. Blue carbon from marine resources includes marine flora and fauna that thrive in a country's economy in a particular area and ultimately benefit the society. India has a long coastline ~7500 km long with a large diversity of marine flora and fauna (Mantri et al., 2020). This chapter discusses the blue carbon source for biomaterial production and its feasible applications.

2. Blue carbon and its source in the marine environment

Carbon-rich materials such as mangroves, tidal marshes, seagrass, macroalgae, phytoplankton, coral reefs, marine fauna, and oyster reefs are considered to be potential sources for ecosystem development (Macreadie et al., 2019; McLeod et al., 2011). These carbon sources are converted into different forms of products for human consumption (Lovelock & Duarte, 2019). In India, different marine environments have various ecosystems with good diversity based on the nature of the water and the climate.

Algae Materials: Applications Benefitting Health. https://doi.org/10.1016/B978-0-443-18816-9.00008-3

India has several coasts as well as the Bay of Bengal and the Arabian sea, with diverse environment and resources. The resources available in Indian marine waters are potentially useful for many commercial applications. The government of India launched research schemes to protect the ecosystem for economic development, such as the prime minister's research fellowship, research parks, the Impacting Research Innovation and Technology initiative, Uchhatar Avishkar Yojana, and Smart India Hackathons. Like India, many other countries have developed sustainable goals and schemes to boost the blue economy by constantly integrating and maintaining the ecosystem.

Renewable resources such as marine flora and fauna have a major role in marine carbon sinks (Fig. 9.1). More than 55% of marine sources sequester CO_2 and mostly convert them into renewable organic carbon. Organic carbon such as hydrocolloids (carbohydrates), lipids, and proteins are major ingredients useful for food and feed applications called blue carbon, used for the development of blu economic growth (Alongi, 2018).

Apart from marine flora, many marine faunas help develop and protect the ecosystem through the food web. Marine fish are use and tolerate marine litter, preventing pollution to an extent. Fish are the main resource for food in many countries and directly reflect the economy (Préat et al., 2018). Living coral breathes in seawater to maintain proper oxygen levels in the ocean's ecosystem; thus, living organisms survive during adverse conditions (Gupta et al., 2013; Steven et al., 2019). Marine resources help develop many useful products for common human use (Enascuta et al., 2018; Ghaly et al., 2013; Vidya et al., 2021).

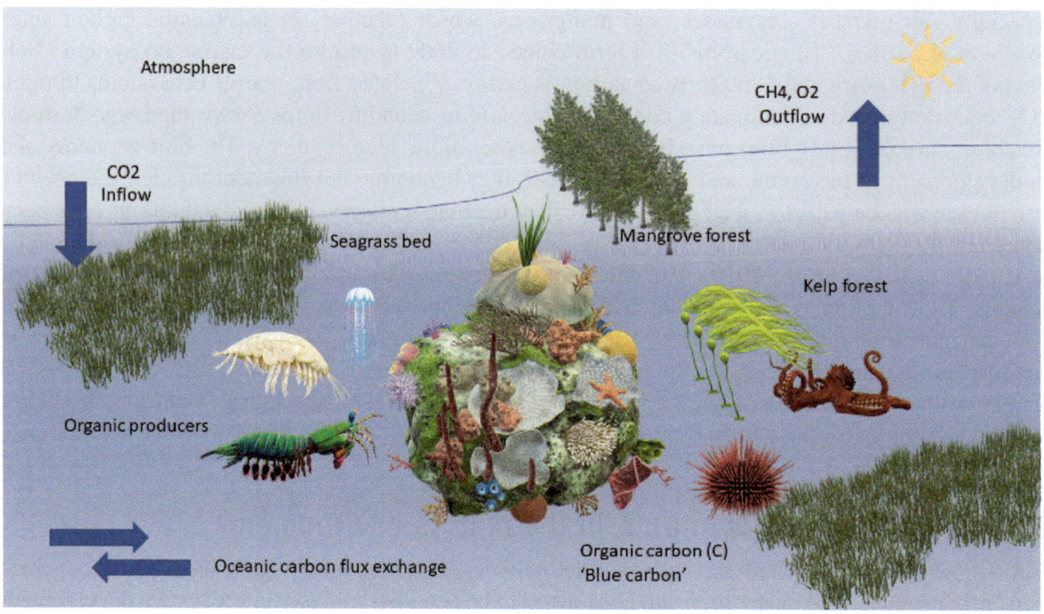

FIGURE 9.1

Graphic conceptual model of marine coastal ecosystem in blue carbon ecology balancing photosynthesis, respiration, and remineralization.

3. Applications of blue carbon in different fields

Blue carbon materials derived from marine sources have been used in many applications such as the food sector, cosmetics, lotions, biofuels (bioethanol, biogas, biohydrogen, etc.), medical applications, colorants, and environmental monitoring (Ali Ahmed et al., 2017; Anil et al., 2017; Holdt & Kraan, 2011; Kilinç et al., 2008; Senthilkumar et al., 2017). Fig. 9.2 shows applications of blue carbon in the biofuel sector.

3.1 Blue carbon biomass as feedstock for bioethanol conversion

Environmental pollution, global warming, and the energy crisis are major concerns throughout the world. To overcome these problems, carbon storage and energy generation are important. Blue carbon, the most powerful resource for producing energy and sequestering carbon, has been available since time immemorial. Sources of carbon feedstock are being depleted daily owing increased demands and the growth in population and consumption. Organic carbon-rich renewable substrates such as cellulose, starch, hemicellulose, and pectin are a few commercial materials widely used as carbon sources. Biomass rich in carbons is obtained from land-based cultivation systems and through afforestation. However, because of certain difficulties in obtaining biomass from land-based cultivation, water scarcity for plantation and cultivation, food to energy crisis are the main reasons to rethink of alternate resources for generating carbon.

Marine biomasses such as seaweed, algae, and seagrass are rich in organic carbon sources, especially sulfated polysaccharides, lipids, and carbohydrates (Howard et al., 2017; Macreadie et al., 2019). For biofuel conversion, these blue carbon sources are useful for developing local energy needs and are an economically inexpensive source. Among the marine resources, seaweeds have an important role in energy production because they are renewable and cultivation is possible onshore and offshore (Ganesan et al., 2019; Mantri et al., 2019). Other types of biomass generated in the marine environment are mostly used for food and feed applications.

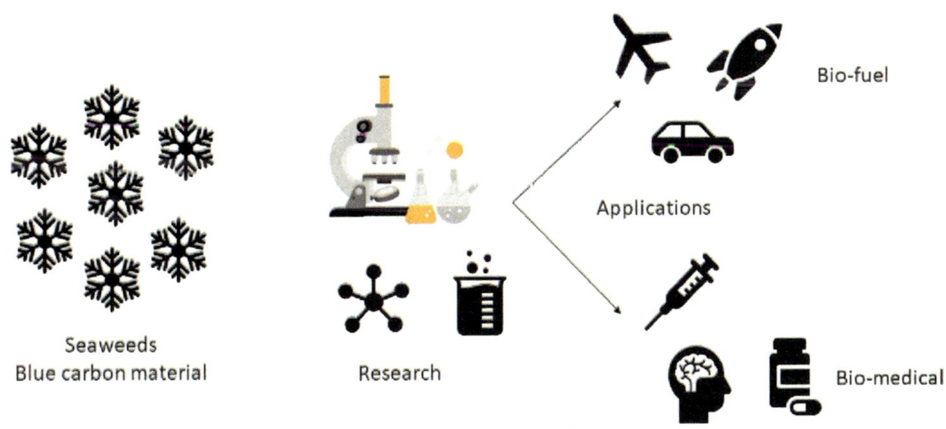

FIGURE 9.2

Applications of blue carbon (seaweed) for commercial use.

Sources of blue carbon material such as seaweed are also called green carbon because of biobased polymers produced through photosynthesis (sequestering CO_2- 173 TgC y^{-1} by seaweed) for tremendous applications, especially in green fuel generation (Krause-Jensen & Duarte, 2016). Blue carbon is a renewable source for various types of energy production such as bioethanol, biogas, biohydrogen, and bioelectricity. The demand for energy is rising daily and increases in the price of crude oil directly affect consumers. The depletion of natural crude oil over time has become critical. Thus, an alternate source for producing fuels through renewable sources is mandatory in certain developing countries to meet the demand. Macroalgal cultivation may directly affect deep sea carbon sequestration from the atmosphere.

Bioethanol can be produced using seaweeds, but the cost of production and pilot-scale demonstration using marine resources is not well-established. The technology for converting seaweed biomass to bioethanol is complicated and the yield of ethanol is small compared with commercial sources of production. However, the use of seaweed to produce bioethanol favors not only blue carbon sinks, it promotes the blue economy. Cultivating seaweed through local fishers has a direct impact on economic development in that area. Concern persons from Government should frame clear policy and legally permit the import and export of seaweed products through "Seaweed Parks" or "Blue Carbon Parks" which ultimately boots country's blue economy. Advantages of blue carbon for ethanol production are:

1. It meets fossil fuel demands.
2. There are fewer carbon emissions.
3. Energy efficiency is higher than for fossil fuel.
4. It improves the blue carbon economy.
5. Small-scale industry for ethanol production will be improved.
6. Integrated Multi-Trophic Aquaculture Systems (IMTA) increase seaweed biomass production for bioethanol production.

3.2 Blue carbon biomass for biomedical applications

Marine biomasses are rich in several bioactive compounds such as biopolymers, secondary metabolites, pigments, fatty acids, and pharmaceuticals. These compounds are used directly or indirectly in the medical field to save precious lives. Several biomaterials have been developed to replace synthetic materials for biomedical use.

Among the marine biomasses, seaweeds are emphasized here owing to their versatile nature for biomedical use. Seaweeds contain several biocompounds, especially secondary metabolites such as steroids, alkaloids, terpenoids, and several hydrocolloid compounds. Secondary metabolites with good activity against many pathogens have been reported for several decades. Seaweed was used for food during the Second World War. Hydrocolloids such as agar, carrageenan, ulvan, fucoidan, and alginate have been widely used for different biomedical applications (Buschmann & Camus, 2019).

Seaweed-derived compounds are widely used in the biomedical fields, esspecially in dental and ortho implants applications (Dilip et al., 2020; Haugen et al., 2020; Sharma et al., 2021), cancer therapy (Martínez Andrade et al., 2018; Zhang et al., 2013), Alzheimer's and Parkinson's disease, cosmetics (Couteau & Coiffard, 2016; Morais et al., 2021; Wang et al., 2015), and skin care applications. Alginate biopolymer is widely used in dental application to take tooth impressions, as injectable hydrogels, for controlled drug release, for guided tissue regeneration, and for guided bone regeneration membrane applications (Ueyama et al., 2006). Agar is a well-known biopolymer mostly used in the food sector and for microbiological applications. Fucoidan is a sulfated polysaccharide used in biomedical applications (Amin et al., 2021). Carrageenan biopolymer is widely used in the food and pharmaceutical sectors as a thickening and gelling agent (Alves et al., 2011; Santo et al., 2009). Several scaffolds and biomembranes were developed using these biopolymers. Every biopolymer has a unique chemical nature. These biopolymers (blue carbon-rich material) can be used in the appropriate fields.

3.3 Marching toward the blue economy through seaweed cultivation

Seaweed is macroalgae, a potential renewable resource available on the Indian coasts. These resources help in local economic growth directly and indirectly through fishing communities by enabling extra income in addition to daily per capita income. The IMTA system (Neori et al., 2017) is an integrated method of cultivating fish and seaweed together with minimal input costs. It favors economic growth, creates job security, increases income to fishers, and creates multiple product development in a single system with automated technology (Kim et al., 2017). This type of IMTA system also favors sustainable development and also protect the environment and ecosystem (Cabrita et al., 2017). IMTA helps in carbon storage and also protects the natural seaweed bed in the ocean, enriching the quantity of blue carbon. Particulate organic carbon and dissolved organic carbon transfer to the deep sea and sedimental areas by seaweed helps the biosphere reserve. This deposition may vary among places and according to the type of seaweed species available (Echem, 2017).

The sequestration of CO_2 and its release back into the atmosphere occurs through a carbon capture method by dissolution and calcification. The reactions are:

$$\text{Calcification } [\text{Ca2} + +\text{HCO3}- \leftarrow \text{CaCO3} + \text{H2O} + \text{CO2}]$$

$$\text{Dissolution } [\text{CaCO3} + \text{H2O} + \text{CO2} \rightarrow \text{Ca2} + + 2\text{HCO3}-]$$

Many techniques are developed for IMTA seaweed cultivation. Among them, long line rope and raft cultivation are predominantly used and are easily harvested manually and mechanically using harvesting vehicles (https://www.springwise.com/innovation/agriculture-energy/sea-tractor-floating-seaweed-farms-sea6). Harvesting vehicles were developed to meet the workforce shortage and for an easy way to harvest (Monagail et al., 2017; Rosentrater, 2014). The genetic development of fast-growing seaweed achieves a good yield within a short time, such as *Kappaphycus* sp. (Qin et al., 2012). The government also plans to collect and develop the healthy germplasm of seaweed to preserve it for future generations and to get the same quality of seaweed with a controlled yield with disease-free cultivation (Wade et al., 2020).

To support the blue economy, the European Union and the United Nations have developed the sustainable goal of improving local economies with the help of farmers and industrialists to promote seaweed cultivation. Various seminars and conferences were conducted to promote the blue economy. In March 2022, the World Ocean Summit was hosted by economists from different countries to strengthen the ocean economy and protect the ocean from overexploitation from local suppliers and industrialists. This offers a way to create products from ocean resources (https://events.economist.com/world-ocean-summit/) (Fig. 9.3).

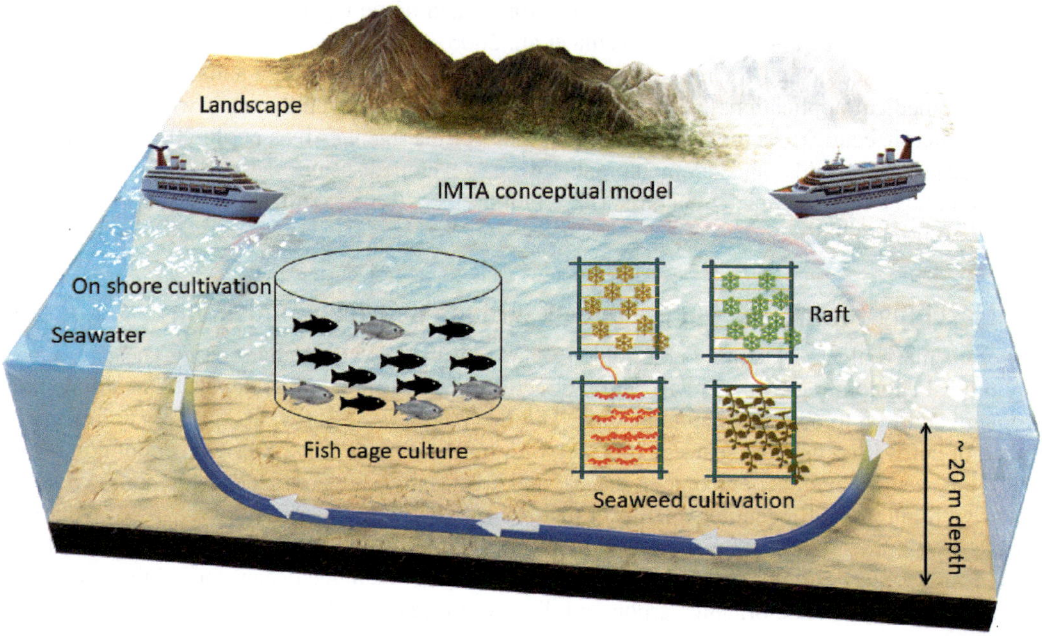

FIGURE 9.3

Conceptual model of Integrated Multi-Trophic Aquaculture Systems (IMTA) for seaweed cultivation in blue carbon economy.

4. Conclusion

Blue carbon and the blue economy are considered as an integrated schemes which ultimately helps in the growth of economy. Carbon sequestration and fixation by seaweed and other supporting organisms in the marine environment have a major role in developing blue carbon. CO_2 uptake by seaweed through photosynthesis and release by respiration and decomposition stabilize the ecosystem and sequester carbon in deep soil in the ocean environment. Socialist involvement and economists advice together help the industrialist to invest in blue carbon economy is possible in future. Different organizations throughout the world are keen to develop the blue economy through seaweed cultivation. The blue economy depends on industrialists, but it is a process in which each individual has the responsibility of improving the ecosystem and economic growth.

References

Ali Ahmed, A. B., Adel, M., Talati, A., Kumar, M. S., Abdulrahim, K., & Abdulhameed, M. M. (2017). Seaweed polysaccharides and their production and applications. *Seaweed polysaccharides: Isolation, biological and biomedical applications*. Elsevier Inc. https://doi.org/10.1016/B978-0-12-809816-5.00020-7

Alongi, D. (2018). *SPRINGER BRIEFS IN CLIMATE STUDIES Blue Carbon Coastal Sequestration for Climate Change Mitigation*. http://www.springer.com/series/11581.

Alves, V. D., Castelló, R., Ferreira, A. R., Costa, N., Fonseca, I. M., & Coelhoso, I. M. (2011). Barrier properties of carrageenan/pectin biodegradable composite films. *Procedia Food Science, 1*, 240−245. https://doi.org/10.1016/j.profoo.2011.09.038

Amin, M. L., Mawad, D., Dokos, S., Koshy, P., Martens, P. J., & Sorrell, C. C. (2021). Fucoidan- and carrageenan-based biosynthetic poly(vinyl alcohol) hydrogels for controlled permeation. *Materials Science and Engineering: C, 121*(October 2020), 111821. https://doi.org/10.1016/j.msec.2020.111821

Anil, S., Venkatesan, J., Chalisserry, E. P., Nam, S. Y., & Kim, S. K. (July 2017). Applications of seaweed polysaccharides in dentistry. *Seaweed polysaccharides: Isolation, biological and biomedical applications* (pp. 331−340). https://doi.org/10.1016/B978-0-12-809816-5.00017-7

Buschmann, A. H., & Camus, C. (2019). An introduction to farming and biomass utilisation of marine macroalgae. *Phycologia* (Vol 58,(5), 443−445. https://doi.org/10.1080/00318884.2019.1638149

Cabrita, A. R. J., Maia, M. R. G., Sousa-Pinto, I., & Fonseca, A. J. M. (2017). Ensilage of seaweeds from an integrated multi-trophic aquaculture system. *Algal Research, 24*(March), 290−298. https://doi.org/10.1016/j.algal.2017.04.024

Couteau, C., & Coiffard, L. (2016). Seaweed application in cosmetics. *Seaweed in health and disease prevention* (pp. 423−441). Elsevier Inc. https://doi.org/10.1016/B978-0-12-802772-1.00014-2

Dilip, A., Gupta, R., & Geiger, Z. (2020). Dental alginate impressions. *StatPearls*, 2−4. http://www.ncbi.nlm.nih.gov/pubmed/29262138.

Echem, R. T. (2017). Enhanced-nutrient cultivation technique of marine macrobenthic algae (Gracilaria arcuata, Kappaphycus alvarezii and Eucheuma denticulatum). *Research Journal, 36*. www.wmsu.edu.ph/re.

Enascuta, C. E., Stepan, E., Bolocan, I., Bombos, D., Calin, C., Oprescu, E. E., & Lavric, V. (2018). Simultaneous production of oil enriched in ω-3 polyunsaturated fatty acids and biodiesel from fish wastes. *Waste Management, 75*, 205−214. https://doi.org/10.1016/j.wasman.2018.01.022

Ganesan, M., Trivedi, N., Gupta, V., Madhav, S. V., Radhakrishna Reddy, C., & Levine, I. A. (2019). Seaweed resources in India - current status of diversity and cultivation: Prospects and challenges. *Botanica Marina, 62*(5), 463−482. https://doi.org/10.1515/BOT-2018-0056/ASSET/GRAPHIC/J_BOT-2018-0056_FIG_015.JPG

Ghaly, A. E., Ramakrishnan, V.v., Brooks, M. S., Budge, S. M., & Dave, D. (2013). Fish processing wastes as a potential source of proteins, amino acids and oils: A critical review. *Journal of Microbial & Biochemical Technology, 5*(4), 107−129. https://doi.org/10.4172/1948-5948.1000110

Gupta, V., Kumar, S., Sood, A., Chaudhary, V., & Prasanna, R. (2013). New insights into the biodiversity and applications of cyanobacteria (blue-green algae) — prospects and challenges. *Algal, 2*(2), 79−97. https://doi.org/10.1016/j.algal.2013.01.006

Haugen, H. J., Basu, P., Sukul, M., Mano, J. F., & Reseland, J. E. (2020). Injectable biomaterials for dental tissue regeneration. *International Journal of Molecular Sciences, 21*(10). https://doi.org/10.3390/ijms21103442

Holdt, S. L., & Kraan, S. (2011). Bioactive compounds in seaweed: Functional food applications and legislation. *Journal of Applied Phycology*, 543−597. https://doi.org/10.1007/s10811-010-9632-5

Howard, J., Sutton-Grier, A., Herr, D., Kleypas, J., Landis, E., Mcleod, E., Pidgeon, E., & Simpson, S. (2017). Clarifying the role of coastal and marine systems in climate mitigation. *Frontiers in Ecology and the Environment, 15*(1), 42−50. https://doi.org/10.1002/fee.1451

Kilinç, B., Cirik, S., Turan, G., Tekogul, H., Koru, E., Berna, K., Semra, C., Gamze, T., Tekogul, H., & Koru, E. (2008). Seaweeds for food and industrial applications. *Parallel Manipulators, New Developments*, 735−748. https://doi.org/10.5772/5372

Kim, J. K., Yarish, C., Hwang, E. K., Park, M., & Kim, Y. (2017). Seaweed aquaculture: Cultivation technologies, challenges and its ecosystem services. *Algae, 32*(1). https://doi.org/10.4490/algae.2017.32.3.3

Krause-Jensen, D., & Duarte, C. M. (2016). Substantial role of macroalgae in marine carbon sequestration. *Nature Geoscience, 9*(10), 737–742. https://doi.org/10.1038/ngeo2790

Lovelock, C. E., & Duarte, C. M. (2019). Dimensions of blue carbon and emerging perspectives. *Biology Letters, 15*(3). https://doi.org/10.1098/rsbl.2018.0781

Macreadie, P. I., Anton, A., Raven, J. A., Beaumont, N., Connolly, R. M., Friess, D. A., Kelleway, J. J., Kennedy, H., Kuwae, T., Lavery, P. S., Lovelock, C. E., Smale, D. A., Apostolaki, E. T., Atwood, T. B., Baldock, J., Bianchi, T. S., Chmura, G. L., Eyre, B. D., Fourqurean, J. W., … Duarte, C. M. (2019). The future of Blue Carbon science. *Nature Communications, 10*(1). https://doi.org/10.1038/s41467-019-11693-w

Mantri, V. A., Ganesan, M., Gupta, V., Krishnan, P., & Siddhanta, A. K. (2019). An overview on agarophyte trade in India and need for policy interventions. *Journal of Applied Phycology*. https://doi.org/10.1007/s10811-019-01791-z

Mantri, V. A., Kavale, M. G., & Kazi, M. A. (2020). Seaweed biodiversity of India: Reviewing current knowledge to identify gaps, challenges, and opportunities. *Diversity, 12*(1). https://doi.org/10.3390/d12010013

Martínez Andrade, K., Lauritano, C., Romano, G., & Ianora, A. (2018). Marine microalgae with anti-cancer properties [JOUR]. *Marine Drugs, 16*(5), 165.

McLeod, E., Chmura, G. L., Bouillon, S., Salm, R., Björk, M., Duarte, C. M., Lovelock, C. E., Schlesinger, W. H., & Silliman, B. R. (2011). A blueprint for blue carbon: Toward an improved understanding of the role of vegetated coastal habitats in sequestering CO2. *Frontiers in Ecology and the Environment, 9*(10), 552–560. https://doi.org/10.1890/110004

Monagail, M. mac, Cornish, L., Morrison, L., Araújo, R., & Critchley, A. T. (2017). Sustainable harvesting of wild seaweed resources Sustainable harvesting of wild seaweed resources. *European Journal of Phycology*. https://doi.org/10.1080/09670262.2017.1365273

Morais, T., Cotas, J., Pacheco, D., & Pereira, L. (2021). Seaweeds compounds: An ecosustainable source of cosmetic ingredients? *Cosmetics, 8*(1), 8. https://doi.org/10.3390/cosmetics8010008

Nellemann, C., & GRID–Arendal. (2009). *Blue carbon: The role of healthy oceans in binding carbon: A rapid response assessment*. GRID-Arendal.

Neori, A., Shpigel, M., Guttman, L., & Israel, A. (2017). Development of Polyculture and Integrated Multi-Trophic Aquaculture (IMTA) in Israel: A Review. *The Israeli Journal of Aquaculture-Bamidgeh*. http://www.siamb.org.il.

Préat, N., de Troch, M., van Leeuwen, S., Taelman, S. E., de Meester, S., Allais, F., & Dewulf, J. (2018). Development of potential yield loss indicators to assess the effect of seaweed farming on fish landings. *Algal Research, 35*(June), 194–205. https://doi.org/10.1016/j.algal.2018.08.030

Qin, S., Lin, H., & Jiang, P. (2012). Advances in genetic engineering of marine algae. *Biotechnology Advances, 30*(6), 1602–1613. https://doi.org/10.1016/j.biotechadv.2012.05.004

Rosentrater, K. A. (2014). *Techno-economic analysis of harvesting, processing, and transporting of seaweed* (pp. 1–27).

Santo, V. E., Frias, A. M., Carida, M., Cancedda, R., Gomes, M. E., Mano, J. F., & Reis, R. L. (2009). Carrageenan-based hydrogels for the controlled delivery of PDGF-BB in bone tissue engineering applications. *Biomacromolecules, 10*(6), 1392–1401. https://doi.org/10.1021/BM8014973

Senthilkumar, K., Ramajayam, G., Venkatesan, J., Kim, S. K., & Ahn, B. C. (2017). Biomedical applications of fucoidan, seaweed polysaccharides. *Seaweed polysaccharides: Isolation, biological and biomedical applications*. Elsevier Inc. https://doi.org/10.1016/B978-0-12-809816-5.00014-1

Sharma, V., Kaur, M., Sehgal, R., & Gupta, R. (2021). Alginate-based bionanocomposites in dental applications. *Bionanocomposites in Tissue Engineering and Regenerative Medicine*, 309–326. https://doi.org/10.1016/B978-0-12-821280-6.00011-8

Steven, A. D. L., Vanderklift, M. A., & Bohler-Muller, N. (2019). A new narrative for the blue economy and blue carbon. *Journal of the Indian Ocean Region, 15*(2), 123–128. https://doi.org/10.1080/19480881.2019.1625215

Ueyama, Y., Koyama, T., Ishikawa, K., Mano, T., Ogawa, Y., Nagatsuka, H., & Suzuki, K. (2006). Comparison of ready-made and self-setting alginate membranes used as a barrier membrane for guided bone regeneration. *Journal of Materials Science: Materials in Medicine, 17*(3), 281–288. https://doi.org/10.1007/s10856-006-7315-1

Vidya, D., Nayana, K., Sreelakshmi, M., Keerthi, K.v., Mohan, K. S., Sudhakar, M. P., & Arunkumar, K. (2021). A sustainable cultivation of microalgae using dairy and fish wastes for enhanced biomass and bio-product production. *Biomass Conversion and Biorefinery*, 0123456789. https://doi.org/10.1007/s13399-021-01817-y

Wade, R., Augyte, S., Harden, M., Nuzhdin, S., Yarish, C., & Alberto, F. (2020). Macroalgal germplasm banking for conservation, food security, and industry. *PLoS Biology, 18*(2). https://doi.org/10.1371/journal.pbio.3000641

Wang, H. M. D., Chen, C. C., Huynh, P., & Chang, J. S. (2015). Exploring the potential of using algae in cosmetics. *Bioresource Technology, 184*, 355–362. https://doi.org/10.1016/j.biortech.2014.12.001

Wylie, L., Sutton-Grier, A. E., & Moore, A. (2016). Keys to successful blue carbon projects: Lessons learned from global case studies. *Marine Policy, 65*, 76–84. https://doi.org/10.1016/j.marpol.2015.12.020

Zhang, Z., Teruya, K., Yoshida, T., Eto, H., & Shirahata, S. (2013). Fucoidan extract enhances the anti-cancer activity of chemotherapeutic agents in MDA-MB-231 and MCF-7 breast cancer cells. *Marine Drugs, 11*(1), 81–98. https://doi.org/10.3390/md11010081

CHAPTER

Applications of bacterial quorum sensing in algal biotechnology

10

Xiaotian Zhao[1], Bharathi Ramalingam[2] and Raju Sekar[1]

[1]*Department of Biological Sciences, School of Science, Xián Jiaotong-Liverpool University, Suzhou, Jiangsu, China;*
[2]*Suzhou Xishan Biotech Inc., VRL Asia, Suzhou, Jiangsu, China*

1. Introduction

Quorum sensing (QS) is a process of cell-to-cell communication mediated by chemical signal molecules produced by the bacteria, which is directly dependent on the local cell density. Bacterial communication is a coordinated behavior exhibited by the bacteria for virulence activity such as biofilm formation, bioluminescence, antibiotic resistance, gene expression, and production of toxins and secondary metabolites (McBrayer et al., 2020; Xiao et al., 2022). Each bacterium produces signal molecules that are released into the environment, and when signal concentration reaches threshold, it triggers the virulence factors indicating that there is enough number of bacteria within the colony. These small diffusible signal molecules are known as autoinducers (AIs), which activates specific receptors for coordinated behavior of the bacterial population (Mieszkin et al., 2013). Genetically, QS requires two sets of genes for communication and gene expression: one for synthase genes which synthesizes small diffusible signal molecules known as acyl-homoserine lactones (AHLs) and other genes for receptors which detects the AHL compounds and cascades the gene expression. The activation of receptor proteins is based on the concentration of autoinducers, and thus, gene expression and physiological process such as biofilm formation, motility, and toxin production are coordinated by the QS systems (Dow, 2021).

QS system was first discovered in a marine bacterium, *Photobacterium fischeri* (Eberhard et al., 1981), and since then, this has been studied widely in most of the well-known bacteria such as *Vibriyo harveyii, Pseudomonas aeruginosa,* and members of *Roseobacter clade* (Dow, 2021). To date, three major QS systems have been identified based on the type of signaling molecules produced. Gram-negative bacteria that produce AHLs have *LuxR/I* system, Gram-positive bacteria that produce auto-inducing peptides (AIPs) have *LuxS*/AI-2 system (McBrayer et al., 2020) (Fig. 10.1), and AI-3/epinephrine/nor epinephrine is a third QS system for interkingdom communication (Zhou et al., 2016).

The bacterial communication can be mediated (1) within the same species called as "intraspecies communication," (2) between different species known as "interspecies communication," and (3) between different kingdoms known as "interkingdom communication." Currently, the most well-known AIs are AHLs produced by most of the Gram-negative bacteria (e.g., *P. aeruginosa*), AIPs produced by

Algae Materials: Applications Benefitting Health. https://doi.org/10.1016/B978-0-443-18816-9.00013-7

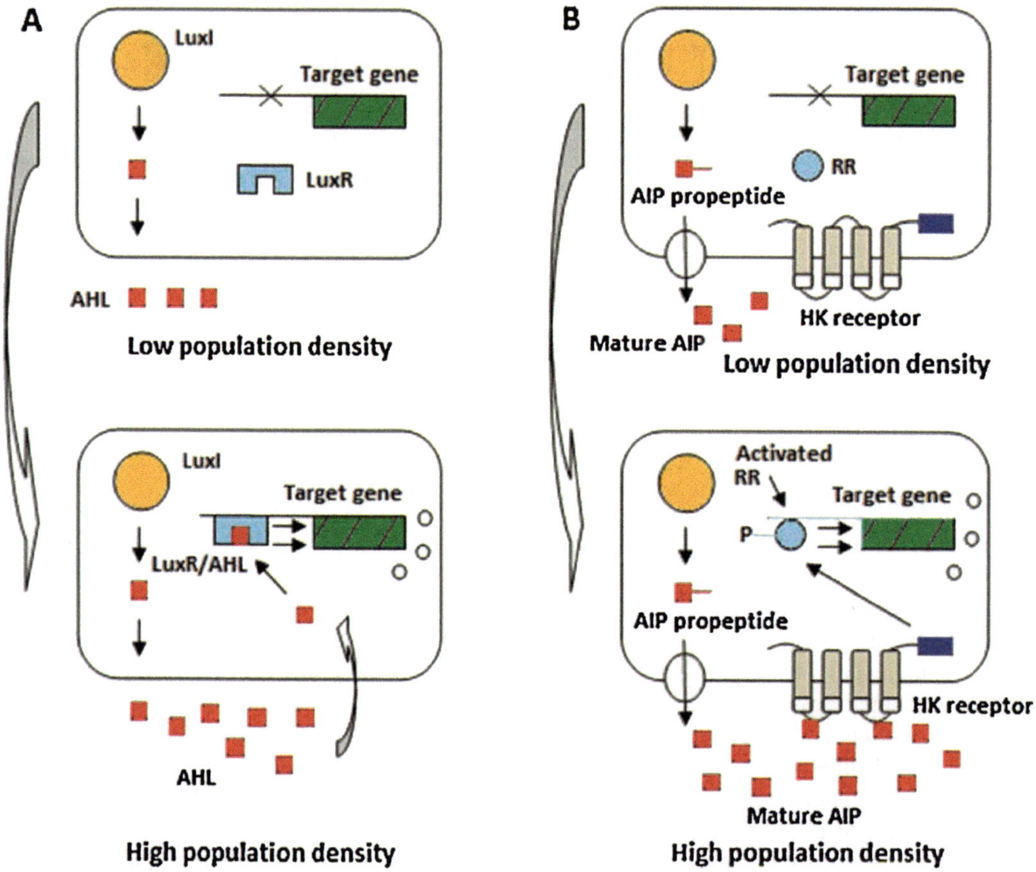

FIGURE 10.1

Quorum sensing systems in bacteria. (A) AHL-mediated QS system in Gram-negative bacteria. (B) Autoinducer peptide (AIP) QS in Gram-positive bacteria.

Reproduced from Siddiqui, M.F., Rzechowicz, M., Winters, H., Zularisam A.W., Fane, A.G. (2015) Quorum sensing based membrane biofouling control for water treatment: A review, Journal of Water Process Engineering 7, 112-122. https://doi.org/10.1016/j.jwpe.2015.06.003 with permission.

the Gram-positive bacteria (e.g., *Staphylococcus aureus*) and 4, 5-dihydroxy-2, 3-pentanedione (DPD) produced by the Gram-negative and Gram-positive bacteria for interspecies commination. QS molecules with various low molecular weights (e.g., AHL, AI-2, DPD, 3-hydoxypalmitic acid methyl ester (3-hydroxy PAME), cis-11-methyl-2-dodecenoic acid (DSF), 2-isocapryloyl-3R-hydroxymethyl-y-butyrolactone (A-factor), diketopiperazines (DKP), 2-heptyl 3-hydroxy-4-quinolone (PQS), and 4-hydroxy-2-heptylquinoline (HHQ) to high molecular weights such as oligopepetides have been identified (Haque et al., 2019).

1.1 Quorum sensing signals and algal–bacterial interactions

The term "phycosphere" in microbial ecology was coined by Bell and Mitchell (1972) to describe the association of algae and bacteria where bacterial growth is stimulated by algal exudates. Phycosphere is a microcosm in which QS plays an important role between microalgae and bacteria (Dow, 2021). The microalgae are microscopic organisms, commonly found in the marine and freshwater systems. The cyanobacteria (commonly known as blue-green algae) are prokaryotes, while all other microalgae (e.g., green algae, diatoms, dinoflagellates) are eukaryotes. All these photosynthetic organisms play important roles in various environments, e.g., from biomass production to role in food webs and biogeochemical cycles (Dow, 2021). The microorganisms in the aquatic and terrestrial environments interact with other microbes and higher organisms such as animals and plants. The outcome of the interaction could be beneficial, detrimental, or neutral. The interaction between the algae and bacteria is known to be mediated by the chemical signal molecules produced by the bacteria, which can be both pervasive and diverse, and the molecules produced endo- or exogenously play a critical role in the interactions. The interaction between the bacteria and algae is primarily mediated by surface colonization followed by production of continuous thin layer of exopolysaccharide comprised of proteins, carbohydrates, lipids, and humic substances. This continuous thin layer formed on algae by bacteria consists of planktonic bacteria from diverse taxa, and the bacteria survive within this environment for nutrients. In this layer, various activities take place particularly the QS. The QS is commonly observed in bacteria, and it has also been found in algae-associated bacteria during the interaction between them. Positive interactions include nutritional contribution, biological matter, and ecological niche for symbiosis, and negative interactions include algicidal activity (Zhou et al., 2016). The most common QS signals regularly isolated from marine bacteria are AHL and 2-alkyl-4-quinolones (AQs) particularly from the members of *Roseobacter clade, Pseudomonas, Alteromonas, Vibrio, and Pseudoalteromonas* (Zhou et al., 2016).

1.2 Positive effects of quorum sensing molecules on algal–bacterial interactions

The algae are associated with specific communities of bacteria, fungi, protozoa, and other eukaryotic organisms. In these communities, biofilm formation is most common, which is regulated by symbiotic QS system (Parsek & Greenberg, 2005). Each algal species hosts and supports specific bacterial communities (Trias et al., 2012). AHLs and other metabolites produced by the algae and bacterium determine the symbiotic association between them. Biofilm promotes the construction of algal–bacterial interaction, and to achieve these, bacteria must locate the suitable substratum for colonization (algae), reversible adhesion followed by irreversible adhesion process leading to formation of biofilm. Overtime, the biofilm matures and coordinates the synergistic behavior of algae and bacteria (Chen, 2013). Some algae reduce the bacterial overpopulation by interfering with bacterial QS system. For example, Australian red alga *Delisea pulchra*, which produces QS molecules known as halogenated furanones, interferes with AHLs through competitive inhibition and selectively inhibits bacterial colonization on algal surfaces (McLean et al., 2004). Another bacterial activity such as motility is also known to be controlled by QS system. For instance, members of *Roseobacter* clade are predominantly responsible for degradation of dimethylsulfoniopropionate (DMSP). Marine bacterium *Silicibacter* sp.

strain TM1040 isolated from *Pfiesteria piscicda* dinoflagellates degrades DMSP and TM1040 bacterium being highly motile is attracted to compounds produced by dinoflagellates through chemotactic response (Miller et al., 2004).

1.3 Negative effects of quorum sensing molecules on algal–bacterial interactions

Competition between algae and associated bacteria exists for the nutrition and space. Bacteria depend and survive on algal exudates for nutrition, and thus to obtain food from algae, bacteria secrete various toxins and compounds that are harmful to algae. Algicidal bacteria secrete chemical substances to kill algae, which are regulated by QS. An interaction of the algicidal bacterium *Kordia algicida* on various diatoms (*Skeletonema costatum, Thalassiosira weissflogii, Phaeodactylum tricornutum, and Chaetoceros didymus*) was studied and found that protease-mediated interactions with bacteria and diatom were dependent on the *K. algicida* cell density, which results in algal cell lysis (Paul & Pohnert, 2011). Marine algicidal bacterium *Ponticoccus* sp. PD-2 isolated from marine alga *Prorocentrum donghaiense* was verified for QS activity, and the study showed that two AHL compounds were produced by PD-2 bacterium, which has algicidal effect on the microalga by inhibiting the algal growth (Xu et al., 2012). This algicidal bacterial activity produced by *Sulfitobacter porphyrae* ZFX1 was used to control harmful algal blooms (HABs) caused by *P. donghaiense* (Zhang F., et al., 2020b). A recent study showed that algicidal bacterium *Microbulbifer* sp. YX04 induces oxidative damage and autophagic cell death in HABs caused by *Phaeocystis globose*, which is regulated by QS system (Zhu et al., 2022). All these studies suggest that the algicidal activity can be an useful bioremediation tool for controlling HABs.

1.4 Applications of algal-bacterial interaction in biotechnology

QS signal molecules and QS systems have been explored widely in bacteria, algae, fungi, and other eukaryotes, and these signal molecules have been applied in different fields of study for biotechnological applications, for example, medical uses of AHL compounds and its analogs, agricultural uses, and scaling up of microbial products using QS compounds (Zhou et al., 2016). This chapter focuses on the biological and environmental applications.

2. Biological applications

With the development of modern biotechnology and the utilization of various energy sources by human beings, algae have emerged as a group of valuable targets for biological applications. In addition to the traditional application of algae in human diet, medical treatment, and the production of livestock feeds, they are potentially recognized as a source of sustainable biofuel for their ability of converting solar energy to bioenergy (Koyande et al., 2019; Zhang B., et al., 2020a). In this regard, the enhancement of biomass accumulation and lipid production of algae have intrigued researchers in the recent studies on algal biotechnology (Kumsiri et al., 2021). Moreover, the applications of algae in the production of certain products using biological methods, such as hydrogen and electricity generation, are also reported (Jiang et al., 2021). Additionally, algae are introduced in the manufacturing of bioproducts including fertilizers and food additives. Nevertheless, several drawbacks concerning the cultivation and utilization of algae substantially limit these promising applications. The presence of algae is normally found in the suspension, resulting in great challenges for biomass harvesting and further applications in large-scale cultivation. The scale-up of algal cultivation is also considered difficult due to the limitations of biomass productivity and extraction of biofuels (Zhang B., et al., 2020a). Therefore, in recent studies, the idea that QS related to bacterial–algal interaction to be exploited in the algal production and related applications are broadly discussed and investigated. This idea is not confined within only the employment of certain bacterial species, but also the presence of QS molecules or potentially activated sludges into the cultivation of algae. In this section, the introduction of QS in the biological applications of algae will be highlighted.

2.1 Algae as biofuel feedstock

Biofuels refer to the fuels that are acquired from biological sources with proper processing to be applied in engineering or industrial situations (Anto et al., 2020). These sources include plants, animals, and microorganisms that either possess biological materials to be directly used for energy supplying or have the ability to produce resulting fuels such as ethanol or biodiesel. Although currently biofuels only contribute to a minor part of the transport fuel consumption across the world, they are promising in future energy structures. The reason is that, compared with traditional fuels, biofuels are more friendly to environment and safer to operate with (Leong et al., 2018). Besides, scientists regard biofuels as sustainable feedstock opposed to the unrenewable fossil fuels that are widely used nowadays.

Based on the development of biofuels, four generations of biofuels are categorized depending on the biomass sources used (Fig. 10.2). The first generation refers to the biofuels produced from oil-bearing food crops. Representative crops in this generation include maize, wheat, and other cereals containing starch and oil. However, concerns have been raised about the first generation of biofuels due to the competition with food production and the potential impact on the biodiversity, resulting in food-versus-fuel debates that facilitate the appearance of the second generation of biofuel. In the second generation, biofuels are manufactured from non-edible feedstocks such as energy crops and forestry and agricultural residues. Yet massively growing such feedstocks could be quite challenging with the requirements of high-standard management practices to ensure consistent oil yields. Besides, low energy conversion rates and lack of stable sources contribute to the limitation and unsustainability of the second-generation biofuel feedstocks as well (Leong et al., 2018).

With the first and second generation to be the only commercial sources of biofuels so far, the applications of biofuels are fairly limited. Consequently, researchers turned to search for other sustainable feedstocks and explored potential microorganisms including algae, fungi, and bacteria. The conversion from the lipid content in such microorganisms to biofuels such as biodiesel could be easily achieved with several steps of processing. These microbes used as biofuels are recognized as the third generation, among which microalgae exhibit most advantages because of rich lipid content and rapid growth and biomass accumulation (Leong et al., 2018). The emerge of the third generation of biofuels makes it possible for mass production of biofuels with tremendously reducing the land occupation and avoiding the food-versus-fuel debates. The production of oil through microalgae is reported to be 20–300 times of that from traditional biomass crops (Shokravi et al., 2021). Additionally, the residual microalgae biomass could be further processed into other valuable bioproducts, resulting in more efficient consumption of the feedstock. Their potential ability of wastewater treatment along with the production of biofuel is also among the advantages of algae compared with other sources (Nguyen et al., 2020).

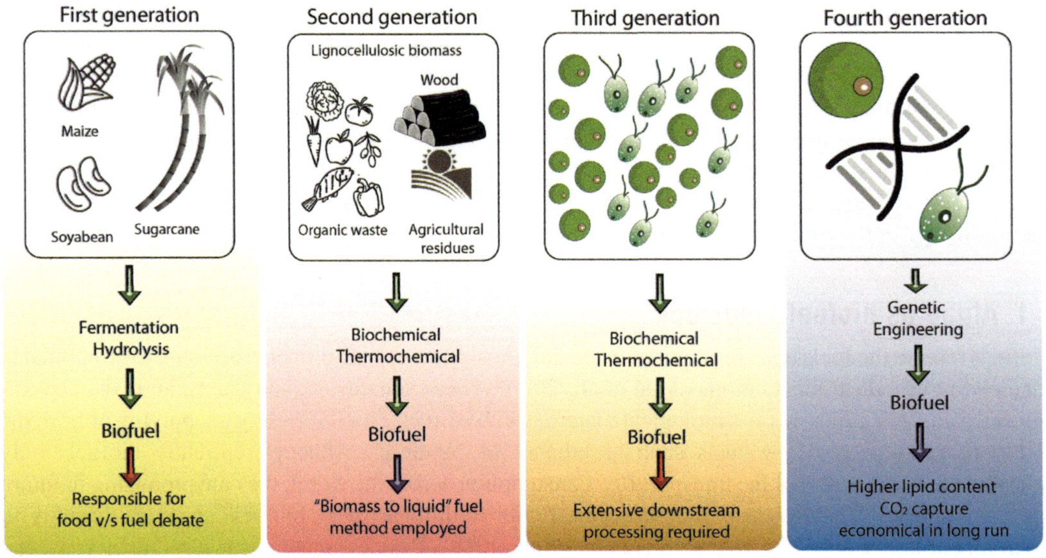

FIGURE 10.2

Generations of biofuels, biomass resources used, and the process involved in the production of biofuels.

Reproduced from Anto, S., Mukherjee, S. S., Muthappa, R., Mathimani, T., Deviram, G., Kumar, S.S., Verma, T. N., & Pugazhendhi, A. (2020). Algae as green energy reserve: Technological outlook on biofuel production, Chemosphere, 242, *125079. https://doi.org/10.1016/j.chemosphere.2019.125079 with permission.*

Although algae are considered as promising feedstock for biofuel generation and successful bench-scale attempts have been established for their production, efficient output of biomass and lipid extraction as well as the scale-up of such systems challenge the development of the third-generation biofuels (Menegazzo & Fonseca, 2019; Shokravi et al., 2021). Naturally, the growth of biomass and the increase of lipid content are two sides of the one scale of algae cultivation, namely, microalgal species with higher lipid content, such as *Botryococcus braunii*, usually demonstrate low growth rate; on the other hand, for species with low lipid content, such as *Scenedesmus* sp. and *Chlorella* sp. (in the 20%–40% range), the growth of biomass could be at a more satisfying level (Aziz et al., 2020). Practically, algal species with lower but acceptable lipid content, for instance, *Chlorella* sp., are chosen for application and research. To facilitate the production of biofuels and related bioproducts, enhancement of either biomass accumulation or lipid content level, or even both simultaneously, has become one of the most discussed topics. Thus, the selection of suitable species and culturing methods is crucial in the optimization of algae for biofuel production.

Following this idea, earlier researchers proposed to exploit the interactions between microalgae and other microorganisms, especially bacteria, for promotion of algal growth and lipid synthesis, consequently benefiting large-scale production of related biofuels and bioproducts (Das et al., 2019). In such interactions, the presence of bacteria could demonstrate both detrimental and beneficial effects to algae, depending on the species employed and environmental conditions. The detrimental influences are usually connected to nutrient competition between bacteria and algae, where bacteria are regarded as contamination, leading to loss of bioresources, low metabolites productivity, and the low quality of biomass (Rashid et al., 2019). Meanwhile, microalgae are capable of responding to bacterial QS molecules (QSMs), negatively or positively under various situations. One group of QSMs, *N*-acyl-homoserine lactones (AHLs), was reported to trigger the self-aggregation of *Chlorophyta* sp. (FACHB-729), hence enhancing their settling ability (Zhou et al., 2017). As microalgae commonly are distributed in the suspension, the stress from AHLs could make the harvest of microalgae from suspension more efficient. Furthermore, it is also believed that QSMs help in the restoration the spore-release capacity of *Acrochaetium* sp., by promoting timely spore liberation (Zhang B., et al., 2020a). Contrarily, in the situation of competition, algae have developed metabolic reactions to counter the threat from bacterial QSMs. Through the release of QS analogs or inhibitors, microalgae use these weapons to either occupy the binding sites or sabotage the connections to interfere the intercellular communication of target bacteria (Manefield & Turner, 2002; Rajamani et al., 2008). Additionally, the interactions between bacteria and algae in their consortia involve the exchange of nutrients and cycling of O_2 and CO_2, as well as providing ecological niche to survive from harsh environment (Zhang B., et al., 2020a) Consequently, the application of QS in algal biotechnology concerning biological production of biofuels is promising, and questions have been raised about how QSMs could be introduced into algal cultivation.

2.2 Algal–bacterial consortia

A popular strategy for biological production of biofuels is the co-culture of bacteria and algae. Nevertheless, as discussed before, the impact of bacteria on algae in the algal–bacterial consortia is not merely beneficial, as the interactions between certain species of bacteria and algae could lead to reduction of biomass accumulation. Therefore, the bacterial species employed in the consortia are usually growth-promoting bacteria, which increase the growth rate of microalgae by 10%–70% (Zhang B., et al., 2020a). One example of growth promotion bacteria is *Rhizobium* sp., which is reported to enhance the growth of *Chlorella sorokiniana* by approximately 13% under desired conditions (Zhou et al., 2021). Metabolites released from these bacteria such as indole-3-acetic acid (IAA) and vitamin B_{12} are considered to crucial in the growth promotion of algae (Fuentes et al., 2016). In addition to the enhancement of biomass growth, bacteria demonstrated the ability of facilitating microalgal bioflocculation. The designated bacterial strain HW001 had the capability of initiating algal aggregation when cocultivated with several algal species (Wang et al., 2016). Further research suggested *Bacillus* sp. strain RP1137 to be more powerful stimulation to the algal aggregation, indicating that certain bacterial strains possess the potential to be exploited for algal biomass harvest (Wang et al., 2016). A recent study on the co-culture of microalgae and actinomycete isolated from biogas effluent with biogas digestate effluent found simultaneous increment of total biomass production and lipid yield, in which the biomass production and lipid productivity of microalga *Tetradesmus obliquus* AARL G022 were increased by 1.30 and 1.55 folds, respectively (Kumsiri et al., 2021). The roles of QSMs in algal–bacterial consortia are not explicated, yet there is a possibility that the participation of these molecules to be one of the mechanisms for algal aggregation, as reported in the related literature (Zhou et al., 2017).

Besides merely introducing growth promoting bacteria into the co-culture, the idea of using activated sludge acquired from wastewater systems was also proved to be feasible. This idea was based on the fact that algae could be applied in wastewater treatment to achieve simultaneous removal of nutrient and increase biomass and lipid content in algal cells. More information about the applications of algae in wastewater treatment and other aspects of environmental issues will also be included in this chapter. Regarding the enhancement of biomass and lipid content, wastewater could act as sources of nutrients such as nitrogen and phosphate that also contain bacteria and related metabolites and molecules (Jiang et al., 2021).

Recently, one of representative studies concerning this strategy was conducted to explore the ability of *Scenedesmus* sp. 336, *Chlorella* sp. 1602, and activated sludge in combination for nutrient removal and lipid accumulation in artificial municipal wastewater systems (Chen et al., 2019). The experiments were set up in several different combinations: aside from the monoculture as control for each species and activated sludge, co-culture systems, namely, *Scenedesmus* sp. 336 with activated sludge, *Chlorella* sp. 1602 with activated sludge, and *Scenedesmus* sp. 336 and *Chlorella* sp. 1602 with activate sludge, were incubated individually for 7 days in controlled conditions of light and dark. Results suggested that ideal removal of nutrients including nitrogen and phosphate was observed in all combinations except the activated sludge only inoculation. However, analysis on the dry weight and lipid production indicates that symbiotic systems demonstrated superior abilities of biomass accumulation and lipid content enrichment. It was also confirmed the participation of IAA via HPLC in this study, demonstrating the beneficial interaction between algae and bacteria. More practically, another research on the co-culture of microalgae-activated sludge for wastewater treatment and biomass production concluded that with an inoculation ratio of 3:1, that is, microalgae (*Chlorella* sp.) to activated sludge from local wastewater treatment plant, nutrient removal efficiency, and biomass production during the operation of photobioreactor could be optimized (Nguyen et al., 2020).

2.3 Quorum sensing molecules and algal co-culture

As discussed, algae can interact with bacteria through the regulation or interference of QSMs (Fig. 10.3). Many signal molecules as QS analogs also mediate the sexuality and reproduction behavior and thus have the capability to increase the algal biomass production (Hombeck & Boland, 1998). Accordingly, instead of inoculations of viable bacteria or activated sludge with complex composition to co-culture systems, direct employment of QSMs in algal cultivation system is considered another strategy to enhance biomass accumulation and lipid content. In this manner, the culturing system is simplified, and it is more practical to control the quality and quantity of added materials. Additionally, investigations on the impact of QSMs on algal growth and lipid production could provide insights into the interactions between bacteria and algae in the microbial consortia.

Recent research suggested that QSMs extracted from activated sludges are capable of triggering lipid synthesis in algae (Zhang et al., 2018). In this study, the cultivation of a strain of *Chlorophyta* sp. (FACHB-729), as model microalgae, was dosed with QSMs extracted from different dry weight of activated sludges sampled from aeration tank. Biomass and lipid content were evaluated during 8 days of culturing. The results indicated that lipid productions were enhanced by around 86%, while the biomass of algae slightly decreased. Therefore, it is concluded that the introduction of QSMs potentially increased the lipid synthesis yet inhibited algal growth and thus produced less biomass. Intriguingly, despite the usage of QSMs extracted from different dry weight of activated sludges, which potentially represented different final concentrations of QSMs in algal cultivations, biomass and lipid production demonstrated no significant variations, implying that the change of QSM concentration in the system may not influence lipid synthesis once it reaches a desired level. Furthermore, the

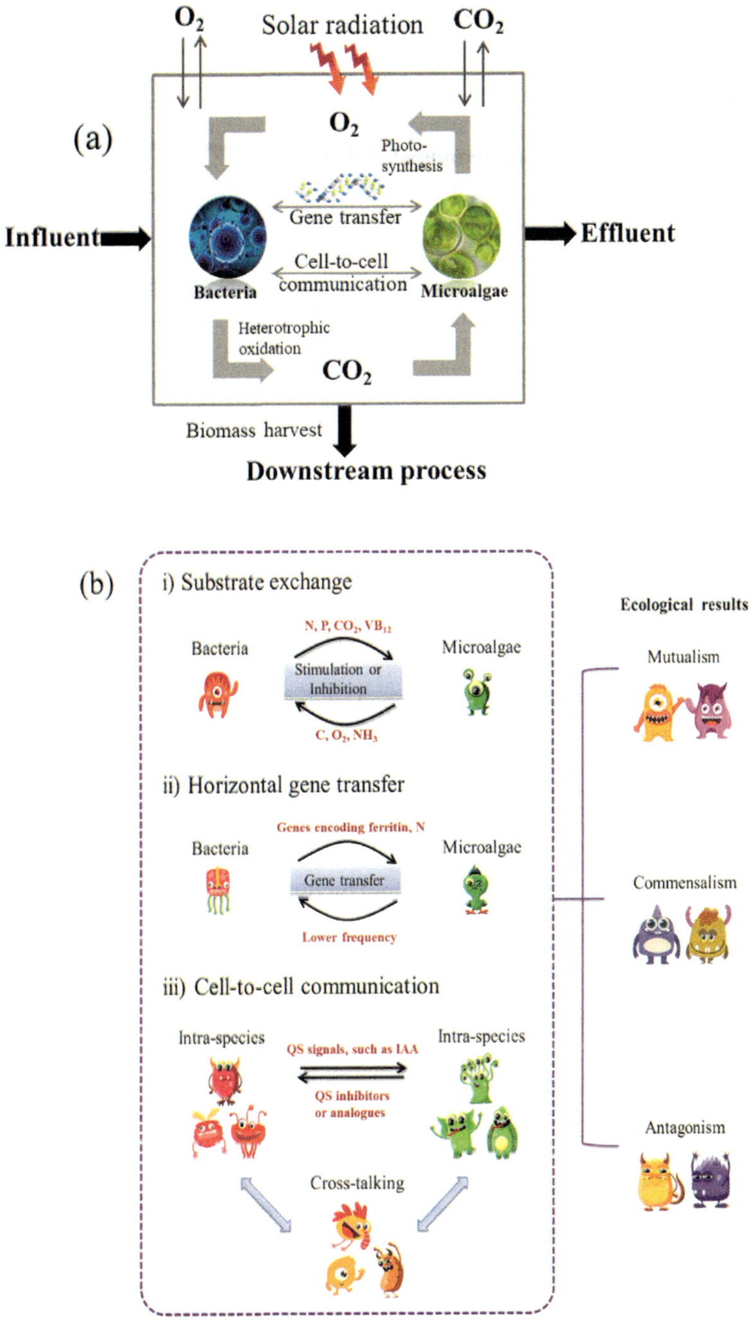

FIGURE 10.3

(A) Schematic diagram of the interaction modes between microalgae and bacteria; (B) The nutrients exchange, horizontal gene transfers, and the cell-to-cell communications between bacteria and microalgae and the interaction results.

Reproduced from Zhang, B., Li, W., Guo, Y., Zhang, Z., Shi, W., Cui, F., Lens, P. N. L., & Tay, J. H. (2020). Microalgal-bacterial consortia: From interspecies interactions to biotechnological applications. Renewable and Sustainable Energy Reviews, 118, 109563. *https://doi.org/10.1016/j.rser.2019.109563 with permission.*

peak of lipid production and productivity occurred at the fourth day from inoculation and fell back to the similar levels of that in control group at the end of the experiment.

However, further study of similar topic indicated that different QSM concentrations in the systems in fact led to different biomass production and lipid content levels (Das et al., 2019). The QSMs extracted from anaerobic bacterial sludge were dosed into the algal culture at different concentrations (0.5 g/L as QS+, 1.0 g/L as QS++, 1.5 g/L as QS+++). Results with 5 days of operation suggested that in terms of biomass accumulation and lipid productivity, algal culture inoculated with the dosage of 1 g/L of QSMs (QS++) exhibited the most optimal output of 2.25 times and 1.28 times increase, respectively. Considering the differences in experimental design and culture methods, the discrepancies between the studies may result from various reasons including aerobic versus anaerobic activated sludge, heterotrophic versus mixotrophic cultivation, *Chlorophyta* sp. (FACHB-729) versus *Chlorella sorokiniana* as model algal species. Therefore, future investigations on the interactions between QSMs and algae, especially the strategies to maintain the increase of lipid content inside such systems in long-term operation, are still in need for the development of biofuel production.

In addition to the monoculture of model algal species, the strategy of coculturing is also explored in the studies on the interactions of QSMs and algae. As discussed previously, AHLs are considered to be a group of common QSMs mediating the interactions between bacteria and algae. The stress provided by the presence of AHLs in the cultivation could trigger the aggregation of algae and make the harvest of biomass less costly (Zhou et al., 2017). It was hypothesized that co-culture can be employed to enhance the abilities of stress toleration from QSMs such as AHLs of algal species (Zhang et al., 2021). In co-culture systems, microbes are reported to communicate and mutually support to deal with environmental stress or biological threats through multiple pathways. Accordingly, the adverse impact of bacterial QSMs on algal cells may be reduced by coculturing several species of algae. For the co-culture of *Chlorella regularis* and *Scenedesmus obliquus* used in this study, the addition of C_6-HSL hardly influenced the algal growth, while the cell growth of each species was reduced by 50% and 35% respectively in the monoculture. Meanwhile, the lipid production of the co-culture system increased by 20%–79%, indicating the feasibility of applying co-culture strategy to synthesis biofuel.

Notably, despite the enhancement of lipid production compared with monocultures, there were no significant differences between the co-culture systems with or without QSMs. It could be implicated that the algal co-culture systems are capable of producing more biomass and lipid than monoculture without involvement QSMs in the process. In fact, the autotrophic co-culture of two freshwater microalgal species, *Ettlia* sp. and *Chlorella* sp. HS-2, was recently reported to improve the biomass accumulation with comparable content composition (carbohydrates, lipids, proteins, and pigments) as the monoculture (Rashid et al., 2019). Although the change of lipid production was not specified in this study, it could be inferred that the lipid production was enhanced along with significant increase in the biomass. Earlier research investigated the co-culture of two species of microalgae without QSMs and also found that the total lipid productivity was improved in the co-culture (Zhao et al., 2014). Therefore, it remains a question whether QSMs are the promotion factor of lipid production in algal co-culture systems. The interactions between QSMs and algae could be altered by the species involved and changes of environmental conditions, leading to discrepancies in microbial behaviors in monoculture and co-culture situations; thus, more investigations into the applications of QSMs in such systems should be proposed in the future for evaluations of related microorganisms and molecules.

Apart from QSMs, other molecules that may interact with algae are also potential promoters for biomass accumulation and lipid production. For instance, one of the molecules introduced in algal cultivation refers to phytohormone, which is regarded as signal molecule that regulates intercellular communications in vascular plants. The application of phytohormones is a promising strategy to improve the performance of algae in biofuel production as well yet related interactions are potentially unsimilar to those of QSMs in that algae themselves are capable of synthesizing a wide range of phytohormones (Stirk & van Staden, 2020). Moreover, with the rapid development of molecular biology and bioinformatics, the idea of gene manipulation in algae has provided insight into the evolution of biofuel production concepts, leading to the emergence of fourth-generation biofuel. Genetically engineered algae demonstrated the abilities of adaption to harsh environment, reduction in nutrient requirement, enhancement of lipid production, and so on and thus are intriguing as exploitation of modern biotechnology (Godbole et al., 2021). Conclusively, algae act as important and promising feedstock of biofuel synthesis, where algal–bacterial interactions, especially the participation of QSMs, could play a vital role in the enhancement of biomass accumulation and lipid production.

2.4 Algae and related bioproducts

In addition to the production of biofuels, algae are involved in the manufacturing of a wide array of other biological or chemical products, including fatty acids, antibiotics, enzymes, biopolymers, algal pigments, and hydrogen (Ahmad et al., 2022; Fakhimi & Tavakoli, 2019). The application of these bioproducts covers various fields such as human nutritional supplements, biomedical treatments, pollution control, and bioenergy for industrial or civil consumptions. Generally, the production of the biochemical compounds starts from the metabolism of algae under favorable conditions; thus, applying algal–bacterial consortium systems usually interferes with the bioproduction due to the introduction of bacterial competition and stress (Aditya et al., 2022). In fact, there are relatively few examples of bioproducts derived from microalgal–bacterial consortia. One of them refers to an earlier study that achieved a 20% of improvement in polyhydroxyalkanoate (PHA) synthesis through the co-culture of bacteria and microalgae (Fradinho et al., 2013). Nevertheless, algae are regarded as promising suppliers of hydrogen production, and the co-culture of algae and bacteria was reported to promote hydrogen yield (Ahmad et al., 2022; Giri et al., 2022). Furthermore, algae also contribute in other forms of energy supplying by improving the performance of microbial fuel cells (MFCs) (Aiyer, 2021; Chia et al., 2022). The following sections will discuss over the biological applications of algae on these energy-related products.

2.5 Algal hydrogen production

Hydrogen is considered as a clean and efficient energy source. The production of hydrogen through biological processes is more friendly to environment and energy conservative than traditional chemical methods and thus attracts scientists as a promising future fuel option. In some literature, biohydrogen released from algae is considered as a form of biofuel comparable with biodiesels (Ahmad et al., 2022). However, the generation of these two products involves in different pathways, where the conversion to biodiesel utilizes the lipid content from algal cells yet hydrogen is usually synthesized by algae through photolysis. The production of photobiological hydrogen is categorized into light-dependent and light-independent types, depending on whether photosystem II is involved. For algae, biohydrogen is generated in the presence of light and water under anaerobic conditions. In this process, the saturation of oxygen in the environment acts as an inhibition factor, suppressing both the metabolism and related gene expression (Mus et al., 2007). When in co-culture systems, the competition and interspecies stress from bacteria and their QSMs could also add to the restrictive conditions, resulting in biomass reduction or loss of cell activity.

A recent study investigated the cocultivation of *Chlamydomonas reinhardtii* with four different species of bacteria, to explore the effects of algal–bacterial consortia on hydrogen bioproduction (Fakhimi & Tavakoli, 2019). *Escherichia coli* K-12 MG1655, *Pseudomonas stutzeri* A1501, *Pseudomonas putida* 12,264, and an unknown bacterial consortium were employed in this experiment and co-cultured with the algae, respectively. In terms of algal growth, the unknown consortium demonstrated no significant impact on the growth rate, while in the cocultivation with bacterial species, the growth of algae was observed to be inhibited at high inoculation concentrations. Nevertheless, hydrogen production was enhanced largely in all co-cultures except that with *E. coli*, which only slightly promoted the hydrogen generation at very low initial concentration. When the concentrations of oxygen and carbon source acetic acid were compared with the footprint of hydrogen production, it could be implied that increase of oxygen in the environment or deficiency of carbon source led to the reduction of hydrogen release. The most significant augment of hydrogen production occurred in the co-culture of algae with the unknown bacteria consortium, in which algae also bore least growth reduction. However, the composition of the bacterial consortium was not specified in the study. Considering the ability of algal co-culture to resist QSM stress discussed in previous sections, hypothesis could be made that the bacterial consortium may contain algal or bacterial species with similar potential, which assisted in the growth of algae and slightly raised the oxygen level in headspace.

Algal biohydrogen is released to the headspace following production, thus the harvesting of algae in systems aiming specifically for wastewater treatment and biohydrogen production. For sustainable production of hydrogen, facilitating granular formation in the reactors was reviewed to be a promising strategy. The involvement of QS in the granulation of algal–bacterial consortia is believed to be critical (Giri et al., 2022). It has been discussed before the presence of bacteria or QSMs stimulating algal aggregation, providing possibilities of their participation in granulation. Algal–bacterial consortia in the forms of granules may lead to altered environmental conditions for algae concerning biohydrogen production. One of the most impactful changes could be the reduction of oxygen levels for imbedded algae, thus improving the hydrogen generation. Similar concepts have been testified by encapsulation of *Chlorella* in polydopamine, laccase, and tannic acid layers, where the hydrogen production was significantly enhanced (Su et al., 2019). Accordingly, the application of algal–bacterial granulation strategy has the potential to play a key role in biohydrogen production.

2.6 Algae and microbial fuel cells

Algal biotechnology also involves other aspects of bioenergy, particularly the generation of bioelectricity. Specifically, bioelectricity herein refers to the electricity produced by MFCs. MFC represents a bioelectrochemical device that can generate electric power along with wastewater treatment. Certain species of electroactive bacteria are exploited in the system for their ability of extracellular transportation of electrons in the presence of a solid electrode (Roy et al., 2017). Microbes cultured in the anodic chamber release electrons via the digestion of wastewater substrates, which are transferred to the cathodic chamber and generate electricity (Sivasankar et al., 2019). At the same time, carbon dioxide is generated in the anodic chamber, and proton moves to cathodic chamber to react with oxygen. For bacteria in anodic chamber, QS is regarded as an important regulatory mechanism due to its ability of increasing microbial diversity, redox activity, and biofilm formation (Das et al., 2022). The participation of algae in this process follows two strategies: acting as mediator or cosubstrate in anodic chamber, or suppling oxygen in the cathodic chamber to reduce or replace air bubbling (Zhang B., et al., 2020a). Oxygen production in the cathodic chamber by algae through photosynthesis demonstrated advantageous effect on electric power generation (Aiyer, 2021). Further research suggested that algal biomass harvested from cathodic chamber could potentially benefit long-term operation of the system as slow release substrate (Gajda et al., 2015).

A step forward has been taken in a recent study toward the application of algal biomass in anodic chamber of MFC (Das et al., 2019). This experiment was coupled with the investigation on the interactions between algae and QSMs extracted from treated anaerobic sludges collected from the anodic chamber of an MFC system. The addition of QSMs into algal culture led to increase in algal productivity by 2.25 times, lipid content by 1.28, and harvesting efficiency by 1.21 times, comparable with the role of QSMs on biomass accumulation and lipid production discussed previously in this chapter (Das et al., 2019). Algal biomass was harvested from algae and QS-induced algae after lipid extraction and then inoculated in the anodic chamber with synthetic wastewater. Results confirmed that the presence of algal biomass from QS-induced algae after lipid production as mediator and substrate in the anodic chamber of MFCs enhanced the electrical power generation as well as the treatment of wastewater. The ideas of reusing biomass harvested from anodic chamber in this study and recycling algal biomass from cathodic chamber previously mentioned share similar concept that it is feasible to apply algal biomass to enhance the performance of MFC and achieve fully or partial self-sustainable system, of which the electricity generation could be largely improved through the participation of QSMs.

In the situation where algae are contained in the cathodic chamber, the system of MFC is specifically termed photosynthetic algae microbial fuel cells (PAMFCs) (Kannan & Donnellan, 2021). Algae in such systems mainly provide photosynthetic oxygenation for electricity recovery, while assisting in wastewater treatment through the consumption of substrates and nutrients (Reddy et al., 2019). Despite that the bioelectricity production could be influenced by a wide range of factors including algal species employed, cathode material, and light density, other environmental parameters, quorum sensing, and related bacterial behaviors are not considered to be crucial in PAMFC (Reddy et al., 2019). Thus, details in this topic are not explicated in this chapter.

3. Environmental applications

3.1 Nutrients removal

The interactions between algae and bacteria are not only exploited in the biological productions and energy generations but also introduced into environmental treatment and ecological management processes. One of the most popular applications of algal–bacterial consortia is to couple the production of biomass and lipid content as well as bioproducts, such as bioelectricity, with wastewater treatment. In fact, algae are considered to be of great value in wastewater treatment strategies with the advantages of low cost, environmental friendly, easy to operate, and generation of commercial products along with their ability to help with the removal of nutrients including nitrogen and phosphorus in industrial or domestic wastewater (Nguyen et al., 2020). Several examples of algal environmental applications are already discussed in previous sections. In the study where algae-activated sludge symbiotic system was investigated for its capability of enhancing lipid accumulation and wastewater treatment, the removal rate of PO_4^{3-}-P and NH^{4+}-N were reported to be 99.82% and 87.13%, respectively (Chen et al., 2019). Meanwhile, algae were widely employed in MFC systems, which aimed to remove nutrients in wastewater along with the generation of bioelectricity. Recent research concerning microalgae—bacteria powered biofuel cell for aquaculture wastewater and energy recovery demonstrated satisfying removal efficiency of nutrients in wastewater and enhanced power output (Sun et al., 2020).

Algal species are capable of directly utilizing nitrogen or phosphorus for their growth and reproduction, and they could participate in the proliferation and nutrient removal processes of denitrifying bacteria or phosphate accumulating organisms (Qixin et al., 2022; Zhang et al., 2021). In this manner, the combination of algae and bacteria in wastewater treatment process is potentially an advanced strategy that leads to enhanced wastewater treatment performances (Zhang et al., 2021). As previously discussed, the interactions between algae and bacteria involve multiple aspects, among which QSMs play important roles in intercellular communications and regulatory interactions between algae and bacteria (Zhang B., et al., 2020a; Zhang et al., 2021). Nevertheless, the processes of nutrient removal from wastewater by microalgal—bacteria consortia largely depend on different strains combined in the systems (Zhang et al., 2021). Algal species communicate and interact with different species of bacteria following different mechanisms that may improve the abilities of these bacteria in nutrient removal (Fig. 10.4). For instance, hydrogen consumer denitrifiers were reported to have enhanced denitrifying performance with the presence of microalgal species *Chlorella vulgaris*, *Ettlia* sp., and *C. reinhardtii,* indicating that hydrogen produced from microalgal metabolism facilitated the removal of nitrogen by the denitrifiers (Rezvani et al., 2020). In phosphorus removal, phosphate accumulating organisms (PAOs) are crucial participants that can uptake phosphate in wastewater and

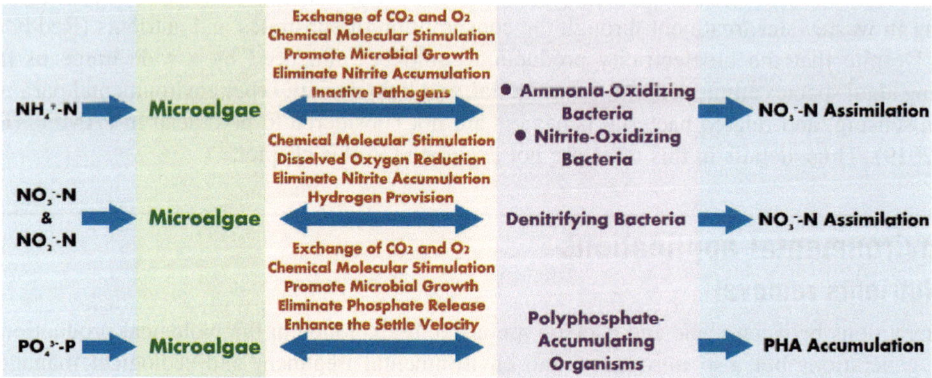

FIGURE 10.4

Potential mechanisms for nutrient removal using microalgae—bacterial consortia.

Reproduced from Zhang, C., Li, S., & Ho, S.H. (2021). Converting nitrogen and phosphorus wastewater into bioenergy using microalgae-bacteria consortia: A critical review. Bioresource Technology 342, 126056 https://doi.org/10.1016/j.biortech.2021.126056 with permission.

accumulate inside the cells. The interactions between PAOs and microalgae were reviewed and proved to be assisting the phosphorus removal processes (Fallahi et al., 2021). Therefore, the application of algal–bacterial consortia to utilize their interactions and quorum sensing regulations is a promising strategy in wastewater treatment.

3.2 Algae and harmful algal blooms

Despite that algae, especially those in algal–bacterial co-culture systems, are widely applied in biological production and environmental management, the unlimited growth of certain algae species such as *Karenia mikimotoi, Karlodinium veneficum, Margalefidinium polykrikoides,* and *Heterocapsa* spp. could severely damage the aquatic systems with the huge depletion of oxygen in the water bodies and massive secretion of dangerous toxins, resulting in a major environmental issue termed as HAB (Gu et al., 2022). Naturally, the HABs occur when such algae are growing in slow-moving water bodies with excessive nutrients and sufficient sunlight. Yet with the eutrophication in aquatic systems caused by increasing human activities, HABs have been recognized as a rising problem threatening the ecological balance in related environments and health of human beings involved (US EPA, 2022). The release of toxins from the overgrown algae could directly harm human bodies and penetrate skin and the blood–brain barrier to cause even lethal effects, due to which over 2000 human poisoning cases were reported annually (Pal et al., 2020). Apart from oxygen consumption and toxins generation to harm organisms in the environment, HABs may also lead to low light penetration into the water bodies and thus impact the growth of submerged aquatic vegetation (Anderson, 2009). Consequently, HAB as an environmental issue worsened by improper human activities deserves our attention and efforts to investigate and mitigate the problem.

To reduce the harmful effects caused by HAB, several methods were proposed in the past decades. Solutions introduced in earlier years involved chemical agents to reverse the impact of HAB as well as control the overgrowth of related algae. Chemicals that were proved to be effective included Triosyn, hydrogen peroxide, and copper sulfate, yet these chemicals could also influence the environment in water bodies (Pal et al., 2020). Recent research on HAB supported more the application of biological treatment, namely, the employment of microorganisms such as bacteria into the aquatic systems affected. These bacteria or microorganisms of other species usually demonstrated algicidal activities to participate in the diminishing of bloomed algae, or the abilities to degrade toxins produced without causing secondary pollution (Pal et al., 2020). Accordingly, the harms resulted from HAB could be under control in a more economical and environmentally friendly way (Yang et al., 2012).

When scientists were exploring further into the mechanisms through which bacteria with algicidal activities inhibit the expanding of HAB, QS emerged to be a crucial participant in the interactions between these bacteria and algae (Fig. 10.5). Importantly, the development of modern molecular biotechnology provides powerful tools for researchers to evaluate the role of QS systems in such interactions. A study concerning the complete genome sequencing of *Sulfitobacter pseudonitzschiae* H46, a marine bacterium known to have algicidal activity against HAB microalgae *Chattonella marina* and *Alexandrium tamarense*, revealed the participation of QS system in the bacterial activities against algae bloom (Hu et al., 2021). The whole-genome sequencing confirmed the existence of two pairs of *LuxI/R* homologous genes in the chromosome: *HtyIA/RA and HtyIB/RB*. Results from thin-layer chromatography (TLC) bioautography indicated C8-HSL and OC8-HSL to be primary QSMs generated. Furthermore, it was reported that through the analysis of downstream genes, QS system of

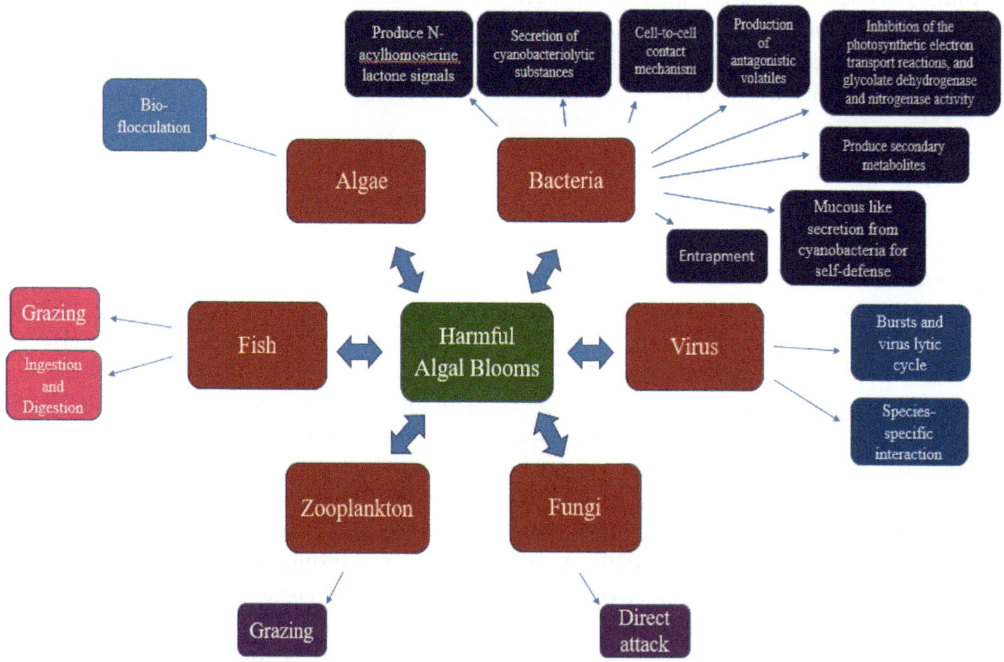

FIGURE 10.5

The interactions of microorganisms with harmful algal blooms for mitigation.

Reproduced from Pal, M., Yesankar, P. J., Dwivedi, A., & Qureshi, A. (2020). Biotic control of harmful algal blooms (HABs): A brief review, Journal of Environmental Management, *268, 110687. https://doi.org/10.1016/j.jenvman.2020.110687 with permission.*

S. pseudonitzschiae H46 may regulate the colonization and membrane formation onto algae cells. Specifically, the ability of this bacterium to use DMSP as its carbon source was implied by the presence of a *dddL* gene; thus, it could be hypothesized that *S. pseudonitzschiae* H46 are capable of attacking microalgal cells, which produce DMSP as intracellular substance for more DMSP uptake. Accordingly, *S. pseudonitzschiae* H46 interacts with target algae via the regulation of QS system and exhibits its algicidal activity driven by its need for DMSP.

A step forward has been taken to further investigate the bacterial species involved in the phycosphere of HAB and their ecological functions. 842 microalgae-associated bacteria isolated from eight typical red tide algae were screened to obtain strains with QS systems, resulting in 113 strains identified (Zhang et al., 2022). Ten different representative strains selected from the 113 identified after removing repetitions were sequenced for 16S rRNA gene. The results suggested that these 10 strains were found in four genera, *Sulfitobacter, Ponticoccus, Leisingera,* and *Nitratireductor*, and most of them (9/10) belonged to *Roseobacter* clade, which was believed to have ecological functions such as biofilm formation, DMSP degradation, and algicidal activities in the interactions with algae. Biofilm forming abilities and DMSP consumption of these 10 strains were tested in the study as well. It was reported that most of the strains were capable of facilitating biofilm formation and utilizing DMSP.

Therefore, the conclusion drawn from this study was consistent with the previously introduced one that the algicidal activities of such bacteria relied on QS systems to regulate the interactions with algae and were supported by their abilities to degrade DMSP.

4. Conclusions

The biological applications of algal biotechnology related to QS and interactions with bacteria are promising in various fields from biofuel production to energy generation. The exploitation of algal–bacterial consortia in suitable conditions demonstrates advantages over monoculture systems in biomass production and accumulation of bioproducts. QSMs and other signal molecules are proved to be effective in promoting algal lipid production as well. The introduction of algae enhances the performance of MFCs, while algal biomass is considered feasible for possibly self-sustainable MFC designs. However, as an emerging research, current studies on biological applications of QS in algal biotechnology extensively focused on the conceptual design and testification of hypothesis and thus lacked explorations for commercialization and scale-up productions, which are critical in their civil and industrial employment and needed to be addressed in the future work.

In general, QS is an important feature of interactions between algae and bacteria, which is widely applied by algal biotechnology to environmental management and hazard control. Wastewater treatment by algal–bacterial consortia coupled with bioproduction lowers the cost of both treatment and production while potentially improving the efficiency of nutrient removal and energy generation. In terms of HAB caused by overgrowing algae, QS systems play vital roles in algicidal activities and has proved the feasibility of exploiting biological methods to mitigate HAB threats. The discussion of various applications of QS in algal biotechnology in this chapter aims to share recent developments and provide insights into the advances in related research.

Acknowledgments

R.S. would like to acknowledge the Key Program Special Fund in Xián Jiaotong-Liverpool University (XJTLU; Grant No. KSF-E-20) and Continuous Support Fund (RDF-SP-88) for financial support. Funding support to X.Z. was provided through the Postgraduate Research Scholarship (PGRS2106047) awarded by XJTLU.

References

Aditya, L., Mahlia, T. M. I., Nguyen, L. N., Vu, H. P., & Nghiem, L. D. (2022). Microalgae-bacteria consortium for wastewater treatment and biomass production. *Science of the Total Environment, 838*, 155871. https://doi.org/10.1016/j.scitotenv.2022.155871

Ahmad, A., Banat, F., Alsafar, H., & Hasan, S. W. (2022). Algae biotechnology for industrial wastewater treatment, bioenergy production, and high-value bioproducts. *Science of the Total Environment, 806*, 150585. https://doi.org/10.1016/j.scitotenv.2021.150585

Aiyer, K. S. (2021). Synergistic effects in a microbial fuel cell between co-cultures and a photosynthetic alga *Chlorella vulgaris* improve performance. *Heliyon, 7*(1), e05935. https://doi.org/10.1016/j.heliyon.2021.e05935

Anderson, D. M. (2009). Approaches to monitoring, control and management of harmful algal blooms (HABs). *Ocean and Coastal Management, 52*(7), 342–347. https://doi.org/10.1016/j.ocecoaman.2009.04.006

Anto, S., Mukherjee, S. S., Muthappa, R., Mathimani, T., Deviram, G., Kumar, S. S., Verma, T. N., & Pugazhendhi, A. (2020). Algae as green energy reserve: Technological outlook on biofuel production. *Chemosphere, 242*, 125079. https://doi.org/10.1016/j.chemosphere.2019.125079

Aziz, M. M. A., Kassim, K. A., Shokravi, Z., Jakarni, F. M., Lieu, H. Y., Zaini, N., Tan, L. S., Islam, S., & Shokravi, H. (2020). Two-stage cultivation strategy for simultaneous increases in growth rate and lipid content of microalgae: A review. *Renewable and Sustainable Energy Reviews, 119*, 109621. https://doi.org/10.1016/j.rser.2019.109621 (April 2019).

Bell, W., & Mitchell, R. (1972). Chemotactic and growth responses of marine bacteria to algal extracellular products. *The Biological Bulletin*, 265−277. https://doi.org/10.2307/1540052. Woods Hole, Mass, Marine Biological Laboratory.

Chen, L. (2013). *Cooperative division of labor between quorum-sensing bacteria within marine self-organized biofilm*. Master Scholar Thesis. Beijing, China: Tsinghua University.

Chen, X., Hu, Z., Qi, Y., Song, C., & Chen, G. (2019). The interactions of algae-activated sludge symbiotic system and its effects on wastewater treatment and lipid accumulation. *Bioresource Technology, 292*(92), 122017. https://doi.org/10.1016/j.biortech.2019.122017

Chia, S. R., Nomanbhay, S. B. H. M., Chew, K. W., Munawaroh, H. S. H., Shamsuddin, A. H., & Show, P. L. (2022). Algae as potential feedstock for various bioenergy production. *Chemosphere, 287*(P1), 131944. https://doi.org/10.1016/j.chemosphere.2021.131944

Das, S., Das, S., & Ghangrekar, M. M. (2019). Quorum-sensing mediated signals: A promising multi-functional modulators for separately enhancing algal yield and power generation in microbial fuel cell. *Bioresource Technology, 294*, 122138. https://doi.org/10.1016/j.biortech.2019.122138

Das, S., Das, S., & Ghangrekar, M. M. (2022). Bacterial signalling mechanism: An innovative microbial intervention with multifaceted applications in microbial electrochemical technologies: A review. *Bioresource Technology, 344*(PB), 126218. https://doi.org/10.1016/j.biortech.2021.126218

Dow, L. (2021). How do quorum-sensing signals mediate algae−bacteria interactions? *Microorganisms, 9*(7). https://doi.org/10.3390/microorganisms9071391

Eberhard, A., Burlingame, A. L., Eberhard, C., Kenyon, G. L., Nealson, K. H., & Oppenheimer, N. J. (1981). Structural identification of autoinducer of Photobacterium fischeri luciferase. *Biochemistry, 20*, 2444−2449. https://doi.org/10.1021/bi00512a013

Fakhimi, N., & Tavakoli, O. (2019). Improving hydrogen production using co-cultivation of bacteria with Chlamydomonas reinhardtii microalga. *Materials Science for Energy Technologies, 2*(1), 1−7. https://doi.org/10.1016/j.mset.2018.09.003

Fallahi, A., Rezvani, F., Asgharnejad, H., Khorshidi, E., Hajinajaf, N., & Higgins, B. (2021). Interactions of microalgae-bacteria consortia for nutrient removal from wastewater: A review. *Chemosphere, 272*, 129878. https://doi.org/10.1016/j.chemosphere.2021.129878

Fradinho, J. C., Domingos, J. M. B., Carvalho, G., Oehmen, A., & Reis, M. A. M. (2013). Polyhydroxyalkanoates production by a mixed photosynthetic consortium of bacteria and algae. *Bioresource Technology, 132*, 146−153. https://doi.org/10.1016/j.biortech.2013.01.050

Fuentes, J. L., Garbayo, I., Cuaresma, M., Montero, Z., González-Del-Valle, M., & Vílchez, C. (2016). Impact of microalgae-bacteria interactions on the production of algal biomass and associated compounds. *Marine Drugs, 14*(5). https://doi.org/10.3390/md14050100

Gajda, I., Greenman, J., Melhuish, C., & Ieropoulos, I. (2015). Self-sustainable electricity production from algae grown in a microbial fuel cell system. *Biomass and Bioenergy, 82*, 87−93. https://doi.org/10.1016/j.biombioe.2015.05.017

Giri, D. D., Dwivedi, H., Khalaf D. Alsukaibi, A., Pal, D. B., Otaibi, A. Al, Areeshi, M. Y., Haque, S., & Gupta, V. K. (2022). Sustainable production of algae-bacteria granular consortia based biological hydrogen: New insights. *Bioresource Technology, 352*, 127036. https://doi.org/10.1016/j.biortech.2022.127036

Godbole, V., Pal, M. K., & Gautam, P. (2021). A critical perspective on the scope of interdisciplinary approaches used in fourth-generation biofuel production. *Algal Research, 58*, 102436. https://doi.org/10.1016/j.algal.2021.102436

Gu, H., Wu, Y., Lü, S., Lu, D., Tang, Y. Z., & Qi, Y. (2022). Emerging harmful algal bloom species over the last four decades in China. *Harmful Algae, 111*, 102059. https://doi.org/10.1016/j.hal.2021.102059 (May 2021).

Haque, S., Yadav, D. K., Bisht, S. C., Yadav, N., Singh, V., Dubey, K. K., Jawed, A., Wahid, M., & Dar, S. A. (2019). Quorum sensing pathways in gram-positive and -negative bacteria: Potential of their interruption in abating drug resistance. *Journal of Chemotherapy, 31*(4), 161–187. https://doi.org/10.1080/1120009X.2019.1599175

Hombeck, M., & Boland, W. (1998). Biosynthesis of the algal pheromone fucoserratene by the freshwater diatom Asterionella formosa (Bacillariophyceae). *Tetrahedron, 54*(37), 11033–11042. https://doi.org/10.1016/S0040-4020(98)00660-7

Hu, T., Wang, S., Shan, Y., Zhang, Y., Zhu, Y., & Zheng, L. (2021). Complete genome of marine microalgae associated algicidal bacterium sulfitobacter pseudonitzschiae H46 with quorum sensing system. *Current Microbiology, 78*(10), 3741–3750. https://doi.org/10.1007/s00284-021-02632-4

Jiang, L., Li, Y., & Pei, H. (2021). Algal–bacterial consortia for bioproduct generation and wastewater treatment. *Renewable and Sustainable Energy Reviews, 149*(17923), 111395. https://doi.org/10.1016/j.rser.2021.111395

Kannan, N., & Donnellan, P. (2021). Algae-assisted microbial fuel cells: A practical overview. *Bioresource Technology Reports, 15*, 100747. https://doi.org/10.1016/j.biteb.2021.100747

Koyande, A. K., Chew, K. W., Rambabu, K., Tao, Y., Chu, D. T., & Show, P. L. (2019). Microalgae: A potential alternative to health supplementation for humans. *Food Science and Human Wellness, 8*(1), 16–24. https://doi.org/10.1016/j.fshw.2019.03.001

Kumsiri, B., Pekkoh, J., Pathom-aree, W., & Lumyong, S. (2021). Enhanced production of microalgal biomass and lipid as an environmentally friendly biodiesel feedstock through actinomycete co-culture in biogas digestate effluent. *Bioresource Technology, 337*, 125446. https://doi.org/10.1016/j.biortech.2021.125446

Leong, W. H., Lim, J. W., Lam, M. K., Uemura, Y., & Ho, Y. C. (2018). Third generation biofuels: A nutritional perspective in enhancing microbial lipid production. *Renewable and Sustainable Energy Reviews, 91*, 950–961. https://doi.org/10.1016/j.rser.2018.04.066

Manefield, M., & Turner, S. L. (2002). Quorum sensing in context: Out of molecular biology and into microbial ecology. *Microbiology, 148*(12), 3762–3764. https://doi.org/10.1099/00221287-148-12-3762

McBrayer, D. N., Cameron, C. D., & Tal-Gan, Y. (2020). Development and utilization of peptide-based quorum sensing modulators in Gram-positive bacteria. *Organic and Biomolecular Chemistry, 18*(37), 7273–7290. https://doi.org/10.1039/d0ob01421d

McLean, R. J. C., Pierson, L. S., 3rd, & Fuqua, C. (2004). A simple screening protocol for the identification of quorum signal antagonists. *Journal of Microbiological Methods, 58*(3), 351–360. https://doi.org/10.1016/j.mimet.2004.04.016

Menegazzo, M. L., & Fonseca, G. G. (2019). Biomass recovery and lipid extraction processes for microalgae biofuels production: A review. *Renewable and Sustainable Energy Reviews, 107*, 87–107. https://doi.org/10.1016/j.rser.2019.01.064 (July 2018).

Mieszkin, S., Callow, M. E., & Callow, J. A. (2013). Interactions between microbial biofilms and marine fouling algae: A mini review. *Biofouling, 29*(9), 1097–1113. https://doi.org/10.1080/08927014.2013.828712

Miller, T. R., Hnilicka, K., Dziedzic, A., Desplats, P., & Belas, R. (2004). Chemotaxis of Silicibacter sp. strain TM1040 toward dinoflagellate products. *Applied and Environmental Microbiology, 70*(8), 4692–4701. https://doi.org/10.1128/AEM.70.8.4692-4701.2004

Mus, F., Dubini, A., Seibert, M., Posewitz, M. C., & Grossman, A. R. (2007). Anaerobic acclimation in Chlamydomonas reinhardtii: Anoxic gene expression, hydrogenase induction, and metabolic pathways. *Journal of Biological Chemistry, 282*(35), 25475–25486. https://doi.org/10.1074/jbc.M701415200

Nguyen, T. D., Nguyen, T., An, Q., Bui, X., Hao, H., Nhat, H., Lin, K. A., Vo, T., Guo, W., Lin, C., & Breider, F. (July 2020). Co-culture of microalgae-activated sludge for wastewater treatment and biomass production : Exploring their role under different inoculation ratios. *Bioresource Technology, 314*, 123754. https://doi.org/10.1016/j.biortech.2020.123754

Pal, M., Yesankar, P. J., Dwivedi, A., & Qureshi, A. (2020). Biotic control of harmful algal blooms (HABs): A brief review. *Journal of Environmental Management, 268*, 110687. https://doi.org/10.1016/j.jenvman.2020.110687

Parsek, M. R., & Greenberg, E. P. (2005). Sociomicrobiology: The connections between quorum sensing and biofilms. *Trends in Microbiology, 13*(1), 27–33. https://doi.org/10.1016/j.tim.2004.11.007

Paul, C., & Pohnert, G. (2011). Interactions of the algicidal bacterium *Kordia algicida* with diatoms: Regulated protease excretion for specific algal lysis. *PLoS One, 6*(6). https://doi.org/10.1371/journal.pone.0021032

Qixin, L., Xuan, F., Zhiya, S., Wenxin, S., Shuo, W., & Ji, L. (2022). Enhanced wastewater treatment performance by understanding the interaction between algae and bacteria based on quorum sensing. *Bioresource Technology, 354*. https://doi.org/10.1016/j.biortech.2022.127161

Rajamani, S., Bauer, W. D., Robinson, J. B., Iii, J. M. F., Pesci, C., Teplitski, M., Gao, M., Sayre, R. T., & Phillips, D. A. (2008). The vitamin riboflavin and its derivative lumichrome activate the LasR bacterial quorum sensing receptor. *Molecular Plant-Microbe Interactions, 21*(9). https://doi.org/10.1094/MPMI-21-9-1184

Rashid, N., Ryu, A. J., Jeong, K. J., Lee, B., & Chang, Y. K. (2019). Co-cultivation of two freshwater microalgae species to improve biomass productivity and biodiesel production. *Energy Conversion and Management, 196*, 640–648. https://doi.org/10.1016/j.enconman.2019.05.106

Reddy, C. Nagendranatha, Nguyen, H. T. H., Noori, M. T., & Min, B. (2019). Potential applications of algae in the cathode of microbial fuel cells for enhanced electricity generation with simultaneous nutrient removal and algae biorefinery: Current status and future perspectives. *Bioresource Technology, 292*, 122010. https://doi.org/10.1016/j.biortech.2019.122010

Rezvani, F., Sarrafzadeh, M. H., & Oh, H. M. (2020). Hydrogen producer microalgae in interaction with hydrogen consumer denitrifiers as a novel strategy for nitrate removal from groundwater and biomass production. *Algal Research, 45*, 101747. https://doi.org/10.1016/j.algal.2019.101747. October 2019).

Roy, S., Marzorati, S., Schievano, A., & Pant, D. (2017). Microbial fuel cells. In M. Abraham (Ed.), *Encyclopedia of sustainable technologies* (pp. 245–259). Oxford: Elsevier.

Siddiqui, M. F, Rzechowicz, M., Winters, H., Zularisam, A. W., & Fane, A. G, (2015). Quorum sensing based membrane biofouling control for water treatment. A review. *Journal of Water Process Engineering, 7*, 112–122. https://doi.org/10.1016/j.jwpe.2015.06.003

Shokravi, H., Shokravi, Z., Heidarrezaei, M., Ong, H. C., Rahimian Koloor, S. S., Petrů, M., Lau, W. J., & Ismail, A. F. (2021). Fourth generation biofuel from genetically modified algal biomass: Challenges and future directions. *Chemosphere, 285*. https://doi.org/10.1016/j.chemosphere.2021.131535

Sivasankar, P., Poongodi, S., Seedevi, P., Sivakumar, M., Murugan, T., & Loganathan, S. (2019). Bioremediation of wastewater through a quorum sensing triggered MFC: A sustainable measure for waste to energy concept. *Journal of Environmental Management, 237*, 84–93. https://doi.org/10.1016/j.jenvman.2019.01.075

Stirk, W. A., & van Staden, J. (2020). Potential of phytohormones as a strategy to improve microalgae productivity for biotechnological applications. *Biotechnology Advances, 44*, 107612. https://doi.org/10.1016/j.biotechadv.2020.107612

Su, D., Qi, J., Liu, X., Wang, L., Zhang, H., Xie, H., & Huang, X. (2019). Enzyme-modulated anaerobic encapsulation of *Chlorella* cells allows switching from O_2 to H_2 production. *Angewandte Chemie - International Edition, 58*(12), 3992–3995. https://doi.org/10.1002/anie.201900255

Sun, J., Li, N., Yang, P., Zhang, Y., Yuan, Y., Lu, X., & Zhang, H. (2020). Simultaneous antibiotic degradation, nitrogen removal and power generation in a microalgae-bacteria powered biofuel cell designed for aquaculture wastewater treatment and energy recovery. *International Journal of Hydrogen Energy, 45*(18), 10871–10881. https://doi.org/10.1016/j.ijhydene.2020.02.029

Trias, R., García-Lledó, A., Sánchez, N., López-Jurado, J. L., Hallin, S., & Bañeras, L. (2012). Abundance and composition of epiphytic bacterial and archaeal ammonia oxidizers of marine red and brown macroalgae. *Applied and Environmental Microbiology, 78*(2), 318–325. https://doi.org/10.1128/AEM.05904-11

US EPA. (2022). *Harmful algal blooms.* Available at: https://www.epa.gov/nutrientpollution/harmful-algal-blooms.

Wang, H., Hill, R. T., Zheng, T., Hu, X., & Wang, B. (2016). Effects of bacterial communities on biofuel-producing microalgae: Stimulation, inhibition and harvesting. *Critical Reviews in Biotechnology, 36*(2), 341–352. https://doi.org/10.3109/07388551.2014.961402

Xiao, Y., Zou, H., Li, J., Song, T., Lv, W., Wang, W., Wang, Z., & Tao, S. (2022). Impact of quorum sensing signaling molecules in gram-negative bacteria on host cells: Current understanding and future perspectives. *Gut Microbes, 14*(1), 1–19. https://doi.org/10.1080/19490976.2022.2039048

Xu, L. Y., Zheng, L., Han, H. T., Cui, Z. S., Guo, X. C., & Li, X. Z. (2012). Screening of microalgae associated bacteria with quorum sensing system and there algicidal activity. *Oceanologia et Limnologia Sinica, 43*, 1149–1155.

Yang, L., Maeda, H., Yoshikawa, T., & Zhou, G. Q. (2012). Algicidal effect of bacterial isolates of *Pedobacter* sp. against cyanobacterium *Microcystis aeruginosa. Water Science and Engineering, 5*(4), 375–382. https://doi.org/10.3882/j.issn.1674-2370.2012.04.002

Zhang, B., Li, W., Guo, Y., Zhang, Z., Shi, W., Cui, F., Lens, P. N. L., & Tay, J. H. (2020a). Microalgal-bacterial consortia: From interspecies interactions to biotechnological applications. *Renewable and Sustainable Energy Reviews, 118.* https://doi.org/10.1016/j.rser.2019.109563 (October 2019).

Zhang, F., Fan, Y., Zhang, D., Chen, S., Bai, X., Ma, X., Xie, Z., & Xu, H. (2020b). Effect and mechanism of the algicidal bacterium Sulfitobacter porphyrae ZFX1 on the mitigation of harmful algal blooms caused by Prorocentrum donghaiense. *Environmental Pollution, 263*, 114475. https://doi.org/10.1016/j.envpol.2020.114475

Zhang, C., Ho, S. H., Li, A., Fu, L., & Zhou, D. (2021). Co-culture of *Chlorella* and *Scenedesmus* could enhance total lipid production under bacteria quorum sensing molecule stress. *Journal of Water Process Engineering, 39*, 101739. https://doi.org/10.1016/j.jwpe.2020.101739

Zhang, C., Li, Q., Fu, L., Zhou, D., & Crittenden, J. C. (2018). Quorum sensing molecules in activated sludge could trigger microalgae lipid synthesis. *Bioresource Technology, 263*, 576–582. https://doi.org/10.1016/j.biortech.2018.05.045

Zhang, C., Li, S., & Ho, S. H. (2021). Converting nitrogen and phosphorus wastewater into bioenergy using microalgae-bacteria consortia: A critical review. *Bioresource Technology, 342*, 126056. https://doi.org/10.1016/j.biortech.2021.126056

Zhang, Y., Zheng, L., Wang, S., Zhao, Y., Xu, X., Han, B., & Hu, T. (2022). Quorum sensing bacteria in the phycosphere of hab microalgae and their ecological functions related to cross-kingdom interactions. *International Journal of Environmental Research and Public Health, 19*(1). https://doi.org/10.3390/ijerph19010163

Zhao, P., Yu, X., Li, J., Tang, X., & Huang, Z. (2014). Enhancing lipid productivity by co-cultivation of *Chlorella* sp. U4341 and *Monoraphidium* sp. FXY-10. *Journal of Bioscience and Bioengineering, 118*(1), 72–77. https://doi.org/10.1016/j.jbiosc.2013.12.014

Zhou, D., Zhang, C., Fu, L., Xu, L., Cui, X., Li, Q., & Crittenden, J. C. (2017). Responses of the microalga *Chlorophyta* sp. to bacterial quorum sensing molecules (N-acylhomoserine lactones): Aromatic protein-induced self-aggregation. *Environmental Science and Technology, 51*(6), 3490–3498. https://doi.org/10.1021/acs.est.7b00355

Zhou, Z., Li, Q., Song, K., Wang, R., Wen, S., Zhang, D., & Cong, W. (2021). Exploration of applying growth-promotion bacteria of *Chlorella sorokiniana* to open cultivation systems. *Bioprocess and Biosystems Engineering, 44*(7), 1567–1576. https://doi.org/10.1007/s00449-021-02542-6

Zhou, J., Lyu, Y., Richlen, M., Anderson, D. M., & Cai, Z. (2016). Quorum sensing is a language of chemical signals and plays an ecological role in algal-bacterial interactions. *Critical Reviews in Plant Sciences, 35*(2), 81–105. https://doi.org/10.1080/07352689.2016.1172461

Zhu, X., Chen, S., Luo, G., Zheng, W., Tian, Y., Lei, Y., Yao, L., Wu, C., & Xu, H. (2022). A novel algicidal bacterium, *Microbulbifer* sp. YX04, triggered oxidative damage and autophagic cell death in *Phaeocystis globosa*, which causes harmful algal blooms. *Microbiology Spectrum, 10*(1), 1–22. https://doi.org/10.1128/spectrum.00934-21

CHAPTER

11 Algae-derived materials and pathways for applications in the automobile industries

Ramesh Desikan[1], Suchitra Rakesh[2] and Karthikeyan Subburamu[1,*]

[1]*Department of Renewable Energy Engineering, Agricultural Engineering College and Research Institute, Tamil Nadu Agricultural University, Coimbatore, Tamil Nadu, India;* [2]*Department of Microbiology, School of Life Sciences, Central University of Tamil Nadu, Thiruvarur, Tamil Nadu, India*

[*]*Karthikeyan Subburamu, skarthy@tnau.ac.in e-mail address: skarthy@tnau.ac.in*

1. Introduction

The petroleum refinery industries are the primary source for supplying transportation fuels to society. The rapid growth of world population, industrialization, and urbanization could increase energy demand, which will cause a shrinking of fossil reserves, increased fuel consumption, greenhouse gas emissions, and associated global warming/climate change issues. Another side, conventional fuels are price varied, and they may be available for a few decades. Searching for alternate fuels reveals that biodiesel and bioethanol biofuels are the potential competitors to mitigate the global warming issues and replace petroleum-based fuels and their pollution. Therefore, the paradigm shift from a petroleum-based economy to a biobased economy can offer a sustainable solution to the aforementioned issues and ensure energy security. Presently, most biodiesels are derived from edible oil feedstocks and reached a commercialized scale in different countries. India focused on nonedible oils as feedstocks for biodiesel production compared with foreign countries. However, the price of plant oil—based biodiesels is still higher than that of commercial petrofuels due to the higher cost of raw oils, seasonal availability, and more demand from food and nonfood industries. Therefore, finding a low-cost feedstock suitable for biodiesel production is essential.

Algae have a more potential resources to fulfill the energy demands in the future. Generally, microalgae generate more lipids than macroalgae and have faster growth in nature, which requires carbon dioxide, sunlight, and nutrients for their growth (Lee et al., 2014). The potential algae strains are collected from various aquatic environments and mimic similar conditions for mass-scale production. The merits of algae feedstocks used as biofuel feedstocks are higher lipid yield, multiple harvesting in a year, and higher productivity. Algae-based wastewater treatment results win—win situation by a simultaneous process of wastewater purification and algae growth utilizing the nutrients (e.g., nitrogen, phosphorus, etc.). Several biofuels/products could be derived from algae feedstocks such as biodiesel, biohydrogen, ethanol, biogas, and bioproducts.

Algae Materials: Applications Benefitting Health. https://doi.org/10.1016/B978-0-443-18816-9.00022-8

2. Algal biomass production

Currently, most carbon dioxide emission globally is due to the transport sector (Balat & Balat, 2010). According to recent studies, microalgal biomass is the best source of renewable and sustainable energy. Hence, to address the severe impact of fossil fuel depletion and hazardous gas emissions, there is a need to address such issues and go for renewable, eco-friendly usage. As microalgae do not compete with food crops or nutrients, they are ideal candidates for carbon dioxide sequestration and effective land utilization. Many studies have reported using microalgae biomass for biofuel and other value-added product generation (Bahadar & Bilal, 2013). Microalgae can utilize the nutrients and organic carbon in wastewater and convert it into biomass. It will further balance respiratory loss, ameliorate the energy budget, and encourage biomass productivity (Wang et al., 2018). Acien-Fernandez et al. (2012) have reported that using water and nutrients from wastewater will help reduce the production cost and make algal biofuel production commercially viable. In addition to biomass production for fuel recovery, we can also harvest various value-added bioproducts. It can compensate for the algae cultivation cost. Hence, wastewater utilization for microalgal biomass production can encourage a carbon-neutral, zero-waste-free, sustainable, eco-friendly approach.

The technoeconomic and life cycle analysis has reported that microalgae-based biofuels are not competitive enough with conventional fossil fuels. As a result, biomass harvesting is expensive (Garcia-Prieto et al., 2017). However, microalgae-based wastewater biorefinery is the best alternative to reduce biofuel production costs. Puyol et al. (2017) reported that wastewater could generate 6.5 MJ/kg of energy, which adds up to 1% of the world's total energy.

3. Recent advances in algal biomass production

The strain must multiply, be robust, and be tolerant of biotic and abiotic stresses during cultivation. There are many algal species with high biomass productivity under several environmental conditions, according to Subramanian et al. (2013). Microalgae can shift the biochemistry of primary energy storage cellular components into high-energy hydrocarbons such as TAG, polyterpenoids, etc. (Breuer et al., 2012). Under stress, carbohydrates are converted to hydrocarbons (Gong et al., 2013). Studies have reported high biomass production and growth rates during nutrient-deprived growth conditions (Sharma et al., 2012). Most algal genetic engineering has focused on altering metabolic pathways for improved biomass and coproducts yield. There has been a two- to fivefold increase in biomass production due to genetic enhancement (Aketo et al., 2020).

Cultivation factors influencing the algal biomass production involve selecting the right production site, strain, cultivation system, seasonal influence, harvesting, nutrient, and water recycling. Fig. 11.1 shows the various factors affecting algal biomass production. The suitable site for algae cultivation includes land selection, sunlight availability, seasonal temperature, carbon dioxide, water sources, power supply, well-connected transport facility, etc. (Xu et al., 2019). The low altitude area with a warm climate and an average of 250 h/month of solar radiation is considered an ideal site (Bhujade et al., 2017). Algal production ideal site selection models have been developed (Sun et al., 2020). Microalgae cultivation under raceway/outdoor depends on temperature, sunlight, and humidity, influencing physiological activity and biomass yield. High light leads to photoinhibition, protein damage, oxidative stress, etc., affecting CO_2 fixation and biomass yield.

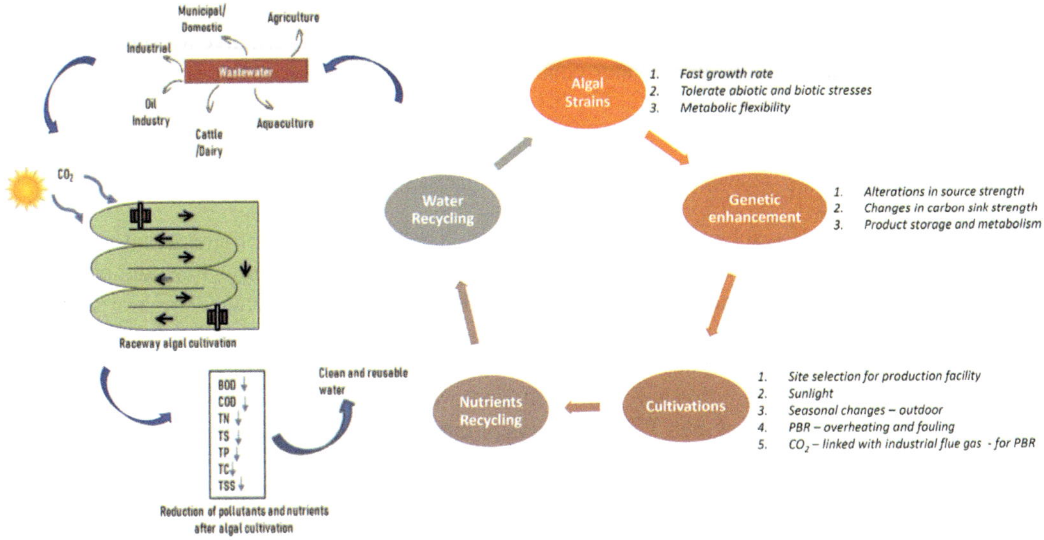

FIGURE 11.1

Factors influencing the production of algal biomass.

Similarly, low light intensity reduces photosynthetic efficiency and biomass yield (Janssen, 2016). The majority of algal strains have optimal growth between 20 and 25°C. The temperature above 35°C leads to photorespiration, affects nutrient availability, increases the concentration of NH_3 in the medium, decreases CO_2 solubility, and increases evaporation losses leading to salinity variations (Posadas et al., 2017). The optimal growth of microalgae for biomass production depends on an adequate supply of nutrients, water, and sunlight. Nitrates and phosphates in agricultural wastewater can be recycled for biomass production (Rajvanshi et al., 2019).

4. Challenges

The high population growth rate and exhaustive use of conventional fossil fuels increase the demand for energy and pose serious environmental problems such as global warming. Hence, researchers worldwide are searching for an alternative solution to overcome these issues. In these contests, microalgae are the best promising candidate for alternative renewable energy resources and serve as an excellent option for carbon sequestration. Microalgae biomass usage for biofuel production faces many challenges and limitations, and we must address these problems by improving the technologies from laboratory to commercialization. The most crucial issues are improving algal biomass productivity, dewatering, pretreatment, and downstream processing. Despite various advanced techniques, microalgal biomass production and harvesting are not cost-efficient.

A cultivation system requires temperature and growth-limiting conditions such as carbon dioxide nutrients, water, etc., whereas dewatering/harvesting is highly energy-intensive. Hence, the cultivation and harvesting of microalgae involve high operating and capital costs. There is a vast technological gap during the commercialization of biofuels. Another major hurdle is the demand for a large quantity of water, carbon dioxide, nutrients, etc. The wastewater can be the best alternative for such issues, but there is a severe concern about wastewater contamination with pathogens, bacteria, and chemicals harmful to algal growth.

5. Perspectives

Algal biofuel production reaches its heights with many interferences and progresses at every step. Metabolic engineering with recombinant DNA technology for genetic modifications and changes in lipid or triglyceride pathways leads to oil-rich algal strains. Setting up large-scale photobioreactors for commercially algae cultivation by utilizing carbon dioxide from flue gas released from industries can be best looked into. Hydrothermal liquefaction for crude oil production directly from wet algal biomass could be well explored; this will further eliminate various cost-intensive steps such as drying, oil extraction, transesterification, etc. This third-generation biofuel with algae as feedstock requires a huge initial investment for setting up various facilities such as cultivation production system optimization, production of genetically modified algae with high lipid producing efficiency, and algae-based biorefinery for wastewater treatment and production of high-value products.

6. Algae mass-scale production

According to a US study, microalgal biomass production costs \$4.92/kg at present technology levels (Kumar & Singh, 2019). Another European study estimated the cost of biomass production to be €4.95, 4.16, and 5.96/kg from open ponds, horizontal tubular systems, and flat panel photobioreactors (Norsker et al., 2011). Although biomass production costs drop to \$0.5 per kg, scaling up of microalgae for stand-alone biofuel production is economically unfeasible due to rapid competition with fossil fuels. To scale up successfully, cost reductions and revenue generation steps must be integrated.

In outdoor pilot-scale raceway ponds for wastewater treatment, *Scenedesmus* sp. produced biomass productivity ranging from 4 g/m^2/d in December when the average temperature was 13°C, and the irradiance was 300−157 w/m^2, 17 g/m^2/d in July when the average temperature was 23°C. The irradiance was 468 ± 292 w/m^2 (Posadas et al., 2015).

Several microalgal species were evaluated in open bioreactors with a 30 L capacity in greenhouse conditions, including *Chlorella vulgaris, Botryococcus braunii, Chlamydomonas reinhardtii, Euglena gracilis*, and *Nannochloropsis oculata*. The growth response of these microalgae was evaluated from March to April, June to July, and October to November. *C. vulgaris* and *B. braunii* resulted in the highest growth during March and April, when the average temperature was 28.5°C and irradiance was 15.9 MJ/m^2/day. *C. reinhardtii* and *N. oculata* grew best in June when the average temperature was 36.1°C, and irradiance was 6.6 MJ/m^2/day. In contrast, *E. gracilis* grew comparably during March and June months. In general, temperature and irradiance were low during October and November for all five microalgae (Metsoviti et al., 2019).

The environmental variations such as seasonal changes have a significant impact on algal biomass productivity. Morales et al. (2017) studied the seasonal growth of *Scenedesmus obtusiusculus* in an airlift-extended loop photobioreactor under outdoor environmental conditions. The study revealed that the biomass productivity was high during the spring season (0.29 g/L/day), followed by autumn. However, under optimum laboratory conditions, the biomass productivity was much higher.

The large-scale cultivation of algae in closed photobioreactor is very expensive and also faces many challenges such as fouling and overheating. Hence, open ponds with mixed raceways are well preferred, not only due to less cost involved to build but can also be easily scaled up to large areas. This makes open raceway ponds an ideal choice for commercial-scale biomass production. According to the literature survey, very few papers have discussed the cultivation designs, yields, operations, and other aspects of commercialization (Zhu et al., 2017).

7. Conversion methods used for biofuel production

Algae have much potential as a biofuel source. Algae can produce various products depending on the species, culture conditions, and biomass processing. Biodiesel, jet fuel, gasoline, ethanol, methanol, hydrogen, and syngas are some of the energy products made from algae (European Biofuels, 2016). Algae may be processed in a variety of ways to provide a variety of energy outputs. After harvesting, algal biomass can be treated via thermochemical, biochemical, and chemical conversion techniques.

8. Biochemical process

Anaerobic digestion, fermentation, and photobiological methods are all used in the biochemical conversion of microalgae. The biological conversion route involves microorganisms hydrolyzing cell walls into fermentable sugars (Raheem et al., 2018). Microalgae are transformed into biofuels using microorganisms and enzymatic reactions in various conversion technologies.

9. Anaerobic digestion

The anaerobic breakdown of sugars into biogas, bioethanol, or biohydrogen is known as fermentation. The production of biogas from anaerobic digestion of microalgae has recently received a lot of attention. For example, the technique is suitable for microalgae with a moisture level of 80−90% and high polysaccharide content but no lignin. All fermentable products are oxidized into acetate, which is then transformed into methane and CO_2 during methanogenesis (Adeniyi et al., 2018). The carbon−nitrogen (C: N) ratio in the feedstock, as well as time, temperature, pH, solids, and feeding rates, all influence biogas generation (Raheem et al., 2018). The biogas yield is poor because of algal sensitivity to bacterial degradation and the low C: N ratio, which results in ammonia generation (inhibitor). Before anaerobic digestion, an effective pretreatment improves biogas yield substantially (Passos et al., 2014). By modifying the cell wall structure, microwave pretreatment of biomass can boost biogas output by 56% (Passos et al., 2013). Enzymatic, mechanical, and thermal pretreatments have improved methane generation (Kandiyoti et al., 2017; Marsolek et al., 2014). Biomass is commonly codigested with waste papers and sewage sludge to compensate for the decreased C: N ratio (Hidaka et al., 2014; Kandiyoti et al., 2017; Suganya et al., 2016). It has been found that codigestion increases CH_4 generation by 26% (Costa et al., 2012). High protein content has been shown to inhibit anaerobic digestion when salt-adapted microorganisms are used (Brennan & Owende, 2010).

10. Fermentation

Bioethanol is made by fermenting hydrolyzed carbohydrates with yeast (Laurens et al., 2015). Pretreatment, enzymatic hydrolysis, and fermentation are all steps in the bioethanol production process. Microalgae are a better feedstock for enzymatic hydrolysis because the cell wall lacks lignin, which minimizes recalcitrance compared with lignocellulosic biomass. Many pretreatment methods have been examined so far, including mechanical, physical, thermal, chemical, and combination procedures, for disrupting the cell wall and removing lipids, favoring enzymatic hydrolysis and bioethanol synthesis (Onumaegbu et al., 2018). Milling, hot water wash, liquefaction, enzymatic hydrolysis, saccharification, and alginate extraction are necessary for successful bioethanol synthesis (Fasahati et al., 2015; Lee & Lee, 2016; Teixeira et al., 2016, pp. 1–7). Bioethanol synthesis by glycogen-containing cyanobacteria yielded 6.5 g/L (Aikawa et al., 2013). Harun et al. (2010) compared untreated intact microalgae to lipids derived from microalgae; they discovered 60% more ethanol.

11. Photobiological hydrogen

Hydrogen (H_2) is a reliable and sustainable energy carrier that can be created from microalgae by using hydrogenase or nitrogenase enzymes in oxygenic photosynthesis. Microalgae convert water into protons, electrons, and oxygen during photosynthesis; a proton is then transformed into H_2 by hydrogenase enzymes (Cantrell et al., 2008). However, the hydrogenase enzyme is inhibited by the synthesis of oxygen, which is the principal hindrance to hydrogen production. This problem can be overcome by properly separating the oxygen generated during photosynthesis. Compared with batch production, continuous flow yielded a significant increase (20-fold) in hydrogen generation (Das et al., 2015).

12. Thermochemical process

Thermochemical conversion is one of the most efficient methods for converting biomass into biofuel (Chan et al., 2019). This approach is more favorable than the biochemical pathway because it is more flexible in processing different types of biomass, and it is still one of the most direct, quick, and simple conversion methods. Thermal decomposition of organic matter happens at very high temperatures in this process, resulting in the decomposition of biomass yield and the formation of biooils and gases.

13. Torrefaction

Torrefaction, also known as slow or mild pyrolysis, high-temperature drying or roasting, or wood cooking, is a pretreatment technique for converting biomass into energy (López Barreiro et al., 2013). This is a type of thermal therapy that takes place in an inert atmosphere with atmospheric pressure and temperatures ranging from 200 to 300 °C. There are four types of torrefaction processes: conventional, microwave, wet, and oxidative torrefaction. Torrefaction can be categorized as light (200–235 °C), mild (235–275 °C), and intense (275–300 °C) depending on the temperature (Chen et al., 2014). The conventional torrefaction process occurs in the presence of inert gas or carrier gas (mostly nitrogen) at a temperature of 200–300°C for 10–60 min at atmospheric pressure (Zhang et al., 2018). The process of oxidative torrefaction is identical to that of conventional torrefaction, except that instead of N_2, the air is used as the carrier gas, and other oxidative gases are utilized in some situations (Ho et al., 2020; Zhang et al., 2019). The biomass is pyrolyzed from the inner material using microwave irradiation in the microwave torrefaction process (Motasemi & Afzal, 2013). The biomass is treated in hydrothermal or hot compressed water at a temperature of 180–260°C for 10–240 min in the wet torrefaction process. Higher reliability and stability, as well as ease of control, are the main benefits of the torrefaction process. The procedure is limited by higher operational costs and a more extended residency period (Ho et al., 2020; Zhang et al., 2019).

14. Pyrolysis

Pyrolysis is carried out at a higher temperature in the absence of oxygen and with or without a catalyst. To recover fuels, biomass can be decomposed at various temperatures and at different times. Pyrolysis produces solid biochar, biooil, or gas as a final product (Sirajunnisa & Surendhiran, 2016; Xiong et al., 2017). Dehydration, devolatilization, and decomposition are the critical stages in converting microalgae to bioenergy (Pourkarimi et al., 2019). *Chlorella protothecoides*, *Chlorella vulgaris*, *Scenedesmus* sp., *Microcystis aeruginosa*, and *Nannochloropsis* sp. residue are among the most critical algae for biofuel conversion. According to heating rates, the pyrolysis process is divided into four categories: slow, fast, microwave, and catalytic. Fast pyrolysis occurs at a faster rate (600–36,000°C/min) with a shorter residence time (0.003–0.05 min), whereas slow pyrolysis occurs at a slower rate (3.5–10°C/min) with a much longer residence time (20–120 min). The temperature range for this process is 300–800°C. Microwave pyrolysis requires a maximum reaction temperature of 800°C, whereas catalytic pyrolysis requires a reaction temperature of just 300°C. Pyrolysis, on the other hand, is generally carried out in the absence of oxygen at a temperature of 400–600°C and at atmospheric pressure (Auersvald et al., 2019; Lee et al., 2017). Overall, in comparison with other processes, fast pyrolysis is better for storing and transporting liquid fuels, whereas slow pyrolysis is better for higher oil production applications (Chen et al., 2015; Campanella & Harold, 2012).

15. Gasification

Gasification is a well-known thermochemical conversion method for converting algal biomass to gaseous fuels. Algal biomass contains carbonaceous material transformed into synthesis gases (syngas). The principal gases developed during the gasification process with other combustible gases such as C_2H_4, C_2H_6, C_3H_8, and others include H_2, CO, CO_2, and CH_4 in partial oxidation of air, steam, or oxygen (Arun et al., 2021; Hirano et al., 1998; Rawat et al., 2011). Gasification can be classified as a conventional or a supercritical water gasification method (SWG). The conventional gasification process takes place at temperatures ranging from 800 to 1000 °C and pressures ranging from 0.1 to 1 MPa. Beyond the critical point of water (373.946°C, 22.064 MPa), algal biomass is immediately transformed into gas in SWG. The process's main benefit is that when the gasification temperature rises, the H_2 yield also rises (Raheem et al., 2017). SWG is an energy-efficient process because biomass can be fed without being dried, reducing the cost of drying to the same level as HTG (hydrothermal gasification). The noted restriction is that the gasification efficiency diminishes as the process progresses and more catalysts are used (Duan et al., 2018; Guan et al., 2012).

16. Hydrothermal process

Hydrothermal processes include hydrothermal carbonization (HTC), hydrothermal liquefaction (HTL), and hydrothermal gasification (HTG). Within 60 min of residence time, HTC reaches a temperature of 200 °C, and a pressure of less than 2 MPa. HTL, on the other hand, requires a temperature of 300–350°C and a pressure of 5–20 MPa, as well as a residence time of 5–60 min. Wet biomass (almost 80% moisture) can be treated hydrothermally without drying in HTG, saving money on drying. Beyond the critical point of water, this method can directly transform algal biomass into a gas (Jena & Das, 2011). Due to the process's working conditions at critical temperature and pressure, solid deposition and corrosion in the reactor are significant drawbacks (Chiaramonti et al., 2015).

17. Chemical process

The extraction of lipids from microalgae and then the conversion of lipids into biodiesel via transesterification are two chemical steps in the algal biofuel production process. Transesterification occurs between triglycerides (lipids) and alcohol in the presence of an acidic, alkaline, or enzyme-based catalyst to produce fatty acid methyl ester (FAME) and glycerol (Kandiyoti et al., 2017; Suganya et al., 2016). The type of alcohol, catalyst, and molar ratio all play a significant role in the reaction. This method is necessary to reduce the viscosity of algal oil and improve its fluidity so that it can be blended with petroleum diesel and directly applied to engines (Adeniyi et al., 2018). Conventional transesterification requires the extraction of algal lipids followed by esterification, whereas direct transesterification is a one-step process that treats wet biomass without extraction. Direct transesterification yields biodiesel at a lower rate than conventional processes, but it saves reagents, energy, and time. It is recommended to combine microwave and ultrasonic irradiation techniques to save energy and boost the yield to 90%. Another option for lowering costs is to use in situ supercritical methanol transesterification (Pati et al., 2011; Wahlen et al., 2011). Palmitic acid (C16:0), stearic acid (C18:0), oleic (C18:1), linoleic acid (C18:2), and linolenic acid (C18:3) are the major fatty acids utilized in making certified biodiesel (Lee et al., 2011). Using an alkaline catalyst, Milano et al. converted up to 90% of FAME from two microalgae species, Spirogyra and Oedigonium (Milano et al., 2016). Salam et al. (2016) suggested that biogas production from residual microalgae from the biodiesel production process could provide the energy needed for microalgae growth or separation from dilute solution prior to *the* in situ process.

18. Algae biodiesel as engine fuel

More options will be available to convert the algae oil or wet/dry algae biomass into biofuels using a suitable conversion process. However, these conversion technologies are not yet commercialized. Most of the biofuel conversion technologies are still in the development stage only. Presently, biodiesel and bioethanol are possible biofuels derived from algae biomass. Generally, vegetable oils have higher kinematic viscosity, specific gravity, lower volatility, and calorific value than that diesel fuels. These characteristics cause problems in diesel engines if vegetable oils are used as a fuel. In addition, the combustion of glycerol releases highly toxic gas. Similar issues are also applicable to algae oils. For example, the algae oil viscosity is 10–20 times higher than diesel fuel, which results in incomplete combustion and leads to the failure of diesel engines (Demirbaş, 2008). Many algal oils contain more saturated and polyunsaturated fatty acids than vegetable oils. Therefore, glycerol removal and improvement of the algae oil fuel properties are essential, which can be achieved through trans-esterification. The important properties of biodiesels derived from different algae strains are presented in Table 11.1. The fuel properties are almost closer to diesel fuel properties and fulfill the international biodiesel standards. Interesting to note that the cetane number for the algal biodiesel is higher than diesel fuel. This property represents the antiknocking quality of algae biodiesel fuel. However, few algae biodiesels have poor cold flow properties and low oxidative stability due to algae oils' higher saturated and polyunsaturated fatty acids (Knothe, 2011).

From an engine performance point of view, the algae biodiesel and diesel—biodiesel blended fuels can be used in compression ignition engines without any modifications. Generally, biodiesel and its blend fuels would release lesser CO_2 content and unburned hydrocarbon emissions than diesel fuel.

Table 11.1 Comparison of algal fuel quality with diesel fuel standard.

S. No	Fuel source	Kinematic viscosity, mm^2/s	Specific gravity	Cetane number	Calorific value, MJ/kg	References
1.	*Thraustochytrid*	1.28	0.940	47.44	42.14	Byreddy et al., (2015)
2.	*Crypthecodinium cohnii*	5.06	0.912	47.99	39.86	Islam et al., (2015)
3.	*Botryococcus braunii*	3.60–5.40	0.855	42.00	36.00 MJ/L	Prathima and Karthikeyan, (2017)
4.	*Chaetoceros calcitrans*	4.38	0.870	59.41	93.18	Joseph, (2017)
5.	*Thalassiosira sp.*	4.13	0.880	55.49	39.01	Joseph, (2017)
6.	*Nitzschia sp.*	4.70	0.870	58.00	39.8	Jebali et al., (2019)
7.	*Tetraselmis sp.*	4.40	0.870	54.00	40.8	Jebali et al., (2019)
8.	*Chlamydomonas sp. BTA 9032*	4.31	–	58.56	41.7	Mondal et al., (2019)
9.	*Chlorella sp. BTA 9031*	2.50–3.91	0.840–0.848	51.17–53.72	44.04	Fiveriati et al., (2020)
10.	*Cylindrotheca sp.*	5.20	0.881	56.20	41.20	Khoobbakht et al., (2021)
11.	Diesel fuel (ASTM standards)	1.3–4.1	820–860 kg/m^3	40.00 min.	–	Demirbas, 2009a
12.	Biodiesel fuel (ASTM standards)	1.9–6.0	880 kg/m^3	47.00 min.	–	Demirbas, 2009b

A similar trend is observed for algae biodiesels. The reason may be oxygen presence in the biodiesel fuel. Similar to plant oil–based biodiesels, the NOx emission is higher for algae biodiesel and their blends due to the higher temperature observed in the engine (Sharif et al., 2020).

19. Conclusions

Fossil fuels are fast diminishing resources, and research on alternate fuels is intensively done worldwide. Biodiesel from vegetable oils is already commercialized across the world. However, low-cost biodiesel feedstock availability is a big issue, and biodiesel synthesized from conventional plant oils feedstocks is still higher than diesel fuel price. Another side, the area required to meet huge demand is also for growing oil crops. These hindrances increased oil crop cultivation and led to food versus fuel conflict. Recently, algae have been identified as an alternative to existing biodiesel feedstocks due to higher oil production per unit area. It results in gaining algae biofuel research to find alternate fuels. Among the various possibilities of biofuels production, algae biodiesel reached up to engine performance testing. The result shows that the performance of algae biodiesel is similar to other biodiesels in compression ignition engines without any modifications. The technology readiness level reached up to four for biofuel productions, whereas it was three for algae cultivation and harvesting. Therefore, optimizing several stages of algae cultivation and biofuel production would generate a low-cost feedstock for algae biofuels. This step will help compete for the algae-derived biofuels with existing biodiesel/conventional fuels and retain them in the fuel market.

References

Acién Fernández, F. G., GonzálezLópez, C. V., Fernández Sevilla, J. M., & Molina Grima, E. (2012). Conversion of CO_2 into biomass by microalgae: How realistic a contribution may it be to significant CO_2 removal? *Applied Microbiology and Biotechnology, 96*(3), 577–586.

Adeniyi, O. M., Azimov, U., & Burluka, A. (2018). Algae biofuel: Current status and future applications. *Renewable and Sustainable Energy Reviews, 90*, 316–335.

Aikawa, S., Joseph, A., Yamada, R., Izumi, Y., Yamagishi, T., Matsuda, F., Kawai, H., Chang, J. S., Hasunuma, T., & Kondo, A. (2013). Direct conversion of Spirulina to ethanol without pretreatment or enzymatic hydrolysis processes. *Energy & Environmental Science, 6*, 1844–1849.

Aketo, T., Hashizume, R., Yabu, Y., Hoshikawa, Y., Nojima, D., Maeda, Y., et al. (2020). Characterization of a novel marine unicellular alga, *Pseudoneochloris* sp. strain NKY372003 as a high carbohydrate producer. *Journal of Bioscience and Bioengineering, 129*(6), 687–692.

Arun, J., Gopinath, K. P., Sivaramakrishnan, R., SundarRajan, P., Malolan, R., & Pugazhendhi, A. (2021). Technical insights into the production of green fuel from CO2 sequestered algal biomass: A conceptual review on green energy. *Science of The Total Environment, 755*, 142636.

Auersvald, M., Shumeiko, B., Vrtiška, D., Straka, P., Staš, M., Šimáček, P., Blažek, J., & Kubička, D. (2019). Hydrotreatment of straw bio-oil from ablative fast pyrolysis to produce suitable refinery intermediates. *Fuel, 238*, 98–110.

Bahadar, A., & Bilal Khan, M. (2013). Progress in energy from microalgae: A review. *Renewable and Sustainable Energy Reviews, 27*, 128–148.

Balat, M., & Balat, H. (2010). Progress in biodiesel processing. *Applied Energy, 87*(6), 1815–1835.

Bhujade, R., Chidambaram, M., Kumar, A., & Sapre, A. (2017). Algae to economically viable low-carbon footprint oil. *Annual Review of Chemical and Biomolecular Engineering, 8*, 335–357.

Brennan, L., & Owende, P. (2010). Biofuels from microalgae—A review of technologies for production, processing, and extractions of biofuels and co-products. *Renewable and Sustainable Energy Reviews, 14*, 557–577.

Breuer, G., Lamers, P. P., Martens, D. E., Draaisma, R. B., & Wijffels, R. H. (2012). The impact of nitrogen starvation on the dynamics of triacylglycerol accumulation in nine microalgae strains. *Bioresource Technology, 124*, 217–226.

Byreddy, A. R., Gupta, A., Barrow, C. J., & Puri, M. (2015). Comparison of cell disruption methods for improving lipid extraction from thraustochytrid strains. *Marine Drugs, 13*(8), 5111–5127.

Campanella, A., & Harold, M. P. (2012). Fast pyrolysis of microalgae in a falling solids reactor: Effects of process variables and zeolite catalysts. *Biomass and Bioenergy, 46*, 218–232.

Cantrell, K. B., Ducey, T., Ro, K. S., & Hunt, P. G. (2008). Livestock waste-to-bioenergy generation opportunities. *Bioresource Technology, 99*, 7941–7953.

Chan, Y. H., Cheah, K. W., How, B. S., Loy, A. C. M., Shahbaz, M., Singh, H. K. G., Yusuf, N. R., Shuhaili, A. F. A., Yusup, S., Ghani, W. A. W. A. K., Rambli, J., Kansha, Y., Lam, H. L., Hong, B. H., & Ngan, S. L. (2019). An overview of biomass thermochemical conversion technologies in Malaysia. *The Science of the Total Environment, 680*, 105–123.

Chen, W. H., Wu, Z. Y., & Chang, J. S. (2014). Isothermal and non-isothermal torrefaction characteristics and kinetics of microalga *Scenedesmus obliquus* CNW-N. *Bioresource Technology, 155*, 245–251.

Chen, W. H., Lin, B. J., Huang, M. Y., & Chang, J. S. (2015). Thermochemical conversion of microalgal biomass into biofuels: A review. *Bioresource Technology, 184*, 314–327.

Chiaramonti, D., Prussi, M., Buffi, M., Casini, D., & Rizzo, A. M. (2015). Thermochemical conversion of microalgae: Challenges and opportunities. *Energy Procedia*, 819–826.

Costa, J. C., Gonçalves, P. R., Nobre, A., & Alves, M. M. (2012). Biomethanation potential of *macroalgae Ulva spp. and Gracilaria spp.* and in co-digestion with waste activated sludge. *Bioresource Technology, 114*, 320–326.

Das, A. A., Esfahani, M. M., Velev, O. D., Pamme, N., & Paunov, V. N. (2015). Artificial leaf device for hydrogen generation from immobilised *C. reinhardtii* microalgae. *Journal of Materials Chemistry, 3*, 20698–20707.

Demirbaş, A. (2008). Production of biodiesel from algae oils. *Energy Sources, Part A: Recovery, Utilization, and Environmental Effects, 31*(2), 163–168.

Demirbas, A. (2009a). Progress and recent trends in bio-diesel fuels. *Energy Conversion and Management, 50*, 14–34.

Demirbas, A. (2009b). Political, economic and environmental impacts of biofuels: A review. *Applied Energy, 86*, S108–S117.

Duan, P. G., Li, S. C., Jiao, J. L., Wang, F., & Xu, Y. (2018). Supercritical water gasification of microalgae over a two-component catalyst mixture. *The Science of the Total Environment, 630*, 243–253.

European Biofuels Technology Platform. (2016). *Strategic research and innovation agenda 2016*. London, UK: European Biofuels Technology Platform.

Fasahati, P., Woo, H. C., & Liu, J. (2015). Industrial-scale bioethanol production frombrown algae: E_ects of pretreatment processes on plant economics. *Applied Energy, 1139*, 175–187.

Fiveriati, A., Yonatan, Y., Anne, O., & Adinurani, P. G. (2020). Characteristic biofuel microalgae Chlorella sp. as renewable energy source. In *E3S web of conferences: 1st International Conference on Renewable Energy Research and Challenge (ICoRER 2019): Banyuwangi, East Java, Indonesia, November 12–13, 2019*, (vol. *190* pp. 1–7). EDP Sciences.

García Prieto, C. V., Ramos, F. D., Estrada, V., Villar, M. A., & Diaz, M. S. (2017). Optimization of an integrated algaebased biorefinery for the production of biodiesel, astaxanthin and PHB. *Energy, 139*, 1159–1172.

Gong, Y., Guo, X., Wan, X., Liang, Z., & Jiang, M. (2013). Triacylglycerol accumulation and change in fatty acid content of four marine oleaginous microalgae under nutrient limitation and at different culture ages. *Journal of Basic Microbiology, 53*(1), 29–36, 2013.

Guan, Q., Savage, P. E., & Wei, C. (2012). Gasification of alga *Nannochloropsis sp.* in supercritical water. *The Journal of Supercritical Fluids, 61*, 139–145.

Harun, R., Danquah, K., & Forde, G. M. (2010). Microalgal biomass as a fermentation feedstock for bioethanol production. *Journal of Chemical Technology and Biotechnology, 85*, 199–203.

Hidaka, T., Inoue, K., Suzuki, Y., & Tsumori, J. (2014). Growth and anaerobic digestion characteristics of microalgae cultivated using various types of sewage. *Bioresource Technology, 170*, 83–89.

Hirano, A., Hon-Nami, K., Kunito, S., Hada, M., & Ogushi, Y. (1998). Temperature effect on continuous gasification of microalgal biomass: theoretical yield of methanol production and its energy balance. *Catal Today, 45*(1–4), 399–404. https://doi.org/10.1016/s0920-5861(98)00275-2

Ho, S. H., Zhang, C., Tao, F., Zhang, C., & Chen, W. H. (2020). Microalgal torrefaction for solid biofuel production. *Trends in Biotechnology, 38*(9), 1023–1033.

Islam, M. A., Rahman, M. M., Heimann, K., Nabi, M. N., Ristovski, Z. D., Dowell, A., ... Brown, R. J. (2015). Combustion analysis of microalgae methyl ester in a common rail direct injection diesel engine. *Fuel, 143*, 351–360.

Janssen, M. (2016). Microalgal photosynthesis and growth in mass culture. In J. Legrand (Ed.), *Advances in chemical engineering* (pp. 185–256). Academic Press.

Jebali, A., Acién, F. G., Jiménez-Ruiz, N., Gómez, C., Fernández-Sevilla, J. M., Mhiri, N., ... Molina-Grima, E. (2019). Evaluation of native microalgae from Tunisia using the pulse-amplitude-modulation measurement of chlorophyll fluorescence and a performance study in semi-continuous mode for biofuel production. *Biotechnology for Biofuels, 12*(1), 1–17.

Jena, U., & Das, K. C. (2011). Comparative evaluation of thermochemical liquefaction and pyrolysis for bio-oil production from microalgae. *Energy Fuels, 25*, 5472–5482.

Joseph, S. (2017). Algal culture methods. In *Winter School on Structure and Function of the Marine Ecosystem: Fisheries, 1–21 December 2017, Kochi*. http://eprints.cmfri.org.in/id/eprint/12753.

Kandiyoti, R., Herod, A., & Bartle, K. (2017). *Solid fuels and heavy hydrocarbon liquids: Thermal characterization and analysis* (2nd ed.). Oxford, United Kingdom: Elsevier.

Khoobbakht, G., Kheiralipour, K., & Karimi, M. (2021). Optimization of Chlamydomonas alga biodiesel percentage for reducing exhaust emission of diesel engine. *Process Safety and Environmental Protection, 152*, 25–36.

Knothe, G. (2011). A technical evaluation of biodiesel from vegetable oils vs. algae. Will algae-derived biodiesel perform? *Green Chemistry, 13*(11), 3048–3065.

Kumar, D., & Singh, B. (2019). Algal biorefinery: An integrated approach for sustainable biodiesel production. *Biomass and Bioenergy, 131*, 105398.

Laurens, L. M. L., Nagle, N., Davis, R., Sweeney, N., Van Wychen, S., Lowell, A., & Pienkos, P. T. (2015). Acid-catalyzed algal biomass pretreatment for integrated lipid and carbohydrate-based biofuels production. *Green Chemistry, 17*, 1145–1158.

Lee, K., Eisterhold, M. L., Rindi, F., Palanisami, S., & Nam, P. K. (2014). Isolation and screening of microalgae from natural habitats in the Midwestern United States of America for biomass and biodiesel sources. *Journal of Natural Science, Biology and Medicine, 5*, 333–339.

Lee, J. H., Kim, S. B., Kang, S. W., Song, Y. S., Park, C., Han, S. O., & Kim, S. W. (2011). Biodiesel production by a mixture of Candida rugosa and Rhizopus oryzae lipases using a supercritical carbon dioxide process. *Bioresource Technology, 102*, 2105–2108.

Lee, O. K., & Lee, E. Y. (2016). Sustainable production of bioethanol from renewable brown algae biomass September. *Biomass and Bioenergy, 92*, 70–75.

Lee, J., Yang, X., Cho, S. H., Kim, J. K., Lee, S. S., Tsang, D. C. W., Ok, Y. S., & Kwon, E. E. (2017). Pyrolysis process of agricultural waste using CO_2 for waste management, energy recovery, and biochar fabrication. *Applied Energy, 185*, 214–222.

López Barreiro, D., Prins, W., Ronsse, F., & Brilman, W. (2013). Hydrothermal liquefaction (HTL) of microalgae for biofuel production: state of the art review and future prospects. *Biomass and Bioenergy, 53*, 113–127.

Marsolek, M. D., Kendall, E., Thompson, P. L., & Shuman, T. R. (2014). Thermal pretreatment of algae for anaerobic digestion. *Bioresource Technology, 151*, 373–377.

Metsoviti, M. N., Papapolymerou, G., Karapanagiotidis, I. T., & Katsoulas, N. (2019). Comparison of growth rate and nutrient content of five microalgae species cultivated in greenhouses. *Plants, 8*(8), 279.

Milano, J., Chyuan, H., Masjuki, H. H., Chong, W. T., & Kee. (2016). Microalgae biofuels as an alternative to fossil fuel for power generation. *Renewable and Sustainable Energy Reviews, 58*, 180–197.

Mondal, M., Khan, A. A., & Halder, G. (2019). Estimation of biodiesel properties based on fatty acid profiles of Chlamydomonas sp. BTA 9032 and Chlorella sp. BTA 9031 obtained under mixotrophic cultivation conditions. *Biofuels*.

Morales, M., Sánchez, L., & Revah, S. (2017). The impact of environmental factors on carbon dioxide fixation by microalgae. *FEMS Microbiology Letters, 365*(3).

Motasemi, F., & Afzal, M. T. (2013). A review on the microwave-assisted pyrolysis technique. *Renewable and Sustainable Energy Reviews, 28*, 317–330.

Norsker, N.-H., Barbosa, M. J., Vermuë, M. H., & Wijffels, R. H. (2011). Microalgal production — a close look at the economics. *Biotechnology Advances, 29*(1), 24–27.

Onumaegbu, C., Mooney, J., Alaswad, A., & Olabi, A. G. (2018). Pretreatment methods for production of biofuel from microalgae biomass. *Renewable and Sustainable Energy Reviews, 93*, 16–26, 2018.

Passos, F., Solé, M., García, J., & Ferrer, I. (2013). Biogas production from microalgae grown in wastewater: Effect of microwave pretreatment. *Applied Energy, 108*, 168–175, 2013.

Passos, F., Uggetti, E., Carre're, H., & Ferrer, I. (2014). Pretreatment of microalgae to improve biogas production: A review. *Bioresource Technology, 172*, 403–412.

Patil, P. D., Gude, V. G., Mannarswamy, A., Deng, S., Cooke, P., Munson-McGee, S., Rhodes, I., Lammers, P., & Nirmalakhandan, N. (2011). Optimization of direct conversion of wet algae to biodiesel under supercritical methanol conditions. *Bioresource Technology, 102*, 118–122.

Posadas, E., Alcántara, C., García-Encina, P. A., Gouveia, L., Guieysse, B., Norvill, Z., Acién, F. G., Markou, G., Congestri, R., Koreiviene, J., & Muñoz, R. (2017). Microalgae cultivation in wastewater. In C. Gonzalez Fernandez, & R. Muñoz (Eds.), *Microalgae-based biofuels and bioproducts* (pp. 67–91). Woodhead Publishing.

Posadas, E., MdM, M., Gomez, C., Acién, F. G., & Muñoz, R. (2015). Influence of pH and CO_2 source on the performance of microalgae-based secondary domestic wastewater treatment in outdoors pilot raceways. *Chemical Engineering Journal, 265*, 239–248.

Pourkarimi, S., Hallajisani, A., Alizadehdakhel, A., & Nouralishahi, A. (2019). Biofuel production through micro- and macroalgae pyrolysis—A review of pyrolysis methods and process parameters. *Journal of Analytical and Applied Pyrolysis, 142*, 104599.

Prathima, A., & Karthikeyan, S. (2017). Characteristics of micro-algal biofuel from Botryococcus braunii. *Energy Sources, Part A: Recovery, Utilization, and Environmental Effects, 39*(2), 206–212.

Puyol, D., Batstone, D. J., Hülsen, T., Astals, S., Peces, M., & Krömer, J. O. (2017). Resource recovery from wastewater by biological technologies: Opportunities, challenges, and prospects. *Frontiers in Microbiology, 7*, 2106.

Raheem, A., Dupont, V., Channa, A. Q., Zhao, X., Vuppaladadiyam, A. K., Taufiq-Yap, Y. H., Zhao, M., & Harun, R. (2017). Parametric characterization of air gasification of *Chlorella vulgaris* biomass. *Energy Fuels, 31*, 2959–2969.

Raheem, A., Prinsen, P., Vuppaladadiyam, A. K., Zhao, M., & Luque, R. (2018). A review on sustainable microalgae based biofuel and bioenergy production: Recent developments. *Journal of Cleaner Production, 181*, 42–59.

Rajvanshi, M., Gautam, K., Manjre, S., Krishna Kumar, G. R., Kumar, C., Govindachary, S., et al. (2019). Stoichiometrically balanced nutrient management using a newly designed nutrient medium for large scale cultivation of *Cyanobacterium aponinum. Journal of Applied Phycology, 31*(5), 2779–2789.

Rawat, I., Kumar, R. R., Mutanda, T., & Bux, F. (2011). Dual role of microalgae: phycoremediation of domestic wastewater and biomass production for sustainable biofuels production. *Applied Energy, 88*(10), 3411–3424.

Salam, K. A., Velasquez-orta, S. B., & Harvey, A. P. (2016). A sustainable integrated in situ transesterification of microalgae for biodiesel production and associated co-product-a review. *Renewable and Sustainable Energy Reviews, 65*, 1179–1198.

Sharif, S. K., Nageswara Rao, B., & Jagadish, D. (2020). Comparative performance and emission studies of the CI engine with *Nodularia Spumigena* microalgae biodiesel versus different vegetable oil derived biodiesel. *SN Applied Sciences, 2*, 858. https://doi.org/10.1007/s42452-020-2697-0

Sharma, K. K., Schuhmann, H., & Schenk, P. M. (2012). High lipid induction in microalgae for biodiesel production. *Energies, 5*(5), 1532–1553.

Sirajunnisa, A. R., & Surendhiran, D. (2016). Algae—A quintessential and positive resource of bioethanol production: A comprehensive review. *Renewable and Sustainable Energy Reviews, 66*, 248–267.

Subramanian, S., Barry, A. N., Pieris, S., & Sayre, R. T. (2013). Comparative energetics and kinetics of autotrophic lipid and starch metabolism in chlorophytic microalgae: Implications for biomass and biofuel production. *Biotechnology for Biofuels, 6*(1), 150.

Suganya, T., Varman, M., Masjuki, H. H., & Renganathan, S. (2016). Macroalgae and microalgae as a potential source for commercial applications along with biofuels production : A biorefinery approach. *Renewable and Sustainable Energy Reviews, 55*, 909–941.

Sun, N., Skaggs, R. L., Wigmosta, M. S., Coleman, A. M., Huesemann, M. H., & Edmundson, S. J. (2020). Growth modeling to evaluate alternative cultivation strategies to enhance national microalgal biomass production. *Algal Research, 49*, 101939.

Teixeira, A. C. R., Sodré, J. R., Guarieiro, L. L. N., Vieira, E. D., de Medeiros, F. F., & Alves, C. T. (2016). *A review on second and third generation bioethanol production*. Warrendale, PA, USA: SAE International.

Wahlen, B. D., Willis, R. M., & Seefeldt, L. C. (2011). Biodiesel production by simultaneous extraction and conversion of total lipids from microalgae, cyanobacteria, and wild mixed-cultures. *Bioresource Technology, 102*, 2724–2730.

Wang, S. K., Wang, X., Tao, H.-H., Sun, X.-S., & Tian, Y.-T. (2018). Heterotrophic culture of *Chlorella pyrenoidosa* using sucrose as the sole carbon source by co-culture with immobilized yeast. *Bioresource Technology, 249*, 425–430.

Xiong, X., Iris, K. M., Cao, L., Tsang, D. C., Zhang, S., & Ok, Y. S. (2017). A review of biochar-based catalysts for chemical synthesis, biofuel production, and pollution control. *Bioresource Technology, 246*, 254–270.

Xu, H., Lee, U., Coleman, A. M., Wigmosta, M. S., & Wang, M. (2019). Assessment of algal biofuel resource potential in the United States with consideration of regional water stress. *Algal Research, 37*, 30–39.

Zhang, C., Ho, S. H., Chen, W. H., Xie, Y., Liu, Z., & Chang, J. S. (2018). Torrefaction performance and energy usage of biomass wastes and their correlations with torrefaction severity index. *Applied Energy, 220*, 598–604.

Zhang, C., Wang, C., Cao, G., Chen, W. H., & Ho, S. H. (2019). Comparison and characterization of property variation of microalgal biomass with non-oxidative and oxidative torrefaction. *Fuel, 246*, 375–385.

Zhu, L., Nugroho, Y. K., Shakeel, S. R., Li, Z., Martinkauppi, B., & Hiltunen, E. (2017). Using microalgae to produce liquid transportation biodiesel: What is next? *Renewable and Sustainable Energy Reviews, 78*, 391–400.

CHAPTER

Algae-derived biosensor materials and their applications 12

Rehab H. Mahmoud[1] and Rabeay Y.A. Hassan[2]

[1]*Water Pollution Research Department, National Research Centre (NRC), Giza, Egypt;* [2]*Nanoscience Program, University of Science and Technology (UST), Zewail City of Science and Technology, Giza, Egypt*

1. Introduction

Due to the extensive industrialization of the world associated with improper waste management policies, a wide range of toxic contaminants have been released into aquatic ecosystems, causing a rapid decline in the world's freshwater resources, thus requiring the implementation of legislation to define water remediation and protection strategies (Lagarde & Jaffrezic-Renault, 2011). Therefore, all water bodies need to be restored to good qualitative and quantitative conditions. Monitoring programs will be required, as well as methods for monitoring and assessing the toxicological effects of water pollutants that are rapid, reliable, and cost-effective. In this context, biosensors appear as excellent alternatives to or techniques complementary to conventional chemical methods.

Algae are easy to cultivate and manipulate, contain several types of enzymes and biomaterials that can catalyze a wide range of bioreactions, and can be paired with different transducers. Besides, they act as indicator for the toxicity of the pollutants toward eukaryotic or prokaryotic cells (Lagarde & Jaffrezic-Renault, 2011). Subsequently, algae-based biosensors are proper apparatus for the online monitoring of the pesticides, volatile organic compounds (VOCs), and heavy metals and security concern in a sensitive, long-term, and multiplexed manner (Altamirano et al., 2004). In the 1990s, algae-based biosensors were first used. A brief story of the history of the algae-based biosensors was recited in Fig. 12.1. The preliminary algal biosensors were based on the photosynthetic activity. In such biosensors, the photosynthetic activity of the living algal cells is affected by the presence of toxic pollutants. This effect can be transformed into electrical or optical signals. Optical sensors are based on the fluorescence of the chlorophyll present in chloroplasts (Grieshaber et al., 2008; Mota et al., 2022), whereas electric (amperometric) sensors depend on the evolution of photosynthetic oxygen with a Clark electrode (Stefan et al., 1999). Several researchers have already described the development of algal biosensors to detect pesticides, VOCs, and heavy metals (Campanella et al., 2001; Chouteau et al., 2005; Shitanda et al., 2009). Among them, algal optical biosensors based on photosynthesis disturbance (targeting chlorophyll fluorescence disturbance) were widely studied for pesticide analysis in contaminated liquid media, with low detection limits (Ferro et al., 2012; Naessens et al., 2000; Védrine et al., 2003a). In contrast to algal optical biosensors, electrochemical biosensors are not sensitive to light; hence, they can be used for turbid samples analysis and do not require labeling.

Algae Materials: Applications Benefitting Health. https://doi.org/10.1016/B978-0-443-18816-9.00016-2

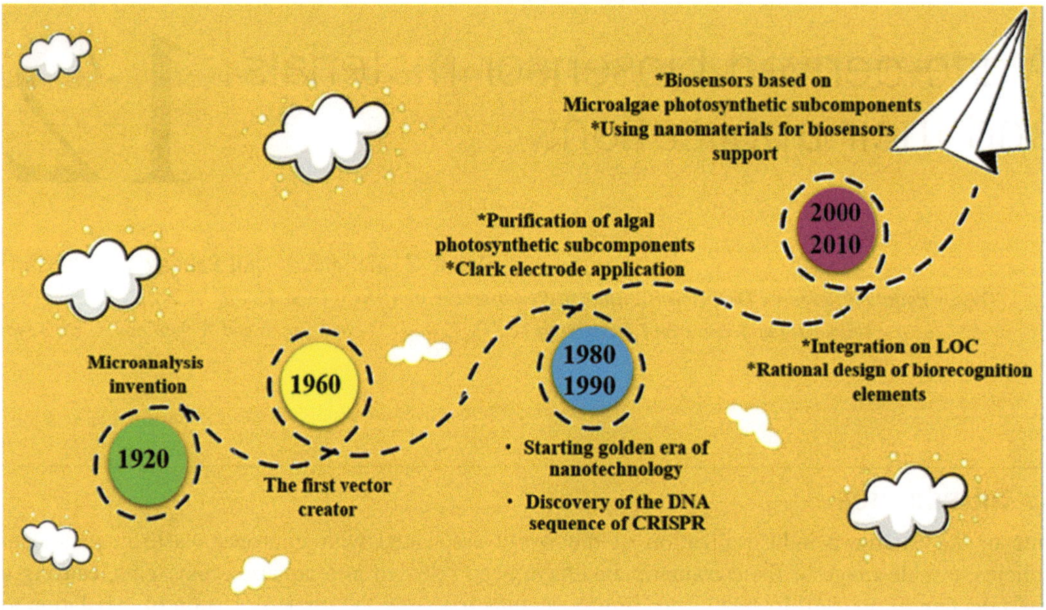

FIGURE 12.1

Timeline recites the story of the microalgae-based biosensors.

In some cases, it is possible to achieve higher selectivity, even without cell modification, by operating at specific potentials where interferences are limited (Lagarde & Jaffrezic-Renault, 2011).

One of the most significant advantages of algae-based biosensors is that they allow for accurate measurements without requiring extensive sample preparation. However, they are low selective for separate analyses. There is only one global signal obtained that represents a variety of toxins. Nonetheless, for water quality assessment, this global response is frequently more useful than individual concentration measurements (Mota et al., 2022). As the green alga *Chlorella vulgaris* has several enzymes surrounding its extracellular membrane and can as indicators for the presence of pesticides, VOCs, and heavy metals, this algal species is commonly used for algae-based biosensors fabrication (Naessens et al., 2000).

This chapter outlines the different microalgal biomaterials (whole cell, enzymes, thylakoid membranes, and biomass) used for biosensors fabrication. Furthermore, it mentions the various applications of such microalgal biomaterials for the detection of pesticides, heavy metals, and VOCs).

1.1 Mechanisms of the algae-based biosensors

A biosensor is an integrated self-contained device that merges a biological recognition element, known as bioreceptor (e.g., enzymes, antibodies, DNA, or microorganisms), immobilized with a physical transducer (e.g., electrochemical, optical, thermal, or piezoelectric), which turns the biological recognition event into a signal proportional to the analyte concentration. The mechanism of biosensors was shown in Fig. 12.2. These microelectronic devices provide accurate, fast, and low-concentration detection of a wide range of substances present in body fluids, soil, air, or water. Unlike the bioassays, transducer in the biosensors is not an integral part of the analytical device. Among the various types of biosensors, the electrochemical one is commonly applied due to its high sensitivity and specificity, and the possibility to reduce the required hardware, providing compact and portable monitoring tools (Grieshaber et al., 2008).

Several algal biosensors have been developed to detect pesticides, VOCs, heavy metals, and even chemical warfare agents (Brayner et al., 2011). In these biosensors, these toxic substances have a significant effect on the metabolic activity of the living algal cells, and this effect can be transformed into electrical or optical signals. One of the main target analytes of algae-based biosensors is pesticides, which includes herbicides, insecticides, and fungicides. Herbicides that target the enzyme acetolactate synthase (ALS) or acetohydroxy acid synthase (AHAS) in algae are widely used in the world. However, the most published scientific papers within the past decade deal with PSII inhibitors. The crucial parameters for an algal biosensor are the cell concentration, temperature, carbon dioxide

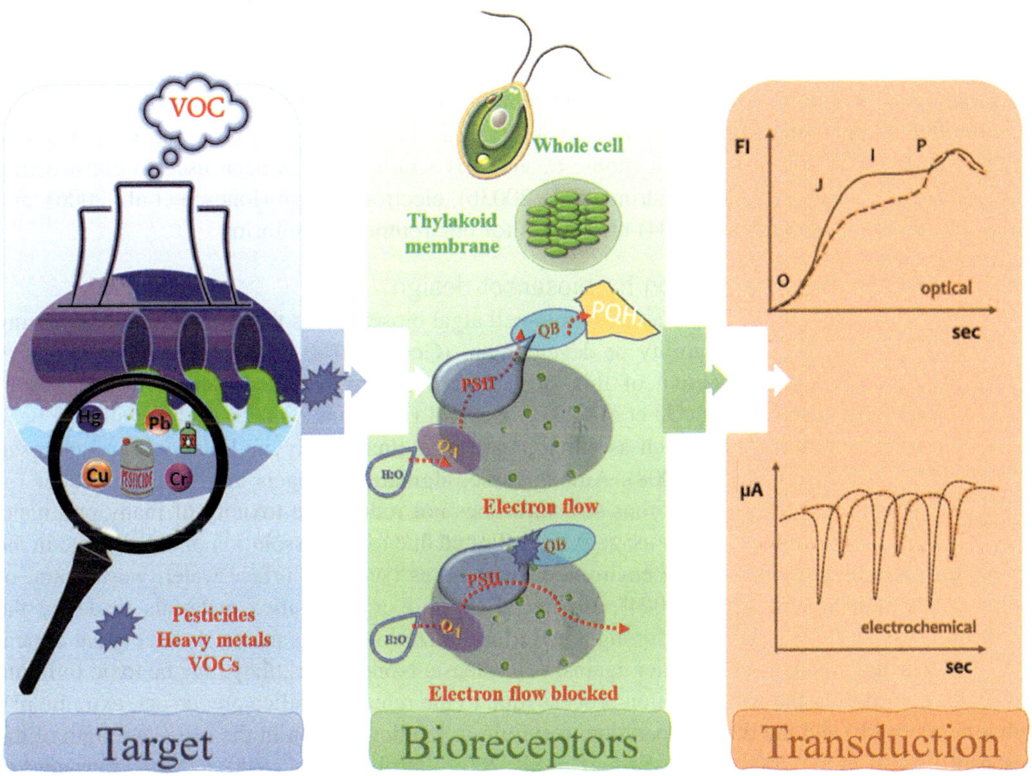

FIGURE 12.2

Schematic for the mechanisms of the algae-based biosensors.

consumption, and photosynthetic dissolved oxygen production (Tsopela et al., 2016). In addition, detecting the molecules that participate in the algal metabolism and photosynthesis, such as hydrogen peroxide, oxygen, and H+ ions, can monitor the presence of inhibitors, such as herbicides. Amperometric measuring techniques are the methods of choice for biosensors that detect dissolved oxygen concentration as an indicator of pollutant-induced photosynthesis inhibition (Tsopela et al., 2014). The high sensitivity and specificity of electrochemical sensors have been documented (Stefan et al., 1999).

1.2 Different algal materials applied for electrochemical biosensors

There are many algal bioreceptors applied in for biosensors fabrication, like the whole algal cells and their photosynthetic subcomponents owing to their ability to be combined into dual transduction miniature devices and monitoring the environmental pollution. Many efforts are exerted for the synthesis of new materials suitable for the screening and monitoring of biologic samples (Immink et al., 2019; Muti et al., 2010; Yumak et al., 2011). Therefore, researchers have tried to synthesize eco-friendly, low-cost, highly stable, and renewable materials; however, biomass has almost all these properties and is abundant (Sethi et al., 2019; Yu et al., 2019).

1.2.1 Algal whole cell

Whole algal cell provides highly sensitive pesticides, heavy metals, and VOCs analysis; hence, it appears to be an attractive candidate biosensor. The green alga *C. vulgaris* is an ideal choice as a biosensing element because it is ubiquitous and grows year-round in all climatic conditions. Its wide availability in nature, simple cell structure, and ability to acclimatize to low nutrient levels make it an excellent choice for biosensing applications. *C. vulgaris* whole cell has been used in constructing optical (Rodriguez Jr et al., 2002; Védrine et al., 2003b), electrochemical (Ionescu et al., 2006), and conductometric (Chouteau et al., 2004) biosensors for environment monitoring.

1.2.1.1 Algal cell immobilization for biosensor design

One of the important steps for constructing whole-cell algal biosensors is the immobilization of living algal cells to study the cells' viability or development (Gosset et al., 2016; Lagarde & Jaffrezic-Renault, 2011). The main objective of this step is to maintain biosensor sensitivity (Eltzov & Marks, 2011) and reproducibility (Gu et al., 2004). Several (in-)organic materials have already been applied for algal immobilization, such as silica, agarose, alginate, polyvinyl alcohol, and chitosan (Mallick, 2020; Moreno-Garrido, 2008). Among these, alginate hydrogels are the most proper for biosensor fabrication. Their macroporous structure does not reduce the toxicity of many pollutants (Abdel-Hamid, 1996). However, their usage is time-limited due to the dissolution of the alginate in the phosphate solutions present in many environmental samples (such as surface water, wastewater, or stormwater) (Taqieddin & Amiji, 2004). Advanced (hybrid or non)-silica materials made by the sol–gel process have been modified for algal cell immobilization. Silica materials have great optical characteristics and create a solid barrier against cell release. Nonetheless, they may be toxic to living algal cells, both during the encapsulation process (due to the release of silica precursors, extreme pH, high ionic strength) and the device operation, resulting in rapid reduction in the signal strength of the biosensor (Nassif et al., 2003; Peña-Vázquez et al., 2009; Perullini et al., 2007). A previous studies implemented a technique to fabricate a new generation of biosensors by double encapsulation of algae in alginate and silica (Perullini et al., 2014). This work was reported that this two-step encapsulation process did not inhibit algal growth and was not toxic to encapsulated algal cells. This technique has been replicated by several researchers, even achieving the coencapsulation of different microorganisms (Pannier et al., 2014; Perullini et al., 2014). An electrochemical biosensor based on *C. vulgaris* immobilized on a platinum electrode was applied for monitoring the inhibition of alkaline phosphatase activity (APA) caused by the pesticide p-nitrophenol (Ionescu et al., 2006). However, the low stability and specificity of whole cells, as well as the low repeatability of the measurements, are critical factors. Furthermore, the whole algal cell biosensors are affected by entire biochemical processes such as the photosynthesis, and respiration mediated by intricate macromolecular assemblies that are embedded in complex surroundings that can alter the results. For these reasons, purified photosynthetic subcomponents (e.g., PSII complexes, thylakoid membranes, and reaction center proteins) are valid alternatives, with higher selectivity due to the direct target analyte recognition. Thus, the drawbacks of these complicated cell structure are avoided, for example, the target does not enter unexpected metabolic flow, altering its structure; and diffusion of the target into cell and thylakoid membranes does not prevent the target from reaching the binding niche. Nonetheless, the extraction procedures for the photosynthetic subcomponents are relatively laborious, expensive, and time-consuming. Moreover, thylakoid membranes, PSII, and reaction center proteins, once isolated from their biological surroundings, are particularly unstable. All these aspects represent serious obstacles to marketability (Antonacci et al., 2018; Antonacci & Scognamiglio, 2019; Veltman & Eltzov, 2019).

1.2.1.2 Microbial fuel cell—based biosensor

Microbial fuel cells (MFCs) are bioelectrochemical systems (BESs) that exploit biological catalytic processes for the generation of electric power or the accumulation of useful compounds. In the algae-based MFCs, algae produce electrons during photosynthesis, some of which are exported outside the cell where they can provide electric current to power devices (Mahmoud et al., 2020, 2022). Many studies observed a decrease in the light-dependent electrical response in pMFCs that was proportional to the concentration of toxicants, thus demonstrating the potential of BESs as environmental

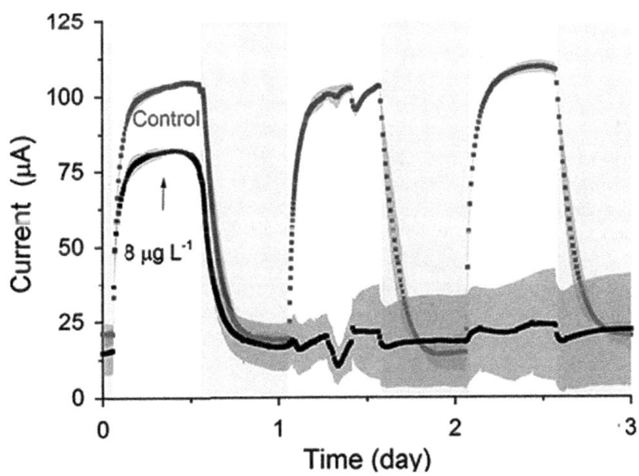

FIGURE 12.3

Response of p-MFC to atrazine. The *arrow* indicates the point of atrazine injection. Shadowed areas indicate the operation during the dark cycle (Gonzalez Olias et al., 2019).

biosensors (Labro et al., 2017; Pescheck et al., 2022). Hence, algae-based MFCs are widely used as biosensors for environmental pollutants. In a previous study, a photosynthetic cathode enriched with the alga *Scenedesmus obliquus* in a photo microbial fuel cell (p-MFC) used for the herbicide atrazine monitoring in water via the detection of dissolved oxygen changes in the catholyte. A detectable change in the photosynthetic pattern was observed after exposure to atrazine (Gonzalez Olias et al., 2019).

1.2.1.3 Brown algae–based biosensors

Brown algae (diatom) frustules have shown great potential as biosensor platforms because of their ability to accept functional groups and because of their variable photoluminescence (PL), Raman spectroscopic, and optoelectric properties in the presence of target molecules and surface modifications (Ren et al., 2013). The structural and material characteristics of diatom frustules even have potential as optical microsystems, such as a micromonochromator (Ferrara et al., 2014). In recent studies, various biosensors with ultrahigh sensitivity using diatom biosilica, peptides, and nanoparticles have been used in proteins self-assembly on Ag and Au NP on pores in diatom surface, which under the laser beam undergo heating and thus release of bioenergy (Gupta et al., 2018; Leonardo et al., 2016). Its uniform porous architecture not only gives it a porous sponge with wide surface area but is also photoluminescent, which is an analog of crystalline silicon (Fig. 12.4). Researchers have developed photoluminescence-based diatom biosensors for sensing the nitroaromatic compounds such as 2, 4, 6-trinitrotoluene (TNT) (Zhen et al., 2016). Thus, adsorbents with essential parameters such as quick adsorption, easy separation, high adsorption capacity, equilibrium time, and with high stability are generally used for treatment of wastewater (Lu et al., 2019). Moreover, diatoms can be affected by

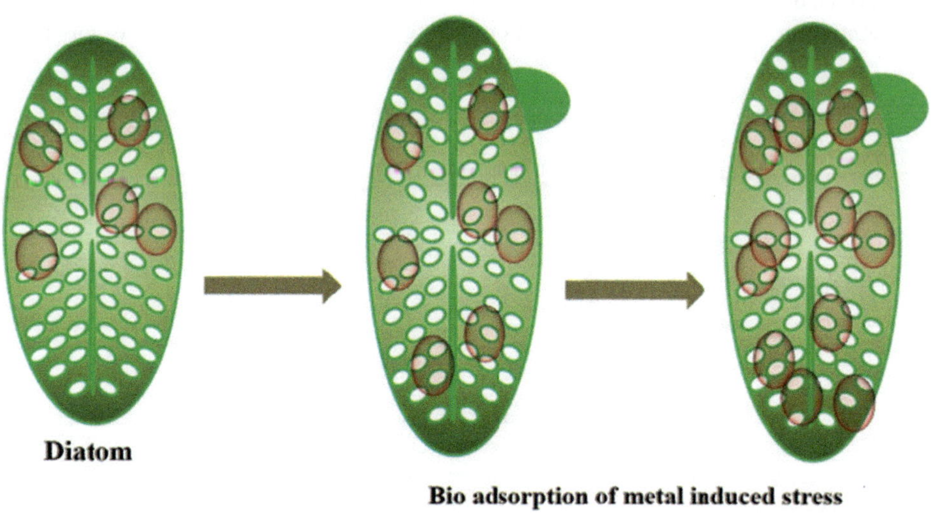

FIGURE 12.4

Diatoms undergoing valve deformation and increase lipid accumulation under stress conditions in presence of heavy metals.

metal contamination such as Ge, Cu, and Zn, which intervene silicate pathway in the diatoms (Pandey & Bergey, 2016). For example, it was seen that *Gomphonema pseudoagur* found in Saraswati River Kurukshetra showed teratological deformation due to heavy metals (Gautam et al., 2017). The inductively coupled plasma mass spectrometry (ICPMS) analysis of the site showed contamination by heavy metals, mainly Pb and Se. The deformity was further associated with high lipid content. Hence, the heavy metal pollutants result into valve or teratological deformation in diatom frustule. The morphological deformations not only make identification difficult but also put the diatom under metabolic stress by increasing the lipid bioaccumulation (Fig. 12.5).

2. Enzymes

Using enzymes present on the external membrane of the algal cells for biosensors fabricating biosensors has potential advantages for detecting the toxic compounds, which act as inhibitors for these enzymes. These enzymes remain in their natural environment which favors long-term stability and reflects the mechanism of toxic inhibition, being therefore of ecological interest. Algal AP is mostly sensitive to heavy metal ions, while algal esterase is commonly affected by organophosphorus pesticides (Brayner et al., 2011). The PSII activity is inhibited by certain pesticides, endocrine disruptive compounds (Rodriguez-Mozaz et al., 2004), heavy metals (Rouillon et al., 2007), explosives such as TNT (Bhalla et al., 2011), or ionizing radiation (Esposito et al., 2006). PSII sensitivity to these toxic compounds depends on a specific interaction between a particular toxic compound and the photosynthetic enzymes in addition to the type of photosynthetic element and its preparation. Many studies reported that the isolated PSII systems are more sensitive than thylakoid membranes or whole cells, as cells possess, in addition to the cell membrane acting as a diffusion barrier, protective intracellular mechanisms that help preventing to a certain extent the effect of herbicides. Phosphatase enzyme is

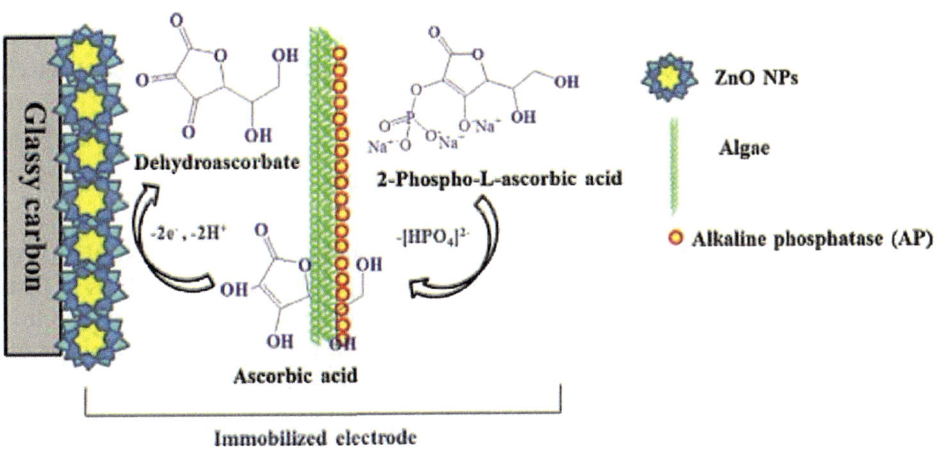

FIGURE 12.5

Mechanism of reaction occurring at AP-algae/ZnO/GC electrode (Pabbi et al., 2018).

one of the enzymes, which can be found abundantly on the *C. vulgaris* cell wall. Under alkaline conditions, the activity of phosphatase enzyme, also known as alkaline phosphatase activity (APA), is inhibited by heavy metals ions. The APA activity is responsible for the dephosphorylation of p-nitrophenyl phosphate into p-nitrophenol and phosphate ions.

A voltammetric detection of another pesticide, chlorpyrifos, was carried out using algal ALP-based biosensor (Pabbi et al., 2018). The detection principle of this biosensor was based on the inhibition of AP algae in presence of chlorpyrifos. For the construction of the biosensor, a glassy carbon electrode was used, onto which were placed ZnO nanoparticles that had the role of increasing the conductivity and/or to increase the electron transfer rate between the electrode surface and the immobilized algae. Algae BSA was deposited on the newly obtained electrode surface and allowed to react for 15 min, and thereafter, the electrode was immersed in glutaraldehyde to cross-link the amines provided by BSA (Fig. 12.5). The obtained biosensor was used for selective voltammetric detection of chlorpyrifos without interferences from malathion, acephate, triazophos, and alkali metals (Pabbi et al., 2018).

2.1 Algal biomass

Algal biomasses are one of cost-effective sources for carbon-based material especially the biochar production (Sindhu et al., 2019). Owing to the salty seawater habitat, macroalgae are rich with metals comparing with microforms. Owing to these characteristics, macroalgae biomass has a high electrical conductivity (42–53 mS cm^{-1}), which makes it an excellent choice as an electrode material (Amoah et al., 2019; Bird et al., 2011; Reddy et al., 2019). A new algae biomass-based electrochemical sensor was designed for detecting vardenafil (VRF) as a model drug-active substance (Bilge et al., 2020). Moreover, algal biomass has strong binding capacities for various metal ions due to the biopolymer's accumulation in their cell walls. As a result, algae biomass can be used for the rapid screening and detection of a wide range of metal ions from aqueous solutions; however, different metal ions show different affinities for algae. Therefore, algal biomass is a cost-effective and efficient material for selective detection of heavy metals in different wastewaters (Yılmaz et al., 2005). Biomass of green microalga *Tetraselmis chuii* was fabricated in a biosorption-based biosensor for the electrochemical detection of copper (II) (Alpat et al., 2007). The combination of biosorption with electrochemical methods results in biosorption-based biosensors, which provides easy, precise, and cheap detection of heavy metals.

2.2 Different applications of algae-based biosensors

2.2.1 Detection of pesticides

Pesticides are widely used in the world and have a significant effect on acetolactate synthase (ALS) or acetohydroxy acid synthase (AHAS) enzymes. Conventional physicochemical analysis of pesticides needs a sample preparation, followed by liquid or gas chromatography and tandem mass spectrometric detection. The use of microalgae-based biosensors opens new avenues for the development of automatic, rapid, and direct analytical methods. They prevent or need minimal sample preparation, allowing for on-site field monitoring. To detect herbicides such as diuron, which inhibit the photosynthetic electron flow, the whole cells of blue green microalga *Synechococcus* sp. PCC 6301 and PCC 7002 were immobilized on silica gel. Both chlorophyll *a* and phycocyanin can be detected via characteristic reflectance bands. Chlorophyll *a* was detectable after 12 weeks of incubation for

Synechococcus PCC 6301 yet the phycocyanin component had disappeared. For *Synechococcus* PCC 7002, this trend was reversed with the water-soluble phycocyanin remaining longer. These hybrid materials could be used as biosensors to detect traces of metals and herbicides by monitoring the photosynthetic activity of the cyanobacteria (Léonard et al., 2010; Meunier et al., 2010; Rooke et al., 2008). In these photoelectrochemical cells, microalgal cells capture light energy and convert it into reduced species. The PSII acceptor DCBQ could be used as an electroactive mediator to transport electrons from the reduced species in the photosynthetic membrane to the working platinum electrode (Marty et al., 1992). Different examples of microalgae-based biosensors applied for pesticides detection were summarized in Table 12.1.

2.2.2 Detection of heavy metals

Several algae-based biosensors showed high sensitivity toward different heavy metals with the lowest limit of detection (LLD) fall within the range of 0.025–1.195 μg/L (Teo, 2014). The linear detection ranges reported that the biosensor is capable of being used for qualitative and quantitative detections. The blue green microalga *Anabaena torulosa*–based biosensor was used for the detection of heavy metals (Cu, Pb, and Cd). The whole algal cells were entrapped on a cellulose membrane through filtration. Then, the membrane was dried and fixed into a cylindrical well, which was designed to be attached to an optical probe. The probe was connected to fluorescence spectrometer with optical fiber. The presence of the heavy metals was indicated by the change of fluorescence emission, before and after the exposure (Wong et al., 2013). In addition, microalgae A. *torulosa* and *Anabaena flos-aquae* had been successfully coupled with optical and amperometric transducers, respectively, for the detection of Cu, Pb, and Cd (Chay et al., 2005; Shing et al., 2013; Tay et al., 2003). In addition, three heavy metals (Cd, Zn, and Pb) were reported to inhibit APA in *C. vulgaris* after exposure depending on

Table 12.1 Examples of microalgae-based biosensors for pesticides detection.

Pesticide/Real sample	Microalgae material affected by pesticide	Type of biosensor	References
Diuron	PSII particles from *Synechococcus bigranulatus*	Amperometry	(Maly et al., 2005)
Atrazine Simazine Ioxynil Bromoxynil Dinoseb	PSII particles from *Synechococcus elongatus*	Amperometry	(Koblizek et al., 1998)
Prometryne Terbuthylazine	Mutant strains of *Chlamydomonas reinhardtii* with engineered D1 protein	Fluorescence	(Giardi et al., 2009)
Linuron Simazine	*C. reinhardtii* whole cell	Amperometry	(Husu et al., 2013)
Diuron, Simazine Irgarol	*Chlorella mirabilis* whole cell	Fluorescence	(Zamaleeva et al., 2011)

the ion concentration (Chouteau et al., 2005). Examples of the different algae-based biosensors developed for the detection of heavy metals in environment were summarized in Table 12.2.

2.2.3 Detection of volatile organic compounds

Bringing together the advantages of biosensors with the survival under different environmental conditions, microalgae-based biosensors for monitoring the VOCs were developed. VOCs are usually used in municipal treatment plants vapor or aerosol. They have a significant effect on the photosynthetic activity specially in the green microalga *C. vulgaris*. Hence, microalgae-based biosensor was designed for rapid detection of methanol using the photosynthetic rate as response signal (Naessens & Tran-Minh, 1998). This biosensor was modified to detect aerosolic perchloroethylene (Naessens & Tran-Minh, 1999). In the modified biosensor, *C. vulgaris* cells were immobilized on the membrane of an oxygen Clark electrode. It measures the photosynthetic oxygen produced by microalgae under illumination conditions in the presence of atmospheric CO_2. As the microalgal membrane is fabricated to work under atmospheric conditions, the biosensor is kept inside a controlled atmosphere chamber, and a water reservoir was fixed to the electrode body to provide humidity for microalgae. This method was based on modification of the photosynthetic activity after spraying the perchloroethylene into the controlled atmosphere chamber, which led to a modification of the photosynthetic oxygen produced under light which can be detected with a Clark electrode (Fig. 12.6). Another microalgae-based biosensor for chemical warfare agents (another type of VOCs), responding within few minutes and employing chlorophyll fluorescence parameters to measure algal responses, was designed (Sanders et al., 2001). These studies successfully demonstrated that, in principle, algal biosensors can be used for monitoring different types of VOCs.

Table 12.2 Examples of the algae-based biosensors developed for the detection of heavy metals in environment.

Algal species	Type of transducer	Heavy metals and media	References
Chlorella vulgaris	Electrochemical—conductometry	Cd, Co, Ni, Pb, and Zn in water	(Berezhetskyy et al., 2007)
Anabaena torulosa	Electrochemical—amperometry	Cu in water	(Chay et al., 2005)
A. torulosa	Optical—fluorescence	Cu, Pb, and Cd in water	(Wong et al., 2013)
Synechocystis sp.	Optical—bioluminescent	Cobalt (Co), Zinc (Zn), Ni in water	(Peca et al., 2008)
Anabaena flos-aquae	Electrochemical—amperometry	Cu and Pb in water	(Tay et al., 2003)

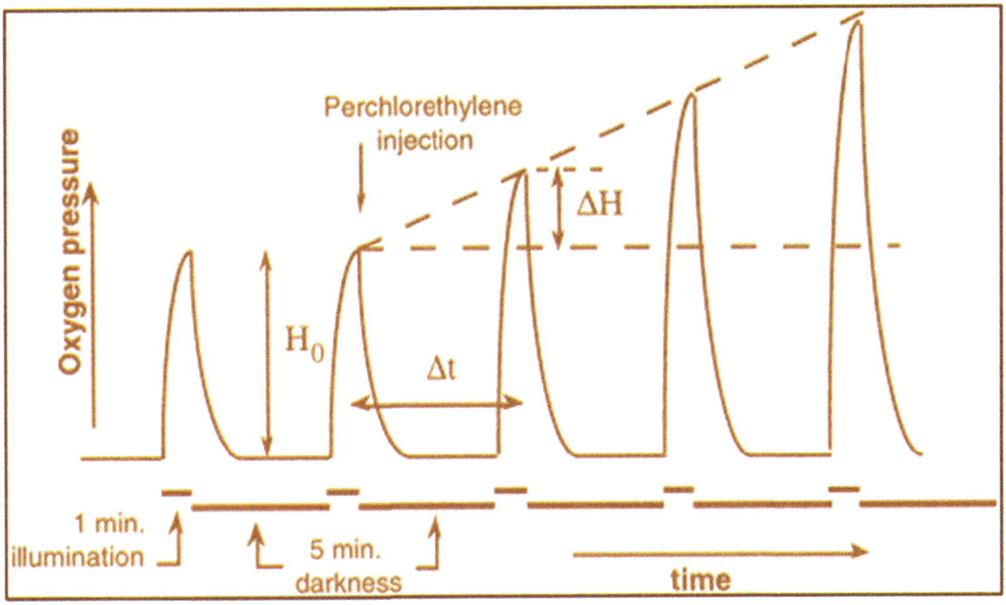

FIGURE 12.6

Typical responses of the algal biosensor to perchloroethylene aerosols.

Reprinted from Ref Naessens, M., & Tran-Minh, C. (1999). Biosensor using immobilized Chlorella microalgae for determination of volatile organic compounds. Sensors and Actuators, B: Chemical, 59*(2). https://doi.org/10.1016/S0925-4005(99)00203-8*

2.3 Conclusions

Microalgae are commonly regarded as a platform for valuable materials such as enzymes, and biomass for different biotechnological applications. Algae-based biosensors have emerged as a promising tool for environmental toxicity monitoring. Their ease of culture, low cost, and potential of mass production have made them a good complement to chemical analyses; they report on global toxicity and, most importantly, on pollutant bioavailability. However, whole algae cell immobilization has appeared to reduce the efficiency of these whole-cell biosensors for some of heavy metals such as Pb^{2+}. Further optimization of biofunctionalization will have to be investigated. Although these biosensors have remarkable sensitivity and accuracy, the stability and storability of these bioanalytical tools are yet to be improved. Hence, further research and studies have to be carried out to further enhance the practicality of algae-based biosensors.

References

Abdel-Hamid, M. I. (1996). Development and application of a simple procedure for toxicity testing using immobilized algae. *Water Science and Technology, 33*(6). https://doi.org/10.1016/0273-1223(96)00289-2

Alpat, S. K., Alpat, Ş., Kutlu, B., Özbayrak, Ö., & Büyükışık, H. B. (2007). Development of biosorption-based algal biosensor for Cu(II) using Tetraselmis chuii. *Sensors and Actuators B: Chemical, 128*(1), 273–278. https://doi.org/10.1016/j.snb.2007.06.011

Altamirano, M., García-Villada, L., Agrelo, M., Sánchez-Martín, L., Martín-Otero, L., Flores-Moya, A., Rico, M., López-Rodas, V., & Costas, E. (2004). A novel approach to improve specificity of algal biosensors using wild-type and resistant mutants: An application to detect TNT. *Biosensors and Bioelectronics, 19*(10). https://doi.org/10.1016/j.bios.2003.11.001

Amoah, J., Hasunuma, T., Ogino, C., & Kondo, A. (2019). 5-Hydroxymethylfurfural production from salt-induced photoautotrophically cultivated Chlorella sorokiniana. *Biochemical Engineering Journal, 142*, 117–123.

Antonacci, A., Lambreva, M. D., Arduini, F., Moscone, D., Palleschi, G., & Scognamiglio, V. (2018). A whole cell optical bioassay for the detection of chemical warfare mustard agent simulants. *Sensors and Actuators, B: Chemical, 257*. https://doi.org/10.1016/j.snb.2017.11.020

Antonacci, A., & Scognamiglio, V. (2019). Photosynthesis-based hybrid nanostructures: Electrochemical sensors and photovoltaic cells as case studies. *TrAC, Trends in Analytical Chemistry, 115*. https://doi.org/10.1016/j.trac.2019.04.001

Berezhetskyy, A. L., Durrieu, C., Nguyen-Ngoc, H., Chovelon, J. M., Dzyadevych, S.v., & Tran-Minh, C. (2007). Conductometric biosensor based on whole-cell microalgae for assessment of heavy metals in wastewater. *Biopolymers and Cell, 23*(6). https://doi.org/10.7124/bc.000786

Bhalla, V., Zhao, X., & Zazubovich, V. (2011). Detection of explosive compounds using Photosystem II-based biosensor. *Journal of Electroanalytical Chemistry, 657*(1–2). https://doi.org/10.1016/j.jelechem.2011.03.026

Bilge, S., Bakirhan, N. K., Donar, Y. O., Sinag, A., Ozkan, S. A., & Okudan, E.Ş. (2020). Green synthesis of carbon based biosensor materials from algal biomass for the sensitive detection of vardenafil. *Journal of Electroanalytical Chemistry, 871*, 114286.

Bird, M. I., Wurster, C. M., de Paula Silva, P. H., Bass, A. M., & De Nys, R. (2011). Algal biochar—production and properties. *Bioresource Technology, 102*(2), 1886–1891.

Brayner, R., Couté, A., Livage, J., Perrette, C., & Sicard, C. (2011). Micro-algal biosensors. *Analytical and Bioanalytical Chemistry, 401*(2). https://doi.org/10.1007/s00216-011-5107-z

Campanella, L., Cubadda, F., Sammartino, M. P., & Saoncella, A. (2001). An algal biosensor for the monitoring of water toxicity in estuarine environments. *Water Research, 35*(1). https://doi.org/10.1016/S0043-1354(00)00223-2

Chay, T. C., Surif, S., & Heng, L. Y. (2005). A copper toxicity biosensor using immobilized cyanobacteria, Anabaena torulosa. *Sensor Letters, 3*(1). https://doi.org/10.1166/sl.2005.006

Chouteau, C., Dzyadevych, S., Chovelon, J.-M., & Durrieu, C. (2004). Development of novel conductometric biosensors based on immobilised whole cell Chlorella vulgaris microalgae. *Biosensors and Bioelectronics, 19*(9), 1089–1096.

Chouteau, C., Dzyadevych, S., Durrieu, C., & Chovelon, J. M. (2005). A bi-enzymatic whole cell conductometric biosensor for heavy metal ions and pesticides detection in water samples. *Biosensors and Bioelectronics, 21*(2). https://doi.org/10.1016/j.bios.2004.09.032

Eltzov, E., & Marks, R. S. (2011). Whole-cell aquatic biosensors. *Analytical and Bioanalytical Chemistry, 400*(4). https://doi.org/10.1007/s00216-010-4084-y

Esposito, D., Faraloni, C., Margonelli, A., Pace, E., Torzillo, G., Zanini, A., & Giardi, M. T. (2006). The effect of ionising radiation on photosynthetic oxygenic microorganisms for survival in space flight revealed by automatic photosystem II-based biosensors. *Microgravity Science and Technology, 18*(3–4). https://doi.org/10.1007/BF02870412

Ferrara, M. A., Dardano, P., de Stefano, L., Rea, I., Coppola, G., Rendina, I., Congestri, R., Antonucci, A., de Stefano, M., & de Tommasi, E. (2014). Optical properties of diatom nanostructured biosilica in Arachnoidiscus sp: Micro-optics from mother nature. *PLoS One, 9*(7). https://doi.org/10.1371/journal.pone.0103750

Ferro, Y., Perullini, M., Jobbagy, M., Bilmes, S. A., & Durrieu, C. (2012). Development of a biosensor for environmental monitoring based on microalgae immobilized in silica hydrogels. *Sensors, 12*(12). https://doi.org/10.3390/s121216879

Gautam, S., Pandey, L. K., Vinayak, V., & Arya, A. (2017). Morphological and physiological alterations in the diatom Gomphonema pseudoaugur due to heavy metal stress. *Ecological Indicators, 72*. https://doi.org/10.1016/j.ecolind.2016.08.002

Giardi, M. T., Scognamiglio, V., Rea, G., Rodio, G., Antonacci, A., Lambreva, M., Pezzotti, G., & Johanningmeier, U. (2009). Optical biosensors for environmental monitoring based on computational and biotechnological tools for engineering the photosynthetic D1 protein of Chlamydomonas reinhardtii. *Biosensors and Bioelectronics, 25*(2). https://doi.org/10.1016/j.bios.2009.07.003

Gonzalez Olias, L., Cameron, P. J., & di Lorenzo, M. (2019). Effect of electrode properties on the performance of a photosynthetic microbial fuel cell for atrazine detection. *Frontiers in Energy Research, 7*. https://doi.org/10.3389/fenrg.2019.00105

Gosset, A., Ferro, Y., & Durrieu, C. (2016). Methods for evaluating the pollution impact of urban wet weather discharges on biocenosis: A review. *Water Research, 89*. https://doi.org/10.1016/j.watres.2015.11.020

Grieshaber, D., MacKenzie, R., Vörös, J., & Reimhult, E. (2008). Electrochemical biosensors—Sensor principles and architectures. *Sensors, 8*(3). https://doi.org/10.3390/s8031400

Gu, M. B., Mitchell, R. J., & Kim, B. C. (2004). Whole-cell-based biosensors for environmental biomonitoring and application. In , *Vol. 87. Advances in biochemical engineering/biotechnology*. https://doi.org/10.1007/b13533

Gupta, S., Kashyap, M., Kumar, V., Jain, P., Vinayak, V., & Joshi, K. B. (2018). Peptide mediated facile fabrication of silver nanoparticles over living diatom surface and its application. *Journal of Molecular Liquids, 249*. https://doi.org/10.1016/j.molliq.2017.11.086

Husu, I., Rodio, G., Touloupakis, E., Lambreva, M. D., Buonasera, K., Litescu, S. C., Giardi, M. T., & Rea, G. (2013). Insights into photo-electrochemical sensing of herbicides driven by Chlamydomonas reinhardtii cells. *Sensors and Actuators, B: Chemical, 185*. https://doi.org/10.1016/j.snb.2013.05.013

Immink, A. H. J., Nieuwenhuis, J. H., Sabatte, G., & Hardeman, W. M. (2019). Biosensor for the detection of target components in a sample. *Google Patents*.

Ionescu, R. E., Abu-Rabeah, K., Cosnier, S., Durrieu, C., Chovelon, J., & Marks, R. S. (2006). Amperometric algal *Chlorella vulgaris* cell biosensors based on alginate and polypyrrole-alginate gels. *Electroanalysis: An International Journal Devoted to Fundamental and Practical Aspects of Electroanalysis, 18*(11), 1041–1046.

Koblizek, M., Masojidek, J., Komenda, J., Kucera, T., Pilloton, R., Mattoo, A. K., & Giardi, M. T. (1998). A sensitive photosystem II-based biosensor for detection of a class of herbicides. *Biotechnology and Bioengineering, 60*(6). https://doi.org/10.1002/(SICI)1097-0290(19981220)60:6<664::AID-BIT3>3.0.CO;2-B

Labro, J., Craig, T., Wood, S. A., & Packer, M. A. (2017). Demonstration of the use of a photosynthetic microbial fuel cell as an environmental biosensor. *International Journal of Nanotechnology, 14*(1–6). https://doi.org/10.1504/IJNT.2017.082467

Lagarde, F., & Jaffrezic-Renault, N. (2011). Cell-based electrochemical biosensors for water quality assessment. *Analytical and Bioanalytical Chemistry, 400*(Issue 4). https://doi.org/10.1007/s00216-011-4816-7

Leonardo, S., Prieto-Simón, B., & Campàs, M. (2016). Past, present and future of diatoms in biosensing. *TrAC, Trends in Analytical Chemistry, 79*. https://doi.org/10.1016/j.trac.2015.11.022

Léonard, A., Rooke, J. C., Meunier, C. F., Sarmento, H., Descy, J. P., & Su, B. L. (2010). Cyanobacteria immobilised in porous silica gels: Exploring biocompatible synthesis routes for the development of photobioreactors. *Energy & Environmental Science, 3*(3). https://doi.org/10.1039/b923859j

Lu, M., Li, L., Shen, S., Chen, D., & Han, W. (2019). Highly efficient removal of Pb2+ by a sandwich structure of metal-organic framework/GO composite with enhanced stability. *New Journal of Chemistry, 43*(2). https://doi.org/10.1039/c8nj05091k

Mahmoud, R. H., Abdo, S. M., Samhan, F. A., Ibrahim, M. K., Ali, G. H., & Hassan, R. Y. A. (2020). Biosensing of algal-photosynthetic productivity using nanostructured bioelectrochemical systems. *Journal of Chemical Technology and Biotechnology, 95*(4). https://doi.org/10.1002/jctb.6282

Mahmoud, R. H., Gomaa, O. M., & Hassan, R. Y. A. (2022). Bio-electrochemical frameworks governing microbial fuel cell performance: Technical bottlenecks and proposed solutions. *RSC Advances, 12*(10). https://doi.org/10.1039/d1ra08487a

Mallick, N. (2020). Immobilization of microalgae. In , *Vol. 2100. Methods in molecular biology*. https://doi.org/10.1007/978-1-0716-0215-7_31

Maly, J., Masojidek, J., Masci, A., Ilie, M., Cianci, E., Foglietti, V., Vastarella, W., & Pilloton, R. (2005). Direct mediatorless electron transport between the monolayer of photosystem II and poly(mercapto-p-benzoquinone) modified gold electrode—New design of biosensor for herbicide detection. *Biosensors and Bioelectronics, 21*(6). https://doi.org/10.1016/j.bios.2005.02.013

Marty, J. L., Mionetto, N., & Roui lion, R. (1992). Entrapped enzymes in photocrosslinkable gel for enzyme electrodes. *Analytical Letters, 25*(8). https://doi.org/10.1080/00032719208017123

Meunier, C. F., Dandoy, P., & Su, B. L. (2010). Encapsulation of cells within silica matrixes: Towards a new advance in the conception of living hybrid materials. *Journal of Colloid and Interface Science, 342*(2). https://doi.org/10.1016/j.jcis.2009.10.050

Moreno-Garrido, I. (2008). Microalgae immobilization: Current techniques and uses. *Bioresource Technology, 99*(10). https://doi.org/10.1016/j.biortech.2007.05.040

Mota, G. F., de Sousa, I. G., de Oliveira, A. L. B., Cavalcante, A. L. G., da Silva Moreira, K., Cavalcante, F. T. T., da Silva Souza, J. E., de Aguiar Falcão, Í. R., Rocha, T. G., & Valério, R. B. R. (2022). Biodiesel production from microalgae using lipase-based catalysts: Current challenges and prospects. *Algal Research, 62*, 102616.

Muti, M., Kuralay, F., Erdem, A., Abaci, S., Yumak, T., & Sinağ, A. (2010). Tin oxide nanoparticles-polymer modified single-use sensors for electrochemical monitoring of label-free DNA hybridization. *Talanta, 82*(5), 1680–1686.

Naessens, M., Leclerc, J. C., & Tran-Minh, C. (2000). Fiber optic biosensor using Chlorella vulgaris for determination of toxic compounds. *Ecotoxicology and Environmental Safety, 46*(2). https://doi.org/10.1006/eesa.1999.1904

Naessens, M., & Tran-Minh, C. (1998). Whole-cell biosensor for direct determination of solvent vapours. *Biosensors and Bioelectronics, 13*(3–4). https://doi.org/10.1016/S0956-5663(97)00121-8

Naessens, M., & Tran-Minh, C. (1999). Biosensor using immobilized Chlorella microalgae for determination of volatile organic compounds. *Sensors and Actuators, B: Chemical, 59*(2), 100–102. https://doi.org/10.1016/S0925-4005(99)00203-8

Nassif, N., Roux, C., Coradin, T., Rager, M. N., Bouvet, O. M. M., & Livage, J. (2003). A sol-gel matrix to preserve the viability of encapsulated bacteria. *Journal of Materials Chemistry, 13*(2). https://doi.org/10.1039/b210167j

Pabbi, M., Kaur, A., Mittal, S. K., & Jindal, R. (2018). A surface expressed alkaline phosphatase biosensor modified with flower shaped ZnO for the detection of chlorpyrifos. *Sensors and Actuators, B: Chemical, 258*. https://doi.org/10.1016/j.snb.2017.11.079

Pandey, L. K., & Bergey, E. A. (2016). Exploring the status of motility, lipid bodies, deformities and size reduction in periphytic diatom community from chronically metal (Cu, Zn) polluted waterbodies as a biomonitoring tool. *Science of the Total Environment, 550*. https://doi.org/10.1016/j.scitotenv.2015.11.151

Pannier, A., Soltmann, U., Soltmann, B., Altenburger, R., & Schmitt-Jansen, M. (2014). Alginate/silica hybrid materials for immobilization of green microalgae *Chlorella vulgaris* for cell-based sensor arrays. *Journal of Materials Chemistry B, 2*(45). https://doi.org/10.1039/c4tb00944d

Peca, L., Kós, P. B., Máté, Z., Farsang, A., & Vass, I. (2008). Construction of bioluminescent cyanobacterial reporter strains for detection of nickel, cobalt and zinc. *FEMS Microbiology Letters, 289*(2). https://doi.org/10.1111/j.1574-6968.2008.01393.x

Peña-Vázquez, E., Maneiro, E., Pérez-Conde, C., Moreno-Bondi, M. C., & Costas, E. (2009). Microalgae fiber optic biosensors for herbicide monitoring using sol-gel technology. *Biosensors and Bioelectronics, 24*(12). https://doi.org/10.1016/j.bios.2009.05.013

Perullini, M., Orias, F., Durrieu, C., Jobbágy, M., & Bilmes, S. A. (2014). Co-encapsulation of Daphnia magna and microalgae in silica matrices, a stepping stone toward a portable microcosm. *Biotechnology Reports, 4*(1). https://doi.org/10.1016/j.btre.2014.10.002

Perullini, M., Rivero, M. M., Jobbágy, M., Mentaberry, A., & Bilmes, S. A. (2007). Plant cell proliferation inside an inorganic host. *Journal of Biotechnology, 127*(3). https://doi.org/10.1016/j.jbiotec.2006.07.024

Pescheck, M., Schweizer, A., & Bláha, L. (2022). Innovative electrochemical biosensor for toxicological investigations on algae and cyanobacteria. *Bioelectrochemistry, 143*. https://doi.org/10.1016/j.bioelechem.2021.107926

Reddy, C. N., Nguyen, H. T. H., Noori, M. T., & Min, B. (2019). Potential applications of algae in the cathode of microbial fuel cells for enhanced electricity generation with simultaneous nutrient removal and algae biorefinery: Current status and future perspectives. *Bioresource Technology, 292*, 122010.

Ren, F., Campbell, J., Wang, X., Rorrer, G. L., & Wang, A. X. (2013). Enhancing surface plasmon resonances of metallic nanoparticles by diatom biosilica. *Optics Express, 21*(13). https://doi.org/10.1364/oe.21.015308

Rodriguez-Mozaz, S., Marco, M. P., Lopez De Alda, M. J., & Barceló, D. (2004). Biosensors for environmental monitoring of endocrine disruptors: A review article. *Analytical and Bioanalytical Chemistry, 378*(3). https://doi.org/10.1007/s00216-003-2385-0

Rodriguez, M., Jr., Sanders, C. A., & Greenbaum, E. (2002). Biosensors for rapid monitoring of primary-source drinking water using naturally occurring photosynthesis. *Biosensors and Bioelectronics, 17*(10), 843–849.

Rooke, J. C., Meunier, C., Léonard, A., & Su, B. L. (2008). Energy from photobioreactors: Bioencapsulation of photosynthetically active molecules, organelles, and whole cells within biologically inert matrices. *Pure and Applied Chemistry, 80*(11). https://doi.org/10.1351/pac200880112345

Rouillon, R., Piletsky, S. A., Breton, F., Piletska, E.v., & Carpentier, R. (2007). Photosystem II biosensors for heavy metals monitoring. In *Biotechnological applications of photosynthetic proteins: Biochips, biosensors and biodevices*. https://doi.org/10.1007/978-0-387-36672-2_15

Sanders, C. A., Rodriguez, M., & Greenbaum, E. (2001). Stand-off tissue-based biosensors for the detection of chemical warfare agents using photosynthetic fluorescence induction. *Biosensors and Bioelectronics, 16*(7–8). https://doi.org/10.1016/S0956-5663(01)00158-0

Sethi, M., Bantawal, H., Shenoy, U. S., & Bhat, D. K. (2019). Eco-friendly synthesis of porous graphene and its utilization as high performance supercapacitor electrode material. *Journal of Alloys and Compounds, 799*, 256–266.

Shing, W. L., Heng, L. Y., & Surif, S. (2013). Performance of a cyanobacteria whole cell-based fluorescence biosensor for heavy metal and pesticide detection. *Sensors, 13*(5). https://doi.org/10.3390/s130506394

Shitanda, I., Takamatsu, S., Watanabe, K., & Itagaki, M. (2009). Amperometric screen-printed algal biosensor with flow injection analysis system for detection of environmental toxic compounds. *Electrochimica Acta, 54*(21). https://doi.org/10.1016/j.electacta.2009.04.005

Sindhu, R., Binod, P., Pandey, A., Ankaram, S., Duan, Y., & Awasthi, M. K. (2019). Biofuel production from biomass: Toward sustainable development. In *Current developments in biotechnology and bioengineering* (pp. 79–92). Elsevier.

Stefan, R. I., van Staden, J. F., & Aboul-Enein, H. Y. (1999). Electrochemical sensor arrays. *Critical Reviews in Analytical Chemistry, 29*(2). https://doi.org/10.1080/10408349891199293

Taqieddin, E., & Amiji, M. (2004). Enzyme immobilization in novel alginate-chitosan core-shell microcapsules. *Biomaterials, 25*(10). https://doi.org/10.1016/j.biomaterials.2003.08.034

Tay, C. C., Surif, S., & Lee, Y. H. (2003). *Detection of metals toxicity biosensor using immobilized cyanobacteria Anabaena flos-aquae*. https://doi.org/10.1109/ASENSE.2003.1225018, 2003 Asian Conference on Sensors, AsiaSENSE 2003.

Teo, S. (2014). Whole cell-based biosensors for environmental heavy metals detection. *Annual Research & Review in Biology, 4*(17). https://doi.org/10.9734/arrb/2014/9472

Tsopela, A., Laborde, A., Salvagnac, L., Ventalon, V., Bedel-Pereira, E., Séguy, I., Temple-Boyer, P., Juneau, P., Izquierdo, R., & Launay, J. (2016). Development of a lab-on-chip electrochemical biosensor for water quality analysis based on microalgal photosynthesis. *Biosensors and Bioelectronics, 79*. https://doi.org/10.1016/j.bios.2015.12.050

Tsopela, A., Lale, A., Vanhove, E., Reynes, O., Séguy, I., Temple-Boyer, P., Juneau, P., Izquierdo, R., & Launay, J. (2014). Integrated electrochemical biosensor based on algal metabolism for water toxicity analysis. *Biosensors and Bioelectronics, 61*. https://doi.org/10.1016/j.bios.2014.05.004

Védrine, C., Leclerc, J. C., Durrieu, C., & Tran-Minh, C. (2003a). Optical whole-cell biosensor using Chlorella vulgaris designed for monitoring herbicides. *Biosensors and Bioelectronics, 18*(4). https://doi.org/10.1016/S0956-5663(02)00157-4

Védrine, C., Leclerc, J.-C., Durrieu, C., & Tran-Minh, C. (2003b). Optical whole-cell biosensor using Chlorella vulgaris designed for monitoring herbicides. *Biosensors and Bioelectronics, 18*(4), 457–463.

Veltman, B., & Eltzov, E. (2019). Whole-cell-based fiber-optic biosensors. In *Handbook of cell biosensors*. https://doi.org/10.1007/978-3-319-47405-2_126-1

Wong, L. S., Lee, Y. H., & Surif, S. (2013). Whole cell biosensor using Anabaena torulosa with optical transduction for environmental toxicity evaluation. *Journal of Sensors, 2013*, 1–8. https://doi.org/10.1155/2013/567272

Yılmaz, A. B., Işık, O., & Sayın, S. (2005). Bioaccumulation and toxicity of different copper concentrations in Tetraselmis chuii. *Su Urunleri Dergisi, 22*(3), 297–301.

Yu, F., Li, S., Chen, W., Wu, T., & Peng, C. (2019). Biomass-derived materials for electrochemical energy storage and conversion: Overview and perspectives. *Energy & Environmental Materials, 2*(1), 55–67.

Yumak, T., Kuralay, F., Muti, M., Sinag, A., Erdem, A., & Abaci, S. (2011). Preparation and characterization of zinc oxide nanoparticles and their sensor applications for electrochemical monitoring of nucleic acid hybridization. *Colloids and Surfaces B: Biointerfaces, 86*(2), 397–403.

Zamaleeva, A. I., Sharipova, I. R., Shamagsumova, R.v., Ivanov, A. N., Evtugyn, G. A., Ishmuchametova, D. G., & Fakhrullin, R. F. (2011). A whole-cell amperometric herbicide biosensor based on magnetically functionalised microalgae and screen-printed electrodes. *Analytical Methods, 3*(3). https://doi.org/10.1039/c0ay00627k

Zhen, L., Ford, N., Gale, D. K., Roesijadi, G., & Rorrer, G. L. (2016). Photoluminescence detection of 2,4,6-trinitrotoluene (TNT) binding on diatom frustule biosilica functionalized with an anti-TNT monoclonal antibody fragment. *Biosensors and Bioelectronics, 79*. https://doi.org/10.1016/j.bios.2016.01.002

CHAPTER 13

Algae materials for bionanopesticides: nanoparticles and composites

Ahmed A. El-Refaey and Salem S. Salem

Botany and Microbiology Department, Faculty of Science, Al-Azhar University, Nasr City, Cairo, Egypt

1. Introduction

Nanotechnology is a branch of science concerned with particles smaller than 100 nm in at least one of their three dimensions (Salem et al., 2023). These particles bridge the gap between bulk materials and atomic structures and have a high ratio of surface area to mass. Thus, they have unique physicochemical properties (Shaheen et al., 2021; Abdelaziz et al., 2022; Baker et al., 2013; Abu-Elghait et al., 2021). These attributes lead to significant changes in their mechanical properties, melting point, optical absorption, thermal, electrical conductivity, biological, and catalytic activities (Al-Zahrani et al., 2022; Al-Rajhi et al., 2022; Hammad et al., 2022; Shah et al., 2015; Salem, 2022). Because of the unique physicochemical properties of nanoparticles (NPs), developments and extensive research related to NPs expand their potential application in wide areas of science and technology, such as electronics, energy, biomedicine, cosmetics, the automotive industry, and agriculture (Bayda et al., 2019; Lewisoscar et al., 2016; Verma et al., 2022; Mohamed et al., 2021; Salem, 2022). The global nanotechnology industry is a rapidly growing field of science and technology that was estimated to reach about \$ 41.8 billion by 2020 (BCC, 2015). By 2024, it is expected to exceed US \$125 billion (Global nanotechnology market report, 2018). NPs are usually fabricated through several physical and chemical routes such as mechanical milling, laser ablation, microemulsion, and sol-gel (Elakraa et al., 2022; Shaheen et al., 2019; Doghish et al., 2022; Hashem et al., 2022; Cheng et al., 2016; Abdel-moneim et al., 2022). However, traditional physical and chemical procedures are costly, the yield is low, and they have poor size dispersion. In addition, the chemicals used in these procedures cause threats to the ecosystem and the techniques are too expensive to be applied on a large scale (Agarwal et al., 2019). To overcome these hazards, sustainable biological routes to synthesizing NPs are needed. An eco-friendly, economical, high-yielding, expeditious, and energy-efficient approach is the biosynthesis of stable metal NPs (Salem et al., 2022; Saied et al., 2022; Hashem et al., 2022; Hammad et al., 2022; Guan et al., 2022). Researchers are concentrating efforts on fabricating NPs based on plants, bacteria, and fungi because they contain biomolecules such as proteins, amines, amino acids, phenols, sugars, ketones, and aldehydes, which act as reducing agents, capping agents, and stabilizers for NPs (Hashem & Salem, 2022; Aref & Salem, 2020; Fouda et al., 2018; Salem et al., 2022; Siddiqi & Husen, 2016). Of all the organisms, algae are called bionanofactories because both live and dead

Algae Materials: Applications Benefitting Health. https://doi.org/10.1016/B978-0-443-18816-9.00004-6

dried biomasses are used to synthesize metallic NPs. Algae have long been employed as food, feed, additives, cosmetics, pharmaceuticals, and fertilizer both commercially and industrially, but the attention is switching to the green synthesis of NPs. Algae are a good source of metabolites; they are easy to culture and grow quickly. Metabolites are used to reduce, cap, and stabilize metal precursors, yielding metal, metal oxide, or bimetallic NPs, and demonstrate promising results that are cost-effective and environmentally friendly. The term "algae" includes a wide range of photosynthetic organisms that inhabit fresh and marine aquatic habitats, although some algae can thrive in a terrestrial environment. Algae are divided into microalgae (visible only by light microscope) and macroalgae (visible to the naked eye). Cyanobacteria (blue-green algae) represent prokaryotic microalgae, whereas seaweed and other groups of algae represent eukaryotic algae (Barsanti & Gualtieri, 2014; Gärtner et al., 2021; Specht et al., 2017). Algae are ideal candidates for the biosynthesis of NPs because of their fast growth rate, ease of harvesting, and cost-effective scale-up. Algae are the most primitive organisms inhabiting diverse ecosystems. They serve as primary photosynthetic organisms (Fawcett et al., 2017). Because of their biocompatibility, safety, and unique physicochemical properties, algal biosynthesized NPs have been investigated for biomedical applications that are antibacterial, antifungal, anticancerous, antifouling, bioremediating, and biosensing (Chaudhary et al., 2020). Algae have the ability to accumulate metals and convert them into NPs. This makes them the greatest alternative for environmentally friendly synthesis, and hence they are called bionanofactories, because live and dead dried biomass is used to synthesize metallic NPs (Mukherjee et al., 2021). NP synthesis may be intracellular or extracellular, depending on the site of NP synthesis and reducing agents: The extracellular synthesis of NPs is carried out as a result of the activities of reducing agents present in the algal aqueous phase (dead dried biomass), such as polysaccharides, reducing sugars, peptides, proteins, pigments, and other reducing factors that may reduce metal ions and precipitate metallic ions in the form of NPs (Khanna et al., 2019). The intracellular formation of NPs is completed (live cells) owing to algal metabolism (i.e., photosynthesis and respiration), which might be responsible for reducing metallic ions (Arya et al., 2018; Jeffryes et al., 2015). Algae with different divisions are employed to synthesize different types of metal NPs (Chan et al., 2022). Among the various types of algae, the most extensively studied ones to biosynthesize NPs are brown, red, blue-green, and micro and macro green algae (El-Belely et al., 2021; Elrefaey et al., 2022). The green synthesis of NPs mainly depends on various factors such as the pH, the ratio of the metal precursor and bioextract, the duration of the reaction mixture, and the temperature. These parameters influence the nucleation, growth, aging, agglomeration, and stabilization of NPs, which can change their size, shape, and crystallinity (Aboelfetoh et al., 2017; Elrefaey, 2022). After NPs are synthesized, they are subjected to various techniques to characterize the morphology (size, shape, and size distribution), crystal structure, elemental composition, quantity, and purity of particles. The most widely used techniques for determining the unique features of NPs include UV-visible spectroscopy, high-resolution transmission electron microscopy, Fourier transmission infrared spectroscopy, X-ray diffraction, dynamic light scattering, energy-dispersive spectroscopy, and selected area electron diffraction (Salem & Fouda, 2020). Crop diseases triggered by plant pathogens and pests are the main causes of crop production loss and restrictions to agricultural development. Over the past century, pesticides have successfully controlled disease, reduced agricultural productivity loss, and improved the quality and quantity of food, clothing, and animal health, as well as fought certain diseases transferred to humans (Krishnamoorthy, 2012). On the other hand, excessive pesticide use and misuse leads to the evolution of drug resistance and makes agrochemicals problematic. Moreover, pesticides result in environmental pollution by accumulating in other organisms such as fish and threaten beneficial microorganisms in the soil as well as honeybees and earthworms. Pesticide accumulation endangers biodiversity by disrupting the structure and function of the ecosystem and causing broad environmental contamination through the persistent use of insecticides. Eventually, it threatens human health through the food chain (Krishnamoorthy, 2012). Therefore, alternative environmental approaches to controlling disease and pests are critical to developing sustainable and intensive agriculture (Lindsey et al., 2020; Saratale et al., 2018). With progress in nanotechnology, biosynthetic NPs have proven to be an alternative approach to agrochemicals in the heavy metal remediation of soil, for the promotion of plant growth, and for managing plant disease with less toxicity and higher performance. Therefore, the bioinspired synthesis of NPs is an inevitable trend for sustainable development in the agricultural field (Xu et al., 2021). NPs such as copper (Cu-NPs) magnesium oxide, gold (Au-NPs), iron oxide, palladium, sele nium (Se-NPs), silver (Ag-NPs), titanium dioxide, and zinc oxide (ZnO-NPs) have a dual effect: they protect the plant from bacterial, fungal, and pest infection; and biosynthesized NPs can become promoters for plant growth, improving the crop yield (Lade et al., 2017). Algae and their metabolites also have antimicrobial, nematocidal, herbicidal, and insecticidal/acaricidal activities against crop pathogens and can be used as biopesticides (Asimakis et al., 2022). Moreover, Cyanobacteria (blue-green algae) are regarded as one of the most important biological pest management agents for controlling plant disease. They produces a variety of biomolecules that are effective against fungi, bacteria, nematodes, and other diseases. They destroys the organism by deactivating vital enzymes, disrupting cytoplasmic membranes, and inhibiting protein synthesis. They can also boost the immune system of plants and promote the production of antioxidant compounds for disease resistance. They destroy organisms by inhibiting protein synthesis, inactivating vital enzymes, and disrupting cytoplasmic membranes. They can also help plants' immune systems and promote the production of antioxidant enzymes for disease resistance (Costa et al., 2019; Mol & Purushothaman, 2020).

2. Nanomaterials as pesticides

The efficacy and environmental safety of nanopesticides are thought to enhance as well as the pesticides' ability to penetrate pests more effectively. (Nuruzzaman et al., 2016). NPs have a large surface area, which leads to increased activity, reactivity, effectiveness, and absorption (Elrefaey et al., 2022). Thus, nanomaterials (NMs) could help to produce new pesticides and insect repellants.

Iron as a trace element has a vital role in forming chlorophyll and hormones (such as gibberellic acid and ethylene), photosynthesis, reactive oxygen species generation and scavenging, mitochondrial respiration, and osmotic regulator in plants (Hänsch et al., 2009). The green synthesis of iron oxide NP showed significant antifungal activity against *Aspergillus niger*. It cause "black mold" and *Mucor piriformis*, which causes *Mucor* rot (Devi et al., 2019). Ashraf et al. (2022) showed that the biosynthesized of iron oxide NP had powerful antifungal activity by inhibiting the growth of *Fusarium oxysporum* during in vitro and in vivo assays. There was also a decline in the disease incidence and severity. This made it more efficient than commercially available fungicides. In addition to acting as a pesticide, iron oxide NPs at a low concentration have positive effects on the growth and development of plants by promoting nutrient absorption and enhancing photosynthetic efficiency (El-Temsah & Joner, 2012; Lau et al., 2020; Sekhon, 2014).

Silicon dioxide NPs have the potential for an efficient insecticidal effect on Colorado potato beetle larvae and crucifer flea beetles (Shatalova et al., 2022). Also, plant growth can be influenced by silica NPs. There is a positive effect of silicon NPs on plant productivity and some physiologic characteristics: the proline content and antioxidant enzyme activity were increased to overcome environmental stress on plants (Elshayb et al., 2021; Thabet et al., 2021).

Ag-NPs are a type of nanopesticide that is active against a variety of pests. Biosynthesized Ag-NPs were efficacious as a potent larvicidal agent. *Aedes aegypti*, *Anopheles stephensi*, and *Culex quinquefasciatus* were all destroyed in a larvicidal bioassay with Ag-NPs at 10 mg/L. These results suggest that biosynthesized Ag-NPs have the potential to be used as an ideal eco-friendly agent to control of mosquito larvae (Elumalai et al., 2017). Ag-NPs are used to control the cowpea seed beetle *Callosobruchus maculatus* (Rouhani, et al.,2013), the black cutworm *Agrotis ipsilon*, and the land snail *Monacha obstructa* (Morsy & Mohamed, 2022), and controlled *Spodoptera littoralis* in cotton fields (Ahmed et al., 2018). *Ulva fasciata*–based bionanoparticles of Ag-NPs inhibited the growth of *Xanthomonas campestris* pv. Malvacearum (Rajesh et al., 2012). Ag-NPs increasing the antifungal activity of fluconazole against *Phoma glomerata*, *Phoma herbarum*, *Fusarium semitectum*, and *Trichoderma* sp. (Gajbhiye et al., 2009). A composite of Ag-NPs@copper chlorophyllin controlled *Thrips tabaci* on onion fields (Merghany et al., 2019).

Copper oxide NPs (CuO-NPs) have been greatly explored for use in agriculture. Copper is an essential micronutrient involved in chlorophyll synthesis as well as protein and carbohydrate metabolism (Pudake et al., 2019). Cu-NPs are effective against fungal disease pathogens including *Fusarium* sp. and *Phytophtora infestans* (Cioffi et al., 2005). Biosynthesized CuO-NPs exhibited high efficacy against *Sitophilus granarius* and *Rhyzopertha dominica* insects, which damage stored wheat grain (Badawy et al., 2021), and against pathogens of apple orchard pests *Alternaria mali*, *Diplodia seriata*, and *Botryosphaeria dothidea* (Ahmad et al., 2020). CuO-NPs biosynthesized from an aquatic plant, *Eichhornia crassipes*, act as a growth inhibitor of *A. niger* and *Fusarium culmorum* (Vanathi et al., 2016). Ag-NPs and CuO-NPs had dose-dependent inhibitory effects on the radial growth of *Alternaria alternata*, *Pleomorphomonas oryzae*, and *Sclerotinia sclerotiorum* sclerotia formation, indicating their potential for plant pathogen control (Consolo et al., 2020). Generally, Cu-NPs are applied as a bactericide against various plant pathogenic bacteria (*Ralstonia* and *Xanthomonas*) and a fungicide against various plant pathogenic fungi (*Alternaria*, *Botrytis*, *Colletotrichum*, *Fusarium*, *Macrophomina*, *Penicillium*, *Phytophthora*, and *Rhizoctonia*). They are also used as an insecticide and larvicide against plant insect pests such as leaf worms, mealybugs, thrips, stored grain pests, spider mites, and various mosquito vectors and insects, and as nematicides. The biocidal effect of copper nanopesticides is related to the production of reactive oxygen species, oxidative stress, membrane damage, protein and enzyme inactivation, and DNA damage (Kora, 2022). It is not just about combating the disease; rather, it goes beyond diagnosing it before the situation worsens. An electrochemical sensor composed of Cu-NPs was developed to detect the fungus *S. sclerotiorum* in oilseeds (Wang et al., 2010).

ZnO-NPs are considered to be as one of the most promising NPs in agriculture. ZnO-NPs are effective as a nanofertilizer as well as an antifungal agent. These NPs are used effectively against *Penicillium expansum*, *Botrytis cinerea*, *Aspergillus flavus*, and more (He et al., 2011; Jayaseelan et al., 2012). Kalaba et al. (2021) found that biosynthesized ZnO-NPs have antimicrobial activity against plant pathogenic *Erwinia amylovora*, *A. flavus*, *A. niger*, *F. oxysporum*, *Fusarium moniliform*, and *A. alternata*. The researchers also demonstrated that ZnO-NP at low concentrations promote seed germination and seedling development. Elrefaey et al. (2022) revealed that biosynthesized ZnO-NPs using brown alga *Cystoseira crinita* showed antifungal action against plant pathogenic *A. niger*. ZnO-NPs were exploited as a green insecticide to combat the postharvest disease rice weevil *Sitophilus oryzae* (L.) and Angoumois grain moth *Sitotroga cerealella* (Olivier) (Ibrahim et al., 2022). They can also be used efficiently to manage pulse beetles in chickpea to inhibit egg laying, the emergence of adults, and seed damage, and to increase the mortality of adults (Wazid et al., 2018). ZnO-NPs have the potential to be employed as an alternative to more harmful insecticides in the control of insects such as *Spodoptera frugiperda* (Pittarate et al., 2021) and *B. cinerea* (Bayat et al., 2019). Thiamethoxam's insecticidal activity can be increased by ZnO-NPs (Jameel et al., 2020). Chhipa (2017) reported that zinc, silicon, titanium dioxide, and Ag-NPs could be used in plant protection as nanopesticides for pest and disease control.

Se-NPs have attracted much attention and offer wide applications. The plant benefits alleviates several abiotic and biotic stressors such as heavy metals, salinity, drought, and high temperatures, and it has the ability to control plant pathogenic microbes (Zohra et al., 2021). Biosynthesized Se-NPs control the growth of fungal strains *A. niger* and *A. flavus* (Abdel-Moneim et al., 2022). Arunthirumeni et al. (2021) reported that biosynthesized Se-NPs are effective as larvicidal and antifeedant agents and can be used to control *S. litura* larvae.

3. Nanocomposite to control plant pathogens

Compared with pure Ag-NPs and Graphene oxide (GO) suspensions, this nanocomposite showed threefold and sevenfold improvement in inhibitory efficiency, respectively, against *Fusarium graminearum*, which causes leaf spot disease (Chen et al. 2016). The remarkable synergetic characteristics of silver–titanate NPs indicate that this composite is a promising green fungicide capable of photo-killing pathogenic fungi such as *B. cinerea* (gray mold disease) in minutes (Rodríguez-González et al., 2016). The antimicrobial action of this nanocomposite synergized with the fungicide Antracol against *Phytophthora capsici*, which causes phytophthora blight in peppers, showed promising results (Chen et al., 2016). These nanocomposites demonstrated maximum antifungal activity against *Rhizoctonia solani*, which causes damping off in cotton (Abd-Elsalam et al., 2018). In the presence of such a silver nanocomposite, fungal growth of *A. niger*, which causes black mold disease, was inhibited. Fungal spores were disrupted owing to Ag-NP dispersed as fillers in pullulan in the presence of the nanocomposites (Pinto et al., 2013). The presence of cytoplasmic coagulation, abnormal mycelia morphology, and hyphae destruction was proved in plant pathogens *Sclerotium rolfsii* and *R. solani* (Rubina et al. (2017). Intercalation of the composite with hexaconazole improved the thermal stability of the moiety, and the nanodelivery system had better inhibition toward *Ganoderma boninense* growth compared with the free hexaconazole (Mustafa et al., 2018).

4. Conclusion

Nanoscience and nanotechnology have witnessed great advances that paved the way to developing efficient, economical, and eco-friendly approaches to environmental remediation. A variety of NMs exhibit physicochemical properties that make them a promising candidate for bionanopesticides. NMs can be made inexpensively using biological methods and employed immediately as pesticides in agriculture. Algae are the fastest biological sources for creating NMs that may be used in a variety of applications. With a reduction in the use of solvents and synthetic chemicals, the use of algae to synthesize NMs has fostered the development of green, simple, and time-effective techniques. As a result, nanotechnology provides a green, effective, and environmentally friendly alternative to pest and insect management, which is becoming an issue in agriculture. The promotion of tools and strategies using nanotechnology might change how people view agriculture. NMs containing fungicides, herbicides, composts, and such have great promise as “magic bullets” that may target plant cells and discharge their charge to achieve the desired results. The strength of NMs, along with biotechnological advancements and the availability of rapid and precise diagnostic instruments, offers an outstanding and bright future for innovative horticultural operations.

References

Abdelmoneim, H. E. M., Wassel, M. A., Elfeky, A. S., Bendary, S. H., Salem, S. S., & Mahmoud, S. A. (2022). Multiple applications of CdS/TiO_2 nanocomposites synthesized via microwave-assisted Sol–Gel. *Journal of Cluster Science, 33*(3), 1119–1128. https://doi.org/10.1007/s10876-021-02041-4

Abd-Elsalam, K. A., Vasil'kov, A. Y., Said-Galiev, E. E., Rubina, M. S., Khokhlov, A. R., Naumkin, A. V., Shtykova, E. V., & Alghuthaymi, M. A. (2018). Bimetallic blends and chitosan nanocomposites: Novel antifungal agents against cotton seedling damping-off. *European Journal of Plant Pathology, 151*(1), 57–72.

Abdelaziz, A. M., Salem, S. S., Khalil, A. M. A., El-Wakil, D. A., Fouda, H. M., & Hashem, A. H. (2022). Potential of biosynthesized zinc oxide nanoparticles to control Fusarium wilt disease in eggplant (Solanum melongena) and promote plant growth. *BioMetals, 35*(3), 601–616. https://doi.org/10.1007/s10534-022-00391-8

Abdel-Moneim, A. M. E., El-Saadony, M. T., Shehata, A. M., Saad, A. M., Aldhumri, S. A., Ouda, S. M., & Mesalam, N. M. (2022). Antioxidant and antimicrobial activities of Spirulina platensis extracts and biogenic selenium nanoparticles against selected pathogenic bacteria and fungi. *Saudi Journal of Biological Sciences, 29*(2), 1197–1209.

Aboelfetoh, E. F., El-Shenody, R. A., & Ghobara, M. M. (2017). Eco-friendly synthesis of silver nanoparticles using green algae (*Caulerpa serrulata*): Reaction optimization, catalytic and antibacterial activities. *Environmental Monitoring and Assessment, 189*(7), 1–15.

Abu-Elghait, M., Hasanin, M., Hashem, A. H., & Salem, S. S. (2021). Ecofriendly novel synthesis of tertiary composite based on cellulose and myco-synthesized selenium nanoparticles: characterization, antibiofilm and biocompatibility. *International Journal of Biological Macromolecules, 175*, 294–303. https://doi.org/10.1016/j.ijbiomac.2021.02.040

Agarwal, P., Gupta, R., & Agarwal, N. (2019). Advances in synthesis and applications of microalgal nanoparticles for wastewater treatment. *Journal of Nanotechnology, 2019*, 1–9.

Ahmad, H., Venugopal, K., Bhat, A. H., Kavitha, K., Ramanan, A., Rajagopal, K., Srinivasan, R., & Manikandan, E. (2020). Enhanced biosynthesis synthesis of copper oxide nanoparticles (CuO-NPs) for their antifungal activity toxicity against major phyto-pathogens of apple orchards. *Pharmaceutical Research, 37*(12), 1–12.

Ahmed, S. S., Abd El-Rahman, F. S., & Abdel Kader, H. M. (2018). Field evaluation of some photosensitizers and nanocomposites against cotton leafworm, *Spodoptera littoralis* (Boisd.) (Lepidoptera: Noctuidae). *Middle East Journal of Applied Sciences, 8*(4), 1471–1479.

Al-Rajhi, A. M. H., Salem, S. S., Alharbi, A. A., & Abdelghany, T. M. (2022). Ecofriendly synthesis of silver nanoparticles using Kei-apple (Dovyalis caffra) fruit and their efficacy against cancer cells and clinical pathogenic microorganisms. *Arabian Journal of Chemistry, 15*(7). https://doi.org/10.1016/j.arabjc.2022.103927

Al-Zahrani, F. A. M., Salem, S. S., Al-Ghamdi, H. A., Nhari, L. M., Lin, L., & El-Shishtawy, R. M. (2022). Green synthesis and antibacterial activity of Ag/Fe_2O_3 nanocomposite using *Buddleja lindleyana* extract. *Bioengineering, 9*(9), Article 452. https://doi.org/10.3390/bioengineering9090452

Aref, M. S., & Salem, S. S. (2020). Bio-callus synthesis of silver nanoparticles, characterization, and antibacterial activities via Cinnamomum camphora callus culture. *Biocatalysis and Agricultural Biotechnology, 27*, Article 101689. https://doi.org/10.1016/j.bcab.2020.101689

Arunthirumeni, M., Veerammal, V., & Shivakumar, M. S. (2021). Biocontrol efficacy of mycosynthesized selenium nanoparticle using Trichoderma sp. on insect pest Spodoptera litura. *Journal of Cluster Science, 33*, 1−9.

Arya, A., Gupta, K., Chundawat, T. S., & Vaya, D. (2018). Biogenic synthesis of copper and silver nanoparticles using green alga *Botryococcus braunii* and its antimicrobial activity. *Bioinorganic Chemistry and Applications, 2018*, 7879403.

Ashraf, H., Batool, T., Anjum, T., Illyas, A., Riaz, S., & Naseem, S. (2022). Antifungal potential of green synthesized Magnetite nanoparticles (BC-Fe3O4 NPs) against wilt infection by ameliorating enzymatic activity and gene expression in *Solanum lycopersicum. L.* Frontiers in Microbiology, *31*, 754292.

Asimakis, E., Shehata, A. A., Eisenreich, W., Acheuk, F., Lasram, S., Basiouni, S., Emekci, M., Ntougias, S., Taner, G., May-Simera, H., & Yilmaz, M. (2022). Algae and their metabolites as potential bio-pesticides. *Microorganisms, 10*(2), 307.

Badawy, A. A., Abdelfattah, N. A., Salem, S. S., Awad, M. F., & Fouda, A. (2021). Efficacy assessment of biosynthesized copper oxide nanoparticles (CuO-NPs) on stored grain insects and their impacts on morphological and physiological traits of wheat (*Triticum aestivum* L.) plant. *Biology, 10*(3), 233.

Baker, S., Harini, B. P., Rakshith, D., & Satish, S. (2013). Marine microbes: Invisible nanofactories. *Journal of Pharmacy Research, 6*(3), 383−388.

Barsanti, L., & Gualtieri, P. (2014). *Algae: anatomy, biochemistry, and biotechnology* (2nd ed.). CRC Press.

Bayat, M., Chudinova, E., Zargar, M., Lyashko, M., Louis, K., & Adenew, F. K. (2019). Phytoassisted green synthesis of zinc oxide nanoparticles and its antibacterial and antifungal activity. *Research on Crops, 20*, 725−730.

Bayda, S., Adeel, M., Tuccinardi, T., Cordani, M., & Rizzolio, F. (2019). The history of nanoscience and nanotechnology: From chemical−physical applications to nanomedicine. *Molecules, 25*(1), 112.

BCC. (2015). *Nanotechnology in environmental applications: The global market, market research report*. www.bccresearch.com/market research/nanotechnology/nanotechnology-environmental-applications market.html.

Chan, S. S., Low, S. S., Chew, K. W., Ling, T. C., Rinklebe, J., Juan, J. C., Ng, E. P., & Show, P. L. (2022). Prospects and environmental sustainability of phyconanotechnology: A review on algae-mediated metal nanoparticles synthesis and mechanism. *Environmental Research, 212*, 113140.

Chaudhary, R., Nawaz, K., Khan, A. K., Hano, C., Abbasi, B. H., & Anjum, S. (2020). An overview of the algae-mediated biosynthesis of nanoparticles and their biomedical applications. *Biomolecules, 10*(11), 1498.

Cheng, W., Zhang, W., Hu, L., Ding, W., Wu, F., & Li, J. (2016). Etching synthesis of iron oxide nanoparticles for adsorption of arsenic from water. *RSC Advances, 6*(19), 15900−15910.

Chen, J., Sun, L., Cheng, Y., Lu, Z., Shao, K., Li, T., … Han, H. (2016). Graphene oxide-silver nanocomposite: Novel agricultural antifungal agent against Fusarium graminearum for crop disease prevention. *ACS Applied Materials and Interfaces, 8*(36), 24057−24070.

Chhipa, H. (2017). Nanofertilizers and nanopesticides for agriculture. *Environmental Chemistry Letters, 15*(1), 15−22.

Cioffi, N., Torsi, L., Ditaranto, N., Tantillo, G., Ghibelli, L., Sabbatini, L., Bleve-Zacheo, T., D'Alessio, M., Zambonin, P. G., & Traversa, E. (2005). Copper nanoparticle/polymer composites with antifungal and bacteriostatic properties. *Chemistry of Materials, 17*(21), 5255−5262.

Consolo, V. F., Torres-Nicolini, A., & Alvarez, V. A. (2020). Mycosinthetized Ag, CuO and ZnO nanoparticles from a promising Trichoderma harzianum strain and their antifungal potential against important phytopathogens. *Scientific Reports, 10*(1), 1–9.

Costa, J. A. V., Freitas, B. C. B., Cruz, C. G., Silveira, J., & Morais, M. G. (2019). Potential of microalgae as biopesticides to contribute to sustainable agriculture and environmental development. *Journal of Environmental Science and Health, Part B, 54*(5), 366–375.

Devi, H. S., Boda, M. A., Shah, M. A., Parveen, S., & Wani, A. H. (2019). Green synthesis of iron oxide nanoparticles using Platanus orientalis leaf extract for antifungal activity. *Green Processing and Synthesis, 8*(1), 38–45.

Doghish, A. S., Hashem, A. H., Shehabeldine, A. M., Sallam, Al. A., El-Sayyad, G. S., & Salem, S. S. (2022). Nanocomposite based on gold nanoparticles and carboxymethyl cellulose: synthesis, characterization, antimicrobial, and anticancer activities. *Journal of Drug Delivery Science and Technology, 77*, Article 103874. https://doi.org/10.1016/j.jddst.2022.103874

Elakraa, A. A., Salem, S. S., El-Sayyad, G. S., & Attia, M. S. (2022). Cefotaxime incorporated bimetallic silver-selenium nanoparticles: promising antimicrobial synergism, antibiofilm activity, and bacterial membrane leakage reaction mechanism. *RSC Advances, 12*, 26603–26619. https://doi.org/10.1039/D2RA04717A

El-Belely, E. F., Farag, M. M., Said, H. A., Amin, A. S., Azab, E., Gobouri, A. A., & Fouda, A. (2021). Green synthesis of zinc oxide nanoparticles (ZnO-NPs) using Arthrospira platensis (Class: Cyanophyceae) and evaluation of their biomedical activities. *Nanomaterials, 11*(1), 95.

Elrefaey, A. A. K., El-Gamal, A. D., Hamed, S. M., & El-belely, E. F. (2022). Algae-mediated biosynthesis of zinc oxide nanoparticles from *Cystoseira crinite* (Fucales; Sargassaceae) and it's antimicrobial and antioxidant activities. *Egyptian Journal of Chemistry, 65*(4), 231–240.

Elshayb, O. M., Nada, A. M., Ibrahim, H. M., Amin, H. E., & Atta, A. M. (2021). Application of silica nanoparticles for improving growth, yield, and enzymatic antioxidant for the hybrid rice ehr1 growing under water regime conditions. *Materials, 14*(5), 1150.

Elumalai, D., Hemavathi, M., Deepaa, C. V., & Kaleena, P. K. (2017). Evaluation of phytosynthesised silver nanoparticles from leaf extracts of *Leucas aspera* and *Hyptis suaveolens* and their larvicidal activity against malaria, dengue and filariasis vectors. *Parasite epidemiology and control, 2*(4), 15–26.

El-Temsah, Y. S., & Joner, E. J. (2012). Impact of Fe and Ag nanoparticles on seed germination and differences in bioavailability during exposure in aqueous suspension and soil. *Environmental Toxicology, 27*(1), 42–49.

Fawcett, D., Verduin, J. J., Shah, M., Sharma, S. B., & Poinern, G. E. J. (2017). A review of current research into the biogenic synthesis of metal and metal oxide nanoparticles via marine algae and seagrasses. *Journal of Nanoscience, 2017*, 1–15.

Fouda, A., EL-Din Hassan, S., Salem, S. S., & Shaheen, T. (2018). In-vitro cytotoxicity, antibacterial, and UV protection properties of the biosynthesized zinc oxide nanoparticles for medical textile applications. *Microbial Pathogenesis, 125*, 252–261. https://doi.org/10.1016/j.micpath.2018.09.030

Gajbhiye, M., Kesharwani, J., Ingle, A., Gade, A., & Rai, M. (2009). Fungus-mediated synthesis of silver nanoparticles and their activity against pathogenic fungi in combination with fluconazole. *Nanomedicine: Nanotechnology, Biology and Medicine, 5*(4), 382–386.

Gärtner, G., Stoyneva-Gärtner, M., & Uzunov, B. (2021). Algal toxic compounds and their aeroterrestrial, airborne and other extremophilic producers with attention to soil and plant contamination: A review. *Toxins, 13*(5), 322.

Guan, Z., Ying, S., Ofoegbu, P. C., Clubb, P., Rico, C., He, F., & Hong, J. (2022). *Green synthesis of nanoparticles: Current developments and limitations. 26* p. 102336). Environmental Technology & Innovation.

Hammad, E. N., Salem, S. S., Mohamed, A. A., & El-Dougdoug, W. (2022). Environmental impacts of ecofriendly iron oxide nanoparticles on dyes removal and antibacterial activity. *Applied Biochemistry and Biotechnology.* https://doi.org/10.1007/s12010-022-04105-1

Hammad, E. N., Salem, S. S., Zohair, M. M., Mohamed, A. A., & El-Dougdoug, W. (2022). Purpureocillium lilacinum mediated biosynthesis copper oxide nanoparticles with promising removal of dyes. *Biointerface Research in Applied Chemistry, 12*(2), 1397−1404. https://doi.org/10.33263/BRIAC122.13971404

Hänsch, R., & Mendel, R. R. (2009). Physiological functions of mineral micronutrients (cu, Zn, Mn, Fe, Ni, Mo, B, cl). *Current Opinion in Plant Biology, 12*(3), 259−266.

Hashem, A. H., & Salem, S. S. (2022). Green and ecofriendly biosynthesis of selenium nanoparticles using Urtica dioica (Stinging nettle) leaf extract: antimicrobial and anticancer activity. *Biotechnology Journal, 17*(2), Article 2100432. https://doi.org/10.1002/biot.202100432

Hashem, A. H., Selim, T. A., Alruhaili, M. H., Selim, S., Alkhalifah, D. H. M., Al Jaouni, S. K., & Salem, S. S. (2022). Unveiling antimicrobial and insecticidal activities of biosynthesized selenium nanoparticles using Prickly Pear Peel Waste. *Journal of Functional Biomaterials, 13*(3), Article 112. https://doi.org/10.3390/jfb13030112

Hashem, A. H., Shehabeldine, A. M., Ali, O. M., & Salem, S. S. (2022). Synthesis of chitosan-based gold nanoparticles: antimicrobial and wound-healing activities. *Polymers, 14*(11), Article 2293. https://doi.org/10.3390/polym14112293

He, L., Liu, Y., Mustapha, A., & Lin, M. (2011). Antifungal activity of zinc oxide nanoparticles against *Botrytis cinerea* and *Penicillium expansum. Microbiological Research, 166*(3), 207−215.

Ibrahim, S., Elbehery, H., & Samy, A. (2022). Insecticidal activity of ZnO NPs synthesized by green method using pomegranate peels extract on stored product insects. *Egyptian Journal of Chemistry, 65*(4), 135−145.

Jameel, M., Shoeb, M., Khan, M. T., Ullah, R., Mobin, M., Farooqi, M. K., & Adnan, S. M. (2020). Enhanced insecticidal activity of thiamethoxam by zinc oxide nanoparticles: A novel nanotechnology approach for pest control. *ACS Omega, 5*(3), 1607−1615.

Jayaseelan, C., Rahuman, A. A., Kirthi, A. V., Marimuthu, S., Santhoshkumar, T., Bagavan, A., Gaurav, K., Karthik, L., & Rao, K. B. (2012). Novel microbial route to synthesize ZnO nanoparticles using Aeromonas hydrophila and their activity against pathogenic bacteria and fungi. *Spectrochimica Acta Part A: Molecular and Biomolecular Spectroscopy, 90*, 78−84.

Jeffryes, C., Agathos, S. N., & Rorrer, G. (2015). Biogenic nanomaterials from photosynthetic microorganisms. *Current Opinion in Biotechnology, 33*, 23−31.

Kalaba, M. H., Moghannem, S. A., El-Hawary, A. S., Radwan, A. A., Sharaf, M. H., & Shaban, A. S. (2021). Green synthesized ZnO nanoparticles mediated by *Streptomyces plicatus*: Characterizations, antimicrobial and nematicidal activities and cytogenetic effects. *Plants, 10*(9), 1760.

Khanna, P., Kaur, A., & Goyal, D. (2019). Algae-based metallic nanoparticles: Synthesis, characterization, and applications. *Journal of Microbiological Methods, 163*, 105656.

Kora, A. J. (2022). Copper-based nanopesticides. *Copper Nanostructures: Next-Generation of Agrochemicals for Sustainable Agroecosystems, Nanobiotechnology for Plant Protection*, 133−153. https://doi.org/10.1016/B978-0-12-823833-2.00019-2

Krishnamoorthy, V. (2012). *Insecticide microbiology*. Springer Science & Business Media.

Lade, B. D., Gogle, D. P., & Nandeshwar, S. B. (2017). Nano bio pesticide to constraint plant destructive pests. *Journal of Nano Research, 6*(3), 1−9.

Lau, E. C., Carvalho, L. B., Pereira, A. E., Montanha, G. S., Corrêa, C. G., Carvalho, H. W., Ganin, A. Y., Fraceto, L. F., & Yiu, H. H. (2020). Localization of coated iron oxide (Fe3O4) nanoparticles on tomato seeds and their effects on growth. *ACS Applied Bio Materials, 3*(7), 4109−4117.

LewisOscar, F., Vismaya, S., Arunkumar, M., Thajuddin, N., Dhanasekaran, D., & Nithya, C. (2016). Algal nanoparticles: Synthesis and biotechnological potentials. *Algae-Organisms for Imminent Biotechnology, 7*, 157−182.

Lindsey, A. P. J., Murugan, S., & Renitta, R. E. (2020). Microbial disease management in agriculture: Current status and future prospects. *Biocatalysis and Agricultural Biotechnology, 23*, 101468.

Merghany, M. M., Abdelgawad, K. F., Tawfic, G. A., & Ahmed, S. S. (2019). Yield, quality and leaves anatomy structure of spring onion sprayed by nanocomposite to control *Thrips tabaci*. *Plant Archives, 19*(1), 1839–1849.

Mohamed, A. A., Abu-Elghait, M., Ahmed, N. E., & Salem, S. S. (2021). Eco-friendly mycogenic synthesis of ZnO and CuO nanoparticles for in vitro antibacterial, antibiofilm, and antifungal applications. *Biological Trace Element Research, 199*, 2788–2799. https://doi.org/10.1007/s12011-020-02369-4

Mol, I., & Purushothaman, T. (2020). Micro-algae as bio-pesticides for the development of sustainable agriculture. *Wide Spectrum, 8*(6), 5–22.

Morsy, A. R., & Mohamed, G. R. (2022). The efficiency of chitosan & silver and their nano-particles on histological and enzymatic activities of land snail, Monacha obstructa and cutworm, *Agrotis ipsilon*. *Egyptian Academic Journal of Biological Sciences, F. Toxicology & Pest Control, 14*(1), 159–178.

Mukherjee, A., Sarkar, D., & Sasmal, S. (2021). A review of green synthesis of metal nanoparticles using algae. *Frontiers in Microbiology, 12*, 693899.

Mustafa, I. F., Hussein, M. Z., Saifullah, B., Idris, A. S., Hilmi, N. H. Z., & Fakurazi, S. (2018). Synthesis of (hexaconazole-zinc/aluminum-layered double hydroxide nanocomposite) fungicide nanodelivery system for controlling Ganoderma disease in oil palm. *Journal of Agricultural and Food Chemistry, 66*(4), 806–813.

Nuruzzaman, M. D., Rahman, M. M., Liu, Y., & Naidu, R. (2016). Nanoencapsulation, nano-guard for pesticides: A new window for safe application. *Journal of Agricultural and Food Chemistry, 64*(7), 1447–1483.

Pinto, R. J., Almeida, A., Fernandes, S. C., Freire, C. S., Silvestre, A. J., Neto, C. P., & Trindade, T. (2013). Antifungal activity of transparent nanocomposite thin films of pullulan and silver against Aspergillus Niger. *Colloids and Surfaces B: Biointerfaces, 103*, 143–148.

Pittarate, S., Rajula, J., Rahman, A., Vivekanandhan, P., Thungrabeab, M., Mekchay, S., & Krutmuang, P. (2021). Insecticidal effect of zinc oxide nanoparticles against Spodoptera frugiperda under laboratory conditions. *Insects, 12*(11), 1017.

Pudake, R. N., Chauhan, N., & Kole, C. (Eds.). (2019). *Nanoscience for sustainable agriculture* (p. 711). Springer International Publishing.

Rajesh, S., Raja, D. P., Rathi, J. M., & Sahayaraj, K. (2012). Biosynthesis of silver nanoparticles using Ulva fasciata (Delile) ethyl acetate extract and its activity against Xanthomonas campestris pv. malvacearum. *Journal of Biopesticides, 5*, 119.

Rodríguez-González, V., Domínguez-Espíndola, R. B., Casas-Flores, S., Patrón-Soberano, O. A., Camposeco-Solis, R., & Lee, S. W. (2016). Antifungal nanocomposites inspired by titanate nanotubes for complete inactivation of Botrytis cinerea isolated from tomato infection. *ACS Applied Materials & Interfaces, 8*(46), 31625–31637.

Rouhani, M., Samih, M. A., & Kalantari, S. (2013). Insecticidal effect of silica and silver nanoparticles on the cowpea seed beetle, *Callosobruchus maculatus F.* (*Col.: Bruchidae*). *Journal of Entomological Research, 4*(4), 297–305.

Rubina, M. S., Vasil'kov, A. Y., Naumkin, A. V., Shtykova, E. V., Abramchuk, S. S., Alghuthaymi, M. A., & Abd-Elsalam, K. A. (2017). Synthesis and characterization of chitosan–copper nanocomposites and their fungicidal activity against two sclerotia-forming plant pathogenic fungi. *Journal of Nanostructure in Chemistry, 7*(3), 249–258.

Saied, E., Salem, S. S., Al-Askar, A. A., Elkady, F. M., Arishi, A. A., & Hashem, A. H. (2022). Mycosynthesis of hematite (α-Fe_2O_3) nanoparticles using aspergillus niger and their antimicrobial and photocatalytic activities. *Bioengineering, 9*(8), Article 397. https://doi.org/10.3390/bioengineering9080397

Salem, S. S. (2022). Baker's yeast-mediated silver nanoparticles: characterisation and antimicrobial biogenic tool for suppressing pathogenic microbes. *BioNanoScience*. https://doi.org/10.1007/s12668-022-01026-5

Salem, S. S. (2022). Bio-fabrication of selenium nanoparticles using baker's yeast extract and its antimicrobial efficacy on food borne pathogens. *Applied Biochemistry and Biotechnology, 194*, 1898–1910. https://doi.org/10.1007/s12010-022-03809-8

Salem, S. S., Ali, O. M., Reyad, A. M., Abd-Elsalam, K. A., & Hashem, A. H. (2022). Pseudomonas indica-mediated silver nanoparticles: antifungal and antioxidant biogenic tool for suppressing mucormycosis fungi. *Journal of Fungi, 8*(2), Article 126. https://doi.org/10.3390/jof8020126

Salem, S. S., Badawy, M. S. E. M., Al-Askar, A. A., Arishi, A. A., Elkady, F. M., & Hashem, A. H. (2022). Green biosynthesis of selenium nanoparticles using orange peel waste: characterization, antibacterial and antibiofilm activities against multidrug-resistant bacteria. *Life, 12*(6), Article 893. https://doi.org/10.3390/life12060893

Salem, S. S., & Fouda, A. (2020). Green synthesis of metallic nanoparticles and their prospective biotechnological applications: An overview. *Biological Trace Element Research, 199*(1), 344–370.

Salem, S. S., Hammad, E. N., Mohamed, A. A., & El-Dougdoug, W. (2023). A comprehensive review of nanomaterials: types, synthesis, characterization, and applications. *Biointerface Research in Applied Chemistry, 13*(1), Article 41. https://doi.org/10.33263/BRIAC131.041

Saratale, R. G., Saratale, G. D., Shin, H. S., Jacob, J. M., Pugazhendhi, A., Bhaisare, M., & Kumar, G. (2018). New insights on the green synthesis of metallic nanoparticles using plant and waste biomaterials: Current knowledge, their agricultural and environmental applications. *Environmental Science and Pollution Research, 25*(11), 10164–10183.

Sekhon, B. S. (2014). Nanotechnology in agri-food production: An overview. *Nanotechnology, Science and Applications, 7*, 31.

Shah, M., Fawcett, D., Sharma, S., Tripathy, S. K., & Poinern, G. E. J. (2015). Green synthesis of metallic nanoparticles via biological entities. *Materials, 8*(11), 7278–7308.

Shaheen, T. I., Fouda, A., & Salem, S. S. (2021). Integration of cotton fabrics with biosynthesized CuO nanoparticles for bactericidal activity in the terms of their cytotoxicity assessment. *Industrial & Engineering Chemistry Research, 60*(4), 1553–1563. https://doi.org/10.1021/acs.iecr.0c04880

Shaheen, T. I., Salem, S. S., & Zaghloul, S. (2019). A new facile strategy for multifunctional textiles development through in situ deposition of SiO_2/TiO_2 nanosols hybrid. *Industrial & Engineering Chemistry Research, 58*(44), 20203–20212. https://doi.org/10.1021/acs.iecr.9b04655

Shatalova, E. I., Grizanova, E. V., & Dubovskiy, I. M. (2022). The effect of silicon dioxide nanoparticles combined with entomopathogenic bacteria or fungus on the survival of Colorado potato beetle and cabbage beetles. *Nanomaterials, 12*(9), 1558.

Siddiqi, K. S., & Husen, A. (2016). Fabrication of metal and metal oxide nanoparticles by algae and their toxic effects. *Nanoscale Research Letters, 11*(1), 1–11.

Specht, E. A., Karunanithi, P. S., Gimpel, J. A., Ansari, W. S., & Mayfield, S. P. (2017). Host organisms: Algae. *Industrial biotechnology* (pp. 605–641). Hoboken, NJ, USA: John Wiley & Sons, Ltd., ISBN 978-3-527-80779-6.

Thabet, A. F., Boraei, H. A., Galal, O. A., El-Samahy, M. F., Mousa, K. M., Zhang, Y. Z., Tuda, M., Helmy, E. A., Wen, J., & Nozaki, T. (2021). Silica nanoparticles as pesticide against insects of different feeding types and their non-target attraction of predators. *Scientific Reports, 11*(1), 1–13.

Vanathi, P., Rajiv, P., & Sivaraj, R. (2016). Synthesis and characterization of Eichhornia-mediated copper oxide nanoparticles and assessing their antifungal activity against plant pathogens. *Bulletin of Materials Science, 39*(5), 1165–1170.

Verma, K. K., Song, X. P., Joshi, A., Tian, D. D., Rajput, V. D., Singh, M., Arora, J., Minkina, T., & Li, Y. R. (2022). Recent trends in nano-fertilizers for sustainable agriculture under climate change for global food security. *Nanomaterials, 12*(1), 173.

Wang, Z., Wei, F., Liu, S. Y., Xu, Q., Huang, J. Y., Dong, X. Y., Yu, J. H., Yang, Q., Zhao, Y. D., & Chen, H. (2010). Electrocatalytic oxidation of phytohormone salicylic acid at copper nanoparticles-modified gold electrode and its detection in oilseed rape infected with fungal pathogen *Sclerotinia sclerotiorum*. *Talanta, 80*(3), 1277–1281.

Wazid, S. N., Prabhuraj, A., Naik, R. H., Shakuntala, N. M., & Sharanagouda, H. (2018). Effect of biosynthesized zinc oxide green nanoparticles on pulse beetle, Callosobruchus analis (Coleoptera: Chrysomelidae). *International Journal of Current Microbiology Applied Science, 7*(09), 503–512.

Xu, L., Zhu, Z., & Sun, D. W. (2021). Bioinspired nanomodification strategies: Moving from chemical-based agrosystems to sustainable agriculture. *ACS Nano, 15*(8), 12655–12686.

Zohra, E., Ikram, M., Omar, A. A., Hussain, M., Satti, S. H., Raja, N. I., & Ehsan, M. (2021). Potential applications of biogenic selenium nanoparticles in alleviating biotic and abiotic stresses in plants: A comprehensive insight on the mechanistic approach and future perspectives. *Green Processing and Synthesis, 10*(1), 456–475.

CHAPTER 14

Algae materials for textile industries

Ishwarya Ayyanar, Sunandha Jeeva Bharathi G., Akila Ravindran, Rajapriya Rajendran, Ajucarmelpricilla Arulprakasam and Balasubramanian Vellaisamy

Department of Microbiology, Alagappa University, Karaikudi, Tamil Nadu, India

1. Introduction

Consumption has exploded due to the unprecedented emergence of the globalization economy in recent years. Although some of this increased consumption is required for humans to exist, the majority is not. Consumption, like wealth, is likewise unevenly distributed. Even if per capita consumption has increased in many parts of the globe, these gains are not uniform, and the disparities between them are significant (Toprak & Anis, 2017). Through various industrial activities, the textile business converts fibers into yarn, then changes the yarn into textiles or associated commodities, and dyes, and finishes these materials. During wet processing in the textile industry, a variety of dyes, chemical substances, additives, and sizing ingredients are utilized. The use of dyes to create color is the most pressing issue in textile industry effluent. Detergents, salt, toxic metals, enzymes, oxidizing and reducing agents, and other pollutants can be found in effluents. Due to the emission of highly contaminated effluents, the condition of both groundwater and surface water in the proximity of textile factories degrades over time. Untreated wastewater effluents from the use of toxic, dangerous, and persistent chemical agents in wet processing affect the ecosystem (Madhav et al., 2018). By 2050, the textile industry will account for a quarter of the world carbon budget, or 26%, if current global trends continue. By 2050, the textile and apparel industries will consume 300 million tons of nonrenewable raw materials and spew 22 million tons of microplastics into the oceans if current trends continue (Ütebay et al., 2020). Global demand for textile goods is steadily increasing as a result of population growth and economic development. In the meanwhile, the textile industry has substantial environmental and resource constraints. Petrochemicals make about 63% of textile fibers, and their production and disposal generate enormous pollution. This thirsty plant is linked to water depletion and hazardous contamination as a result of pesticide usage. The majority of environmental consequences are exacerbated by subsequent phases in the textile manufacturing process (Sandin & Peters, 2018). The textile industry is a group of interconnected businesses that produce fabric from natural (cotton, silk, wool, etc.) and/or synthetic fibers. Around 60% of total energy is consumed by dyeing and finishing procedures. One of the most common environmental challenges associated with the textile sector is water contamination. Natural impurities removed from the fibers being processed, as well as the chemicals used in the processing, are the main causes of contamination. Effluents are frequently hot, alkaline,

Algae Materials: Applications Benefitting Health. https://doi.org/10.1016/B978-0-443-18816-9.00018-6

odorous, and colored with coloring agents. Some of the compounds that were released are potentially dangerous. The textile industry is increasingly seeing air emissions, notably volatile organic compounds (VOCs), as equally important and relevant (VOC) (Moses & Ammayappan, 2006).

1.1 Chemical usage in the textile industry

High exploitation of natural resources, particularly water, goes hand in hand with industrial growth. A wide range of hazardous chemicals are used in the manufacture of various products, posing a major threat to water quality and harming human health by producing infectious and chronic illnesses. The fashion industry is one of the most complex industrial sectors, with over 8000 different chemicals used in various textile production processes, as well as vast amounts of raw water (Tounsadi et al., 2020). Chemicals found in textiles come from a variety of places. Chemical products might be introduced willingly or inadvertently throughout the production process for specialized objectives. The coloring process, where dyes are introduced, is one of the main toxic sources in textile manufacture. The leftover chemical impurities of the dye's composition present in the fabric may pose a health concern if taken up by and passed through the human skin, depending on the condition of the chemicals used for coloring as well as the coloring/washing techniques (Iadaresta et al., 2018).

1.2 Water consumption in textile industry

The amount of water needed for textile wet processing varies based on the type of process material, colors used, finishing chemicals used, and processing equipment employed. To process 1 kg of textile material, 50–100 L of water is required on average (global average). Almost every stage of textile wet processing necessitates the use of water as well as the requisite chemicals. The textile industry consumes a lot of water in various processes, and nearly as much water is released at the end of the processing activity as was utilized as a processing medium (Samanta et al., 2019).

1.3 Energy consumption in textile industry

Due to the dusty air within the mill and the loud noise from the fabric machines, manpower is also in short supply in textile companies. As a result of these circumstances, workers are less interested in working in textile mills, requiring mill management to install more automated machinery. As a result, fabric machinery manufacturers are introducing automation into their machines, which raises the indirect power requirements of the machines (khude, 2017).

1.4 Air emissions in textile industry

Pollutants are released into the atmosphere by the majority of textile mill processes. Gaseous emissions have been recognized as the textile industry's second-largest source of pollution. There has been significant conjecture regarding the amount and types of air pollutants generated by textile activities, but data on fabric manufacturing processes are sometimes lacking. Air emissions include dirt, oil mists, acid vapors, odors, and boiler exhausts. Cleaning and manufacturing adjustments result in sludge from tanks and discarded processing chemicals, which may contain dangerous organics and metals (Karthik & Gopalakrishnan, 2014, pp. 153–88).

1.5 Desizing

There are so many processes involved before and after the dyeing process (Fig. 14.1). The process of removing the sizing substances from the cloth is known as desizing. Desizing is the process in the wet processing or pre-dyeing did after the singeing process. It is a chemical process, and the pace of this process, i.e., the rate of hydrolysis of starch, may be adjusted. The primary goal of textile sizing and desizing is to boost weaving production. This is accomplished by reducing yarn breakages, which allows for faster running rates. Indeed, contemporary weaving methods could not have achieved such high speeds without equivalent advancements in sizing technology (Hasan et al., 2022).

1.6 Scouring

Dust, oil, wax, minerals, and other contaminants are common in natural fiber. Scouring is the process of eliminating such contaminants. Scouring is the most common wet procedure used on textiles before

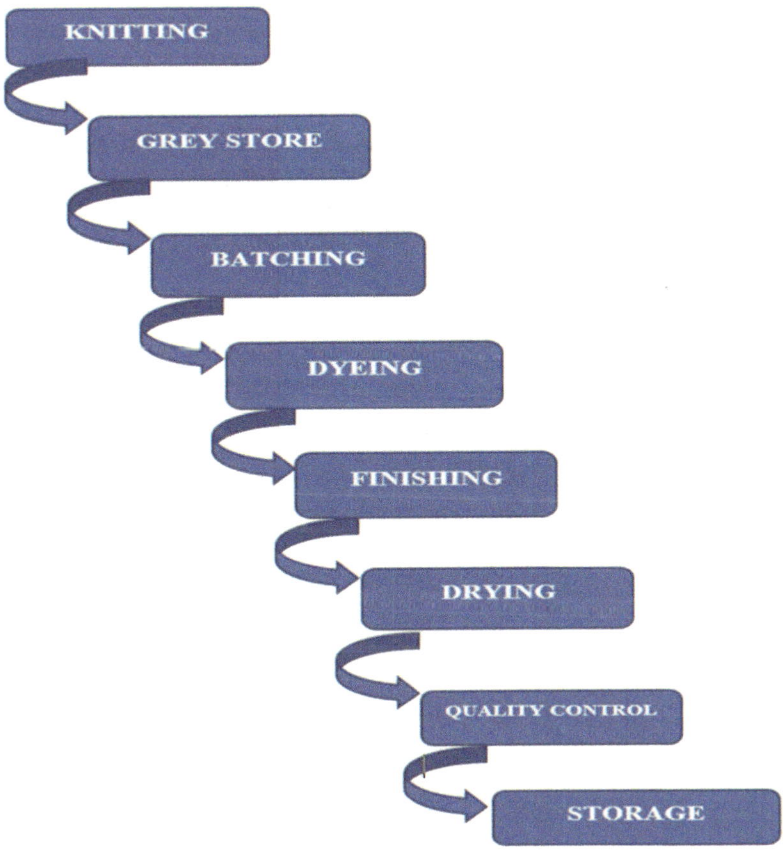

FIGURE 14.1

Steps involved in the dyeing process in textile industries.

dyeing or printing. Cotton scouring's main goal is to remove natural and synthetic contaminants from the cloth, making it sustainable and more absorbent. Natural cotton scouring is a chemical cleaning process that eliminates natural wax, nonfibrous impurities, as well as any extra soiling or debris from the materials. Cotton cloth is often scrubbed in kiers, big iron containers (Mazharul Islam Kiron, 2021).

1.7 Dyeing process in textile industry

Textile industries use a wide spectrum of synthetic dyes to suit the ever-increasing demands for quality, fastness, diversity, and color depth. There are almost 10,000 different dyes and pigments utilized. In the textile business, a wide range of dyes are employed, including sulfur dyes, acid dyes, basic dyes, dispersion dyes, direct dyes, reactive dyes, and vat dyes (Fig. 14.2). Because of their vivid color, fastness, and simplicity of application, reactive dyes are popular in the textile garment sector. Since the reactive dyes are widely utilized in the textile industry, this analysis focuses primarily on environmentally friendly reactive dyeing techniques (Varadarajan & Venkatachalam, 2016).

1.8 Exhaust dyeing method

In the presence of heat input, synthetic chemicals, aqueous or solvent, and mechanical force of the dye bath or material, the exhaust dyeing process gives color to the fabric in a dyeing bath. Soft stream, jigger, and turbo dyeing machines are popular exhaust dyeing equipment. It is a batch dyeing method that can handle small to huge batches. To keep the dyeing procedure going, research has been done to lower the dyeing temperatures and dyeing liquid (MLR) (Muthu et al., 2016; Shamey & Zhao, 2014).

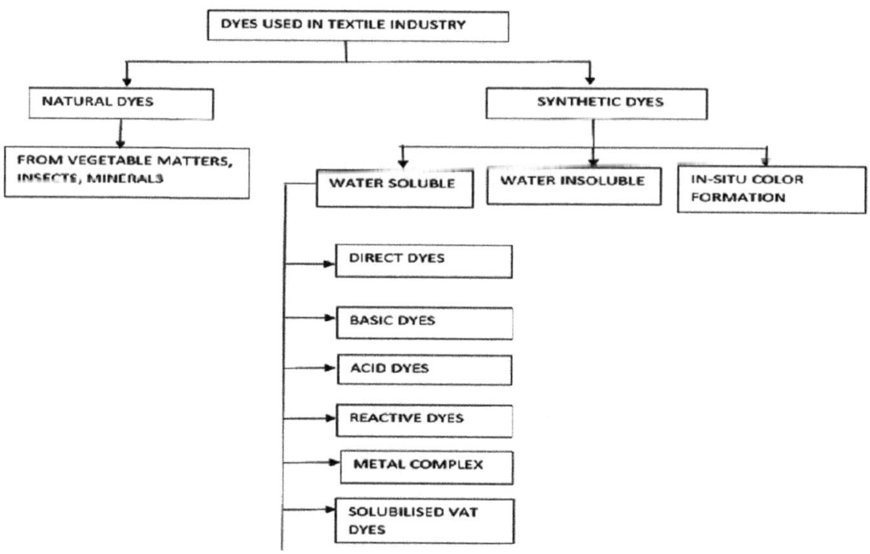

FIGURE 14.2

Classification of dyes used in the textile industries.

1.9 Pad-stream dyeing method

The exhaust dyeing process consumes time, water, and energy for some reactive and vat dyes. To save energy, the pad dry steam technique was developed, in which woven fabric in open width form is padded through a 50–300 gpL dyeing solution with 100% expression, and then steamed with saturated steam at 105°C for 5–10 min. During steaming, dye molecules on the fiber's surface travel to the expanded fiber structure and are fixed. It is a continuous process using a tiny volume of liquor, such as MLR of 1:1 or 1:2. However, the fastness properties are inferior, and the effluent discharge is larger as compared with exhaust dyeing (Muthu et al., 2016).

1.10 Dyes used in textile industry

Dyes are becoming more widely used as a result of their wide range of applications in numerous industries. Dyes are more stable to light, heat, and oxidizing chemicals because of their aromatic molecular structure, making them more difficult to remove from wastewater. A dye is a pigment that adds color to a product or material, such as paper, leather, textiles, rubber, plastic, and cosmetics. A product or substance with color applied to it has a more appealing appearance and adds social value to an esthetic. Natural or human-made chemicals can be used to make coloring compounds (synthetic). Synthetic dyes are aromatic, water-soluble, and dispersible organic colorants (Ayele et al., 2021). Dyes are classified according to their intended application and chemical structural features, and are made up of chromophores, which are responsible for the dye's color. azo, anthraquinone, methane, nitro, arylmethane, carbonyl, and other functional groups, which make up these chromophore-containing centers. Auxochromes are electrons that remove or donate substituents from chromophores to create or amplify their color. The most common auxochromes are amine, carboxyl, sulfonate, and hydroxyl. Industry is predicted to use almost 10,000 different dyes and pigments, with over 7 × 105 tons of synthetic dyes produced each year. Textile dyeing procedures include batch, continuous, and semicontinuous dyeing. A variety of elements influence the technique used, including the type of material (fiber, yarn, fabric, fabric structure, and garment), as well as the general type of fiber, dye lot size, and colored fabric quality criteria (Drumond Chequer et al., 2013). Natural dyes and colorants are environmentally friendly and have grown in popularity in recent years as a result of their positive effects on the environment and human health.

2. Classification of dyes

2.1 Natural dyes

Natural dyes are organic compounds obtained from naturally occurring sources such as plants (e.g., indigo and saffron), insects (e.g., cochineal beetles and lac scale insects), animals (e.g., some types of mollusks or shellfish), and minerals (e.g., indigo and saffron) (e.g., ferrous sulfate, ochre, and clay). The aforementioned sources offer a wide range of beautiful natural colors, from yellow to black. A variety of organic and inorganic substances absorb light in the visible range of 400–800 nm, resulting in these colors (pigments). The structure or contents of the coloring pigment/molecules, which contain varied chromophores present in the dye-producing plant to exhibit a wide spectrum of colors, determine the absorption of light (Affat, 2021).

2.2 Azoic dyes

Azo dyes now make up the majority of dye chemistry, and their importance is likely to expand in the future. They play an important role in the governance of the dye and printing markets. A simple diazotization and coupling procedure is used to create these colors. Many procedures and changes are employed to acquire the requisite color properties, yield, and molecular size of the dyes for greater dispersibility. Azo dyes are by far the most popular, accounting for more than 60% of all dyes used. Azo dyes make up over 70% of all dyes used in industry. These compounds are distinguished by the functional group (-N=N-) formed by two symmetrical and/or asymmetrical identical or nonazo alkyl or aryl radicals (Benkhaya et al., 2020). On the other hand, these compounds meet human needs while also causing ecological and sanitary alterations in water systems, soil, and the environment. The use of dyes in aquatic bodies is an esthetic issue but could also be hazardous to human health. Many textile industry liquid and solid effluents, on the other hand, are treated before being discharged into the environment, decreasing their impact on aquatic life (de Campos Ventura-Camargo & Marin-Morales, 2013).

2.3 Acidic dyes

Organic sulfonic acid sodium salts make up the majority of acid dyes. It is always a sulfonic acid salt with an aromatic structure that includes chromogen and solubilizing groups (D-SO3Na). The color comes from the anionic component of the soluble ionic molecules in acid dyes. Acid dyes got their name from the fact that an acid was required for their application. The moniker "acid dye" fits their chemical structure because they are salts of organic acids. Anionic dyes are sometimes known as dies. Acid dyes are water-soluble anionic dyes that are used to color fabrics such as silk, wool, nylon, and modified acrylic fibers in neutral to acid dye baths (Upadhye., 2021).

2.4 Basic dye

Basic dyes are also known as cationic dyes because they ionize in solution, causing their colored component to become a positively charged radical. Basic dyes are used to color wool, silk, acrylic, and mod-acrylic fibers (Mia et al., 2018).

2.5 Direct dyes

Direct dyes are sulfonic acid sodium salts with an azo group as the principal chromophore. This dyeing process is done at a low temperature. It tends to leak colors after absorbing them, resulting in uneven hues that are resistant to various types of aftertreatment. Direct dyes are anionic dyes that are substantive to cellulose fibers when applied from an aqueous solution containing electrolytes (Rahman & Islam, 2019).

2.6 Brief description of algae

The Greek word "phyco," which means "phycophyte," is the source of the word "algae." These algae, which range in size from single cells to fully formed plants, are eukaryotic and aquatic in nature. The world's population is expected to surpass 10 billion people by 2050. Agriculture is practically fully exploited; the majority of arable land is already in use, and factors such as climate change and urbanization offer significant obstacles to agriculture's future. To fulfill future demands, just increasing the intensity of agricultural, farming, fisheries, and fossil energy production will not be enough. Rising global temperatures, harsh weather, shifting climatic patterns, and the loss of cultivable land will necessitate significant modifications in present agro technology (Fabris et al., 2020; Wurtzel et al.) to minimize the environmental impact through sustainable source of commodities including food, bio-products, and bulk chemicals. Using high-tech engineering and molecular genetic technologies, such as phenomics and genetic engineering, agricultural crops including soy, corn, wheat, and rice have improved their overall efficiency, cost-effectiveness, and environmental impact (Fabris et al., 2020). From animal feed and aquaculture to human nutrition and health products, algal biomass and algae-derived chemicals have a wide variety of potential applications. The basic characteristics of algae are given in Table 14.1.

Although macroalgae have garnered a lot of interest as possible natural antioxidants, certain algae are believed to be rich providers of natural antioxidants (Shalaby, 2011). Algae are photosynthetic creatures with a wide range of biological variety and metabolic flexibility. They are significantly abler to change their biochemical metabolic pathways and cellular composition in response to environmental stimuli

Table 14.1 Basic characteristics of algae.

S.No.	Algae basics characteristics	Characteristics description
1.	Algae reproduce quickly	Algae can quadruple their population in a matter of hours, can be harvested on a daily basis, and have the potential to provide a volume of biomass and ethanol many times that of our most prolific crops.
2.	Algae consumes CO_2	When algae are produced in the sun, it consumes (or absorbs) carbon dioxide (CO_2) as it grows, releasing oxygen (O_2) for the rest of us to breathe. Algae require additional CO_2 for high production, which may be provided by emissions sources such as power stations, ethanol factories, and other sources.
3.	Agriculture and algae do not compete	Algae farming makes use of terrain that is not suited for traditional agriculture and water that is not ideal for other crops, such as seawater, brackish water, and wastewater. As a result, algae-based fuels are a good complement to biofuels produced through traditional agricultural methods.
4.	Wastewater can be screened with algae	Algae flourish in nutrient-rich environments such as sewage, animal waste, and some industrial discharges, cleaning the wastes while creating a sufficient biomass.
5.	Biomass from algae can be used as a source of energy	The residual algal biomass can be dried and "pelletized" and utilized as fuel in blast furnaces and other power production sources after oil extraction.
6.	Algae can be used to make a variety of products	Plastics, chemical feedstocks, lubricants, fertilizers, and even cosmetics may all be made from algae, which can be grown for big or local markets. Other uses for algae may be found here.
7.	The algae industry is a job generator	Algae may be grown in a number of climates and using a variety of production methods, such as ponds, photobioreactors, and fermenters, and will therefore provide a wide range of occupations across the United States, from research to engineering, building to farming, marketing to financial services.

than terrestrial plants. According to several scientific articles, the effectiveness of CO_2 fixation by microalgae is 10–50 times higher than that of terrestrial plants. The maximal photosynthetic efficiency of microalgae is predicted to be 4.5% using a practical methodology (Mironiuk & Chojnacka, 2018). Algae are divided into two classes depending on their physical characteristics and size: microalgae and macroalgae. The number of algae species is estimated to be between 1 and 10 million, with microalgae accounting for the majority. Because they may live in a variety of biological settings and severe circumstances such as high temperatures and pH. Microalgae are the most abundant living creatures on the planet because of these qualities (Rajkumar & Takriff, 2015).

2.7 Microalgae

For biofuels, health supplements, pharmaceuticals, and cosmetics, microalgae can be a rich source of carbon compounds. They can also be used to purify wastewater and minimize CO_2 emissions. Microalgae generate polysaccharides, lipids, pigments, proteins, vitamins, bioactive chemicals, and antioxidants, among other bioproducts. The use of microalgae as a renewable and sustainable feedstock for biofuel production has reignited interest in biorefinery. Growth augmentation technologies and genetic engineering may be used to increase their potential as a future source of renewable bio products (Khan et al., 2018).

2.8 Algae as a biofabric material

Algae include thallophytes, or plants without roots, stalks, or leaves. Algae are categorized as either aquatic or freshwater plants, depending on their carbon cutoff and photosynthetic efficiency. Chlorophyll is the most abundant photosynthetic pigment in these species. On the contrary, they do not have a sterile cell layer surrounding the reproductive cells. Other photosynthetic microorganisms, such as green microalgae, dinoflagellates, and cyanobacteria, have a considerable advantage over terrestrial ecosystems in terms of a diversified supply of biomolecules for food, feed, and fuel. Their main advantages include a faster rate of growth, higher biomass production, and the ability to synthesize complex metabolites with less inputs. Microalgal species have a wide range of taxonomic and inherent biochemical diversity, making them a useful source of a wide range of biomolecules with industrial and medicinal uses. As a result, microalgae have been utilized to manufacture biomolecules such as lipids, proteins, and carbohydrates. To explain the importance of algal fabrics, simplified table is given (Table 14.2).

From animal feed and fisheries to adult nutrition and health products, algal biomass and algal blooms chemicals offer a wide variety of potential applications. Although macroalgae have attracted

Table 14.2 Brief table explains why should consider algae fabric.

	Why algae is used for fabric
Algaeing process	When mixed with cellulose, a plant fiber that clothing manufacturers can make themselves using algaeing's exclusive recipe, the algal is converted into a liquid mixture that may be used as a dye or transformed into a textile.
Algal-based fibres	Natural fire resistance is a feature of algae-based fabrics, which might reduce the need for hazardous flame retardants in clothing.
Biodegradation	Algae degrade quicker than cotton, the most popular natural clothing material, and do not require pesticides or a lot of space to produce.
Sustainability of the algae-based fabric	Clothes created form algae are high-quality materials that can be produced sustainably and are fully biodegradable.

great attention as possible natural antioxidants, certain algae are considered rich providers of natural antioxidants (Shalaby, 2011). Algae are divided into two types: microalgae and macroalgae (seaweeds), both macro- and microalgae are used to make biofuel (oil, biodiesel, bioethanol, biogas, and biogas), as binding agents (agar, carrageen), as a biostimulant, as a source of nutrients, and for pollution management (wastewater treatment, reduction of CO_2 emissions) (Mironiuk & Chojnacka, 2018). Microalgae are naturally occurring biofactories that are fast gaining popularity due to its long viability. The microalgal biotechnology industry includes both upstream (bioprospecting of superior microalgal strains, isolation and strain purification, strain selection for specific purposes, cultivation systems, biomass harvesting) and downstream (cultivation systems, biomass harvesting) processes (microalgal disruption techniques and intracellular metabolite extraction, biochemical transformations, and product recovery).

It is necessary to conduct high-throughput screening of microalgal strains that produce substantial amounts of the compounds (Molitor & Schnoor, 2020; Mutanda et al., 2020). Green microalgae, dinoflagellates, and cyanobacteria, among other photosynthetic microorganisms, have a significant advantage over terrestrial ecosystems as a rich supply of diverse biomolecules for food, feed, and fuel purposes. Their primary benefits are a quicker growth rate, increased biomass yield, and the capacity to synthesis complex metabolites with minimum inputs. Microalgal species have a vast taxonomic and intrinsic biochemical variety, making them a good source of numerous biomolecules with industrial and pharmaceutical applications. As a result, microalgae have been used to produce biomolecules such as lipids, proteins, and carbohydrates for a long time (Kumar et al., 2020).

The textile sector benefits greatly from algae-based dye since it is nontoxic and allergen free. Algae products would assist the textile industry since they are energy and water efficient, and they help to preserve good working conditions by eliminating the use of dangerous pollutants (Bose & Rajasekaran, 2016) (Fig. 14.3). Biotextiles like these might help reshape the modern fashion business, which is now a disaster for the environment. Clothing production for ever-changing low-cost fashion emits more carbon than aircraft and shipping combined. It also consumes a lot of water and creates harmful effluent as well as microplastic contamination. Plus, after a few years, nearly three-fifths of clothing winds up in landfills or incinerators. Algae take up no valuable land, grow extremely quickly, and consume far less water and resources than traditional natural textiles such as cotton, wool, or silk. However, it is not a durable substance that can survive the weather for an extended period of time on its own (Patel, 2021).

2.9 Natural fabric dyeing using algae

Natural pigments are derived from mineral, animal, or plant sources and can be found in any part of the plant, such as leaves, fruits, twigs, seeds, flowers, bark root, and so on. Plant pigment production has two primary drawbacks: a large quantity of land is required for cultivation, and CO_2 emissions are produced (Mona et al., 2019). Because they avoid the drawbacks of synthetic dyes and the typically inferior performance of plant-based dyes, algae-based dyes offer an environmentally safe and long-lasting solution for textile dyeing and printing operations. Green color's objective is to replace synthetic dyes with algae-based alternatives, resulting in more environmentally friendly dyeing and printing technologies (Rabie et al.) (Fig. 14.4). Classification of natural dye was divided on the basis of three factors, which is mentioned in Table 14.3.

Algae are a rich source of various photosynthetic pigments such as chlorophylls, carotenes, phycobilins, phycocyanin, and phycoerythrin, which are found in both fresh and saline water. Chlorophyll b is found in green algae, fucoxanthin and carotenoids are found in brown algae, while carotene and zeaxanthin

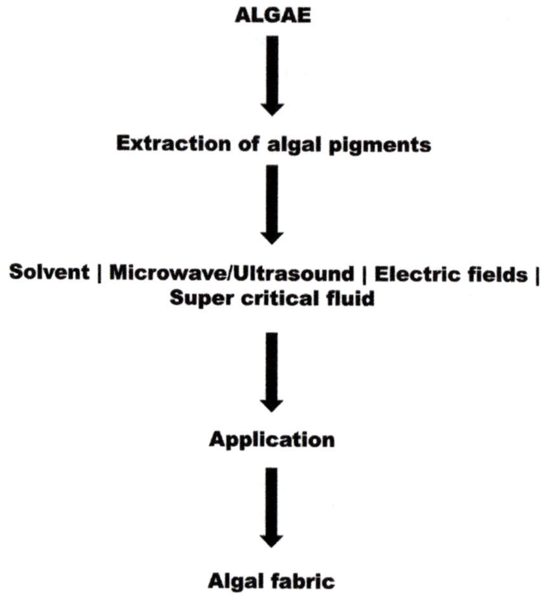

FIGURE 14.3

Extraction process of fabric from algae.

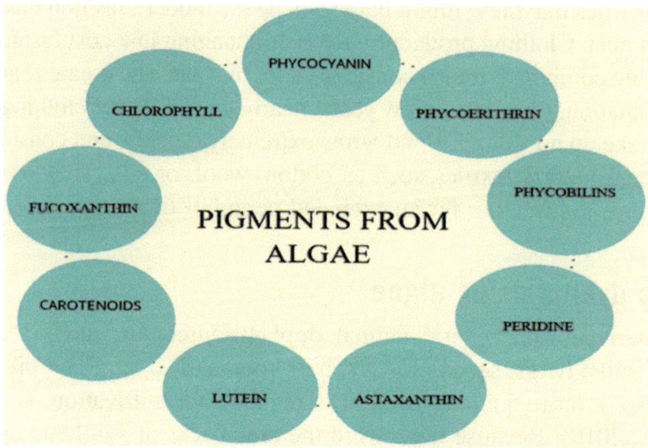

FIGURE 14.4

Pigments produced from algae.

are found in red algae. Due to a scarcity of publications on algal-based colorants in fabric processing and information on algal dye extraction, the usage of algal pigments in the textile industry is minimal (Azeem et al., 2019). Many modern garments are woven in factories with acrylic, nylon, or polyester threads and then cut and sewed. These are all humanmade and nonbiodegradable materials. These specialists predict

Table 14.3 Classification of natural colorants using in textile industry.

Natural colorants classification		
On the basis of chemical nature	**On the basis of application**	**On the basis of color**
Lac dye	Disperse dye	Blue dye
Naphthoquinone dye	Vat dye	Red dye
Carotenoid dye	Mordant dye	Brown dye
Tannins	Acid dye	Yellow dye
Flavonoid dye	Basic dye	Black dye
Chromene dye		
Vegetable tannins		

that some of tomorrow's clothing may be bioengineered that is made from live bacteria, algae, yeast, animal cells, or fungi and that, when discarded, it will decompose into harmless compounds. This type of technology has the potential to reduce waste and pollution. Azeem et al. (2019) concluded that *Sargassum muticum* generated more color strength on cotton textiles when dyed for 30 min in the presence of 4.0 g per 100 mL NaCl as an exhausting agent at 75°C temperature and basic pH of medium. Also, cotton fabric dyed with *Iyengaria stellata* extract at 75°C, pH 5.0, 4.0 g per 100 mL NaCl concentration, and 75-min time interval produced a stronger color. Their study results further showed that cotton fabric colored with *Colpomenia sinuosa* and *Lithomyrtus obtusa* dye has produced the best color strength at pH 11, dyeing for 60 min at 60°C, and 8.0 g per100mL NaCl and at 30°C dyeing temperature, 45 min duration, pH 7.0, and 6.0 g per 100 mL concentration of NaCl, respectively (Azeem et al., 2019). Mona et al. (2019) presented their study findings, which included pigments taken from red, green, and brown algae, and evaluated for their capacity to resist the growth of common pathogenic microorganisms as well as how well they stick to textiles and keep color. Using several natural mordants, the adhesion and coloring qualities of these were examined and compared (Mona et al., 2019).

2.10 Scope of algae in textile industries

Algae may also be used to color our clothes in an environmentally responsible way (Fig. 14.5). British scientists pioneered the use of algae to add color to textiles and paintings. Diatoms are single-celled organisms that live in glistening shells. These shells provide the algae a feature that permits them to

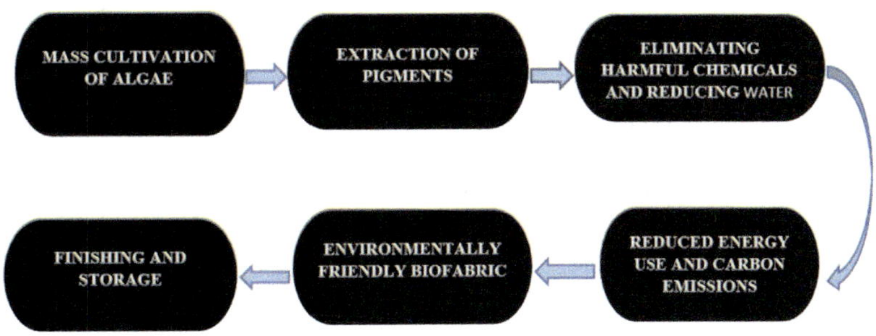

FIGURE 14.5

Scope of algal dyes and fabrics.

Table 14.4 Scope of algae in environmental and fashion industries.

S.No.	Algaeing impact	Description
1.	Scope in biodiversity	Toxins and chemicals are being replaced with environmentally friendly algal formulations across the production chain.
2.	Scope in fashion industry	Developing a new category of skin care qualities for clothing and textiles.
3.	Scope in water conservation	Compared with traditional textile, algaeing's solution saved millions of liters of contaminated water. The amount of water used in the manufacture of 1 million cotton t-shirts is comparable.

behave like crystals, changing color depending on how the holes in their shell are arranged. Color sensitivity is maintained without changing the chemical composition of the cloth. When the paint dries, the algae arrange themselves horizontally on the surface, reflecting light evenly. As a result, even translucent paint and transparent silica shells have the potential to cast a strong shadow. This could result in a dyeing solution that is both safe and cost-effective. Algae are a prime example of how environmentally friendly resources may be put to good use. New applications for algal fiber production will likely make it a valuable resource in the future (Table 14.4) (Objective, 2018).

2.11 Growth rate of wastes produced by textile industry in various countries

The textile industry accounts for around 5% of India's gross domestic product (GDP). As a result, annual consumer spending on apparel and footwear amounts to about 68 billion dollars. The successful algal fabric producing companies are listed in Table 14.5. According to the Indian Textile Journal, more than 1 million tons of textiles are discarded each year, with the majority of this originating from

Table 14.5 Companies involved in production of algal fabric.

Competitors identified in this market		
S. No.	**Companies**	**Country**
1.	Algix LLC	(The United States)
2.	Corbio N.V.	The Netherlands
3.	Diversified energy corporation	The United States
4.	DSM nutritional products AG	Switzerland
5.	Kent Bio Energy corporation	(United States)
6.	Pond Technologies inc	Canada
7.	Sapphire Engineering (NE) ltd	The United Kingdom
8.	Algenol biotech	The United States

home sources. This equates to 0.79 kg of textile waste per person. Let us look at how this compares with other countries' textile waste. In 2011, Americans dumped 18.76 million tons of textile trash, or around 60.2 kg per individual. United Kingdom, Italy, Portugal, EU-28, Austria, Germany, Norway, and Sweden dispose of 19, 7, 6, 6, 5, 4, 1, and 1 kg of textile waste per capita, which is shown as histogram (Aggarwal, 2021).

Conventional cotton production utilizes a lot of nitrogen and phosphorus-based fertilizers, as well as herbicides and insecticides, to protect crops from insects like the boll weevil and mealybugs. Pesticides, insecticides, and herbicides have all been used less in recent decades, but the decline appears to be uneven. A wide range of chemicals are used in the dyeing, processing, and finishing of products, many of which are extremely hazardous and should be avoided. Prior to dyeing, fabrics are scrubbed with chlorinated solvents such as trichloroethane (TCE) and nonylphenol ethoxylates (NPEs), which are both potentially harmful to human health (Objective, 2018).

3. Conclusion

Algae, which include pond scum and seaweed, are an important source of food, cosmetics, and fertilizers. They could potentially be used to make biofuels, bioplastics, and now ecological clothing. Advanced functional materials are composed of living, breathing algae, and clothing constructed from it would be photosynthetic, which means it would absorb carbon dioxide from the air and exhale oxygen like plants. Furthermore, the material is simple to mass-produce and biodegradable. These biotextiles might help change the modern fashion industry, which is now a disaster for the environment. Clothing manufacturing for ever-changing low-cost fashion generates more CO_2 than airplanes and shipping combined. It also consumes a lot of water, produces toxic effluent, and causes microplastic pollution. Nearly three-fifths of apparel ends up in landfills or incinerators after a few years. Algae consume significantly less water and resources than traditional natural fabrics such as cotton, wool, or silk and take up no precious land. Algal materials will be the dominating source for textile businesses in the future; however, extensive study in this subject to create more durable clothing materials is required.

References

Affat, S. S. (2021). Classifications, advantages, disadvantages, toxicity effects of natural and synthetic dyes: A review [internet]. *University of Thi-Qar Journal of Science (UTsci), 8*(1), 130–135 [cited 2022 Jun 20]. Available from: https://www.researchgate.net/publication/351885086_Classifications_Advantages_Disadvantages Toxicity_Effects_of_Natural_and_Synthetic_Dyes_A_review.

Aggarwal, A. (2021). Circular economy for textiles as engrained in the traditional Indian life [cited 2022 Jun 20] *Environmental Science: An Indian Journal Case Reports [Internet], 17*(1), 180. Available from: https://www.researchgate.net/profile/Anupriya-Aggarwal/publication/349287424_Circular_economy_for_textiles_as_engrained_in_the_traditional_Indian_life/links/6027db90299bf1cc26c28f20/Circular-economy-for-textiles-as-engrained-in-the-traditional-Indian-life.

Ayele, A., Getachew, D., Kamaraj, M., & Suresh, A. (2021). Phycoremediation of synthetic dyes: An effective and eco-friendly algal technology for the dye abatement [internet]. *Journal of Chemistry, 2021* [cited 2022 Jun 20]. Available from: https://www.hindawi.com/journals/jchem/2021/9923643/.

Azeem, M., Iqbal, N., Mir, R. A., Adeel, S., Batool, F., Khan, A. A., & Gul, S. (December 1, 2019). Harnessing natural colorants from algal species for fabric dyeing: A sustainable eco-friendly approach for textile processing. *Journal of Applied Phycology, 31*(6), 3941–3948.

Benkhaya, S., M'rabet, S., & El Harfi, A. (2020). *Classifications, properties, recent synthesis and applications of azo dyes* (Vol 6, p. e03271). Heliyon. Elsevier.

Bose, P., & Rajasekaran, A. (2016). Decolourization of textile dyes using bacterial consortium. *International Journal of Environmental Science and Technology, 4*(1), 17–22.

de Campos Ventura-Camargo, B., & Marin-Morales, M. A. (2013). U. Azo dyes: Characterization and toxicity-a review [cited 2022 Jun 20] *Textiles and Light Industrial Science and Technology, 4*(1), 85–103. Available from: https://www.airitilibrary.com/Publication/alDetailedMesh?docid=P20150611018-201304-201508180026-201508180026-85-103.

Drumond Chequer, F. M., de Oliveira, G. A. R., Anastacio Ferraz, E. R., Carvalho, J., Boldrin Zanoni, M. V., & de Oliveir, D. P. (2013). Textile dyes: Dyeing process and environmental impact. In *Eco-friendly textile dyeing and finishing [internet]* (6, pp. 151–176). IntechOpen [cited 2022 Jun 20]. Available from: undefined/state.item.id.

Fabris, M., Abbriano, R. M., Pernice, M., Sutherland, D. L., Commault, A. S., Hall, C. C., Labeeuw, L., McCauley, J. I., Kuzhiuparambil, U., Ray, Parijat, Kahlke, T., & Ralph, P. J. (2020). Emerging technologies in algal biotechnology: Toward the establishment of a sustainable, algae-based bioeconomy. *Frontiers of Plant Science, 11*, 279. Frontiers Media S.A.

Hasan, M., Al Biruni, M. T., Azad, S., & Tanvir, A. (2022). Adsorptive removal of dye from textile wastewater employing Moringa oleifera leaves biochar as a natural biosorbent. *Biomass Conversion and Biorefinery*, 1–17.

Iadaresta, F., Manniello, M. D., Östman, C., Crescenzi, C., Holmbäck, J., & Russo, P. (September 1, 2018). Chemicals from textiles to skin: An in vitro permeation study of benzothiazole. *Environmental Science & Pollution Research, 25*(25), 24629–24638.

Karthik, T., & Gopalakrishnan, D. (2014). *Environmental analysis of textile value chain: An overview* [cited 2022 Jun 20] (p. 1). Singapore: Springer. Available from: https://link.springer.com/chapter/10.1007/978-981-287-110-7_6.

Khan, M. I., Shin, J. H., & Kim, J. D. (March 5, 2018). The promising future of microalgae: Current status, challenges, and optimization of a sustainable and renewable industry for biofuels, feed, and other products [cited 2022 Jun 20] *Microbial Cell Factories [Internet], 17*(1), 1–21. Available from: https://microbialcellfactories.biomedcentral.com/articles/10.1186/s12934-018-0879-x.

khude, P. (September 6, 2017). A review on energy management in textile industry [cited 2022 Jun 20] *Innovative Energy Research [Internet], 6*(2). Available from: https://www.omicsonline.org/open-access/a-review-on-energy-management-in-textile-industry.php.

Kiron, M. I. (2021). Quality Control System in Garment Industry, Quality control system in Apparel Industry, Quality control system for Knitted Garment Industry. https://textilelearner.net/quality-control-system-in-garmentsindustry Accessed 04 January 2021.

Kumar, G., Shekh, A., Jakhu, S., Sharma, Y., Kapoor, R., & Sharma, T. R. (2020). Bioengineering of microalgae: Recent advances, perspectives, and regulatory challenges for industrial application. *Frontiers in Bioengineering and Biotechnology, 8.* . Frontiers Media S.A.

Madhav, S., Ahamad, A., Singh, P., & Mishra, P. K. (March 1, 2018). A review of textile industry: Wet processing, environmental impacts, and effluent treatment methods [cited 2022 Jun 20] *Environmental Quality Management [Internet], 27*(3), 31–41. Available from: https://onlinelibrary.wiley.com/doi/full/10.1002/tqem.21538.

Mia, S., Amanullah, S., & Rahman, M. M. (2018). *Faculty of engineering department of textile engineering study on garments inspection [internet]* [cited 2022 Jun 20]. Available from: https://www.researchgate.net/profile/

Asif-Safwan/publication/340465120_Competitive_study_between_Natural_dye_and_Synthetic_dye_in_case_of_dying_of_100_Nylon_fabric/links/5e8baf8f299bf1307983da43/Competitive-study-between-Natural-dye-and-Synthetic-dye-in-c.

Mironiuk, M., & Chojnacka, K. (2018). The environmental benefits arising from the use of algae biomass in industry. In *Algae biomass: Characteristics and applications* (pp. 7–16). Springer International Publishing.

Molitor, H. R., & Schnoor, J. L. (October 27, 2020). Using simulated flue gas to rapidly grow nutritious microalgae with enhanced settleability [cited 2022 Jun 20] *ACS Omega [Internet, 5*(42), 27269–27277. Available from: https://pubs.acs.org/doi/full/10.1021/acsomega.0c03492.

Mona, S., Yazhini, M., Fakhruddin Shaukat, P., Chandra Sekarenthiran, S., & Maya, S. (2019). Extraction of algal pigments and their suitability as natural dyes. *Journal of Algal Biomass Utilization, 10*(2), 1–8.

Moses, J. J., & Ammayappan, L. (2006). Growth of textile industry and their issues on environment with [cited 2022 Jun 20] *Asian Dye [Internet], 3*(2), 61–67. Available from: https://www.researchgate.net/profile/Arvind-Singh-21/post/Research_articles_on_textile_industry_especially_its_impacts_on_environment_and_mitigation/attachment/59d650a379197b80779a965f/AS%3A503972612263936%401497167842561/download/growth+and+environmental.

Mutanda, T., Naidoo, D., Bwapwa, J. K., & Anandraj, A. (December 17, 2020). Biotechnological applications of microalgal oleaginous compounds: Current trends on microalgal bioprocessing of products. *Frontiers in Energy Research, 8*, 299.

Muthu, S. S., & Gardetti, M. A. (2016). Green fashion [internet] [cited 2022 Jun 20]. In S. S. Muthu, & M. A. Gardetti (Eds.), *Vol. 1. Environmental footprints and eco-design of products and processes* (p. 256). Singapore: Springer Singapore. Available from: http://link.springer.com/10.1007/978-981-10-0245-8.

Objective, Common (2018). *Fashion industry pollution | sustainable chemicals [internet]* [cited 2022 Jun 20]. Available from: https://www.commonobjective.co/article/the-issues-chemicals.

Patel, P. (2021). In a big step towards sustainable fashion, scientists create a biodegradable, carbon-capturing textile from algae. https://www.anthropocenemagazine.org/2021/05/algae-could-help-make-the-fashion-industry-green/ Accessed 06 May 2021.

Rabie, A. M., Abukhadra, M. R., Rady, A. M., Ahmed, S. A., Labena, A., Mohamed, H. S. H., Betiha, M. A., & Shim, J.-J. (March 1, 2020). Instantaneous photocatalytic degradation of malachite green dye under visible light using novel green Co–ZnO/algae composites [cited 2022 Jun 20] *Research on Chemical Intermediates [Internet], 46*(3), 1955–1973. Available from: https://link.springer.com/article/10.1007/s11164-019-04074-x.

Rahman, P. T., & Islam, T. (2019). *Study on investigation of fabric wastage in cutting section of a knit garments industry* [cited 2022 Jun 20]. Available from: http://dspace.daffodilvarsity.edu.bd:8080/handle/123456789/4276.

Rajkumar, R., & Takriff, M. S. (December 12, 2015). Prospects of algae and their environmental applications in Malaysia: A case study [cited 2022 Jun 20] *Journal of Bioremediation and Biodegradation [Internet], 7*(1), 1–12. Available from: https://www.omicsonline.org/open-access/prospects-of-algae-and-their-environmental-applications-in-malaysia-acase-study-2155 6199-1000321.php?aid=64731.

Samanta, K. K., Pandit, P., Samanta, P., & Basak, S. (January 1, 2019). Water consumption in textile processing and sustainable approaches for its conservation. *Water Text Fash*, 41–59.

Sandin, G., & Peters, G. M. (May 20, 2018). Environmental impact of textile reuse and recycling—A review. *Journal of Cleaner Production, 184*, 353–365.

Shalaby, E. A. (2011). Algae as promising organisms for environment and health [Internet] [cited 2022 Jun 20] *Plant Signaling and Behavior, 6*, 1338–1350. pmc/articles/PMC3258063/.

Shamey, R., & Zhao, X. (2014). Modelling, simulation and control of the dyeing process [Internet] [cited 2022 Jun 20] *Modelling, Simulation and Control of the Dyeing Process*, 1–230. Available from: https://books.google.co.in/books?hl=en&lr=&id=l2B7AwAAQBAJ&oi=fnd&pg=PP1&dq=Shamey+et+al.,+2014&ots=hVw0rb5XkA&sig=ETolhP7aTRDLqdTsB_IJMwNedmk.

Toprak, T., & Anis, P. (August 10, 2017). Textile industry's environmental effects and approaching cleaner production and sustainability, an overview [cited 2022 Jun 20] *Journal of Textile Engineering and Fashion Technology [Internet], 2*(4). Available from: https://medcraveonline.com/JTEFT/JTEFT-02-00066.php.

Tounsadi, H., Metarfi, Y., Taleb, M., El Rhazi, K., & Rais, Z. (July 1, 2020). Impact of chemical substances used in textile industry on the employee's health: Epidemiological study [cited 2022 Jun 20] *Ecotoxicology and Environmental Safety [Internet]*, 197. Available from: https://pubmed.ncbi.nlm.nih.gov/32335392/.

Upadhye, N. Y. (2021). Synthetic Dyes in Textiles. https://textilelearner.net/synthetic-dyes-in-textiles/#more-11532 Accessed 08 September 2021.

Ütebay, B., Çelik, P., & Çay, A. (2020). Textile wastes: Status and perspectives. In *Waste in textile and leather sectors*. IntechOpen.

Varadarajan, G., & Venkatachalam, P. (March 1, 2016). Sustainable textile dyeing processes [cited 2022 Jun 20] *Environmental Chemistry Letters [Internet], 14*(1), 113–122. Available from: https://link.springer.com/article/10.1007/s10311-015-0533-3.

Wurtzel, E. T., Vickers, C. E., Hanson, A. D., Millar, A. H., Cooper, M., Voss-Fels, K. P., Nikel, Pablo I., & Erb, Tobias J. (December 1, 2019). Revolutionizing agriculture with synthetic biology [cited 2022 Jun 20] *Nature Plants [Internet], 5*(12), 1207–1210. Available from: https://pubmed.ncbi.nlm.nih.gov/31740769/.

CHAPTER

15 Algae nanomaterials and nanostructures

Xu Zhou[1,2], Shiyu Yin[1,2], Zhongqi He[1,2] and Wenbiao Jin[1,2]

[1]*State Key Lab of Urban Water Resource and Environment, School of Civil and Environmental Engineering, Harbin Institute of Technology Shenzhen, Shenzhen, Guangdong, China;* [2]*Shenzhen Engineering Laboratory of Microalgae Bioenergy, Harbin Institute of Technology Shenzhen, Shenzhen, Guangdong, China*

1. Introduction

Nowadays, nanomaterials (NMs) are widely used in antibacterial agents, gene therapy, drug delivery, catalysts, biosensors, and pollutants removal (Li et al., 2011). NMs are controlled and manipulated by nanotechnology to keep their size in a range of 1—100 nm (Ahmed et al., 2017). The synthesis approaches of NMs include physical, chemical, and biological methods. Among them, physical and chemical synthesis is expensive and prone to toxic by-products Kumari et al. (2019), while biosynthesis offers advantages of low cost, ease of synthesis, and reduced environmental load. The synthesis process involved in biological NMs is known as green nanotechnology (Rana et al., 2020). Green nanotechnology refers to the use of nanotechnology to improve the external environment so that the environment can be sustainable, and also refers to the use of nanotechnology to increase the sustainability of products.

Algae are economically and ecologically important in the field of green nanotechnology because of their high metal bioaccumulation and reduction capacities and low toxicity. Besides that, the photosynthesis of algae can act as a light-driven redox reaction in which carbon dioxide (CO_2) and H_2O are converted to energy-rich organic compounds ($(CH_2O)n$) and oxygen (Damergi et al., 2021). Thus, algae can also achieve both CO_2 capture and value-added by-product production. It is estimated that 1 kg dry weight of microalgae can recover approximately 1.88 kg of CO_2 (Adamczyk et al., 2016). Algal biomass contains approximately 15%—20% lipid, 20%—40% carbohydrate, and 30%—40% protein based on its dry weight (Kannah et al., 2021). Very recently, algae as a living cell factory or algae extracts as the natural reducing agent can be used for the green synthesis of various metal NMs, thus achieving clean and sustainable development (Li et al., 2021). Among them, algae extracts include carbohydrates, polyunsaturated fatty acids (PUFA), proteins, pigments (e.g., carotenoids and chlorophyll), etc. (Uzair et al., 2020). The photosynthetic pigment fucoxanthin of microalgae can reduce silver ions to NPs (Brar et al., 2022).

In this chapter, we first provide a short overview of the algal biosynthesis of nanoparticles (NPs) and then focus on algae-mediated AgNPs, AuNPs, ZnO NPs, CuO NPs, CuNPs, Fe_3O_4 NPs, MIONPs,

Algae Materials: Applications Benefitting Health. https://doi.org/10.1016/B978-0-443-18816-9.00015-0

and PtNPs and their applications. Finally, the current challenges of algal applications for NMs synthesis are discussed, and research trends are prospected.

2. Algae-based nanoparticles and nanostructures

Generally, metal NMs can be synthesized through in vivo/in vitro biological systems. In the vivo biological systems, algae acting as cell factories absorb and transport metal ions from the aqueous environment to the cell or cell surface utilizing enzymes with high bioreductive capacity (as natural catalysts) for further conversion to metal NMs (Li et al., 2021). Components such as carboxyl groups, amino groups, binding proteins, and extracellular polysaccharides on the cell surface allow metal ions to be effectively adsorbed (Chandel et al., 2022). In the vitro biological systems, the substances used for synthesizing metal NMs contain (1) extracellular enzymes or biomolecules on the cell surface or secreted outside the cell and (2) intracellular enzymes or biomolecules extracted from the cell (Dahoumane et al., 2016). Algae have attracted much attention on in vitro synthesis of NMs because of their richness in bioactive molecules.

The process of synthesizing metal NMs using algae intracellularly/extracellularly is different. In the case of intracellular synthesis of metal NMs, the general steps are (1) collection of algal biomass followed by rinsing with distilled water, (2) interaction of a specific metal solution with algal biomass (live algae), and (3) sonication and centrifugation to harvest stable NMs. In contrast, for extracellular synthesis, algal biomass is first collected and then rinsed with distilled water, followed by different treatments to obtain algal extracts, which then interact with specific metal solutions to synthesize NMs (Uzair et al., 2020). Table 15.1 shows the algae-based NPs and nanostructures.

2.1 Silver

Among the many metal NPs, silver nanoparticles (AgNPs) are used in medical fields due to their unique biocidal activity. AgNPs can control the growth of bacteria by destroying their cell walls (Aritonang et al., 2019; Wu et al., 2020). In addition, AgNPs can be used in wastewater treatment, electronic components, pharmaceuticals, and cosmetics, among others (Marimuthu et al., 2020; Shanmuganathan et al., 2019).

Currently, AgNPs can be synthesized by different species of algae, such as *Chlorophyceae*, *Cyanophyceae*, *Bacillariophyceae*, *Rhodophyceae*, and *Phaeophyceae*. Sayed Ahmed et al. (2021) used synthesized biosynthetic AgNPs using *Chlorella* sp. MF1 and $AgNO_3$ solution, which was applied as a biofortifier to enhance the glucosinolates and kaempferol content of *Eruca sativa* Mill. The treatment methods of algal species affect the synthesis and stability of AgNPs. Velgosova et al. (2020) confirmed that different treatments (heat treatment, centrifugation, and combination of methods) affect the composition of *Parachlorella kessleri* extract and thus the reducing capacity of the extract, which further has an impact on the synthesis of AgNPs. Alzahrani et al. (2020) studied the production potentials of AgNPs extracellularly using *Hydroclathrus clathratus* crude methanol extract and aqueous solution, respectively, and found that the produced particle size and shape of the AgNPs varied. The particles produced using *H. clathratus* aqueous solution were much smaller. Lashgarian et al. (2021) synthesized Ag/AgCl NPs using ethanol extract of *Ulva fasciata*. The XRD patterns of Ag/AgCl NPs showed an irregular crystalline phase, indicating an amorphous structure. Kusumaningrum et al.

Table 15.1 Synthesis of algae-based nanomaterials.

Strain	Types of nanoparticle	Size (nm)	Shape	Extracellular/ Intracellular	Application	References
Caulerpa racemosa	AgNPs	5−25	Spherical, triangular	Extracellular	Antimicrobial agents	Kathiraven et al. (2014)
Chlorella sp. MF1	AgNPs	1.45−5.08	Spherical	Extracellular	Biofortification agents	Sayed Ahmed et al. (2021)
Parachlorella kessleri (extract)	AgNPs	~20	Near spherical	Extracellular	Antibiofilm activity	Velgosova et al. (2020)
Sargassum wightii (extract)	AgNPs	18.45−41.59	Spherical	Extracellular	Pharmacological agents	Deepak et al. (2018)
Chaetoceros sp.	AgNPs	149.03 ± 3.04	Rectangular or square	Extracellular	Environmental, pharmaceutical, medical, and biotechnological	Mishra et al. (2020)
Skeletonema sp.		186.73 ± 4.90	Nonspherical			
Thalassiosira sp.		239.46 ± 44.30	Nonspherical			
Chlorella pyrenoidosa (extract)	AgNPs	/	Spherical and cuboidal	Extracellular	Antimicrobial agents	Kusumaningrum et al. (2019)
Neochloris aquatica RDS02 (extract)	AgNPs	15−20	Spherical	Extracellular	Drug development	Tamil Selvan et al. (2018)
Gelidiella acerosa (extract)	AgNPs	/	/	Extracellular	Antibacterial agents	Thiruchelvi et al. (2021)
Hydroclathrus clathratus (crude methanol extract)	AgNPs	7−31	Spherical and polygonal	Extracellular	Antibacterial/ antibiofilm drugs	Alzahrani et al., (2020)
Hydroclathrus clathratus (aqueous solution)		11−49	Spherical, triangle, quadrangular, and rod shapes			
Natural diatoms	AgNPs	/	Spherical	Extracellular	Antimicrobial activity	Sherief et al. (2021)
Sargassum myriocystum (extract)	AgNPs	20 ± 2.20	Hexagonal	Extracellular	Nanodrug for anticancer	Balaraman et al. (2020)
Sargassum angustifolium	AgNPs	32 ± 10	Spherical	Extracellular	/	Bakhshan et al. (2018)
Chlorella vulgaris	AgNPs	1−50	Spherical	Extracellular	Antibacterial activity	Torabfam and Yüce (2020)

Continued

Table 15.1 Synthesis of algae-based nanomaterials.—cont'd

Strain	Types of nanoparticle	Size (nm)	Shape	Extracellular/ Intracellular	Application	References
Navicula cincta	AgNPs	32	Polydisperse spherical	Extracellular	Antimicrobial activity	Boobalakrishnan et al. (2020)
Sargassum sp.	AuNPs	50–100	Triangular, hexagonal, and truncated triangular	Extracellular	/	Liu et al. (2005)
Chondrus crispus *Spirogyra insignis*	AuNPs	~30, ~50	Triangular, hexagonal, and spherical	Extracellular	/	Castro et al. (2013)
Egregia sp.	AuNPs	8	Spherical	Extracellular	/	Colin et al. (2018)
Stephanopyxis turris	AuNPs	≥50	/	Intracellular	/	Pytlik et al. (2019)
Callithamnion corymbosum sp.	AuNPs	60–90	/	Extracellular	Medical applications	Lucaci et al. (2021)
Chlorella vulgaris	Au/cellulose nanocomposite	113.33–203.50	Spherical	Extracellular	Cancer therapy	Hamouda et al. (2021)
	Cellulose nanofibrils	28.43–82.01	Rod-shaped			
	AuNPs	13–16	Spherical			
Dunaliella salina (extract)	AuNPs	5–36	/	Extracellular	Anticancer activity	Chumakov et al. (2020)
Spirulina platensis and *Oscillatoria* sp. (algal supernatant)	AuNPs	/	/	/	Antibiotics and fungicides	El-Sheekh, Hassan et al. (2021)
Spirulina platensis and *Oscillatoria* sp. (algal aqueous extracts)		/	/	Extracellular		
Spirulina platensis and *Oscillatoria* sp. (ffresh biomass)		15.49–55.08	Octahedral, pentagonal, and triangular shaped	Intracellular		

Sargassum serratifolium (extract)	AuNPs	17.47 ± 0.13	Spherical, triangles, pentagons, and narrow square	Extracellular	Catalytic activity	Kim et al. (2021)
Cystoseira tamariscifolia (extract)	AuNPs	7.6 ± 2.2	Spherical and polycrystalline	Extracellular	Biomedical applications	Machado et al. (2021)
Spirulina platensis *Oscillatoria* sp.	AuNPs	15.60–77.13	Octahedral, pentagonal, and the triangular	Intracellular	Therapeutic antiviral agents	El-Sheekh et al. (2022)
Sargassum muticum (extract)	ZnO NPs	30–57	Hexagonal wurtzite	Extracellular	Cancer treatments	Azizi et al. (2014), Sanaeimehr et al. (2018)
Arthrospira platensis	ZnO NPs	30–55	Spherical	Extracellular	Antimicrobial agents	El-Belely et al. (2021)
Limnothrix sp. KO05 (C-phycocyanin)	Phycocyanin–zinc oxide nanorods	around 33	Rod-like	Extracellular	Nanodrug	Davaeifar et al. (2019)
Scenedesmus dimorphus (extract)	ZnO nanorods	27.37	Crystalline hexagonal nanorod	Extracellular	Catalytic activity	Mahana and Mehta (2021)
Turbinaria ornata (extract)	ZnO NPs	15–52	Rod, spherical, and hexagonal	Extracellular	Nanoagriculture	Itroutwar et al. (2019)
Chlorella (extract)	ZnO NPs	20 ± 2.20	Hexagonal	Extracellular	Degradation of ddibenzothiophene (DBT) pollutant	Khalafi et al. (2019)
Tetraselmis indica (extract)	ZnO NPs	20–40	Crystalline spherical	Extracellular	Antibacterial, antioxidant, and hemolytic activity	Thirumoorthy et al. (2021)
Ulva fasciata (aqueous extract)	ZnO NPs	10–14	Rod	Extracellular	Nanomedicine	Hamouda et al. (2020)
Sargassum vulgare (extract)	ZnO NPs	50–150	Spherical	Extracellular	Antioxidant and photocatalytic activity	Karkhane et al. (2020)
Ulva lactuca (extract)	ZnO NPs	10–50	Triangles, hexagons, rods, and rectangles	Extracellular	Photocatalytic, antibiofilm, and insecticidal activity	Ishwarya et al. (2018)
Anabaena cylindrica (extract)	CuO NPs	40–60	Rod	Extracellular	Disinfection of drinking water	Bhattacharya et al. (2019)

Continu

Table 15.1 Synthesis of algae-based nanomaterials.—cont'd

Strain	Types of nanoparticle	Size (nm)	Shape	Extracellular/ Intracellular	Application	References
Macrocystis pyrifera (extract)	CuO NPs	2–50	Spherical	Extracellular	/	Araya-Castro et al. (2020)
Cystoseira trinodis (extract)	CuO NPs	5–7.8	Spherical	Extracellular	Catalytic, antioxidant, and antibacterial	Gu et al. (2018)
Chlamydomonas reinhardtii	CuNPs	Sub-10	Spherical	/	Bioremediation	Žvab et al. (2021)
Haematococcus pluvialis, *Sargassum angustifolium* (aqueous extract)	Perlite nanoparticles (CuO/Per-NPs)	13–24	Spherical	Extracellular	/	Ali Babazadeh et al. (2021)
Pterocladia capillacea (extract)	CuO NPs	62 ± 17.7	/	Extracellular	Anticancer activity	Aboeita et al. (2022)
Chaetomorpha antennina	Nanorods of cuprous oxide	90	Polyhedral rod	/	Biomedical, packaging, biotechnological, textile, water treatment, and pharmaceutical	Bhutiya et al. (2018)
Petalonia fascia	Fe_3O_4 NPs	6.54–13.46	Spherical	Extracellular	Bioremediation of wastewaters	El-Sheekh, El-Kassas, et al. (2021)
Colpomenia sinuosa		10.56–19.91	Cubic			
Padina pavonica		26.50–34.01	Spherical			
Chlorella K01 (extract)	Fe_3O_4 NPs	76.5	Spheroid	Extracellular	Plant growth stimulants and antifungal	Win et al. (2021)
Sargassum crassifolium (extract)	Magnetic iron oxide nanoparticles (MIONPs)	40–215	Quasi-spherical	Extracellular	Environmental, biomedical, and industrial	Budlayan et al. (2021)
Halymenia dilatata (extract)	PtNPs	15 ± 1.7	Spherical	Extracellular	Pharmaceutical and biomedical	Sathiyaraj et al. (2021)

(2018) compared the effect of agitation/no agitation on the synthesis of AgNPs using *Chlorella pyrenoidosa* and found that the agitation treatment improves performance in particle insertion and cell stability. AgNPs and silver-doped formulations are safer and less toxic than other heavy metal formulations. For example, Ag-modified TiO_2 NPs have higher photocatalytic properties and therefore have higher bacterial activity than unmodified TiO_2 NPs. As a semiconductor material with photocatalytic properties, TiO_2 shows antibacterial activity. Meanwhile, silver also exhibits antibacterial activity against a variety of microorganisms (Durango-Giraldo et al., 2019). In addition, pH affects the synthesis of AgNPs. It was found that pH in the range of 6.0–8.0 is suitable for the synthesis of AgNPs due to the acceleration of electrostatic repulsion between metal ions and extracellular polysaccharides (EPSs) (Li et al., 2017).

2.2 Aurum

Aurum nanoparticles (AuNPs) are used widely in biomedicine, catalysis, cancer therapy, electronics manufacturing, and numerous fields (Khan et al., 2019; Samanta et al., 2019). AuNPs bound to oligonucleotides can be used as probes for detecting target DNA (Park et al., 2002). The fabricated and heparin-conjugated AuNPs can be used for early tumor detection and destruction of cancer cells by photothermal activity (Kah et al., 2007; Lee et al., 2010). AuNPs have been reported for infrared absorbing optical coatings due to their ability to absorb infrared radiation (Shankar et al., 2005). AuNPs are used in microreflectors because of their optical properties (Kim & Osterloh, 2006).

Liu et al. (2005) synthesized 50–100 nm AuNPs with triangular, hexagonal, and truncated triangular morphologies by using *Sargassum* sp. The pH significantly affects the shape and size of metal NMs. Among them, the shape of AuNPs is influenced by the pH during the synthesis. It was found that polygon NPs are easily generated in acidic media, while nanospheres are easily generated in alkaline media. For instance, Castro et al. observed that most of the AuNPs formed in acidic media were triangular and hexagonal (polygonal) by transmission electron microscopy (TEM). As the pH increased to 4, the AuNPs started to appear spherical (Castro et al., 2013). Colin et al. (2018) found that algae extract from *Egregia* sp. has a dual role in the synthesis of AuNPs, including the reduction of gold ions and the provision of biomolecules (for the stability of the ligand shell around AuNPs). In addition to the aforementioned, intracellular synthesis of AuNPs has also been reported. In vivo biosynthesis of AuNPs was detected in diatom cells using surface-enhanced Raman spectroscopy (SERS) by the addition of gold salts to the growth medium (Pytlik et al., 2017). Pytlik et al. (2019) detected AuNPs $\geq$50 nm inside the cells, and smaller AuNPs were found near the cell walls. In addition to AuNPs, algae can also be used to synthesize nanocomposites. Nanocomposites are the combination of two or more independent constituents in one material. Hamouda et al. synthesized nanocellulose, AuNPs, and Au/cellulose nanocomposite with enhanced anticancer activity by using *Chlorella vulgaris* biosynthesis. They found that nanocellulose and Au/cellulose nanocomposite treatments were able to increase the relative expression of tumor suppressor 53 significantly (Hamouda et al., 2021).

2.3 Platinum

Currently, platinum nanoparticles (PtNPs) have been used in water–gas exchange reactions, fuel cells, petrochemicals, biosensors, and organocatalysis (Jeyaraj et al., 2019). PtNPs are also used in the biomedical field because of various applications such as anticancer, antibacterial, bioimaging, and drug delivery (Bloch et al., 2021). Sathiyaraj et al. biosynthesized PtNPs with an average particle size of 15 ± 1.7 nm from *Halymenia dilatata*. PtNPs showed promising performance in terms of free radical scavenging, reducing ability, and antioxidant properties (Sathiyaraj et al., 2021).

2.4 Zinc

Zinc oxide nanoparticles (ZnO NPs) are the most common NMs in practice and research. ZnO and ZnO NPs are commonly added to glass, ceramics, cement, plastics, rubber, pigments, food additives, and paints because of their excellent properties including easy availability, low price, and high electrical and thermal conductivity (Mahmoud et al., 2020). In addition to this, ZnO NPs can be used as sunscreen with their photostability and ability to absorb UV radiation (Al Jabri et al., 2022; Bhat et al., 2022; Gebre & Sendeku, 2019).

El-Belely et al. (2021) synthesized spherical ZnO-NPs with approximately 30–55 nm diameter using *Arthrospira platensis* isolated from aqueous samples, which show good antimicrobial activity and toxicity against cancer cells and are promising for biomedical applications. NPs can also be synthesized from pigments extracted from algae. Davaeifar et al. (2019) purified C-phycocyanin pigment from *Limnothrix* sp. KO05 and synthesized phycocyanin–zinc oxide nanorods (PHY-ZnONRs) using C-phycocyanin pigment. The PHY-ZnONRs can be used to develop low-toxicity antimicrobial agents and combination therapy. ZnO NPs prepared from *Turbinaria ornata* extract were able to improve the agronomical traits of rice. Among them, rice seeds have improved germination, seedling vigor, and dry matter yield. Therefore, ZnO NPs can be applied to sustainable nanoagriculture (Itroutwar et al., 2019). The ZnO NPs synthesized using zinc nitrate and microalgae *Chlorella* extract were recoverable and durable for dibenzothiophene (DBT) pollutant degradation (Khalafi et al., 2019).

2.5 Copper

Copper nanoparticles (CuNPs) have good catalytic activity, antioxidant potential, and antibacterial properties and are widely used in pollutants removal and biomedical applications.

CuO NPs with an average particle size of 3.6 nm were biosynthesized by *Anabaena cylindrica* extracts and exhibited antibacterial activity against both *Escherichia coli* and gram-negative bacteria. Thus, CuO NPs can realize potential applications for disinfection in drinking water (Bhattacharya et al., 2019). Araya-Castro et al. (2020) obtained protein fractions from aqueous extracts of *Macrocystis pyrifera* and also successfully synthesized CuO-NPs. Bhutiya et al. (2018) extracted a bird's nest–type architectural network of cellulosic nanofibers from *Chaetomorpha antennina*. The nanorods of cuprous oxide were subsequently grown over algal cellulose by single-step hydrothermal method. In this case, algal cellulose acts as a reducing agent to reduce CuO to Cu_2O (Bhutiya et al., 2018). Temperature is the key factor affecting the rate, size, and stability of NPs synthesis. Akintelu et al. (2020) stated that the recommended temperature for the synthesis of CuO NPs or other metal oxide NPs using plant extracts is 25–100°C.

2.6 Iron

Iron-based NPs have been widely studied and applied due to their excellent properties. For example, zero-valent iron nanoparticles (nZVIs) have been used in soil and groundwater remediation due to their ability to effective removal of contaminants (Lei et al., 2018; Yan et al., 2013). Magnetic iron oxide nanoparticles (MIONPs) have superparamagnetic properties. As a result, MIONPs can be used for magnetic resonance imaging, targeted cancer therapy, magnetic storage, and wastewater treatment (Nitnavare et al., 2022). El-Sheekh, El-Kassas, et al. (2021) used *Petalonia fascia*, *Colpomenia sinuosa*, and *Padina pavonica* to biosynthesize Fe_3O_4 NPs with average diameters of 9.42, 17.78, and 31.41 nm, respectively. The algae-mediated Fe_3O_4 NPs had the advantage in nitrogen and phosphorus removal. For instance, the biological removal capacities of TN and TP were 89.8% and 93.96%, respectively, using 4 μg/mL of Fe_3O_4 NPs synthesized by *P. pavonica* extracts (El-Sheekh, El-Kassas, et al., 2021). Fe_3O_4 NPs have also been used in agriculture. Fe_3O_4 NPs synthesized using *Chlorella* K01 extracts were able to stimulate plant growth and improve plant germination. The antifungal activity of Fe_3O_4 NPs has some potential applications in the control of fungal diseases (Win et al., 2021). Budlayan et al. (2021) showed the effect of different iron precursor concentrations during the synthesis of MIONPs using *Sargassum crassifolium* extract. High temperature completely inhibits the activity of iron reductase, resulting in the poor synthesis of NPs (Rajendran & Sen, 2016).

3. Limitations and challenges

- Biosynthetic NMs have the advantages of simplicity, low cost, and environmental friendliness. Among them, algae-based NMs are widely used in medicine, agriculture, catalysis, and environmental remediation. However, most of the research on algae-based NMs is still at the laboratory scale, where algal growth and biosynthesis parameters are relatively easy to control. However, at the industrial scale, the control of relevant parameters is more complex and requires more resources, thus increasing the cost of algae-based NMs biosynthesis. Therefore, how to achieve large-scale production applications is a direction to be considered for future research.
- In addition, it is difficult to ensure good homogeneity of algae-based NMs (e.g., size and shape) during biosynthesis. There is still a long way to go to achieve industrial-scale mass production. Because of the aforementioned problems, future research on the biosynthesis of algae-based NMs may consider (1) simplifying the biosynthesis process to reduce the technical difficulty and (2) improving the homogeneity and efficiency of the large-scale synthesis of algae-based NMs to achieve industrial-scale production applications.
- Most of the research techniques still focus on the extracellular synthesis of NPs, which leads to more dispersion of NPs and even causes size changes, so downstream isolation techniques are also a key consideration for current research. The dispersion of NPs is related to their fluidic behavior, and techniques such as centrifugation and freeze-thawing are currently used to control the synthesis of intracellular and extracellular NPs, thus improving dispersion (Brar et al., 2022). In addition, the use of genetic modifications and metabolic engineering can improve the catalytic activity of biological substrates or facilitate the production of glutathione, metallothionein, and membrane proteins as a stabilizing agent (Ramya et al., 2015).

4. Summary points

- This chapter describes algae-based NMs and nanostructures, as well as applications and relevant factors affecting biosynthesis.

- Compared with the synthesis of NMs by physical or/and chemical methods, the biosynthesis of NMs using algae is a safe, economical, and sustainable approach. Algae-mediated synthesis of NPs is divided into intracellular and extracellular.
- Bioactive compounds of algae (e.g., pigments, carbohydrates, alkaloids, proteins, etc.) are capable of converting metal ions into NPs, and the synthetic techniques are used in agricultural, aquacultural, pharmaceutical, cosmetical, biotechnological, energetical, and nanotechnological applications.
- The metal NPs that can be biosynthesized using algae include AgNPs, AuNPs, ZnO NPs, CuO NPs, CuNPs, Fe_3O_4 NPs, MIONPs, and PtNPs. Among them, most of the studies focus on AgNPs and AuNPs.
- Most of the algal-mediated metal NPs are spherical, but due to different synthesis conditions, they may also appear triangular, quadrangular, hexagonal, and rod. The algae-mediated metal NPs generally have various excellent properties such as antibiofilm activity, catalytic activity, antibiofilm, and insecticidal activity. Therefore, they have potential applications in antimicrobial agents, biofortification agents, cancer therapy, bioremediation, and plant growth stimulants.

References

Aboeita, N. M., Fahmy, S. A., El-Sayed, M. M. H., Azzazy, H. M. E.-S., & Shoeib, T. (2022). Enhanced anticancer activity of nedaplatin loaded onto copper nanoparticles synthesized using red algae. *Pharmaceutics, 14*(2). https://doi.org/10.3390/pharmaceutics14020418

Adamczyk, M., Lasek, J., & Skawińska, A. (2016). CO2 biofixation and growth kinetics of Chlorella vulgaris and nannochloropsis gaditana. *Applied Biochemistry and Biotechnology, 179*(7), 1248–1261. https://doi.org/10.1007/s12010-016-2062-3

Ahmed, S., Chaudhry, S. A., & Ikram, S. (2017). A review on biogenic synthesis of ZnO nanoparticles using plant extracts and microbes: A prospect towards green chemistry. *Journal of Photochemistry and Photobiology B, 166*, 272–284. https://doi.org/10.1016/j.jphotobiol.2016.12.011

Akintelu, S. A., Folorunso, A. S., Folorunso, F. A., & Oyebamiji, A. K. (2020). Green synthesis of copper oxide nanoparticles for biomedical application and environmental remediation. *Heliyon, 6*(7), e04508. https://doi.org/10.1016/j.heliyon.2020.e04508

Al Jabri, H., Saleem, M. H., Rizwan, M., Hussain, I., Usman, K., & Alsafran, M. (2022). Zinc oxide nanoparticles and their biosynthesis: Overview. *Life (Basel), 12*(4). https://doi.org/10.3390/life12040594

Ali Babazadeh, B., Razeghi, J., Jafarirad, S., & Motafakkerazad, R. (2021). Are biosynthesized nanomaterials toxic for the environment? Effects of perlite and CuO/perlite nanoparticles on unicellular algae haematococcus pluvialis. *Ecotoxicology, 30*(5), 899–913. https://doi.org/10.1007/s10646-021-02406-5

Alzahrani, R. R., Alkhulaifi, M. M., & Al-Enazi, N. M. (2020). In vitro biological activity of Hydroclathrus clathratus and its use as an extracellular bioreductant for silver nanoparticle formation. *Green Processing and Synthesis, 9*(1), 416–428. https://doi.org/10.1515/gps-2020-0043

Araya-Castro, K., Chao, T.-C., Durán-Vinet, B., Cisternas, C., Ciudad, G., & Rubilar, O. (2020). Green synthesis of copper oxide nanoparticles using protein fractions from an aqueous extract of Brown algae Macrocystis pyrifera. *Processes, 9*(1). https://doi.org/10.3390/pr9010078

Aritonang, H. F., Koleangan, H., & Wuntu, A. D. (2019). Synthesis of silver nanoparticles using aqueous extract of medicinal plants' (impatiens balsamina and lantana camara) fresh leaves and analysis of antimicrobial activity. *International Journal of Microbiology, 8*. https://doi.org/10.1155/2019/8642303

Azizi, S., Ahmad, M. B., Namvar, F., & Mohamad, R. (2014). Green biosynthesis and characterization of zinc oxide nanoparticles using brown marine macroalga Sargassum muticum aqueous extract. *Materials Letters, 116*, 275–277. https://doi.org/10.1016/j.matlet.2013.11.038

Bakhshan, Y., Samari, F., Ghaemi, M., Ghafarigousheh, S., & Kakoee, A. (2018). Experimental study on the thermal conductivity of silver nanoparticles synthesized using Sargassum angostifolium. *Iranian Journal of Science and Technology, Transactions of Mechanical Engineering, 43*(S1), 251–257. https://doi.org/10.1007/s40997-018-0153-1

Balaraman, P., Balasubramanian, B., Kaliannan, D., Durai, M., Kamyab, H., Park, S., Chelliapan, S., Lee, C. T., Maluventhen, V., & Maruthupandian, A. (2020). Phyco-synthesis of silver nanoparticles mediated from marine algae Sargassum myriocystum and its potential biological and environmental applications. *Waste and Biomass Valorization, 11*(10), 5255–5271. https://doi.org/10.1007/s12649-020-01083-5

Bhat, J. A., Faizan, M., Bhat, M. A., Huang, F., Yu, D. Y., Ahmad, A., Bajguz, A., & Ahmad, P. (2022). Defense interplay of the zinc-oxide nanoparticles and melatonin in alleviating the arsenic stress in soybean (Glycine max L.). *Chemosphere, 288*, 14. https://doi.org/10.1016/j.chemosphere.2021.132471

Bhattacharya, P., Swarnakar, S., Ghosh, S., Majumdar, S., & Banerjee, S. (2019). Disinfection of drinking water via algae mediated green synthesized copper oxide nanoparticles and its toxicity evaluation. *Journal of Environmental Chemical Engineering, 7*(1). https://doi.org/10.1016/j.jece.2018.102867

Bhutiya, P. L., Misra, N., Abdul Rasheed, M., & Zaheer Hasan, S. (2018). Nested seaweed cellulose fiber deposited with cuprous oxide nanorods for antimicrobial activity. *International Journal of Biological Macromolecules, 117*, 435–444. https://doi.org/10.1016/j.ijbiomac.2018.05.210

Bloch, K., Pardesi, K., Satriano, C., & Ghosh, S. (2021). Bacteriogenic platinum nanoparticles for application in nanomedicine. *Frontiers of Chemistry, 9*, 624344. https://doi.org/10.3389/fchem.2021.624344

Boobalakrishnan, S., Mani, P., Pottail, L., & Sundararajan, S. K. (2020). Synthesis of nano-silver from Navicula cincta and the evaluation of its antimicrobial activity. *Micro & Nano Letters, 15*(2), 119–124. https://doi.org/10.1049/mnl.2019.0365

Brar, K. K., Magdouli, S., Othmani, A., Ghanei, J., Narisetty, V., Sindhu, R., Binod, P., Pugazhendhi, A., Awasthi, M. K., & Pandey, A. (2022). Green route for recycling of low-cost waste resources for the biosynthesis of nanoparticles (NPs) and nanomaterials (NMs)-A review. *Environmental Research, 207*, 112202. https://doi.org/10.1016/j.envres.2021.112202

Budlayan, M. L. M., Patricio, J. N., Arco, S. D., Capangpangan, R. Y., & Alguno, A. C. (2021). Effects of precursor concentration on the properties of magnetic iron oxide nanoparticles synthesized using brown seaweed (Sargassum crassifolium) extract. *Materials Today Proceedings, 46*, 1608–1612. https://doi.org/10.1016/j.matpr.2020.07.251

Castro, L., Blazquez, M. L., Munoz, J. A., Gonzalez, F., & Ballester, A. (2013). Biological synthesis of metallic nanoparticles using algae. *IET Nanobiotechnology, 7*(3), 109–116. https://doi.org/10.1049/iet-nbt.2012.0041

Chandel, N., Ahuja, V., Gurav, R., Kumar, V., Tyagi, V. K., Pugazhendhi, A., Kumar, G., Kumar, D., Yang, Y. H., & Bhatia, S. K. (2022). Progress in microalgal mediated bioremediation systems for the removal of antibiotics and pharmaceuticals from wastewater. *Science of the Total Environment, 825*, 21. https://doi.org/10.1016/j.scitotenv.2022.153895

Chumakov, D., Pylaev, T., Avdeeva, E., Dykman, L., Khlebtsov, N., Bogatyrev, V., Tuchin, V. V., & Genina, E. A. (2020). Anticancer properties of gold nanoparticles biosynthesized by reducing of chloroaurate ions with Dunaliella salina aqueous extract. In *Saratov fall meeting 2019: Optical and nano-technologies for biology and medicine*.

Colin, J. A., Pech-Pech, I. E., Oviedo, M., Águila, S. A., Romo-Herrera, J. M., & Contreras, O. E. (2018). Gold nanoparticles synthesis assisted by marine algae extract: Biomolecules shells from a green chemistry approach. *Chemical Physics Letters, 708*, 210–215. https://doi.org/10.1016/j.cplett.2018.08.022

Dahoumane, S. A., Wujcik, E. K., & Jeffryes, C. (2016). Noble metal, oxide and chalcogenide-based nanomaterials from scalable phototrophic culture systems. *Enzyme and Microbial Technology, 95*, 13–27. https://doi.org/10.1016/j.enzmictec.2016.06.008

Damergi, E., Qin, P., Sharma, S., Nazeeruddin, M. K., & Ludwig, C. (2021). Enhancing algae biomass production by using dye-sensitized solar cells as filters. *ACS Sustainable Chemistry & Engineering, 9*(43), 14353–14364. https://doi.org/10.1021/acssuschemeng.1c03780

Davaeifar, S., Modarresi, M. H., Mohammadi, M., Hashemi, E., Shafiei, M., Maleki, H., Vali, H., Zahiri, H. S., & Noghabi, K. A. (2019). Synthesizing, characterizing, and toxicity evaluating of phycocyanin-ZnO nanorod composites: A back to nature approaches. *Colloids and Surfaces B: Biointerfaces, 175*, 221–230. https://doi.org/10.1016/j.colsurfb.2018.12.002

Deepak, P., Amutha, V., Birundha, R., Sowmiya, R., Kamaraj, C., Balasubramanian, V., Balasubramani, G., Aiswarya, D., Arul, D., & Perumal, P. (2018). Facile green synthesis of nanoparticles from brown seaweed Sargassum wightii and its biological application potential. *Advances in Natural Sciences: Nanoscience and Nanotechnology, 9*(3), 035019. https://doi.org/10.1088/2043-6254/aadc4a (10 pp.)-035019 (10 pp.).

Durango-Giraldo, G., Cardona, A., Zapata, J. F., Santa, J. F., & Buitrago-Sierra, R. (2019). Titanium dioxide modified with silver by two methods for bactericidal applications. *Heliyon, 5*(5). https://doi.org/10.1016/j.heliyon.2019.e01608

El-Belely, E. F., Farag, M. M. S., Said, H. A., Amin, A. S., Azab, E., Gobouri, A. A., & Fouda, A. (2021). Green synthesis of zinc oxide nanoparticles (ZnO-NPs) using Arthrospira platensis (class: Cyanophyceae) and evaluation of their biomedical activities. *Nanomaterials, 11*(1). https://doi.org/10.3390/nano11010095

El-Sheekh, M. M., El-Kassas, H. Y., Shams El-Din, N. G., Eissa, D. I., & El-Sherbiny, B. A. (2021). Green synthesis, characterization applications of iron oxide nanoparticles for antialgal and wastewater bioremediation using three brown algae. *International Journal of Phytoremediation, 23*(14), 1538–1552. https://doi.org/10.1080/15226514.2021.1915957

El-Sheekh, M. M., Hassan, L. H. S., & Morsi, H. H. (2021). Evaluation of antimicrobial activities of blue-green algae-mediated silver and gold nanoparticles. *Rendiconti Lincei. Scienze Fisiche e Naturali, 32*(4), 747–759. https://doi.org/10.1007/s12210-021-01016-x

El-Sheekh, M. M., Shabaan, M. T., Hassan, L., & Morsi, H. H. (2022). Antiviral activity of algae biosynthesized silver and gold nanoparticles against Herps Simplex (HSV-1) virus in vitro using cell-line culture technique. *International Journal of Environmental Health Research, 32*(3), 616–627. https://doi.org/10.1080/09603123.2020.1789946

Gebre, S. H., & Sendeku, M. G. (2019). New frontiers in the biosynthesis of metal oxide nanoparticles and their environmental applications: An overview. *SN Applied Sciences, 1*(8). https://doi.org/10.1007/s42452-019-0931-4

Gu, H., Chen, X., Chen, F., Zhou, X., & Parsaee, Z. (2018). Ultrasound-assisted biosynthesis of CuO-NPs using brown alga Cystoseira trinodis: Characterization, photocatalytic AOP, DPPH scavenging and antibacterial investigations. *Ultrasonics Sonochemistry, 41*, 109–119. https://doi.org/10.1016/j.ultsonch.2017.09.006

Hamouda, R. A., Abd El Maksoud, A. I., Wageed, M., Alotaibi, A. S., Elebeedy, D., Khalil, H., Hassan, A., & Abdella, A. (2021). Characterization and anticancer activity of biosynthesized Au/cellulose nanocomposite from Chlorella vulgaris. *Polymers, 13*(19). https://doi.org/10.3390/polym13193340

Hamouda, R. A., Yousuf, W. E., Mohammed, A. B. A., Mohammed, R. S., Darwish, D. B., & Abdeen, E. E. (2020). Comparative study between zinc oxide nanoparticles synthesis by biogenic and wet chemical methods in vivo and in vitro against *Staphylococcus aureus*. *Microbial Pathogenesis, 147*, 104384. https://doi.org/10.1016/j.micpath.2020.104384

Ishwarya, R., Vaseeharan, B., Kalyani, S., Banumathi, B., Govindarajan, M., Alharbi, N. S., Kadaikunnan, S., Al-Anbr, M. N., Khaled, J. M., & Benelli, G. (2018). Facile green synthesis of zinc oxide nanoparticles using Ulva lactuca seaweed extract and evaluation of their photocatalytic, antibiofilm and insecticidal activity. *Journal of Photochemistry and Photobiology B, 178*, 249–258. https://doi.org/10.1016/j.jphotobiol.2017.11.006

Itroutwar, P. D., Govindaraju, K., Tamilselvan, S., Kannan, M., Raja, K., & Subramanian, K. S. (2019). Seaweed-based biogenic ZnO nanoparticles for improving agro-morphological characteristics of rice (oryza sativa L.). *Journal of Plant Growth Regulation, 39*(2), 717–728. https://doi.org/10.1007/s00344-019-10012-3

Jeyaraj, M., Gurunathan, S., Qasim, M., Kang, M. H., & Kim, J. H. (2019). A comprehensive review on the synthesis, characterization, and biomedical application of platinum nanoparticles. *Nanomaterials, 9*(12), 41. https://doi.org/10.3390/nano9121719

Kah, J. C. Y., Kho, K. W., Lee, C. G. L., Sheppard, C. J. R., Shen, Z. X., Soo, K. C., & Olivo, M. C. (2007). Early diagnosis of oral cancer based on the surface plasmon resonance of gold nanoparticles. *International Journal of Nanomedicine, 2*(4), 785–798.

Kannah, R. Y., Kavitha, S., Banu, J. R., Sivashanmugam, P., Gunasekaran, M., & Kumar, G. (2021). A mini review of biochemical conversion of algal biorefinery. *Energy & Fuels, 35*(21), 16995–17007. https://doi.org/10.1021/acs.energyfuels.1c02294

Karkhane, M., Lashgarian, H. E., Mirzaei, S. Z., Ghaffarizadeh, A., cherghipour, K., Sepahvand, A., & Marzban, A. (2020). Antifungal, antioxidant and photocatalytic activities of zinc nanoparticles synthesized by Sargassum vulgare extract. *Biocatalysis and Agricultural Biotechnology, 29*. https://doi.org/10.1016/j.bcab.2020.101791

Kathiraven, T., Sundaramanickam, A., Shanmugam, N., & Balasubramanian, T. (2014). Green synthesis of silver nanoparticles using marine algae Caulerpa racemosa and their antibacterial activity against some human pathogens. *Applied Nanoscience, 5*(4), 499–504. https://doi.org/10.1007/s13204-014-0341-2

Khalafi, T., Buazar, F., & Ghanemi, K. (2019). Phycosynthesis and enhanced photocatalytic activity of zinc oxide nanoparticles toward organosulfur pollutants. *Scientific Reports, 9*(1), 6866. https://doi.org/10.1038/s41598-019-43368-3

Khan, A. U., Khan, M., Malik, N., Cho, M. H., & Khan, M. M. (2019). Recent progress of algae and blue-green algae-assisted synthesis of gold nanoparticles for various applications. *Bioprocess and Biosystems Engineering, 42*(1), 1–15. https://doi.org/10.1007/s00449-018-2012-2

Kim, J. Y., & Osterloh, F. E. (2006). Planar gold nanoparticle clusters as microscale mirrors. *Journal of the American Chemical Society, 128*(12), 3868–3869. https://doi.org/10.1021/ja057958k

Kim, B., Song, W. C., Park, S. Y., & Park, G. (2021). Green synthesis of silver and gold nanoparticles via Sargassum serratifolium extract for catalytic reduction of organic dyes. *Catalysts, 11*(3). https://doi.org/10.3390/catal11030347

Kumari, S., Tyagi, M., & Jagadevan, S. (2019). Mechanistic removal of environmental contaminants using biogenic nano-materials. *International journal of Environmental Science and Technology, 16*(11), 7591–7606. https://doi.org/10.1007/s13762-019-02468-3

Kusumaningrum, H. P., Zainuri, M., Marhaendrajaya, I., & Subagio, A. (2018). Nanosilver microalgae biosynthesis: Cell appearance based on SEM and EDX methods. *Journal of Physics: Conference Series, 1025*. https://doi.org/10.1088/1742-6596/1025/1/012084

Lashgarian, H. E., Karkhane, M., Alhameedawi, A. K., & Marzban, A. (2021). Phyco-mediated synthesis of Ag/AgCl nanoparticles using ethanol extract of a marine green algae, Ulva fasciata delile with biological activity. *Biointerface Research in Applied Chemistry, 11*(6), 14545–14554. https://doi.org/10.33263/Briac116.1454514554

Lee, K., Lee, H., Bae, K. H., & Park, T. G. (2010). Heparin immobilized gold nanoparticles for targeted detection and apoptotic death of metastatic cancer cells. *Biomaterials, 31*(25), 6530–6536. https://doi.org/10.1016/j.biomaterials.2010.04.046

Lei, C., Sun, Y., Tsang, D. C. W., & Lin, D. (2018). Environmental transformations and ecological effects of iron-based nanoparticles. *Environment Pollution, 232*, 10–30. https://doi.org/10.1016/j.envpol.2017.09.052

Li, Y., Li, Y., Li, Q., & Bao, J. (2017). Rapid biosynthesis of silver nanoparticles based on flocculation and reduction of an exopolysaccharide from arthrobacter sp B4: Its antimicrobial activity and phytotoxicity. *Journal of Nanomaterials, 8*. https://doi.org/10.1155/2017/9703614

Liu, B., Xie, J., Lee, J. Y., Ting, Y. P., & Chen, J. P. (2005). Optimization of high-yield biological synthesis of single-crystalline gold nanoplates. *Journal of Physical Chemistry B, 109*(32), 15256–15263. https://doi.org/10.1021/jp051449n

Li, S. N., Wang, R., & Ho, S. H. (2021). Algae-mediated biosystems for metallic nanoparticle production: From synthetic mechanisms to aquatic environmental applications. *Journal of Hazardous Materials, 420*, 126625. https://doi.org/10.1016/j.jhazmat.2021.126625

Li, X., Xu, H., Chen, Z.-S., & Chen, G. (2011). Biosynthesis of nanoparticles by microorganisms and their applications. *Journal of Nanomaterials*, 1–16. https://doi.org/10.1155/2011/270974

Lucaci, A. A.-R., Bulgariu, D., & Bulgariu, L. (2021). Green synthesis of gold nanoparticles using marine red algae biomass. In *2021 international conference on e-health and bioengineering (EHB)* (pp. 1–4).

Machado, S., Gonzalez-Ballesteros, N., Goncalves, A., Magalhaes, L., Sarria Pereira de Passos, M., Rodriguez-Arguelles, M. C., & Castro Gomes, A. (2021). Toxicity in vitro and in zebrafish embryonic development of gold nanoparticles biosynthesized using cystoseira macroalgae extracts. *International Journal of Nanomedicine, 16*, 5017–5036. https://doi.org/10.2147/IJN.S300674

Mahana, A., & Mehta, S. K. (2021). Potential of scenedesmus-fabricated ZnO nanorods in photocatalytic reduction of methylene blue under direct sunlight: Kinetics and mechanism. *Environmental Science and Pollution Research International, 28*(22), 28234–28250. https://doi.org/10.1007/s11356-021-12682-7

Mahmoud, A. W. M., Abdeldaym, E. A., Abdelaziz, S. M., El-Sawy, M. B. I., & Mottaleb, S. A. (2020). Synergetic effects of zinc, boron, silicon, and zeolite nanoparticles on confer tolerance in potato plants subjected to salinity. *Agronomy-Basel, 10*(1), 23. https://doi.org/10.3390/agronomy10010019

Marimuthu, S., Antonisamy, A. J., Malayandi, S., Rajendran, K., Tsai, P. C., Pugazhendhi, A., & Ponnusamy, V. K. (2020). Silver nanoparticles in dye effluent treatment: A review on synthesis, treatment methods, mechanisms, photocatalytic degradation, toxic effects and mitigation of toxicity. *Journal of Photochemistry and Photobiology B-Biology, 205*, 13. https://doi.org/10.1016/j.jphotobiol.2020.111823

Mishra, B., Saxena, A., & Tiwari, A. (2020). Biosynthesis of silver nanoparticles from marine diatoms Chaetoceros sp., Skeletonema sp., Thalassiosira sp., and their antibacterial study. *Biotechnology Reports (Amsterdam), 28*, e00571. https://doi.org/10.1016/j.btre.2020.e00571

Nitnavare, R., Bhattacharya, J., Thongmee, S., & Ghosh, S. (2022). Utilizing photosynthetic microbes in nanobiotechnology: Applications and perspectives. *Science of the Total Environment*, 156457. https://doi.org/10.1016/j.scitotenv.2022.156457

Park, S. J., Taton, T. A., & Mirkin, C. A. (2002). Array-based electrical detection of DNA with nanoparticle probes. *Science, 295*(5559), 1503–1506.

Pytlik, N., Kaden, J., Finger, M., Naumann, J., Wanke, S., Machill, S., & Brunner, E. (2017). Biological synthesis of gold nanoparticles by the diatom Stephanopyxis turris and in vivo SERS analyses. *Algal Research-Biomass Biofuels and Bioproducts, 28*, 9–15. https://doi.org/10.1016/j.algal.2017.10.004

Pytlik, N., Klemmed, B., Machill, S., Eychmüller, A., & Brunner, E. (2019). In vivo uptake of gold nanoparticles by the diatom Stephanopyxis turris. *Algal Research, 39*. https://doi.org/10.1016/j.algal.2019.101447

Rajendran, K., & Sen, S. (2016). Optimization of process parameters for the rapid biosynthesis of hematite nanoparticles. *Journal of Photochemistry and Photobiology B-Biology, 159*, 82–87. https://doi.org/10.1016/j.jphotobiol.2016.03.023

Ramya, S., Shanmugasundaram, T., & Balagurunathan, R. (2015). Biomedical potential of actinobacterially synthesized selenium nanoparticles with special reference to anti-biofilm, anti-oxidant, wound healing, cytotoxic and anti-viral activities. *Journal of Trace Elements in Medicine & Biology, 32*, 30–39. https://doi.org/10.1016/j.jtemb.2015.05.005

Rana, A., Yadav, K., & Jagadevan, S. (2020). A comprehensive review on green synthesis of nature-inspired metal nanoparticles: Mechanism, application and toxicity. *Journal of Cleaner Production, 272*. https://doi.org/10.1016/j.jclepro.2020.122880

Samanta, S., Agarwal, S., Nair, K. K., Harris, R. A., & Swart, H. (2019). Biomolecular assisted synthesis and mechanism of silver and gold nanoparticles. *Materials Research Express, 6*(8). https://doi.org/10.1088/2053-1591/ab296b

Sanaeimehr, Z., Javadi, I., & Namvar, F. (2018). Antiangiogenic and antiapoptotic effects of green-synthesized zinc oxide nanoparticles using Sargassum muticum algae extraction. *Cancer Nanotechnology, 9*(1), 3. https://doi.org/10.1186/s12645-018-0037-5

Sathiyaraj, G., Vinosha, M., Sangeetha, D., Manikandakrishnan, M., Palanisamy, S., Sonaimuthu, M., Manikandan, R., You, S., & Prabhu, N. M. (2021). Bio-directed synthesis of Pt-nanoparticles from aqueous extract of red algae Halymenia dilatata and their biomedical applications. *Colloids and Surfaces A: Physicochemical and Engineering Aspects, 618*. https://doi.org/10.1016/j.colsurfa.2021.126434

Sayed Ahmed, H. I., Elsherif, D. E., El-Shanshory, A. R., Haider, A. S., & Gaafar, R. M. (2021). Silver nanoparticles and Chlorella treatments induced glucosinolates and kaempferol key biosynthetic genes in Eruca sativa. *Beni-Suef University Journal of Basic and Applied Sciences, 10*(1). https://doi.org/10.1186/s43088-021-00139-2

Shankar, S. S., Rai, A., Ahmad, A., & Sastry, M. (2005). Controlling the optical properties of lemongrass extract synthesized gold nanotriangles and potential application in infrared-absorbing optical coatings. *Chemistry of Materials, 17*(3), 566–572. https://doi.org/10.1021/cm048292g

Shanmuganathan, R., Karuppusamy, I., Saravanan, M., Muthukumar, H., Ponnuchamy, K., Ramkumar, V. S., & Pugazhendhi, A. (2019). Synthesis of silver nanoparticles and their biomedical applications - a comprehensive review. *Current Pharmaceutical Design, 25*(24), 2650–2660. https://doi.org/10.2174/1381612825666190708185506

Sherief, M. A., El-Bassyouni, G. T., Gamal, A. A., & Esawy, M. A. (2021). Modification of diatom using silver nanoparticle to improve antimicrobial activity. *Materials Today Proceedings, 43*, 3369–3374. https://doi.org/10.1016/j.matpr.2020.05.391

Tamil Selvan, S., Balasubramani, G., Narayanasamy, S., & Ramamurthy, D. (2018). Evaluation of multitudinous potentials of photosynthetic microalga, neochloris aquatica RDS02 derived silver nanoparticles. *Smart Science, 7*(2), 116–129. https://doi.org/10.1080/23080477.2018.1491743

Thiruchelvi, R., Jayashree, P., & Mirunaalini, K. (2021). Synthesis of silver nanoparticle using marine red seaweed Gelidiella acerosa -A complete study on its biological activity and its characterisation. *Materials Today Proceedings, 37*, 1693–1698. https://doi.org/10.1016/j.matpr.2020.07.242

Thirumoorthy, G. S., Balasubramaniam, O., Kumaresan, P., Muthusamy, P., & Subramani, K. (2021). Tetraselmis indica mediated green synthesis of zinc oxide (ZnO) nanoparticles and evaluating its antibacterial, antioxidant, and hemolytic activity. *BioNanoScience, 11*(1), 172–181. https://doi.org/10.1007/s12668-020-00817-y

Torabfam, M., & Yüce, M. (2020). Microwave-assisted green synthesis of silver nanoparticles using dried extracts of Chlorella vulgaris and antibacterial activity studies. *Green Processing and Synthesis, 9*(1), 283–293 https://doi.org/10.1515/gps-2020-0024

Uzair, B., Liaqat, A., Iqbal, H., Menaa, B., Razzaq, A., Thiripuranathar, G., Fatima Rana, N., & Menaa, F. (2020). Green and cost-effective synthesis of metallic nanoparticles by algae: Safe methods for translational medicine. *Bioengineering, 7*(4). https://doi.org/10.3390/bioengineering7040129

Velgosova, O., Dolinská, S., Mražíková, A., & Briančin, J. (2020). Effect of P. kessleri extracts treatment on AgNPs synthesis. *Inorganic and Nano-Metal Chemistry, 50*(9), 842–852. https://doi.org/10.1080/24701556.2020.1726388

Win, T. T., Khan, S., Bo, B., Zada, S., & Fu, P. (2021). Green synthesis and characterization of Fe3O4 nanoparticles using Chlorella-K01 extract for potential enhancement of plant growth stimulating and antifungal activity. *Scientific Reports, 11*(1), 21996. https://doi.org/10.1038/s41598-021-01538-2

Wu, L. L., Zhu, G. S., Zhang, X. X., & Si, Y. B. (2020). Silver nanoparticles inhibit denitrification by altering the viability and metabolic activity of Pseudomonas stutzeri. *Science of the Total Environment, 706*, 9. https://doi.org/10.1016/j.scitotenv.2019.135711

Yan, W. L., Lien, H. L., Koel, B. E., & Zhang, W. X. (2013). Iron nanoparticles for environmental clean-up: Recent developments and future outlook. *Environmental Science-Processes & Impacts, 15*(1), 63–77. https://doi.org/10.1039/c2em30691c

Žvab, U., Kukulin, D. S., Fanetti, M., & Valant, M. (2021). Bioremediation of copper polluted wastewater-like nutrient media and simultaneous synthesis of stable copper nanoparticles by a viable green alga. *Journal of Water Process Engineering, 42*. https://doi.org/10.1016/j.jwpe.2021.102123

CHAPTER

16 Algae materials for drugs and drug delivery

Soraya Paz-Montelongo[1,2], Daniel Niebla-Canelo[1,2], Samuel Alejandro-Vega[1,2], Carmen Rubio-Armendáriz[1,2], Ángel Gutiérrez-Fernández[1,2] and Arturo Hardisson[1,2]

[1]*Área de Toxicología, Universidad de La Laguna, Tenerife, Islas Canarias, Spain;* [2]*Grupo Interuniversitario de Toxicología Alimentaria y Ambiental, Universidad de La Laguna, Tenerife, Islas Canarias, Spain*

1. Introduction

Algae are protist organisms mainly found in marine ecosystems and are the base of their trophic chains. Most of them are photosynthetic; however, recently, it has been discovered that some algae are also able to have heterotrophic behavior.

Algae have different colorings, which depend on their main pigment. They are classified accordingly as brown algae, red algae, or green algae (Paz et al., 2022). Blue algae, also known as cyanobacteria, are also remarkable as genera as *Spirulina* is under this classification. It should be noted that cyanobacteria are photoautotrophs that carry out oxygenic photosynthesis; meanwhile the microalgae are microscopic algae both eukaryotic and prokaryotic (Singh & Saxena, 2015).

Algae and microalgae have been classified according to the research by Ruggiero et al. (2015) and Levasseur et al. (2020) (Table 16.1). Red algae belong to the Rhodophyta phylum; meanwhile, green algae and brown algae are from the Chlorophyta phylum and Phaeophyceae class, respectively (Quitral et al., 2012).

Even though algae are vulgarly called as "sea plants," the reality is that there are key traits that clearly differ from algae and plants. Algae morphology is based on a rhizoid that allows them to attach themselves in the substratum, but it is not used to take nutrients. Stipites are similar to a plant stalk, and laminas are the plain parts of the algae.

Algae are key organisms in the production of organic matter from inorganic one. The presence of high algae abundance could be synonymous with eutrophication in an area. Therefore, the presence of certain algae's species or algae's blooming in some zones may indicate water pollution. The production of biotoxins and "red tides" by the algae's blooming is also remarkable.

Algae have been used in food in many Asian countries such as Japan, China, or Vietnam since ancient times. However, their nutritional qualities and their beneficial properties have contributed to their inclusion in the western countries diet (Paz et al., 2019). Kombu (*Laminaria)*, nori (*Porphyra*), hijiki (*Hizikia fusiformis*), thongweed (*Himanthalia elongata*), and cochayuyo (*Durvillaea antarctica*) are the most frequently found algae species. There are multiple dishes based on algae such as sushi, algae snacks, soups, and stews, among others. Therefore, algae are a valuable food which stand of by

Algae Materials: Applications Benefitting Health. https://doi.org/10.1016/B978-0-443-18816-9.00017-4

Table 16.1 Distribution and classification of Algae used in biotechnology.

Domain	Kingdom	Phylum	Class
Prokaryota	Eubacteria	Cyanobacteria	Cyanophyceae
			Gloeobacterophyceae
	Archaebacteria	———	———
Eukaryota	Protozoa	Euglenozoa	Euglenophyceae
	Chromista	Cryptista	Cryptophyceae
		Haptophyta	Coccolithophyceae
			Pavlovophyceae
		Heterokontophyta	Bacillariophyceae
			Chrysophyceae
			Eustigmatophyceae
			Dictyochophyceae
			Phaeophyceae
			Xanthophyceae
	Plantae	Glaucophyta	Glaucophyceae
		Rhodophyta	Cyanidiophyceae
		Chlorophyta	Chlorophyceae
			Trebouxiophyceae
	Fungi	———	———
	Animalia	———	———

their great protein content rich in glutamic acid, alanine, or arginine. They also have vitamins such as vitamin A or D, and their trace element content (zinc (Zn), potassium (K), iodine (I), etc.) is not falling behind, same as the fiber one that could reach the 30%–60% of the total dry weight. Furthermore, the presence of bioactive compounds such as carotenoids and polyphenols, antioxidant substances, may suggest that algae should be considered as a functional food (Darias-Rosales et al., 2020; Martín León et al., 2021; Paz et al., 2018, 2019b).

Marine algae have been used in traditional medicine since antiquity in many civilizations, as it could be seen in China where different dynasties registered their use in many diseases' treatment. They are even registered in the "Chinese Materia Medica" (Chapman, 1950). *Saccharina japonica* algae have been used to treat the menstrual symptoms. Even Confusio, the famous Chinese philosopher, wrote about the algae.

Japanese population has used the seaweeds for the treatment of glandular problems and gout since 300 BCE. *Laminaria saccharina* algae have been used to treat syphilis and goiter (Rao & Periyasami, 2020). Roman civilization used seaweed in the treatment of wound, rash, and burn. Meanwhile the British used the red algae *Porphyra* to prevent scurvy and *Chondrus* for the treatment of internal discomfort.

In recent decades, the interest of society, and especially the scientific community, in the study of the composition of algae and their various applications has increased. This is why the use of algae for purposes other than food is frequent, and due to their chemical composition, algae play an important role in numerous biotechnological applications. Algae and their extracts are present in the pharmaceutical industry, cosmetic industry, chemical industry, etc. Fig. 16.1 summarizes the most important applications of algae.

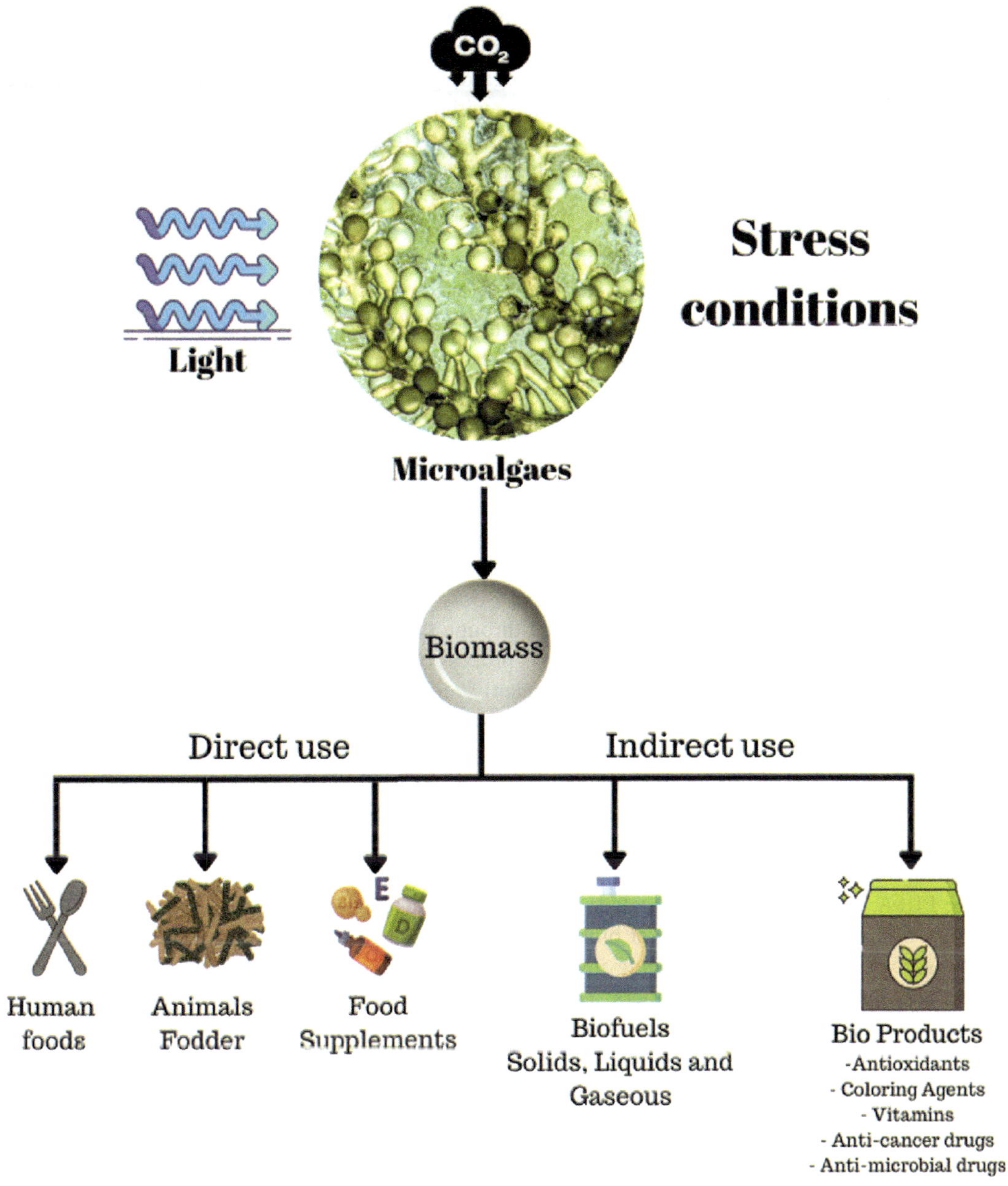

FIGURE 16.1

Direct and indirect uses of microalgae.

Adapted from Khan, M. I., Shin, J., Kim, J.-D. (2018). The promising future of microalgae: Current status, challenges, and optimization of a sustainable and renewable industry for biofuels, feed, and other products. Microbial Cell Factories, *17. 10.1186/s12934-018-0879-x.*

The use of algae or their derivatives in the pharmaceutical industry represents a new research field, which offers more sustainable and natural alternatives for the population. However, the biggest challenges faced by the pharmaceutical industry regarding the use of algae for medical purposes are the difficulty of obtaining an adequate algal biomass without disturbing the environment and isolating the bioactive substances present in the different species of algae. For this reason, in this chapter, we will highlight the bioactive substances and their usefulness in obtaining pharmaceuticals, as well as the procedures for obtaining and their use.

2. Algae's bioactive substances

Algae contain a variety of bioactive chemicals. The bioactive substances present in algae can be classified into polyunsaturated fatty acids (PUFAs), essential amino acids, vitamins, antioxidants, fiber, minerals, and proteins.

2.1 Polyunsaturated fatty acids

Fatty acids are biological molecules whose main characteristics are their insolubility in water and their high energy value due to the high number of carbon–hydrogen bonds. Although some animal lipids, such as cholesterol, are related to cardiovascular diseases, among others, plant or marine lipids are different since they are mainly unsaturated lipids (Li-Beisson et al., 2019; Thompson, 1996). Humans cannot generate PUFA, so they must be ingested through the diet.

PUFAs are key compounds in the biosynthesis process of cellular hormones (eicosanoids). Due to their chemical structure, PUFAs present predominantly "cis" orientation and rigid nonlinear structure; this structure improves the fluidity of cell membranes and therefore intracellular communication and maintenance of homeostasis (Aluko, 2012; Duhan et al., 2020).

PUFA consumption has been epidemiologically related to a lower incidence of cardiovascular diseases associated with elevated blood cholesterol levels, obesity, and diabetes. Likewise, it has been related to improved immunity, decreased asthma symptoms, and decreased levels of proinflammatory compounds such as interleukin-6 and prostaglandins (Duhan et al., 2020).

The PUFA content differs among algae species. Some species of algae stand out for their PUFA content and have been used mainly for biofuel production.

The species that are currently being investigated are *Botryococcus, Chlorella, Dunaliella, Neochloris,* and *Scenedesmus*, among others. These algae species all belong to the Chlorophyceae class, that is, they are green algae. However, other authors are investigating the PUFA content in brown and red algae. *Botryococcus* is used to obtain biofuel. The PUFA content of *Botryococcus braunii* ranges from oleic acid, linoleic acid, and palmitic acid (Fig. 16.2) (Cheng et al., 2013).

Abd El-Baky et al. (2004) studied the composition of polyunsaturated fatty acids in *Dunaliella salina* species under different circumstances and concluded that growth of the alga under low-nitrogen conditions increases the final content of fatty acids, especially alpha-linoleic acid (C18) and palmitic acid (C16). It is also shown that the *D. salina* alga contains omega-3 fatty acids.

Studies carried out by Tejeda-Benítez et al. (2015) in which the content and quality of PUFAs from *Chlorella* sp. and *Dunaliella* sp. microalgae were analyzed concluded that linoleic acid and oleic acid (Fig. 16.2) were the main compounds in the lipid phase extracted from both species (Tejeda-Benítez et al., 2015).

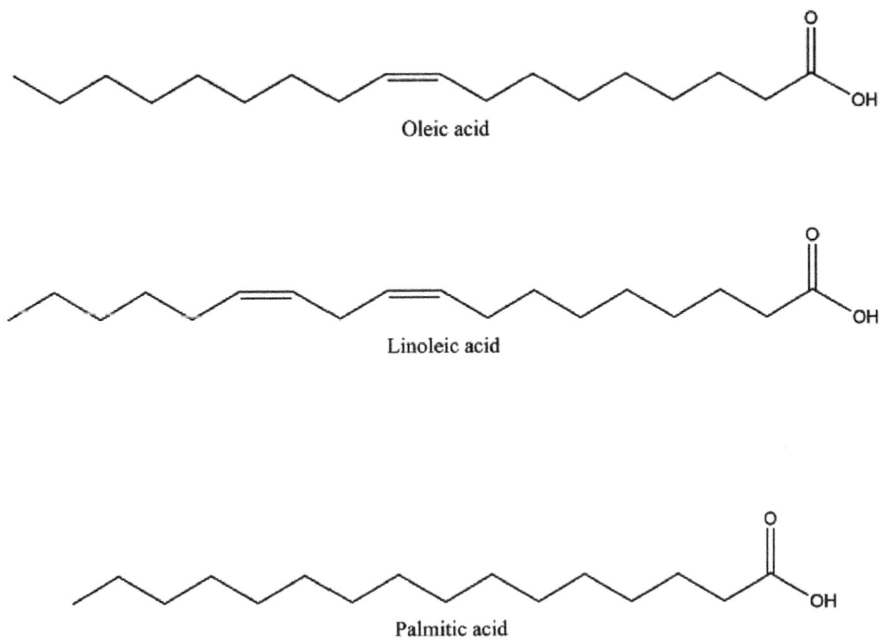

FIGURE 16.2

Chemical structure of oleic acid, linoleic acid and palmitic acid.

2.2 Vitamins and antioxidants

Algae stand out for containing a large number of vitamins. However, the interest between vitamins and algae lies in the presence of provitamin A, vitamin C, and vitamin E. The type of algae influences the presence of some of these vitamins, and there are differences between red, brown, and green algae. Likewise, the content of group B vitamins in algae is remarkable, a feature that gives a great benefit compared with vegetables, which lack this type of vitamins (Table 16.2) (Helliwell et al., 2016).

Table 16.2 lists the main vitamins in some edible algae and other foods (Pereira et al., 2013; Slavin & Lloyd, 2012; Wells et al., 2017). The alga *Dunaliella* sp. is coveted because it is an optimal source for obtaining β-carotene (Hosseini et al., 2009). In addition, *Dunaliella* is noted for its high content of B vitamins (Fabregas & Herrero, 1990). Other edible algae, such as *Himanthalia elongata* or *Palmaria palmata*, stand out for their vitamin E and C and beta-carotene content (Ferraces-Casais et al., 2012). Meanwhile, spirulina (*Arthrospira platensis* and *Arthrospira maxima*), a widespread cyanobacterium, stands out as an important source of B vitamins, vitamin C, vitamins D and E, and photosynthetic pigments such as chlorophyll, xanthophyll, and beta-carotene (Ali & Saleh, 2012; Rubio et al., 2021).

Other studies confirm that the *Haematococcus* species, a freshwater green alga, contains a high concentration of astaxanthin, a ketocarotenoid that is a natural nutritional component (Lorenz & Cysewski, 2000). This ketocarotenoid is found in about 2%–3% of the dry weight of the alga and constitutes almost the totality of the carotenoids contained in this species. It should be noted that this substance is the natural pigment of other aquatic organisms such as salmon and shrimp. The

Table 16.2 Main vitamins in algae and other foods.

Category	Subcategory	Main vitamins
Algae	*Himanthalia elongata*	β-Carotene
		Vitamin E
		Vitamin C
	Palmaria palmata	β-Carotene
	Dunaliella tertiolecta	Vitamin B6
		Vitamin B5
		Vitamin B2
		Vitamin B1
Fruit/vegetable	Fruit	Vitamin C
		Vitamin K
	Legumes	B Vitamins
		Vitamin C
		Vitamin K
	Leafy vegetables	Vitamin C
		Vitamin B9
	Roots and tubers	Vitamin E
Meat	Red meat	Vitamin B2
		Vitamin B3
		Vitamin B6
		Vitamin B5
		Vitamin B12
	Poultry	Vitamin B3
		Vitamin B6

antioxidant activity of astaxanthin is very potent, surpassing beta-carotene, alpha-tocopherol, or lycopene. Astaxanthin has been found to be 10 to 100 times more potent as an antioxidant than vitamin E and beta-carotene (Cysewski & Lorenz, 2007, p. 281; Guerin et al., 2003).

2.3 Proteins, fibers, and minerals

Table 16.3 summarizes the protein content of different algae species. Some edible algae are an important source of protein, especially algae belonging to the Chlorophyceae (green algae) and Rhodophyceae (brown algae) phylum. The value of the proteins present in these algae is based on their amino acids balance.

Taking soybean protein as an example of vegetable protein, proteins extracted from algae are found in the same or even higher proportions than those from soybeans.

As for the amino acids present in algae, their proportion varies among species. For example, the green alga *Ulva pertusa* stands out for its arginine, alanine, glutamic acid, leucine, and aspartic acid content. Meanwhile the red seaweed *Palmaria palmata* has a great content of valine and methionine (Wells et al., 2017).

Table 16.3 Protein content in percentage (%) in different algae species (Wells et al., 2017).

Cyanobacteria	Protein content (% dry weight)
Spirulina (*Arthrospira platensis*)	60–70
Chlorophyta	
Chlorella spp.	40–70
Ulva spp.	7–44
Phaeophyceae	
Alaria esculenta	9–20
Himanthalia elongata	5–15
Saccharina latissima	6–26
Undaria pinnatifida	12–23
Rhodophyta	
Gracilaria tikvahiae	10.7–19.6
Palmaria palmata (dulse)	9.7–25.5
Porphyra spp. (laver)	29–39
Porphyridium	56
Pyropia spp. (nori)	28–44

Algae contain various essential minerals, including iodine (I) (González et al., 2020). Iodine is a necessary element due to its role in the production of thyroid hormones and, therefore, in thyroid regulation. Some algae such as kelp (*Laminaria*) contain iodine levels ranging from 130 to 900 mg per 100 g of dehydrated kelp (Ho & Redan, 2021).

Another essential element whose deficiency directly affects human health is iron (Fe), which is also prominent in several kelp species. Studies conducted by Paz et al. (2019) showed that the *U. pinnatifida* algae contained 58.8 mg/kg dry weight of Fe.

The mineral content of algae makes it possible to obtain sustainable and environmentally friendly pharmaceutical and nutraceutical preparations aimed at preventing deficiencies of essential trace elements in humans (Mahadevan, 2015; Vilcanqui et al., 2021).

3. Algae and their uses in the pharmaceutical industry

Nowadays, some drugs and other products such as supplements or food supplements containing substances extracted from natural products such as algae can be found in the market.

As previously reported, the high vitamin and antioxidant content of algae, microalgae, and cyanobacteria has been exploited by the pharmaceutical industry, which has commercialized some of these species, such as spirulina, in capsules, pills, and powder form, claiming their beneficial properties as a food supplement.

In recent years, research in the pharmaceutical industry has focused on the use of natural resources such as algae to obtain drugs. In China, a research group has succeeded in obtaining a brown algae–based drug for the treatment of mild and moderate Alzheimer's disease, which, in 2019, received approval from the National Medical Products Administration of China for sale (Wang et al., 2019).

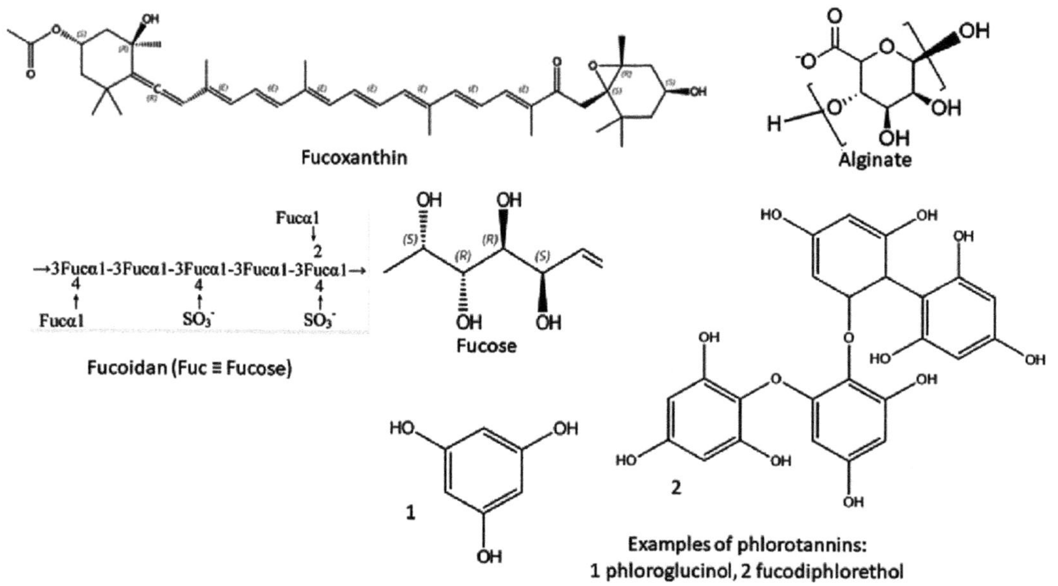

FIGURE 16.3

Chemical structure of fucoxanthin, alginates, fucoidans, and florotannins

Another efficient application of algae and algae derivatives is for the treatment of obesity, a major global epidemic with great repercussions for human health as it is related to metabolic syndrome, type 2 diabetes, and cardiovascular diseases. Antiobesity drugs are currently being evaluated as some algae contain bioactive compounds that have potential as antiobesity agents. These compounds are fucoxanthin, alginates, fucoidans, and florotannins (Fig. 16.3) (Wan-Loy & Siew-Moi, 2016). The anti-obesity effects of these compounds are due to several mechanisms, including inhibition of lipid absorption and metabolism (fucoxanthin and fucoidans), effect on satiety (alginates), and inhibition of adipocyte differentiation (fucoxanthin).

Several authors have studied the application of compounds derived from algae against lung cancer (Saadaoui et al., 2020). Table 16.4 lists the different species of algae and the bioactive substance studied that have shown effects on the inhibition of cancerous processes. Thus, for example, fucoidans extracted from algae such as *U. pinnatifida* or *Fucus vesiculosus* have shown activity inhibiting the pathways that produce cancer cells at the lung level (Boo et al., 2011; Huang et al., 2015).

The research team of Cano-Europa et al. (2012) studied the use of the algae *Chlamydomonas gloeopara* and *Chlorella* against oxidative stress and kidney damage caused by mercury intoxications, improving the symptoms and effects caused by such intoxication, since *C. gloeopara* and *Chlorella* decrease lipid peroxidation and reactive oxygen species.

Other studies indicate the possibility of using the seaweed *Laminaria japonica* for its fucoidan content as an alternative to current antibiotics since it has been shown to inhibit the proliferation of *Staphylococcus aureus* and *Escherichia coli*. In addition, extracts of this algae act as inhibitors of bacterial efflux pumps, thus preventing resistance by blocking drug efflux mechanisms (Liu et al., 2017; Lu et al., 2019).

Table 16.4 Different species of algae and the bioactive substance studied that have shown effects on the inhibition of cancerous processes.

Algae species	Active molecule	Study model	Ref.
Undaria pinnatifida	Fucoidan	Human lung cancer A549 cells	Boo et al. (2011)
Fucus vesiculosus			Huamg et al. (2015)
U. pinnatifida		Tumor cells: PC-3 Normal cells In vivo	Boo et al. (2013)
Sargassum mcclurei		Tumor cells: DLD-1 Normal cells In vivo	Thinh et al. (2013)
F. vesiculosus		Tumor cells Normal cells In vivo	Jeong et al. (2012)
Codium decorticatum	Glycoprotein	Human lung cancer A549 cells	Senthilkumar el al. (2016)
Bryopsis	Kahalalide F		Smit el al. (2004)
Hypnea musciformis	Chloroform extract	NCI−H292 (human lung mucoepidermoid carcinoma)	Guedes et al. (2013)
	Ethanol extract	NCI−H292	
Gracilariopsis lemaneiformis	Polysaccharide	NSCLC cell line, A549	Kang et al. (2017)
Sargassum fusiforme		lung adenocarcinoma SPC-A-1 cells and its xenograft model	Chen et al. (2017)
Sargassum macrocarpum	Tuberatolide B (TTB, $C_{27}H_{34}O_4$)	(A549 and H1299)	Choi et al. (2017)
Sargassum stenophyllum	Sarg A, sulfated polysaccharide	Tumor cells: B16F10 Normal cells In vivo: angiogenesis assays: CAM and gel foam plug implanted SC in mice flank. Tumor growth delay: B16F10 melanoma cells injected SC into back of Male Swiss-Tecpar mice	Dias et al. (2005)
Bostrychia montagnei *Porphyra columbina* *Laminaria brasiliensis* *Sargassum stenophyllum*	Sulfated galactans, alginic acid, fucoidan	Tumor cells: HeLa Normal cells: In vivo	Stevan et al. (2001)
Laminaria digitata	Laminarin	Tumor cells: HT-29 Normal cells In vivo	Park et al. (2013)

Continued

Table 16.4 Different species of algae and the bioactive substance studied that have shown effects on the inhibition of cancerous processes.—cont'd

Algae species	Active molecule	Study model	Ref.
Gracilaria lemaneiformis	Sulfated polysaccharide	Tumor cells Normal cells In vivo: H22 cells implanted in the right thigh of the lower limb in female ICR mice	Fan et al. (2012)
Chondrus ocellatus	λ-Carrageenan	Tumor cells Normal cells In vivo: H-22 cells implanted SC into the right hind groin of ICR mice	Zhou et al. (2006)
Enteromorpha intestinalis	Sulfated polysaccharide	Tumor cells: S-180 Normal cells In vivo: S-180 implanted into the right armpits in male ICR mice	Jiao et al. (2009)

Adapted from and Murphy, C., Hotchkiss, S., Worthington, J., McKeown, S. R. 2014. The potential of seaweed as a source of drugs for use in cancer chemotherapy. Journal of Applied Phycology, 26, *2211–2264.*

4. Drug delivery and algaes

Algae, in addition to being a source of high-value bioactive substances, can be used as carriers or drug delivery systems (Zhong et al., 2021). This is why, in recent years, several researchers have studied the possibility of using some bioactive compounds of different structures for these purposes.

The cell walls of algae contain sulfated polysaccharides such as carrageenan (zinc acetate-1-(5-cyano-2-pyridinyl)-3-[(1S,2S)-2-(6-fluoro-2-hydroxy-3-propionylphenyl) cyclopropyl] urea (1:2:1)) in red algae, ulvan (composed mainly by rhamnose, glucuronic acid, iduronic acid and xylose) in green algae or fucoidan ((6R)-2,6-anhydro-1-deoxy-6-methyl-5-O-methyl-3-O-sulfo-D-galactitol) in brown algae (Cunha & Grenha, 2016; Kigdell et al., 2019; Peso-Echarri et al., 2012) (Fig. 16.4).

Sulfated polysaccharides are polymers that contain a high amount of sulfate ester groups in their structure, mostly present in bacteria, algae, and animals. Sulfated polysaccharides can be used in the design of drug delivery systems.

Rodrigues et al. (2020) demonstrated the potential application of carrageenan from red algae in the treatment of tuberculosis. This is because polysaccharides could be good candidates to compose the matrix of macrophage-targeted drug carriers, as they can use the same recognition pathways of infecting organisms. Carrageenan (CRG), a polysaccharide extracted from edible red algae, is an interesting possibility for the approach of directly targeting alveolar macrophages, as its composition is recognized by several macrophage lectin receptors. This research group successfully obtained appropriately sized inhalable microparticles of starch/CRG associated with isoniazid and rifabutin simultaneously. This study showed that the antituberculosis activity of the drugs was not altered after encapsulation. Rodrigues et al. (2020) have been able to demonstrate that the use of these compounds

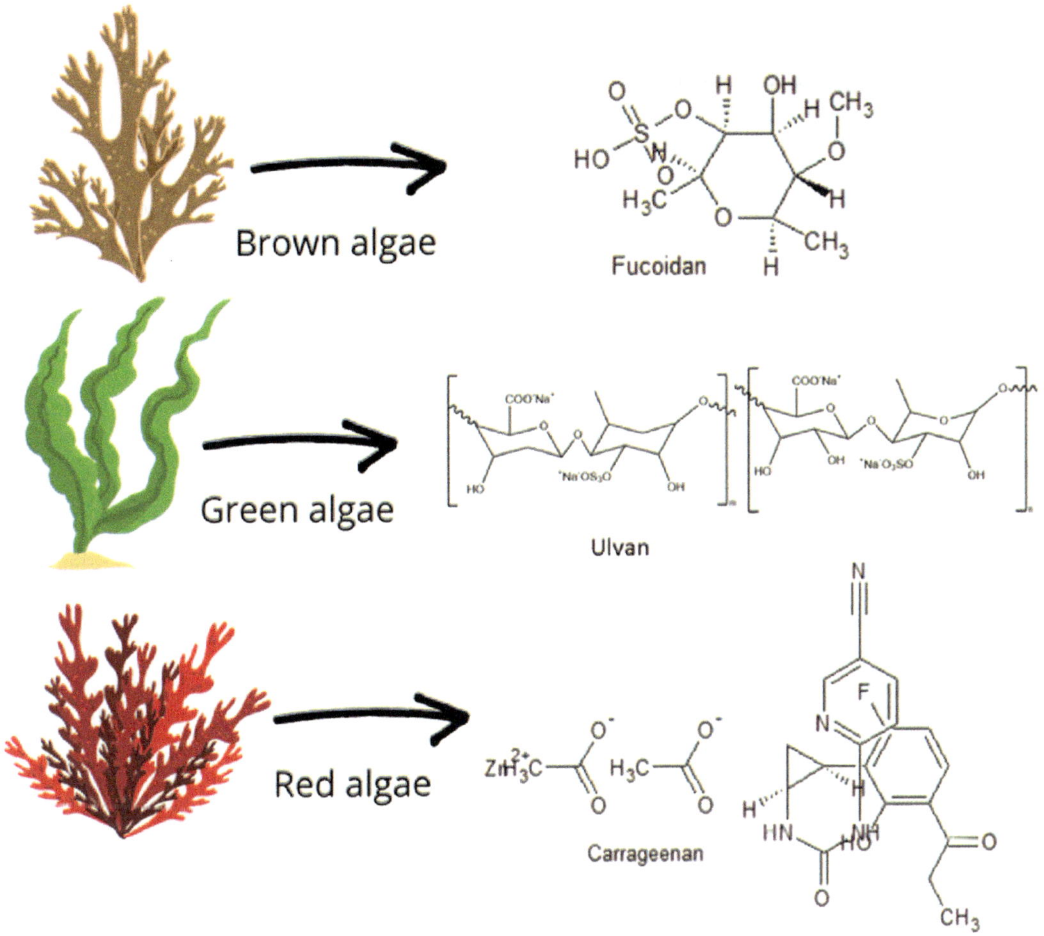

FIGURE 16.4

Mainly polysaccharides contained in the brown, green, and red algae.

from red algae opens the door to their application as drug carriers, especially for treating respiratory diseases such as tuberculosis due to their ability to be inhaled.

Shchelik et al. (2020) succeeded in developing a new drug delivery system based on the green algae *Chlamydomonas reinhardtii*, with the aim of combating skin diseases. They chemically modified the algae with the antibiotic vancomycin, discovered that *C. reinhardtii*-masked vancomycin, and therefore used the insertion of a photoessential linker for the release of the antibiotic. This live drug vehicle was evaluated in the presence of *Bacillus subtilis*, and only after UVA1-mediated release, inhibition of bacterial growth was observed. These results represent one of the first examples of a living organism used as a drug delivery system for antibiotic release by UVA1 irradiation.

5. Conclusions

Algae contain many bioactive chemicals, notably PUFAs, essential amino acids, vitamins, antioxidants, fiber, minerals, and proteins. Drugs and other products such as food supplements or complements containing substances derived from natural products such as those found in algae are now available.

The high vitamin and antioxidant content of algae, microalgae, and cyanobacteria has been exploited by the pharmaceutical industry, which has commercialized some of these species in capsules, tablets, and powders.

The pharmaceutical industry has also discovered the use of algae and their extracts as carriers or drug delivery systems. Beneficial effects have been found from their use in reducing the adverse effects of some drugs in various therapies.

References

Abd El-Baky, H. H., El Baz, F. K., & El-Baroty, G. S. (2004). Production of lipids rich in omega 3 fatty acids from the halotolerant alga Dunaliella salina. *Biotechnology, 3*(1), 102–108.

Ali, S. K., & Saleh, A. M. (2012). Spirulina – an overview. *International Journal of Pharmacy and Pharmaceutical Sciences, 4*(3), 9–15.

Aluko, R. (2012). Bioactive lipids. In *Functional foods and nutraceuticals*. New York, USA: Springer. https://doi.org/10.1007/978-1-4614-3480-1

Boo, H.-J., Hong, J.-Y., Kim, S.-C., Kang, J.-I., Kim, M.-K., Kim, E.-J., Hyun, J.-W., Koh, Y.-S., Yoo, E.-S., Kwon, J.-M., & Kang, H.-K. (2013). The anticancer effect of fucoidan in PC-3 prostate cancer cells. *Marine Drugs, 11*, 2982–2999.

Boo, H.-J., Hyun, J.-H., Kim, S.-C., Kang, J.-I., Kim, M.-K., Kim, S.-Y., Cho, H., Yoo, E.-S., & Kang, H.-K. (2011). Fucoidan from Undaria pinnatifida induces apoptosis in A549 human lung carcinoma cells. *Phytotherapy Research, 25*, 1082–1086.

Cano-Europa, E., Blas-Valdivia, V., Rodríguez-Sánchez, R., Torres-Manzo, P., Franco-Colín, M., Hernández-García, A., & Ortiz-Butrón, R. (2012). Uso terapéutico de algunos microorganismos, microalgas, algas y hongos. *Revista Mexicana de Ciencias Farmacéuticas, 43*(4), 22–30.

Chapman, V. S. (1950). *Seaweed and their uses*. London, UK: Methuen & Co., Ltd.

Cheng, P., Ji, B., Gao, L., Zhang, W., Wang, J., & Liu, T. (2013). The growth, lipid and hydrocarbon production of Botryococcus braunii with attached cultivation. *Bioresource Technology, 138*, 95–100.

Chen, H., Zhang, L., Long, X., Li, P., Chen, S., Kuang, W., & Guo, J. (2017). Sargassum fusiforme polysaccharides inhibit VEGF-A-related angiogenesis and proliferation of lung cancer in vitro and in vivo. *Biomedicine & Pharmacotherapy, 85*, 22–27.

Choi, Y. K., Kim, J., Lee, K. M., Choi, Y., Ye, B., Kim, M., Ko, S., Lee, S., Kang, D., & Heo, S. (2017). Tuberatolide B suppresses cancer progression by promoting ROS-mediated inhibition of STAT3 signaling. *Marine Drugs, 15*.

Cunha, L., & Grenha, A. (2016). Sulfated seaweed polysaccharides as multifunctional materials in drug delivery applications. *Marine Drugs, 14*(3), 42.

Cysewski, R., & Lorenz, T. (2007). *Industrial production of microalgal cell-mass and secondary products - species of high potential: Haematococcus. Handbook of handbook of microalgal culture: Biotechnology and applied phycology* (p. P281). USA: Richmond A.

Darias-Rosales, J., Rubio, C., Gutiérrez, Á. J., Paz, S., & Hardisson, A. (2020). Risk assessment of iodine intake from the consumption of red seaweeds (Palmaria palmata and Chondrus crispus). *Environmental Science and Pollution Research, 27*(36), 45737–45741.

Dias, P. F., Siqueira, J. M., Vendruscolo, L. F., de Jesus, N. T., Gagliardi, A. R., Maraschin, M., & Ribeiro-do-Valle, R. M. (2005). Antiangiogenic and antitumoral properties of a polysaccharide isolated from the seaweed Sargassum stenophyllum. *Cancer Chemotherapy and Pharmacology, 56*, 436–446.

Duhan, N., Barak, S., & Mudgil, D. (2020). Bioactive lipids: Chemistry & health benefits. *Biointerface Research in Applied Chemistry, 10*(6), 6676–6687.

Fabregas, J., & Herrero, C. (1990). Vitamin content of four marine microalgae. Potential use as source of vitamins in nutrition. *Journal of Industrial Microbiology and Biotechnology, 5*(4), 259–263.

Fan, Y., Wang, W., Song, W., Chen, H., Teng, A., & Liu, A. (2012). Partial characterization and anti-tumor activity of an acidic polysaccharide from Gracilaria lemaneiformis. *Carbohydrate Polymers, 88*, 1313–1318.

Ferraces-Casais, P., Lage-Yusty, M. A., Rodríguez-Bernaldo de Quirós, A., & López-Hernández, J. (2012). Evaluation of bioactive compounds in fresh edible seaweeds. *Food Analytical Methods, 5*(4), 828–834.

González, A., Paz, S., Rubio, C., Gutiérrez, Á. J., & Hardisson, A. (2020). Human exposure to iodine from the consumption of edible seaweeds. *Biological Trace Element Research, 197*(2), 361–366.

Guedes, É. A. C., da Silva, T. G., Aguiar, J. S., de Barros, L. D., Pinotti, L. M., & Sant'Ana, A. E. G. (2013). Cytotoxic activity of marine algae against cancerous cells. *Revista Brasileira de Farmacognosia, 23*, 668–673.

Guerin, M., Huntley, M. E., & Olaizola, M. (2003). Haematococcus astaxanthin: Applications for human health and nutrition. *Trends in Biotechnology, 21*(5), 210–216.

Helliwell, K. E., Lawrence, A. D., Holzer, A., Kudahl, U. J., Sasso, S., Kräutler, B., … Smith, A. G. (2016). Cyanobacteria and eukaryotic algae use different chemical variants of vitamin B12. *Current Biology, 26*(8), 999–1008.

Ho, K. K., & Redan, B. W. (2021). Impact of thermal processing on the nutrients, phytochemicals, and metal contaminants in edible algae. *Critical Reviews in Food Science and Nutrition, 62*(2), 508–526.

Hosseini Tafreshi, A., & Shariati, M. (2009). Dunaliella biotechnology: Methods and applications. *Journal of Applied Microbiology, 107*(1), 14–35.

Huang, T. H., Chiu, Y. H., Chan, Y. L., Chiu, Y. H., Wang, H., Huang, K. C., Li, T. L., Hsu, K. H., & Wu, C. J. (2015). Prophylactic administration of fucoidan represses cancer metastasis by inhibiting vascular endotelial growth factor (VEGF) and matrix metalloproteinases (MMPs) in Lewis tumor-bearing mice. *Marine Drugs, 13*, 1882–1900.

Jeong, B.-E., Ko, E.-J., & Joo, H.-G. (2012). Cytoprotective effects of fucoidan, an algae-derived polysaccharide on 5-fluorouracil-treated dendritic cells. *Food and Chemical Toxicology, 50*, 1480–1484.

Jiao, L., Li, X., Li, T., Jiang, P., Zhang, L., Wu, M., & Zhang, L. (2009). Characterization and anti-tumor activity of alkali-extracted polysaccharide from Enteromorpha intestinalis. *International Immunopharmacology, 9*, 324–329.

Kang, Y., Wang, Z., Xie, D., Sun, X., Yang, W., Zhao, X., & Xu, N. (2017). Characterization and potential antitumor activity of polysaccharide from gracilariopsis lemaneiformis. *Marine Drugs, 15*.

Khan, M. I., Shin, J., & Kim, J.-D. (2018). The promising future of microalgae: Current status, challenges, and optimization of a sustainable and renewable industry for biofuels, feed, and other products. *Microbial Cell Factories, 17*. https://doi.org/10.1186/s12934-018-0879-x

Kidgell, J. T., Magnusson, M., de Nys, R., & Glasson, C. R. (2019). Ulvan: A systematic review of extraction, composition and function. *Algal Research, 39*, 101422.

Levasseur, W., Perre, P., & Pozzobon, V. (2020). A review of high value-added molecules production by microalgae in light of the classification. *Biotechnology Advances, 41*, 107545. https://doi.org/10.1016/j.biotechadv.2020.107545

Li-Beisson, Y., Thelen, J. J., Fedosejevs, E., & Harwood, J. L. (2019). The lipid biochemistry of eukaryotic algae. *Progress in Lipid Research, 74*, 31−68.

Liu, M., Liu, Y., Cao, M. J., Liu, G. M., Chen, Q., Sun, L., & Chen, H. (2017). Antibacterial activity and mechanisms of depolymerized fucoidans isolated from Laminaria japonica. *Carbohydrate Polymers, 172*, 294−305.

Lorenz, R. T., & Cysewski, G. R. (2000). Commercial potential for Haematococcus microalgae as a natural source of astaxanthin. *Trends in Biotechnology, 18*(4), 160−167.

Lu, W. J., Lin, H. J., Hsu, P. H., Lai, M., Chiu, J. Y., & Lin, H. T. V. (2019). Brown and red seaweeds serve as potential efflux pump inhibitors for drug-resistant *Escherichia coli*. *Evidence-based Complementary and Alternative Medicine, 2019*, 2019.

Mahadevan, K. (2015). Seaweeds: A sustainable food source. In *Seaweed sustainability* (pp. 347−364). Academic Press.

Martín-León, V., Paz, S., D'Eufemia, P. A., Plasencia, J. J., Sagratini, G., Marcantoni, G., … Rubio-Armendáriz, C. (2021). Human exposure to toxic metals (Cd, Pb, Hg) and nitrates (NO3−) from seaweed consumption. *Applied Sciences, 11*(15), 6934.

Murphy, C., Hotchkiss, S., Worthington, J., & McKeown, S. R. (2014). The potential of seaweed as a source of drugs for use in cancer chemotherapy. *Journal of Applied Phycology, 26*, 2211−2264.

Park, H.-K., Kim, I.-H., Kim, J., & Nam, T.-J. (2013). Induction of apoptosis and the regulation of ErbB signaling by laminarin inHT-29 human colon cancer cells. *International Journal of Molecular Medicine, 32*, 291−295.

Paz, S., Rubio-Armendáriz, C., Frías, I., Guillén-Pino, F., Niebla-Canelo, D., Alejandro-Vega, S., Gutiérrez, Á. J., Hardisson, A., & González-Weller, D. (2022). Toxic and trace elements in seaweeds from a north Atlantic ocean region (tenerife, canary Islands). *Sustainability, 14*(10), 5967. https://doi.org/10.3390/su14105967. MDPI AG. Retrieved from.

Paz, S., Rubio, C., Frías, I., Gutiérrez, Á. J., González-Weller, D., Martín, V., … Hardisson, A. (2019). Toxic metals (Al, Cd, Pb and Hg) in the most consumed edible seaweeds in Europe. *Chemosphere, 218*, 879−884.

Paz, S., Rubio, C., Frías, I., Gutiérrez, A. J., González-Weller, D., Revert, C., & Hardisson, A. (2018). Metal concentrations in wild-harvested phaeophyta seaweed from the Atlantic ocean (canary Islands, Spain). *Journal of Food Protection, 81*(7), 1165−1170.

Paz, S., Rubio, C., Frías, I., Luis-González, G., Gutiérrez, Á. J., González-Weller, D., & Hardisson, A. (2019). Human exposure assessment to macro-and trace elements in the most consumed edible seaweeds in Europe. *Environmental Science and Pollution Research, 26*(36), 36478−36485.

Pereira, Manuela de Castro Cardoso, Paula, Vicente, & Filipa dos Reis Baltazar, Ana (2013). Meat nutritional composition and nutritive role in the human diet. *Meat Science, 93*(3), 586−592. https://doi.org/10.1016/j.meatsci.2012.09.018

Peso-Echarri, P., Frontela-Saseta, C., González-Bermúdez, C. A., Ros-Berruezo, G. F., & Martínez-Graciá, C. (2012). Polisacáridos de algas como ingredientes funcionales en acuicultura marina: Alginato, carragenato y ulvano. *Revista de Biología Marina y Oceanografía, 47*(3), 373−381.

Quitral, V., G, C., L, M., & Schwartz, M. (2012). Propiedades nutritivas y saludables de algas marinas y su potencialidad como ingrediente funcional. *Revista Chilena de Nutricion, 39*, 196−202.

Rao, P. V., & Periyasamy, C. (2020). Biodiversity, conservation and medicinal uses of seaweeds: the glimpses. *Medicinal Plants: Biodiversity, Sustainable Utilization and Conservation* (pp. 21−32). Singapore: Springer.

Rodrigues, S., Cunha, L., Rico, J., Rosa da Costa, A. M., Almeida, A. J., Faleiro, M. L., … Grenha, A. (2020). Carrageenan from red algae: An application in the development of inhalable tuberculosis therapy targeting the macrophages. *Drug Delivery and Translational Research, 10*(6), 1675−1687.

Rubio, C., Dominik-Jakubiec, M., Paz, S., Gutiérrez, Á. J., González-Weller, D., & Hardisson, A. (2021). Dietary exposure to trace elements (B, Ba, Li, Ni, Sr, and V) and toxic metals (Al, Cd, and Pb) from the consumption of commercial preparations of Spirulina platensis. *Environmental Science and Pollution Research, 28*(17), 22146–22155.

Ruggiero, M. A., Gordon, D. P., Orrell, T. M., Bailly, N., Bourgoin, T., Brusca, R. C., Cavalier-Smith, T., Guiry, M. D., & Kirk, P. M. (2015). A higher level classification of all living organisms. *PLoS One, 10*(4), e0119248. https://doi.org/10.1371/journal.pone.0119248

Saadaoui, I., Rasheed, R., Abdulrahman, N., Bounnit, T., Cherif, M., Al Jabri, H., & Mraiche, F. (2020). Algae-derived bioactive compounds with anti-lung cancer potential. *Marine Drugs, 18*, 197. https://doi.org/10.3390/md18040197

Senthilkumar, D., & Jayanthi, S. (2016). Partial characterization and anticancer activities of purified glycoprotein extracted from green seaweed Codium decorticatum. *Journal of Functional Foods, 25*, 323–332.

Shchelik, I. S., Sieber, S., & Gademann, K. (2020). Green algae as a drug delivery system for the controlled release of antibiotics. *Chemistry–A European Journal, 26*(70), 16644–16648.

Singh, J., & Saxena, R. C. (2015). In chapter 2—An introduction to microalgae: Diversity and significance. In S. Kim (Ed.), *Handbook of marine microalgae* (pp. 11–24). Boston: Academic Press.

Slavin, J. L., & Lloyd, B. (2012). Health benefits of fruits and vegetables. *Advances in Nutrition, 3*(4), 506–516. https://doi.org/10.3945/an.112.002154

Smit, A. J. (2004). Medicinal and pharmaceutical uses of seaweed natural products: A review. *Journal of Applied Phycology, 16*, 245–262.

Stevan, F., Oliveira, M., Bucchi, D., Noseda, M., Iacomini, M., & Duarte, M. (2001). Cytotoxic effects against HeLa cells of polysaccharides from seaweeds. *Journal of Submicroscopic Cytology and Pathology, 33*, 477–484.

Tejeda-Benítez, L., Henao-Argumedo, D., Alvear-Alayón, M., & Castillo-Saldarriaga, C. R. (2015). Caracterização e perfil lipídico de azeites de microalgas. *Revista Facultad de Ingeniería, 24*(39), 43–54.

Thinh, P. D., Menshova, R. V., Ermakova, S. P., Anastyuk, S. D., Ly, B. M., & Zvyagintseva, T. N. (2013). Structural characteristics and anticancer activity of fucoidan from the brown alga Sargassum mcclurei. *Marine Drugs, 11*, 1456–1476.

Thompson, G. A., Jr. (1996). Lipids and membrane function in green algae. *Biochimica et Biophysica Acta (BBA) - Lipids and Lipid Metabolism, 1302*(1), 17–45.

Vilcanqui, Y., Mamani-Apaza, L. O., Flores, M., Ortiz-Viedma, J., Romero, N., Mariotti-Celis, M. S., & Huamán-Castilla, N. L. (2021). Chemical characterization of Brown and red seaweed from southern Peru, a sustainable source of bioactive and nutraceutical compounds. *Agronomy, 11*(8), 1669.

Wan-Loy, C., & Siew-Moi, P. (2016). Marine algae as a potential source for anti-obesity agents. *Marine Drugs, 14*(12), 222.

Wang, X., Sun, G., Feng, T., Zhang, J., Huang, X., Wang, T., Xie, Z., Chu, X., Yang, J., Wang, H., Chang, S., Gong, Y., Ruan, L., Zhang, G., Yan, S., Lian, W., Du, C., Yang, D., Zhang, Q., Lin, F., Zhang, H., Ge, C., Xiao, S., Ding, J., & Geng, M. (2019). Sodium oligomannate therapeutically remodels gut microbiota and suppresses gut bacterial amino acids-shaped neuroinflammation to inhibit Alzheimer's disease progression. *Cell Research, 29*(10), 787–803.

Wells, M. L., Potin, P., Craigie, J. S., Raven, J. A., Merchant, S. S., Helliwell, K. E., Smith, A. G., Camire, M. E., Brawley, S. H., et al. (2017). Algae as nutritional and functional food sources: Revisiting our understanding. *Journal of Applied Phycology, 29*(2), 949–982. https://doi.org/10.1007/s10811-016-0974-5

Zhong, D., Du, Z., & Zhou, M. (2021). Algae: A natural active material for biomedical applications. *View, 2*(4), 20200189.

Zhou, G., Sheng, W., Yao, W., & Wang, C. (2006). Effect of low molecular lambda-carrageenan from *Chondrus ocellatus* on antitumor H-22 activity of 5-Fu. *Pharmacological Research, 53*, 129–134.

Further reading

Boo, H., Hyun, J., Kim, S., Kang, J., Kim, M., Kim, S., Cho, H., Yoo, E., & Kang, H. (2022). Fucoidan from Undaria pinnatifida induces apoptosis in A549 human lung carcinoma cells. *Phytotherapy Research, 25*, 1082–1086.

Ferraces-Casais, P., Lage-Yusty, M., Rodriguez, A., & Lopez-Hernandez, J. (2011). Evaluation of bioactive compounds in fresh edible seaweeds. *Food Analytical Methods, 5*.

Hannan, M., Dash, R., Haque, M., Mohibbullah, M., Sohag, A. A. M., Rahman, M., ... Moon, I. S. (2020). Neuroprotective potentials of marine algae and their bioactive metabolites: Pharmacological insights and therapeutic advances. *Marine Drugs, 18*(7), 347.

Sabry, O. M. M., Goeger, D. E., Valeriote, F. A., & Gerwick, W. H. (2017). Cytotoxic halogenated monoterpenes from Plocamium cartilagineum. *Null, 31*, 261–267.

Sakthivel, R., Muniasamy, S., Archunan, G., & Devi, K. P. (2016). Gracilaria edulis exhibit antiproliferative activity against human lung adenocarcinoma cell line A549 without causing adverse toxic effect in vitro and in vivo. *Food & Function, 7*, 1155–1165.

Soares, C., Sousa, S., Machado, S., Vieira, E., Carvalho, A. P., Ramalhosa, M. J., ... Delerue-Matos, C. (2021). Bioactive lipids of seaweeds from the Portuguese north coast: Health benefits versus potential contamination. *Foods, 10*(6), 1366.

CHAPTER 17

Algae material for clinical applications

Kelvii Guo

Department of Mechanical and Biomedical Engineering, City University of Hong Kong, Kowloon Tong, Hong Kong, China

1. Introduction

Nowadays, biomaterials are successfully applied in medical field not only as imaging, drug carrier, and biosensors, but also as devices or implants to replace or support damaged tissues and improve the correlated biofunctions or as an accelerator for tissues regeneration in clinical applications (Anyanwu et al., 2022; Kumar et al., 2021; Wang et al., 2017).

Algae materials, as natural materials, are biocompatible and biodegradable and have outstanding properties of minimum adverse effects and avoiding the second surgical operation, which are superior to the traditional implant materials (such as stainless steel, cobalt-based alloys, commercially pure titanium or titanium–based alloys, etc.). Meanwhile, algae materials are nontoxic, safe, and friendly to human environment (Arokiarajan et al., 2022; Pérez-Alva et al., 2022; Wan et al., 2021; Zargarzadeh et al., 2020).

To date, algae materials have been tailored to meet with the requirements of clinical applications, especially in the field of TE (tissue engineering), drug delivery, and wound dressing.

2. Tissue engineering

TE is taken as a method to make defective organs or tissues restored or improved with created function tissues by encapsulating in or seeding onto the tailored materials to accelerate cell growth, differentiation, and proliferation. For bone construction, algae materials can be used as scaffolding materials, and for bone tissue regeneration, it can be used to deliver cells. The results show that the algae material–encapsulated cells can elaborate and mineralize type I collagen and type III collagen and fibronectin. In addition, it can also support murine embryonic stem cells differentiation to bone cells. It should be noted that algae materials are lack of mechanical strength results in its unsuitable for load-bearing orthopedic applications. Therefore, the research on enhancing the strength of the relevant algae materials has been exploring, for example, mixing other natural or artificial materials with algae materials, to realize the ultimate goal-improving algae material strength (Kerney et al., 2022; Rajeshkumar et al., 2020).

As a bone protector, cartilage prevents bone ends from wear or friction due to the unique characteristic. It distributes in the body everywhere such as bronchial tubes, nose, and ear. It is well known

Algae Materials: Applications Benefitting Health. https://doi.org/10.1016/B978-0-443-18816-9.00010-1

that there are no blood vessels and nerves in cartilage. The repairing speed is obviously so slow that causes inflammation, pain, and unmovement. Promising results show that algae materials grafted to the defective sites of cartilage can stimulate the growth of cartilage to restore the relevant damage. Research reveals that algae materials can be beneficial to ameliorate the growth of chondrocytes without any harmful effect on the correlated properties. Also, it should be pointed out that the scaffold made with algae materials cannot possess high mechanical property to agree with the issues of articular cartilage repairing because of the peculiarity of algae materials. To overcome such drawbacks, adding other materials will be an effective alternative method to enhance algae material mechanical performance (Bal-Ozturk et al., 2021).

Recently, the investigation on hepatocytes seeding onto or encapsulating in algae material shows that hepatocytes are still viable, and the correlated functional phenotype is holding stably, which indicates that algae materials can improve or enhance the viability and proliferability of hepatocyte in clinical applications. Algae materials are also applied in other tissues and organs such as eye, heart, muscle, nerve, etc. to accelerate or improve cell adhesion, growth, and differentiation (Aguero et al., 2021; Beaumont et al., 2021; Cheng et al., 2019).

3. Drug delivery

As the final purpose of drug delivery, sending the relevant drugs to the needed therapeutic tissues or organs realizes the maximum effectiveness of drug treatment. Due to the effect of the gentle gelation of algae materials, it is impressive to take algae materials as drug delivery to sustain the related medicine release (e.g., antimicrobial drugs, antibiotic drugs) (Kesavan et al., 2021; Phogat et al., 2021; Thorn et al., 2021; Hernández-Ramírez et al., 2020).

Up to now, owing to the development of nanotechnology, algae materials used for bioapplications can be made into smaller size nanoscale, which increases effective surface areas to enhance the required performance for drug delivery. Moreover, the results show that combining algae materials with magnetic materials can strengthen or enhance precise drug target, increase the drug content load, and mitigate or alleviate the toxicity to the required therapeutic cells. Also, as a natural material, algae material is effectively biocompatible and has the relatively low side effects on human health. To date, algae materials used for clinical applications are still in primary stage, and the correlated research is rare and deficient (Ahirwar et al., 2022; Ashaolu et al., 2021; Salvioni et al., 2021; Yugay et al., 2020; Kaur et al., 2019; Korsunsky et al., 2019).

4. Wound dressing

In general, the wound dressing usually adopts cloth with a sterile pad and cloth gauze as the traditional therapeutic method. As the lowest cost method, it easily sticks to the wound and is prone to bacterial infection that resulted in delaying wound healing.

Due to the unique characteristics of algae materials such as high biocompatibility and water solubility, algae materials can ameliorate and accelerate the period of wound recovery by supporting the environment under the optimal pH. Algae materials should be sufficiently tailored by the relevant techniques exploring for hindering bacterial infection, especially for the chronic wound healing. To realize such targets, excellent cell attachment, adhesion, viability, differentiation, proliferation, and growth, algae materials possessed fine bioactivities, such as protein adsorption, antibacterial activity, and ability to stop and prevent bleeding, that is, blood clotting and hemostasis, should be deeply explored. Also, the correlated mechanical properties should be spontaneously investigated such as elongation, swelling ratio, thermal stability, and tensile strength under the clinical situation (Ashraf et al., 2022; Eldeeb et al., 2022; Vijayan et al., 2022; Miguel et al., 2021; Savencu et al., 2021; Alavi & Nokhodchi, 2020; Batista et al., 2020; Filho et al., 2020; Jain et al., 2019; Shahzad et al., 2019; Türe, 2019).

Although the research on the efficient inhibition of bacterial growth of algae materials is relatively insufficient, it is promising to have an attractive scenario that more and more researchers are focusing on the relevant algae materials R&D to effectively stimulate angiogenesis and inhibit bacterial growth for the wound healing enhancement. Recently, the interesting results indicate that the tailored algae materials stimulate not only cell proliferation but also cell migration. In addition, for wound dressing, the high long-term durability and excellent sustained drugs release are also required on the basis of the practical clinical applications. At the same time, algae materials for wound dressing with the prolonged and sustained release of antimicrobial drug must be biocompatible and nontoxic to human body. Therefore, a systematic investigation on the correlated algae materials should be performed to realize the ideal wound healing process, especially for the self-healing materials. Before the wound dressings are commercialized and used by patients, clinical studies are important and necessary that the relevant performance aspects of wound dressing on the patients including adverse effects must be evaluated (Cooperman et al., 2022; Mal et al., 2022; Reynolds et al., 2021; Abdo & Ortman, 2020; Gaudio et al., 2020; Taemeh et al., 2020).

It should be noted that since 2019, numerous clinical trials have been carried out on Corona Virus Disease-2019 (COVID-19). However, there is currently no one-enough vaccine or entirely effective antiviral treatments for it. Hans et al. (2021) explore the potential of the sulfated polysaccharides as antiviral agents in preventing recent COVID-19 because of their antiviral effects, which can effectively disrupt the interaction between virus and host cell receptors, thus blocking the internalization of virus particles.

5. Conclusion

Due to the complicated in vivo environment, it is crucial to tailor algae materials for intricate scaffold structure to promote cell bioactivities. Up to now, some algae materials (such as algae polysaccharides, algae sulfated polysaccharides, etc.) are though investigated in the clinical situation, the breakthrough technology tends to ameliorate healing capability and promote tissue regeneration, drug release and cell viability without any harmful effects should be further explored.

References

Abdo, J., & Ortman, H. (2020). Biologic and synthetic cellular and/or tissue-based products and smart wound dressings/coverings. *Surgical Clinics of North America, 100*(4), 741–756.

Aguero, L., Alpdagtas, S., Ihan, E., Zaldivar-Silva, D., & Gunduz, O. (2021). Functional role of crosslinking in alginate scaffold for drug delivery and tissue engineering: A review. *European Polymer Journal, 160*, 110807.

Ahirwar, A., Kesharwani, K., Deka, R., Muthukumar, S., Khan, M. J., Rai, A., Vinayak, V., Varjani, S., Joshi, K. B., & Morjaria, S. (2022). Microalgal drugs: A promising therapeutic reserve for the future. *Journal of Biotechnology, 349*, 32–46.

Alavi, M., & Nokhodchi, A. (2020). An overview on antimicrobial and wound healing properties of ZnO nanobiofilms, hydrogels, and bionanocomposites based on cellulose, chitosan, and alginate polymers. *Carbohydrate Polymers, 227*, 115349.

Anyanwu, R. C., Rodriguez, C., & Durrant, A. (2022). Micro-macroalgae properties and applications. *Encyclopedia of Smart Materials, 2*, 732–758.

Arokiarajan, M. S., Thirunavukkarasu, R., Joseph, J., Ekaterina, O., & Aruni, W. (2022). Advance research in biomedical applications on marine sulfated polysaccharide. *International Journal of Biological Macromolecules, 194*, 870–881.

Ashaolu, T. J., Samborska, K., Lee, C. C., Tomas, M., Capanoglu, E., Tarhan, Ö., Taze, B., & MahdiJafari, S. (2021). Phycocyanin, a super functional ingredient from algae; properties, purification characterization, and applications. *International Journal of Biological Macromolecules, 193*(B), 2320–2331.

Ashraf, S. S., Parivar, K., Roodbari, N. H., Mashayekhan, S., & Amini, N. (2022). Fabrication and characterization of biaxially electrospun collagen/alginate nanofibers, improved with *Rhodotorula mucilaginosa* sp. GUMS16 produced exopolysaccharides for wound healing applications. *International Journal of Biological Macromolecules, 196*, 194–203.

Bal-Ozturk, A., Cecen, B., Avci-Adali, M., Topkaya, S. N., Alarcin, E., Yasayan, G., Li, Y. C. E., Bulkurcuoglu, B., Akpek, A., Avci, H., Shi, K., Shin, S. R., & Hassan, S. (2021). Tissue adhesives: From research to clinical translation. *Nanotechnology, 36*, 101049.

Batista, M. P., Gonçalves, V. S. S., Gaspar, F. B., Nogueira, I. D., Matias, A. A., & Gurikov, P. (2020). Novel alginate-chitosan aerogel fibres for potential wound healing applications. *International Journal of Biological Macromolecules, 156*, 773–782.

Beaumont, M., Tran, R., Vera, G., Niedrist, D., Rousset, A., Pierre, R., Shastri, V. P., & Forget, A. (2021). Hydrogel-forming algae polysaccharides: From seaweed to biomedical applications. *Biomacromolecules, 22*(3), 1027–1052.

Cheng, S. Y., Show, P. L., Lau, B. F., Chang, J. S., & Ling, T. C. (2019). New prospects for modified algae in heavy metal adsorption. *Trends in Biotechnology, 37*(11), 1255–1268.

Cooperman, S., Jhaveri, V., Kruse, D., & Sachs, B. (2022). First reported case of non-coastal *Shewanella algae* in the United States: A case report and literature review. *Foot and Ankle Surgery: Techniques, Reports & Cases, 2*(1), 100126.

Eldeeb, A. E., Salah, S., Amer, M. S., & Elkasabgy, N. A. (2022). 3D nanocomposite alginate hydrogel loaded with pitavastatin nanovesicles as a functional wound dressing with controlled drug release; preparation, in-vitro and in-vivo evaluation. *Journal of Drug Delivery Science and Technology, 3*, 103292.

Filho, R. N. F. M., Vasconcelos, N. F., Andrade, F. K., de FreitasRosa, M., & Vieira, R. S. (2020). Papain immobilized on alginate membrane for wound dressing application. *Colloids and Surfaces B: Biointerfaces, 194*, 111222.

Gaudio, P. D., Amante, C., Civale, R., Bizzarro, V., Petrella, A., Pepe, G., Campiglia, P., Russo, P., & Aquino, R. P. (2020). In situ gelling alginate-pectin blend particles loaded with Ac2-26: A new weapon to improve wound care armamentarium. *Carbohydrate Polymers, 227*, 115305.

Hans, N., Malik, A., & Naik, S. (2021). Antiviral activity of sulfated polysaccharides from marine algae and its application in combating COVID-19: Mini review. *Bioresource Technology Reports, 13*, 100623.

Hernández-Ramírez, J., Wong-Arce, A., González-Ortega, O., & Rosales-Mendoza, S. (2020). Expression in algae of a chimeric protein carrying several epitopes from tumor associated antigens. *International Journal of Biological Macromolecules, 147*, 46–52.

Jain, R., & Wairkar, S. (2019). Recent developments and clinical applications of surgical glues: An overview. *International Journal of Biological Macromolecules, 137*, 95–106.

Kaur, P. K. A., & Goyal, D. (2019). Algae-based metallic nanoparticles: Synthesis, characterization and applications. *Journal of Microbiological Methods, 163*, 105656.

Kerney, R., Cherubino, M., & Bender, Z. (2022). Algae on the brain in bioengineering. *Trends in Biotechnology, 40*(3), 259–260.

Kesavan, S., Meena, K. S., Sharmili, S. A., Govindarajan, M., Alharbi, N. S., Kadaikunnan, S., Khaled, J. M., Alobaidi, A. S., Alanzi, K. F., & Vaseeharan, B. (2021). Ulvan loaded graphene oxide nanoparticle fabricated with chitosan and d-mannose for targeted anticancer drug delivery. *Journal of Drug Delivery Science and Technology, 65*, 102760.

Korsunsky, A. M., Sapozhnikov, P. V., Everaerts, J., & Salimon, A. I. (2019). Nature's neat nanostructuration: The fascinating frustules of diatom algae. *Materials Today, 22*, 159–160.

Kumar, B. R., Mathimani, T., Sudhakar, M. P., Rajendran, K., Nizami, A. S., Brindhadevi, K., & Pugazhendhi, A. (2021). A state of the art review on the cultivation of algae for energy and other valuable products: Application, challenges, and opportunities. *Renewable and Sustainable Energy Reviews, 138*, 110649.

Mal, N., Satpati, G. G., Raghunathan, S., & Davoodbasha, M. A. (2022). Current strategies on algae-based biopolymer production and scale-up. *Chemosphere, 289*, 133178.

Miguel, S. P., Ribeiro, M. P., Otero, A., & Coutinho, P. (2021). Application of microalgae and microalgal bioactive compounds in skin regeneration. *Algal Research, 58*, 102395.

Pérez-Alva, A., MacIntosh, A. J., Baigts-Allende, D. K., García-Torres, R., & Ramírez-Rodrigues, M. M. (2022). Fermentation of algae to enhance their bioactive activity: A review. *Algal Research, 64*, 102684.

Phogat, S., Saxena, A., Kapoor, N., Aggarwal, C., & Tiwari, A. (2021). Diatom mediated smart drug delivery system. *Journal of Drug Delivery Science and Technology, 63*, 102433.

Rajeshkumar, G., & Selvadeepak, R. V. S. (2020). Tissue mimicking material an idealized tissue model for clinical applications: A review. *Materials Today Proceedings, 22*(4), 2696–2703.

Reynolds, D., Huesemann, M., Edmundson, S., Sims, A., Hurst, B., Cady, S., Beirne, N., Freeman, J., Berger, A., & Gao, S. (2021). Viral inhibitors derived from macroalgae, microalgae, and cyanobacteria: A review of antiviral potential throughout pathogenesis. *Algal Research, 57*, 102331.

Salvioni, L., Morelli, L., Ochoa, E., Labra, M., Fiandra, L., Palugan, L., Prosperi, D., & Colombo, M. (2021). The emerging role of nanotechnology in skincare. *Advances in Colloid and Interface Science, 293*, 102437.

Savencu, I., Iurian, S., Porfire, A., Bogdan, C., & Tomuţă, I. (2021). Review of advances in polymeric wound dressing films. *Reactive and Functional Polymers, 168*, 105059.

Shahzad, A., Khan, A., Afzal, Z., Umer, M. F., Khan, J., & Khan, G. M. (2019). Formulation development and characterization of cefazolin nanoparticles-loaded cross-linked films of sodium alginate and pectin as wound dressings. *International Journal of Biological Macromolecules, 124*, 255–269.

Taemeh, M. A., Shiravandi, A., Korayem, M. A., & Daemi, H. (2020). Fabrication challenges and trends in biomedical applications of alginate electrospun nanofibers. *Carbohydrate Polymers, 228*, 115419.

Thorn, C. R., Howell, P. L., Wozniak, D. J., Prestidge, C. A., & Thomas, N. (2021). Enhancing the therapeutic use of biofilm-dispersing enzymes with smart drug delivery systems. *Advanced Drug Delivery Reviews, 179*, 113916.

Türe, H. (2019). Characterization of hydroxyapatite-containing alginate-gelatin composite films as a potential wound dressing. *International Journal of Biological Macromolecules, 123*, 878–888.

Vijayan, A., Vipin, C. L., & Kumar, G. S. V. (2022). Dual growth factor entrapped nanoparticle enriched alginate wafer-based delivery system for suppurating wounds. *International Journal of Biological Macromolecules, 208*, 172–181.

Wang, H. M. D., Li, X. C., Lee, D. J., & Chang, J. S. (2017). Potential biomedical applications of marine algae. *Bioresource Technology, 244*(2), 1407–1415.

Wan, M. C., Qin, W., Lei, C., Li, Q. H., Meng, M., Fang, M., Song, W., Chen, J. H., Tay, F., & Niu, L. N. (2021). Biomaterials from the sea: Future building blocks for biomedical applications. *Bioactive Materials, 6*(12), 4255–4285.

Yugay, Y. A., Usoltseva, R. V., Silant'ev, V. E., Egorov, A. E., Karabtsov, A. A., Kumeiko, V. V., Ermakova, S. P., Bulgakov, V. P., & Shkryl, Y. N. (2020). Synthesis of bioactive silver nanoparticles using alginate, fucoidan and laminaran from brown algae as a reducing and stabilizing agent. *Carbohydrate Polymers, 245*, 116547.
Zargarzadeh, M., Amaral, A. J. R., Custódio, C. A., & Mano, J. F. (2020). Biomedical applications of laminarin. *Carbohydrate Polymers, 232*, 115774.

CHAPTER

Algae materials for cosmetics and cosmeceuticals 18

Bruna Bertoloni dos Santos[1], Ana Lucía Morocho-Jácome[1], Zulita Adriana Prieto[2], Mohammad Hossein Morowvat[3], Fabiana Vieira Lima[4], Maria Valéria Robles Velasco[1], João Carlos Monteiro de Carvalho[1] and André Rolim Baby[1]

[1]*Faculty of Pharmaceutical Sciences, University of São Paulo, São Paulo, Brazil;* [2]*Faculty of Biological Sciences, National University of Trujillo, Trujillo, Peru;* [3]*Pharmaceutical Sciences Research Center, Shiraz University of Medical Sciences, Shiraz, Iran;* [4]*Division of Faculty of Pharmacy, University of Espírito Santo, Espírito Santo, Brazil*

1. Introduction

Human skin is constantly exposed to ultraviolet (UV) radiation. This happens to many organisms. Such a stressor induces various types of acute and chronic disorders (De Jager et al., 2017). In excess, this stressor can promote lipid peroxidation, damaging cell membranes. They can activate or inactivate enzymes, causing DNA damage that can lead to cancer, among other pathologies (Avadhani et al., 2017; Broinizi et al., 2007; Costa, 2010; Hirata et al., 2004; Goldstein, 2008; Gomes & Damazio, 2009; Mambro & Fonseca, 2007; Montagner & Costa, 2009; Steiner, 2011). To reduce damage caused by radiation, the use of sunscreens is highly recommended. However, there has been an increase in the demand for dermocosmetics developed with natural and sustainable ingredients. This offers the opportunity for investigations into natural, sustainable, and renewable sources of biocompounds and bioactive molecules to develop products that can be considered eco-friendly (Chrapusta et al., 2017; McIntosh et al., 2018; Show et al., 2017).

In search of their survival, many organisms, such as microalgae, develop strategies and mechanisms to defend themselves against challenges of the environment. The production of secondary metabolites by various organisms has a fundamental role in survival, acting as a defense in response to environmental challenges (Delgoda & Murray, 2017). Such difficulties faced by organisms include predation, competition with other plants and organisms for resources, forms of occupation, establishment and propagation in the environment, and solar radiation. In the case of exposure to UV radiation, its effects can cause damage to the physiologic and biochemical processes of these organisms, such as growth, survival, buoyancy, photosynthesis, pigmentation, CO_2 absorption, and structural changes in proteins and DNA (Bacova et al., 2019; Delgoda & Murray, 2017; Kumar et al., 2016; Rastogi & Incharoensakdi, 2013).

Chlorella vulgaris, a species of green microalga, is a eukaryotic microorganism with high photosynthetic ability. *C. vulgaris* is found in different environments, predominantly in fresh and salt water, where there is exposure to high levels of UV radiation and extreme survival conditions. In

Algae Materials: Applications Benefitting Health. https://doi.org/10.1016/B978-0-443-18816-9.00003-4

response to these challenges, *C. vulgaris* produces several secondary metabolites that may have antioxidant properties (Cha et al., 2010; Hynstova et al., 2017; Jiang & Kalodimos, 2017; Khan et al., 2018; Priyadarshani & Rath, 2012; Putri et al., 2021; Wang et al., 2015).

Cultivation of *C. vulgaris* can be performed in different production systems, both in the open and in more sophisticated and specific systems. When the goal is to achieve high productivity, strains are grown in a system called closed-system photobioreactor, where it is possible to control cultivation conditions such as temperature, light, and pH value. This implies a high production scale, enabling the commercial production of a series of desired compounds and greater biomass yield, in addition to the advantage of its unicellular nature to ensure biomass with a unique composition (Derner, 2006; Duran et al., 2018).

Considering the field of Cosmetology, knowledge about ingredients derived from microalgae is scarce and needs further investigations. However, through studies carried out so far, *C. vulgaris* showed the potential to develop attributes including an ability to prevent skin blemishes, repair skin damage, treat seborrhea, condition hair, repair collagen, inhibit inflammation, hydrate, and protect against damage caused by solar radiation (Bhalamurugan et al., 2018; Cunha et al., 2022; Mourelle et al., 2017).

According to Mourelle and colleagues (2017), microalgae cultivation is a modern biotechnology. Moreover, the use of microalgae biomass is ecologically correct because of the possibility of using it in various fields of application, such as in foods, as a source of biodiesel, and in pharmaceuticals and cosmetics (Ariede et al., 2017; Khan et al., 2018; McIntosh et al., 2018; Show et al., 2017). Another advantage is the possibility of directing the production of substances of interest, according to the intended objective, through different cultivation strategies (Show et al., 2017). Thus, microalgae extracts can be relevant types of constituents in products considered natural for sun protection (Bhalamurugan et al., 2018; Fonseca, 2016; Liu & Chen, 2014; Stages, 2010).

2. Literature review

2.1 Microalgae

Microalgae are eukaryotic and photoautotrophic beings consisting of a simple structure. They require water, CO_2, light, and inorganic nutrients for growth and reproduction. Microalgae can be found in all ecosystems, including marine waters. Generally, they live in environments with limiting and stressful conditions such as exposure to heat, cold, high salinity, osmotic pressure, different light sources, and solar radiation (Markou et al., 2014; Morocho Jácome et al., 2012; Panahi et al., 2019; Skjånes et al., 2013).

They are identified as photosynthetic microorganisms that convert CO_2 into energy and bioactive compounds, such as pigments, carbohydrates, proteins, polyunsaturated fatty acids, and vitamins. Microalgae are mostly autotrophic, but they can also be heterotrophic or mixotrophic. Autotrophs carry out photosynthesis as the main metabolic process and use sunlight and CO_2 as a carbon source. Heterotrophics need an organic carbon source and do not need light; this group does not carry out photosynthesis. Mixotrophs have both strategies (Show et al., 2017).

Compared with superior plants, microalgae have the advantages of fast growth, a low consumption of water for production, easy harvesting, and high CO_2 fixation. They help to reduce the concentration of greenhouse gases in the atmosphere, produce O_2 and nonseasonal biomass, may be cultivated in waste or nonpotable water (Cezare-Gomes et al., 2019; Marino, 2018; Pohndorf et al., 2016). However, it is essential to control the cultivation conditions to obtain uniform biomass composition (Ariede et al., 2017). Therefore, there is wide commercial interest in photosynthetic microorganisms in several areas, such as to develop nutraceutical, pharmaceutical, and cosmetic products, because they are able to convert CO_2 into biomass, which is a source of natural bioactive compounds. Moreover, they present factors that aid in balancing the earth's ecosystem. Finally, they contribute to the demand for eco-friendly, renewable, and sustainable ingredients and products (Ariede et al., 2017; Cezare-Gomes et al., 2019; Show et al., 2017).

2.1.1 *Chlorella vulgaris*

The taxonomic classification of *C. vulgaris* (Van Den Hoek et al., 1995) is phylum Chlorophyta > class Trebouxiophyceae > order Chlorellades > family Chlorellaceae > genus *Chlorella* > specie *C. vulgaris*.

C. vulgaris is a spherical microalgae 2–10 μm in diameter. There is no flagellum; it is an immobile cell. Its cytoplasm is composed of water, proteins, and minerals. Only one chloroplast has photosynthetic pigments, such as chlorophyll *a* and *b*, in its interior. It is greenish and its reproduction occurs by binary partition through asexual spores that grow quickly, originating an average of one cell every 24 h under optimal cultivation conditions (Araújo, 2015b; Marino, 2018; Raven et al., 2007). Its wall is rigid owing to the presence of glucosamine in its basic composition. However, its thickness and composition vary according to the growth phase (Marino, 2018; Safi et al., 2014).

C. vulgaris, as well as other species of microalgae, has unique culture characteristics, with adequate growth in a nutrient-rich medium. However, studies are needed to determine its metabolism, so necessary adjustments should be adopted regarding cultivation conditions to achieve the high productivity of biomass and biocompounds of interest (Araújo, 2015b).

2.2 Microalgae composition

The composition of microalgae can vary according to the cultivation conditions to which they were subjected. However, microalgae biomass is composed in greater amounts of proteins and carbohydrates, and in smaller amounts of vitamins and antioxidants (Panahi et al., 2019).

Proteins are essential for the composition of microalgae and are involved in various physiologic processes of the cell, such as growth, repair, and maintenance. They also act as chemical messengers and regulators of cellular activities, and even as defense against invaders (Safi et al., 2014). Amino acids found in species of *C. vulgaris* at greater amounts are alanine and glutamic acid (Ursu et al., 2014).

Lipids are a group of compounds defined as being insoluble in water. They mainly include glycolipids, waxes, hydrocarbons, phospholipids, and fatty acids. These compounds can be modulated according to the amount of nitrate and light intensity available during their cultivation (Liao et al., 2017). They are synthesized by chloroplasts and are also found in cell walls and in some organelles, such as mitochondria (Marino, 2018; Safi et al., 2014). Carbohydrates represent a group of reduced sugars and polysaccharides, such as starch and cellulose. Starch is the most abundant polysaccharide in the microalga *C. vulgaris*. It is located in the chloroplast. In addition to sugars, it acts as a store of energy for the cell. Cellulose is located in the cell wall and is a highly resistant structural polysaccharide that works as a protective barrier. The amount of carbohydrates in *C. vulgaris* species can vary from 12% to 55% of its total composition, depending on growing conditions (Lordan et al., 2011; Safi et al., 2014).

The most abundant pigments in *C. vulgaris* are chlorophylls and carotenoids. Chlorophylls are located in the chloroplast thylakoids in about 1–2% of the dry biomass of the microalgae. Carotenoids such as β-carotene perform some vital cell functions, such as capturing light and transferring it to the photosystem. They are complex proteins involved in cellular photosynthesis; they even protect chlorophyll molecules from degradation during exposure to intense solar radiation (Cezare Gomes et al., 2019; Fernández Linares et al., 2017; Safi et al., 2014; Solomon & Martin et al., 2011).

2.3 Types of cultivation

The cultivation of microalgae can occur in open or closed systems according to the intended cultivation objectives. Both have advantages and disadvantages. When cultivation is carried out in an open system, the main advantage is the lower cost of the process. However, there is no contamination control. In closed system cultivation, on the other hand, there is the advantage of less contamination and the possibility of controlling conditions, favoring greater biomass production. However, the cost of this process is higher because the necessary infrastructure is expensive (Borowitzka, 1999; Brennan & Owende, 2010; Carvalho et al., 2014; Duran et al., 2018; Singh & Sharma, 2012; Ugwu et al., 2008).

2.3.1 Closed cultivation system

The closed cultivation system was developed with the intention of reducing influences and limitations that affect open system cultivation. It promotes the opportunity of control the conditions of the environment and obtaining a high cell concentration as well as the substances of interest produced by microalgae that are widely used to produce pharmaceuticals, nutraceuticals, and cosmetics (Cezare-Gomes et al., 2019; Safi et al., 2014). The low risk for contamination in these systems is an advantage that allows the commercialization of microalgae biomass, called a clean algae culture. It is a type of crop suitable for species more sensitive to competition with other algae and growing conditions (Safi et al., 2014). With this, several models of closed system were created, such as flat plate, tubular, and column photobioreactors (Cezare-Gomes et al., 2019; Show et al., 2017). Furthermore, a comparative study carried out on *C. vulgaris* by Lam and Lee (2012), which showed its most efficient growth when it was cultivated in a closed system.

2.3.2 Tubular photobioreactor

Among the photobioreactor models, the tubular one is the most suitable for large-scale biomass production because it enables the greatest productivity, permitting commercialization (Cezare-Gomes et al., 2019). The tubular photobioreactor can be built with plastic or glass tubes with a maximum diameter to ensure greater light penetration and absorption by cells. The length must also be optimized. Homogenization of the medium is usually performed by air pumps, promoting agitation and preventing cell sedimentation (Marino, 2018; Show et al., 2017).

2.4 Factors related to cultivation

For cell growth, microalgae cultures of the *Chlorella* genus basically need macronutrients, micronutrients, and factors such as light intensity, pH, and CO_2, which are regulated and optimized according to the selected species (Araújo, 2015b; Liu & Chen, 2014; Show et al., 2017).

Among main macronutrients needed to cultivate microalgae of the *Chlorella* genus are carbon, nitrogen, and phosphorus (Liu & Chen, 2014). Carbon constitutes about 40–50% of the total content of cells. It is also responsible for their central metabolism, in which carbon dioxide is the main source of carbon for *Chlorella*. Right after carbon, we have nitrogen as the second main macronutrient, which represents about 1–10% of the total content of cells. It is the main constituent of amino acids and nucleic acids and it can be used by the *Chlorella* genus in the form of nitrate, ammonium, or urea. Micronutrients include elements such as Fe, Mg, Na, K, Ca, Cl, B, Cu, Mn, Zn, V, Co, and Se, which ensure cell function, acting as cofactors, whereas others are important in biosynthesis processes. Vitamins are also added to culture (Araújo, 2015b; Liu & Chen, 2014; Rebolloso-Fuentes et al., 2001; Show et al., 2017).

2.4.1 Light intensity

The light intensity and photoperiod directly influence the growth rate of microalgae. They are important factors in cultivating photosynthetic organisms (Ahluwalia et al., 2016; Andrade, 2014). Light can affect cell growth and metabolism under three possible conditions: light limitation, saturation, and inhibition (Show et al., 2017). According to Andrade (2013) and Baiee (2016), limiting light in cultivation causes an increase in protein production, whereas the adequate availability of light according to the need of each microalga can cause a greater accumulation of polysaccharides, carbon fixation for energy reserves, and greater growth and cell concentration (Hamedi et al., 2016). The light saturation constant is the luminous intensity that corresponds to half of the maximum specific growth rate. The increase in light intensity from a certain value reduces the cellular photosynthetic capacity, a phenomenon called photoinhibition, inducing the production of reactive oxygen species and causing irreversible damage to cellular structures (Andrade, 2014; Chisti, 2007).

2.4.2 pH

The pH value is related to the concentration of H^+ ions and the solubility of CO_2 and minerals present in the medium that can directly or indirectly affect the cellular metabolism of microalgae. In photoautotrophic cultivation, a change in pH medium is directly related to the CO_2 that acidifies it, so an acidic medium has greater availability of CO_2 (Ahluwalia et al., 2016). However, because of the low solubility of CO_2 in aqueous media, CO_2 supply must occur throughout the entire cultivation (Klasson et al., 1991). During cultivation in response to the biological activities of the cells, the pH value tends to increase. This can be considered an indicator of microalgal growth. The growth of microalgae is natural in an alkaline medium (Andrade, 2014). However, in the specialized literature, *C. vulgaris* precipitated when the pH value of the culture was above 9.5, consequently causing a decrease in biomass production. The ideal pH range can vary among microalgae species. For *C. vulgaris*, the pH interval could be 7–8, when the specific growth rate can reach 2.03/d. Control of the pH can be performed by injecting CO_2 into the medium or buffering agents (Ahluwalia et al., 2016).

2.4.3 Temperature

In microalgae growth, temperature is also an important factor. Raising the temperature can increase biomass production and even double antioxidant activity, as shown by Miranda and Sato (2001) using *C. vulgaris*. However, the temperature limit varies according to each species. It can cause growth inhibition by inactivation and heat shock in cell proteins (Andrade, 2014). For *C. vulgaris*, Chinnasamy et al. (2009) verified that the optimal temperature for high growth was around 30°C. On the other hand, Serra-Maia et al. (2016) noticed that the highest specific growth velocity occurred at 25°C. Bashan et al. (2015) worked with *C. vulgaris* UTEX 2714 and verified that the temperature of greatest cell growth was 27°C, with a decrease in growth at 35°C. Considering this information, the condition for cultivating *Chlorella* at 27 ± 2°C would be adequate. However, this value can vary for each species, with tolerance for growth of some strains at temperatures of up to 37°C. The temperature can also vary according to the light intensity received. That can influence other parameters, such as CO_2 solubility and salts present in the culture medium.

2.4.4 Nitrogen

Nitrogen is one of the most important macronutrients in the culture medium because it is part of the chemical structure of various molecules, such as proteins, photosynthetic pigments, amino acids, vitamins, and nucleic acids (Acioli, 2014; Araújo, 2015b). The main sources of nitrogen added to cultivations are inorganic, such as nitrate, nitrite, and ammonium, as well as other organic nitrogen compounds, such as urea (Araújo, 2015b; lúcio, 2013). Nitrogen sources in microalgae cultures, including those of *Chlorella*, are able to influence the biomass composition and growth of microalgae at greater or lesser concentrations. According to the literature, for *C. vulgaris*, a high concentration of nitrogen can induce an increase in proteins and chlorophyll in the cells, raising their photosynthetic capacity. However, there will be a decrease in the concentration of carbohydrates (Lourenço, 2006; Marino, 2018; Venckus et al., 2017). At low concentrations, it can induce lipid accumulation in up to 70% of dry biomass and carbohydrate accumulation. However, it inhibits cell division, causing a decrease in protein levels, biomass productivity, the photosynthetic capacity of cells, the number of photosynthetic pigments, and chlorosis (Adamakis et al., 2018; Li et al., 2016; Liu & Chen, 2014; Venckus et al., 2017). According to Li and colleagues (2016), nitrogen limitation affects the photosynthetic capacity, which in turn compromises the lipid yield. It is essential to maintain a balance among these factors so that an optimal yield is achieved.

2.5 Application of microalgae in dermocosmetics

In cosmetology, microalgae have been used as a raw material in product development owing to their ability to produce biologically bioactive compounds of interest to the industry. Generally, they are used as antioxidants, antiinflammatory agents, photoprotectors, moisturizers, thickening agents, actives against abrasion, tanning products, hair care products, antiaging ingredients, antiwrinkle agents, and skin and whitening products, among others (Joshi et al., 2018; Mourelle et al., 2017).

Among the different species of photosynthetic microorganisms, *Spirulina* and *Chlorella* have metabolites and pigments that perform various cosmetic functions. Amino acids and proteins have moisturizing capacity and are used in skin and hair care products. Other biocompounds, such as lipids, phycobiliproteins, terpenoids, and pigments, contain antiinflammatory and antioxidant activities and can act as stabilizers (Cunha et al., 2022; Joshi et al., 2018; Mourelle et al., 2017). Table 18.1 lists biocompounds derived from microalgae and some of their functions.

Microalgae extracts have a wide variety of biocompounds recognized to add value to cosmetic products. Therefore, studies aim to optimize their production and cultivation so that the biomass will be optimized in the production of compounds of interest. These studies are relevant for the cosmetic industry because the composition of biomass directly influences the effect of the final product. It also represents a promising and profitable source of raw materials for industries as a new biotechnology to be included on the market (Ariede et al., 2017; Mourelle et al., 2017; Wang et al., 2015).

Table 18.1 Different species of microalgae and their biocompounds, uses, and activities in dermocosmetics and some pharmaceuticals.

Microalga	Biocompounds	Uses and activities in cosmetics	References
Chlorella sp.	Polysaccharides	Moisturizing and thickening action	Mourelle et al. (2017) Joshi et al. (2018)
Chlorella sp.	Chlorophyll	Mask odors in toothpastes and deodorants, pigment, anticarcinogenic property	Mourelle et al. (2017)
Chlorella vulgaris	Extract of *C. vulgaris*	Collagen repair, tissue regeneration, and antiaging	Skjånes et al. (2013) Koller et al. (2014) Wang et al. (2015) Fonseca. (2016) Ariede et al. (2017)
C. vulgaris	Chlorella	Antimicrobial action	Fonseca. (2016)
C. vulgaris	Astaxanthin	Suppresses skin hyperpigmentation, inhibits melanin production	Skjånes et al. (2013) Wang et al. (2015)
Chlorella sp.	Lutein	Photoprotective and antioxidant action	Cha et al. (2010) Wang et al. (2015) Li et al. (2016) Hynstova et al. (2017) Muszynska et al. (2018)
C. vulgaris	Chlorophyll *a*, chlorophyll *b*, and β-carotene	Antiaging, moisturizing, thickening and depigmenting action	Cha et al. (2010) Li et al. (2016) Hynstova et al. (2017) Joshi et al. (2018)
C. vulgaris	Vitamins	Toning skin, curing dark circles, hair growth and dandruff treatment	Joshi et al. (2018)
C. vulgaris	α-Carotene	Antioxidant and anticarcinogenic action	Hynstova et al. (2017)
C. vulgaris	Palmitic acid, palmitoleic acid	Antiaging, depigmentation, moisturizing and thickening agent	Joshi et al. 2018
Chlorella sp.	β-1,3-Glucan	Antiinflammatory action, immune system booster, and free radical scavenger	Mourelle et al. (2017)
C. vulgaris	Complex B vitamins	Helps maintain skin and hair health	Safi et al. (2014)
C. vulgaris	Xanthophylls	Antioxidant action, helps stabilize membrane and protects cells from intense radiation	Panahi et al. (2019)

2.5.1 Sunscreens

UV filters available on the market are divided into chemical (oxybenzone, octyl methoxycinnamate, aminobenzoic acid, avobenzone, etc.) and physical (titanium dioxide and zinc oxide). Sunscreens are widely used to protect against damage caused by exposure to UV radiation. However, they can cause skin irritation, photosensitivity, and contact dermatitis. Thus, there is a need to search for compounds that can be as effective as the protectors already in use. These compounds need to be from natural, sustainable sources and cause fewer adverse events with chronic use (Avadhani et al., 2017).

Among the compounds and substances available on the market for formulating cosmetics are biocompounds produced by microalgae. Some are able to synthesize substances that absorb UV radiation and promote other benefits to the skin. Some species of microalgae that have potential as a raw material for sunscreens are *Chlorella* spp., which are already recognized for having benefits such as antiaging and antioxidant qualities, moisturizing action, and the ability to prevent wrinkles and blemishes from forming on skin, among others (Ariede et al., 2017).

3. Objectives

The main objective of this research was to evaluate the employability of *C. vulgaris* as a photoprotective ingredient.

4. Experimental

4.1 Microalgae

Strains of *C. vulgaris* were used in the cultures. They were identified as UTEX 2714 and originated from the wastewater treatment pond in Santa Fé de Bogotá, Colombia. These were kept in 500-mL Erlenmeyer flasks on a rotary shaker at 100 rpm at 25°C ± 2°C (Vonshak, 1997) and a luminous intensity of 6 klx (72 μmol photons/m^2s) to obtain the inoculum (Carvalho et al., 2004).

4.2 Culture medium

The culture medium used for cell growth was the basal Bold (Silva, 2020; Silva et al., 2019). It was also used at higher $NaNO_3$ concentrations, with values three (Bold 3N medium) and six times higher (Bold 6N medium). Culture media were autoclaved at 121°C for 30 min before use.

4.3 Evaluation of cell growth and concentration

A concentrated suspension sample was collected for the calibration curves of each strain. Ten different dilutions of the suspension were made and analyzed in a spectrophotometer (FEMTO 600 PLμS, version 2.4) at the wavelength of 720 nm (DO_{720}) with a 1 cm optical path cuvette (Li et al., 2018). Filtration membranes with a porosity of 0.45 μm (Millipore) were placed in a Petri dish, dried in an oven at 60°C for 1 h, left for another 1 h in a desiccator to cool, and then weighed. Twenty milliliters of a concentrated suspension was vacuum filtered on one of the 0.45 μm membranes that had been previously dried and weighed. After filtering, the membrane containing the cells was placed in a Petri dish, left in an oven at 60°C for 24 h to remove all humidity, left for another hour in a desiccator to cool, and then weighed. The cell mass was obtained through the difference in membrane weight with and without cells (Pereira, 2013).

4.4 Inoculum

An aliquot of 350 mL (equivalent to 10% of the photobioreactor volume) of a cell suspension (cell fraction and culture medium) of each strain was kept under the conditions described in Section 4.1 (Carvalho et al., 2004; Vonshak, 1997) until the cellular concentration to obtain the inoculum was reached, which was 0.1 g/L (Li et al., 2018).

4.5 Cultivation conditions

The strains were cultivated in a tubular airlift photobioreactor developed at the Microalgal Biotechnology Laboratory of the Department of Biochemical-Pharmaceutical Technology, Faculty of Pharmaceutical Sciences, University of São Paulo (Fig. 18.1). The photobioreactor consisted of transparent

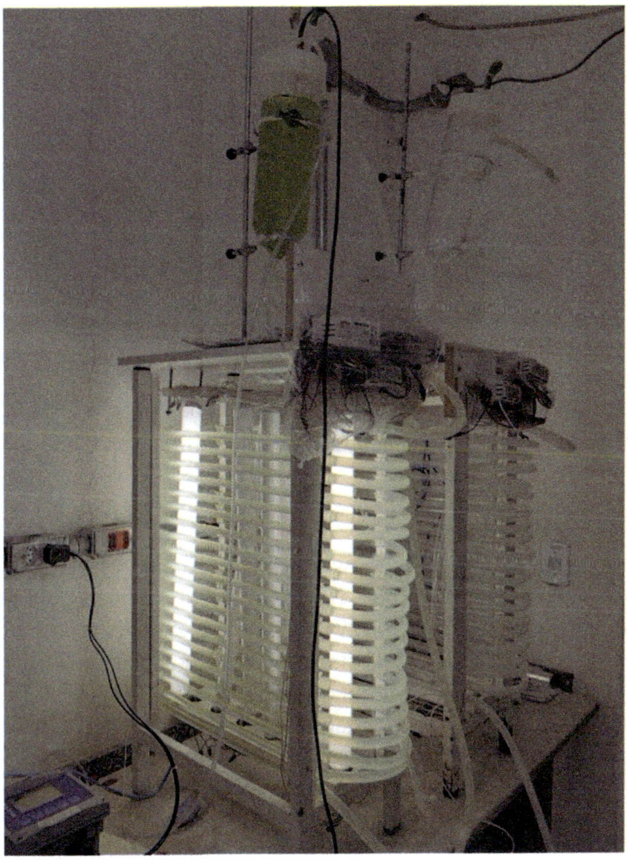

FIGURE 18.1

Photograph of the photobioreactor (3.5L).

tubes with an internal diameter volume of 1.0 cm and a total system volume of 3.5 L. The illuminated volume corresponded to 2.0 L, and the culture flow of 40 ± 1 L/h (Jácome, 2014). The photobioreactors underwent chemical asepsis before the experiments began, using sodium hypochlorite with an active chlorine content of 2.0–2.5% w/w, sodium thiosulfate at a concentration of 12.5 g/L to neutralize the action of chlorine, and deionized water (Andersen, 2005; Ahluwalia et al., 2016).

The temperature was maintained at 27°C ± 2°C and the pH value of the medium was monitored and kept at 7.5 ± 0.3 (Montoya et al., 2014). The light intensity was maintained at 60–70 μmol photons.m/s (Adamakis et al., 2018).

The UTEX 2714 strain of the microalga *C. vulgaris* was cultivated under three different concentrations of sodium nitrate: 1N (2.94 mM), 3N (8.82 mM), and 6N (17.64 mM). The medium used in the cultures was the Bold basal medium, varying in the concentrations of sodium nitrate, proposed to be the main source of nitrogen for the culture.

4.6 Biomass productivity and specific growth rate

Biomass productivity (mg.L /d) was obtained from the dried biomass concentration (mg/L) divided by the specified time range (days) based on Eq. (18.1) (Santos, 2013; Toledo-Cervantes et al., 2013)

$$P = \frac{X_m - X_i}{Tc} \tag{18.1}$$

In Eq. (18.1), biomass productivity P (mg/L/d), X_i is the inoculum cell concentration; X_m is the maximum cell concentration; and Tc is time needed to obtain X_m.

The specific growth rate (d^{-1}) of *C. vulgaris* was calculated using Eq. (18.2) (Clesceri et al., 1989; Santos, 2013):

$$\mu_{MÁX} = \frac{\ln\left(\frac{X}{X_0}\right)}{(t - t_0)} \tag{18.2}$$

In Eq. (18.2), specific growth rate, $\mu_{MÁX}$ (d^{-1}),

$\mu_{MÁX}$ is the slope of the lnX line as a function of time in its linear stretch; X is the final biomass concentration; X_0 is the initial biomass concentration; t is the final time; and t_0 is the initial time.

4.7 Dry biomass and extract

At the end of cultures, the microalgal biomass was centrifuged and washed with deionized water to remove excess salts from the culture medium. To obtain the dry biomass, samples were dried with ventilation at 65°C for 12 h (Pelizer et al., 1999), macerated, and frozen for storage. To obtain the extract, samples of wet biomass of *C. vulgaris* strains were suspended and shaken with glass spheres under agitation at 1000 rpm for 55 min, with microscopic observation to verify cell lysis. After lysis, the liquid fraction was separated from the glass spheres by filtration and centrifuged at 4500 rpm for 30 min to obtain the extract, which was the supernatant (Buono et al., 2012). The extract was lyophilized for better storage conditions. Dry biomass and extract were subjected to the following techniques.

4.8 Total lipid content

A mixture of apolar and polar solvent was used to determine the total lipid fraction (Pelizer et al., 1999). Approximately, 1.0 g of the samples was placed in a Soxhlet extractor and refluxed with chloroform and methanol at 2:1 v/v until the extracting liquid became clear (Olguín et al., 2001). Then, the solvent mixture containing the total lipid fraction was recovered in a rota-evaporator. The lipid content of the samples was calculated according to the ratio of the mass of the total lipid fraction and the mass of the samples used in extraction. Calculation of the percentage of total lipids was performed according to Eq. (18.3) (Pelizer et al., 1999):

$$\text{Total lipids } (\%) = \left[(M_f - M_i)/M_a\right] \cdot 100 \tag{18.3}$$

In Eq. (18.3), calculation of the percentage of total lipids, M_f is the weight of the balloon with the sample; M_i is the empty balloon weight; and M_a is the amount of sample used to perform the test.

4.9 Fatty acid composition

The determination was performed after converting fatty acids into their corresponding methyl esters (Hartman & Lago, 1973). Aliquots of 200–250 mg of the lipid fraction (item 5.10) were weighed and then 3.0 mL of methanol solution of 0.5M KOH was added. The mixture was heated to reflux for 5 min. Afterward, 9.0 mL of the esterification reagent (2 g NH_4Cl + 3 mL H_2SO_4 in 60 mL methanol) was added and the mixture was left to rest for 10 min until the reaction occurred. Then, it was heated to reflux for another 5 min and transferred to a separatory funnel together with 15.0 mL of petroleum ether and 30.0 mL of deionized water. After the phases were stirred and separated, the aqueous phase was discarded. In the organic phase, 15.0 mL of deionized water was added. After stirring and phase separation, the aqueous phase was discarded and the procedure was repeated. The organic phase was collected, the solvent was evaporated in a rotary evaporator in a water bath at 40°C, and the residues were removed under nitrogen flow. Finally, an analysis of fatty acid methyl esters was performed according to Rodrigues-Ract (2008), with adaptations, for later injection into the gas chromatograph (Model 7890 CX, Agilent, EUA). Identification of fatty acid methyl esters was performed according to a comparison of the retention time of sample constituents with individual standards.

4.10 Total protein content

To determine the protein content of the biomass and the extract, the Kjeldahl method was used with 6.25 as the nitrogen-to-protein conversion factor (AOAC, 1984). About 20 ± 0.5 mg of the defatted samples (resulted from the lipid determination) were sprayed, weighed, and placed in the test tubes, to which 2.0 g of the catalyst (1.9 g of potassium sulfate and 10.0 mg of copper sulfate) and 3.0 mL of sulfuric acid were added and subjected to digestion at 350°C. In this process, organic nitrogen in the sample was converted to ammonia, which was kept in the system in the form of ammonium sulfate. This sample, which went through acid digestion, was then alkalinized with 10.0 mL of 60% sodium hydroxide and heated in a still with steam drag. The ammonia was received in a saturated solution of boric acid. This was displaced in titration in solution with 0.02N HCl, with a mixture of methyl red and methylene blue as an indicator. The percentage of total protein was calculated according to Eq. (18.4) (AOAC, 1984):

$$\text{PT } (\%) = [(V \cdot f \cdot \text{mEq} \cdot 6.25)/P(g)] \cdot 100 \tag{18.4}$$

In Eq. (18.4), calculation of the percentage of total protein, V is the HCl spent volume; f is the HCl factor; mEq is milliequivalent grams of nitrogen; 6.25 is the general conversion factor of nitrogen to protein; and P is the sample weight in grams.

4.11 Photoprotective formulation

Semisolid formulations, oil-in-water (O/W) emulsions, were prepared because this type of vehicle is widely used in cosmetic and dermatologic products (Eccleston, 1997). To evaluate the potential of the extract of *C. vulgaris* as an ingredient for sunscreens, it was examined at different concentrations (1%, 5%, and 10% w/w) in association with a mixture of UV filters. Octyl *p*-methoxycinnamate and avobenzone were used to achieve a sunscreen/photostability model (Ruscinc, 2018). The formulations are described in Table 18.2.

To handle the photoprotective formulations, the raw materials were separated into two phases: aqueous and oily. The UV filters and Crodafos CES were mixed with isopropyl myristate, composing the oil phase. Aristoflex AVC, a polymer with stabilizing properties used at a wide pH range, was mixed with glycerin and the preservative system in water at 60°C, forming the aqueous phase. Then, the bioactive compound was incorporated into the purified water. The oil phase was heated to 55°C, and then the samples were stored so that the oil phase reached 35°C, and the aqueous phase 40°C. Then, the oil phase was added to the aqueous phase to obtain an emulsion. The pH value of the formulations was adjusted to 5.0–6.0 with triethanolamine or citric acid at room temperature (24.0°C ± 2.0°C). The pH value of the samples was defined to obtain a pH close to that of skin (4.6–5.8) (Leonardi et al., 2002). The samples were stored in glass containers at room temperature.

4.12 Physicochemical characterization and organoleptic properties

The samples were characterized by their pH value and organoleptic properties. The pH value was estimated with a universal pH strip (Sigma–Aldrich) by direct immersion into the samples at room

Table 18.2 Photoprotective formulations: qualitative and quantitative composition.

Composition	Function	Proportion % (w/w)			
		F0	F1	F2	F3
Crodafos CES	Emulsifier	4.0	4.0	4.0	4.0
Isopropyl myristate	Emollient	5.0	5.0	5.0	5.0
Octyl *p*-methoxycinnamate	Ultraviolet B filter	10.0	10.0	10.0	10.0
Avobenzone	Ultraviolet A filter	5.0	5.0	5.0	5.0
Glycerin	Humectant	5.0	5.0	5.0	5.0
EUXYL K 300	Preservative	0.75	0.75	0.75	0.75
Aristoflex AVC	Emulsifier polymer	1.0	1.0	1.0	1.0
Extract of *Chlorella vulgaris*	Active	–	1.0	5.0	10.0
Citric acid	pH corrector	q.s.p.	q.s.p.	q.s.p.	q.s.p.
Triethanolamine	pH corrector	q.s.p.	q.s.p.	q.s.p.	q.s.p.
Purified aqua	Vehicle	q.s.p.	q.s.p.	q.s.p.	q.s.p.

temperature (24°C + 2°C). Estimations of pH were performed in replicates of three (Baby, 2007). Appearance, color, and odor (organoleptic properties) of the samples were evaluated macroscopically and subjectively (Anvisa, 2004).

4.13 In vitro photoprotective activity

The evaluation of in vitro efficacy against UVB and broad-spectrum protection of the formulations was carried out using diffuse reflectance spectrophotometry with an integration sphere (Labsphere UV2000S Ultraviolet Transmittance Analyzer). Samples were weighed and applied as a uniform thin film at 0.75 mg/cm^2 (Baby, 2007; Peres et al., 2017; Velasco et al., 2008) over the substrate (polymethylmethacrylate [PMMA] plate) (Heliplates). After application, the plates were kept at rest and protected from light for 15 min and at room temperature (24°C ± 1°C). Subsequently, the plates were submitted to a spectrophotometric reading using a PMMA plate treated with glycerin as a blank, at a spectral wavelength interval (λ) of 290–400 nm and a progression rate of 1.0 nm. The following parameters were obtained from the UV2000 equipment software: in vitro sun protection factor (SPF) and critical wavelength (nm) (Velasco et al., 2008). The experiment was performed in triplicate (Springsteen et al., 1999; United States, 2011).

4.14 Evaluation of photostability

Evaluating photostability in photoprotective formulations obtains information about possible degradation that UV filters can undergo when exposed to artificial UV radiation emitted by a controlled source of radiation (United States, 2011; Brasil, 2012). After the absorbance of samples in the in vitro photoprotective activity assay was determined, the same PMMA plates were irradiated for 2 h using a UV irradiation simulator (CPS+, Atlas Suntest) equipped with a xenon lamp and an irradiance of 580 W/m^2 (Diffey et al., 2000; United States, 2011). After irradiation, the in vitro SPF and critical wavelength were obtained and compared with those before irradiation. The test was performed in triplicate.

4.15 Statistical analysis

Data were treated using the Minitab program (version 19), with a significance level of 5.0% ($P \leq .05$). In comparisons of the two variables, results were treated according to the Student *t* test. In comparisons among three or more variables, one-way analysis of variance was used, followed by Tukey test.

5. Results and discussion

5.1 Cell concentration

To determine the growth and cellular concentration, a calibration curve was constructed from a concentrated suspension of UTEX 2714 strain. The result is illustrated in Fig. 18.2. The calibration curve was used to maintain the inoculum, through a reading of the absorbance (DO_{720nm}), until the appropriate cell concentration fixed at 0.1 g/L was reached. After, this inoculum was prepared to initiate microalgal cultivation (Li et al., 2018).

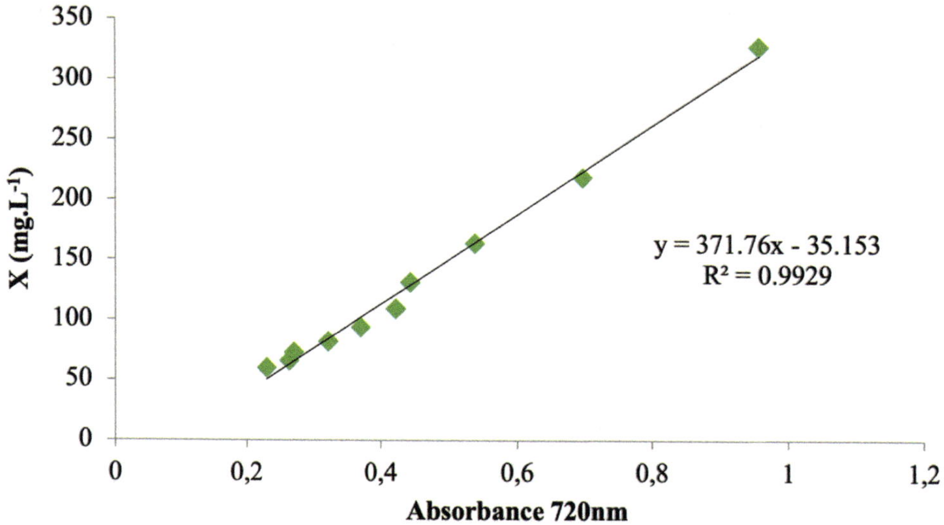

FIGURE 18.2

Calibration curve and cell concentration (x) of UTEX 2714 strain (absorbance at 720nm). In the equation, y is coefficient and x is the sample absorbance.

5.2 Cultivation

Cultivation of the strains started after acquisition of the inoculum with a duration of 6–19 days, varying according to the growth of each strain. The growth of strains was monitored and cultivations were interrupted when they stabilized, preventing cell death and biochemical cell alteration (Martins et al., 2010). The biomass harvest was performed by centrifugation, an effective method in which 95% of biomass is recovered without damaging the cell structure, It is an inexpensive and fast process (Safi et al., 2014).

5.3 Growth curve

Daily absorbance readings (DO_{720nm}) were taken to follow strain growth. They were used to monitor the growth phases indicating the best time to interrupt cultivations without cells entering the death phase. Microalgae growth are composed of five phases: an adaptation or lag phase, an exponential or log phase, the end of the exponential phase or phase of decreased cell growth, the stationary phase, and the senescence phase or cellular death (Martins et al., 2010). According to Santos (2013), the exponential phase is the range of interest for establishing biomass production and productivity. Fig. 18.3 shows the growth curve of the UTEX 2714 strain grown in Bold culture medium with a sodium nitrate concentration of 2.94 mM (1N), 8.82 mM (3N), and 17.64 mM (6N).

Nitrate is the main source of nitrogen and an essential nutrient for microalgae. Their cellular structure is composed of nitrogen and it is part of their physiologic and biological processes (Acioli, 2014; Araújo, 2015b). Under limiting conditions of this nutrient, microalgal growth will be proportional to the amount of nitrate available, especially when for a discontinuous type of cultivation in

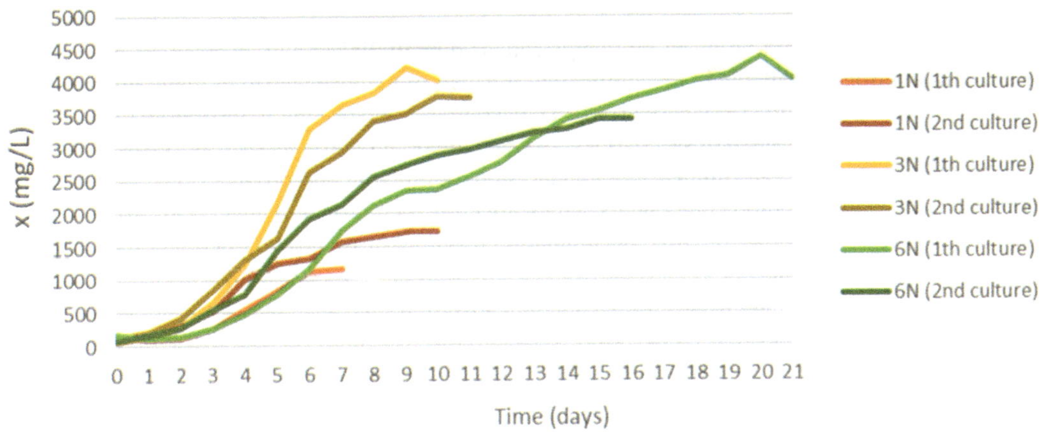

FIGURE 18.3

Cell concentration (X) of C. vulgaris UTEX 2714 versus cultivation time in different sodium nitrate concentration.

which cells consume nutrients available in the medium and are not replenished (Adamakis et al., 2018; Li et al., 2016; Marino, 2018).

In the 6N condition, the maximum cell concentration was 4366.48 mg/L in 19 days, whereas the maximum cell concentration of 3N was 4207.62 mg/L, which was reached in 8 days. From an industrial point of view, the 3N cultivation would be the best option regarding the yield and cultivation time. Also, the parameters for cell productivity and the maximum specific growth rate were obtained. As the productivity results show, a high cell density and biomass productivity could not be reached under a nitrogen limitation, whereas the content of chlorophyll, an important photosynthetic pigment, was be negatively affected under such conditions because of the decreased biosynthesis of crucial chlorophyll proteins complexes (Adamakis et al., 2018). The results were statistically treated and there was a difference between the 1N and 3N/6N cultivations ($P < .5$). There was no difference between the 3N and 6N conditions.

The maximum specific growth rate (μ_{MAX}) represents the number of cell divisions per unit of time. It is constant during the exponential phase (Andrade, 2014). This parameter was calculated for the cultivations. The results for the maximum specific growth rate from cultivations 1N, 3N, and 6N were compared and demonstrated no statistical differences ($P < .5$) when these were compared.

In our experiment condition, nitrogen was able to limit cell growth and cell productivity at a low concentration, but it did not influence the maximum specific growth rate. The medium and high concentrations (3N and 6N) of nitrogen demonstrated similar behavior among the analyzed parameters, but with a difference in the duration of cultivation. The 3N cultivation obtained the same results as for 6N, but in the shortest cultivation period. To prepare the cosmetic formulation, we choose to use the extract from the 3N cultivation, according to its performance in the last tests.

5.4 Biomass and extract composition

The biomass and extract of *C. vulgaris* is fundamentally composed of proteins, carbohydrates, and lipids. According to the cultivation conditions, the proportions of this composition can vary.

5.4.1 Lipid

The amount of accumulated lipids on microalga cells is directly linked and inversely proportional to the amount of nitrogen available in the cultivation. It is part of the mechanism of many microalgae to deal with nutrient deprivation (Skjånese al., 2013). In contrast to plants, microalgae store their primary energy in lipid materials instead of saccharides, so when important nutrients such as nitrogen, which is essential for cell structure, are limited, microalgae stop multiplying and produce some proteins, carbohydrates, and other structural components, which affecting their growth (Panahi et al., 2019; Skjånes et al., 2013). Table 18.3 shows the lipid accumulation potential of *C. vulgaris* in relation to available nitrogen during cultivation.

The accumulation of lipids widely varies among cultivations, as listed in Table 18.3, however, our results were within the expected range. They corroborated with the previous literature because we observed that stress caused by N-deprivation raised the lipid content. As shown in Table 18.4, the lipid content of the 3N extract was higher than the biomass under 3N cultivation conditions. The process of lipid extraction has a large influence on the range of lipids recovered from a biomass or extract. The difference in the process of extracting lipids between dry biomass and the extract was in the breakage of the cell walls, which, according to Dvoretsky et al. (2016), is a variable that could influence the recovery of the lipid rate. To extract the dry biomass lipid, the biomass was macerated before extraction, and to obtain the extract, the biomass went through the mechanical process of breaking the cell wall, which may have helped release lipids from the membranes and cytoplasm.

5.4.2 Fatty acids

The total lipid content of a microalga can reach 1%–70% of its dry weight and can basically be divided into two groups: storage (nonpolar lipids, mainly triacyl glycerides) and structural lipids (polar lipids such as phospholipids and sterols) (Hamed, 2016; Sharma and Schuhmann, 2012). The profile of

Table 18.3 Lipid accumulation (% dry cell weight) of *Chlorella vulgaris*, according to the specialized literature.

Condition	Lipid % (dry cell weight)	References
N-replete	18	Illman et al. (2000)
N-depriv	40	
N-replete	5–40	Safia et al. (2014)
N-deprived	68	
N-replete	16.5	Scarsella et al. (2009)
N-deprived	51.8	
N-replete	25	Widjaja et al. (2009)
N-deprived	52	
N-replete	5.90	Converti et al. (2009)
N-deprived	16.41	
N-replete	21.67	Adamakis (2018)
N-deprived	36.60	

Source: Authors.

fatty acids of *C. vulgaris* cultivation under three different nitrogen concentrations and the 3N extract was determined by CG-MS. The results are shown in Table 18.4 and Fig. 18.4.

According to the literature, major components of the lipid content of *C. vulgaris* are oleic, palmitic, and linolenic acid, which agrees with our results (Hamed, 2016). Palmitic acid can be used as a coemulsifier and opacifying agent in the cosmetic industry. It is also used to improve the texture of products and maintain skin moisture (Khan et al., 2018; Is Palmitic Acid Bad for the Environment and Your Skin?, 2020). Oleic acid is similar to lipids found in the hair and epidermis. Is use in cosmetic formulations could help to improve the compatibility of formulations with cutaneous tissue. It can also promote the maintenance of skin hydration (Dario et al., 2018; Monte, 2013; Viola & Viola, 2009). Used in cosmetics, linoleic acid accumulates on the stratum corneum, contributing to a better skin barrier, cutaneous regeneration, and moisture. It is mainly used in products for the aging population (Dario et al., 2018). In cosmetic products, it functions as an emollient, thickening agent and has antiinflammatory, acne reductive, and skin-lightning properties (Linoleic Acid, 2021).

5.4.3 Proteins

The accumulation of proteins is directly proportional to available nitrogen in the culture medium, as well as the growth rate. Our results for the protein content for the *C. vulgaris* cultivation were corroborated by Venckus et al. (2017).

Proteins are considered to be an important product of microalgae and are widely used for human and animal nutrition (Chew et al., 2017). The literature has shown other uses of proteins extracted from microalgae, such as *C. vulgaris,* especially in fields, such as in the cosmetic industry (Cunha et al., 2022; Hamed, 2016; Mourelle et al., 2017; Ursu et al., 2014). Urso and colleagues (2014) showed that proteins extracted from *C. vulgaris* had an emulsifying property and could be used as a functional ingredient. Another study pointed out that a protein-rich extract from microalgae could repair signs of early skin aging, exerting a tightening effect and preventing stria formation (Spolaore et al., 2006).

Table 18.4 Fatty acid concentration (% GC-MS peak area) of lipid/oil fraction of *Chlorella vulgaris* samples grown under different nitrogen levels (1N, 3N, and 6N) and 3N extract.

Fatty acids	Classification	1N	3N	6N	3N extract
C16:0 (palmitic acid)	SFA	23.47	25.02	21.98	27.33
C16:1 (palmitoleic acid)	MUFA	4.23	7.57	5.98	5.69
ND	–	0.57	0.90	0.80	0.91
ND	–	4.40	3.23	4.15	2.24
C17:1 (cis-10-heptadecenoic acid)	Unsaturated fatty acid	9.00	6.87	10.86	4.90
C18:0 (stearic acid)	SFA	3.52	4.06	1.70	8.27
C18:1n9C (oleic acid)	MUFA	17.98	20.79	13.20	–
C18:2n6c (linoleic acid)	PUFA	24.10	21.19	23.36	25.24
C18:3n6 (γ-linolenic)	PUFA	13.10	10.66	17.97	8.03

MUFA, *monounsaturated fatty acid;* ND, *not detected;* PUFA, *polyunsaturated fatty acid;* SFA, *saturated fatty acid.*
1N = 2.94 mM; 3N = 8.82 mM; 6N = 17.64 mM. Values are the average of three replicates.

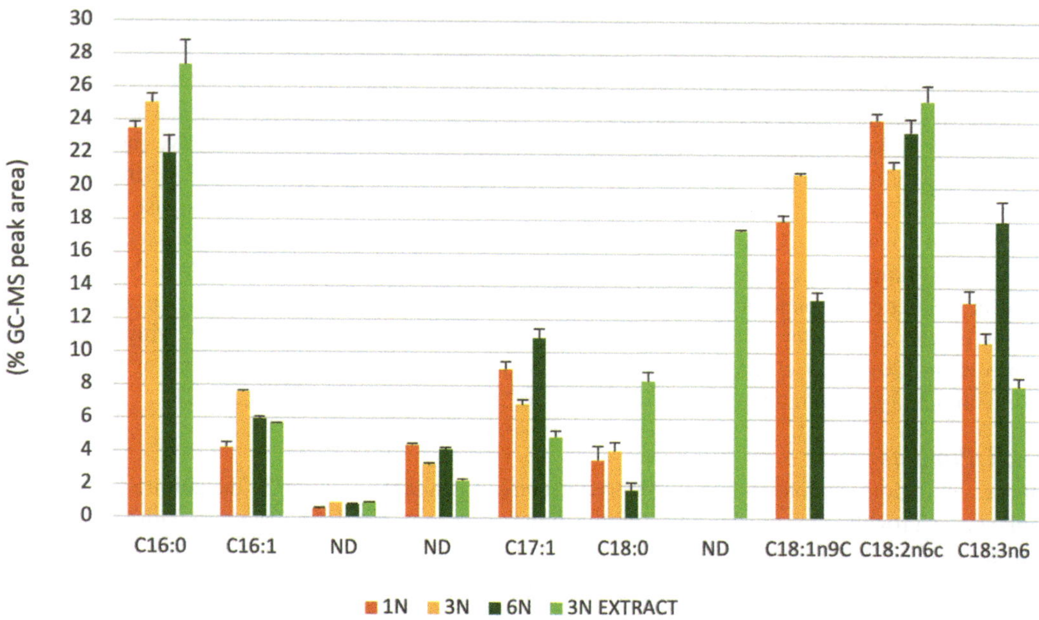

FIGURE 18.4

Fatty acid concentration of the cultivations (%) and the 3N extract.

C. vulgaris extract can reach 50–60% protein content of the biomass dry weight (Venckus et al., 2017). It has been shown to develop similar effects as other protein-rich extracts (antiaging, thickening, and antiwrinkle effects, for instance) (Ariede et al., 2017; Fonseca, 2016; Koller et al., 2014; Skjånes et al., 2013; Wang et al., 2015).

5.5 Sunscreen formulations

The O/W emulsified system containing UV filters was macroscopically compatible with the 3N extract. UV active ingredients of sunscreen (avobenzone and octyl *p*-methoxycinnamate) were selected to obtain a photounstable prototype product with the aim of combining attributes from the *C. vulgaris* extract with this system. All samples were macroscopically evaluated with regard to their organoleptic properties. They were considered suitable and their pH values (6–7) were biocompatible with skin (Velasco et al., 2012). The odor of samples was characteristic. It became more intense as the concentration of the extract increased (Fig. 18.5). The green color of formulations resulted from the color of the extract, which also became more intense according to the concentration. Those properties remained unchanged until the end of the studies.

An extract of *C. vulgaris* was incorporated into the formulation at different concentrations (1.0%, 5.0%, and 10.0% w/w) to verify the ideal concentration that would provide the highest in vitro SPF and increased photostability (Table 18.5). The minimum suitable SPF value was established as at least 6 with a critical wavelength of at least 370 nm, according to Brazilian legislation (Brasil, 2012; Ruscinc, 2018).

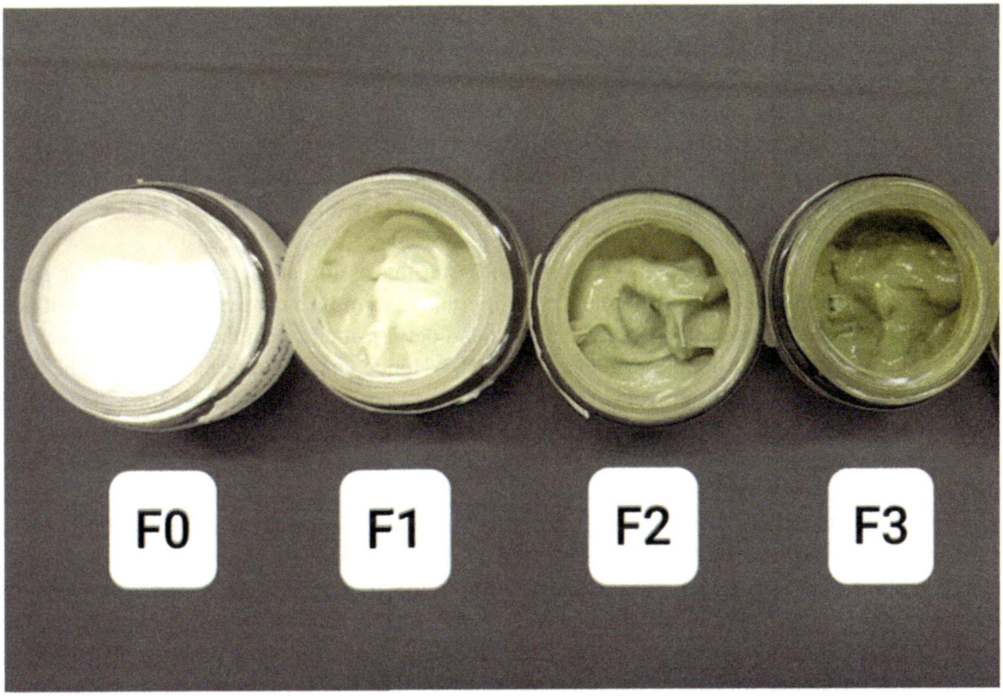

FIGURE 18.5

Photograph of photoprotective formulations.

Table 18.5 In vitro sun protection factor (SPF) and critical wavelength (nm) of sunscreen samples (n = 3).

Samples	In vitro SPF	Standard deviation	Critical wavelength (nm)	Standard deviation
F0	54.33^{A}	7.37	382.0^{C}	0.00
F1	45.67^{A}	2.08	381.3^{C}	0.57
F2	43.33^{A}	6.11	381.3^{C}	0.57
F3	80.33^{B}	0.57	381.0^{C}	0.00

F0 = formulation without extract; F1 = 1.0% extract; F2 = 5.0% extract; F3 = 10.0% extract. Results that share the same letter are statistically equal ($P < .5$).

The extract of *C. vulgaris* was able to promote an increase in the in vitro SPF value at a concentration of 10.0%. For all other concentrations, the extract did not interfere with the performance of this parameter. The critical wavelength did not experience interference from the extract, and all values were above the minimum established by Brazilian legislation. Secondary metabolites could exert some mechanisms to increase the SPF and consequently modify skin–UV interaction, improving cutaneous protection. The mechanism may involve absorbing UV radiation, inhibiting UV-induced free radical and skin surface-protecting lipids, and attenuating inflammatory responses, among other actions (Kostyuk et al., 2021). According to the specialized literature, extracts of *C. vulgaris* can have lutein, α-carotene, and xanthophylls; and β-1,3-glucan and linoleic acid (a compound identified in our investigation), which also have antiinflammatory properties; and other compounds, which would similarly act to provide skin protection against UV radiation and consequently increase the SPF of the formulation (Ando et al., 1998; Barham et al., 2000; Cha et al., 2010; Hynstova et al., 2017; Li et al., 2016; Muszynska et al., 2018; Panahi et al., 2019; Skjånes et al., 2013; Wang et al., 2015).

The photostability of sunscreens is an important issue that is directly correlated to their performance. The association of different UV filters is an alternative to improving the photostability of novel sunscreens. However, the association of avobenzone and octyl *p*-methoxycinnamate was shown to be photounstable, as expected (Lhiaubet-Vallet et al., 2010). Hence, another alternative would be to add photostabilizers and natural antioxidants to the formulations to maintain or enhance the efficacy of the product (Morocho-Jácome et al., 2020). F0, F1, F2, and F3 were irradiated for 2 h, because it is recommended to reapply sunscreens every 2 h (United States, 2021). The in vitro SPF was read every 30 min to monitor photostability.

Unfortunately, the extract derived from *C. vulgaris* biomass did not work as a photostabilizer at any concentration tested and the active ingredients degraded at less than 2 h of artificial irradiation. Our results suggested that the extract was unable to reduce the formation of free radical by-products produced by photodegradation of the avobenzone. It was also unable to balance the photochemical reaction of the degradation of octyl *p*-methoxycinnamate (Ariede et al., 2017). However, the *C. vulgaris* extract should not be discarded as a photostabilizing agent. Further tests are required with an extract obtained under new conditions of microalga culture. As observed, cultivation conditions directly interfered with secondary metabolites produced by this microalga, which are bioactives that could have important functions in several types of topical formulations.

There has been increasing demand for dermocosmetics developed with natural and sustainable products. This creates the opportunity and need for new searches for natural, sustainable, and renewable sources of biocompounds and bioactive molecules to develop formulations that can be considered eco-friendly products (Chrapusta et al., 2017; McIntosh et al., 2018; Show et al., 2017).

6. Conclusion

Considering the experimental conditions adopted in this research, we conclude that (1) the best growing condition for *C. vulgaris* cultivation was at a 3N concentration of nitrogen (8.82 mM), particularly regarding the yield and cultivation time to obtain the biomass. This process generated cell productivity of 3978.67 m/L d; a growth rate of 0.71 μ_{MAX} (d^{-1}); 23.50%, and 19.31% of lipid content and protein content, respectively. (2) The extract prepared from the biomass cultivated in the 3N condition had 43.17% lipid content and 19.78% protein content. (3) A sunscreen formulation was prepared as an O/W emulsion containing octyl *p*-methoxycinnamate and avobenzone as UV filters. The extract of *C. vulgaris* was incorporated at 0.0%, 1.0%, 5.0%, and 10.0%. (4) The sunscreen sample with the best performance was the one with a 10.0% extract of *C. vulgaris*, with an in vitro SPF of 80.33 ± 0.57.

Acknowledgments

The authors gratefully acknowledge FAPESP, CAPES, and CNPq (305250/2019-1).

References

Acioli, T. S. (2014). *Cultivo da microalga Chlorella spp. em pequena escala visando ao aumento da produção de biomassa e óleos para biodiesel.*

Adamakis, I. D., et al. (2018). Cultivation, characterization, and properties of Chlorella vulgaris microalgae with different lipid contents and effect on fast pyrolysis oil composition. *Environmental Science and Pollution Research, 25*(23), 23018–23032.

Ahluwalia, M. S., et al. (2016). Cultivo de Chlorella sp. em fotobiorreatores enriquecidos com CO_2 e avaliação da biomassa para produção de biocombustíveis. *IOSR Journal of Economics and Finance, 3*(1), 56. www.bertelsmann-stiftung.de/fileadmin/files/BSt/Publikationen/GrauePublikationen/MT_Globalization_Report_2018.pdf%0A. http://eprints.lse.ac.uk/43447/1/India_globalisation%2C society and inequalities%28lsero%29.pdf%0Ahttps://www.quora.com/What-is-the.

Andersen, R. A. (2005). *Algal culturing techniques.*

Ando, H., Ryu, A., Hashimoto, A., Oka, M., & Ichihashi, M. (1998). Linoleic acid and alpha-linolenic acid lightens ultraviolet-induced hyperpigmentation of the skin. *Archives of Dermatological Research, 290*, 375–381. https://doi.org/10.1007/s004030050320

Andrade, L. M. D. E. (2014). Produção de microalgas E Caracterização de Sua Composição Protêica E Lipídica via Espectrometria de Massas. *Tese de Doutorado, 141.*

Andrade, L. M. De, Mendes, M. A., & Nascimento, C. A. O. (2013). *Avaliação da intensidade luminosa nos cultivos em shaker da Chlorella vulgaris.*

Anvisa (Agência Nacional De Vigilância Sanitária). (2004). Guia de estabilidade de produtos cosméticos. *Brasilia: Anvisa, 1*, 45 (Séries temáticas da Anvisa. Qualidade em cosméticos, 1).

AOAC. (1984). In *AOAC association of Official Analytical Chemists* (Vol 14) ([s.l: s.n.]).

Araújo, F. O. (2015b). *Avaliação do crescimento e composição de micro- organismos fotossintetizantes para uso como matéria-prima em fotoprotetor.* Universidade de São Paulo.

Ariede, M. B., et al. (2017). *Cosmetic attributes of algae - a reviewAlgal Research.*

Avadhani, K. S., et al. (2017). *Skin delivery of epigallocatechin-3-gallate (EGCG) and hyaluronic acid loaded nano-transfersomes for antioxidant and anti-aging effects in UV radiation induced skin damage loaded nano-transfersomes for antioxidant and anti-aging effects, 7544.*

Baby, A. R. (2007). *Avaliação in vitro da permeabilidade cutânea da rutina em emulsões cosméticas.* Universidade de São Paulo.

Bacova, R., et al. (2019). The effects of 5-azacytidine and cadmium on global 5-methylcytosine content and secondary metabolites in the freshwater microalgae Chlamydomonas reinhardtii and Scenedesmus quadricauda. *Journal of Phycology, 55*(2), 329–342.

Baiee, M. A., & Salman, J. M. (2016). Effect of phosphorus concentration and light intensity on protein content of microalga Chlorella vulgaris. *Mesopotamia Environmental Journal, 2*(2), 75–86.

Barham, J. B., Edens, M. B., Fonteh, A. N., Johnson, M. M., Easter, L., & Chilton, F. H. (2000). Addition of eicosapentaenoic acid to gamma-linolenic acid- supplemented diets prevents serum arachidonic acid accumulation in humans. *Journal of Nutrition, 130*, 1925–1931. https://doi.org/10.1093/jn/130.8.1925

Bashan, Y., Lopez, B. R., Huss, V. A. R., Amavizca, E., & de-Bashan, L. E. (2015). Chlorella Sorokiniana (formerly *C. vulgaris*) UTEX 2714, non-thermotolerant microalga useful for biotechnological application and as a reference strain. *Journal of Applied Phycology.*

Bhalamurugan, G. L., Valerie, O., & Mark, L. (2018). Valuable bioproducts obtained from microalgal biomass and their commercial applications: A review. *Environmental Engineering Research, 23*(3), 229–241.

Borowitzka, M. (1999). Commercial production ofmicroalgae: Pond, tanks, tubes and fermenters. *Journal of Biotechnology, 70*(1–3), 313–321.

Brasil. (2012). *Guia para Avaliação de Segurança de Produtos Cosméticos* [s.l: s.n.]. http://portal.anvisa.gov.br/documents/106351/107910/Guia+para+Avaliaçáo+de+Segurança+de+Produtos+Cosméticos/ab0c660d-3a8c-4698-853a-096501c1dc7c. Acesso em: 29 out. 2019

Brennan, L., & Owende, P. (2010). Biofuels from microalgae — a review of technologies for production, processing, and extractions of biofuels and co-products. *Renewable and Sustainable Energy Reviews, 14*(2), 557–577. fev https://linkinghub.elsevier.com/retrieve/pii/S1364032109002408.

Broinizi, P. R. B., et al. (2007). Avaliação da atividade antioxidante dos compostos fenólicos naturalmente presentes em subprodutos do pseudofruto de caju (Anacardium occidentale L.). *Ciência e Tecnologia de Alimentos, 27*(4), 902–908.

Buono, S., Langellotti, A. L., Martello, A., Bimonte, M., Tito, A., Carola, A., Apone, F., Colucci, G., & Fogliano, V. (2012). Biological activities of dermatological interest by the water extract of the microalga *Botryococcus braunii. Archives of Dermatological Research, 304*, 755–764.

Carvalho, J. C. M., Francisco, F. R., almeida, K. A., Sato, S., & Converti, A. (2004). Cultivation of Arthrospira (Spirulina) platensis by fed-batch addition of ammonium chloride at exponentially-increasing feeding rate. *Journal of Phycology, 40*, 589–597.

Carvalho, J., Matsudo, M., Bezerra, R., Ferreira-Camargo, L., & Sato, S. (2014). Microalgae bioreactors. In R. Bajpai, A. Prokop, & M. Z. (Eds.), *Algal biorefineries* (pp. 83–126) [s.l: s.n.].

Cezare-Gomes, E. A., et al. (2019). Potential of microalgae carotenoids for industrial application. *Applied Biochemistry and Biotechnology, 188*(3), 602–634.

Cha, K. H., et al. (2010). Effect of pressurized liquids on extraction of antioxidants from Chlorella vulgaris. *Journal of Agricultural and Food Chemistry, 58*(8), 4756–4761.

Chew, K. W., et al. (2017). Microalgae biorefinery: High value products perspectives. *Bioresource Technology, 229*, 53–62. https://doi.org/10.1016/j.biortech.2017.01.006

Chien, A. C., & Ju, Y. (2009). Study of increasing lipid production from fresh water microalgae *Chlorella vulgaris. Journal of the Taiwan Institute of Chemical Engineers, 40*, 13–20.

Chinnasamy, S., Ramakrishnan, B., Bhatnagar, A., & Das, K. C. (2009). Biomass production potential of a wastewater alga chlorella vulgaris ARC 1 under elevated levels of CO_2 and temperature. *International Journal of Molecular Sciences*, 518–532.

Chisti, Y. (2007). Biodiesel from microalgae. *Biotechnology Advances, 25*(3), 294–306. https://doi.org/10.1016/j.biotechadv.2007.02.001.

Chrapusta, E., et al. (2017). Mycosporine-like amino acids: Potential health and beauty ingredients. *Marine Drugs, 15*(10), 1–29.

Clesceri, L. S., Greenberg, A. E., & R. T, R. (1989). *Standard methods for the examination of water and wastewater.* American Public Health Association.

Costa, D. C. M. C. W. S. C. (2010). Constituintes químicos, fenóis totais e atividade antioxidante de Sterculia striata St. Hil. et Naudin Chemical constituents, total phenolics and antioxidant activity of Sterculia striata. *St. Hil. et Naudin, 40*(1), 207–212. http://www.scielo.br/pdf/aa/v40n1/v40n1a26.pdf.

Cunha, S. A., Coscueta, E. R., Nova, P., Silva, J. L., & Pintado, M. M. (2022). *Bioactive Hydrolysates from Chlorella vulgaris: Optimal process and bioactive properties, 27 (8)* p. 2505). https://doi.org/10.3390/molecules27082505.

Dario, M. F., et al. (2018). *Synergistic photoprotective activity of nanocarrier containing oil of Acrocomia aculeata (Jacq.) Lodd. Ex. Martius—Arecaceae. Industrial crops and products.*

De Jager, T. L., Cockrell, A. E., & Du Plessis, S. S. (2017). Ultraviolet light induced generation of reactive oxygen species. *Advances in Experimental Medicine and Biology, 996*, 15–23.

Delgoda, R., & Murray, J. E. (2017). Evolutionary perspectives on the role of plant secondary metabolites. In *Pharmacognosy* (pp. 93–100). Elsevier [s.l: s.n.].

Derner. (2006). *Microalgas, produtos e aplicações.*

Diffey, B. L., et al. (2000). In vitro assessment of the broad-spectrum ultraviolet protection of suscreen products. *Journal of the American Academy of Dermatology, 43*(6), 1024–1035.

Duran, S. K., Kumar, P., & Sandhu, S. S. (2018). A review on microalgae strains, cultivation, harvesting, biodiesel conversion and engine implementation. *Biofuels, 7269*, 1–12. https://doi.org/10.1080/17597269.2018.1457314.

Dvoretsky, D., et al. (2016). Enhanced lipid extraction from microalgae chlorella vulgaris biomass: Experiments, modelling, optimization. *Chemical Engineering Transactions, 49*, 175–180.

Eccleston, G. M. (1997). Functions of mixed emulsifiers and emulsifying waxes in dermatological lotions and creams. *Colloids Surfaces A: Physicochemical and Engineering Aspects, 123–124*, 169–182.

Fernández-Linares, L. C., et al. (2017). Assessment of *Chlorella vulgaris* and indigenous microalgae biomass with treated wastewater as growth culture medium. *Bioresource Technology, 244*, 400–406. https://doi.org/10.1016/j.biortech.2017.07.141

Fonseca, J. A. (2016). *Aplicação de Algas na Indústria Alimentar e Farmacêutica.*

Goldstein, M. (2008). Recuperação do DNA e fotoenvelhecimento. *Cosmetcs & Toiletries, 20*(1), 48–51.

Gomes, R. K., & Damazio, M. G. (2009). In *Cosmetologia: Descomplicando os princípios ativos, 3* ([s.l: s.n.]).

Hamed, I. (2016). The evolution and versatility of microalgal biotechnology: A review. *Comprehensive Reviews in Food Science and Food Safety, 15*(6), 1104–1123.

Hamedi, S., Mahdavi, M. A., & Gheshlaghi, R. (2016). Improved lipid and biomass productivities in Chlorella vulgaris by differing the inoculation medium from the production medium. *Biofuel Research Journal, 3*(2), 410–416.

Hartman, L., & Lago, R. C. (1973). Rapid preparation of fatty acid methyl esters from lipids. *Lab Prat, 22*, 475–476.

Hirata, L. L., Sato, M. E. O., & Santos, C. A. D. M. (2004). Radicals livres e o envelhecimento cutâneo. *Acta Farmaceutica Bonaerense, 23*(3), 418–424.

Hynstova, V., et al. (2017). Separation, identification and quantification of carotenoids and chlorophylls in dietary supplements containing Chlorella vulgaris and Spirulina platensis using High Performance Thin Layer Chromatography. *Journal of Pharmaceutical and Biomedical Analysis*. https://doi.org/10.1016/j.jpba.2017.09.018

Is palmitic acid Bad for the environment and Your skin? Disponível em.(2021). Acesso em: 15 ago https://thedermreview.com/palmitic-acid/.

jácome, A. L. M. (2014). *Estudo do reaproveitamento de meio no cultivo de Arthrospira (Spirulina) platensis.*

Jiang, Y., & Kalodimos, C. G. (2017). NMR studies of large proteins. *Journal of Molecular Biology, 429*(17), 2667–2676. https://doi.org/10.1016/j.jmb.2017.07.007

Joshi, S., Kumari, R., & Upasani, V. N. (2018). Applications of algae in cosmetics: An overview. *International Journal of Innovative Research in Science, Engineering and Technology, 7*(March), 1269–1278.

Khan, N. R., Gawas, Sarita, D., Rathod, & Virendra, K. (2018). Enzyme-catalysed production of n-butyl palmitate using ultrasoundassisted esterification of palmitic acid in a solvent-free system. *Bioprocess and Biosystems Engineering*.

Khan, M. I., Shin, J. H., & Kim, J. D. (2018). The promising future of microalgae: Current status, challenges, and optimization of a sustainable and renewable industry for biofuels, feed, and other products. *Microbial Cell Factories, 17*(1), 1–21. https://doi.org/10.1186/s12934-018-0879-x

Klasson, K. T., et al. (1991). *Bioreactor Design for sirjthesis gas fermentations* ([s.l: s.n.]).

Koller, M., Muhr, A., & Braunegg, G. (2014). Microalgae as versatile cellular factories for valued products. *Algal Research, 6*(PA), 52–63. out https://linkinghub.elsevier.com/retrieve/pii/S2211926414000782.

Kostyuk, V., et al. (2021). Natural substances for prevention of skin photoaging: Screening systems in the development of sunscreen and rejuvenation cosmetics. *Rejuvenation Research, 21*(2), 91–101, 1 abr. 2018. www.liebertpub.com/doi/abs/10.1089/rej.2017.1931. Acesso em: 22 ago

Kumar, J., et al. (2016). UV-B radiation stress causes alterations in whole cell protein profile and expression of certain genes in the rice phyllospheric bacterium *Enterobacter cloacae. Frontiers in Microbiology, 7*(SEP), 1–14.

Lam, M. K., & Lee, K. T. (2012). Potential of using organic fertilizer to cultivate *Chlorella vulgaris* for biodiesel production. *Applied Energy, 94*, 303–308. https://doi.org/10.1016/j.apenergy.2012.01.075

Lhiaubet-Vallet, V., et al. (2010). Filter-filter interactions. Photostabilization, triplet quenching and reactivity with singlet oxygen. *Photochemical and Photobiological Sciences, 9*, 552–558.

Liao, Q., et al. (2017). Simultaneous enhancement of Chlorella vulgaris growth and lipid accumulation through the synergy effect between light and nitrate in a planar waveguide flat-plate photobioreactor. *Bioresource Technology, 243*, 528–538. https://doi.org/10.1016/j.biortech.2017.06.091

Li, T., et al. (2016). Morphology, growth, biochemical composition and photosynthetic performance of *Chlorella vulgaris* (Trebouxiophyceae) under low and high nitrogen supplies. *Algal, 16*, 481–491. https://doi.org/10.1016/j.algal.2016.04.008

Li, J., et al. (2018). Effects of sodium bicarbonate on cell growth, lipid accumulation, and morphology of *Chlorella vulgaris. Microbial Cell Factories, 17*(1), 1–10. https://doi.org/10.1186/s12934-018-0953-4

Liu, J., & Chen, F. (2014). *Biology and industrial applications of* Chlorella*: Advances and prospects*. December.

Linoleic acid. Disponível em. (2021). Acesso em: 15 ago https://thedermreview.com/linoleic-acid/.

Lordan, S., Ross, R. P., & Stanton, C. (jun. 2011). *Marine bioactives as functional food ingredients: Potential to reduce the incidence of chronic diseasesMarine Drugs*.

Lourenço, S. (2006). *Cultivo de microalgas marinhas: Princípios e aplicações*. http://www.sidalc.net/cgi-bin/wxis.exe/?IsisScript=AGB.xis&method=post&formato=2&cantidad=1&expresion=mfn=242602. Acesso em: 28 nov. 2019.

Lúcio, M. J. (2013). *Cultivo de microalgas* Chlorella vulgaris *com efluente doméstico como meio de cultura alternativo. Trabalho de Conclusão de Curso do Curso de Engenharia Sanitária e Ambiental da UFCS. 71*.

Mambro, V. M. D., & Fonseca, M. J. V. (2007). Mudanças por radicais livres e ação tópica de antioxidantes. *Cosmetcs & Toiletries, 19*(5), 74–76.

Marino, V. F. (2018). *Análises dos parâmetros de cultivo da microalga Chlorella vulgaris*. Universidade de São Paulo, 2018.

Markou, G., Vandamme, D., & Muylaert, K. (2014). Microalgal and cyanobacterial cultivation: The supply of nutrients. *Water Research, 65*, 186–202. https://doi.org/10.1016/j.watres.2014.07.025

Martins, A., Caetano, N. S., & Mata, T. M. (2010). *Microalgae for biodiesel production and other applications : A review. 14* pp. 217–232).

McIntosh, K., et al. (2018). Alkenones as a promising green alternative for waxes in cosmetics and personal care products. *Cosmetics, 5*(2).

Miranda, M. S., Sato, S., & M.-f, J. (2001). Antioxidant activity of the microalga *Chlorella vulgaris* cultered on special conditions. *Bollettino Chimico Farmaceutico, 3*(140), 165–168.

Montagner, S., & Costa, A. (2009). Bases biomoleculares do fotoenvelhecimento. *Anais Brasileiros de Dermatologia, 84*(3), 263–269.

Monte, G. C. D. O. (2013). *Desenvolvimento e avaliação da estabilidade física de condicionador capilar contendo óleo de pequi (Caryocar brasiliense Camb.)*.

Montoya, E. Y. O., et al. (2014). *Production of* Chlorella vulgaris *as a source of essential fatty acids in a tubular photobioreactor continuously fed with air enriched with CO_2 at different concentrations*.

Morocho Jácome, A. L., et al. (2012). Kinetic and thermodynamic investigation of *Arthrospira* (Spirulina) platensis fed-batch cultivation in a tubular photobioreactor using urea as nitrogen source. *Journal of Chemical Technology and Biotechnology, 87*(11), 1574–1583.

Morocho-Jácome, A. L., Freire, T. B., Oliveira, A. C., Santos de Almeida, T., Rosado, C., Velasco, M. V. R., & Baby, A. R. (2020). In vivo SPF from multifunctional sunscreen systems developed with natural compounds—a review. *Journal of Cosmetic Dermatology, 20*, 729–737. https://doi.org/10.1111/jocd.13609

Mourelle, M. L., Gómez, C. P., & Legido, J. L. (2017). The potential use of marine microalgae and cyanobacteria in cosmetics and thalassotherapy. *Cosmetics, 4*(4).

Muszyńska, B., et al. (2018). Bioaccessibility of phenolic compounds, lutein, and bioelements of preparations containing Chlorella vulgaris in artificial digestive juices. *Journal of Applied Phycology, 30*(3), 1629–1640.

Olguín, E. J., Galicia, S., Angulo-Guerrero, O., & Hernández, E. (2001). The effect of low light flux and nitrogen deficiency on the chemical composition of *Spirulina* sp. (*Arthrospira*) grown on digested pig waste. *Bioresource Technology, 77*(1), 19–24.

Panahi, Y., et al. (1 jun. 2019). Impact of cultivation condition and media content on chlorella vulgaris composition. *Advanced Pharmaceutical Bulletin, 9*(2), 182–194. https://doi.org/10.15171/jcvtr.2015.24

Pelizer, L. H., Sassano, C. E., Carvalho, J. C. M., Sato, S., Gioielli, L. A., & Moraes, I. O. (1999). Padronização do método de secagem da biomassa de *Spirulina platensis*. *Farmácia e Química, 32*(1), 37–40.

Pereira, A. B. (2013). *Produção De biomassa e de lipídeos por cultivo misto de microalgas suplementado com* co_2. Universidade federal do paraná.

Peres, D. D., et al. (2017). Quality by design (QbD), Process Analytical Technology (PAT), and design of experimented applied to the development of multifunctional sunscreens. *Drug Development and Industrial Pharmacy, 43*(2), 246–256.

Pohndorf, R. S., Camara, A. S., Larrosa, A. P. Q., Pinheiro, C. P., Strieder, M. M., & Pinto, L. A. A. (2016). Production of lipids from microalgae *Spirulina* sp.: Influence of drying, cell disruption and extraction methods. *Biomass and Bioenergy, 93*, 25–32.

Priyadarshani, I., & Rath, B. (2012). *Commercial and industrial applications of micro algae – a review, 3 (4)* pp. 89–100).

Putri, T. W., Nursida, N. F., & Raya, I. (2021). Antioxidant activity of *Chlorella vulgaris* used as an antioxidant cream. *Journal of Physics: Conference Series*. 1899012030. https://iopscience.iop.org/article/10.1088/1742-6596/1899/1/012030

Rastogi, R. P., & Incharoensakdi, A. (2013). UV radiation-induced accumulation of photoprotective compounds in the green alga Tetraspora sp. CU2551. *Plant Physiology and Biochemistry, 70*, 7–13. https://doi.org/10.1016/j.plaphy.2013.04.021

Raven, P. H., Evert, R. F., & Eichhorn, S. E. (2007). Biologia vegetal. In *Biologia Vegetal. 7. Ed. Rio de Janeiro: Editora Guanabara Koogan* (p. 906).

Rebolloso-Fuentes, M. M., et al. (2001). Biomass nutrient profiles of the microalga Nannochloropsis. *Journal of Agricultural and Food Chemistry, 49*(6), 2966–2972.

Rodrigues-Ract, J., & L. A, G. (2008). Lipídios modificados obtidos a partir de gordura do leite, óleo de girassol e ésteres de fitosteróis para aplicação em spreads. *Quimica Nova, 31*, 1960–1965.

Ruscinc, N. (2018). *Fotoprotetores bioativos contendo extrato de mirtilo (*Vaccinium myrtillus *L .): Caracterização físico-química e funcional*. Faculdade de Ciências Farmacêuticas da Universidade de São Paulo, 2018

Safi, C., et al. (2014). *Morphology, composition, production, processing and applications of* Chlorella vulgaris*: A review*.

Santos, P. V. D. O. S. (2013). *Seleção de cepas oleaginosas da microalga* Chlorella *spp. e otimização do seu cultivo em meio sintetico e vinhaça visando à produção sustentável de biodiesel*. Universidade de são paulo escola de engenharia de são carlos.

Scarsella, M., et al. (2009). Achievements and perspectives in hetero- and mixotrophic culturing of microalgae. *Chemical Engineering Transactions, 17*, 1065–1070.

Serra-Maia, R., et al. (2016). Influence of temperature on *Chlorella vulgaris* growth and mortality rates in a photobioreactor. *Algal Research, 18*, 352–359.

Sharma, K. K., & Schuhmann, H.,S. P. (2012). High lipid induction in microalgae for biodiesel production. *Energies, 5*, 1532–1553.

Show, P. L., et al. (2017). A holistic approach to managing microalgae for biofuel applications. *International Journal of Molecular Sciences, 18*(1).

Silva, L. C. M. (2020). *Avaliação do crescimento de micro-organismos fotossintetizantes em fotobiorreator tubular empregando nutrientes provenientes da vinhaça tratada.*

Silva, L. N., Leite, M. G. A., & Maia Campos, P. M. B. G. (2019). Development of hair care formulations containing *Spirulina platensis* and *Ascophyllum nodosum* extracts. *International Journal of Phytocosmetics and Natural Ingredients, 6*(1). pp. 13–13.

Singh, R. N., & Sharma, S. (2012). Development of suitable photobioreactor for algae production—a review. *Renewable and Sustainable Energy Reviews, 16*(4), 2347–2353.

Skjånes, K., Rebours, C., & Lindblad, P. (2013). Potential for green microalgae to produce hydrogen, pharmaceuticals and other high value products in a combined process. *Critical Reviews in Biotechnology, 33*(2), 172–215.

Solomon, Berg, & Martin. (2011). *Biology. 9.* Fort Worth: Saunders College Publishing [s.l.].

Spolaore, P., et al. (2006). Commercial applications of microalgae. *Journal of Bioscience and Bioengineering, 101*(2), 87–96.

Springsteen, A., et al. (1999). In vitro measurement of sun protection factor of sunscreens by diffuse transmittance. *Analytica Chimica Acta, 380*, 155–164.

Stages, E. (2010). *Effects of marine toxins on the reproduction and early stages development of aquatic organisms.*

States, U. Food and Drug Administration. Department of Health and Human Services. (2021). *Sunscreen: How to Help Protect Your Skin from the Sun.* Available in: https://www.fda.gov/drugs/understanding-over-counter-medicines/sunscreen-how-help-protect-your-skin-sun.

States, U. Food and Drug Administration. Department of Health and Human Services 21 CFR Parts 201 and 310. (2011). Labeling and effectiveness testing; sunscreens drug products for over-the-counter human use. *Federal Register, 76*(117), 35620–35665.

Steiner, D. (2011). *Antioxidantes no tratamento do envelhecimento cutâneo.* http://www.cosmeticsonline.com.br/noticias/detalhes-colunas1/714/antioxidantes+no+tratamento+do+envelhecimento+cutáneo. Acesso em: 22 jan. 2020.

Toledo-Cervantes, A., Morales, M., & E Novelo, S. R. (2013). Carbon dioxide fixation and lipid storage by Scenedesmus obtusiusculus. *Bioresource Technology, 130*, 652–658.

Ugwu, C. U., Aoyagi, H., & Uchiyama, H. (jul. 2008). Photobioreactors for mass cultivation of algae. *Bioresource Technology, 99*(10), 4021–4028. https://linkinghub.elsevier.com/retrieve/pii/S0960852407001368.

Ursu, A. V., et al. (2014). Extraction, fractionation and functional properties of proteins from the microalgae *Chlorella vulgaris. Bioresource Technology, 157*, 134–139. https://doi.org/10.1016/j.biortech.2014.01.071

Van Den Hoek, C., Mann, D.G., & H. M. J. Algae: An introduction to phycology. ([s.l: s.n.]), 1995

Velasco, M. V. R., et al. (2008). Associação da rutina com p-metoxicinamato de octila e benzofenona-3: Avaliação in vitro da eficácia fotoprotetora por espectrofotometria de refletância. *Latin American Journal of Pharmacy, 27*(1), 23–27.

Velasco, M. V., et al. (2012). Influences of bioactive substances on the physicochemical and functional stability of sunscreen emulsions. *Biomedical and Biopharmaceutical Research, 1*(9), 119–130.

Venckus, P., Kostkevičienė, J., & Bendikienė, V. (2017). Green algae *Chlorella vulgaris* cultivation in municipal wastewater and biomass composition. *Journal of Environmental Engineering and Landscape Management, 25*(1), 56–63.

Viola, P., & Viola, M. (2009). Virgin olive oil as a fundamental nutritional component and skin protector. *Clinical Dermatology, 27*, 159–165.

Vonshak, A. (1997). Spirulina platensis *(Arthrospira) physiology, cell-biology and biotechnology.* London: Taylor & Francis.

Wang, H. D., et al. (2015). Exploring the potential of using algae in cosmetics. *Bioresource Technology, 184*, 355–362. https://doi.org/10.1016/j.biortech.2014.12.001

Further reading

De Araújo, L. A. Desenvolvimento de formulações cosméticas contendo óleos vegetais para a proteção e reparação capilar. 2015a. Universidade de são paulo faculdade de ciências farmacêuticas de ribeirão preto, 2015.

Dario, M. F. (2016). *Desenvolvimento e avaliação de eficácia de nanoemulsão catiônica bioativa na proteção capilar aos danos foto-oxidativos*. Faculdade de Ciências Farmacêuticas, 2016.

Converti, A., Casazza, A. A., Ortiz, E. Y., Perego, P., & Borghi, M. D. (2009). Effect of temperature and nitrogen concentration on the growth and lipid content of Nannochloropsis oculata and Chlorella vulgaris for biodiesel production. *Chemical Engineering and Processing: Process Intensification, 48*(6), 1146–1151.

Dario, M. F., et al. (2013). Efficacy of *Punica granatum* L. hydroalcoholic extract on properties of dyed hair exposed to UVA radiation. *Journal of Photochemistry and Photobiology B: Biology, 120*, 142–147. https://doi.org/10.1016/j.jphotobiol.2012.12.011

Dias, M. F. R. G., Almeida, A. M., & Pichler, J. (2014). The Shampoo pH can affect the hair: Myth or Reality? *Internatinal Journal of Trichology, 6*(3), 95–99.

Evans, T. (2011). Evaluating hair conditioning with instrumental combing. *Cosmetics & Toiletries, 126*(8).

Evans, A., Marsh, J., & Wickett, R. (2011). The structural implications of water hardness metal uptake by human hair. *International Journal of Cosmetic Science, 33*(5), 477–482.

Fregonesi, A., et al. (2009). Brazilian oils and butters: The effect of different fatty acid chain composition on human hair physiochemical properties. *Journal of Cosmetic Science, 60*(2), 273–280.

Freitas, L. F. de L. e (2016). *Estudo da solubilidade de sais inorgânicos em misturas de solventes*. Universidade de Tiradentes.

Goshiyama, A. M. (2019). *Avaliação das propriedades das fibras capilares tratadas com alisante ácido com diferentes valores de pH*. http://www.teses.usp.br/teses/disponiveis/9/9139/tde-07052019-154247/pt-br.php%0Ahttp://fi-admin.bvsalud.org/document/view/87855.

Griffiths, M. J., & Harrison, S. T. L. (2009). Lipid productivity as a key characteristic for choosing algal species for biodiesel production. *Journal of Applied Phycology, 21*, 493–507.

Illman, A. M., Scragg, A. H., & Shales, S. W. (2000). Increase in Chlorella strains calorific values when grown in low nitrogen medium. *Enzyme and Microbial Technology*, 631–635.

Kockler, J., et al. (2012). Photostability of sunscreens. *Journal of Photochemistry and Photobiology C: Photochemistry Reviews, 13*(1), 91–110. https://doi.org/10.1016/j.jphotochemrev.2011.12.001

Korte, M., et al. (2014). Distribution and localization of hydrophobic and ionic chemical groups at the surface of bleached human hair fibers. *Langmuir, 30*(41), 12124–12129.

Lautenschlager, S., Wulf, H. C., & Pittelkow, M. R. (2007). Photoprotection. *Lancet, 370*(9586), 528–537.

Leite, M. G. A., & MAIA Campos, P. M. B. G. (2018). Photoprotective effects of a multifunctional hair care formulation containing botanical extracts, vitamins, and UV filters. *Photochemistry and Photobiology, 94*(5), 1010–1016.

Leonardi, G. R., Gaspar, L. R., & Campos, P. M. B. G. M. (2002). Estudo da variação do pH da pele humana exposta à formulação cosmética acrescida ou não das vitaminas A, E ou de ceramida, por metodologia não invasiva. *Anais Brasileiros de Dermatologia, 77*(5), 563–569.

Lv, J. M., et al. (2010). Enhanced lipid production of *Chlorella vulgaris* by adjustment of cultivation conditions. *Bioresource Technology, 101*(17), 6797–6804.

Mobin, S. M. A., Chowdhury, H., & Alam, F. (2019). Commercially important bioproducts from microalgae and their current applications-A review. *Energy Procedia, 160*(2018), 752–760. https://doi.org/10.1016/j.egypro.2019.02.183

Moraes, C. A. P. (2012). *Síntese e avaliação da segurança in vitro da rutina e do succinato de rutina visando sua incorporação em formulações fotoprotetoras eficazes associados a filtros químicos e físico.*

Rebenfeld, L., Weigmann, H., & Dansizer, C. (1966). *Temperature dependence of the mechanical properties of human hair in relation to structure.*

Robbins, C. R. Chemical and physical behavior of human hair. In: Springer-verlaq. Third Edit ed. (s.l: s.n).

Robbins, C., & Reich, C. (1986). Prediction of hair assembly characteristics from single-fiber properties. Part II. The relationship of fiber curvature, friction, stiffness, and diameter to combing behavior. *Journal of the Society of Cosmetic Chemists, 37*(3), 141–158.

Schoeller, H. L. E. S. (1962). *Hydrologie Dynamique et Chimique, Recherche, Exploitation et Évaluation des Ressources* (p. 642). Paris: Masson Et Cie.

Sorokin, C.,K. R. W. (1958). The effects of light intensity on the growth rates of green algae. *Plant Physiology, 2*(33), 109–113.

CHAPTER

Algal lipids, lipidomics, and biosurfactants

19

Mounika Munagala Srinivas Reddy[1,2], Sonal Tiwari[1] and Vikas Singh Chauhan[1,2]

[1]*Plant Cell Biotechnology (PCBT) Department, CSIR-Central Food Technological Research Institute (CFTRI), Mysuru, Karnataka, India;* [2]*Academy of Scientific and Innovative Research (AcSIR), Ghazibad, Uttar Pradesh, India*

1. Introduction

Algae are known to be the world's oldest plants. They first appeared 3.5 billion years ago. These photosynthetic species are commonly categorised as macroalgae with lengths up to 60 m and microalgae with sizes as small as 0.2 μm (Koyande et al., 2019). Microalgae are a varied category of organisms with an estimated 200,000 to 800,000 species. Some of these species are grown commercially to extract bioactive chemicals for human and animal consumption, cosmetics, and for the biofuel (Bleakley et al., 2017). Macroalgae have received a lot of attention in recent years because of their various health-promoting properties that reduce the risk of many chronic diseases and even help people live longer (Ganesan et al., 2019). The lipid content of macroalgae ranges from 2% to 13% of dry weight (Kazir et al., 2019). Microalgae have high lipid content ranging from 8.2% to 37.8% of dry weight (Huerlimann et al., 2010). Microalgae cultivation has shifted its focus in recent years to the production of lipids for commercial use. Lipids are a diverse group of substances for which there is no universally accepted definition. They are chemically heterogenous substances that are soluble in nonpolar solvents but insoluble in water. As hydrophobic building blocks, all lipids contain fatty acyl/alkyl, sphingosine, or isoprene fractions (De Luca et al., 2021). Two major lipid types in algae are storage lipids (mostly triacylglycerol [TAG]) and membrane components (glycolipids, betaine lipids, and phospholipids) (Li-Beisson et al., 2019). Nonpolar lipids such as sterols, free fatty acids, wax and steryl esters are found in algae (Gunstone et al., 2010). Lipids are necessary components of all living cells and have a vital role in cell membranes. Lipids were once thought to be merely structural components, but now they have been found to have prominent role in cell signaling and the production of bioactive metabolites (Cockcroft et al., 2021). Signal transduction, protein processing, and membrane turnover are all facilitated by lipid-driven membrane domains (Aureli et al., 2015). Lipids make up the majority of cellular membranes. They have a wide range of structural characteristics, and the distribution of different lipids and their species (membrane lipid composition) differs by organism, cell type, organelle, membrane, membrane subdomain, and bilayer leaflet (Harayama and Howard, 2018).

Algae Materials: Applications Benefitting Health. https://doi.org/10.1016/B978-0-443-18816-9.00019-8

2. Algal lipids

2.1 Polar algal lipids

Glycolipids and phospholipids (PLs) make up the majority of polar lipids in most algae. Betaine lipids (BLs) may also be found in certain species or orders. Polar lipids are both structural components of membranes and lipid synthesis intermediates (Mimouni et al., 2018).

2.1.1 Glycolipids

Glycolipids (glycosylglycerides, GLs) have a 1,2-diacyl-stereospecifically numbered (sn)-glycerol moiety with a mono- or oligosaccharide linked to the glycerol backbone at the sn-3 position. Galactosylglycerides are uncharged, polar lipids that constitute the majority of plastid lipids (Mizusawa & Wada, 2012). They contain one or two galactose molecules coupled to the glycerol at the sn-3 position, related to 1,2-diacyl-3-O-(b-D-galactopyranosyl)-sn-glycerol (or MGDG, monogalactosyldiacylglycerol) and 1,2-diacyl-3-O-(a-D-galactopyranosyl)-(1 → 6)-O-b-D-galactopyranosyl-sn-glycerol or (DGDG, digalactocyldiacylglycerol). In plant thylakoid membranes, monogalactosyldiacylglycerol and digalactocyldiacylglycerol make up 40%−55% and 15%−35% of the overall lipids, respectively. Algae also contain sulfolipid sulfoquinovosyl diacylglycerol (SQDG), which is another variety of glycosylglyceride. The sulfonic acid linkage in SQDG makes it unique. At physiological pH, the sulfonoglucosidic moiety (6-deoxy-6-sulfono-glucoside) is referred to as sulfoquinovosyl, and the sulfonic residue carries a full negative charge. The structures of MGDG, DGDG, and SQDG are given in Fig. 19.1. The two main galactolipids, DGDG and MGDG, have a varying ratio in microalgae because environmental and dietary factors affect them. MGDG might not be the primary membrane lipid in diatoms because their GL composition differs from that of plants and chlorophytes (Guschina & Harwood, 2013; Lepetit et al., 2012).

The anionic lipid SQDG, which is present in equal or even more concentrations than MGDG, is the most abundant lipid in diatom thylakoids. As a result, it has been anticipated that the high concentration of SQDG and phosphatidylglycerol in diatom thylakoids has important implications on the thylakoid membrane's composition and the diatom xanthophyll cycle's functions (Garab et al., 2016; Lepetit et al., 2012). The high n-3 PUFA (polyunsaturated fatty acid) content of microalgal glycolipid (GL), which is comparable with that of higher plants, is a distinguishing feature. Many diatoms (*Phaeodactylum tricornutum, Fistulifera solaris, Skeletonema costatum, Stephanodiscus* sp.) have 20:5/C16 and C16/C16-MGDG as their main MGDG species. However, in *P. tricornutum*, an acylated form of SQDG with 20:5 attached to the sugar moiety was recently discovered (Abida et al., 2015). Eukaryotic-kind galactolipids such as 18:4/18:4-MGDG and 18:3/18:4-DGDG are found in some diatoms, such as *Cyclotella meneghiniana* (Guschina & Harwood, 2013). Interestingly, galactolipids are abundant in the nonphotosynthetic diatom *Nitzschia alba*. The plastidic compartment is where galactolipids are made. Microalgae have galactolipids built into their membranes, which are a component of the chloroplast compartment. Diacylglycerol (DAG) is used to make all glycolipids. Sulfolipid synthase and monogalactosyl diacylglycerol synthase, respectively, are involved in the production of SQDG and MGDG from this molecule. The digalactosyl diacylglycerol synthase uses the MGDG produced in this last reaction as a substrate to synthesize DGDG (Liang et al., 2014; Xu et al., 2010).

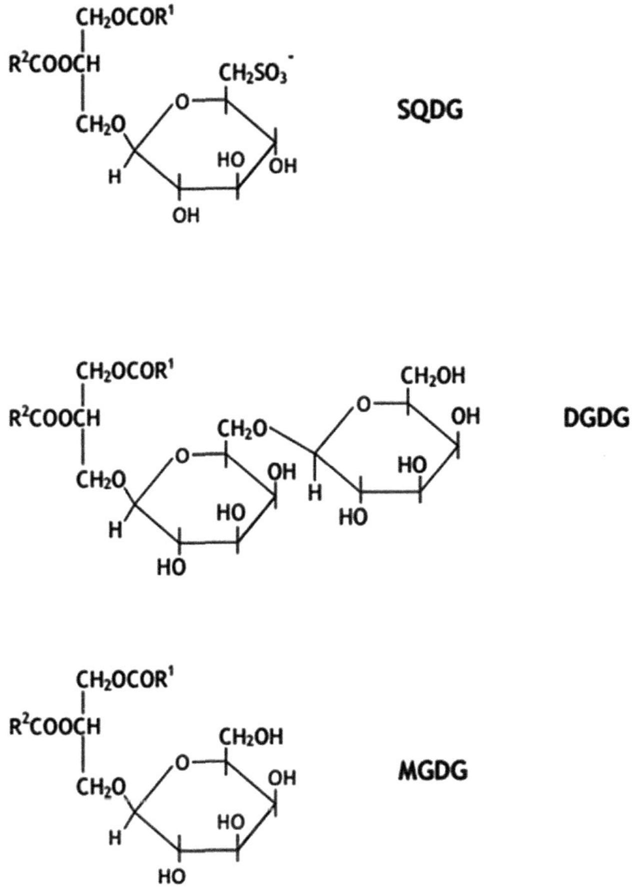

FIGURE 19.1

Structures of SQDG, DGDG, and MGDG.

2.1.2 Phospholipids

Phospholipids (Phosphoglycerides, PLs) have a glycerol backbone with esterified nonpolar acyl groups at 1 and 2 positions, and phosphate is esterified to the sn-3 position attached ahead to a hydrophilic base group. The major phosphoglycerides found in many algae species are phosphatidylcholine (PC), phosphatidylethanolamine (PE) (see Fig. 19.2), and phosphatidylglycerol (PG). Diphosphatidylglycerol (DPG), phosphatidylinositol (PI), and phosphatidylserine (PS) are present in significant amounts in various algal cells. Although phosphatidic acid is typically a relatively small part, it is an important metabolite and may also function as a signaling substance. With the exception of PG, the PLs are found in the extrachloroplast membranes (Guschina & Harwood, 2013). Thylakoid membranes contain a significant amount of PL. In eukaryotic green algae, PG makes up between 10% and 20% of total polar glycerolipids. D3-trans-hexadecenoic acid (16:1, 3 t) is an unusual fatty acid

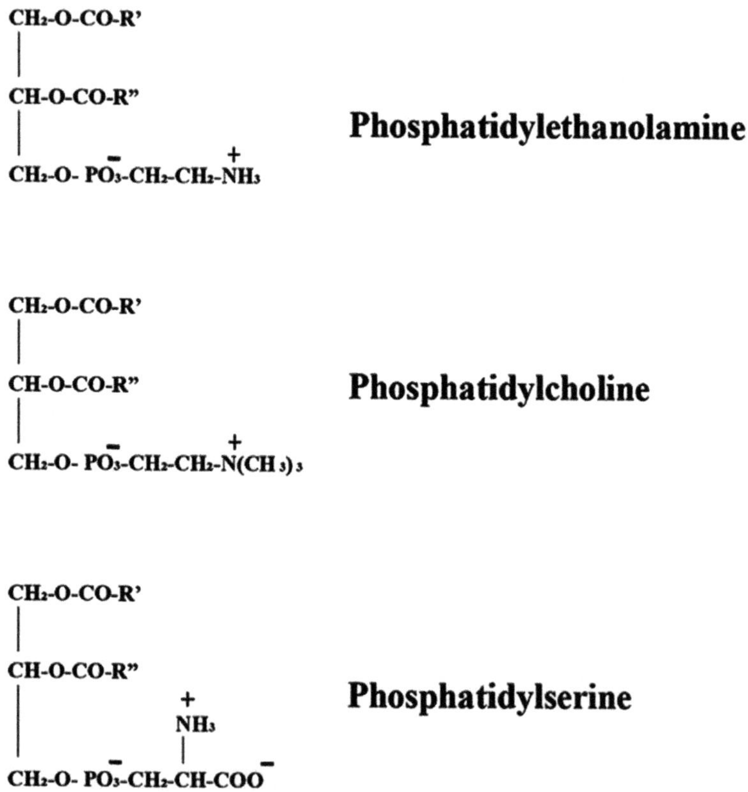

FIGURE 19.2

Structures of phosphatidylethanolamine, phosphatidylcholine, and phosphatidylserine.

found in all photosynthetic eukaryotes, and it is greatly enhanced at the sn-2 position of phosphatidylglycerol. Phosphatidic acid (PA) is also a component of PL, though in smaller amounts. It is a metabolic intermediate as well as a signaling compound (Guschina & Harwood, 2013). Except for PG, which is found in significant amounts in thylakoid membranes, PL is found in extrachloroplast membranes. PLs contribute to extrachloroplast membrane structural stability and selective permeability, while phosphatidylglycerol assists glycolipids in upholding photosynthetic apparatus stability. In most of the algae, PLs are present in a low percentage (10%–20%) of total lipids (Guschina & Harwood, 2013) except in some dinophytes where the amount is 78%–95%, such as *Kraenia, Takayama*, and *Karlodinium* species (Mooney et al., 2007) and *Polarella glacialis* (Guschina & Harwood, 2013). In diatoms, PC plays an important part in the biosynthesis of galactolipids, as evidenced by *Phaeodactylum* or *Fistulifera solaris* (Liang et al., 2014). When phosphate levels are low, plant plastids produce more DGDG and SQDG, and DGDG is transported to extraplastidial cellular membrane to substitute PLs. BLs are frequently used by algae in place of PLs (Guschina & Harwood, 2013).

2.1.3 Betaine lipids

Betaine lipids (BLs) are a diverse group of functional and structural polar lipids that have been studied since their discovery in the initial 1960s. BLs are complex lipids that contain a betaine moiety as a polar group instead of phosphorus or carbohydrate and are attached to the sn-3 position of glycerol by an ether bond with fatty acid (FA) esterified in the sn-1 and sn-2 positions. 1,2-Diacylglyceryltrimethylhomoserine (DGTS), 1,2-diacylglycerylcarboxyhydroxymethylcholine (DGCC), and 1,2-diacylglycerylhydroxymethyltrimethyl-β-alanine (DGTA) (see Fig. 19.3) are three types of BLs identified in algae. Because their molecules have a negatively charged carboxyl group and positively charged trimethylammonium group, they are all zwitterionic at neutral pH. BLs are abundant in algae, but not in higher plants such as gymnosperms or angiosperms (Guschina et al., 2013). Freshwater and marine species have significantly different fatty acid compositions in DGTS.

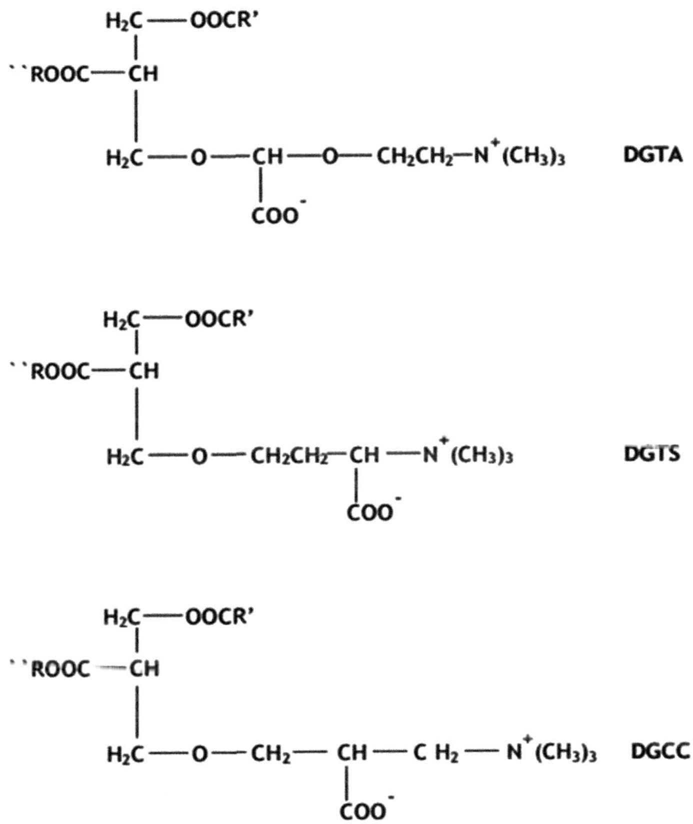

FIGURE 19.3

Structures of DGTA, DGTS, and DGCC.

Saturated fatty acids have been noticed at sn-1 position of the glycerol backbone in freshwater algae, while C18 acids were found at sn-2 position. Marine algae often have very long-chain PUFAs in their DGTS at both the sn-1 and sn-2 positions. For example, DGTS was exceptionally rich in EPA (eicosapentaenoic acid) in the marine eustigmatophyte UTEX 2341 (previously identified as *Chlorella minutissima*), which produced DGTS at unusually high levels (about 44% of overall lipids). In this alga, the EPA accounted for more than 90% of total DGTS fatty acids. It has been proposed that BLs, particularly DGTS, are more evolutionarily primitive lipids that perform the same functions in membranes in lower plants as PC does in higher plants and animals because of the structural similarities among BLs and phosphatidylcholine (and also with their taxonomic allocation, which shows a direct correlation between BLs as well as PC in several microalgae species) (Guschina and Harwood, 2013).

2.2 Nonpolar lipids

Nonpolar triacylglycerols (TAGs), distinct from PLs present in biological membranes, are not a part of a cell's structural framework but act as storage energy. After being created, TAGs are stored in distinct globular structures called lipid droplets (LDs) in cytoplasmic cytosol or chloroplasts. TAG accumulation varies greatly (for example, from 2% of total lipids in *Fucus serratus* to 77% in stationary phase in *Parietochloris incisa*) and is influenced with a variety of external conditions. Fatty acids are converted into TAG molecules by algal cells once growth is slow and new membrane components are not needed. This process continues until things improve and additional growth is required. TAG is primarily synthesized in the light, stored in cytosolic lipid bodies, and then repurposed in the dark for polar lipid synthesis. TAG synthesis is greatly influenced by nitrogen deficiency, as numerous algae exhibit a two- to threefold rise in the amount of lipid, primarily TAG, when N (nitrogen) is limited. Saturated and monounsaturated fatty acids make up the majority of algal TAG. On the other hand, some oleaginous species have demonstrated the capacity to obtain substantial amounts of long-chain polyunsaturated fatty acids in TAG (Guschina et al., 2009). Some algal species have been found to have alternative storage lipids that are not glycerolipids. When the marine cryptomonad *Chroomonas salina* was aged and *Euglena gracilis* was started growing in the shadow, wax esters were collected (Guschina et al., 2009). *Isochrysis galbana* and *Emiliania huxleyi*, and haptophyte algae, produce a variety of polyunsaturated long-chain (C37 − C39) alkenes, alkenones, and alkenoates (Guschina et al., 2009). A physiological function for these polyunsaturated fatty acids as an energy reserve has been proposed based on their metabolic behavior and cellular localization (Guschina et al., 2009). When adapting to abrupt climatic changes, new PUFA production could be extremely slow, and also PUFA-rich TAG may donate particular acyl groups to polar lipids and MGDG to ensure fast adaptive membrane reorganization (Makewicz et al., 1997).

3. Biosynthesis of triacylglycerol in algae

Eukaryotic algae are a diverse group of organisms that play an important role in a wide range of ecosystems. They produce more than half of the primary food at the bottom of the food chain.

Eukaryotic algae have a varied structural and functional properties of acyl lipids and uncommon fatty acids that are not present in other phyla due to their variable morphology and habitats. Many microalgae are known to accumulate TAG in lipid droplets under physical environmental stresses like unfavourable light intensity, temperature, high salinity, alkaline pH and dehydration (Zienkiewicz et al., 2016). Indeed, some authors have indicated that the ability of few algae to accumulate VLC-PUFAs (very-long-chain polyunsaturated fatty acids) in TAG could serve as a reserve, allowing the organisms to adapt to any future rapid changes in their environment, such as increased light intensity, UV radiation, or cold temperatures. Based on their lipid diversity, thousands of species of algae and cyanobacteria have been scanned for high lipid content, leading to the identification and purification of about a hundred oleaginous species. These species can be found in a variety of taxonomic groups, and the total lipid content of individual species or strains can differ significantly within and between taxonomic groups (Zienkiewicz et al., 2016). In microalgae, fatty acid (FA) synthesis takes place in plastid, same as in plant cells. Following their transport to the cytosol, fatty acids enter the TAG assembly pathway, which is catalyzed by the enzymes of endoplasmic reticulum (ER) (Liu et al., 2013). Furthermore, de novo TAG assembly in the chloroplast has recently been suggested in the *Chlamydomonas reinhardtii* after nitrogen deficiency of a mutant deficient in starch, increasing the possibility of new TAG assembly biosynthetic pathways in microalgae unobserved in general plants (Zienkiewicz et al., 2016). Until this hypothesis is further confirmed or refuted, similar TAG assembly pathways in the ER and plastidic envelopes must be considered. Indeed, the discovery of a growing amount of presumed acyltransferase targeting plastid in various genes of microalgae could generate evidence for TAG assembly pathway in chloroplast (Zienkiewicz et al., 2016; Li et al., 2014). Atleast two major TAG biosynthetic pathways in microalgae have been proposed, the Kennedy or glycerol phosphate pathway and the monoacylglycerol (MAG) pathway (Zienkiewicz et al., 2016). TAG is made by esterifying an acyl-CoA and a glycerol hydroxyl set in both pathways. The kennedy pathway begins with the acylation of glycerol-3-phosphate (G-3-P). Acyl-CoA:glycerol-3-phosphate acyltransferase (GPAT) acylates glycerol-3-phosphate (G-3-P) to produce lysophosphatidic acid (LPA), which may then be further acylated to make phosphatidic acid (PA) by acyl-CoA-dependent acyl-CoA:LPA acyltransferase (LPAAT). Phosphatidic acid phosphatase (PAP) then dephosphorylates the PA to produce diacylglycerol (DAG). A mechanism called acyl-CoA:DAG acyltransferase catalyses the conversion of DAG into TAG by integrating the third acyl-CoA (DGAT). A pathway from G-3-P to DAG in chloroplasts has been reported, although it is unknown whether DAG produced in chloroplasts contributes to the production of TAG (Chen & Wang, 2021). In MAG pathway MAG acyltransferase (MGAT) converts 2-monoacylglycerol (sn-2-MAG) to DAG, which is the first step in the monoacylglycerol pathway. GPAT4 and GPAT6 enzymes, which use dicarboxylic and hydroxy acyl-CoA as acyl donors to produce sn-2-MAG, have previously been shown to catalyze de novo synthesis of MAG in plant cells (Yang et al., 2010; Reddy et al., 2010). After that, the DAG produced by either pathway is converted to TAG. An acyl-CoA:DAG acyltransferase (DGAT) catalyzes this enzymatic reaction, which is thought to be the committed step in the synthesis of TAG pathway. Phospholipid:diacylglycerol acyltransferase (PDAT) catalyzes the third pathway of TAG biosynthesis, in which PLs or galactolipids act as an acyl donor (Dahlqvist et al., 2000; Zhang et al., 2009; Yhoon et al., 2012). TAG formation via the PDAT pathway has been demonstrated in yeast and plants (Dahlqvist et al., 2000;

Zhang et al., 2009). Plants with PDAT activity were found in a limited oil seed species, but their insights into TAG biosynthetic pathway differed. All microalgae studied had PDAT-encoding genes, implying that acyl-CoA-independent mechanisms of TAG biosynthesis occur in microalgae as well (Zienkiewicz et al., 2016).

4. Enhancement of algal lipids

Production of lipids is significantly affected by cultivation techniques. Continuous, semicontinuous, batch, fed-batch, and two-stage cultivation strategies are the five main divisions of microalgal cultivation methods (Ryu et al., 2019). Due to the lack of stressful conditions present throughout the cultivation process, fed-batch and continuous cultivation strategies do not support high lipid productivity. While producing consistent biomass, two-stage cultivation systems and semicontinuous strategies promote lipid accumulation (Aziz et al., 2020). A more effective strategy is required, focusing on lowering the overall production cost and improving scalability, to enhance the production of lipids from microalgae. Combining abiotic stress and nutrients, chemical additives, using EDTA (ethylenediaminetetraacetic acid), and plant hormones and cocultivating bacteria and microalgae are just a few of the novel approaches (Singh et al., 2016). Limiting the amount of nitrogen (N) and phosphorus (P) in the cells is one of the most popular methods for increasing lipid concentration (Behera et al., 2021). In an effort to make microalgae cultivation systems more effective, numerous research papers have been published. There are two main groups that these studies fall under. The first group focuses on improving bioreactor design and other methods to boost microalgae's lipid productivity. By using an appropriate cultivation strategy, the other category's main goal is to increase biomass and lipid yields (Aziz et al., 2020). To increase the productivities of microalgal bioproducts, metabolic engineering strategies have been used. The development of genetic engineering techniques and multi-omics analyses has resulted in significant advancements in microalgae metabolic engineering over the past ten years (Kang et al., 2021). Table 19.1 gives few examples of techniques used to enhance various algal lipids.

5. Environmental influences on algal lipids

5.1 Effect of temperature

Temperature has been shown to impact FA proportions and lipid class content in various algae in previous studies. Algal growth is best at around 25°C, but each species has its own optimum temperature. It has significant consequences for outdoor cultivation systems (Li-Beisson et al., 2019). Naturally, some strains of algae can survive in temperatures exceeding 40°C, while those that thrive in thermal springs can survive in temperatures as high as 80°C. Given the variety of algal species and the range of temperatures found in their habitats, it is not surprising that lipid analyses have found differences in their fatty acid profiles. Dinoflagellates from various freshwater habitats at low temperature have been studied for their lipid classes. The molecular species of their lipid classes could be used to categorize them. The glycosylglycerides (SQDG, MGDG, and DGDG) appeared to be especially helpful for categorization. The range of dinoflagellate tolerance was found to be best reflected in

Table 19.1 Various strategies used for the enhancement of algal lipids.

Sr No	Strategy used for the Enhancement of algal lipids	Microalgae used	Lipid yield W/W (%)	References
1	**Cultural conditions/ nutrient stress**			
	Phosphorus	*Chlorella sp.*	23.6%	(Liang et al., 2013)
		Chlorella vulgaris	37.73%	(Chu et al., 2013)
	Nitrogen	*Scenedesmus deserticola*	54.4%	(Li et al., 2013)
		Chlorococcum ellipsoideum	40.5%	(Li et al., 2013)
		Nannochloropsis oculata	15.51%	(Converti et al., 2009)
		C. vulgaris	16.51%	(Converti et al., 2009)
	Magnesium	*Monoraphidium sp.*	58%	(Huang et al., 2014)
	Iron	*C. vulgaris*	56.6%	(Liu et al., 2008)
2	**Different cultivation conditions**			
	Temperature	*N. oculata*	7.42%–14.9%	(Converti et al., 2009)
	Salinity	*Dunaliella salina*	67%	(Takagi et al., 2006)
	Light	*Nannochloropsis sp.*	31.3%	(Atta et al., 2013)
	Carbon dioxide	*Scenedesmus obliquus*	33%	(Abd El Baky et al., 2012)
3	**Chemical inducers**			
	Auxin	*Dunaliella tertiolecta*	24%–70%	(El Arroussi et al., 2015)
	Gibberellic acid, zeatin, indole acetic acid	*Acutodesmus obliquus*	6.61%	(Renuka et al., 2018)
	Fulvic acid	*Monoraphidium sp. FXY-10*	12.07%	(Che et al., 2016)
	Ethylene-diamine-tetraacetic acid (EDTA)	*Scenedesmus obliquus*	29.7%	(Ren et al., 2014)
4	**Various wastewater used as culture media**			
	Industrial brewery	*C. vulgaris*	6.04%	(Farooq et al., 2013)
	Palm oil mill effluent	*Chlorella sorokiniana CY-1*	15.07%	(Cheah et al., 2018)
	Farm wastewater	*C. sorokiniana*	83.5%	(Hena et al., 2015)
5	**Cultivation methods**			
	Two-stage cultivation method	*C. vulgaris*	7.11%	(Mujtaba et al., 2012)
	Three-stage cultivation method	*C. sorokiniana*	83.5%	(Hena et al., 2015)

Continued

Table 19.1 Various strategies used for the enhancement of algal lipids.—cont'd

Sr No	Strategy used for the Enhancement of algal lipids	Microalgae used	Lipid yield W/W (%)	References
6	**Combined cultivation methods**			
	Nutrient combination	*Ankistrodesmus falcatus KJ671624*	59.6%	(Singh et al., 2015)
	Yeast–microalgae cocultivation	*S. obliquus Rhodotorula glutinis*	60%–70%	(Yen et al., 2015)
	Microalgae–bacteria cocultivation	*Rhizobiumsp., Ankistrodesmus sp.*	11.2%	(Do Nascimento et al., 2013)
7	**Molecular approach**			
	Genetic engineering tool	*Chlamydomonas reinhardtii*	18.76%	(Ahmad et al., 2015)
	Functional gene overexpression	*Phaeodactylum tricornutum*	57.8%	(Xue et al., 2015)
		Chlorella minutissima	44%	(Hsieh et al., 2012)
		Chlorella pyrenoidosa	50.4%	(Hsieh et al., 2012)

thylakoid glycolipids, whereas the distribution of nonthylakoid lipids and their species revealed phylogeny better (Anesi et al., 2016). In several species of algae, including *Chlamydomonas*, increased growth temperature can be used to increase lipid content. With the importance of algae in biotechnology, commercially important species have become a recent focus. Thus, the thraustochytrid *Aurantiochytrium*, which forms appreciable docosahexaenoic acid (DHA), was studied at temperatures ranging from 10 to 30°C and found to be optimal at 10°C. In addition, the combination of nitrogen stress and varying growth temperature in *C. reinhardtii* has been investigated (James et al., 2013). With increased global warming, some recent studies on heat stress (in *Chlamydomonas*) and climate warming in *Scenedesmus obliquus* have been conducted (Schroda et al., 2015).

5.2 Effects of pH

Lipids are also affected when algae grow at extreme pH levels. As a result, basic pH stress raised TAG amount accumulation while decreasing comparative levels of membrane lipids in *Chlorella* sp. (Guschina et al., 2009). The influence of pH on the lipid and fatty acid components of *Chlamydomonas* sp. isolated from a volcanic acidic lake and *C. reinhardtii* isolated from an algal collection was investigated and equated. When grown under the same conditions, *Chlamydomonas* sp. had more saturated FAs in the polar lipids than *C. reinhardtii*. The TAG content (as a percentage of total lipids) in *Chlamydomonas* species grown at pH 1 was also higher as compared with cells grown at higher pH (Wilhelm et al., 2014).

5.3 Effect of light

Light is a major controlling factor for algal growth and different algal species have different needs (Khan et al., 2018). Both biochemical composition and algal mass yield are affected by its duration and intensity (Krzemińska et al., 2014). Light intensity is among the most crucial elements that affects algal growth and lipid composition. GLs (which include MGDG, DGDG, and SQDG) and PLs (which include phosphatidylglycerol and phosphatidylcholine) constitute the typical lipid composition of the thylakoid membrane of the diatom *P. tricornutum* (Wilhelm et al., 2014). Even though MGDG is still the most abundant lipid, SQDG concentration increases when this diatom is cultured under low light intensity. However, when grown under intense lighting, the SQDG concentration exceeds the MGDG amounts (Lepetit et al., 2012). Because MGDG and SQDG are linked to thylakoid membranes, many studies on their role in photosynthesis have been conducted. When diatoms are exposed to saturating light intensities, MGDG acts as a photoprotector (Wilhelm et al., 2014). Increased light intensity boosts rate of photosynthesis and growth until the latter reaches a plateau because photosynthesis is balanced by photorespiration and photoinhibition. For a number of algal species, maximum growth (and lipid production) intensities and also the impact of varying light intensities in combination with carbon dioxide levels or salinity and nitrogen have recently been reported (Daliry et al., 2017). Both qualitative and quantitative aspects of lipid metabolism can be influenced by these factors. The amount of eicosapentaenoic acid and decosahexaenoic acid in polar lipids in *Pavlova lutheri* was maximum at low intensity light, whereas their synthesis is quite high at intermediate intensity of light (19 w/m^2) (Li-Beisson et al., 2019). Reduced light intensity improved membrane components, particularly SQDG, phosphatidylglycerol, and phosphatidylcholine in the red alga *Trichocarpus crinitus* (Li-Beisson et al., 2019). Saturating light causes TAG to accumulate in LDs in *Chlamydomonas reinhardtii* for a long time. Extensive (24-hour) dark periods also cause TAG accumulation in *Chlamydomonas* sp. (Hemschemeier et al., 2013). The impact of UV-B irradiation has also been investigated on different algae. Whether the algae were UV sensitive or tolerant determined the extent of the changes. There was a decrease in VLC PUFAs such as EPA and DHA, as well as the fact that nutrient-depleted cells were more sensitive (Li-Beisson et al., 2019).

5.4 Effect of nutrients

The basic macronutrients for all algal species are carbon, nitrogen, and phosphorus. Silicon is required by some microalgae (e.g., many diatoms) (Zulu et al., 2018). Nitrogen deficiency below 0.5 mg/L induces lipid production optimally in *Chlorella vulgaris* (Daliry et al., 2017). N (nitrogen) is, in fact, the most common stress that causes TAG production. Similarly, phosphorus deficiency has a significant impact on lipid metabolism and, as a result, on the content of algal oil (Abida et al., 2015). In *P. tricornutum*, under both the N and P limiting conditions decrease in MGDG and PtdGro (Phosphatidylglycerol) and enhancement of accumulation of palmitic (16:0) and palmitoleic acid (16:1) were observed (Abida et al., 2015). The increase in TAG in *Nannochloropsis oceanica* under nitrogen deprivation was accompanied by reduction in MGDG as well as the extrachloroplastic lipids DGTS and phosphatidylcholine (Han et al., 2017). Also, there is a scarcity of information about the mechanisms that control TAG accumulation during N deficiency. N-starvation induced TAG production in *Chlamydomonas* sp. was accompanied by increased expression of a nitrogen response regulator (Boyle et al., 2012). Furthermore, in *P. tricornutum*, RNAi knockdown of nitrate reductase can improve lipid biosynthesis (Li-Beisson et al., 2019). On N-deprivation, the transcription factor ROC40, which affects circadian rhythm, was discovered to significantly rise. Further evidence from mutant analysis supported ROC40's involvement in N-starvation-induced TAG synthesis (Goncalves et al., 2016).

It has been reported that on P-deprivation, *C. reinhardtii* had lower amounts of all phosphoglycerides. Phosphatidylglycerol, a crucial thylakoid (photosynthetic) membrane lipid was significantly reduced, but it can be substituted by SQDG, an anionic membrane lipid (Riekhof et al., 2003). Normal responses to P-deprivation (and S-deprivation) were observed in a low phosphate bleaching mutant of *C. reinhardtii* (lpb1), but the lpb1 mutant lacked critical acclimation ability (Li-Beisson et al., 2019). P-deficiency caused increases in DGDG and DGTS (and TAG) in the freshwater eustigmatophyte *Monodus subterraneus*, which occurred due to the loss of phosphoglycerides. The increase in DGDG (but not MGDG) resembled how higher plants react to a phosphorus (P) deficiency. FA contents of different lipid classes also changed. Similar to N-deprivation, but to a lesser extent, P-deprivation results in lipid (TAG) accumulation (Abida et al., 2015). It was discovered that there was a step-wise adaptive response in *P. tricornutum* when looking at the time course of changes. The authors proposed that phosphoglycerides provide alternative phosphorus after catabolism and that non-phosphorus lipids, such as SQDG for phosphotidylglycerol and DGTA for phosphotidylcholoine, were used to replace some of the phosphoglycerides (Abida et al., 2015). The impact of P deficiency on diatoms is thoroughly discussed (Zulu et al., 2018). Other macronutrients that can affect lipid metabolism include carbon dioxide and sulfur (Devi & Mohan, 2012). S-deficiency could increase nonpolar lipids in *Chlorella ellipsoidea* and *Chlamydomonas reinhardtii* (Li-Beisson et al., 2019). Silicon is a macronutrient for many diatoms, including oleaginous species and its reduction has been shown to increases TAG biosynthesis in diatoms (Jiang et al., 2015; Zulu et al., 2018). The *Thalassiosira pseudonana*, for example, accumulates 24% more lipids under silicon deprivation than it does when N is scarce (Zendejas et al., 2012). Carbon is clearly a macronutrient, despite the fact that it is typically obtained from the atmosphere. However, different concentrations of CO_2 or different regimes have been studied with culture conditions in mind. CO_2 concentrations in *Chlorella kessleri* have a significant impact on lipid metabolism. Glycerol feeding in batch cultures of *Schizochytrium* sp. and changes in inorganic carbon regimes supplied to *C. reinhardtii* altered FA patterns (Li-Beisson et al., 2019; Lohman et al., 2014). The impact of environmental factors on carbon metabolism in diatoms has recently been discussed (Li-Beisson et al., 2019). A general commentary on growth and lipid accumulation in *C. reinhardtii* by nutrient depletion and supplementation was recently published (Yang et al., 2018). Micronutrients (e.g., Co, Cu, Fe, Mg, Mn, Mo, Zn) that are required in trace amounts may have a significant impact on algal growth because they alter normal enzyme activity (Khan et al., 2018). TAG accumulation in *Chlamydomonas* has been shown to be influenced by elements such as Fe and Zn (Urzica et al., 2013). There is a delicate line between nutrient deficit, adequacy, and harmfulness for many of these elements. Iron appears to be a main regulator of phytoplankton biomass. $FeCl_3$ added to fresh medium at 1.2×10^5 mol/L increased biomass and percent lipid composition in late exponential phase cultures of *C. vulgaris*, whereas lower concentrations resulted in lower lipid levels (Li-Beisson et al., 2019). LDs and TAG accumulated in Fe-starved *C. reinhardtii.* It was discovered that the saturation index had increased, implying that desaturation activity had been conceded. The FA profiles of DGDG and DGTS (but not MGDG) have been altered, and gene expression of enzymes or proteins involved in TAG accumulation was elevated (Urzica et al., 2013). Copper, like iron, is required for certain enzyme functions. However, toxic effects, as well as changes in lipid metabolism, associated with copper have been observed in a variety of algae (Hemschemeier et al., 2013). When chromium effects were investigated in *Euglena gracilis* acquired from a culture or obtained from a polluted river, PUFA levels were most significantly impacted. In treated cells, electron microscopy revealed thylakoid disorganization. Cadmium is another toxic metal that affects lipid metabolism. Vulnerability to heavy metal toxicity differs greatly among algae species, and *P. tricornutum* is comparatively resistant to cadmium, likely due to protective transport and detoxification processes (Li-Beisson et al., 2019).

6. Extraction of algal lipids

Microalgae have thick cell walls, and the stiffness of the microalgae matrix restricts the release of their bioactive chemicals. The production of microalgal biofuel and nutraceuticals is not energy-efficient due to the pretreatment stage that involves cell destruction and lipid extraction. Therefore, choosing the right pretreatment and extraction techniques to extract lipids from microalgae while preserving their bioactivity is crucial. In general, mechanical (homogenizer, sonication, microwave, pulse electric field) and nonmechanical or chemical procedures are used to disturb cells (acid, surfactant, enzymes) (George et al., 2015). Additionally, there is a disparity between mechanical and chemical approaches because chemical methods may be scaled up, whereas mechanical methods cannot (De Carvalho et al., 2020). Table 19.2 shows a variety of cell disruption techniques for microalgae lipid extraction. While the traditional techniques, such as Soxhlet extraction, Folch extraction, and Bligh—Dyer extraction, are typically easy to use and inexpensive, which encourages their use, they also have drawbacks such as high organic reagent consumption and environmental pollution, in addition to the prolonged extraction time. A difficult area of research is the extraction of lipids using appropriate environmentally acceptable solvents. Finding substitute solvents that may be utilized for lipid extraction without endangering the environment or human health is urgently needed. To attain great efficiency and superior outcomes, single-step integration technology that combines cell disruption and lipid extraction procedures can be applied (Lee et al., 2021). Because solvents are used to extract the lipids from the microalgae cells, recovering the same solvents from the extracted lipids takes additional energy, making the process of extracting lipids or oil from microalgae cells an expensive and energy-intensive operation. In addition, the method used should be fast and should not damage the extracted lipids (Iqbal et al., 2013).

Many cutting-edge techniques have been investigated recently for the extraction of lipids, including supercritical fluid extraction (SFE), ultrasound-assisted extraction (UAE), microwave-assisted extraction (MAE), and others. The supercritical fluid extraction method replaces harmful

Table 19.2 Various extraction methods of lipid from microalgae.

Extraction methods	Microalgae	Solvent and procedure	Remarks	References
Traditional Soxhlet extraction method	*Nannochloropsis oculata*	n-Hexane	5.79% lipid yield with n-hexane	(Liau et al., 2010)
		Ethanol	40.90% lipid yield with ethanol	
Ultrasound-assisted extraction	*Nannochloropsis oculata*	Petroleum ether, frequency 40 kHz, 1 h	3.3% lipid yield	(Converti et al., 2009)
	Scenedesmus sp.	Hexane, Ultrasonic power: 0 to 50 W, Time interval: 0 to 5 sec., Time: 0 ~ 10 min	Lipid yield increased from 0.76 g/L to 1.31 g/L.	(Sivaramakrishnan and Incharoensakdi, 2019)
Microwave-assisted extraction	*Chlorella PY-ZU1*	Chloroform:methanol Temperature: 80−120°C, Time: 5−10 min	8.45% lipid yield	(Cheng et al., 2013)
Supercritical carbon dioxide (SC−CO_2)	*Schizochytrium limacinum*	35 MPa, 40°C, 95% v/v ethanol as a cosolvent	33.9% lipid yield with high DHA purity	(Tang et al., 2011)
	Nannochloropsis oculata, Chlorella vulgaris	Ethanol (10 %, used as cosolvent), Temperature: 50°C Pressure: 250/450/750 bar, CO_2 flow: 25 g/min	97% of neutral lipids were extracted from *C. vulgaris,* and 83% neutral lipids from *N. oculata*	(Obeid et al., 2018)
Subcritical fluid extraction	*Tetradesmus Obliquus*	Dimethyl ether at 20°C temperature and 0.55 MPa pressure	Subcritical dimethyl ether could recover the extraction rate of lipid by altering the permeability of cell wall	(Wang et al., 2021)
Pressurized fluid extraction	*Nannochloropsis oculata*	n-hexane 60°C, 10−12 MPa, 10 min	6.1 (w/w) % lipid yield	(Pieber et al., 2012)
		n-hexane/propan-2-ol (2:1 vol.%),	20 (w/w) % lipid yield	
		Ethanol : 96 vol.%	36 (w/w) % lipid yield	
Ionic liquids extraction	*Spirulina platensis*	Reflux at 95°C for 2 h using pure caprolactam-based ionic liquids with methanol (as cosolvent)	Lipids' yield enhancement of 14%	(Naiyl et al., 2022)

Enzyme-assisted extraction	*Nannochloropsis* sp.	Multienzyme, pretreatment with cellulase (CEL) and mannanase (GMA) temperature ($T = 15-75°C$), pH (2–8), pretreatment time ($P = 30-270$ min), CEL dosage ($D_1 = 0-20$ mg/g), and GMA dosage ($D_2 = 0-2$ mg/g)	90% of lipids were recovered from *Nannochloropsis* cells pretreated under optimal process conditions ($T = 53°C$, pH = 4.4, $P = 210$ min, $D_1 = 13.8$ mg/g and $D_2 = 1.5$ mg/g)	(Zuorro et al., 2016)
Photoelectrochemical system	*Chlorella* sp.	Photoelectrochemical (PEC) with TiO_2-based photoanode utilized UV–Vi's light to produce •OH at a phosphate–palladium cathode by using hexane, isopropanol	The electrons are taken up by a palladium–phosphorus cathode, which produces hydroxyl radicals (•OH). To improve lipid extraction and purity, the •OH targets microalgal cell walls and membranes, leading to an oil extraction efficiency of 96%	(Wu et al., 2021)
Osmotic shock	*Chaetoceros muelleri*	Silica starvation	Lipid recovery efficiency of 72%	(Astals et al., 2019)
Wet extraction	*Chlorella* and *Scenedesmus* sp.	Hexane, 90°C, 0.1 MPa	Lipid yield 59.3%	(Sathish et al., 2012)

solvents with nontoxic CO_2 gas as the solvent. These cutting-edge methods partially address the drawbacks of traditional methods while also helping to increase the rate at which lipids are extracted and keep the extracts' bioactive properties (Iqbal and Chandra, 2013).

7. Algal lipidomics

Lipidomics allows to examine the quality and quantity of lipid bodies in cells. The term "lipidomics" refers to the study of the structure and function of a cell's or organism's entire lipidome, as well as their interactions with other lipids, metabolites, and proteins. To increase lipid productivity, researchers must understand the pathways and factors that affect lipid biosynthesis, which can now be discovered through lipidomic research (Anand et al., 2019). By systematic quantification of cell lipidome, cell protein interactions, and lipid metabolism kinetics, lipidomics has also provided new information on health and disease (Bisht et al., 2021). Lipid composition and exact knowledge are required for the production of a variety of products, including biofuels and PUFAs. Table 19.3 shows various lipidomic methods used in the past 10 years.

Table 19.3 Various lipidomic studies with different algae.

Sr No	Methods	Algae used	Remarks/Features	References
1	Chloroform:methanol (2:1)	*Chlorella vulgaris*	After ultrasonication, a mixture of chloroform–methanol (2:1) yielded 19% total lipids.	(dos Santos et al., 2015)
2	High-pressure liquid chromatography coupled with mass spectrometry (HPLC-MS)	*Chlorella* sp. (saltwater)	The HPLC-MS method has been used to analyze triacyl glyceride.	(Jones et al., 2012)
3	High-pressure liquid chromatography coupled with evaporative light scattering detection (HPLC-ELSD)	*Chlorella lobophora*	Extensively used technique to analyze neutral lipids from microalgal extracts.	(Sathoshkumar et al., 2015)
4	Gravimetric methods	Various microalgae	Lower concentration of lipids even from wet algae can be detected by using the gravimetric methods such as sulfophosphovanillin.	(Byreddy et al., 2016)
5	Fluorescent technique	*Auxenochlorella protothecoides*	Noninvasive technique used to visualize lipid bodies inside algal cells.	(Gusbeth et al., 2016)

Table 19.3 Various lipidomic studies with different algae.—cont'd

Sr No	Methods	Algae used	Remarks/Features	References
6	Raman spectroscopy	*Chlamydomonas reinhardtii*	A noninvasive technique that uses Raman spectroscopy to analyze cells in vivo. It allows to measure lipid bodies, unsaturation, chain length, Tm, and other properties.	(Sharma, 2015)
7	Ultrahigh-performance liquid chromatography (UHPLC)	*Cyclotella cryptica, Nannochloropsis salina, Thalassiosira weissflogii, Alexandrium minutum, Alexandrium tamutum*	Targeted analysis of the main lipid classes such as phospholipids, glycolipids, and triacylglycerol can be achieved by using UHPLC.	(Cutignano et al., 2016)
8	Hydrophilic interaction chromatography coupled with high-resolution mass spectrometry (HILIC)–LC-(HRMS)	*Fucus vesiculosus*	Identification of the lipidome at the molecular level has become possible due to (HILIC)–LC-(HRMS). A total of 187 molecular species distributed by phospholipids, glycolipids, and betaine lipids have been identified.	(Da Costa et al., 2019)
9	Gas chromatography–mass spectrometry (GC-MS) and liquid chromatography–mass spectrometry (LC-MS).	*Chlorella vulgaris*	48 lipids in the major clusters of *C. vulgaris* were identified and putatively annotated using mass spectrometry-based molecular networking.	(Pantami et al., 2020)
10	Nuclear magnetic resonance (NMR)	Various microalgae	NMR procedures have enormous potential for rapid screening of microalgae strains with the required metabolite and lipid profiles, which are more difficult to obtain using LC-MS and GC-MS.	(Bisht et al., 2021)

In a study, column chromatography was used to find out the lipid fractions (polar and nonpolar lipids) in the extracted lipid. Lipidomic analysis found that non-polar lipids were more abundant in *Chlorella vulgaris* (57%), *Scenedesmus* sp. (54%) whereas in *Synechococcus* sp. (69%) polar lipids were present (Shanmugam et al., 2020). Two closely related strains of microalga *Parietochloris* (*Trebouxiophyceae, Chlorophyta*) were distinguished based on the lipidomic analysis using both shotgun lipidomics and hydrophilic interaction liquid chromatography-tandem mass spectrometry with electrospray ionization. The most significant variations were seen in three types of phospholipids: phosphatidylethanolamine, phosphatidylinositol and phosphatidylcholine (Rezanka et al., 2017). High-resolution mass spectrometry (HR-MS) lipidomic studies are required to resolve lipid regulation and metabolic fractionation of lipids in microalgae. HR-MS is useful for providing information at the cellular level as well as studying the composition of complex extracts. In a study, researchers have used a liquid chromatography quadrupole time of flight mass spectrometry (LC-QToF MS) technique developed in their lab to examine changes in lipid composition in the microalga *Ettlia oleoabundans* (Matich et al., 2018). The polar lipidome of the haptophyte microalga *Chrysotila pseudoroscoffensis* was discovered for the first time by hydrophilic interaction liquid chromatography-high resolution mass spectrometry (HILIC-HRMS and MS/MS) and gas chromatography-mass spectrometry (GC-MS). The polar lipid species identified by HILIC-MS/MS included phospholipids, betaine lipids, and glycolipids, some of which have recognised bioactive properties (Moreira et al., 2022). To determine how stress conditions during cultivation affect lipid composition, the lipidomic pattern of algae is frequently investigated. In a series of studies, the effects of NaCl stress and stress caused by nitrate or phosphate marginalization on the alga *Chlamydomonas nivalis* were investigated (Lu et al., 2013). Aside from identifying common PLs, they used UHPLC/ESI-MS in positive and negative ion modes to detect lipid biomarkers such as DGTS, MGDG, DGDG, and SQDG. By using multivariate statistical analysis, lipids obtained under various culture conditions (control, nitrate-, and phosphate-deprived) were clearly detected (Lu et al., 2013).

The differences in the representation of polar lipids such as DGTS, MGDG, DGDG, and SQDG were identified by processing data statistically (for example, PCA), which were obtained by positive and negative ESI-MS. DGTS (20:4–20:5), MGDG (20:5–14:0), DGDG (20:5–16:1), and, for example, TAG (16:1–20:5–20:5) were among the molecular species identified. The lipidomic profile of lipids from geographical varieties of the snow alga *Chloromonas pichinchae* was explored by using silver-LC/APCI-MS and NARP-LC/APCI-MS (Řezanka et al., 2014). The analysis of TAG-containing PUFA was the main focus of the study. The extremophilic red alga *Galdieria sulphuraria* was studied using lipidomic analysis and because this alga grows finest at lower pH, the lipidomic profile of cells grown at pH 1 to 4 was investigated. Depending on the pH of the culture medium, the alga uses a variety of mechanisms to adapt to low pH, such as substituting BLs for PLs or changing the regioisomers ratio of polar lipids, such as PLs, glycolipids, and BLs (Vítová et al., 2016). The alga *Chlamydomonas reinhardtii* was used to study polar lipids using a nanoelectrospray direct infusion MS method. Positive and negative mode electrospray ions were used to identify polar lipids found in alga (DGTS, MGDG, DGDG, and SQDG), as well as their molecular species (DGTS 16:0/18:4, SQDG 16:0/18:3, PtdGro (Phosphatidylglycerol) 16:1/18:3, and so on) (Yang et al., 2015). TAG profiles in four oleaginous saltwater microalgae were investigated using MALDI-TOF/MS, ESI linear ion trap–orbitrap MS, and 1H NMR spectroscopy: *Phaeodactylum tricornutum, Nannochloropsis salina, N. oculi*, and *Tetraselmis suecica* (Danielewicz et al., 2011). As a result, the method provides a quick and accurate analysis that can be used in the production of biofuels. The relative abundance of sole classes of lipids and their molecular species in the lipidomic profile of algal crude oils from *N. salina* evaluated without LC but by using ESI (electrospray ionization), APCI (atmospheric pressure chemical ionization), APPI (atmospheric pressure photoionization), and MALDI (matrix-assisted laser desorption/ionization) disclosed large variations in different classes of lipids and their molecular species (Lee et al., 2013). More than 100 molecular species of TAG were identified using UHPLC-HR-MS in six algae from the genera *Phaeodactylum, Botryococcus, Porphyridium, Neochloris, Scenedesmus,* and *Nannochloropsis* (MacDougall et al., 2011).

The lipidome of *Chlorococcum amblystomatis* was identified using HILIC−HR−ESI−MS/MS (hydrophilic interaction liquid chromatography high resolution electrospray ionization- tandem mass spectrometry). The lipid content of the *Chlorococcum amblystomatis* strain was 20.77%, and the fatty acid profile evaluated by gas chromatography-mass spectrometry revealed that this microalga includes a large amount of omega-3 polyunsaturated fatty acids (PUFAs). The reported lipidome included 245 molecular ions and 350 lipid species from 15 different classes of glycolipids, phospholipids, and betaine lipids (Conde et al., 2021). The cytotoxicity of three herbicides (atrazine, clomazone, and norflurazon) on *Chlamydomonas reinhardtii* was examined by assessing lipid component variations following herbicide exposure. Lipidomic analysis demonstrates that atrazine-exposed cells have significantly altered lipid content at $>EC_{50}$. The microchip-based matrix-assisted laser desorption ionisation mass spectrometry (MALDI-MS) approach demonstrates lipidomic profiling of aquatic species subjected to various stressors, showing usefulness for high-throughput screening and single-cell analysis in ecotoxicity research (Shanta et al., 2021).

Low amounts of lipids are present in macroalgae (1−8% of dry weight of lipids) (Ibañez and Cifuentes, 2013). However, the majority of them contain PUFAs, such as docosahexaenoic acid, eicosapentaenoic acid, and arachidonic acid, all of which have high nutritional value for humans and animals (Burri et al., 2012; Ginzberg et al., 2000; Küllenberg et al., 2012; Rossmeisl et al., 2012). Also, it is well known that a high ratio of n-6/n-3 essential fatty acids, such as 15/1, as found in today's Western diets, promotes the pathogenesis of many diseases (Simopoulos, 2002) and that this ratio is low (below 7.0) in seaweeds (Pereira et al., 2012). Glycolipids are the most abundant lipids in macroalgae, but there are also neutral lipids, PLs, BLs (Klug & Benning, 2001), and soluble pigments (Fichtali & Senanayake, 2010; Holdt et al., 2011). The fatty acid composition of several algae species has been extensively revealed over the past decade (Kumari et al., 2013; Sanina et al., 2004; Tasende et al., 2000), but the polar lipid components are still unknown (Mendes et al., 2013; He et al., 2011). Understanding the lipid profile of algae is essential for developing high-quality, sustainable commercial goods from marine sources as well as for manipulating lipid biomass and characteristics (Stengel et al., 2011; Ibanez et al., 2013). To speed up the characteristics of lipid profile and support the strategic choices of algal producers, effective methods for molecular analyses of algae must be created. However the application of lipidomic techniques like liquid chromatography−mass spectrometry (LC-MS) have been successfully applied in recent years to research the lipidome of a variety of biological materials, but their use in seaweed research is still in its early stages (He et al., 2011; Herrero et al., 2007). For the first time, a lipidomic approach using hydrophilic interaction liquid chromatography−electrospray ionization−mass spectrometry (HILIC−ESI−MS) was used to characterize red seaweed *C. crispus,* specifically its FA and polar lipid components (Melo et al., 2015). There are fewer lipidomic bottlenecks caused by analytical methods due to the rapid development of analytical techniques. In biological sample, almost all lipid species can be profiled and quantified using analytical methods that are optimized. However, sensitivity is still an issue that must be considered when performing low-abundance lipid profiling in small samples. There are two aspects to the future goals in lipidomics. One goal is to improve the accuracy of lipid compound profiling in cellular or subcellular structures. The other goal is to elucidate the possible mechanisms of lipid fluctuations at the gene and protein levels by exploring entire lipid metabolism pathways.

8. Algae as biosurfactant

Biosurfactants are a diverse group of lipids with amphiphilic properties, meaning they have both hydrophilic and hydrophobic domains. A hydrocarbon chain is frequently found in the nonpolar (hydrophobic) region. Biosurfactants have a wide range of biological and physicochemical properties due to their structural diversity. The hydrophilic polar portion can be amphoteric, nonionic, or ionic, which allows them to have various advantageous characteristics, such as low surface tension, antimicrobial activity, and efficient hydrocarbon emulsion (Walter et al., 2010). Biosurfactants have important uses in various sectors of industries, including hygiene, bioremediation, and healthcare due to these desirable characteristics. Surfactants are currently used in a variety of products in the cosmetics and health care industries, including insect repellants, lens solutions, antacids, acne pads, deodorants, toothpaste, baby products, and many others. In the last two decades, research on biosurfactants has exploded, with an exponential increase in the last ten years (Marcelino et al., 2020). As these compounds are secreted extracellularly, biosurfactants are commonly found on the surface of the organisms that produce them, and they act as a layer of defense for microorganisms in hostile environments. Glycolipids, FAs, PLs, polymers, and lipopeptides are the five types of biosurfactants. The commercialization of biosurfactants is currently the focus of research; however, it faces the challenge of low productivity and high economic costs. Different microorganisms produce a wide range of biosurfactants (Markande et al., 2021). Microbial aggregates and assemblages known as biofilms result from microbial natural preference for interfaces and their ability to produce amphipathic surface-active agents (or biosurfactants) for dominating these niches. These amphipathic biomolecules form a "conditioning film" on the surfaces, altering their original properties and impacting microbe–interface interactions. Low-molecular-weight biosurfactants and high-molecular-weight bioemulsifiers have traditionally been classified as biosurfactants (Ron & Rosenberg, 1997). Microbial biosurfactants and bioemulsifiers are good substitutes for their chemical counterparts (Sałek & Euston, 2019). Microbial biosurfactants have been found in a variety of environments, including terrestrial, freshwater, and marine (Markande et al., 2021). Although genetic engineering can increase biosurfactant production, consumers do not favor the use of genetically modified organisms (GMOs) (Sałek & Euston, 2019). Biosurfactants have already been used in industry as commercially available or green surfactants synthesized using microbes and chemical transformation/synthesis processes and applied individually or in combination with biosurfactants (Schultz & Soares Rosado, 2020). Bioinformatics and biosynthesis/chemical synthesis tools, are needed to explore new spheres for biosurfactant research (Schultz & Soares Rosado, 2020). Chemical and biological engineering is a rapidly developing field that can help expand the repertoire of usable biosurfactants while also increasing production (Sałek & Euston, 2019).

One study investigated the use of cyanobacteria *Arthrospira* sp. and *Synechococcus nidulans* and of chlorophytes *Chlorella minutissima*, *Chlorella vulgaris* and *Chlorella homosphaera* for the production of biosurfactants using autotrophic and mixotrophic cultivation (Elisngela et al., 2015). A study found that ruptured microalgae contain a variety of surface-active chemicals with varying properties. This study's fractionation approach is a first step toward understanding the emulsifying capabilities of these surface-active substances. It may be interesting in the future to isolate particular surface-active molecules for in-depth study. In this perspective, microalgae may be viewed as a source of biosurfactants and bioemulsifiers, which may have greater commercial potential than biofuel production. Food, cosmetic, and pharmaceutical industries are increasingly replacing synthetic surfactants with naturally derived alternatives. Microalgae can be grown on a large scale as a source of surface-active fractions with unique features (Law et al., 2018). Table 19.4 gives various applications of algal lipids as a biosurfactant.

Table 19.4 Various applications of algal lipid as a biosurfactant.

Sr No	Type of algal lipid	Algal species	Properties/ application as biosurfactant	References
1	Glycolipid	*Chlorella vulgaris, Chlorella pyrenoidosa, Chlorella sorokiniana*	Cosmetics, medicinal	(Kosaric et al., 1992)
2	Total lipids (polar: phospholipids, glycolipids) (Nonpolar: TAG)	*Nannochloropsis salina*	Various commercial applications	(Law et al., 2018)
3	Lipid and nonlipid Biomass	*Chlorella* sp.	For the production of biosurfactant from *Bacillus subtilis* C9 *Chlorella* biomass was used as the sole fermentation substrate	(Yun et al., 2020)
4	Total algal lipid	*Chlorella* sp.	Anticancer Antimicrobial Antioxidant Antiinflammatory	(Rahman et al., 2019)
5	Neutral and polar algal lipid	Different microalgae species	Microbial-based approaches Biotherapeutics Wetting and foaming agents Functional food ingredients Detergent in petroleum Environmental management Cosmetics pharmaceuticals	(Garay et al., 2014)
6	Algal total lipid content	*Porphyridium cruentum, Spirulina platensis,* *Dunaliella bardawil, Chlorella minutissima*	Commercial applications Pharmaceutical industry Bioremediation	(Marcelino et al., 2020)

9. Conclusion

Microalgae are an excellent source of lipids with bioactive activities that include antiviral, anti-inflammatory, and antitumor effects. The structural elucidation of these lipids is essential for comprehending the links between their structure and biological activity since they offer the considerable biotechnological potential for food and health applications. The data provided in this chapter highlights the various major algal lipids present, extraction procedure, algal lipidomics, combined techniques used to extract high-accuracy lipidomic results and various applications of algal lipids including biosurfactant. It has been suggested that mass spectrometry-based methods are effective in identifying and quantifying algal lipids in routine analysis. Algal lipids and their fatty acid concentration and composition are closely correlated with the condition of chloroplast membranes, environmental stress factors, and nutritional availability. In the framework of the bioeconomy, high-value lipids from algae create high hopes for a new wave of goods with improved nutritional and functional attributes. Future challenges include determining the whole molecular profile of algal lipids in order to have a better knowledge of the metabolic routes of synthesis and conducting a comprehensive screening of the bioactive metabolites generated by microalgae.

References

Abd El Baky, H. H., El-Baroty, G. S., Bouaid, A., Martinez, M., & Aracil, J. (2012). Enhancement of lipid accumulation in Scenedesmus obliquus by optimizing CO2 and Fe3+ levels for biodiesel production. *Bioresource Technology, 119*, 429–432.

Abida, H., Dolch, L. J., Meï, C., Villanova, V., Conte, M., Block, M. A., Finazzi, G., Bastien, O., Tirichine, L., Bowler, C., & Rébeillé, F. (2015). Membrane glycerolipid remodeling triggered by nitrogen and phosphorus starvation in Phaeodactylum tricornutum. *Plant Physiology, 167*(1), 118–136.

Ahmad, I., Sharma, A. K., Daniell, H., & Kumar, S. (2015). Altered lipid composition and enhanced lipid production in green microalga by introduction of brassica diacylglycerol acyltransferase 2. *Plant Biotechnology Journal, 13*(4), 540–550.

Anand, V., Kashyap, M., Samadhiya, K., & Kiran, B. (2019). Strategies to unlock lipid production improvement in algae. *International Journal of Environmental Science and Technology, 16*(3), 1829–1838.

Anesi, A., Ulrike, O., Gert, H., Sukenik, A., Flaim, G., & Guella, G. (2016). Comparative analysis of membrane lipids in psychrophilic and mesophilic freshwater dinoflagellates. *Frontiers of Plant Science, 7*, 524.

Astals, S., Pratt, S., Jensen, P. D., & Schenk, P. M. (2019). Impact of osmotic shock pre-treatment on microalgae lipid extraction and subsequent methane production. *Bioresource Technology Reports, 7*, 100214.

Atta, M., Ani, I., Ataullah, B., & Wahidin, S. (2013). Intensity of blue LED light: A potential stimulus for biomass and lipid content in fresh water microalgae chlorella vulgaris. *Bioresource Technology, 148*, 373–378.

Aureli, M., Grassi, S., Prioni, S., & Prinetti, A. (2015). Lipid membrane domains in the brain. *Biochimica et Biophysica Acta (BBA) - Molecular and Cell Biology of Lipids, 1851*(8), 1006–1016.

Aziz, M. M. A., Kassim, K. A., Shokravi, Z., Jakarni, F. M., Liu, H. Y., Zaini, N., Tan, L. S., Islam, A. S., & Shokravi, H. (2020). Two-stage cultivation strategy for simultaneous increases in growth rate and lipid content of microalgae: A review. *Renewable and Sustainable Energy Reviews, 119*, 109621.

Behera, B., Unpaprom, Y., Ramaraj, R., Maniam, G. P., Govindan, N., & Paramasivan, B. (2021). Integrated biomolecular and bioprocess engineering strategies for enhancing the lipid yield from microalgae. *Renewable and Sustainable Energy Reviews, 148*, 111270.

Bisht, B., Kumar, V., Gururani, P., Tomar, M. S., Nanda, M., Vlaskin, M. S., Kumar, S., & Kurbatova, A. (2021). The potential of nuclear magnetic resonance (NMR) in metabolomics and lipidomics of microalgae-a review. *Archives of Biochemistry and Biophysics, 710*, 108987.

Bleakley, S., & Hayes, M. (2017). Algal proteins: Extraction, application, and challenges concerning production. *Foods, 6*(5), 33.

Boyle, N. R., Page, M. D., Liu, B., Blaby, I. K., Casero, D., Kropat, J., Cokus, S. J., Hong-Hermesdorf, A. A., Shaw, J., Karpowicz, S. J., & Gallaher, S. D. (2012). Three acyltransferases and nitrogen-responsive regulator are implicated in nitrogen starvation-induced triacylglycerol accumulation in Chlamydomonas. *Journal of Biological Chemistry, 287*(19), 15811–15825.

Burri, L., Hoem, N., Banni, S., & Berge, K. (2012). Marine omega-3 phospholipids: Metabolism and biological activities. *International Journal of Molecular Sciences, 13*(11), 15401–15419.

Byreddy, A. R., Gupta, A., Barrow, C. J., & Puri, M. (2016). A quick colorimetric method for total lipid quantification in microalgae. *Journal of Microbiological Methods, 125*, 28–32.

Che, R., Ding, K., Huang, L., Zhao, P., Xu, J. W., Li, T., Ma, H., & Yu, X. (2016). Enhancing biomass and oil accumulation of Monoraphidium sp. FXY-10 by combined fulvic acid and two-step cultivation. *Journal of the Taiwan Institute of Chemical Engineers, 67*, 161–165.

Cheah, W. Y., Show, P. L., Juan, J. C., Chang, J. S., & Ling, T. C. (2018). Enhancing biomass and lipid productions of microalgae in palm oil mill effluent using carbon and nutrient supplementation. *Energy Conversion and Management, 164*, 188–197.

Chen, H., & Wang, Q. (2021). Regulatory mechanisms of lipid biosynthesis in microalgae. *Biological Reviews, 96(5)*, 2373–2391.

Cheng, J., Sun, J., Huang, Y., Feng, J., Zhou, J., & Cen, K. (2013). Dynamic microstructures and fractal characterization of cell wall disruption for microwave irradiation-assisted lipid extraction from wet microalgae. *Bioresource Technology, 150*, 67–72.

Chu, F.-F., Chu, P. N., Cai, P. J., Li, W. W., Lam, P. K., & Zeng, R. J. (2013). Phosphorus plays an important role in enhancing biodiesel productivity of Chlorella vulgaris under nitrogen deficiency. *Bioresource Technology, 134*, 341–346.

Cockcroft, S. (2021). Mammalian lipids: Structure, synthesis and function. *Essays in Biochemistry, 65*(5), 813–845.

Conde, T. A., Couto, D., Melo, T., Costa, M., Silva, J., Domingues, M. R., & Domingues, P. (2021). Polar lipidomic profile shows Chlorococcum amblystomatis as a promising source of value-added lipids. *Scientific Reports, 11*(1), 1–23.

Converti, A., Casazza, A. A., Ortiz, E. Y., Perego, P., & Del Borghi, M. (2009). Effect of temperature and nitrogen concentration on the growth and lipid content of Nannochloropsis oculata and Chlorella vulgaris for biodiesel production. *Chemical Engineering and Processing: Process Intensification, 48*(6), 1146–1151.

Cutignano, A., Luongo, E., Nuzzo, G., Pagano, D., Manzo, E., Sardo, A., & Fontana, A. (2016). Profiling of complex lipids in marine microalgae by UHPLC/tandem mass spectrometry. *Algal Research, 17*, 348–358.

Da Costa, E., Domingues, P., Melo, T., Coelho, E., Pereira, R., Calado, R., Abrew, M. H., & Domingues, M. R. (2019). Lipidomic signatures reveal seasonal shifts on the relative abundance of high-valued lipids from the brown algae Fucus vesiculosus. *Marine Drugs, 17*(6), 335.

Dahlqvist, A., Ståhl, U., Lenman, M., Banas, A., Lee, M., Sandager, L., Ronne, H., & Stymne, S. (2000). Phospholipid: Diacylglycerol acyltransferase: An enzyme that catalyzes the acyl-CoA-independent formation of triacylglycerol in yeast and plants. *Proceedings of the National Academy of Sciences, 97*(12), 6487–6492.

Daliry, S., Hallajisani, A., Mohammadi Roshandeh, J., Nouri, H., & Golzary, A. (2017). Investigation of optimal condition for Chlorella vulgaris microalgae growth. *Global Journal of Environmental Science and Management*, 217–230.

Danielewicz, M. A., Anderson, L. A., & Franz, A. K. (2011). Triacylglycerol profiling of marine microalgae by mass spectrometry [S]. *Journal of Lipid Research, 52*(11), 2101–2108.

De Carvalho, J. C., Magalhaes Jr, A. I., de Melo Pereira, G. V., Medeiros, A. B. P., Sydney, E. B., Rodrigues, C., Aulestia, D. T. M., de Souza Vandenberghe, L. P., Soccol, V. T., & Soccol, C. R. (2020). Microalgal biomass pretreatment for integrated processing into biofuels, food, and feed. *Bioresource Technology, 300*, 122719.

De Luca, M., Pappalardo, I., Limongi, A. R., Viviano, E., Radice, R. P., Todisco, S, Martelli, G., Infantino, V., & Vassallo, A. (2021). Lipids from microalgae for cosmetic applications. *Cosmetics, 8*(2), 52.

Devi, M. P., & Mohan, S. V. (2012). CO_2 supplementation to domestic wastewater enhances microalgae lipid accumulation under mixotrophic microenvironment: Effect of sparging period and interval. *Bioresource Technology, 112*, 116–123.

Do Nascimento, M., de los Angeles Dublan, M., Ortiz-Marquez, J. C. F., & Curatti, L. (2013). High lipid productivity of an Ankistrodesmus–Rhizobium artificial consortium. *Bioresource Technology, 146*, 400–407.

dos Santos, K., Moreira, D. M., Kunigami, C. N., Aranda, D. A. G., & Teixeira, C. M. L. L. (2015). Comparison between several methods of total lipid extraction from Chlorella vulgaris biomass. *Ultrasonics Sonochemistry, 22*, 95–99.

El Arroussi, H., Benhima, R., Bennis, I., El Mernissi, N., & Wahby, I. (2015). Improvement of the potential of Dunaliella tertiolecta as a source of biodiesel by auxin treatment coupled to salt stress. *Renewable Energy, 77*, 15–19.

Elisngela, M. R, Etiele, G. D. M, Cibele, F. D. O, Kellen, Z., & Jorge, A. V. C. (2015). Microalgae cultivation for biosurfactant production. *African Journal of Microbiology Research, 9(47)*, 2283–2289.

Farooq, W., Lee, Y. C., Ryu, B. G., Kim, B. H., Kim, H. S., Choi, Y. E., & Yang, J. W. (2013). Two-stage cultivation of two Chlorella sp. strains by simultaneous treatment of brewery wastewater and maximizing lipid productivity. *Bioresource Technology, 132*, 230–238.

Fichtali, J., & Senanayake, S. P. J. N. (2010). Development and commercialization of microalgae-based functional lipids. *Functional Food Product Development*, 206–225.

Ganesan, M., Trivedi, N., Gupta, V., Madhav, S. V., Reddy, C. R., & Levine, I. A. (2019). Seaweed resources in India–current status of diversity and cultivation: Prospects and challenges. *Botanica Marina, 62*(5), 463–482.

Garab, G., Ughy, B., & Goss, R. (2016). Role of MGDG and non-bilayer lipid phases in the structure and dynamics of chloroplast thylakoid membranes. *Lipids in Plant and Algae Development*, 127–157.

Garay, L. A., Boundy-Mills, K. L., & Bruce German, J. (2014). Accumulation of high-value lipids in single-cell microorganisms: A mechanistic approach and future perspectives. *Journal of Agricultural and Food Chemistry, 62*(13), 2709–2727.

George, A., Brandt, A., Tran, K., Zahari, S. M. N. S., Klein-Marcuschamer, D., Sun, N., Sathitsuksanoh, N., Shi, J., Stavila, V., Parthasarathi, R., & Singh, S. (2015). Design of low-cost ionic liquids for lignocellulosic biomass pretreatment. *Green Chemistry, 17*(3), 1728–1734.

Ginzberg, A., Cohen, M., Sod-Moriah, U. A., Shany, S., Rosenshtrauch, A., & Arad, S. M. (2000). Chickens fed with biomass of the red microalga Porphyridium sp. have reduced blood cholesterol level and modified fatty acid composition in egg yolk. *Journal of Applied Phycology, 12*(3), 325–330.

Goncalves, E. C., Wilkie, A. C., Kirst, M., & Rathinasabapathi, B. (2016). Metabolic regulation of triacylglycerol accumulation in the green algae: Identification of potential targets for engineering to improve oil yield. *Plant Biotechnology Journal, 14*(8), 1649–1660.

Goss, R., Nerlich, J., Lepetit, B., Schaller, S., Vieler, A., & Wilhelm, C. (2009). The lipid dependence of diadinoxanthin de-epoxidation presents new evidence for a macrodomain organization of the diatom thylakoid membrane. *Journal of Plant Physiology, 166*(17), 1839–1854.

Gunstone, F. D., Giovanni, M. T., Wing-Keong, N., & Douglas, R. D. (2010). The world's oils and fats. In *Fish oil replacement and alternative lipid sources in aquaculture feeds* (pp. 61–98). Boca Raton, FL, USA: CRC Press, Taylor and Francis.

Gusbeth, C. A., Eing, C., Göttel, M., Sträßner, R., & Frey, W. (2016). Fluorescence diagnostics for lipid status monitoring of microalgae during cultivation. *International Journal of Renewable Energy & Biofuels, 899698*, 2016.

Guschina, I. A., & Harwood, J. L. (2009). Algal lipids and effect of the environment on their biochemistry. *Lipids in Aquatic Ecosystems*, 1–24. Springer, New York, NY.

Guschina, I. A., & Harwood, J. L. (2013). Algal lipids and their metabolism. *Algae for Biofuels and Energy*, 17–36. Springer, Dordrecht.

Han, D., Jia, J., Li, J., Sommerfeld, M., Xu, J., & Hu, Q. (2017). Metabolic remodeling of membrane glycerolipids in the microalga Nannochloropsis oceanica under nitrogen deprivation. *Frontiers in Marine Science, 4*, 242.

Harayama, T., & Howard, R. (2018). Understanding the diversity of membrane lipid composition. *Nature Reviews Molecular Cell Biology, 19*(5), 281–296.

He, H., Rodgers, R. P., Marshall, A. G., & Hsu, C. S. (2011). Algae polar lipids characterized by online liquid chromatography coupled with hybrid linear quadrupole ion trap/fourier transform ion cyclotron resonance mass spectrometry. *Energy & Fuels, 25*(10), 4770–4775.

Hemschemeier, A., Casero, D., Liu, B., Benning, C., Pellegrini, M., Happe, T., & Merchant, S. S. (2013). Copper response regulator1–dependent and–independent responses of the Chlamydomonas reinhardtii transcriptome to dark anoxia. *The Plant Cell Online, 25*(9), 3186–3211.

Hena, S., Fatihah, N., Tabassum, S., & Ismail, N. (2015). Three stage cultivation process of facultative strain of Chlorella sorokiniana for treating dairy farm effluent and lipid enhancement. *Water Research, 80*, 346–356.

Herrero, M., Vicente, M. J., Cifuentes, A., & Ibáñez, E. (2007). Characterization by high-performance liquid chromatography/electrospray ionization quadrupole time-of-flight mass spectrometry of the lipid fraction of Spirulina platensis pressurized ethanol extract. *Rapid Communications in Mass Spectrometry: An International Journal Devoted to the Rapid Dissemination of Up-to-the-Minute Research in Mass Spectrometry, 21*(11), 1729–1738.

Holdt, S. L., & Kraan, S. (2011). Bioactive compounds in seaweed: Functional food applications and legislation. *Journal of Applied Phycology, 23*(3), 543–597.

Hsieh, H.-J., Su, C.-H., & Chien, L.-J. (2012). Accumulation of lipid production in Chlorella minutissima by triacylglycerol biosynthesis-related genes cloned from *Saccharomyces cerevisiae* and Yarrowia lipolytica. *Journal of Microbiology, 50*(3), 526–534.

Huang, L., Xu, J., Li, T., Wang, L., Deng, T., & Yu, X. (2014). Effects of additional Mg2+ on the growth, lipid production, and fatty acid composition of Monoraphidium sp. FXY-10 under different culture conditions. *Annals of Microbiology, 64*(3), 1247–1256.

Huerlimann, R., De Nys, R., & Heimann, K. (2010). Growth, lipid content, productivity, and fatty acid composition of tropical microalgae for scale-up production. *Biotechnology and Bioengineering, 107*(2), 245–257.

Ibañez, E., & Cifuentes, A. (2013). Benefits of using algae as natural sources of functional ingredients. *Journal of the Science of Food and Agriculture, 93*(4), 703–709.

Iqbal, J., & Chandra, T. (2013). Microwave assisted lipid extraction from microalgae using biodiesel as co-solvent. *Algal Research, 2*(1), 34–42.

James, G. O., Hocart, C. H., Hillier, W., Price, J. D., & Djordjevic, M. A. (2013). Temperature modulation of fatty acid profiles for biofuel production in nitrogen deprived Chlamydomonas reinhardtii. *Bioresource Technology, 127*, 441–447.

Jiang, Y., Nunez, M., Laverty, K. S., & Quigg, A. (2015). Coupled effect of silicate and nickel on the growth and lipid production in the diatom Nitzschia perspicua. *Journal of Applied Phycology, 27*(3), 1137–1148.

Jones, J., Manning, S., Montoya, M., Keller, K., & Poenie, M. (2012). Extraction of algal lipids and their analysis by HPLC and mass spectrometry. *Journal of the American Oil Chemists' Society, 89*(8), 1371–1381.

Kang, N. K., Baek, K., Koh, H. G., Atkinson, C. A., Ort, D. R., & Jin, Y. S. (2021). Microalgal metabolic engineering strategies for the production of fuels and chemicals. *Bioresource Technology*, 126529.

Kazir, M., Abuhassira, Y., Robin, A., Nahor, O., Luo, J., Israel, A., Golberg, A., & Livney, Y. D. (2019). Extraction of proteins from two marine macroalgae, Ulva sp. and Gracilaria sp., for food application, and evaluating digestibility, amino acid composition and antioxidant properties of the protein concentrates. *Food Hydrocolloids, 87*, 194–203.

Khan, M. I., Hyuk Shin, J., & Deog Kim, J. (2018). The promising future of microalgae: Current status, challenges, and optimization of a sustainable and renewable industry for biofuels, feed, and other products. *Microbial Cell Factories, 17*(1), 1–21.

Klug, R. M., & Benning, C. (2001). Two enzymes of diacylglyceryl-O-4′-(N, N, N,-trimethyl) homoserine biosynthesis are encoded by btaA and btaB in the purple bacterium Rhodobacter sphaeroides. *Proceedings of the National Academy of Sciences, 98*(10), 5910–5915.

Kosaric, N. (1992). Biosurfactants in industry. *Pure and Applied Chemistry, 64*(11), 1731–1737.

Koyande, A. K., Chew, K. W., Rambabu, K., Tao, Y., Chu, D. T., & Show, P. L. (2019). Microalgae: A potential alternative to health supplementation for humans. *Food Science and Human Wellness, 8*(1), 16–24.

Krzemińska, I., Pawlik-Skowrońska, B., Trzcińska, M., & Tys, J. (2014). Influence of photoperiods on the growth rate and biomass productivity of green microalgae. *Bioprocess and Biosystems Engineering, 37*(4), 735–741.

Küllenberg, D., Taylor, L. A., Schneider, M., & Massing, U. (2012). Health effects of dietary phospholipids. *Lipids in Health and Disease, 11*(1), 1–16.

Kumari, P., Bijo, A. J., Mantri, V. A., Reddy, C. R. K., & Jha, B. (2013). Fatty acid profiling of tropical marine macroalgae: An analysis from chemotaxonomic and nutritional perspectives. *Phytochemistry, 86*, 44–56.

Law, S. Q. K., Mettu, S., Ashokkumar, M., Scales, P. J., & Martin, G. J. (2018). Emulsifying properties of ruptured microalgae cells: Barriers to lipid extraction or promising biosurfactants? *Colloids and Surfaces B: Biointerfaces, 170*, 438–446.

Li, T., Wan, L., & Zhang, C. (2013). Responses in growth, lipid accumulation, and fatty acid composition of four oleaginous microalgae to different nitrogen sources and concentrations. *Chinese Journal of Oceanology and Limnology, 31*(6), 1306–1314.

Li, J., Han, D., Wang, D., Ning, K., Jia, J., Wei, L., Jing, X., Huang, S., Chen, J., Li, Y., & Hu, Q. (2014). Choreography of transcriptomes and lipidomes of Nannochloropsis reveals the mechanisms of oil synthesis in microalgae. *The Plant Cell Online, 26*(4), 1645–1665.

Lee, S. Y., Khoiroh, I., Vo, D. V. N., Senthil Kumar, P., & Show, P. L. (2021). Techniques of lipid extraction from microalgae for biofuel production: A review. *Environmental Chemistry Letters, 19*(1), 231–251.

Lee, Y. J., Leverence, R. C., Smith, E. A., Valenstein, J. S., Kandel, K., & Trewyn, B. G. (2013). High-throughput analysis of algal crude oils using high resolution mass spectrometry. *Lipids, 48*(3), 297–305.

Lepetit, B., Goss, R., Jakob, T., & Wilhelm, C. (2012). Molecular dynamics of the diatom thylakoid membrane under different light conditions. *Photosynthesis Research, 111*(1), 245–257.

Liang, K., Zhang, Q., Gu, M., & Cong, W. (2013). Effect of phosphorus on lipid accumulation in freshwater microalga Chlorella sp. *Journal of Applied Phycology, 25*(1), 311–318.

Liang, Y., Maeda, Y., Yoshino, T., Matsumoto, M., & Tanaka, T. (2014). Profiling of polar lipids in marine oleaginous diatom Fistulifera solaris JPCC DA0580: Prediction of the potential mechanism for eicosapentaenoic acid-incorporation into triacylglycerol. *Marine Drugs, 12*(6), 3218–3230.

Liau, B.-C., Shen, C. T., Liang, F. P., Hong, S. E., Hsu, S. L., Jong, T. T., & Chang, C. M. J. (2010). Supercritical fluids extraction and anti-solvent purification of carotenoids from microalgae and associated bioactivity. *The Journal of Supercritical Fluids, 55*(1), 169–175.

Li-Beisson, Y., Thelen, J. J., Fedosejevs, E., & Harwood, J. L. (2019). The lipid biochemistry of eukaryotic algae. *Progress in Lipid Research, 74*, 31–68.

Liu, B., & Benning, C. (2013). Lipid metabolism in microalgae distinguishes itself. *Current Opinion in Biotechnology, 24*(2), 300–309.

Liu, Z.-Y., Wang, G.-C., & Zhou, B.-C. (2008). Effect of iron on growth and lipid accumulation in Chlorella vulgaris. *Bioresource Technology, 99*(11), 4717–4722.

Lohman, E. J., Gardner, R. D., Halverson, L. D., Peyton, B. M., & Gerlach, R. (2014). Carbon partitioning in lipids synthesized by Chlamydomonas reinhardtii when cultured under three unique inorganic carbon regimes. *Algal Research, 5*, 171–180.

Lu, N., Wei, D., Chen, F., & Yang, S. T. (2013). Lipidomic profiling reveals lipid regulation in the snow alga Chlamydomonas nivalis in response to nitrate or phosphate deprivation. *Process Biochemistry, 48*(4), 605–613.

MacDougall, K. M., McNichol, J., McGinn, P. J., O'Leary, S. J., & Melanson, J. E. (2011). Triacylglycerol profiling of microalgae strains for biofuel feedstock by liquid chromatography–high-resolution mass spectrometry. *Analytical and Bioanalytical Chemistry, 401*(8), 2609–2616.

Makewicz, A., Gribi, C., & Eichenberger, W. (1997). Lipids of Ectocarpus fasciculatus (phaeophyceae). Incorporation of [l-14C] oleate and the role of TAG and MGDG in lipid metabolism. *Plant and Cell Physiology, 38*(8), 952–962.

Marcelino, P. R. F., Gonçalves, F., Jimenez, I. M., Carneiro, B. C., Santos, B. B., & Silva, S. S. (2020). *Sustainable production of biosurfactants and their applications*. Lignocellulosic biorefining technologies.

Markande, A. R., Patel, D., & Varjani, S. (2021). A review on biosurfactants: Properties, applications and current developments. *Bioresource Technology, 330*, 124963.

Matich, E. K., Ghafari, M., Camgoz, E., Caliskan, E., Pfeifer, B. A., Haznedaroglu, B. Z., & Atilla-Gokcumen, G. E. (2018). Time-series lipidomic analysis of the oleaginous green microalga species Ettlia oleoabundans under nutrient stress. *Biotechnology for Biofuels, 11*(1), 1–15.

Melo, T., Alves, E., Azevedo, V., Martins, A. S., Neves, B., Domingues, P., Calado, R., Abreu, M. H., & Domingues, M. R. (2015). Lipidomics as a new approach for the bioprospecting of marine macroalgae—Unraveling the polar lipid and fatty acid composition of Chondrus crispus. *Algal Research, 8*, 181–191.

Mendes, M., Pereira, R., Pinto, I. S., Carvalho, A. P., & Gomes, A. M. (2013). Antimicrobial activity and lipid profile of seaweed extracts from the North Portuguese Coast. *International Food Research Journal, 20*(6), 3337–3345.

Mimouni, V., Couzinet-Mossion, A., Ulmann, L., & Wielgosz-Collin, G. (2018). Lipids from microalgae. *Microalgae in health and disease prevention* (pp. 109–131). Academic Press.

Mizusawa, N., & Wada, H. (2012). The role of lipids in photosystem II. *Biochimica et Biophysica Acta (BBA) - Bioenergetics, 1817*(1), 194–208.

Mooney, B. D., Nichols, P. D., De Salas, M. F., & Hallegraeff, G. M. (2007). LIPID, fatty acid, and sterol composition of eight species of KARENIACEAE (DINOPHYTA): Chemotaxonomy and putative lipid PHYCOTOXINS 1. *Journal of Phycology, 43*(1), 101–111.

Moreira, A. S., Gonçalves, J., Conde, T. A., Couto, D., Melo, T., Maia, I. B., Pereira, H., Silva, J., Domingues, M. R., & Nunes, C. (2022). Chrysotila pseudoroscoffensis as a source of high-value polar lipids with antioxidant activity: A lipidomic approach. *Algal Research, 66*, p.102756.

Mujtaba, G., Hoi, W., Lee, C. G., & Lee, K. (2012). Lipid production by Chlorella vulgaris after a shift from nutrient-rich to nitrogen starvation conditions. *Bioresource Technology, 123*, 279–283.

Naiyl, R. A., Kengara, F. O., Kiriamiti, K. H., & Ragab, Y. A. (2022). Lipid extraction from microalgae using pure caprolactam-based ionic liquids and with organic co-solvent. *PeerJ Analytical Chemistry, 4*, e13.

Obeid, S., Beaufils, N., Camy, S., Takache, H., Ismail, A., & Pontalier, P. Y. (2018). Supercritical carbon dioxide extraction and fractionation of lipids from freeze-dried microalgae Nannochloropsis oculata and Chlorella vulgaris. *Algal Research, 34*, 49–56.

Pantami, H. A., Ahamad Bustamam, M. S., Lee, S. Y., Ismail, I. S., Mohd Faudzi, S. M., Nakakuni, M., & Shaari, K. (2020). Comprehensive GCMS and LC-MS/MS metabolite profiling of chlorella vulgaris. *Marine Drugs, 18*(7), 367.

Pereira, H., Barreira, L., Figueiredo, F., Custódio, L., Vizetto-Duarte, C., Polo, C., Rešek, E., Engelen, A., & Varela, J. (2012). Polyunsaturated fatty acids of marine macroalgae: Potential for nutritional and pharmaceutical applications. *Marine Drugs, 10*(9), 1920–1935.

Pieber, S., Schober, S., & Mittelbach, M. (2012). Pressurized fluid extraction of polyunsaturated fatty acids from the microalga Nannochloropsis oculata. *Biomass and Bioenergy, 47*, 474–482.

Rahman, P. K. S. M., Mayat, A., Harvey, J. G. H., Randhawa, K. S., Relph, L. E., & Armstrong, M. C. (2019). *Biosurfactants and bioemulsifiers from marine algae. The role of microalgae in wastewater treatment* (pp. 169–188). Singapore: Springer.

Reddy, V. S., Rao, D. K. V., & Ram, R. (2010). Functional characterization of lysophosphatidic acid phosphatase from *Arabidopsis thaliana*. *Biochimica et Biophysica Acta (BBA) - Molecular and Cell Biology of Lipids, 1801*(4), 455–461.

Ren, H.-Y., Liu, B. F., Kong, F., Zhao, L., Xie, G. J., & Ren, N. Q. (2014). Enhanced lipid accumulation of green microalga Scenedesmus sp. by metal ions and EDTA addition. *Bioresource Technology, 169*, 763–767.

Renuka, N., Guldhe, A., Singh, P., & Bux, F. (2018). Combined effect of exogenous phytohormones on biomass and lipid production in Acutodesmus obliquus under nitrogen limitation. *Energy Conversion and Management, 168*, 522–528.

Rezanka, T., Nedbalová, L., Lukavský, J., Procházková, L., & Sigler, K. (2017). Lipidomic analysis of two closely related strains of the microalga Parietochloris (Trebouxiophyceae, Chlorophyta). *Algal Research, 25*, 473–482.

Řezanka, T., Nedbalová, L., Procházková, L., & Sigler, K. (2014). Lipidomic profiling of snow algae by ESI-MS and silver-LC/APCI-MS. *Phytochemistry, 100*, 34–42.

Riekhof, W. R., Ruckle, M. E., Lydic, T. A., Sears, B. B., & Benning, C. (2003). The sulfolipids 2′-O-acyl-sulfoquinovosyldiacylglycerol and sulfoquinovosyldiacylglycerol are absent from a Chlamydomonas reinhardtii mutant deleted in SQD1. *Plant Physiology, 133*(2), 864–874.

Rosenberg, E., & Ron, E. Z. (1997). Bioemulsans: Microbial polymeric emulsifiers. *Current Opinion in Biotechnology, 8*, 313–316.

Rossmeisl, M., Macek Jilkova, Z., Kuda, O., Jelenik, T., Medrikova, D., Stankova, B., Kristinsson, B., Haraldsson, G. G., Svensen, H., Stoknes, I., & Sjövall, P. (2012). Metabolic effects of n-3 PUFA as phospholipids are superior to triglycerides in mice fed a high-fat diet: Possible role of endocannabinoids. *PLoS One, 7*(6), e38834.

Ryu, K. H., Kim, B., & Lee, J. H. (2019). A model-based optimization of microalgal cultivation strategies for lipid production under photoautotrophic condition. *Computers & Chemical Engineering, 121*, 57–66.

Sałek, K., & Euston, S. R. (2019). Sustainable microbial biosurfactants and bioemulsifiers for commercial exploitation. *Process Biochemistry, 85*, 143–155.

Sanina, N. M., Goncharova, S. N., & Kostetsky, E. Y. (2004). Fatty acid composition of individual polar lipid classes from marine macrophytes. *Phytochemistry, 65*(6), 721–730.

Santhoshkumar, K., Prasanthkumar, S., & Ray, J. G. (2015). Biomass productivity and fatty acid composition of Chlorella lobophora VM Andreyeva, a potential feed stock for biodiesel production. *American Journal of Plant Sciences, 6*(15), 2453.

Sathish, A., & Sims, R. C. (2012). Biodiesel from mixed culture algae via a wet lipid extraction procedure. *Bioresource Technology, 118*, 643–647.

Schroda, M., Hemme, D., & Mühlhaus, T. (2015). The Chlamydomonas heat stress response. *The Plant Journal, 82*(3), 466–480.

Schultz, J., & Soares Rosado, A. (2020). Extreme environments: A source of biosurfactants for biotechnological applications. *Extremophiles, 24*(2), 189–206.

Shanmugam, S., Mathimani, T., Anto, S., Sudhakar, M. P., Kumar, S. S., & Pugazhendhi, A. (2020). Cell density, Lipidomic profile, and fatty acid characterization as selection criteria in bioprospecting of microalgae and cyanobacterium for biodiesel production. *Bioresource Technology, 304*, 123061.

Shanta, P. V., Li, B., Stuart, D. D., & Cheng, Q. (2021). Lipidomic profiling of algae with microarray MALDI-MS toward ecotoxicological monitoring of herbicide exposure. *Environmental Science & Technology, 55*(15), 10558.

Sharma, S. K., Nelson, D. R., Abdrabu, R., Khraiwesh, B., Jijakli, K., Arnoux, M., O'Connor, M. J., Bahmani, T., Cai, H., Khapli, S., & Jagannathan, R. (2015). An integrative Raman microscopy-based workflow for rapid in situ analysis of microalgal lipid bodies. *Biotechnology for Biofuels, 8*(1), 1–14.

Simopoulos, A. P. (2002). The importance of the ratio of omega-6/omega-3 essential fatty acids. *Biomedicine & Pharmacotherapy, 56*(8), 365–379.

Singh, P., Guldhe, A., Kumari, S., Rawat, I., & Bux, F. (2015). Investigation of combined effect of nitrogen, phosphorus and iron on lipid productivity of microalgae Ankistrodesmus falcatus KJ671624 using response surface methodology. *Biochemical Engineering Journal, 94*, 22–29.

Singh, P., Kumari, S., Guldhe, A., Misra, R., Rawat, I., & Bux, F. (2016). Trends and novel strategies for enhancing lipid accumulation and quality in microalgae. *Renewable and Sustainable Energy Reviews, 55*, 1–16.

Sivaramakrishnan, R., & Incharoensakdi, A. (2019). Low power ultrasound treatment for the enhanced production of microalgae biomass and lipid content. *Biocatalysis and Agricultural Biotechnology, 20*, 101230.

Stengel, D. B., Connan, S., & Popper, Z. A. (2011). Algal chemodiversity and bioactivity: Sources of natural variability and implications for commercial application. *Biotechnology Advances, 29*(5), 483–501.

Takagi, M., & Yoshida, T. (2006). Effect of salt concentration on intracellular accumulation of lipids and triacylglyceride in marine microalgae Dunaliella cells. *Journal of Bioscience and Bioengineering, 101*(3), 223–226.

Tang, S., Qin, C., Wang, H., Li, S., & Tian, S. (2011). Study on supercritical extraction of lipids and enrichment of DHA from oil-rich microalgae. *The Journal of Supercritical Fluids, 57*(1), 44–49.

Tasende, M. G. (2000). Fatty acid and sterol composition of gametophytes and sporophytes of Chondrus crispus (Gigartinaceae, Rhodophyta). *Scientia Marina, 64*(4), 421–426.

Urzica, E. I., Vieler, A., Hong-Hermesdorf, A., Page, M. D., Casero, D., Gallaher, S. D., Kropat, J., Pellegrini, M., Benning, C., & Merchant, S. S. (2013). Remodeling of membrane lipids in iron-starved Chlamydomonas. *Journal of Biological Chemistry, 288*(42), 30246–30258.

Vítová, M., Goecke, F., Sigler, K., & Řezanka, T. (2016). Lipidomic analysis of the extremophilic red alga Galdieria sulphuraria in response to changes in pH. *Algal Research, 13*, 218–226.

Walter, V., Syldatk, C., & Hausmann, R. (2010). Screening concepts for the isolation of biosurfactant producing microorganisms. *Biosurfactants*, 1–13.

Wang, Q., Oshita, K., Takaoka, M., & Shiota, K. (2021). Influence of water content and cell disruption on lipid extraction using subcritical dimethyl ether in wet microalgae. *Bioresource Technology, 329*, 124892.

Wilhelm, C., Jungandreas, A., Jakob, T., & Goss, R. (2014). Light acclimation in diatoms: From phenomenology to mechanisms. *Marine Genomics, 16*, 5–15.

Wu, Y., Xiang, W., Li, L., Liu, H., Zhong, N., Chang, H., & Rittmann, B. E. (2021). A novel photoelectrochemical system to disrupt microalgae for maximizing lipid extraction efficiency. *Chemical Engineering Journal, 420*, 130517.

Xu, J., Chen, D., Yan, X., Chen, J., & Zhou, C. (2010). Global characterization of the photosynthetic glycerolipids from a marine diatom Stephanodiscus sp. by ultra performance liquid chromatography coupled with electrospray ionization-quadrupole-time of flight mass spectrometry. *Analytica Chimica Acta, 663*(1), 60–68.

Xue, J., Niu, Y. F., Huang, T., Yang, W. D., Liu, J. S., & Li, H. Y. (2015). Genetic improvement of the microalga Phaeodactylum tricornutum for boosting neutral lipid accumulation. *Metabolic Engineering, 27*, 1–9.

Yang, L., Chen, J., Qin, S., Zeng, M., Jiang, Y., Hu, L., Xiao, P., Hao, W., Hu, Z., Lei, A., & Wang, J. (2018). Growth and lipid accumulation by different nutrients in the microalga Chlamydomonas reinhardtii. *Biotechnology for Biofuels, 11*(1), 1–12.

Yang, W., Pollard, M., Li-Beisson, Y., Beisson, F., Feig, M., & Ohlrogge, J. (2010). A distinct type of glycerol-3-phosphate acyltransferase with sn-2 preference and phosphatase activity producing 2-monoacylglycerol. *Proceedings of the National Academy of Sciences, 107*(26), 12040–12045.

Yang, D., Song, D., Kind, T., Ma, Y., Hoefkens, J., & Fiehn, O. (2015). Lipidomic analysis of Chlamydomonas reinhardtii under nitrogen and sulfur deprivation. *PLoS One, 10*(9), e0137948.

Yen, H.-W., Chen, P.-W., & Chen, L.-J. (2015). The synergistic effects for the co-cultivation of oleaginous yeast-Rhodotorula glutinis and microalgae-Scenedesmus obliquus on the biomass and total lipids accumulation. *Bioresource Technology, 184*, 148–152.

Yhoon, K., Han, D., Li, Y., Sommerfeld, M., & Hu, Q. (2012). Phospholipid: Diacylglycerol acyltransferase is a multifunctional enzyme involved in membrane lipid turnover and degradation while synthesizing tri-acylglycerol in the unicellular green microalga Chlamydomonas reinhardtii. *The Plant Cell Online, 24*(9), 3708–3724.

Yun, J.-H., Cho, D. H., Lee, B., Lee, Y. J., Choi, D. Y., Kim, H. S., & Chang, Y. K. (2020). Utilization of the acid hydrolysate of defatted Chlorella biomass as a sole fermentation substrate for the production of biosurfactant from Bacillus subtilis C9. *Algal Research, 47*, 101868.

Zendejas, F. J., Benke, P.I, P. I., Lane, P. D., Simmons, B. A., & Lane, T. W. (2012). Characterization of the acylglycerols and resulting biodiesel derived from vegetable oil and microalgae (Thalassiosira pseudonana and Phaeodactylum tricornutum). *Biotechnology and Bioengineering, 109*(5), 1146–1154.

Zhang, M., Fan, J., Taylor, D. C., & Ohlrogge, J. B. (2009). DGAT1 and PDAT1 acyltransferases have overlapping functions in Arabidopsis triacylglycerol biosynthesis and are essential for normal pollen and seed development. *The Plant Cell Online, 21*(12), 3885–3901.

Zienkiewicz, K., Du, Z. Y., Ma, W., Vollheyde, K., & Benning, C. (2016). Stress-induced neutral lipid biosynthesis in microalgae—molecular, cellular and physiological insights. *Biochimica et Biophysica Acta (BBA) - Molecular and Cell Biology of Lipids, 1861*(9), 1269–1281.

Zulu, N. N., Zienkiewicz, K., Vollheyde, K., & Feussner, I. (2018). Current trends to comprehend lipid metabolism in diatoms. *Progress in Lipid Research, 70*, 1–16.

Zuorro, A., Maffei, G., & Lavecchia, R. (2016). Optimization of enzyme-assisted lipid extraction from nanno-chloropsis microalgae. *Journal of the Taiwan Institute of Chemical Engineers, 67*, 106–114.

CHAPTER

Positive and negative environmental impacts on algae

20

H.D. Ambika
Algae Ecologist and Taxonomist, Germany

1. Introduction

Microalgae are a large and diverse group of aquatic organisms that lack the complex cell structures found in higher plants. They are ubiquitous and can be found in diverse environments: some species thriving in freshwater, and others in saline conditions and seawater (Carlsson et al., 2007; Schenk et al., 2008). Estimates suggest that over 800,000 species of microalgae exist, yet of these only 50,000 are documented (Mata et al., 2010). Each species has adapted to a particular environment, for example, extreme climates, salt levels, pH, or light levels (Dragone et al., 2010). Most species are photoautotrophic, converting solar energy into chemical forms through photosynthesis. Algal growth in a water body is a complex activity characterized by a large number of interior physical, chemical, and biological mechanisms and impacted by a series of exterior environmental factors. Total nitrogen (TN), total phosphorus (TP), water temperature, and light intensity have been identified as primary factors influencing algal growth, and their synergic effects contribute to phytoplankton growth (Ni et al., 2012; Bartosh & Banks, 2007).

Microalgae are a natural source of high-value compounds for pharmaceutical and food industry and can also be a source of biofuels. Naturally occurring algae are live completely submerged in aquatic environments, as low-density cellular suspension. To have significant environmental and economic impact on human society, algae must be grown in large-scale systems, to significantly increase the production volume. The process of scale-up is to enlarge the production quantities with similar or higher productivity and product quality.

Expanding microalgal cultivations from a lab-scale unit to a commercial one is a challenge due to the difficulty in assessing the factors affecting the scale-up process during the cultivation. As a result, most large-scale cultivations give a lower yield than is expected in the laboratory (Hsu & Wu, 2002). In addition, large-scale algal growth benefits the environment since algae are a valuable carbon capture source and feed on atmospheric nitrogen oxides, another prominent harmful greenhouse gas (Emeka et al., 2012).

There are two main cultivation methods of photoautotrophic algae: raceway pond systems and photobioreactors (PBRs). A typical raceway pond comprises a closed loop oval channel, open to the air, and mixed with a paddle wheel to circulate the water and prevent sedimentation (ponds are kept shallow as optical absorption, and self-shading by the algal cells limits light penetration through the

Algae Materials: Applications Benefitting Health. https://doi.org/10.1016/B978-0-443-18816-9.00014-9

algal broth). In PBRs, the culture medium is enclosed in a transparent array of tubes or plates, and the microalgal broth is circulated from a central reservoir. PBR systems allow for better control of the algae culture environment but tend to be more expensive than raceway ponds. Auxiliary energy demand may also be higher.

There is a gap between the theoretical biological potential of microalgae and the biomass productivities obtained outdoors, especially in mass culture in PBRs. This chapter will describe the key factors and ways to overcome it. The basic technical (material durability, others) and biological problems (mixing, temperature control, contamination) will be covered and critically discussed.

2. Algal cellular components

The exploitation of microalgae as a protein source has led to increased interest in the use of microalgae (e.g., *Spirulina, Chlorella,* and *Scenedesmus*) in healthy food production (Becker, 2007). Algae such as *Dunaliella* and *Spirulina* are also used for pigment production (Borowitzka & Borowitzka, 1990; Chen & Zhang, 1997). However, algae such as *Chlamydomonas, Nannochloropsis, Schizochytrium, Chlorella,* and *Nitzschia* are high density in lipid content that is attractive for biofuels production (Hu et al., 2008). Lipid content in algae can be as high as 50% (dry weight) with higher percentages of total dry weight (~60%) found as protein and starch (Singh & Gu, 2010). These fractions of microalgae can be used to produce biodiesel (Chisti, 2007; Mata et al., 2010), bioethanol (Harun et al., 2010; John & Anisha, 2011), biohydrogen (Beer et al., 2009; Benemann, 1997), bioplastics (Hempel et al., 2011), and other products, while simultaneously contributing to CO_2 mitigation (Jeong et al., 2003; Keffer & Kleinheinz, 2002). Depending on the species and growth conditions, dry algae weight can contain 20%–50% lipid (Chisti, 2007), which is the major substrate for biofuels production. Total lipid composition includes (but is not limited to) neutral lipids, polar lipids, wax esters, sterols, hydrocarbons, and prenyl derivatives such as tocopherols, carotenoids, terpenes, quinones, and phytylated pyrrole derivatives, including chlorophylls (Hu et al., 2008). For algae, high starch strains including *Chlorella vulgaris* (with 37% dry weight starch) (Hirano et al., 1997) are being studied for potential use and have high-yield feedstocks. Algal starch is known to be readily fermentable by yeast (Nguyen et al., 2009) and, therefore, is being intensely studied for use in ethanol production (Harun et al., 2010; John & Anisha, 2011). Carbohydrates (primarily starch) are another valuable component of the algal cell. Typical dry weight content of carbohydrates in algae ranges from 20% to 40% of total cell mass (Hu, 2004). Certain components, such as starch, can be easily converted to ethanol by hydrolysis and fermentation. Currently, the starch used for ethanol production is obtained from food crops such as corn, wheat, sorghum, and others (Nichols & Bothast, 2008).

Protein and starch content can constitute up to 60% of dry weight of algae (Becker, 2007). This residual protein from the biomass can be used for livestock, poultry, and fish feed additives (Singh & Gu, 2010). It has been reported that algae can replace about 5%–10% of conventional protein sources in poultry feed (Spolaore et al., 2006). Recently, residual algae cake after lipid extraction has been used in large animal feeding trials (Hasan & Chakrabarti, 2009). However high concentrations of nucleic acids in algae can pose challenges for the utilization in animal feed applications.

3. Environmental factors and their impacts on algae

To produce microalgae in large scale with respect to all resources, there are positive and negative impacts on the environment. There has been a broad scope of research based on microalgae, from microalgae as a biological resource to microalgae as a source of industrial and domestic wastewater cleanup, and from microalgae for the production of biofuels and fertilizer to microalgae for the production of food and pharmaceuticals. Fewer studies have looked into environmental impacts that could occur, should cultivation systems be scaled up. Some of the emerging research studies consider greenhouse gas emissions (Fagerstone et al., 2011; Ferrón et al., 2012), water consumption (Guieysse et al., 2013; Yang et al., 2011), and wastewater treatment (Christenson et al., 2011; Woo et al., 2012). There have been several recent reviews discussing the potential environmental impacts of different aspects of microalgal cultivation, for example, the designing of pond-based cultivation systems using ecological principles to reduce environmental impacts (Smith et al., 2010), identification of environmental impacts and their social acceptance, and perceived and actual health impacts (Council, 2012).

For microalgae cultivation, water quality requirements vary depending on alga strains. Using domestic and industrial wastewater sources could be economically and environmentally beneficial for large-scale microalgae cultivation, as this practice could provide low-cost water and nutrients as well as wastewater remediation (Philippa et al., 2014). Microalgae are highly adaptive to their environment and thrive by utilizing nutrients available in the water body. A high surface area-to-volume ratio gives algae the potential to absorb large amounts of nutrients across their surface, enhancing photosynthesis. The demand and rate of uptake of a nutrient depend on the strain and environmental conditions (e.g., temperature, light, limiting nutrients, etc.).

4. Light

The previous research and publications in the past decades clearly showed that if large biomass yields were claimed at the laboratory level, high outputs over a long period were generally not achieved in outside full-scale culture units. As suggested by Richmond (1986), this was likely caused by reduced growth rate related to light limitation. More important than the total light impinging on the culture surface was the quantity of light and energy per cell. High areal biomass densities imply low productivities as consequence of significant light limitation and a high contribution of respiration and endogenous biomass consumption. Low areal biomass densities imply damaging effects of radiation in excess (photoinhibition and photo-oxidation) and lower productivities (Soeder, 1980). This shows that best productivities require a strict control of areal biomass densities over a narrow optimum interval, which should be determined for each species (strain), location, and season of the year.

The phenomenon of light limitation and, to obtain the best light regime, high levels of mixing are necessary in dense cultures to reach a turbulent flow of the culture. If light too high, photoinhibition and photooxidative damage could take place with detrimental effect over the cultures. Apart from the light, other crucial key factors such as volume (S/V) ratio, geometry, orientation, and inclination of reactor, material and thickness of reactor walls, culture depth and density, and mixing should be considered.

Light intensity also affects the cellular composition of algae. *Dunaliella tertiolecta* exhibits a decrease in protein content and an increase in the lipid fraction with increasing light intensities up to saturation (Cuhel, et al., 1984). Similar results were reported by Morris et al. (1974), in a study on the marine diatom, *Phaeodactylum tricornutum*, in which low light (400 lux at the culture surface) led to an increase in the rate of protein synthesis. Low light intensity has been observed to result in higher protein content, while high photon flux density (PFD) results in increased extracellular polysaccharide content (Iqbal & Zafar, 1993). Absence of light was observed to increase the total lipid content of the *Desmarestia viridis* but reduce triglycerides, free fatty acids, free alcohols, and sterols (Smith et al., 1993).

Wu and Merchuk (2001) investigated the effect of light and dark cycles on the growth of algae and observed that with increasing photon flux density (PFD), specific growth rate increases up to a certain threshold PFD value after which a decline in growth rate was observed. Sustained high light intensities have also been reported to cause photoinhibition and reduce light utilization efficiency. Light utilization efficiency may be optimized by prolonging the dark period under conditions of high light intensity. This allows the photosynthesis machinery in the cell to fully utilize captured photons and convert them into chemical energy, thus avoiding the effects of photoinhibition (Long et al., 1994).

5. Temperature

Temperature is one of the most difficult parameters to control and optimize in large-scale outdoor culture systems along with light intensity. Fluctuations in temperature, both daily and seasonally, can lead to significant decreases in productivity. The optimal growth temperature for microalgae is species (strain) specific but seldom is attained outdoors during long periods of time. A high temperature and high irradiance of a low density with a combination of high dissolved oxygen (DO) culture may result in photooxidative damage to algal cells.

Temperature influences algae photosynthesis by changing the photosynthetic rate, or by inducing phenotypic or genotypic changes among algae species (Davison, 1991). During the period of the day when sun is high and in the hot season cooling is very important and crucial step. So many methods have been proposed, such as blocking a part of radiation impinging the reactor, covering it with an opaque material (this technical option affects light and temperature simultaneously), spraying the reactor surface with cool water, spreading cool water through a cascade acting as a heat exchanger, or submerging the reactor in a pool containing cool water acting as a heat dissipator. In contrast, low temperatures in winter may require additional culture heating. Renewable energies, such as geothermal, solar, and photovoltaic, should not be neglected.

One of the most commonly observed changes with temperature shift is the alteration in the level of unsaturation of fatty acids in the lipid membrane (Guschina, 2009; Harwood, 2004; Thompson, 1996). Lower temperatures decrease the fluidity in the cell membrane. Cells then compensate by increasing levels of unsaturated fatty acids to increase fluidity. However, it also makes the membranes more susceptible to damage by free radicals (Raven & Geider, 1988; Nishida & Murata, 1996). For example, in a study involving *Botryococcus braunii*, a green alga that secretes extracellular lipids, differences in lipid composition were observed at three different growth temperatures (18, 25, and 32°C). Intracellular lipid synthesis was found to be inhibited at supraoptimal temperature (32°C); consequently, lipid content decreased to 5% dry weight at 32°C in comparison with 22% at 25°C. The decrease in lipid content led to an accumulation of polysaccharides. However, temperature did not affect the secretion of extracellular lipids (Kalacheva et al., 2002). Increasing temperature beyond the optimum reduces protein synthesis and consequently results in decreased growth rates (Konopka & Brock, 1978). Morris et al. (1974) studied the growth of alga, *P. tricornutum*, a marine diatom, and reported a considerable increase in protein synthesis rates at night with lowering the temperatures, presumably due to the fact that protein synthesis is a significant component of night time algal metabolism (Cuhel et al., 1984). Similarly, Rhee and Gotham (1981) observed an increase in protein concentration in *Scenedesmus* sp. with decreasing temperature.

6. pH

The pH is one of the most critical environmental conditions in microalgal cultivation since it determines the solubility and availability of CO_2 and nutrients, and has a significant influence on microalgal metabolism (Chen & Durbin, 1994). The optimal pH range for biomass and lipid production for each microalgal species is different, which is narrow and strain specific (Goldman et al., 1982; Moheimani, 2013). Specifically, the pH of the relative concentrations of the carbonaceous species in water is critical (Azov, 1982). The pH of microalgal cultures rises gradually during the day due to the uptake of inorganic carbon by microalgae. The availability of CO_2 inhibits the cell growth with higher pH value (Azov, 1982; Chen & Durbin, 1994). On the other hand, algal cultivation at high pH can suppress undesired biological contaminants (Bartley et al., 2014). Alkaline pH increases the flexibility of the cell wall of mother cells, which prevents its rupture and inhibits autospore release, thus increasing the time for cell cycle completion (Guckert & Cooksey, 1990). Alkaline pH indirectly results in an increase in triglyceride accumulation but a decrease in membrane-associated polar lipids because of cell cycle inhibition. Maximum algal growth occurs around neutral pH, although optimum pH is the initial culture pH at which an alga is adapted to grow. Changing pH in media may limit algal growth via metabolic inhibition (Goldman et al., 1982). Pruder and Bolton (1979) observed that *Thalassiosira pseudonana* cells adapted to low pH (6.5) had lower growth rate at suboptimal pH (8.8). Normal growth rate was restored after the pH was lowered by addition of HCl. Similar results were reported by Chen and Durbin, where photosynthetic rate and algal growth were minimal at pH 9.0, but carbon uptake rates were enhanced when the pH was lowered to 8.3 (Chen & Durbin, 1994).

7. Salinity

Salinity is also an important factor that alters the biochemical composition of algal cells. When algae are exposed to lower or higher salinity levels than their natural levels, it can change growth rate and alter composition. For example, higher salinity increases the algae lipid content (Fabregas et al., 1984; Zhila et al., 2011). *Dunaliella*, a marine alga, exhibited an increase in saturated and monounsaturated fatty acids with an increase in NaCl concentration from 0.4 to 4 M (Xu & Beardall, 1997). In another study from Takagi and Yoshida (2006) with *D. tertiolecta*, an increase in intracellular lipids (60%–67%) and triglyceride concentration (40%–56%) with an increase in NaCl concentration from 0.5 (freshwater concentration) to 1.0 M was observed. Ben-Amotz et al. (1985) reported that lipid content of *Botryococcus braunii* grown in 0.50 M NaCl was higher compared with media without NaCl addition, but protein, carbohydrates, and pigments levels were lower. lipid content; however, the greatest biomass concentration was achieved at the lowest salinity level.

8. Nutrients availability

The growth rate of algae is proportional to the uptake rate of the most limiting nutrient under optimal conditions of temperature and p^H. The variation in the biochemical composition depends on conditions of nutrient limitation. Carbon, hydrogen, and oxygen are three essential nonmineral nutrients. Abundance of hydrogen and oxygen in the media for algae cultures means that their availability is not a challenge to cellular growth or metabolism. Carbon is one of the other major nutrients that must be supplied. It is essential for photosynthesis and hence algal growth and reproduction. Nitrogen and phosphate are two important macronutrients for growth and metabolism of algal cells. Nitrogen is a fundamental element for the formation of proteins and nucleic acids. Being an integral part of essential molecules such as ATP, the energy carrier in cells, phosphate is another very important nutrient.

9. Carbon source

Carbon is a vital nutrient that regulates the growth and metabolism of microalgae. Different carbon sources are assimilated by microalgae cells via different pathways. Inorganic carbon sources are mainly used through the CO2-concentrating mechanisms (CCMs), while organic carbon sources are absorbed by microalgae mainly through the pentose phosphate (PPP) pathway and the Embden–Meyerhof–Parnas (EMP) pathway. Therefore, the addition of carbon source has a significant impact on the production of microalgae biomass and lipid accumulation.

One of the essential nutrients required for biomass formation is carbon. It can be acquired by photosynthetic microalgae in an inorganic form from carbon dioxide via carbohydrase activity. However, carbon dioxide has low solubility in water, and the poor net mass transfer from the atmosphere makes it a limiting nutrient for microalgae cultivation. The heterotrophic microalgae cannot assimilate carbon in the same way and requires an organic carbon source. On the other hand, the use of flue gas has been considered as an alternative to overcome that hurdle, but could create additional concerns on aquatic environments due to the dissolution of other pollutants.

10. Nitrogen

Nitrogen is a key nutrient required by microalgae. Nitrogen assimilation is required for the formation of genetic material, energy transfer molecules, proteins, enzymes, chlorophylls, and peptides. Most microalgae will assimilate inorganic nitrogen in the form of ammonium nitrogen (NH_4^+), but when they run out or are not available, they have the ability to utilize other inorganic nitrogen species such as nitrate (NO_3^-) or nitrite (NO_2^-) and are used commonly in culture media (Wu et al., 2012).

We have heard of algae blooms, which is just because an excess of nitrogen in an aquatic environment can lead to uncontrollable microalgae blooms, which could develop toxic conditions (toxins from cyanobacteria, free ammonia which is toxic to fish and low oxygen concentrations during the night due to algal respiration); therefore, nitrogen recovery from wastewater via biological uptake for algal biomass production may contribute to alleviate such negative impacts (Chen et al., 2009; Lardon et al., 2009; Mata et al., 2010). Yang et al. (2008) showed that biomass accumulation in

Chlamydomonas reinhardtii was inhibited under nitrogen deficiency up to 31.7%, with simultaneous increases in total fatty acid yield up to 93%, coupled to enhancement in lipid production up to 113.46 ± 1.78 mg/L. Thus, it can be concluded that nitrogen concentration favors higher biomass productivity, and depletion of nitrogen shifts the flux to lipid production.

Microalgae growth in nitrogen-deficient medium might change their lipid metabolic pathway and accumulate mainly triacyl glycerides (TAGs), which are stored in the cytoplasm of microalgae as a source of carbon and energy. Nitrogen-limiting conditions can decrease the thylakoid membrane cellular content, activate acyl hydrolase, and stimulate phospholipid hydrolysis, resulting in increased intracellular content of fatty acyl-CoA (Chu et al., 2013).

11. Phosphorous

Phosphorous compound plays a significant role in algal growth, lipid production, fatty acid yield, and metabolic processes such as energy transfer, signal transduction, and photosynthesis (Yang et al., 2018; Ota et al., 2016). Phosphorous is an essential nutrient that makes up slightly less than 1% of total algal biomass and is required at approximately 0.03%–0.06% in the medium to sustain algae growth (Hannon, 2010; Ota et al., 2016; Procházková, 2013). Phosphorous is essential to microalgal cells for the production of cellular components such as phospholipids, DNA, RNA, and ATP for the metabolic pathways that involve energy transfer and nucleic acid synthesis (Atiku, 2016). Microalgae can uptake phosphorus in the form of polyphosphate or orthophosphate to enhance their growth and nutritional content in the cells. It is also reported that microalgae growth and phosphate uptake are linearly proportional to the biomass yield. Chu et al. (2013) demonstrated that *Chlorella vulgaris* had been cultivated for biodiesel production under phosphorus-sufficient conditions. The result showed that the lipid yield of *C. vulgaris* for biodiesel production was 58.39 mg/L/day in phosphate-sufficient conditions, which was relatively higher than in a phosphate-deficient regime. Thus, it can be concluded that phosphate is an important macronutrient for microalgal lipid production (Chu et al., 2013). However, the phosphorus uptake by microalgae can reach saturation due to the limitation of light, and reduction of carbon dioxide and oxygen levels in the culture medium.

12. Contamination and leaks

The two most common methods of microalgae cultivation are open cultivation systems, such as open ponds, tanks, and raceway ponds, and controlled closed cultivation systems using different types of bioreactors. Open ponds allow large-scale cultivation at lower cost. However, the open design makes them vulnerable to contamination. This risk can be minimized by altering culture conditions, making them unfavorable to native species. Ponds that are not correctly designed or constructed could pose a threat to the direct environment from leaching of the pond contents into the ground. PBRs are translucent containers that allow light to penetrate to the microalgae. PBRs are closed and therefore are less susceptible to contamination. Depending on the volume, a leak from these containers could also have a significant impact, for example, if located near a natural source of water. However, it would be potentially easier to detect and therefore easier to rectify.

13. Conclusion

For the past four decades, significant progress has been made in algae research toward a better understanding of algae growth. This comprises an improved understanding of the effects of the nutrient availability and environmental factors on algae cell division and composition. In this chapter, it has been tried to explain and give a picture about the effects of various nutritional factors (i.e., carbon, nitrogen, phosphorous), environmental factors (i.e., light, temperature, and pH), and their interactions on algae growth and composition. Along with traditional method, it should take an advantage of advances in biotechnology and bioinformatics, to study more on large volume of genetic information. Continued research will not only lead to a greater understanding of basic algal cell biology but also aid in development of more accurate predictive models for algae growth.

References

Atiku, A., Mohamed, R. M. S. R., Al-Gheethi, A. A., Wurochekke, A. A., & Kassim, A. H. (2016). Harvesting microalgae biomass from the phycoremediation process of greywater. *Environmental Science & Pollution Research, 23*, 24624–24641.

Azov, Y. (1982). Effect of pH on inorganic carbon uptake in algal cultures. *Applied and Environmental Microbiology, 43*, 1300–1306.

Bartley, M. L., Boeing, W. J., Dungan, B. N., Holguin, F. O., & Schaub, T. (2014). pH effects on growth and lipid accumulation of the biofuel microalgae Nannochloropsis salina and invading organisms. *Journal of Applied Phycology, 26*, 1431–1437.

Bartosh, Y., & Banks, C. (2007). Algal growth response in a range of light and temperature conditions: Implications for non-steady state conditions in waste stabilisation ponds. *Water Science and Technology, 55*(11), 211–218.

Becker, E. W. (2007). Micro-algae as a source of protein. *Biotechnology Advances, 25*, 207–210.

Beer, L. L., Boyd, E. S., Peters, J. W., & Posewitz, M. C. (2009). Engineering algae for biohydrogen and biofuel production. *Current Opinion in Biotechnology, 20*, 264–271.

Ben-Amotz, A., Tornabene, T. G., & Thomas, W. H. (1985). Chemical profile of selected species of microalgae with emphasis on lipids. *Journal of Phycology, 21*, 72–81.

Benemann, J. (1997). Feasibility analysis of photobiological hydrogen production. *International Journal of Hydrogen Energy, 22*, 979–987.

Borowitzka, L. J., & Borowitzka, M. A. (1990). Commercial production of-carotene by Dunaliella salina in open ponds. *Bulletin of Marine Science, 47*, 244–252.

Carlsson, A. S., van Beilen, J. B., Möller, R., & Clayton, D. (September 2007). In D. Bowles (Ed.), *Micro- and macro-algae: Utility for industrial applications outputs from the EPOBIO project* (p. 86). Newbury (UK: University of York, CPL Press.

Chen, C. Y., & Durbin, E. G. (1994). Effects of pH on the growth and carbon uptake of marine phytoplankton. *Marine Ecology Progress Series, 109*, 83–94.

Chen, S., Chen, X., Peng, Y., & Peng, K. (2009). A mathematical model of the effect of nitrogen and phosphorus on the growth of blue-green algae population. *Applied Mathematical Modelling, 33*(2), 1097–1106.

Chen, F., & Zhang, Y. (1997). High cell density mixotrophic culture of Spirulina platensis on glucose for phycocyanin production using a fed-batch system. *Enzyme and Microbial Technology, 20*, 221–224.

Chisti, Y. (2007). Biodiesel from microalgae. *Biotechnology Advances, 25*, 294–306.

Christenson, L., & Sims, R. (2011). Production and harvesting of microalgae for wastewater treatment, biofuels, and bioproducts. *Biotechnology Advances, 29*(6), 686–702.

Chu, F. F., Chu, P. N., Cai, P. J., Li, W. W., Lam, P. K. S., & Zeng, R. J. (2013). Phosphorus plays an important role in enhancing biodiesel productivity of Chlorella vulgaris under nitrogen deficiency. *Bioresource Technology, 134*, 341–346.

Council, N. R. (2012). Environmental effects. In *Sustainable development of algal biofuels in the United States* (pp. 139–190). Washington. DC: The National Academies Press.

Cuhel, R. L., Ortner, P. B., & Lean, D. R. S. (1984). Night synthesis of protein by algae. *Limnology & Oceanography, 29*, 731–744.

Davison, I. R. (1991). Environmental effects on algal photosynthesis: Temperature. *Journal of Phycology, 27*(1), 2–8.

Dragone, G., Fernandes, B., Vicente, A. A., & Teixeira, J. A. (2010). Third generation biofuels from microalgae. In A. Mendez-Vilas (Ed.), *Current research, technology and education topics in applied microbiology and microbial biotechnology* (pp. 1355–1366). Spain: Formatex Research Center.

Emeka, U., Ndukwe, G. I., Mustapha, K. G., & Ayo, R. I. (2012). Constrains to large scale algal biomass production and utilization. *Journal Algal Biomass Utilization, 3*, 14–21.

Fabregas, J., Abalde, J., Herrero, C., Cabezas, B., & Veiga, M. (1984). Growth of the marine microalga Tetraselmis suecica in batch cultures with different salinities and nutrient concentrations. *Aquaculture, 42*, 207–215.

Fagerstone, K. D., Quinn, J. C., Bradley, T. H., De Long, S. K., & Marchese, A. J. (2011). Quantitative measurement of direct nitrous oxide emissions from microalgae cultivation. *Environmental Science and Technology, 45*(21), 9449–9456.

Ferrón, S., Ho, D. T., Johnson, Z. I., & Huntley, M. E. (2012). Air-water fluxes of N2O and CH4 during microalgae (Staurosira sp.) cultivation in an open raceway pond. *Environmental Science and Technology, 46*(19), 10842–10848.

Goldman, J. C., Azov, Y., Riley, C. B., & Dennett, M. R. (1982). The effect of pH in intensive microalgal cultures. I. Biomass regulation. *Journal of Experimental Marine Biology and Ecology, 57*, 1–13.

Guckert, J. B., & Cooksey, K. E. (1990). Triglyceride accumulation and fatty acid profile changes in *Chlorella (Chlorophyta)* during high pH induced cell cycle inhibition. *Journal of Phycology, 26*, 72–79.

Guieysse, B., Béchet, Q., & Shilton, A. (2013). Variability and uncertainty in water demand and water footprint assessments of fresh algae cultivation based on case studies from five climatic regions. *Bioresource Technology, 128*, 317–323.

Guschina, I. A., & Harwood, J. L. (2009). Algal lipids and effect of the environment on their biochemistry. In *Lipids in aquatic ecosystems* (pp. 1–24). Berlin, Germany: Springer.

Hannon, M., Gimpel, J., Tran, M., Rasala, B., & Mayfield, S. (2010). Biofuels from algae: Challenges and potential. *Biofuels, 1*, 763–784.

Harun, R., Danquah, M. K., & Forde, G. M. (2010). Microalgal biomass as a fermentation feedstock for bioethanol production. *Journal of Chemical Technology and Biotechnology, 85*, 199–203.

Harwood, J. L. (2004). Involvement of chloroplast lipids in the reaction of plants submitted to stress. In P. A. Siegenthaler, & N. Murata (Eds.), *6. Lipids in photosynthesis: Structure, function and genetics* (pp. 287–302). Berlin, Germany: Springer.

Hasan, M. R., & Chakrabarti, R. (2009). *Use of algae and aquatic macrophytes as feed in small-scale Aquaculture: A review.* Rome, Italy: Food and Agriculture Organization of the United Nations (FAO).

Hempel, F., Bozarth, A. S., Lindenkamp, N., Klingl, A., Zauner, S., Linne, U., Steinbüchel, A., & Maier, U. G. (2011). Microalgae as bioreactors for bioplastic production. *Microbial Cell Factories, 10*, 81. https://doi.org/10.1186/1475-2859-10-81

Hirano, A., Ueda, R., Hirayama, S., & Ogushi, Y. (1997). CO2 fixation and ethanol production with microalgal photosynthesis and intracellular anaerobic fermentation. *Energy, 22*, 137–142.

Hsu, Y., & Wu, W. (2002). A novel approach for scaling-up a fermentation system. *Biochemical Engineering Journal, 11*, 123–130.

Hu, Q. (2004). Environmental effects on cell composition. In A. Richmond (Ed.), *Handbook of microalgal culture: Biotechnology and applied phycology* (pp. 83–93). Oxford, UK: Blackwell.

Hu, Q., Sommerfeld, M., Jarvis, E., Ghirardi, M., Posewitz, M., Seibert, M., & Darzins, A. (2008). Microalgal triacylglycerols as feedstocks for biofuel production: Perspectives and advances. *The Plant Journal, 54*, 621–639.

Iqbal, M., & Zafar, S. (1993). Effects of photon flux density, CO2, aeration rate, and inoculum density on growth and extracellular polysaccharide production by Porphyridium cruentum. *Folia Microbiologica, 38*, 509–514.

Jeong, M. L., Gillis, J. M., & Hwang, J. Y. (2003). Carbon dioxide mitigation by microalgal photosynthesis. *Bulletin of the Korean Chemical Society, 24*, 1763–1766.

John, R. P., Anisha, G., Nampoothiri, K. M., & Pandey, A. (2011). Micro and macroalgal biomass: A renewable source for bioethanol. *Bioresource Technology, 102*, 186–193.

Kalacheva, G., Zhila, N., Volova, T., & Gladyshev, M. (2002). The effect of temperature on the lipid composition of the green alga Botryococcus. *Microbiology, 71*, 286–293.

Keffer, J. E., & Kleinheinz, G. T. (2002). Use of *Chlorella vulgaris* for CO2 mitigation in a photobioreactor. *Journal of Industrial Microbiology and Biotechnology, 29*, 275–280.

Konopka, A., & Brock, T. D. (1978). Effect of temperature on blue-green algae (cyanobacteria) in Lake Mendota. *Applied and Environmental Microbiology, 36*, 572–576.

Lardon, L., Hélias, A., Sialve, B., Steyer, J. P., & Bernard, O. (2009). Life-cycle assessment of biodiesel production from microalgae. *Environmental Science and Technology, 43*(17), 6475–6481.

Long, S., Humphries, S., & Falkowski, P. G. (1994). Photoinhibition of photosynthesis in nature. *Annual Review of Plant Biology, 45*, 633–662.

Mata, T. M., Martins, A. A., & Caetano, N. S. (2010). Microalgae for biodiesel production and other applications: A review. *Renewable and Sustainable Energy Reviews, 14*(1), 217–232.

Moheimani, N. R. (2013). Inorganic carbon and pH effect on growth and lipid productivity of Tetraselmis suecica and Chlorella sp (Chlorophyta) grown outdoors in bag photobioreactors. *Journal of Applied Phycology, 25*, 387–398.

Morris, I., Glover, H., & Yentsch, C. (1974). Products of photosynthesis by marine phytoplankton: The effect of environmental factors on the relative rates of protein synthesis. *Marine Biology, 27*, 1–9.

Nguyen, M. T., Choi, S. P., Lee, J., Lee, J. H., & Sim, S. J. (2009). Hydrothermal acid pre-treatment of *Chlamydomonas reinhardtii* biomass for ethanol production. *Journal of Microbiology and Biotechnology, 19*, 161–166.

Ni, W. M., Zhang, J. Y., Stevenson, R. J., & Zhu, Y. M. (2012). Environmental factors regulating cyanobacteria dominance and microcystin production in a subtropical lake within the Taihu watershed, China. *Journal of Zhejiang University - Science, 13*(4), 311–322.

Nichols, N. N., & Bothast, R. J. (2008). Production of ethanol from grain. In W. Vermerris (Ed.), *Genetic improvement of bioenergy crops* (pp. 75–88). New York, NY, USA: Springer.

Nishida, I., & Murata, N. (1996). Chilling sensitivity in plants and cyanobacteria: The crucial contribution of membrane lipids. *Annual Review of Plant Biology, 47*, 541–568.

Ota, S., Yoshihara, M., Yamazaki, T., Takeshita, T., Hirata, A., Konomi, M., Oshima, K., Hattori, M., Bisova, K., Zachleder, V., et al. (2016). Deciphering the relationship among phosphate dynamics, electron-dense body and lipid accumulation in the green alga Parachlorella kessleri. *Scientific Reports, 6*, Article 25731.

Philippa, K., Ross, B., Camargo-Valero, M. A., & Tomlin, A. S. (2014). An overview of the potential environmental impacts of large-scale microalgae cultivation. *Biofuels, 5*(3), 331–349. https://doi.org/10.1080/17597269.2014.913925

Procházková, G., Brányiková, I., Zachleder, V., & Brányik, T. (2013). Effect of nutrient supply status on biomass composition of eukaryotic green microalgae. *Journal of Applied Phycology, 26*, 1359–1377.

Pruder, G. D., & Bolton, E. T. (1979). The role of CO2 enrichment of aerating gas in the growth of an estuarine diatom. *Aquaculture, 17*, 1–15.

Raven, J. A., & Geider, R. J. (1988). Temperature and algal growth. *New Phytologist, 110*, 441–461.

Rhee, G. Y., & Gotham, I. J. (1981). The effect of environmental factors on phytoplankton growth: Temperature and the interactions of temperature with nutrient limitation. *Limnology & Oceanography, 26*, 635–648.

Richmond, A. (1986). Outdoor mass cultures of microalgae. In A. Richmond (Ed.), *Handbook of microalgal mass culture* (pp. 285–330). Boca Raton: CRC Press.

Schenk, M. P., Thomas-Hall, S. R., Stephens, E., Marx, U. C., Mussgnug, J. H., Posten, C., et al. (2008). Second generation biofuels: High-efficiency microalgae for biodiesel production. *Bioenergy Research, 1*, 20–43.

Singh, J., & Gu, S. (2010). Commercialization potential of microalgae for biofuels production. *Renewable and Sustainable Energy Reviews, 14*, 2596–2610.

Smith, R., Cavaletto, J., Eadie, B., & Gardner, W. (1993). Growth and lipid composition of high Arctic ice algae during the spring bloom at Resolute, Northwest Territories, Canada. *Marine Ecology Progress Series, 97*, 19–29.

Smith, V. H., Sturm, B. S. M., Denoyelles, F. J., & Billings, S.a. (2010). The ecology of algal biodiesel production. *Trends in Ecology & Evolution, 25*(5), 301–309.

Soeder, C. J. (1980). Massive cultivation of microalgae: Results and prospects. *Hydrobiologia, 72*, 197–209.

Spolaore, P., Joannis-Cassan, C., Duran, E., & Isambert, A. (2006). Commercial applications of microalgae. *Journal of Bioscience and Bioengineering, 101*, 87–96.

Takagi, M., & Yoshida, T. (2006). Effect of salt concentration on intracellular accumulation of lipids and triacylglyceride in marine microalgae Dunaliella cells. *Journal of Bioscience and Bioengineering, 101*, 223–226.

Thompson, G. A., Jr. (1996). Lipids and membrane function in green algae. *Biochimica et Biophysica Acta, 1302*, 17–45.

Woo, S. G., Yoo, K., Lee, J., et al. (2012). Comparison of fatty acid analysis methods for assessing biorefinery applicability of wastewater cultivated microalgae. *Talanta*, 1–8.

Wu, L. F., Chen, P. C., Huang, A. P., & Lee, C. M. (2012). The feasibility of biodiesel production by microalgae using industrial wastewater. *Bioresource Technology, 113*, 14–18.

Wu, X., & Merchuk, J. C. (2001). A model integrating fluid dynamics in photosynthesis and photoinhibition processes. *Chemical Engineering and Science, 56*, 3527–3538.

Xu, X. Q., & Beardall, J. (1997). Effect of salinity on fatty acid composition of a green microalga from an antarctic hypersaline lake. *Phytochemistry, 45*, 655–658.

Yang, L., Chen, J., Qin, S., Zeng, M., Jiang, Y., Hu, L., Xiao, P., Hao, W., Hu, Z., Lei, A., et al. (2008). Growth and lipid accumulation by different nutrients in the microalga Chlamydomonas reinhardtii. *Biotechnology for Biofuels, 11*, 40.

Yang, J., Xu, M., Zhang, X., Hu, Q., Sommerfeld, M., & Chen, Y. (2011). Life-cycle analysis on biodiesel production from microalgae: Water footprint and nutrients balance. *Bioresource Technology, 102*(1), 159–165.

Yang, F. F., Xiang, W., Li, T., & Long, L. (2018). Transcriptome analysis for phosphorus starvation-induced lipid accumulation in Scenedesmus sp. *Scientific Reports, 8*, 1–11.

Zhila, N. O., Kalacheva, G. S., & Volova, T. G. (2011). Effect of salinity on the biochemical composition of the alga Botryococcus braunii Kütz IPPAS H-252. *Journal of Applied Phycology, 23*, 47–52.

CHAPTER

21 Genetic engineering of algae material: merits and demerits

Shailaja V.L.[1], Cordelia Mano John[2] and Kalaivani M.K.[3]

[1]*Arham Bioresearch Technology Pvt. Ltd, Chennai, Tamil Nadu, India;* [2]*Department of Biomedical Sciences, Sri Ramachandra Institute of Higher Education and Research, Chennai, Tamil Nadu, India;* [3]*Sree Balaji Medical College and Hospital, Chennai, Tamil Nadu, India*

1. Introduction

For millennia, humans have valued plants as renewable sources of food, medicine, and materials. Agricultural technologies have been consistently modified and improved throughout human history to satisfy the evolving needs of civilization. To solve the practical constraints and major environmental challenges connected with rapidly rising population, greater innovation in current industrial and agricultural practices is the need of the hour. Increased agricultural acreage, crop rotation, and production could fulfill rising food demand; nevertheless, these techniques exacerbate current environmental issues such as deforestation, degradation of soil, and biodiversity loss (FAO, 2002, 2004; Sreeharsha & Venkata Mohan, 2021). As a result, microalgae are gaining popularity in this regard as both a sustainable feedstock and a food supply alternative (Sudhakar et al., 2019).

Microalgae are unicellular, tiny, photosynthetic microorganisms found in freshwater and marine habitats. For decades, microalgae and their components have been used in a variety of industrial applications. High photosynthetic activity, CO_2 sequestration, biomass production, and rapid proliferation rate make algae a good candidate for production of biofuels, proteins, pigments, nutraceuticals, medicines, and other value-added goods (Torres et al., 2020; Venkata Mohan et al., 2020). Their varied nutritional properties, combined with their capacity to develop quickly, make them a promising dietary supplement alternative as well (Paniagua-Michel, 2015). The global market of these value-added biochemical produced from dry algal biomass has an annual market value of 3800−5500 million dollars (Brasil et al., 2017). Since microalgae can be grown in marine water, freshwater, and even industrial wastes, it makes them a good candidate for genetic engineering, nevertheless, with certain limitations.

Genetic engineering is a technique for manipulating genes to produce material for human use. Genetic engineering of algal species is carried out by the following four basic steps (Fig. 21.1):

1. Choosing a host,
2. Gene of interest,
3. Transformation method, and
4. Screening and selection of the transformants.

Algae Materials: Applications Benefitting Health. https://doi.org/10.1016/B978-0-443-18816-9.00012-5

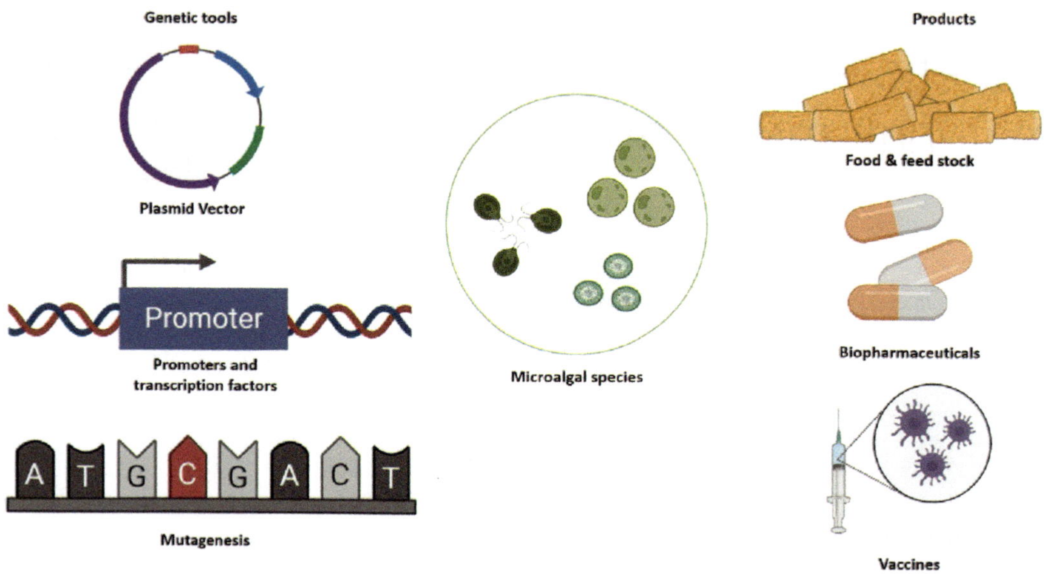

FIGURE 21.1

Genetic Engineering Tools to Develop Microalgae as a Platform for the Manufacturing of Commercial Product.

2. Genetic engineering: a brief history

Genetic engineering, universally termed as recombinant DNA technology, was first originated in 1972 by Paul Berg from Stanford University who successfully engineered a virus by combining genetic material (DNA) from that of a monkey with bacterial virus using genetic transfer methodology. A year later, i.e., in 1973, Herbert Boyer from the University of California and Stanley Cohen from Stanford University constructed the first antibiotic resistance bacterial strain by introducing antibiotic resistance coding genetic element into *Escherichia coli* plasmid. In the following year (1974), Rudolf Jaenisch at the Princeton University produced the first genetically modified animal wherein the genome of the mouse embryo was inserted with a foreign DNA from virus. Apart from this, inventions such as DNA sequencing by Maxam and Gilbert and Sanger and amplification of DNA using polymerase chain reaction (PCR) by Kary Mullis offered a major breakthrough in genetic engineering and also for diagnostic purpose (Blatti et al., 2012).

From the aforementioned findings, it is clear that deoxyribonucleic acid (DNA) is primarily involved in the process of genetic engineering. This is due to the fact that DNA is considered to possess the genetic makeup of an organism, thereby serving as a template for both replication and transcription activities, which in turn helps the organism to thrive. Therefore, genetic engineering involves the modification of an organism to produce desired biological material such as an enzyme or a metabolite which the organism originally is not capable of producing such materials. This process is carried out by extracting the genetic material and isolating the gene of interest from an organism. The obtained DNA is further modified with available molecular techniques and inserted into a vector which is cloned into the organism (host) which lacks the desired gene. Inside the host, the cloned product is subjected to replication and transcription, leading to the production of recombinant DNA. Upon completion of this procedure, not all cells in the host contain the recombinant DNA; hence, screening of the recombinant cells needs to be segregated from the population of wild type (Blatti et al., 2012).

3. Algal biology and genetic engineering

The eukaryotic photosynthetic unicellular organism microalgae are estimated to be around 72,500 in number; they are studied not only for their economic importance but also for their ecological role (Guiry, 2012; Melillo et al., 1993). This organism that belongs to the polyphyletic origin channelizes sunlight, carbon dioxide, and water toward the production of biomass and other natural products (Ortiz-Marquez et al., 2013; Williams & Laurens, 2010). Their capability to thrive in natural environment even with sporadic availability of nutrients has made many microalga species to evolve metabolically to adapt to such conditions and grow luxuriantly. This results in the increased efficiency of photosynthesis than plants, which in turns alleviates the biomass production by algae (Benedetti et al., 2018; Bhola et al., 2014; Litchman, 2007). They also efficiently mitigate the carbon dioxide levels in gas streams, and also the NO_X released from algal biofuel is comparatively less than the fossil fuels (Adeniyi et al., 2018). Apart from this, these organisms are natural producers of biomolecules such as oils and lipids, carbohydrates, various pigments such as phycocyanin, carotenoids, and phycobiliproteins, and many more secondary metabolites, which are of both cosmetic and commercial importance (Koyande et al., 2019). It is clear that microalgae are considered as next-generation, cell-based biofactories for the production of uncountable products following other microbes such as bacteria and yeast (Rasala & Mayfield, 2015; Vavitsas et al., 2018). Therefore, microalgae have been exploited in the manufacture of biofuels and other bioproducts, which includes chemicals and materials used for industrial purpose, pharmaceuticals, and nutraceuticals. These characteristic features gained the attention of researchers to explore microalgae as source of food and energy.

With so much promising benefits from microalgae in the field of pharmaceuticals and biorefinery, its economic viability and commercial availability are hindered due to certain limitations such as slow growth rate with increased contamination, high production and harvesting cost, and complicated extraction of specific secondary metabolite (Li et al., 2020). Implementation of genetic engineering tools such as RNA interference (RNAi), zinc finger nucleases (ZFNs), transcription activator-like effector nucleases (TALENs), and clustered regularly interspaced short palindromic repeats–CRISPR-associated protein 9 (CRISPR-Cas9) in the modification of algal strains is gaining importance for the mass production of proteins required for therapeutic purpose and other carotenoids used in nutraceuticals (Jeon et al., 2017). And also advancement in the genomic- and proteomic-level modifications in engineering an algal strain is widely practiced for biofuel production (Sharma, Puttachary, et al., 2018). With this setting, this chapter focuses on the various methodologies involved in genetic engineering with specific focus on algae along with its application in algal biology. Also, the pros and cons of algal genetic modification is also discussed.

4. Genetic engineering of algal genome for tuning increased metabolite production

Algal genetic engineering has been gaining momentum in recent years due to the advancement in the field of molecular technologies and genome sequencing. With the advent of genome sequencing, many algal nuclear genomes have been sequenced, which has facilitated the development of genetic tools in such organisms namely *Chlamydomonas reinhardtii* (Mussgnug, 2015), *Phaeodactylum tricornutum* (Huang & Daboussi, 2017), *Nannochloropsis gaditana* (Poliner et al., 2018), cyanobacteria *Synechocystis* sp. PCC 6803 (Hagemann et al., 2018), and many more. Apart from this, mitochondrial and chloroplast genomes of the microalgae are also being sequenced.

Genomic data obtained as a result of algal sequencing aided researchers to identify the genetic elements, which need to be engineered and transformed. There are several genetic transformation methodologies developed and optimized for microalgae, which includes electroporation (Fan et al., 2014), biolistic particle bombardment (Dunahay et al., 1995), glass bead–mediated agitation (Blatti et al., 2012), and *Agrobacterium tumefaciens*–mediated transformation (Ahmad et al., 2015). The use of selection markers such as antibiotic resistance genes (bleomycin, hygromycin streptomycin, etc.), green fluorescent protein, and biochemical fluorescent markers (luciferase, β-glucuronidase and β-galactosidase) is efficiently employed in the isolation of transformants (Banerjee et al., 2020). The effective expression of transgene in model organism is demonstrated using constitutive and inducible promoters and bidirectional promoters. Apart from this, these heterologous or synthetic promoter sequences are also described for transgene expression. In cyanobacteria and *C. reinhardtii*, gene expression is regulated by regulatory effectors such as ligand-binding riboswitches (Fabris et al., 2020). Organelle transformation, namely, nuclear and chloroplast transformation is carried out routinely in microalgae species (Mallick et al., 2016). However, nuclear transformation is nonspecific because in this process genes are incorporated randomly, leading to random mutagenesis (Lee et al., 2003).

5. Recent advances in algal genetic engineering

Recently, several genome-editing tools have been developed in the field of microalgae engineering: RNAi, ZFNs, TALENs, and CRISPR/Cas9.

(i) RNA interference

This technique was originally exposed by Fire et al. (1998) in the model organism *Caenorhabditis elegans*. In this, the mRNA molecules are targeted to analyze and study the functions of a particular gene, thereby silencing the gene of interest (Fire et al., 1998; Ketting, 2011). The silencing of genes using RNAi technique has been widely used in microalgae even though there are considerable core elements being lost (Huang et al., 2011; Zhang & Hu, 2014). The knockdown of genes by RNA interference technique is found to be well established in microalgal species, which includes *Nannochloropsis oceanica*, *C. reinhardtii*, and *P. tricornutum*. Though this technique is widely practiced in algal engineering, there are constraints due to this methodology such as unpredictability of target gene activity with unstable phenotypic heritability. Hence, this led to the development of other techniques for more successful mutagenesis (Lu et al., 2021).

(ii) Zinc finger nucleases

ZFNs are synthesized nucleases, which are protein-based nucleases derived from *Flavobacterium okeanokoites* called as FokI endonucleases. This endonuclease consists of DNA-binding domain at the N-terminal region and a nonspecific DNA-binding domain at the C-terminal region. The ZFNs are engineered by fusing the zinc finger proteins at the C-terminal region of the FokI endonucleases, where they identify and bind to 3-bp site on the target gene. Once the FokI enzyme binds to the double stranded, its dimerization activates the mechanism by causing a cleavage or break in the DNA. This is followed by DNA repair system activation by either homologous recombination (HR) or the nonhomologous end joints, leading to DNA repair (Urnov et al., 2010). ZFN is considered as an excellent genetic tool for targeted gene knockout; however, in algal genetic engineering, this method is demonstrated to be efficient in *C. reinhardtii*, wherein introduction of specifically designed ZFN containing cop3 gene within an aphVIII marker into a *Chlamydomonas* cell with a mutation thereby restores the function of the marker (Sizova et al., 2013). Though this method seems to be efficient, the expense involved in the construction of ZFN with accurate specificity toward the target has pushed researchers to search for more feasible methods.

(iii) Transcription activator-like effector nucleases

TALENs are programmed restriction enzymes with three structural domains: DNA binding, transcriptional activation, and nuclear localization signal (NLS) domains. This enzyme is constructed containing 12–27 tandem array repeats with 34 amino acids in each repeat. TALENs are found to be more successful in *P. tricornutum* species, wherein the triacylglycerol buildup was increased to 45% by specific mutagenesis of UDP-glucose pyrophosphorylase gene (Daboussi et al., 2014). More work related with TALENs need to be optimized with new strategies in algal genetic engineering domain.

(iv) CRISPR/Cas9

The advent of CRISPR/Cas9 modification tool greatly influenced the genetic engineers with its flexible nature. This mechanism is mediated by two molecules: Cas9 (endonuclease DNA) and the guide RNA (gRNA), and hence, this method unlike ZFN and TALENs is an RNA-guided nuclease mechanism. Using this system, multiple genes are disrupted at the same time, leading to CRISPR mutant library generation. Apart from the general application of this tool, it can also be used in knock-in of genes, and also the CRISPR interference (CRISPRi) induces transcription of sequence-specific interference of target genes. This method helps in the upregulation of endogenous genes leading to ample opportunities to alter the metabolic construct of an organism. Hence, using this CRISPR/Cas9 system, many microalgal species have been engineered, namely, *C. reinhardtii*, *Nannochloropsis* sp., and *P. tricornutum*. This method is not only used for gene editing, but functional characterization of novel genes is also carried out using CRISPR/Cas9 technology. It is also involved in the metabolic engineering of algae, which helps in the increased production of biohydrogen, pigments, lipids, and other biomolecules (Banerjee et al., 2020; Fayyaz et al., 2020; Lu et al., 2021).

6. Applications of genetically engineered algae

Genetic engineering to alter metabolic pathways achieved via the up- and downregulation of transcriptional and translational genes is accomplished by gene knockout or knock-in. This method is superior to random mutagenesis in that it is more accurate and faster and more accurately targets the gene. This improves production and lowers harvesting costs, making it more practical for daily use. This section will try to elucidate the uses of microalgae metabolites in terms of nutritional supplements, food, feedstock, fuel, and biopolymers.

6.1 Microalgae as source of nutraceuticals

The term "nutraceutical," first coined by DeFelice in 1989, describes "any substance conceived as a food, or part of a food which provides medical or health benefits, as well as the prevention and treatment of a disease." As a result, health-promoting nutrients obtained from food or food products to aid in the prevention or treatment of disease and/or dysfunction are also termed nutraceuticals (Bishop & Zubeck, 2012). Microalgae are labeled "GRAS" (generally recognized as safe) (U.S. Food and Drug Administration (HHS), 2016) and sold as nutraceuticals that provide health benefits. Because algae can produce a variety of biomolecules with bioactive properties, such as beta carotene, lutein, astaxanthin, chlorophyll, phycobiliprotein, and polyunsaturated fatty acids, which are beneficial to human and animal health and development; microalgae are particularly interesting as a source for nutraceutical production. *Nostoc* sp. have been utilized as food in China about 2000 years ago; recently, *Chlorella* and *Spirulina* sp. have been used as food in Japan, Taiwan, and Mexico (Sathasivam et al., 2019). Other species such as *Chlamydomonas*, *Scenedesmus*, *Synechococcus*, *Dunaliella*, *Haematococcus*, and *Porphyridium* are employed in producing nutraceuticals. Furthermore, some microalgae with the ability to store critical mineral can be packaged effectively for the nutraceutical formulations to contain. Many microalgae have been currently established for genetic engineering approaches such as *Synechococcus*, *Synechocystis*, *Anabaena*, *C. reinhardtii*, *N. gaditana*, *Ostreococcus tauri*, and *Porphyridium cruentum* (Doron et al., 2016). The CRISPR/Cas9 technology was recently employed to efficiently induce stable targeted gene alterations in model organisms such as the marine diatom *Phaeodactylum tricornutum* (Nymark et al., 2016) and the oleaginous microalgae *Nannochloropsis* (Wang et al., 2016). But the question remains whether organism transformed utilizing editing technology will be termed genetically modified organisms (GMOs) is now being debated.

Carotenoids are pigments found in many foods, cosmetics, and medications that have a variety of beneficial characteristics. Carotenoids are highly valued commercially for their antioxidant ability, vibrant hues, and health-promoting effects (Chakdar & Pabbi, 2017). Astaxanthin, lutein, and β-carotene are the main carotenoids found in microalgae and are most important because of their potential application in the food industry. Other substances with high antioxidant properties include polyketides, phenolic compounds, polyhydroxyalkanoates, proteins, and peptides. These are created in large quantities to protect cells from oxidative stress–induced hydrogen peroxide and oxygen radicals (Vuppaladadiyam et al., 2018). The unicellular microalga *C. reinhardtii* has been engineered to express the Bkt (β-carotene ketolase) gene from *Haematococcus pluvialis* to produce the ketocarotenoid, astaxanthin, and cantaxanthin (Vila et al., 2012). This organism is normally grown on a large scale using a two-stage cultivation approach, in which the cells are grown under ideal conditions in the first stage and then subjected to a combination of stress conditions in the second stage to increase pigment accumulation (Markou & Nerantzis, 2013). Donor gene organism, the green microalga *H. pluvialis*, is able to accumulate around 4% of astaxanthin and has become the first artificial source of natural ketocarotenoids. Endogenous nuclear phytoene desaturase (*pds*) gene expression in the chloroplast of *H. pluvialis* resulted in up to a 90% increase in astaxanthin accumulation per culture volume (Galarza et al., 2018). This is the first time a microalgae natural *pds* nuclear gene has been used to induce change in a plastid. Baek et al. (2018) found that preassembled DNA-free CRISPR-Cas9 generated a knockout mutant of the zeaxanthin epoxidase encoding gene in *C. reinhardtii* strain CC-4349, which has higher zeaxanthin content than the wild type. Exogenous genes encoding proteins are overexpressed. Anila et al. (2016) described the metabolic engineering of the carotenogenic pathway in *Dunaliella salina* for the synthesis of astaxanthin by similar introduction of a *bkt* gene from *H. pluvialis*. Nuclear overexpression of the phytoene synthase (PSY) encoding gene from *D. salina* in *C. reinhardtii* showed a 2.6 times rise in generation of lutein (Couso et al., 2011). An increase in esterified astaxanthin by *H. pluvialis* was also reported by using the upregulated expression technique (Cui et al., 2021). Natural β-carotene is a high-value product than its synthetic counterpart. In comparison with the synthetic version, β-carotene improves the rate of apoptosis in prostrate malignant cells (Jayappriyan et al., 2013). Therefore, microalgae in the manufacturing of natural carotenoids are becoming more popular. Moreover, β-carotene, also known as orange carotenoid, can act as a precursor for vitamin A that is involved in important physiological processes such as lung development, tissue maintenance, and regeneration and, if deficient, can cause serious problems such as disrupting normal lung function and tissue dysfunction (Timoneda et al., 2018).

The microalgae are a rich source of saturated fatty acids, unsaturated fatty acids, and proteins. Their diverse biochemistry suggests that they have enormous promise as immune-boosting supplements (Bishop & Zubeck, 2012). Many biochemical studies of microalgae have revealed a diverse range of metabolites, including omega-3 fatty acids, sulfated polysaccharides, and carotenoids, all of which have antiinflammatory, antiviral, and antibacterial activities (Barkia et al., 2019). Microalgal biomass is increasingly being used to produce medicines, nutraceuticals, and functional foods that carry health advantages. *Chlorella* and *Arthrospira* species, for example, are dominant nutritional supplements with global production of USD 40 million per year (Levasseur et al., 2020). The global *Spirulina* market is worth USD 629.6 million, and it is expected to grow at a 9.4% CAGR by 2025 (Barkia et al., 2019). With a compounded annual growth rate of 6.9%, the microalgal market (biomass or algal-based products) might earn up to $76.68 million USD by 2028 (Data Bridge Market Research, 2021).

Algal proteins have been discovered to be effective antiviral agents with a broad spectrum of activity, blocking viral entry and replication in cells (Ahmadi et al., 2015). Their extracts also aid in the prevention of heart disease by decreasing cholesterol levels (Reyna-Martinez et al., 2018; Ryu et al., 2014). Angiotensin-converting enzyme inhibitory peptides have been developed using microalgal derived peptides, which aid to reduce blood pressure (Sheih et al., 2009). Protein makes up about 40%–60% of the microalgal body weight (depending on the species) (Shim et al., 2008). Microalgal proteins can compete with conventional sources in terms of both quality and quantity. Most commercially available nonalgal protein sources are low in a few key amino acids, whereas market-ready *Chlorella*-based products include substantial levels of all necessary amino acids (Williams & Laurens, 2010). Bioactive peptides from *Nannochloropsis oculata*, *Chlorella pyrenoidosa*, *Arthrospira maxima*, and *Tetraselmis suecica* have been shown to have antioxidant, antiinflammatory, and anti-tumor activities, as well as lowering blood pressure (Wang & Zhang, 2013).

6.2 Microalgae as source of food

Algae have been consumed as food by various cultures for more than thousands of years (Wells et al., 2017). The oldest recorded use of algae as a food source comes from Chile, where archaeological records suggest ingestion of algae dating back 14,000 years (Dillehay et al., 2008). Furthermore, there are numerous textual records demonstrating the use of algae for several millennia as a food source throughout the world. One of the most important factors in considering algal species as food source is their nutritional content, which varies tremendously even among the same species. Only a few microalgae have been granted GRAS designation by the Food and Drug Administration, namely, *Arthrospira platensis*, *C. reinhardtii*, *Auxenochlorella protothecoides*, *Chlorella vulgaris*, *Dunaliella bardawil*, and *Euglena gracilis*. Due to the development of next-generation sequencing technologies and decreasing cost of sequencing, the list of sequenced genomes is constantly growing as the cost of sequencing decreases (Torres et al., 2020). The genetic modification of microalgal species is possible thanks to the availability of sequenced genomes. Efforts in genetic engineering of genomes, mainly on the nuclear and the chloroplast genomes, have led to complete sequencing of many microalgae. Currently, the nuclear genomes of up to nine algae species have been sequenced, including *C. reinhardtii, Chlorella variabilis, Micromonas pusilla, Ostreococcus lucimarinus, O. tauri, Volvox carteri, and Phaeodactylum tricornis.* Presently, 85 cyanobacterial genomes from over 30 distinct species have been entirely sequenced, with a total of 376 genomes being sequenced whole or partially. Advancement and optimization of genetic transformation processes allow for achieving a desired phenotype or function in the microalgae model. To produce increased biomass yield and desired products, metabolic pathways can be altered by overexpression or silencing of specific genes (Sharma, Saharia et al., 2018). Genome manipulation can also be performed by protein-based nucleases from an endonuclease of *Flavobacterium okeanokoites* (named FokI endonuclease) (Kim et al., 1996). Mutations in Chlamydomonas cell are introduced by ZFNs, which was designed specifically with an insert of cop3 gene into an aminoglycoside 3′-phosphotransferase type VIII (aphVIII) marker (Sizova et al., 2013). ZFNs are widely used alongside TALENs in gene editing with many successes in microalgae like *C. reinhardtii* (Gao et al., 2014). ZFNs and TALENs were the most often used gene editing techniques prior to the discovery of the CRISPR/Cas9 system (Christian et al., 2010; Kim et al., 1996).

Creation of expression vectors in most industrially relevant microorganisms is now possible by Golden Gate Modular Clonings (MoClo) toolkits by offering a library of genetic building blocks as a guide for the design and construction process (Iverson et al., 2016; Moore et al., 2016). Crozet et al. (2018) recently constructed the first MoClo toolkit for *C. reinhardtii*, a microalga with approximately 100 gene components comprising 67 unique genetic elements including promoters, untranslated regions (UTRs), terminators, tags, reporters, antibiotic resistance genes, and introns. Similarly, for *Nannochloropsis*, a first high-capacity gene stacking toolkit was built in a modular structure, consisting of numerous adaptive promoters, resistance indicators, and reporter genes into gateway-compatible vectors that may be coupled in various configurations to achieve multigene expression (Poliner et al., 2020). Following the development and cloning of an optimum expression vector, the plasmid vector DNA must be introduced into the algal genome with high efficiency. Successful transformation procedures for microalgal conversions are biolistic particle bombardment, glass bead agitation, electroporation, *Agrobacterium*-mediated, silicon carbide whiskers, and nanoparticles (Jinkerson & Jonikas, 2015). Although electroporation is still the most common way of transformation for most microalgal species, alternative approaches are also beneficial in some instances. Particle bombardment, polyethylene glycol-mediated, and *Agrobacterium*-mediated methods have all been used to transform haptophytes with particle bombardment providing the best results (Velmurugan & Deka, 2018).

Algae are a source of proteins, lipids, fatty acids, vitamins, and minerals. Protein makes up a large portion of the dry biomass of certain algae. Protein content has been reported to make up to 70% of biomass in species such as *Arthrospira platensis* (Szabo et al., 2013). The remaining GRAS species have an average protein content of about 40%, which is relatively high when compared with other plant sources such as soybean (38%) rice (10%) pea (2.8%), or even animal sources such as milk (4%) or eggs (13%) (Chae et al., 2006). Algae not only have a high protein concentration, but their protein composition is also substantially richer than ordinary plant proteins. Similarly, some algae, such as *A. protothecoides*, can store up to 70% of their dry biomass as lipids (Miao & Wu, 2004). The green alga, *Dunaliella tertiolecta*, has been demonstrated to be a good source of vitamin A, vitamin B1, vitamin B9, and vitamin E (Fabregas & Herrero, 1990). The carotenoid, astaxanthin, naturally by cultivating the microalgae the *H. pluvialis*, has a distinctive red pigment that is currently used in animal feeds to confer a yellow or red in farmed-raised salmon and chicken egg yolks (Koller et al., 2014). In addition to the use in human nutrition, microalgae can be used in the livestock, aquaculture, and pet feed. In fact, 30% of present global algal production is sold for animal feed applications (Becker, 2004), while over 50% of current global *Arthrospira* production is used as a feed additive (Spoloare et al., 2006). For the green algae *C. reinhardtii*, many genetic tools have been created for both constitutive and inducible expression, as well as nuclear and chloroplast expression (Barrera & Mayfield, 2013). Addition of endogenous constitutive promoters psbD, atpA, and rbcL, as well as the endogenous inducible promoter psbA, has resulted in reasonable quantities of recombinant protein accumulation in the chloroplast (Barrera & Mayfield, 2013; Moore et al., 2016). But in *Euglena*, modest success was achieved in for recombinant gene production in chloroplast by using only the natural psbA promoter and 3′UTR (Doetsch et al., 2001). Endogenous promoters from rbcs2, B-2-tub, Nos, hsp70A, nia1, and cyc6 have all been found to result in good recombinant gene expression in *C. reinhardtii* nucleus-encoded genes (Barrera & Mayfield, 2013; Scranton et al., 2016). Microalgae, because of their GRAS status, they can also be used to produce recombinant colostrum and milk proteins, which could be very important as dietary supplements for infants in developing countries. Colostrum proteins can be synthesized in algae and can be orally administered from intact algae without purification, directly delivering the proteins to the gut (Torres et al., 2020). This is one reason why only *Porphyridium* species have been utilized in the biotechnological manufacture of arachidonic acid, pigments (phycocyanin, phycoerythrin), and extracellular polysaccharides (Mata et al., 2010). Engineered *Porphyridium cruentum* can be grown in an artificial seawater medium without the need for vitamin B12, which is required by most other red microalgae. Translational repression of *P. tricornutum* by RNAi introduction has been used to enhance lipid content by the knockdown of the gene-encoding nitrate reductase.

6.3 Microalgae as source of biofuel

The quest for a sustainable alternative to conventional fuel has been driven by the factors such as depletion of fossil fuels, rising prices, escalating demands, and global climate change (Behera et al., 2015). In this scenario, microalgae have emerged as a possible candidate for biofuel since several strains accumulate lipid, grow quickly, and have a higher photosynthetic yield than their plant equivalents. Furthermore, microalgae may be cultured without the use of agricultural land or natural landscapes, and they can help to mitigate global climate change by treating wastewater and sequestering carbon dioxide. Furthermore, compared with solar, tidal, and wind energy, biomass-derived liquid fuels can be easily stored, transported, and used directly in automobiles and other transport engines (Scott et al., 2010). Biofuels derived from photosynthetic organisms provide significant potential for meeting global energy demand while remaining carbon-neutral and allowing CO_2 sequestration from the atmosphere (Ravindran et al., 2017). Despite these advantages, microalgae have a number of drawbacks, including as limited lipid output under restrictive growth circumstances and delayed development in strains with high lipid content. The lack of two qualities in extant microalgal species is growth rate (Chen et al., 2017), as well as inadequate light-harvesting capacity (Stephenson et al., 2011). Understanding the molecular complexities of lipid metabolism, particularly triacylglycerol (TAG) biosynthesis, their genetic and metabolic regulations for triggered metabolic flux, target subcellular storage location, and possible secretion could aid in strain improvement, maximizing the cost-effectiveness of microalgal biofuel production. Biotechnological interventions can lead to significant improvements in these areas for commercial biofuel production. Table 21.1 provides an overview of application of different genome-editing tools in photosynthetic microalgae. Understanding the molecular complexities of lipid metabolism, particularly TAG biosynthesis, their genetic and metabolic regulations for triggered metabolic

Table 21.1 List of different genome-editing tools utilized to enhance photosynthetic microalgae.

Species	Genome-editing tool	Mode of action	Target gene	References
Chlamydomonas reinhardtii	CRISPR/Cas9	Knockout/ knockdown	*Hygro, mGFP,FKB12, Gluc* *MAA7, CpSRP43, chlM* *ZEP, CpFTSY* Zeaxanthin epoxidase gene	Shin et al. (2016) Baek et al. (2016) Baek et al. (2018)
	CRISPRi	Knockdown	*PEPC1* and *RFP*	Kao and Ng (2017)
	CRISPR/Cpf1	Knockout	*CpFTSY, CpSRP43*, and *PHT7*	Ferenczi et al. (2018)
	ZNF and CRISPR/Cas9	Knockout	*COP1/2, COP3* (encoding channel rhodopsin 1 [ChR1]) *COP4* (encoding ChR2) *COP5, PHOT, UVR8, VGCC* *MAT3* and *aCRY*	Greiner et al. (2017)
P. tricornutum	Meganucleases and TALEN-mediated	Knockout	UGPase/NAT gene	Daboussi et al. (2014)
	CRISPR/Cas9	Knockout	*CPSRP54* Urease gene, *Vtc2* *Pho4*	Nymark et al. (2016) Slattery et al. (2018) Stukenberg et al. (2018)
Thalassiosira pseudonana	CRISPR/Cas9	Knockout	Urease, silacidin gene	Hopes et al. (2016) Belshaw et al. (2017)
Nannochloropsis oceanica	CRISPR/Cas9	Knockout	Nitrate reductase, ZnCys	Ajjawi et al. (2017) Poliner et al. (2018)
Nannochloropsis gaditana	CRISPR/Cas9	Knockout	Ble, GFP, Aco1, ZnCys, and seven LHC genes	Verruto et al. (2018)
Phellinus ellipsoidea	TALEN-mediated	Knockout	Uridine monophosphate synthetase	Kasai et al. (2015)

flux, target subcellular storage location, and possible secretion could aid in strain improvement, maximizing the cost-effectiveness of microalgal biofuel production. In this section, we discuss various strategies for increasing lipid accumulation and quality, including using an efficient transformation techniques, to regulate key enzymes involved in lipid production, block competing pathways, enabling high cell biomass under nutrient-depleted conditions and other environmental stresses, and control upstream regulators of targets, transcription factors, and microRNAs.

Oil-rich microalgae are rich and most viable feedstock source for large-scale biofuel production because they do not require freshwater (Stephenson et al., 2011). Many oleaginous microalgal species store a substantial amount of lipids in the form of TAG, a convenient source of biofuel that can account for up to 70% of the dry cell weight in some species (Scott et al., 2010). In microalgal cells, TAG is the predominant energy reserve, accounting for 60%−70% of the dry cell weight (Scott et al., 2010). Normally a pool of varied fatty acid (FA) moieties with different degrees of unsaturation are present within the cell serving various purposes. Biomass derived from microalgae can be used to produce biofuels such as biodiesel, bioethanol, biohydrogen, biomethane, and bioelectricity. Microalgae-derived oils are the most promising for biofuel generation (Shuba & Kifle, 2018). As a result, contemporary research focuses on increasing lipid accumulation in microalgae under various circumstances to increase oil output. TAG production in green algae begins in the chloroplast, where the photosynthetic machinery uses atmospheric carbon to produce starch, which is then catabolized by glycolysis to produce FAs and TAGs. Initiation of FA biosynthesis begins when the precursor molecules are converted to acetyl CoA, which is then used to synthesize malonyl CoA by acetyl-CoA carboxylase (ACC). Elongation of the newly synthesized FA is carried out in the stroma where the prokaryotic type II FA synthase (FAS II) is present. FAS II catabolizes the conversion of malonyl-CoA to malonyl-ACP (Shtaida et al., 2015), and the assimilation of FA with the glycerol backbone to form TAG is carried out by the Kennedy pathway (Li-Beisson et al., 2015).

Most of the key genes and enzymes in TAG biosynthesis have been predicted by microalgal omics in the model organism, *Chlamydomonas*. In the *Chlamydomonas* genome, several copies of genes producing acyl-CoA: diacylglycerol acyltransferases were predicted, albeit in fewer copies compared with higher plants, implying that TAG plays an important role in microalgal cell physiology (Liu & Benning, 2013). Conventional methods of enhancing lipid accumulation include stressors such as nutrient deprivation, change of light intensity, temperature, salinity, and pH (Suyono et al., 2015). Among these nitrogen deficiency is the most effective for increasing and storing lipid content (Belotti et al., 2013). Additionally, dual-stage cultivation (Doan & Obbard, 2015) and coculture techniques with chemical additions (Singh et al., 2016) also help to improve lipid production in culture systems. However, genetically modified microalgae provide greater opportunities for precise regulation of target pathways, resulting in increased cellular lipid accumulation under normal growth circumstances (Lim & Schenk, 2017). Successful nuclear transformation has been documented in more than 40 microalgal species so far (Doron et al., 2016). The first attempt at microalgal metabolic engineering aimed at enhancing FA supply to lipid synthesis by overexpressing the acetyl-CoA carboxylase gene (*ACCase*), which codes for the enzyme that carboxylates acetyl-CoA to malonyl-CoA, the first committed step in FA synthesis, was by Dunahay et al. (1996). Although the altered microalgae had a two- to threefold increase in *ACCase* activity, no increase in FA content was observed (Stukenberg et al., 2018). This demonstrated that *ACCase* overexpression had no direct effect on lipid biosynthesis. However, overexpression of an *ACCase* subunit (accD) in combination with malic enzyme, which converts malate to pyruvate, was successful in increasing total lipid content in microalgae (Talebi et al., 2014). Blocking metabolic pathways that compete with lipogenesis, such as starch production and lipid catabolism, is another option for increasing intracellular lipid concentration. Starch is the major storage molecule in some microalgae strains, and inhibiting starch synthesis can direct carbon flow toward lipid production (Ravindran et al., 2017). Table 21.2 indicates a list of some microalgae and the genes modified to elevate lipid accumulation by redirecting carbon pool toward lipogenesis. Since starch as an energy storage molecule is not ubiquitous in microalgae (León-Saiki et al., 2017), suppressing lipid catabolism is a more viable method for increasing lipid content across microalgal strains.

Table 21.2 List of genes manipulated for oil improvement.

Species	Gene	Mode of action	Observations	References
Chlamydomonas Reinhardtii	Acyl ACP-reductase	Overexpression	2.4 times increase in TAG levels	Rengel et al. (2018)
	Citrate synthase	Suppression	3 times increase in TAG levels	Deng et al. (2013)
	Phosphoenolpyruvate carboxylase	Suppression	20% increase in TAG levels	Deng et al. (2014)
	Ferrodoxins	Overexpression	2.5 times increase in lipid level	Huang et al. (2015)
	Diacylglycerol lipase	Suppression	Delay in TAG hydrolysis	Li et al. (2012)
P. tricornutum	Malic enzyme	Overexpression	2.7 times increase in total lipids	Xu et al. (2015)
	Triglyceride lipase 1	Suppression	Increased TAG level in lipid extracts	Barka et al. (2016)
	DGAT	Overexpression	Increased neutral lipid content	Niu et al. (2013)
	Coexpression of GPAT and DGAT	Overexpression	2.6 times increase in TAG levels	Zou et al. (2018)
	C 12 thioesterase	Overexpression	~1.2 × C 12:0 FA accumulation	Radakovits et al. (2011)
Chlorella minutissima	DGAT, PDAT, GPAT, LPAAT, PAP	Overexpression	>30%–40% increase in TAG content	Hseih et al. (2012)
Dunaliella tertiolecta	C 14 thioesterase	Overexpression	2 times more C 14:0 and C 12:0 FA content	Lin et al. (2018)
Coccomyxa sp.	ADP-glucose pyrophosphorylase (*sta6*)	Suppression	Higher lipid content	Takahashi et al. (2018)
Thalassiosira pseudonana	Phosphofructo-2-kinase	Overexpression	Increased glycolytic activity	Abbriano et al. (2018)
Fistulifera solaris	Glycerol kinase	Overexpression	12% increase in lipid productivity	Muto et al. (2015)
Scenedesmus quadricauda	Glycerol-3-phosphate dehydrogenase	Overexpression	1.9 rise in intracellular G3P level	Gomma et al. (2015)

In *Thalassiosira pseudonana,* suppressing the production of a multifunctional lipase/phospholipase/acyltransferase enzyme enhanced lipid yield without affecting growth (Trentacoste et al., 2013). When exposed to silicon deprivation, the mutant strain accumulated 2.4–3.3 times more lipid than the control strain. This may adversely affect microalgal growth, which can be overcome using RNAi-mediated gene silencing under the control of inducible promoters. Many of these promoters have been discovered in microalgae, including one in *Dunaliella* that has light-responsive components (Baek et al., 2016). Overexpression of acyltransferases has produced interesting results in addition to manipulating carbohydrate metabolism and lipid metabolism-related genes for raising cellular neutral lipids (TAG) (see Table 21.2). Several acyltransferases catalyze steps in the Kennedy pathway for TAG assembly, including acyl-CoA: glycerol-3-phosphate acyltransferase (GPAT), acyl-CoA: lysophosphatidic acyltransferase (LPAAT), and acyl-CoA: DGAT. These TAG assembly genes have been identified as promising candidates in genetically engineering the lipid pathway (Maravi et al., 2016). Simultaneous expression of five acyltransferases (phosphatidic acid phosphatase, LPAAT, glycerol-3-phosphate dehydrogenase, GPAT, and DGAT) resulted in a twofold increase in lipid content in the green microalgae *Chlorella minutissima* (Hsieh et al., 2012). In *N. gaditana*, the knockdown of a single transcription regulator *ZnCys* resulted in a 103% increase in lipid content, indicating a lipid yield to the tune of ~5 g/m^2/day (Ajjawi et al., 2017). The Photofuel project (http://www.photofuel.eu) has thoroughly examined the utilization of GE microalgae strains for the release of biofuel precursors to the culture broth for direct separation without the burden of cell harvesting (Varela Villarreal et al., 2020). Yeast fermentation can convert polysaccharides produced by microalgae to bioethanol fuel (Choi et al., 2015). In the fermentation process, however, glycogen is favored over starch because it is smaller and more water soluble. The model cyanobacterial strains *Synechocystis* sp. PCC 6803, *Synechococcus elongatus* PCC 7942, *Synechococcus* sp. NKBG 15041c, and *Synechococcus* sp. PCC 7002 have been extensively employed to study bioethanol generation by the glycogen metabolism pathway (Badary et al., 2018). *Synechocystis* has been used to generate around 836 mg/L of 1-butanol (can be increased to 4.7 g/L) by metabolic engineering in a flask (Miao et al., 2018). These strains with enhanced or modified metabolic activity show great value for biotechnological exploitation. Microalgal hydrocarbons are also used to produce kerosene, diesel, and gasoline. For instance, *Botryococcus braunii* produces hydrocarbons with a high oil yield (Rangarao et al., 2007). Biosyngas is produced by biomass gasification at high temperature of 800–1200°C in the presence of air, oxygen, or water vapor to yield methane, hydrogen, carbon monoxide, water, and ashes (Buxy et al., 2012, pp. 59–67). Similarly, bioethanol is produced by yeast-fermenting sugar sources and also from cellulose and hemicellulose by transforming them to sugars and ultimately ethanol (Harun et al., 2010). The starch content of certain microalgae has been reported to be higher than 50% (Chen et al., 2013). According to Chen et al. (2017), biotechnology methods could cut the cost of producing microalgal biofuel by 15%–20% than traditional methods. The successful implementation of these strategies can make the production of microalgal biofuels competitive to fossil fuels.

6.4 Microalgae as source of biopharmaceutics

Genetic modification technologies in microalgae have opened up new possibilities for producing and delivering recombinant proteins as biopharmaceuticals. Various biopharmaceuticals are synthesized by in vitro recombinant methods in microalgae (Table 21.3). These include vaccinations, antibodies, enzymes, hormones, antibiotics, cytokines, immune signaling proteins, growth factors, and

Table 21.3 Biopharmaceuticals produced by recombinant technology in microalgae.

Species	Biopharmaceutical	Location	Function	Observations	References
Chlamydomonas reinhardtii	Human interferon-α	Nucleus	Immunity	Anticancer activity in HepG2 cell line	El-Ayouty et al. (2019)
	Human proangiogenic growth factors	Nucleus	Wound healing	Endothelial cell tube formation	Jarquín-Cordero et al. (2020)
	Human lactoferrin	Nucleus	Baby food formulation	In vitro anti-bacterial activity	Pang et al. (2019)
	White spot syndrome Virus protein	Chloroplast	Treatment of white spot disease	Improves survival rate in shrimps	Kiataramgul et al. (2020)
	Viral dsRNA	Chloroplast	Shrimp virus	Improves survival rate in shrimps	Charoonnart et al. (2019)
	Taurine synthesis pathway	Chloroplast	Animal feed supplement	Seven times rise in taurine intermediates	Tevatia et al. (2019)
P. tricornutum	Human PNPLA3	Nucleus	Lipid accumulation	Increased levels of lipids in cells	Wang et al. (2018)
***Schizochytrium** sp.*	Epitope for breast cancer	Nucleus	Breast cancer treatment	Humoral antibody response, IgG antibodies to breast cancer proteins	Hernández-Ramírez et al. (2020)
***Chlorella** sp.*	Antimicrobial peptides	Nucleus	Antimicrobial agent for marine animals	Improved fish survival	He et al. (2018)
Haematococcus pluvialis	Antimicrobial peptide (psicidin-4)	Chloroplast	Antimicrobial agent	Accumulated piscidin-4	Wang et al. (2020)

thrombolytic agents (Rader, 2008; Rosales-Mendoza et al., 2020). Conventionally, these therapeutic agents are produced in in vivo systems such as bacteria, yeast, mammalian, and/or insect cells. These in vivo systems suffer from limitations such as high costs, product range, product efficacy, low yields, contamination risk, fragility, and immunogenic effects (Kesik-Brodacka, 2018). These restrictions can be overcome by using eukaryotic microalgae and are now for biopharmaceutical production and delivery. Both the nucleus and the chloroplast in green microalgae can be genetically modified to create biopharmaceuticals (Walsh, 2018). Most engineered microalgal proteins are designed to be stored in the chloroplast because of advantages such as precise integration via homologous recombination and insensitivity to gene silencing and positional effects. Though the chloroplast lacks mechanisms for posttranslational modification, it is known to accumulate more amounts of protein than the nucleus (Lingg et al., 2012). *C. reinhardtii* is the most studied microalgal species for recombinant biopharmaceutical production (Griesbeck et al., 2006). Other algal systems include *D. salina*, *H. pluvialis*, *Chlorella* sp., *N. oculata*, *Schizochytrium* sp., and *P. tricornutum*.

In the face of increasing antibiotic resistances, alternatives to conventional antibiotics are becoming more popular. Antimicrobial peptides (AMPs) are oligopeptides found in plants, animals, and microorganisms that have wide antimicrobial activity against bacteria, mycobacteria, viruses, and fungus. They can be synthetically produced using recombinant methods in low-cost expression systems, such as microalgae. AMPs produced from *Chlorella* sp. have been orally fed to prevent bacterial infection in fish. A fusion protein with in vitro bactericidal activity was produced in *Chlorella* by converting sequences from two fish AMPs, hepcidin and scygonadin, utilizing a plant expression vector and promoter. In fishes *Sparus macrocephalus* and hybrid grouper fed with transgenic *Chlorella* cells, the survival rate against *Aeromonas hydrophila* infection was over 80% in *Somniosus microcephalus* and ~55% in hybrid grouper fish (He et al., 2018).

Like ToAMP4, an antibacterial and antifungal protein from the flower *Taraxacum officinale* expressed in *Chlorella* sp. driven by the *RBCS2* promoter has shown to have antimicrobial activity against *E. coli, Salmonella enteritidis, Staphylococcus aureus, and Bacillus subtilis* (Xue et al., 2020). Another study conducted by Pang used a plant expression vector backbone with codon optimization to create human lactoferrin (hLF) in *C. reinhardtii*. Because of its wide range of actions, including antiviral and anticancer activity, iron binding, intestinal epithelial cell proliferation, and antioxidizing activity, the hLF protein is of interest as a biopharmaceutical. The isolated hLF showed in vitro antibacterial activity against *E. coli* and *Klebsiella variicola* IV-3 (Pang et al., 2019). The protein extract was also nontoxic to mice and had comparable stability with commercially manufactured hLF. The antimicrobial peptide piscidin 4 gene, which is derived from a family of common peptides found in fish, was successfully expressed in *H. pluvialis* [179,180]. The use of these newly acquired genetic tools in algae will increase pathogen protection for a wide range of organisms, whether through direct intake of recombinant algae or exposure to purified proteins.

Human proteins expressed in algae could be used to treat a range of disorders. In the past, many human proteins, such as cytokines and growth hormones, have been expressed in algae (Rasala et al., 2010). More proteins beneficial to human health have been successfully synthesized in the model organism *C. reinhardtii* since then, including human growth hormone expressed in chloroplasts. The cytokine human interferon-alpha (huIFN-alpha) is well known for its antitumor and antiviral action and is primarily synthesized in mammalian, bacterial, and plant cell systems (Noel et al., 2018). *C. reinhardtii* has been successfully engineered to accumulated huIFN-beta-1 and IFN-alpha-2a mRNA transcript. Using Agrobacterium-mediated nuclear transformation system, the human IFN-alpha-2a gene was amplified, transferred into a plant expression vector, and incorporated into the nuclear genome. This purified protein presented in vitro and in vivo antitumour activity and in vitro antiviral activity against the vesicular stomatitis virus (El-Ayouty et al., 2019). The human proangiogenic growth factors expressed in *C. reinhardtii* are a therapy for chronic tissue hypoxia. Growth factors such as hVEGF-165, hPDGF-B, and hSDF-1 were produced using a set of molecular tools including codon optimization, nuclear transformation, and promoters, psaD, and ar1 in cell wall–defective mutant hosts. In vitro experiment of these recombinant proteins increased endothelial tube production and vascular anastomosis (Jarquín-Cordero et al., 2020).

Vitamin B12 deficiency is most common in children, pregnant women, elderly, vegans, and vegetarians. Absence of intrinsic factor (IF) results in pernicious anemia, as IF is necessary to bind to B12 enabling its uptake and absorption in the small intestine. The intracellular and secreted human IF was successfully produced in *C. reinhardtii* by codon optimization of the gene driven by the psaD promoter combined with the inclusion of the native arylsulfatase secretion signal peptide (Lima et al., 2018).

6.5 Microalgae as source of antigens and vaccines

Expressing antigens and antibodies in microalgae has a number of advantages because of its low cost, capacity to do posttranslational modifications, and a variety of delivery methods. Microalgae, for example, can be used as an oral vaccination carrier for fish. To try this, a delivery system was prepared using *C. reinhardtii* to introduce the oral recombinant vaccines through green fluorescent protein to the bloodstream of zebrafish, *Danio rerio* (Kwon et al., 2019). This method of oral vaccinations in aquaculture has been applied to treat white spot syndrome virus (WSSV) in crustaceans especially in shrimps. The algae *D. salina* was first shown to express WSSV viral envelope protein VP28 in the nucleus (Feng et al., 2014). The contradictory result was observed by Kiatar-amgul et al. (2020) reported that after codon optimization of the gene encoding for WSSV VP28 could be expressed in the chloroplast of a cell wall–deficient strain of *C. reinhardtii*. This protein was efficient at increasing the life span of WSSV-infected shrimp by 13% following seven days (Kiataramgul et al., 2020). The chloroplast of cell wall–deficient strain of *C. reinhardtii* was used to express a fragment of dsRNA from the yellow head virus, another shrimp pathogen by using glass bead transformation method (Charoonnart et al., 2019).

In terms of human disease applications, microalgal-derived biopharmaceuticals can be used to produce tumor antigens or antibodies, which can then be used in cancer immunotherapy. The microalga *Schizochytrium* sp. inserted into the Algevir vector was used to generate a chimeric protein BCB by fusing B cell epitopes from the tumor-associated antigens, human epidermal growth factor receptor-2, mucin-like glycoprotein 1, Wilms' tumor antigen (WT1), and mammaglobin for treating breast cancer (Ramos-Vega et al., 2018). *P. tricornutum* transformed by biolistic method was used to produce the monoclonal antihepatitis B IgG antibody (Hempel et al., 2012). Vaccine for the malarial parasite *Plasmodium falciparum* cell traversal protein for ookinetes and sporozoites antigen (PfCelTOS) has been developed in the chloroplast of *C. reinhardtii* (Shamriz & Ofoghi, 2019). Multiple antigens such as Pfs25/28, Pfs48/45, PfMSP1, PbMSP1, and PbAMA1 have been produced in algae *Chlamydomonas* (Dauvillée et al., 2010; Gregory et al., 2012; Jones et al., 2013). Additionally, *S. aureus* D2 protein is fused to the beta subunit of cholera toxin, a known mucosal adjuvant (Dreesen et al., 2010), and others include against infectious diseases such as human papilloma virus, hepatitis B virus, and HIV (Yan et al., 2016). Genetically altered microalgae have been produced for noninfectious diseases such as type 1 diabetes, atherosclerosis, hypertension, allergies, and malignancies (Rosales-Mendoza, 2016b). A number of microalgal recombinant subunit vaccines against classical swine fever virus, foot-and-mouth disease virus, and other viruses have been created using *C. reinhardtii*, which was used as a host system to express antigenic proteins (Rosales-Mendoza, 2016a; Specht & Mayfield, 2014; Yan et al., 2016). Despite the fact that *C. reinhardtii* is a model organism, nuclear transformation is frequently accompanied by transgenic silencing due to a random integration event, whereas transgenes inserted into the chloroplast genome by homologous recombination are not silenced and provide a platform for the production of recombinant proteins. As a result, chloroplast homologous recombination is the primary means of expression along with nuclear transformation. The use of microalgae for research and development of novel medications or vaccines has significant potential as research in biotechnology for algal active substance.

7. Merits of genetic engineering in microalgae

Cultivation of microalgae neither needs high-quality land as in the case of higher plant crops or nor requires a high-quality fermenter as in case of bacteria or yeast. This organism only requires a light and micronutrients, and the production of biomass was very high in the low cost that is one of the advantages of algal culture (Beacham et al., 2017). For genetic modification, microalgae are the best option for various reasons including quick life span, very high photosynthetic rates, high productivity rate, and high adaptogenic activity (Allison et al., 2013).

8. Demerits of genetic engineering in microalgae

In traditional methods, algae are cultured in three different methods, namely, natural locations, open ponds, and closed system using bioreactor. But the genetically modified microalgae cannot be cultivated in the natural location or open ponds because the release of GMOs into environment possesses various challenges. Thus, closed-system cultivation grown in open building may be considered as safe. Various countries framed different regulation to cultivate and process the GMOs. In the Netherlands, for using GMOs in industrial setup for large-scale production, several criteria were fixed in the Good Industrial Large-Scale Practice (GILSP). The important rules in the GLISP are as follows: the host organism should be nonpathogenic, which should have the long-time history for being used in the history without any safety issues, And the next rule is that the GMOs were synthesized using a safe vector and a safe insert (Wijffels, 2015, p. 6071). The European Commission in 2002 had listed the possible adverse effects of GMOs including any toxic effect or allergic to humans, plants, and animals, changing the susceptibility organism, which may lead to formation of new infectious disease or which may able to act as a new vector to carry the disease. GMOs may alter the carbon and nitrogen cycle through varying the rate of decomposition of organic material in the soil (European commission, 2002). Apart from this problem, the release of GM algae in to environment may eradicate the natural population, through the natural selection process (Flynn et al., 2013). Through horizontal gene transfer method, the inserted foreign can easily transfer to the other organism. Horizontal gene transfer method is one of the important tools for the evolution, and in cyanobacterial, 50% of the evolution occurred through horizontal gene transfer method. Thus, handling of GMOs should be careful to prevent the gene transfer. In vertical transfer method, gene transfer occurs via sexual reproduction; therefore, the sexual reproduction cycle through wild-type partner and GMOs results in the transfer of genetically modified gene into the environment.

Transformation of foreign gene into the microalgae is difficult because of the nature of the cell wall composition of algae. All the transformation methods will not be suitable in microalgae due to the rigidity of the cell wall. To increase the cell permeability, different parameter should be standardized, but the success rate in terms of product recovery is very minimal (Sreenikethanam et al., 2022). Though various techniques have been listed to do the genetic modifications in microalgae, the genome sequences were available for very few species only (Chakdar et al., 2021).

9. Conclusion

The versatile nature of microalgae makes them as a good candidate for production of biofuels, proteins, pigments, nutraceuticals, medicines, and other value-added goods. This chapter lists the various techniques in genetic engineering and how these techniques help tailor the microalgae in terms of valuable products. Though genetic engineering can increase the production, every technique will have the pros and cons. This chapter also explains the merits and demerits of genetically modified microalgae.

References

2002/623/EC: Commission decision of 24 July 2002 establishing guidance notes supplementing Annex II to Directive 2001/18/EC of the European Parliament and of the Council on the deliberate release into the environment of genetically modified organisms and repealing Council Directive 90/220/EEC (Text with EEA relevance) (notified under document number C(2002) 2715). Official Journal L, 200, 22–33. https://eur-lex.europa.eu/legal-content/EN/TXT/?uri=CELEX:32002D0623.

Abbriano, R., Vardar, N., Yee, D., & Hildebrand, M. (2018). Manipulation of a glycolytic regulator alters growth and carbon partitioning in the marine diatom Thalassiosira pseudonana. *Algal Research, 32*, 250–258.

Adeniyi, O. M., Azimov, U., & Burluka, A. (2018). Algae biofuel: Current status and future applications. *Renewable and Sustainable Energy Reviews, 90*, 316–335.

Ahmadi, A., Moghadamtousi, S. Z., Abubakar, S., & Zandi, K. (2015). Antiviral potential of algae polysaccharides isolated from marine sources: A review. *BioMed Research International*, 1–10.

Ahmad, I., Sharma, A. K., Daniell, H., & Kumar, S. (2015). Altered lipid composition and enhanced lipid production in green microalga by introduction of brassica diacylglycerol acyltransferase 2. *Plant biotechnology journal, 13*(4), 540–550.

Ajjawi, I., Verruto, J., Aqui, M., Soriaga, L. B., Coppersmith, J., Kwok, K., Peach, L., Orchard, E., Kalb, R., Xu, W., & Carlson, T. J. (2017). Lipid production in *Nannochloropsis gaditana* is doubled by decreasing expression of a single transcriptional regulator. *Nature Biotechnology, 35*(7), 647–652.

Allison, J. G., Wagner, M., McAllister, M., Ren, A., & Snyder, R. (2013). Sand bottom microalgal production and benthic nutrient fluxes on the Northeastern Gulf of Mexico Nearshore Shelf. *Gulf and Caribbean Research, 25*(1), 1–8.

Anila, N., Simon, D. P., Chandrashekar, A., Ravishankar, G. A., & Sarada, R. (2016). Metabolic engineering of Dunaliella salina for production of ketocarotenoids. *Photosynthesis Research, 127*(3), 321–333.

Badary, A., Takamatsu, S., Nakajima, M., Ferri, S., Lindblad, P., & Sode, K. (2018). glycogen production in marine cyanobacterial strain synechococcus sp. NKBG 15041c. *Marine Biotechnology, 20*(2), 109–117. https://doi.org/10.1007/s10126-017-9792-2

Baek, K., Lee, Y., Nam, O., Park, S., Sim, S. J., & Jin, E. (2016). Introducing *Dunaliella* LIP promoter containing light-inducible motifs improves transgenic expression in chlamydomonas reinhardtii. *Biotechnology Journal, 11*, 384–392. https://doi.org/10.1002/biot.201500269

Baek, K., Yu, J., Jeong, J., Sim, S. J., Bae, S., & Jin, E. (2018). Photoautotrophic production of macular pigment in a chlamydomonas reinhardtii strain generated by using DNA-free CRISPR-Cas9 RNP-mediated mutagenesis. *Biotechnology and Bioengineering, 115*(3), 719–728.

Banerjee, S., Banerjee, S., Ghosh, A. K., & Das, D. (2020). Maneuvering the genetic and metabolic pathway for improving biofuel production in algae: Present status and future prospective. *Renewable and Sustainable Energy Reviews, 133*, 110155.

Barka, F., Angstenberger, M., Ahrendt, T., Lorenzen, W., Bode, H. B., & Büchel, C. (2016). Identification of a triacylglycerol lipase in the diatom Phaeodactylum tricornutum. *Biochimica et Biophysica Acta (BBA)-Molecular and Cell Biology of Lipids, 1861*(3), 239–248.

Barkia, I., Saari, N., & Manning, S. R. (2019). Microalgae for high-value products towards human health and nutrition. *Marine Drugs, 17*, 304.

Barrera, D. J., & Mayfield, S. P. (2013). High-value recombinant protein production in microalgae. In A. Richmond, & Hu Q. Emeritus (Eds.), *Handbook of microalgal culture: Applied phycology and biotechnology* (2nd ed., pp. 532–544). Hoboken: John Wiley & Sons, Ltd.

Beacham, T. A., Sweet, J. B., & Allen, M. J. (2017). Large scale cultivation of genetically modified microalgae: A new era for environmental risk assessment. *Algal Research, 25*, 90–100.

Becker, W. (2004). Microalgae in human and animal nutrition. In , *Vol. 312. Handbook of microalgal culture: Biotechnology and applied phycology*. Hoboken, NJ, USA: Wiley Online Library.

Behera, S., Singh, R., Arora, R., Sharma, N. K., Shukla, M., & Kumar, S. (2015). Scope of algae as third generation biofuels. *Frontiers in Bioengineering and Biotechnology, 2*, 90.

Belotti, G., Bravi, M., de Caprariis, B., de Filippis, P., & Scarsella, M. (2013). Effect of nitrogen and phosphorus starvations on Chlorella vulgaris lipids productivity and quality under different trophic regimens for biodiesel production. *American Journal of Plant Sciences, 4*, 2013.

Belshaw, N., Grouneva, I., Aram, L., Gal, A., Hopes, A., & Mock, T. (2017). *Efficient CRISPR/Cas-mediated homologous recombination in the model diatom Thalassiosira pseudonana*. bioRxiv [Preprint]. https://doi.org/10.1101/215582

Benedetti, M., Vecchi, V., Barera, S., & Dall'Osto, L. (2018). Biomass from microalgae: The potential of domestication towards sustainable biofactories. *Microbial Cell Factories, 17*(1), 1–18.

Bhola, V., Swalaha, F., Ranjith Kumar, R., Singh, M., & Bux, F. (2014). Overview of the potential of microalgae for CO2 sequestration. *International Journal of Environmental Science and Technology, 11*(7), 2103–2118.

Bishop, W. M., & Zubeck, H. M. (2012). Evaluation of microalgae for use as nutraceuticals and nutritional supplements. *Journal of Nutrition & Food Sciences, 2*(5), 1–6.

Blatti, J. L., Beld, J., Behnke, C. A., Mendez, M., Mayfield, S. P., & Burkart, M. D. (2012). Manipulating fatty acid biosynthesis in microalgae for biofuel through protein-protein interactions. *PLOS One, 7*.

Brasil, B. D. S. A. F., de Siqueira, F. G., Salum, T. F. C., Zanette, C. M., & Spier, M. R. (2017). Microalgae and cyanobacteria as enzyme biofactories. *Algal Research, 25*, 76–89.

Buxy, S., Diltz, R., & Pullammanappallil, P. (2012). Biogasification of marine algae Nannochloropsis oculata. In *Materials challenges in alternative and renewable energy II: Ceramic transactions* (pp. 59–67).

Chae, S. R., Hwang, E. J., & Shin, H. S. (2006). Single cell protein production of Euglena gracilis and carbon dioxide fixation in an innovative photo-bioreactor. *Bioresource Technology, 97*(2), 322–329.

Chakdar, H., Hasan, M., Pabbi, S., Nevalainen, H., & Shukla, P. (2021). High-throughput proteomics and metabolomic studies guide re-engineering of metabolic pathways in eukaryotic microalgae: A review. *Bioresource Technology, 321*, 124495.

Chakdar, H., & Pabbi, S. (2017). Algal pigments for human health and cosmeceuticals. In *Algal green chemistry* (pp. 171–188). Elsevier.

Charoonnart, P., Worakajit, N., Zedler, J. A., Meetam, M., Robinson, C., & Saksmerprome, V. (2019). Generation of microalga Chlamydomonas reinhardtii expressing shrimp antiviral dsRNA without supplementation of antibiotics. *Scientific Reports, 9*(1), 1–8.

Chen, B., Wan, C., Mehmood, M. A., Chang, J. S., Bai, F., & Zhao, X. (2017). Manipulating environmental stresses and stress tolerance of microalgae for enhanced production of lipids and value-added products—a review. *Bioresource Technology, 244*, 1198–1206.

Chen, C. Y., Zhao, X. Q., Yen, H. W., Ho, S. H., Cheng, C. L., Lee, D. J., & Chang, J. S. (2013). Microalgae-based carbohydrates for biofuel production. *Biochemical Engineering Journal, 78*, 1–10.

Choi, S., Song, C. W., Shin, J. H., & Lee, S. Y. (2015). Biorefineries for the production of top building block chemicals and their derivatives. *Metabolic Engineering, 28*, 223–239. https://doi.org/10.1016/j.ymben.2014.12.007

Christian, M., Cermak, T., Doyle, E. L., Schmidt, C., Zhang, F., Hummel, A., Bogdanove, A. J., & Voytas, D. F. (2010). Targeting DNA double-strand breaks with TAL effector nucleases. *Genetics, 186*(2), 757–761. https://doi.org/10.1534/genetics.110.120717

Couso, I., Vila, M., Rodriguez, H., Vargas, M. A., & León, R. (2011). Overexpression of an exogenous phytoene synthase gene in the unicellular alga Chlamydomonas reinhardtii leads to an increase in the content of carotenoids. *Biotechnology Progress, 27*(1), 54–60.

Crozet, P., Navarro, F. J., Willmund, F., Mehrshahi, P., Bakowski, K., Lauersen, K. J., & Lemaire, S. D. (2018). Birth of a photosynthetic chassis: A MoClo toolkit enabling synthetic biology in the microalga Chlamydomonas reinhardtii. *ACS Synthetic Biology, 7*(9), 2074–2086.

Cui, H., Xu, W., Zhu, X., Zhao, C., Cui, Y., Ji, C., & Li, R. (2021). Characterization of a Haematococcus pluvialis diacylglycerol acyltransferase 1 and its potential in unsaturated fatty acid-rich triacylglycerol production. *Frontiers of Plant Science, 12*, 771300-771300.

Daboussi, F., Leduc, S., Maréchal, A., Dubois, G., Guyot, V., Perez-Michaut, C., Amato, A., Falciatore, A., Juillerat, A., Beurdeley, M., & Voytas, D. F. (2014). Genome engineering empowers the diatom *Phaeodactylum tricornutum* for biotechnology. *Nature Communications, 5*(1), 1–7. https://doi.org/10.1038/ncomms4831

Data Bridge Market Research. (2021). *Global microalgae market – industry trends and forecast to 2028* Accessed on: 23 April 2021, from https://www.databridgemarketresearch.com/reports/global-microalgae-market.

Dauvillée, D., Delhaye, S., Gruyer, S., Slomianny, C., Moretz, S. E., D'Hulst, C., Long, C. A., Ball, S. G., & Tomavo, S. (2010). Engineering the chloroplast targeted malarial vaccine antigens in *Chlamydomonas* starch granules. *PLoS One, 5*(12), e15424. https://doi.org/10.1371/journal.pone.0015424

Deng, X., Cai, J., & Fei, X. (2013). Effect of the expression and knockdown of citrate synthase gene on carbon flux during triacylglycerol biosynthesis by green algae Chlamydomonas reinhardtii. *BMC Biochemistry, 14*(1), 1–11.

Deng, X., Cai, J., Li, Y., & Fei, X. (2014). Expression and knockdown of the PEPC1 gene affect carbon flux in the biosynthesis of triacylglycerols by the green alga Chlamydomonas reinhardtii. *Biotechnology Letters, 36*(11), 2199–2208.

Dillehay, T. D., Ramírez, C., Pino, M., Collins, M. B., Rossen, J., & Pino Navarro, J. D. (2008). Monte verde: Seaweed, food, medicine, and the peopling of South America. *Science, 320*(5877), 784–786.

Doan, Y. T. T., & Obbard, J. P. (2015). Two-stage cultivation of a Nannochloropsis mutant for biodiesel feedstock. *Journal of Applied Phycology, 27*(6), 2203–2208.

Doetsch, N. A., Favreau, M. R., Kuscuoglu, N., Thompson, M. D., & Hallick, R. B. (2001). Chloroplast transformation in Euglena gracilis: Splicing of a group III twintron transcribed from a transgenic psbK operon. *Current Genetics, 39*(1), 49–60.

Doron, L., Segal, N., & Shapira, M. (2016). Transgene expression in microalgae-from tools to applications. *Frontiers of Plant Science, 7*, 505. https://doi.org/10.3389/fpls.2016.00505

Dreesen, I. A., Charpin-El Hamri, G., & Fussenegger, M. (2010). Heat-stable oral alga-based vaccine protects mice from Staphylococcus aureus infection. *Journal of Biotechnology, 145*, 273–280. https://doi.org/10.1016/j.jbiotec.2009.12.006

Dunahay, T. G., Jarvis, E. E., Dais, S. S., & Roessler, P. G. (1996). Manipulation of microalgal lipid production using genetic engineering. *Applied Biochemistry and Biotechnology, 57*(1), 223–231.

Dunahay, T. G., Jarvis, E. E., & Roessler, P. G. (1995). Genetic transformation of the diatoms Cyclotella cryptica and Navicula saprophila. *Journal of Phycology, 31*(6), 1004–1012.

El-Ayouty, Y., El-Manawy, I., Nasih, S., Hamdy, E., & Kebeish, R. (2019). Engineering Chlamydomonas reinhardtii for expression of functionally active human interferon-α. *Molecular Biotechnology, 61*(2), 134–144.

Fabregas, J., & Herrero, C. (1990). Vitamin content of four marine microalgae potential use as source of vitamins in nutrition. *Journal of Industrial Microbiology and Biotechnology, 5*(4), 259–263.

Fabris, M., Abbriano, R. M., Pernice, M., Sutherland, D. L., Commault, A. S., Hall, C. C., … Ralph, P. J. (2020). Emerging technologies in algal biotechnology: Toward the establishment of a sustainable, algae-based bioeconomy. *Frontiers of Plant Science, 11*, 279.

Fan, J., Cui, Y., Wan, M., Wang, W., & Li, Y. (2014). Lipid accumulation and biosynthesis genes response of the oleaginous Chlorella pyrenoidosa under three nutrition stressors. *Biotechnology for Biofuels, 7*(1), 1–14.

FAO. (2002). *World agriculture: Towards 2015/2030, summary Report*. Available online: http://www.fao.org/docrep/004/Y3557E/y3557e03.htm (accessed on 14 May 2022).

FAO. (2004). *Protein sources for the animal feed industry.* Available online: http://www.fao.org/docrep/007/y5019e/y5019e03.htm (accessed on 14 May 2022).

Fayyaz, M., Chew, K. W., Show, P. L., Ling, T. C., Ng, I. S., & Chang, J. S. (2020). Genetic engineering of microalgae for enhanced biorefinery capabilities. *Biotechnology Advances, 43*, 107554.

Feng, S., Feng, W., Zhao, L., Gu, H., Li, Q., Shi, K., & Zhang, N. (2014). Preparation of transgenic Dunaliella salina for immunization against white spot syndrome virus in crayfish. *Archives of Virology, 159*(3), 519–525.

Ferenczi, A., Pyott, D. E., Xipnitou, A., & Molnar, A. (2018). Efficient targeted DNA editing and replacement in Chlamydomonas reinhardtii using Cpf1 ribonucleoproteins and single-stranded DNA. *Proceeding of the National Academy of Sciences of the United States of America, 114*, 13567–13572. https://doi.org/10.1073/pnas.1710597114

Fire, A., Xu, S., Montgomery, M. K., Kostas, S. A., Driver, S. E., & Mello, C. C. (1998). Potent and specific genetic interference by double-stranded RNA in Caenorhabditis elegans. *Nature, 391*(6669), 806–811.

Flynn, K. J., Mitra, A., Greenwell, H. C., & Sui, J. (2013). Monster potential meets potential monster: Pros and cons of deploying genetically modified microalgae for biofuels production. *Interface focus, 3*(1), 20120037.

Galarza, J. I., Gimpel, J. A., Rojas, V., Arredondo-Vega, B. O., & Henríquez, V. (2018). Over-accumulation of astaxanthin in *Haematococcus pluvialis* through chloroplast genetic engineering. *Algal Research, 31*, 291–297.

Gao, H., Wright, D. A., Li, T., Wang, Y., Horken, K., Weeks, D. P., & Spalding, M. H. (2014). TALE activation of endogenous genes in Chlamydomonas reinhardtii. *Algal Research, 5*, 52–60.

Gomma, A. E., Lee, S. K., Sun, S. M., Yang, S. H., & Chung, G. (2015). Improvement in oil production by increasing Malonyl-CoA and Glycerol-3-phosphate pools in Scenedesmus quadricauda. *55*, 447–455. https://doi.org/10.1007/s12088-015-0546-4

Gregory, J. A., Li, F., Tomosada, L. M., Cox, C. J., Topol, A. B., Vinetz, J. M., & Mayfield, S. (2012). Algae-produced Pfs25 elicits antibodies that inhibit malaria transmission. *PLoS One, 7*(5), e37179.

Greiner, A., Kelterborn, S., Evers, H., Kreimer, G., Sizova, I., & Hegemann, P. (2017). Targeting of photoreceptor genes in Chlamydomonas reinhardtii via zinc-finger nucleases and CRISPR/Cas9. *The Plant Cell, 29*, 2498–2518. https://doi.org/10.1105/tpc.17.00659

Griesbeck, C., Kobl, I., & Heitzer, M. (2006). Chlamydomonas reinhardtii. *Molecular Biotechnology, 34*(2), 213–223.

Guiry, M. D. (2012). How many species of algae are there? *Journal of Phycology, 48*(5), 1057–1063.

Hagemann, M., Gärtner, K., Scharnagl, M., Bolay, P., Lott, S. C., Fuss, J., Huettel, B., Reinhardt, R., Klähn, S., & Hess, W. R. (2018). Identification of the DNA methyltransferases establishing the methylome of the cyanobacterium Synechocystis sp. PCC 6803. *DNA Research, 25*(4), 343–352.

Harun, R., Danquah, M. K., & Forde, G. M. (2010). Microalgal biomass as a fermentation feedstock for bioethanol production. *Journal of Chemical Technology & Biotechnology, 85*(2), 199–203.

Hempel, F., & Maier, U. G. (2012). An engineered diatom acting like a plasma cell secreting human IgG antibodies with high efficiency. *Microbial Cell Factories, 11*(1), 1–6.

He, Y., Peng, H., Liu, J., Chen, F., Zhou, Y., Ma, X., & Wang, K. (2018). Chlorella sp. transgenic with Scy-hepc enhancing the survival of Sparus macrocephalus and hybrid grouper challenged with Aeromonas hydrophila. *Fish & Shellfish Immunology, 73*, 22–29.

Hernández-Ramírez, J., Wong-Arce, A., González-Ortega, O., & Rosales-Mendoza, S. (2020). Expression in algae of a chimeric protein carrying several epitopes from tumor associated antigens. *International Journal of Biological Macromolecules, 147*, 46–52.

Hopes, A., Nekrasov, V., Kamoun, S., & Mock, T. (2016). Editing of the urease gene by CRISPR-Cas in the diatom Thalassiosira pseudonana. *Plant Methods, 12*, 49. https://doi.org/10.1186/s13007-016-0148-0

Hsieh, H. J., Su, C. H., & Chien, L. J. (2012). Accumulation of lipid production in Chlorella minutissima by triacylglycerol biosynthesis-related genes cloned from Saccharomyces cerevisiae and Yarrowia lipolytica. *Journal of Microbiology, 50*(3), 526–534.

Huang, W., & Daboussi, F. (2017). Genetic and metabolic engineering in diatoms. *Philosophical Transactions of the Royal Society*. B3722016041120160411.

Huang, A., He, L., & Wang, G. (2011). Identification and characterization of microRNAs from Phaeodactylum tricornutum by high-throughput sequencing and bioinformatics analysis. *BMC Genomics, 12*(1), 1–11.

Huang, L. F., Lin, J. Y., Pan, K. Y., Huang, C. K., & Chu, Y. K. (2015). Overexpressing ferredoxins in Chlamydomonas reinhardtii increase starch and oil yields and enhance electric power production in a photo microbial fuel cell. *International Journal of Molecular Sciences, 16*(8), 19308–19325.

Iverson, S. V., Haddock, T. L., Beal, J., & Densmore, D. M. (2016). CIDAR MoClo: Improved MoClo assembly standard and new E. coli part library enable rapid combinatorial design for synthetic and traditional biology. *ACS Synthetic Biology, 5*(1), 99–103.

Jarquín-Cordero, M., Chávez, M. N., Centeno-Cerdas, C., Bohne, A. V., Hopfner, U., Machens, H. G., & Nickelsen, J. (2020). Towards a biotechnological platform for the production of human pro-angiogenic growth factors in the green alga Chlamydomonas reinhardtii. *Applied Microbiology and Biotechnology, 104*(2), 725–739.

Jayappriyan, K. R., Rajkumar, R., Venkatakrishnan, V., Nagaraj, S., & Rengasamy, R. (2013). In vitro anticancer activity of natural β-carotene from *Dunaliella salina* EU5891199 in PC-3 cells. *Biomedicine and Preventive Nutrition, 3*(2), 99–105.

Jeon, S., Lim, J. M., Lee, H. G., Shin, S. E., Kang, N. K., Park, Y. I., & Chang, Y. K. (2017). Current status and perspectives of genome editing technology for microalgae. *Biotechnology for Biofuels, 10*(1), 1–18.

Jinkerson, R. E., & Jonikas, M. C. (2015). Molecular techniques to interrogate and edit the Chlamydomonas nuclear genome. *The Plant Journal, 82*(3), 393–412.

Jones, C. S., Luong, T., Hannon, M., Tran, M., Gregory, J. A., Shen, Z., Briggs, S. P., & Mayfield, S. P. (2013). Heterologous expression of the C-terminal antigenic domain of the malaria vaccine candidate Pfs48/45 in the green algae *Chlamydomonas reinhardtii. Applied Microbiology and Biotechnology, 97*(5), 1987–1995. https://doi.org/10.1007/s00253-012-4071-7

Kao, P. H., & Ng, I. S. (2017). CRISPRi mediated phosphoenolpyruvate carboxylase regulation to enhance the production of lipid in Chlamydomonas reinhardtii. *Bioresource Technology, 245*, 1527–1537. https://doi.org/10.1016/j.biortech.2017.04.111

Kasai, Y., Oshima, K., Ikeda, F., Abe, J., Yoshimitsu, Y., & Harayama, S. (2015). Construction of a self-cloning system in the unicellular green alga Pseudochoricystis ellipsoidea. *Biotechnology Biofuel, 8*, 94. https://doi.org/10.1186/s13068-015-0277-0z

Kesik-Brodacka, M. (2018). Progress in biopharmaceutical development. *Biotechnology and Applied Biochemistry, 65*(3), 306–322.

Ketting, R. F. (2011). The many faces of RNAi. *Developmental Cell, 20*(2), 148–161.

Kiataramgul, A., Maneenin, S., Purton, S., Areechon, N., Hirono, I., Brocklehurst, T. W., & Unajak, S. (2020). An oral delivery system for controlling white spot syndrome virus infection in shrimp using transgenic microalgae. *Aquaculture, 521*, 735022.

Kim, Y. G., Cha, J., & Chandrasegaran, S. (1996). Hybrid restriction enzymes: Zinc finger fusions to Fok I cleavage domain. *Proceedings of the National Academy of Sciences, 93*(3), 1156–1160.

Koyande, A. K., Chew, K. W., Rambabu, K., Tao, Y., Chu, D. T., & Show, P. L. (2019). Microalgae: A potential alternative to health supplementation for humans. *Food Science and Human Wellness, 8*(1), 16–24.

Kwon, K. C., Lamb, A., Fox, D., & Jegathese, S. J. P. (2019). An evaluation of microalgae as a recombinant protein oral delivery platform for fish using green fluorescent protein (GFP). *Fish & Shellfish Immunology, 87*, 414–420.

Lee, J. W., & Greenbaum, E. (2003). A new oxygen sensitivity and its potential application in photosynthetic H 2 production. In *Biotechnology for fuels and chemicals* (pp. 303–313). Totowa, NJ: Humana Press.

León-Saiki, G. M., Remmers, I. M., Martens, D. E., Lamers, P. P., Wijffels, R. H., & van der Veen, D. (2017). The role of starch as transient energy buffer in synchronized microalgal growth in Acutodesmus obliquus. *Algal Research, 25*, 160–167.

Levasseur, W., Perré, P., & Pozzobon, V. (2020). A review of high value-added molecules production by microalgae in light of the classification. *Biotechnology Advances, 41*, 107545. https://doi.org/10.1016/j.biotechadv.2020.107545

LiBeisson, Y., Beisson, F., & Riekhof, W. (2015). Metabolism of acyl-lipids in Chlamydomonas reinhardtii. *The Plant Journal, 82*(3), 504–522.

Li, X., Benning, C., & Kuo, M. H. (2012). Rapid triacylglycerol turnover in Chlamydomonas reinhardtii requires a lipase with broad substrate specificity. *Eukaryotic Cell, 11*(12), 1451–1462.

Lima, S., Webb, C. L., Deery, E., Robinson, C., & Zedler, J. A. (2018). Human intrinsic factor expression for bioavailable vitamin B12 enrichment in microalgae. *Biology, 7*(1), 19.

Lim, D. K., & Schenk, P. M. (2017). Microalgae selection and improvement as oil crops: GM vs non-GM strain engineering. *AIMS Bioengineering, 4*(1), 151–161.

Lingg, N., Zhang, P., Song, Z., & Bardor, M. (2012). The sweet tooth of biopharmaceuticals: Importance of recombinant protein glycosylation analysis. *Biotechnology Journal, 7*(12), 1462–1472.

Lin, H., Shen, H., & Lee, Y. K. (2018). Cellular and molecular responses of Dunaliella tertiolecta by expression of a plant medium chain length fatty acid specific Acyl-ACP thioesterase. *Frontiers in Microbiology, 9*, 619.

Litchman, E. (2007). Resource competition and the ecological success of phytoplankton. In *Evolution of primary producers in the sea* (pp. 351–375). Academic Press.

Liu, B., & Benning, C. (2013). Lipid metabolism in microalgae distinguishes itself. *Current Opinion in Biotechnology, 24*, 300–309. https://doi.org/10.1016/j.copbio.2012.08.008

Li, X., Wang, X., Duan, C., Yi, S., Gao, Z., Xiao, C., … Li, J. (2020). Biotechnological production of astaxanthin from the microalga Haematococcus pluvialis. *Biotechnology Advances, 43*, 107602.

Lu, Y., Zhang, X., Gu, X., Lin, H., & Melis, A. (2021). Engineering microalgae: Transition from empirical design to programmable cells. *Critical Reviews in Biotechnology, 41*(8), 1233–1256.

Mallick, N., Bagchi, S. K., Koley, S., & Singh, A. K. (2016). Progress and challenges in microalgal biodiesel production. *Frontiers in Microbiology, 7*, 1019.

Maravi, D. K., Kumar, S., Sharma, P. K., Kobayashi, Y., Goud, V. V., Sakurai, N., Koyama, H., & Sahoo, L. (2016). Ectopic expression of AtDGAT1, encoding diacylglycerol O-acyltransferase exclusively committed to TAG biosynthesis, enhances oil accumulation in seeds and leaves of Jatropha. *Biotechnology for Biofuels, 9*(1), 1–13. https://doi.org/10.1186/s13068-016-0642-7

Markou, G., & Nerantzis, E. (2013). Microalgae for high-value compounds and biofuels production: A review with focus on cultivation under stress conditions. *Biotechnology Advances, 31*(8), 1532–1542.

Mata, T. M., Martins, A. A., & Caetano, N. S. (2010). Microalgae for biodiesel production and other applications: A review. *Renewable and Sustainable Energy Reviews, 14*(1), 217–232.

Melillo, J. M., McGuire, A. D., Kicklighter, D. W., Moore, B., Vorosmarty, C. J., & Schloss, A. L. (1993). Global climate change and terrestrial net primary production. *Nature, 363*(6426), 234–240.

Miao, X., & Wu, Q. (2004). High yield bio-oil production from fast pyrolysis by metabolic controlling of Chlorella protothecoides. *Journal of Biotechnology, 110*(1), 85–93.

Miao, R., Xie, H., Ho, F. M., & Lindblad, P. (2018). Protein engineering of α-ketoisovalerate decarboxylase for improved isobutanol production in Synechocystis PCC 6803. *Metabolic Engineering, 47*, 42–48.

Moore, S. J., Lai, H. E., Kelwick, R. J., Chee, S. M., Bell, D. J., Polizzi, K. M., & Freemont, P. S. (2016). EcoFlex: A multifunctional MoClo kit for E. coli synthetic biology. *ACS Synthetic Biology, 5*(10), 1059–1069.

Mussgnug, J. H. (2015). Genetic tools and techniques for Chlamydomonas reinhardtii. *Applied Microbiology and Biotechnology, 99*(13), 5407–5418.

Muto, M., Tanaka, M., Liang, Y., Yoshino, T., Matsumoto, M., & Tanaka, T. (2015). Enhancement of glycerol metabolism in the oleaginous marine diatom Fistulifera solaris JPCC DA0580 to improve triacylglycerol productivity. *Biotechnology for Biofuels, 8*, 4. https://doi.org/10.1186/s13068-014-0184-9

Niu, Y. F., Zhang, M. H., Li, D. W., Yang, W. D., Liu, J. S., Bai, W. B., & Li, H. Y. (2013). Improvement of neutral lipid and polyunsaturated fatty acid biosynthesis by overexpressing a type 2 diacylglycerol acyltransferase in marine diatom Phaeodactylum tricornutum. *Marine Drugs, 11*(11), 4558–4569.

Noel, N., Jacquelin, B., Huot, N., Goujard, C., Lambotte, O., & Müller-Trutwin, M. (2018). Interferon-associated therapies toward HIV control: The back and forth. *Cytokine & Growth Factor Reviews, 40*, 99–112.

Nymark, M., Sharma, A. K., Sparstad, T., Bones, A. M., & Winge, P. (2016). A CRISPR/Cas9 system adapted for gene editing in marine algae. *Scientific Reports, 6*, 24951. https://doi.org/10.1038/srep24951

Ortiz-Marquez, J. C. F., Do Nascimento, M., Zehr, J. P., & Curatti, L. (2013). Genetic engineering of multispecies microbial cell factories as an alternative for bioenergy production. *Trends in Biotechnology, 31*(9), 521–529.

Pang, X., Tong, Y., Xue, W., Yang, Y. F., Chen, X., Liu, J., & Chen, D. (2019). Expression and characterization of recombinant human lactoferrin in edible alga Chlamydomonas reinhardtii. *Bioscience, Biotechnology, and Biochemistry, 83*(5), 851–859.

Paniagua-Michel, J. (2015). Microalgal nutraceuticals. In *Handbook of marine microalgae* (pp. 255–267). Academic Press.

Poliner, E., Clark, E., Cummings, C., Benning, C., & Farre, E. M. (2020). A high-capacity gene stacking toolkit for the oleaginous microalga, Nannochloropsis oceanica CCMP1779. *Algal Research, 45*, 101664.

Poliner, E., Pulman, J. A., Zienkiewicz, K., Childs, K., Benning, C., & Farré, E. M. (2018). A toolkit for Nannochloropsis oceanica CCMP1779 enables gene stacking and genetic engineering of the eicosapentaenoic acid pathway for enhanced long-chain polyunsaturated fatty acid production. *Plant Biotechnology Journal, 16*(1), 298–309.

Radakovits, R., Eduafo, P. M., & Posewitz, M. C. (2011). Genetic engineering of fatty acid chain length in Phaeodactylum tricornutum. *Metabolic Engineering, 13*(1), 89–95.

Rader, R. A. (2008). (Re) defining biopharmaceutical. *Nature Biotechnology, 26*(7), 743–751.

Ramos-Vega, A., Rosales-Mendoza, S., Bañuelos-Hernández, B., & Angulo, C. (2018). Prospects on the use of Schizochytrium sp. to develop oral vaccines. *Frontiers in Microbiology*, 2506.

Ranga, R., Sarada, R., & Ravishankar, G. A. (2007). Influence of CO_2 on growth and hydrocarbon production in Botryococcus braunii. *Journal of Microbiology and Biotechnology, 17*(3), 414–419.

Rasala, B. A., & Mayfield, S. P. (2015). Photosynthetic biomanufacturing in green algae; production of recombinant proteins for industrial, nutritional, and medical uses. *Photosynthesis Research, 123*(3), 227–239.

Rasala, B. A., Muto, M., Lee, P. A., Jager, M., Cardoso, R. M., Behnke, C. A., & Mayfield, S. P. (2010). Production of therapeutic proteins in algae, analysis of expression of seven human proteins in the chloroplast of Chlamydomonas reinhardtii. *Plant Biotechnology Journal, 8*(6), 719–733.

Ravindran, B., Kurade, M. B., Kabra, A. N., Jeon, B. H., & Gupta, S. K. (2017). Recent advances and future prospects of microalgal lipid biotechnology. In *Algal biofuels* (pp. 1–37). Cham: Springer.

Rengel, R., Smith, R. T., Haslam, R. P., Sayanova, O., Vila, M., & Leon, R. (2018). Overexpression of acetyl-CoA synthetase (ACS) enhances the biosynthesis of neutral lipids and starch in the green microalga Chlamydomonas reinhardtii. *Algal Research, 31*, 183–193.

Reyna-Martinez, R., Gomez-Flores, R., López-Chuken, U., Quintanilla-Licea, R., Caballero-Hernandez, D., Rodríguez-Padilla, C., & Tamez-Guerra, P. (2018). Antitumor activity of Chlorella sorokiniana and Scenedesmus sp. microalgae native of Nuevo León state. *México Peer Journal, 6*, e4358.

Rosales-Mendoza, S. (2016a). Genetic engineering approaches for algae. In S. Rosales-Mendoza (Ed.), *Algae-based biopharmaceuticals* (pp. 15–40). Berlin: Springer International.

Rosales-Mendoza, S. (2016b). Perspectives for the algae-made biopharmaceuticals field. In S. Rosales-Mendoza (Ed.), *Algae-based biopharmaceuticals* (pp. 143–163). Berlin: Springer International.

Rosales-Mendoza, S., Solís-Andrade, K. I., Márquez-Escobar, V. A., González-Ortega, O., & Bañuelos-Hernandez, B. (2020). Current advances in the algae-made biopharmaceuticals field. *Expert Opinion on Biological Therapy, 20*(7), 751–766.

Ryu, N. H., Lim, Y., Park, J. E., Kim, J., Kim, J. Y., Kwon, S. W., & Kwon, O. (2014). Impact of daily Chlorella consumption on serum lipid and carotenoid profiles in mildly hypercholesterolemic adults: A double-blinded, randomized, placebo-controlled study. *Nutrition Journal, 13*, 1–8.

Sathasivam, R., Radhakrishnan, R., Hashem, A., & Abd Allah, E. F. (2019). Microalgae metabolites: A rich source for food and medicine. *Saudi Journal of Biological Sciences, 26*(4), 709–722. https://doi.org/10.1016/j.sjbs.2017.11.003

Scott, S. A., Davey, M. P., Dennis, J. S., Horst, I., Howe, C. J., Lea-Smith, D. J., & Smith, A. G. (2010). Biodiesel from algae: Challenges and prospects. *Current Opinion in Biotechnology, 21*(3), 277–286.

Scranton, M. A., Ostrand, J. T., Georgianna, D. R., Lofgren, S. M., Li, D., Ellis, R. C., & Mayfield, S. P. (2016). Synthetic promoters capable of driving robust nuclear gene expression in the green alga Chlamydomonas reinhardtii. *Algal Research, 15*, 135–142.

Shamriz, S., & Ofoghi, H. (2019). Expression of recombinant PfCelTOS antigen in the chloroplast of Chlamydomonas reinhardtii and its potential use in detection of malaria. *Molecular Biotechnology, 61*(2), 102–110.

Sharma, S., Puttachary, S., Thippeswamy, A., Kanthasamy, A. G., & Thippeswamy, T. (2018). Status epilepticus: Behavioral and electroencephalography seizure correlates in kainate experimental models. *Frontiers in Neurology, 9*, 7.

Sharma, P. K., Saharia, M., Srivstava, R., Kumar, S., & Sahoo, L. (2018). Tailoring microalgae for efficient biofuel production. *Frontiers in Marine Science, 5*, 382.

Sheih, I. C., Fang, T. J., & Wu, T. K. (2009). Isolation and characterisation of a novel angiotensin I-converting enzyme (ACE) inhibitory peptide from the algae protein waste. *Food Chemistry, 115*, 279–284.

Shin, S. E., Lim, J. M., Koh, H. G., Kim, E. K., Kang, N. K., Jeon, S., Kwon, S., Shin, W. S., Lee, B., Hwangbo, K., & Kim, J. (2016). CRISPR/Cas9-induced knockout and knock-in mutations in *Chlamydomonas reinhardtii*. *Scientific Reports, 6*(1), 1–15.

Shtaida, N., Khozin-Goldberg, I., & Boussiba, S. (2015). The role of pyruvate hub enzymes in supplying carbon precursors for fatty acid synthesis in photosynthetic microalgae. *Photosynthesis Research, 125*(3), 407–422.

Shuba, E. S., & Kifle, D. (2018). Microalgae to biofuels: Promising alternative and renewable energy, review. *Renewable and Sustainable Energy Reviews, 81*, 743–755.

Singh, S. K., Sundaram, S., Sinha, S., Rahman, M. A., & Kapur, S. (2016). Recent advances in CO2 uptake and fixation mechanism of cyanobacteria and microalgae. *Critical Reviews in Environmental Science and Technology, 46*(16), 1297–1323.

Sizova, I., Greiner, A., Awasthi, M., Kateriya, S., & Hegemann, P. (2013). Nuclear gene targeting in Chlamydomonas using engineered zinc-finger nucleases. *The Plant Journal, 73*(5), 873–882. https://doi.org/10.1111/tpj.12066

Slattery, S. S., Diamond, A., Wang, H., Therrien, J. A., Lant, J. T., Jazey, T., Lee, K., Klassen, Z., Desgagné-Penix, I., Karas, B. J., & Edgell, D. R. (2018). An expanded plasmid-based genetic toolbox enables Cas9 genome editing and stable maintenance of synthetic pathways in Phaeodactylum tricornutum. *ACS Synthetic Biology, 7*(2), 328–338. https://doi.org/10.1021/acssynbio.7b00191

Specht, E. A., & Mayfield, S. P. (2014). Algae-based oral recombinant vaccines. *Frontiers in Microbiology, 5*, 60. https://doi.org/10.3389/fmicb.2014.00060

Spolaore, P., Joannis-Cassan, C., Duran, E., & Isambert, A. (2006). Commercial applications of microalgae. *Journal of Bioscience and Bioengineering, 101*(2), 87–96.

Sreeharsha, R. V., & Venkata Mohan, S. (2021). Symbiotic integration of bioprocesses to design a self-sustainable life supporting ecosystem in a circular economy framework. *Bioresource Technology, 326*, 124712. https://doi.org/10.1016/j.biortech.2021.124712

Sreenikethanam, A., Raj, S., Rajesh Banu, J., Gugulothu, P., & Bajhaiya, A. K. (2022). Genetic engineering of microalgae for secondary metabolite production: Recent developments, challenges, and future prospects. *Frontiers in Bioengineering and Biotechnology, 10.*

Stephenson, P. G., Moore, C. M., Terry, M. J., Zubkov, M. V., & Bibby, T. S. (2011). Improving photosynthesis for algal biofuels: Toward a green revolution. *Trends in Biotechnology, 29*(12), 615–623.

Stukenberg, D., Zauner, S., Dell'Aquila, G., & Maier, U. G. (2018). Optimizing CRISPR/Cas9 for the diatom Phaeodactylum tricornutum. *Frontiers of Plant Science, 9*, 740. https://doi.org/10.3389/fpls.2018.00740

Sudhakar, M. P., Kumar, B. R., Mathimani, T., & Arunkumar, K. (2019). A review on bioenergy and bioactive compounds from microalgae and macroalgae-sustainable energy perspective. *Journal of Cleaner Production, 228*, 1320–1333.

Suyono, E. A., Haryadi, W., Zusron, M., Nuhamunada, M., Rahayu, S., & Nugroho, A. P. (2015). The effect of salinity on growth, dry weight and lipid content of the mixed microalgae culture isolated from Glagah as biodiesel substrate. *Journal of Life Sciences, 9*, 229–233.

Szabo, N. J., Matulka, R. A., & Chan, T. (2013). Safety evaluation of whole algalin protein (*wap*) from *Chlorella protothecoides*. *Food and Chemical Toxicology, 59*, 34–45.

Takahashi, K., Ide, Y., Hayakawa, J., Yoshimitsu, Y., Fukuhara, I., Abe, J., & Harayama, S. (2018). Lipid productivity in TALEN-induced starchless mutants of the unicellular green alga Coccomyxa sp. strain Obi. *Algal Research, 32*, 300–307.

Talebi, A. F., Tohidfar, M., Bagheri, A., Lyon, S. R., Salehi-Ashtiani, K., & Tabatabaei, M. (2014). Manipulation of carbon flux into fatty acid biosynthesis pathway in *Dunaliella salina* using AccD and ME genes to enhance lipid content and to improve produced biodiesel quality. *Biofuel Research Journal, 1*(3), 91–97.

Tevatia, R., Payne, S., Allen, J., White, D., Clemente, T. E., Cerutti, H., & Blum, P. (2019). A synthetic cdo/csad taurine pathway in the green unicellular alga Chlamydomonas reinhardtii. *Algal Research, 40*, 101491.

Timoneda, J., Rodríguez-Fernández, L., Zaragozá, R., Marín, M. P., Cabezuelo, M. T., Torres, L., & Barber, T. (2018). Vitamin A deficiency and the lung. *Nutrients, 10*(9), 1132.

Torres-Tiji, Y., Fields, F. J., & Mayfield, S. P. (2020). Microalgae as a future food source. *Biotechnology Advances, 41*, 107536. https://doi.org/10.1016/j.biotechadv.2020.107536

Trentacoste, E. M., Shrestha, R. P., Smith, S. R., Glé, C., Hartmann, A. C., Hildebrand, M., & Gerwick, W. H. (2013). Metabolic engineering of lipid catabolism increases microalgal lipid accumulation without compromising growth. *Proceedings of the National Academy of Sciences, 110*(49), 19748–19753.

Urnov, F. D., Rebar, E. J., Holmes, M. C., Zhang, H. S., & Gregory, P. D. (2010). Genome editing with engineered zinc finger nucleases. *Nature Reviews Genetics, 11*(9), 636–646.

U.S. Food and Drug Administration (HHS). (2016). *FDA summary: Substances generally regarded as safe (final rule)*. Available online: http://www.fda.gov/AboutFDA/ReportsManualsForms/Reports/EconomicAnalyses/ucm517103.htm (accessed on 14 May 2022).

Varela Villarreal, J., Burgués, C., & Rösch, C. (2020). Acceptability of genetically engineered algae biofuels in Europe: Opinions of experts and stakeholders. *Biotechnology for Biofuels, 13*(1), 1–21.

Vavitsas, K., Fabris, M., & Vickers, C. E. (2018). Terpenoid metabolic engineering in photosynthetic microorganisms. *Genes, 9*(11), 520.

Velmurugan, N., & Deka, D. (2018). Transformation techniques for metabolic engineering of diatoms and haptophytes: Current state and prospects. *Applied Microbiology and Biotechnology, 102*(10), 4255–4267.

Venkata Mohan, S., Hemalatha, M., Chakraborty, D., Chatterjee, S., Ranadheer, P., & Kona, R. (2020). Algal biorefinery models with self-sustainable closed loop approach: Trends and prospective for blue-bioeconomy. *Bioresource Technology, 295*, 122128. https://doi.org/10.1016/j.biortech.2019.122128

Verruto, J., Francis, K., Wang, Y., Low, M. C., Greiner, J., Tacke, S., Kuzminov, F., Lambert, W., McCarren, J., Ajjawi, I., & Bauman, N. (2018). Unrestrained markerless trait stacking in *Nannochloropsis gaditana* through combined genome editing and marker recycling technologies. *Proceedings of the National Academy of Sciences, 115*(30), E7015–E7022.

Vila, M., Galván, A., Fernández, E., & León, R. (2012). Ketocarotenoid biosynthesis in transgenic microalgae expressing a foreign β-C-4-carotene oxygenase gene. *Methods in Molecular Biology, 892*, 283–295. https://doi.org/10.1007/978-1-61779-879-5_17

Vuppaladadiyam, A. K., Prinsen, P., Raheem, A., Luque, R., & Zhao, M. (2018). Microalgae cultivation and metabolites production: A comprehensive review. *Biofuels, Bioproducts and Biorefining, 12*(2), 304–324.

Walsh, G. (2018). Biopharmaceutical benchmarks 2018. *Nature Biotechnology, 36*(12), 1136–1145.

Wang, K., Cui, Y., Wang, Y., Gao, Z., Liu, T., Meng, C., & Qin, S. (2020). Chloroplast genetic engineering of a unicellular green alga Haematococcus pluvialis with expression of an antimicrobial peptide. *Marine Biotechnology, 22*(4), 572–580.

Wang, Q., Lu, Y., Xin, Y., Wei, L., Huang, S., & Xu, J. (2016). Genome editing of model oleaginous microalgae Nannochloropsis spp. by CRISPR/Cas9. *The Plant Journal, 88*(6), 1071–1081. https://doi.org/10.1111/tpj.13307

Wang, X., Wei, W., Li, N. J., Yuan, W., Ding, Y., Yang, W. D., & Li, H. Y. (2018). Heterogeneous expression of human PNPLA3 triggers algal lipid accumulation and lipid droplet enlargement. *Algal Research, 31*, 276–281.

Wang, X., & Zhang, X. (2013). Separation, antitumor activities, and encapsulation of polypeptide from Chlorella pyrenoidosa. *Biotechnology Progress, 29*(3), 681–687.

Wells, M. L., Potin, P., Craigie, J. S., Raven, J. A., Merchant, S. S., Helliwell, K. E., … Brawley, S. H. (2017). Algae as nutritional and functional food sources: Revisiting our understanding. *Journal of Applied Phycology, 29*(2), 949–982.

Wijffels, R. H. (2015). *The need and risks of using transgenic microalgae for the production of food, feed, chemicals and fuels*. Netherlands: Biosafety and the environmental uses of micro-organisms.

Williams, P. J. L. B., & Laurens, L. M. (2010). Microalgae as biodiesel & biomass feedstocks: Review & analysis of the biochemistry, energetic & economics. *Energy & Environmental Science, 3*(5), 554–590.

Xue, B., Dong, C. M., Hu, H. H., Dong, B., & Fan, Z. C. (2020). Chlamydomonas reinhardtii-expressed multimer of ToAMP4 inhibits the growth of bacteria of both Gram-positive and Gram-negative. *Process Biochemistry, 91*, 311–318.

Xue, J., Niu, Y. F., Huang, T., Yang, W. D., Liu, J. S., & Li, H. Y. (2015). Genetic improvement of the microalga Phaeodactylum tricornutum for boosting neutral lipid accumulation. *Metabolic Engineering, 27*, 1–9.

Yan, N., Fan, C., Chen, Y., & Hu, Z. (2016). The potential for microalgae as bioreactors to produce pharmaceuticals. *International Journal of Molecular Sciences, 17*, 962. https://doi.org/10.3390/ijms17060962

Zou, L. G., Chen, J. W., Zheng, D. L., Balamurugan, S., Li, D. W., Yang, W. D., ... Li, H. Y. (2018). High-efficiency promoter-driven coordinated regulation of multiple metabolic nodes elevates lipid accumulation in the model microalga Phaeodactylum tricornutum. *Microbial Cell Factories, 17*(1), 1–8.

Zhang, C., & Hu, H. (2014). High-efficiency nuclear transformation of the diatom Phaeodactylum tricornutum by electroporation. *Marine genomics, 16*, 63–66.

Further reading

Noga, E. J., Silphaduang, U., Park, N. G., Seo, J. K., Stephenson, J., & Kozlowicz, S. (2009). Piscidin 4, a novel member of the piscidin family of antimicrobial peptides. *Comparative Biochemistry and Physiology Part B: Biochemistry and Molecular Biology, 152*(4), 299–305.

Snow, A. A., & Smith, V. H. (2012). Genetically engineered algae for biofuels: A key role for ecologists. *BioScience, 62*(8), 765–768.

Venkata Mohan, S., Rohit, M. V., Chiranjeevi, P., Hariprasad, R., & Arora, S. (2015). Integrating microalgae cultivation with wastewater treatment for biodiesel production. In *Algal Biorefinery: An integrated approach* (pp. 321–337). Cham, Switzerland: Springer.

Yan, J., Cheng, R., Lin, X., You, S., Li, K., Rong, H., & Ma, Y. (2013). Overexpression of acetyl-CoA synthetase increased the biomass and fatty acid proportion in microalga Schizochytrium. *Applied Microbiology and Biotechnology, 97*(5), 1933–1939.

CHAPTER 22

Perspective of algae materials 2.0

R. Srinithi[1], P. Sangavi[1,a], K.T. Nachammai[3,a], S. Gowtham Kumar[2,a] and K. Langeswaran[3]

[1]*Department of Bioinformatics, Science Campus, Alagappa University, Karaikudi, Tamil Nadu, India;* [2]*Faculty of Allied Health Sciences, Chettinad Hospital & Research Institute, Chettinad Academy of Research and Education, Kelambakkam, Tamil Nadu, India;* [3]*Department of Biotechnology, Science Campus, Alagappa University, Karaikudi, Tamil Nadu, India*

1. Introduction

The microalgal biomass is composed of carbohydrates and lipids produced by sunlight light, CO_2, and some of the micronutrients such as nitrogen, phosphorus, and potassium. The main components of carbohydrates and lipids produce biofuels and biomaterials. Algae can be cultivated in ponds, lakes, rivers, oceans, waterfalls, puddles, and streams. Some biofertilizers, biodiesel, biopolymer, human nutrients, biogas, bioplastics, and animal feed are produced using algal biomass (Barkia et al., 2019). Algal biomass is oleaginous in nature. Microalgae and *cyanobacteria* are microorganisms used for the production of biomaterials sustainably. The microalgae consist of mitochondria, lysozyme, and endoplasmic reticulum. The practical way is to disrupt the cells' significant components of lipid, protein, and carbohydrates from the cytoplasm. The carbohydrates present in microalgae can be used to produce the chemicals. Several bioactive compounds are synthesized as secondary metabolites in microalgae. Some bioactive compounds such as toxins, terpenes, and indoles can be obtained from microalgae. They are used in pharmaceuticals due to their antioxidant, anticoagulation, antioxidant, antitumor, antiviral, antiinflammatory, and anticancer properties (Qamar et al., 2021). Beta-carotene, lutein, and polysaccharides are some of the value-added products produced from microalgae. Algae can be cultivated to produce various products used in small to large industries, especially plastics, chemical feedstocks, fertilizers, tablets, and even cosmetics.

More than 300,000 species are present, and they live in complex natural habitats and adapt to extreme weather conditions. Algae occur in various environments such as marine, freshwater, hot and cold springs, etc. Unicellular and multicellular algae groups are present in several shapes and forms. Phenolic compounds, lipids, polysaccharides, alkaloids, steroids, lipids, and bioactive peptides are the constituents of algae. This ability of the microalgae can be capable of producing some secondary metabolites.

[a]Equal First author.

Algae Materials: Applications Benefitting Health. https://doi.org/10.1016/B978-0-443-18816-9.00009-5

Microalgae grow fast and can produce large biomass with high protein content. Microalgae can produce useful bioproducts such as polysaccharides, antioxidants, bioactive and functional pigments, natural dyes, β-carotene, astaxanthin, docosahexaenoic acid (DHA), eicosapentaenoic acid (EPA), and algal extracts. Microalgae produce many bioproducts such as lipids, pigments, proteins, polysaccharides, bioactive compounds, vitamins, and antioxidants. Microalgae are a renewable and best feedstock for the production of large amounts of biomass and biofuel production. Algae are the natural producer of bioactive pigment. Algae are classified based on the pigments. The algal pigment is highly valued in the nutraceutical, pharmaceutical, and cosmetic industries (Caporgno & Mathys, 2018). In the pharmaceutical industry, the extensive cultivation of algae used to produce *spirulina* tablet is now highly prescribed as healthy food to improve the vitality of a human being. Algae are rich in sources of nitrogen, iodine, iron, calcium, potassium, silica, phosphorous, and vitamins. Seaweed, microalgae, and *cyanobacteria* are commonly used for agriculture production (Sorrenti et al., 2021). Therapeutic properties present in algae are antiviral, anticancer, and antioxidant. Algae are rich in dietary fiber and provide valuable nutrients to the human diet.

2. Algae-based materials

2.1 Bioplastics

Bioplastics are plastic made from algae in seawater and form gigantic molecules called polymers. These polymers are then used to make bioplastics. The polymer formed by the algae is 100% vegetable and biodegradable, and renewable. The polymer breaks down into small pieces; it does not release any toxic substances and does not affect the environment. The main advantage of bioplastics is degraded into small pieces after a few weeks. These trim pieces are utilized for plant feed. The bioplastics produced from algae are nonhazardous, and they solve the pollution problem in the ocean and ecosystems. It is considered an environmentally and economically viable alternative to conventional petrochemical plastics. Most of the environmental problems related to plastics can be eradicated by using biopolymers and biobased fibers generated from renewable wastes and bioresources. Biobased plastics are predominantly made up of renewable resources through the activity of living organisms. Biobased plastics are a notable revolution that can promote and enlarge the bioeconomy. Biobased products are industrial or commercial products (other than food or feed) composed partly or mostly of biological products. The bioplastics biocompatibility and biodegradability properties make leading superior polymers compared with conventional synthetic plastics. This possession makes the extraordinary in applying packaging, biomedical, and value-added industrial (Devadas et al., 2021). Algae materials are used in various fields shown in Fig. 22.1.

2.1.1 Uses of bioplastic in agriculture

Bioplastics can be molded in any shape, size, and dimension; it is used in food industries for food packaging. Some of the bioplastics such as greenhouse film and protection nets are used in agriculture to protect the crops from physical and biological damage and maintain them under inappropriate conditions. Bioplastics are used to farm mushrooms, secure them freshly, and protect them from erosion. Bioplastics degrade quickly and naturally, do not affect the environment, and preserve soil nutrients.

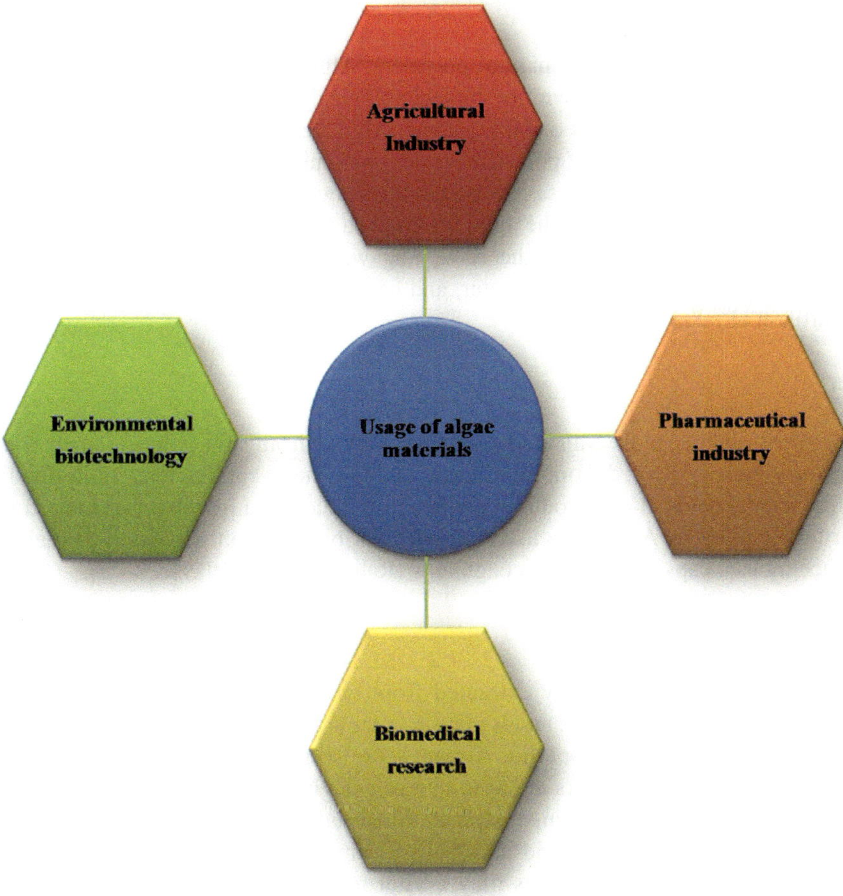

FIGURE 22.1

Usage of algae-based materials.

3. Bioplastics in the pharmaceutical industry

Medical professionals in hospitals and surgeries use nontoxic biodegradable bioplastics. Some therapeutic devices such as pins and screws used in bone healing and surgery are produced with biodegradable plastics. Bioplastics use tablet containers; they can renew easily and are eco-friendly. Biopolymers are widely used in tissue engineering and drug delivery systems. It does not cause a severe immune reaction when introduced into the host organisms, and it is nontoxic. The drugs such as antitumor, antiinflammatory, vaccines, steroids, and antibiotics are encapsulated with the biopolymer for the controlled release of drugs. Its biocompatibility nature does not affect host organisms. Bioplastics are used in pharmaceutical industries by making bioplastic materials such as a screw and staplers. These are used in drug delivery, bone treatment, and other medical uses.

4. Biopolymers

The polymers obtained from algae are natural polymers. Alginate and agar are polysaccharides obtained from algae and are well known for biotechnological applications. The polysaccharide of alginate is derived from the cell walls of brown algae. Alginates are used as a dissolver for the rapid release of tablets. Agar is a mixture of polysaccharides containing both agarose and agaropectin. The extraction of agar from algae contains several steps: the seaweeds are washed to remove the foreign matters, boiled using water to dissolve the agar, filtered and cooled to obtain jelly structure, and then various techniques are used to remove the remaining water to obtain the final product. Agar is insoluble in cold water but soluble in hot water, and agar is used in many pharmaceutical and food industries (Kartik et al., 2021).

4.1 Agar

Agar and alginates are the more valuable products derived from algae due to their vicious and thickening nature. Agar is derived from microalgae and has many applications in the food industry, textile, and biomedical fields. It can produce candies, fruit juices, and frozen foods in food industries, also used for textile printing. Agarose is used in molecular biology. In the biomedical field, it can be used to produce tablets, capsules, and anticoagulants (Rahmati et al., 2019).

4.2 Alginates

Alginate is a natural polymer typically extracted from brown algae by treating with sodium hydroxide. Alginate is gelling or viscous, and it plays a potential role in biomedical applications for its biocompatibility, nontoxic and hydrophilic. Alginates are converted into biomaterials such as biofilms, fibers, etc. Alginate biopolymers are used in tissue engineering and biomedical products with better quality wound dressing components. Alginates are used as a by-product of viral protection and used in food industries to coat fruits and as a gelling, emulsifying, and stabilizing agent in food factories (Kothale et al., 2020). Alginates play a vital role in the release of drug delivery in the pharmaceutical industry, and the biofilm is an algae material derived from microalgae, which can be used in drug delivery systems. Alginate gel is used to deliver drug molecules in the drug delivery system. Alginates play a significant role in biomedical applications for human healthcare. In biotechnology, they can be used in drug delivery, tissue engineering, and gene transformation (Severino et al., 2019).

5. Algal-based fuels

In the modern world, to decrease the energy demand in transportation and protect our environment, algal-based fuels were utilized. The biofuels are produced by using the algal biomass. The algae are cultured in a large pool. The microorganism converts sunlight into energy. The energy is stored as oil, the oil extracted by using a chemical solvent or mechanical process (Ananthi et al., 2021).

5.1 Biofuels

Biofuels are derived from renewable raw materials; biofuel is referred to as solid, liquid, or gaseous fuels. Biofuel is derived from plants or algae. Microalgae have been used as a feedstock for biofuels; they do not compete with the land-based food crops; they grow in water (Eswaran et al., 2021). Different types of techniques were used for biorefinery depending upon the species selection. If the microalgae species does not contain a cell wall, the cell disruption technique is effortless and low energy-intensive. Microalgae species contain a cell wall consisting of mitochondria, lysosomes, and endoplasmic reticulum. It is more practical to disrupt the cell wall to release the protein and carbohydrates from the cytoplasm. Proteins and lipids are primarily present in microalgae (El Zokm et al., 2021). Lipids can be used for the production of biofuels. The proteins and carbohydrates present in microalgae were purified and used as feed for animals. It is also used to produce bioethanol, biomethane, green diesel, biobutanol, biodiesel, biogas, and biohydrogen (Xia et al., 2016). Algae-based materials are highly used in biofuel production, as shown in Fig. 22.2.

5.1.1 Bioethanol

Bioethanol can be produced from algae based on the fermentation of algal polysaccharides. It can be produced by algal biomass using several groups of microbes such as yeast, fungi, and bacteria under the condition of anaerobic fermentation. Commonly, two processes are used for bioethanol production from algae: fermentation and gasification. Agar can be used as a feedstock for bioethanol production; it is a polymer present in a few groups of marine algae. Some green algae that worked for bioethanol production are *Chlorella, Spirulina, Chlamydomonas, and Prymnesium parvum*. Among all the three types of algae (green, brown, and red algae), brown algae are considered a potential feedstock for ethanol production due to their high carbohydrate content. Some countries produce commercial ethanol from sugary and starchy crops on a large scale (Kumar et al., 2018).

5.1.2 Biomethane production

Biomethane is used mainly as both chemical feedstock and fuel. Biomethane is produced from biogas by removing carbon dioxide from biogas. Anaerobic digestion is one of the practical approaches of the algae to generate biomethane. Pretreatment and codigestion methods are the best methods to yield methane from the algae biodegradation. Some other different feedstocks are used for methane production. They are solid waste, wood, and grass, but the algae biomass are potential feedstock for biomethane production. Algae biomass productivity is higher than that of land plants (Zheng et al., 2021).

5.1.3 Biobutanol

Butanol is produced from both microalgae and macroalgae by fermentation of algae. Butanol is used as a fuel for transportation. Butanol is such a biofuel like petroleum to decrease the energy demand in transportation and industries. Microalgae are quickly grown photosynthetic organisms that can increase butanol production without affecting crop fields and ecology. The fermentation of carbohydrates produces butanol. This process is known as ABE fermentation (acetone, butanol, and ethanol). Algal biomass does not require freshwater to produce biobutanol (Zhang, Hong, et al., 2021; Zhang, Sun, et al., 2021).

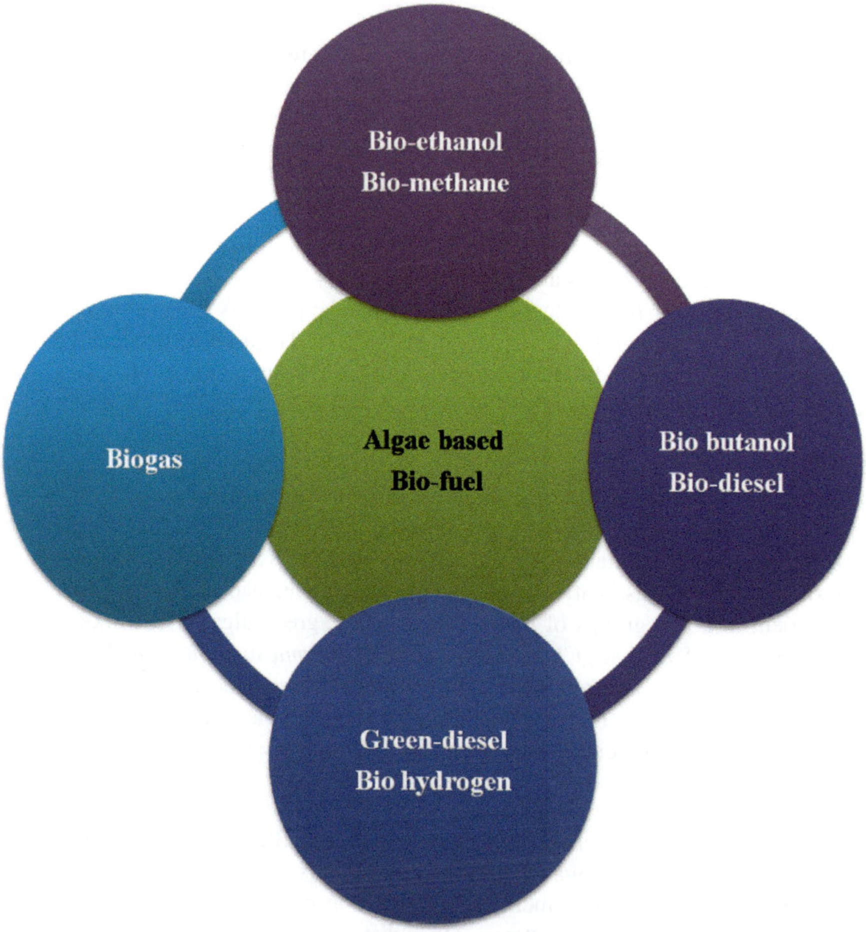

FIGURE 22.2

Algae-based materials used in the production of biofuel.

5.1.4 Biodiesel production

The raw materials for the production of biodiesel are oils or lipids. The oil extracted from the microalgae, the part of residual biomass, can be used to produce methane by the anaerobic digestion method. The energy generated from the methane is further used to produce biodiesel. The algal biodiesel contains 10%–20% viscosity higher than the conventional diesel. The production of diesel from algae is easy. *Tribonema* and *euglena* are some of the promising potential species used to produce biodiesel. Both microalgae and macroalgae are used for the production of biodiesel. Some of the steps involved in biodiesel production are cultivating algae by providing required nutrients, Co_2, and sunlight. Biofuel-contented algae are cultivated and harvested through the proper harvesting equipment. After selecting specific algae, it is dried under the sunlight or some other thermal process. The oil is extracted from the species by disrupting the cell under the physical or mechanical method. Biodiesel is produced by separating lipid and fatty acid contents (Biller et al., 2013).

5.1.5 Green diesel

Algae can be used to produce green diesel, otherwise known as renewable diesel. Green diesel is biodegradable and nontoxic. A large amount of fat can be produced from single-celled microalgae, which can be converted into biodiesel. Typically, algae contain 20%–80% of oil contents. Some of the species such as *Tribonema* (yellow-green algae) and *Ulothrix* (green algae) have a good potential for producing biodiesel. When biofuels are burned, they produce low carbon output and pollutants. When compared with petroleum diesel, biodiesel produces fewer pollutants. Usually, aquatic unicellular green algae can be used for biodiesel production.

5.1.6 Biohydrogen

The oil and pigments are extracted from the microalgae; the primary substance produced from the oil is biodiesel, and pigments and biomass are coproducts. Biohydrogen and carotenoids are produced by utilizing leftover biomass. Biohydrogen production from microalgae is a renewable process.

5.1.7 Biogas

Biogas is produced by the biochemical process of the anaerobic digestion of microalgae. Anaerobic digestion is a well-known technique using high energy. Microorganisms are broken down into organic materials in the absence of air. Biogas is produced with a small amount of water vapor, methane, carbon dioxide, and other gases. The primary component of natural gas is methane. The yield of biogas depends upon the choice of algal strain. Biogas is used as a cooking fuel, lightning fuel, and vehicle fuel (De Obeso Fernandez Del Valle & Scheckhuber, 2021) (Fig. 22.3).

6. Algae pigments and pharmaceutical use

The primary photosynthetic pigments in algae are chlorophylls, carotenoids, and phycobilins (Pereira et al., 2021). Different types of algae have different types of pigments. Pigments are usually present in all types of algae except *Cyanophyceae* (a type of blue-green algae). Chlorophyll is a green pigment present in *cyanobacteria*. Phycoerythrin pigment occurs in red algae. Brown algae are widely used as edible seaweeds. The extraction of brown algae is used as a thickener in food industries. Blue-green algae are highly used to treat high blood pressure and protein supplements (Bobescu et al., 2020). Generally, red algae have been used in traditional medicine. Medical science has identified that extraction of algae contains antiviral and antitumor properties (Kim et al., 2011).

7. Electricity production from algae

The demand for fossil fuel energy is a massive problem in the energy sector. The production of energy should be renewable in future. The algal microbial fuel cell (AMFC) is the best renewable energy technology. The photosynthetic organisms are used for the production of direct electricity from wastewater. AMFCs are an abiotic system that can produce electricity continuously from wastewater simultaneously with algal biomass production. Algae are an excellent renewable energy source for energy management in the future (Inohana et al., 2020).

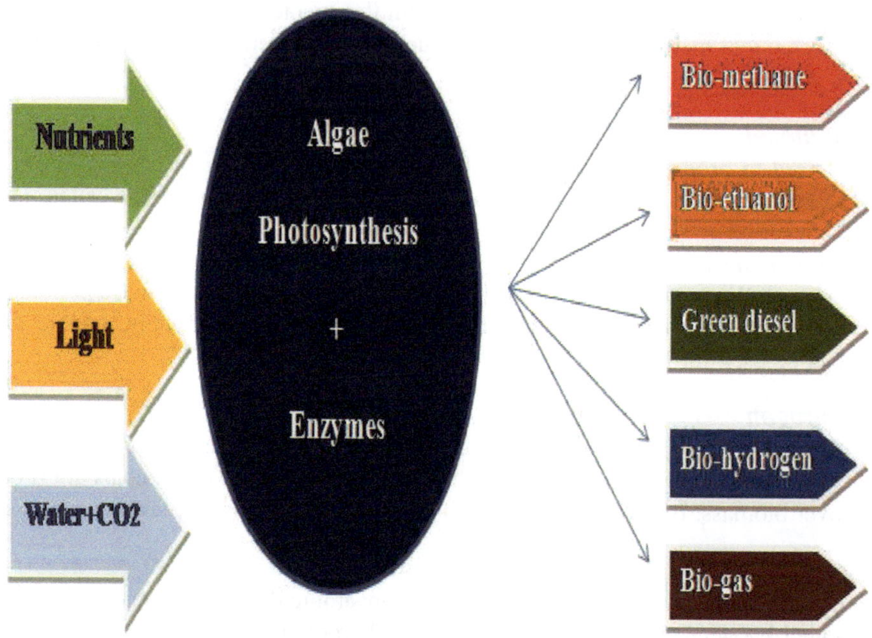

FIGURE 22.3

Algae produce biofuel using various energy sources.

8. Algae-based bioactive compounds

A wide variety of bioactive chemicals are synthesized from microalgae, which is depicted in Fig. 22.4. It contains the substance of carbohydrates, growth substances, vitamins, antibiotics, amino acids, and some toxic substances that show the biological activities, including antiviral, anticancer, and antioxidant effects (Xu et al., 2019). The new drug leads are discovered from the algae metabolites. Natural compounds are more important in the search to produce new pharmacologically active compounds; they play a prominent role in discovering a new drug for the treatment of human diseases (Baranowska & Bartoszek, 2016). Many antitumor and antiviral drugs are discovered from natural algae materials. Microalgae produce various biomaterials or compounds for clinical therapeutics (Zhang et al., 2015).

8.1 Perspectives of microalgae in environmental biotechnology

In environmental biotechnology, microalgae have many applications for bioremediation and bioassay of environmental toxicants such as heavy metals, pharmaceuticals, and pesticides. The most common microalgae used for the bioassays of the toxicant include *Dunaliella tertiolecta* and *Isochrysis galbana.* The algae-based natural polymers are biodegradable and biocompatibility products; they are

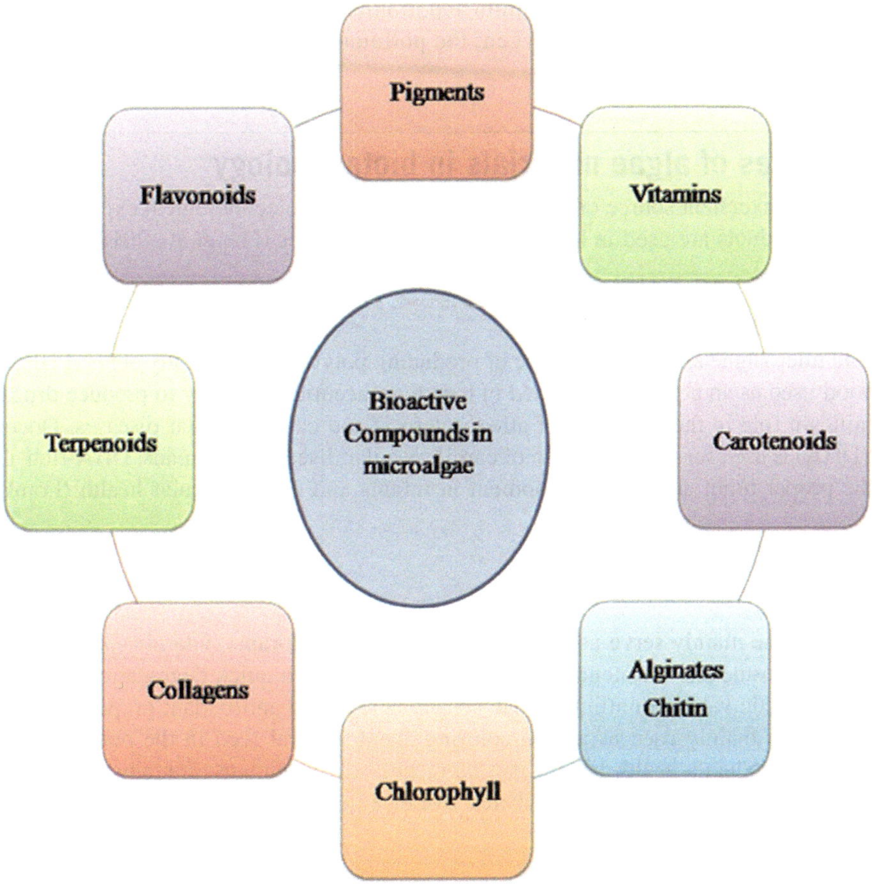

FIGURE 22.4

Bioactive compounds are present in the microalgae.

used for everyday life despite nonbiodegradable products such as plastic bags. Using algae-based materials protects our environment from water and land pollution without releasing toxic substances (Kong et al., 2018).

8.2 Bioremediation of radioactive compounds

Algae materials are used in environmental biotechnology to protect our environment from the pollutant of heavy metals, toxins, pesticides, radioactive compounds, etc. Radioactive compounds are life-threatening because of their high toxicity and radioactivity. The contamination of radioactive compounds affects the groundwater. The algae materials produced from marine algae can absorb radioactive compounds such as radium and uranium. The microalgae can grow in pollutant water under light conditions without oxygen; the pollutant is taken up and digested (Znad et al., 2022).

9. Perspectives of algae materials in biotechnology

Microalgae are an excellent source of high-value products for biotechnological exploitation. Some of the microalgae products are used in biotechnological applications (Fan et al., 2018).

9.1 Polyunsaturated fatty acids

Some marine microalgae are a prime source of producing polyunsaturated fatty acids (PUFAs). PUFA is a superfood used as an aquafeed and used in the pharmaceutical industry to produce drugs. PUFAs play a significant role in the treatment of atherosclerosis and cardiovascular diseases. Docosahexaenoic acid (DHA) is used for the prevention of cardiovascular disease. In humans, DHA-rich algal oil is vital for the proper brain and eye development in infants and cardiovascular health (Frankel et al., 2002).

9.2 Carotenoids

Carotenoids in algae mainly serve as pigments, natural food colorants, vitamin-like (beta-carotene), animal feed, and cosmetics. Carotenoids are insoluble and have a quick degradation capacity. It also contains a therapeutic value of antiinflammatory, anticancer, and antioxidant properties. *Astaxanthin* from microalgae is mainly used as a food-coloring agent, natural feed in the poultry industry, and aquaculture. The product's health benefit is a nutraceutical for humans, contains strong antioxidant properties, and protects against cancer, diabetes, eye disease, and metabolic syndrome (Eggersdorfer & Wyss, 2018).

9.3 Sterols

Algae are rich in the source phytosterol. The structure of the sterol is similar to cholesterol. Some microalgae have a high content of sterols. It is a bioactive compound that contains beneficial health effects such as anticancer, antiviral, antifungal, antibacterial, and antiinflammatory. It is used in the pharmaceutical industry and neutraceuticals as food ingredients. Different sterols are present in different types of algae species, and they differ in genera and chemical structure. Phytosterols are used in cancer treatment significantly since they can inhibit colon, prostate, and breast cancer (Geng et al., 2017).

9.4 Polyhydroxyalkanoates

Polyhydroxyalkanoates (PHAs) are biopolymers synthesized from microorganisms. Microalgae are the specific microorganisms that produce PHAs by the photosynthesis process. Microalgae are the unique organisms to produce PHA at a low cost with the lowest nutritional requirement. More than 100 strains of microalgae are tested for the production of PHA. PHA polyesters are produced through the photosynthesis process. Microalgae use sunlight, water, carbon dioxide, and nutrients to produce the PHAs. Then, PHAs were extracted and purified using physical, chemical, and biological methods. In the chemical method for PHA extraction, the most common solvents are chloroform, acetone, methyl isobutyl ketone, methylene chloride, propylene carbonate, ethyl acetate, and isoamyl alcohol. The most common physical method for PHA extraction is ultrasound. PHA is extracted from the biological process using lysozymes, nucleases, and proteases (Fradinho et al., 2013).

10. The perspective of phycobiliprotein algae material for biomedical research

Phycobiliproteins are oligomeric proteins and water-soluble photosynthetic pigments present in algae such as *cyanobacteria*, *rhodophytes* (a type of red algae), and some specific algae. Three types of phycobiliproteins are present, and they differ in their properties. In microalgae, phycobiliprotein pigments differ due to various environmental conditions. The main application for phycobiliproteins is dyed for food, and cosmetics, as a fluorescent marker for biomedical research and used as a tool for immunology laboratories (Zuorro et al., 2021).

11. Algae products

Algae are rich in protein, carbohydrates, vitamins, minerals, and fats. Essential minerals present in the algae are iron, calcium, zinc, magnesium, etc. Algae can act as an essential ingredient for the thickener of food. A type of algae can be used in dairy items such as cheese, milkshakes, and household products such as toothpaste. Gelatin is a substance made up of a type of algae used to produce numerous edible products. Algae are also used for the production of jelly and bread items. Used as a fertilizer, red and brown algae are the most common algae used to manufacture fertilizer. Seaweed extract is used to produce liquid fertilizer (Hosseini et al., 2013). Algae are used to feed farm animals, and seaweed is the most common type to produce fodder. Red algae are widely used to produce semisolid products such as medicines and cosmetics etc.

12. Synthetic polymers

Synthetic polymers are human-made polymers. Scientist artificially derives synthetic polymers from oil and petroleum. The production of synthetic polymers or nonbiodegradable polymers has become an issue with the production and utilization of these synthetic polymers. The uses of synthetic polymer create unmanageable pollution in the environment. It can affect our environment through persistent organic pollutants (POPs) secretion, food imitation, and landfill accumulation. Synthetic polymers create water pollution and affect oceans by filling synthetic polymers, and there is a leakage of toxic substances after a few years. Polymers such as polyethene and polypropene are nonbiodegradable; it cannot break down into microorganisms. So they cause toxic problems after a few years in the landfills. To solve these problems, there was rapid progress in producing natural or biobased polymers to protect our environment. Synthetic polymers are lightweight, hard to break, and easy to form shapes. One of the most common synthetic polymers is polyethene derived from ethylene. Essential advantages of synthetic polymers are low cost, elasticity, strength, and chemical resistance. Some examples of synthetic polymers are polyethene, polyvinyl chloride (PVC), polypropylene, nylon, etc. Synthetic polymers are used in bottles, toys, electrical, pipes, and flooring. Biodegradable synthetic polymers are used in drug delivery systems. Pins, rods, and screws are some of the devices produced using biodegradable synthetic polymers. Biodegradable synthetic polymers are also used in tissue engineering (Jung et al., 2022).

13. Future perspectives of algae-based materials

There is a demand for plastic products in this generation; plastics recycling and chemical treatment are not enough to reduce plastic pollution. It affects marine pollution and human health. Microalgae secretion helps to degrade the plastic. On the other hand, bioplastics are derived from microalgae. Biobased feedstocks are used for the production of biobased materials due to their biodegradability and biocompatibility (Zhang, Sun, et al., 2021). Nowadays, genetically engineered microalgae are used for the production of starch, cellulose, PHAs, polyvinyl butyral (PHB), and PVC polymers. Algal polymers are highly used in food industries due to their viscosifying and gelling abilities and tremendous application in biological and pharmaceutical activities such as antitumor, anticancer, antioxidant anticoagulant, and antimicrobial anticancer activities (Kizhakkekalam & Chakraborty, 2021). Fig. 22.5 depicted the future perspectives of algae-based materials.

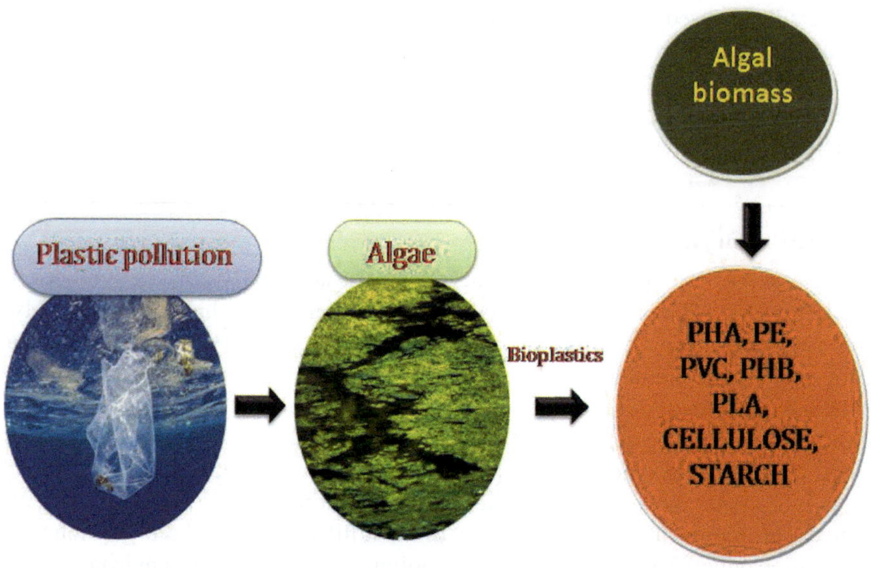

FIGURE 22.5

Schematic representation of production of algal biomass.

14. Conclusion

Algae are used in various fields as a base material. Algae-based materials, including bioplastics, biopolymers, biofuel, bioremediation of radioactive compounds, and algae pigments, are used in pharmaceutical, bioactive compounds, biotechnology, and environmental technology. Algae-based products are fully biodegradable and renewable resources. Algae-based materials are free of toxicity. It is a compostable one while comparing with present conventional products. It helps to reduce CO_2 emissions and increase the efficiency of resources. Algae materials are environmentally friendly as well as sustainable. In future, algae-based material is going to rule the entire universe.

Acknowledgment

The author Dr K. Langeswaran mercifully acknowledges Alagappa University for the infrastructure facility and RUSA phase 2.0 grant sanctioned vide letter No. F.24-51/2014-U policy (TN Multi-Gen), Dept of Govt of India Dt, 09.10.2018 for financial assistance.

References

Ananthi, V., Balaji, P., Sindhu, R., Kim, S. H., Pugazhendhi, A., & Arun, A. (2021). A critical review on different harvesting techniques for algal based biodiesel production. *Science Total Environment, 780*, 146467. https://doi.org/10.1016/j.scitotenv.2021.146467

Baranowska, M., & Bartoszek, A. (2016). Antioxidant and anti-microbial properties of bioactive phytochemicals from cranberry. *Postepy Higieny I Medycyny Doswladczalnej, 70*(0), 1460–1468. https://doi.org/10.5604/17322693.1227896. Published 2016 Dec 31.

Barkia, I., Saari, N., & Manning, S. R. (2019). Microalgae for high-value products towards human health and nutrition. *Marine Drugs, 17*(5), 304. https://doi.org/10.3390/md17050304. Published 2019 May 24.

Biller, P., Friedman, C., & Ross, A. B. (2013). Hydrothermal microwave processing of microalgae as a pre-treatment and extraction technique for bio-fuels and bio-products. *Bioresource Technology, 136*, 188–195. https://doi.org/10.1016/j.biortech.2013.02.088

Bobescu, E., Bălan, A., Moga, M. A., Teodorescu, A., Mitrică, M., & Dima, L. (2020). Are there any beneficial effects of *spirulina* supplementation for metabolic syndrome components in postmenopausal women? *Marine Drugs, 18*(12), 651. https://doi.org/10.3390/md18120651. Published 2020 Dec 17.

Caporgno, M. P., & Mathys, A. (2018). Trends in microalgae incorporation into innovative food products with potential health benefits. *Frontiers in Nutrition, 5*, 58. https://doi.org/10.3389/fnut.2018.00058. Published 2018 Jul 31.

De Obeso Fernandez Del Valle, A., & Scheckhuber, C. Q. (2021). From past to present: Biotechnology in Mexico using algae and fungi. *Plants, 10*(11), 2530. https://doi.org/10.3390/plants10112530. Published 2021 Nov 20.

Devadas, V. V., Khoo, K. S., Chia, W. Y., et al. (2021). Algae biopolymer towards sustainable circular economy. *Bioresource Technology, 325*, 124702. https://doi.org/10.1016/j.biortech.2021.124702

Eggersdorfer, M., & Wyss, A. (2018). Carotenoids in human nutrition and health. *Archives of Biochemistry and Biophysics, 652*, 18–26. https://doi.org/10.1016/j.abb.2018.06.001

El Zokm, G. M., Ismail, M. M., & El-Said, G. F. (2021). Halogen content relative to the chemical and biochemical composition of fifteen marine macro and micro algae: Nutritional value, energy supply, antioxidant potency, and health risk assessment. *Environmental Science and Pollution Research International, 28*(12), 14893–14908. https://doi.org/10.1007/s11356-020-11596-0

Eswaran, N., Parameswaran, S., & Johnson, T. S. (2021). Biofuels and sustainability. *Methods in Molecular Biology, 2290*, 317–342. https://doi.org/10.1007/978-1-0716-1323-8_20

Fan, M., Nath, A. K., Tang, Y., et al. (2018). Investigation of the anti-prostate cancer properties of marine-derived compounds. *Marine Drugs, 16*(5), 160. https://doi.org/10.3390/md16050160. Published 2018 May 12.

Fradinho, J. C., Domingos, J. M., Carvalho, G., Oehmen, A., & Reis, M. A. (2013). Polyhydroxyalkanoates production by a mixed photosynthetic consortium of bacteria and algae. *Bioresource Technology, 132*, 146–153. https://doi.org/10.1016/j.biortech.2013.01.050

Frankel, E. N., Satué-Gracia, T., Meyer, A. S., & German, J. B. (2002). Oxidative stability of fish and algae oils containing long-chain polyunsaturated fatty acids in bulk and in oil-in-water emulsions. *Journal of Agricultural and Food Chemistry, 50*(7), 2094–2099. https://doi.org/10.1021/jf0111458

Geng, H. X., Yu, R. C., Chen, Z. F., Peng, Q. C., Yan, T., & Zhou, M. J. (2017). Analysis of sterols in selected bloom-forming algae in China. *Harmful Algae, 66*, 29–39. https://doi.org/10.1016/j.hal.2017.04.008

Hosseini, S. M., Khosravi-Darani, K., & Mozafari, M. R. (2013). Nutritional and medical applications of spirulina microalgae. *Mini Reviews in Medicinal Chemistry, 13*(8), 1231–1237. https://doi.org/10.2174/1389557511313080009

Inohana, Y., Katsuya, S., Koga, R., Kouzuma, A., & Watanabe, K. (2020). Shewanella algae relatives capable of generating electricity from acetate contribute to coastal-sediment microbial fuel cells treating complex organic matter. *Microbes Environment, 35*(2), ME19161. https://doi.org/10.1264/jsme2.ME19161

Jung, K., Corrigan, N., Wong, E. H. H., & Boyer, C. (2022). Bioactive synthetic polymers. *Advance Materials, 34*(2), e2105063. https://doi.org/10.1002/adma.202105063

Kartik, A., Akhil, D., Lakshmi, D., et al. (2021). A critical review on production of biopolymers from algae biomass and their applications. *Bioresource Technology, 329*, 124868. https://doi.org/10.1016/j.biortech.2021.124868

Kim, S. K., Thomas, N. V., & Li, X. (2011). Anticancer compounds from marine macroalgae and their application as medicinal foods. *Advances in Food & Nutrition Research, 64*, 213–224. https://doi.org/10.1016/B978-0-12-387669-0.00016-8

Kizhakkekalam, V. K., & Chakraborty, K. (2021). Seaweed-associated heterotrophic bacteria: New paradigm of prospective anti-infective and anti-cancer agents. *Archives of Microbiology, 203*(3), 1241–1250. https://doi.org/10.1007/s00203-020-02106-4

Kong, F., Romero, I. T., Warakanont, J., & Li-Beisson, Y. (2018). Lipid catabolism in microalgae. *New Phytologist, 218*(4), 1340–1348. https://doi.org/10.1111/nph.15047

Kothale, D., Verma, U., Dewangan, N., Jana, P., Jain, A., & Jain, D. (2020). Alginate as promising natural polymer for pharmaceutical, food, and biomedical applications. *Current Drug Delivery, 17*(9), 755–775. https://doi.org/10.2174/1567201817666200810110226

Kumar, A., Kaushal, S., Saraf, S. A., & Singh, J. S. (2018). Microbial bio-fuels: A solution to carbon emissions and energy crisis. *Frontiers in Bioscience, 23*(10), 1789–1802. https://doi.org/10.2741/4673. Published 2018 Jun 1.

Pereira, A. G., Otero, P., Echave, J., et al. (2021). Xanthophylls from the sea: Algae as source of bioactive carotenoids. *Marine Drugs, 19*(4), 188. https://doi.org/10.3390/md19040188. Published 2021 Mar 27.

Qamar, H., Hussain, K., Soni, A., Khan, A., Hussain, T., & Chénais, B. (2021). Cyanobacteria as natural therapeutics and pharmaceutical potential: Role in antitumor activity and as nanovectors. *Molecules, 26*(1), 247. https://doi.org/10.3390/molecules26010247. Published 2021 Jan 5.

Rahmati, M., Alipanahi, Z., & Mozafari, M. (2019). Emerging biomedical applications of algal polysaccharides. *Current Pharmaceutical Design, 25*(11), 1335–1344. https://doi.org/10.2174/1381612825666190423160357

Severino, P., da Silva, C. F., Andrade, L. N., de Lima Oliveira, D., Campos, J., & Souto, E. B. (2019). Alginate nanoparticles for drug delivery and targeting. *Current Pharmaceutical Design, 25*(11), 1312–1334. https://doi.org/10.2174/1381612825666190425163424

Sorrenti, V., Castagna, D. A., Fortinguerra, S., Buriani, A., Scapagnini, G., & Willcox, D. C. (2021). Spirulina microalgae and brain health: A scoping review of experimental and clinical evidence. *Marine Drugs, 19*(6), 293. https://doi.org/10.3390/md19060293. Published 2021 May 22.

Xia, A., Jacob, A., Tabassum, M. R., Herrmann, C., & Murphy, J. D. (2016). Production of hydrogen, ethanol and volatile fatty acids through co-fermentation of macro- and micro-algae. *Bioresource Technology, 205*, 118–125. https://doi.org/10.1016/j.biortech.2016.01.025

Xu, D., Hu, M. J., Wang, Y. Q., & Cui, Y. L. (2019). Antioxidant activities of quercetin and its complexes for medicinal application. *Molecules, 24*(6), 1123. https://doi.org/10.3390/molecules24061123. Published 2019 Mar 21.

Zhang, K., Hong, Y., Chen, C., & Wu, Y. R. (2021). Unraveling the unique butyrate re-assimilation mechanism of Clostridium sp. strain WK and the application of butanol production from red seaweed Gelidium amansii through a distinct acidolytic pretreatment. *Bioresource Technology, 342*, 125939. https://doi.org/10.1016/j.biortech.2021.125939

Zhang, W., Oda, T., Yu, Q., & Jin, J. O. (2015). Fucoidan from Macrocystis pyrifera has powerful immune-modulatory effects compared to three other fucoidans. *Marine Drugs, 13*(3), 1084–1104. https://doi.org/10.3390/md13031084. Published 2015 Feb 19.

Zhang, Q. Y., Sun, W., Du, Y. X., & Gong, X. L. (2021). Biodegradation of algae-derived organic matter (I-DOM) from LAKE taihu. *Huanjing Kexue, 42*(4), 1870–1878. https://doi.org/10.13227/j.hjkx.202007227

Zheng, X., Wang, H., Yan, Q., Zhang, G., & Chen, C. (2021). Simultaneous nitrogen removal and methane production from Taihu blue algae against ammonia inhibition using integrated bioelectrochemical system (BES). *Science Total Environment, 777*, 146144. https://doi.org/10.1016/j.scitotenv.2021.146144

Znad, H., Awual, M. R., & Martini, S. (2022). The utilization of algae and seaweed biomass for bioremediation of heavy metal-contaminated wastewater. *Molecules, 27*(4), 1275. https://doi.org/10.3390/molecules27041275. Published 2022 Feb 14.

Zuorro, A., Leal-Jerez, A. G., Morales-Rivas, L. K., et al. (2021). Enhancement of phycobiliprotein accumulation in thermotolerant *oscillatoria* sp. through media optimization. *ACS Omega, 6*(16), 10527–10536. https://doi.org/10.1021/acsomega.0c04665. Published 2021 Apr 12.

CHAPTER

Algae biorefinery: strategies for a sustainable industry

23

Joana Assunção[1,2,a], Fernando Pagels[1,3,a] and A. Catarina Guedes[1]

[1]*CIIMAR – Interdisciplinary Centre of Marine and Environmental Research, University of Porto, Novo Edifício Do Terminal de Cruzeiros de Leixões, Matosinhos, Portugal;* [2]*LEPABE – Laboratory of Engineering of Processes, Environment, Biotechnology and Energy, Faculty of Engineering of University of Porto, Porto, Portugal;* [3]*FCUP – Faculty of Science, University of Porto, Porto, Portugal*

1. Introduction

New health lifestyle trends and more awareness about environmental sustainability have been translating into the increasing demand for natural products. Consumers have recently become more concerned about the potential negative impact that some synthetic compounds may have on their health (Wang et al., 2015). Furthermore, environmental issues are becoming more of public concern, and consumers are becoming more aware of the direct (and indirect) negative impacts caused to the environment by their actions and choices (Puh, 2016). For this reason, the industry of food, nutraceutical, cosmetics, and pharmaceuticals is dramatically diversifying its management and marketing orientation toward new customer trends. Therefore, the exploration of nonconventional sources of natural products such as algae has been increasing.

Algae comprise different and diverse taxonomic groups regarding the physiology and morphology of photosynthetic organisms—cyanobacteria, microalgae, and seaweed (Raven & Giordano, 2014). Seaweeds (macroalgae) are macroscopic and multicellular organisms, while microalgae are unicellular eukaryotic organisms and cyanobacteria are not considered truly algae, because they are prokaryotic organisms but they present the same functional photosynthetic capacity and the same biotechnological importance (Levasseur et al., 2020) They can also be typically divided into (1) cyanobacteria (phylum Cyanophyta); (2) red algae (phylum Rhodophyta); (3) brown algae (phylum Heterokontophyta, class Phaeophyta); and (4) green algae (phylum Chlorophyta), mainly related with the composition of their pigments (De Jesus Raposo et al., 2015).

Being photosynthetic organisms, they are able to grow photoautotrophically by using energy from sunlight and inorganic carbon sources to generate oxygen, biomass, and several organic molecules and distinct metabolites with commercial relevance; however, depending on the carbon source, they can also grow mixotrophically or heterotrophically (Guedes & Malcata, 2012). Heterotrophic species obtain their nutrients from external source of organic compounds (e.g., glucose); on the other hand,

[a]Both authors contributed equally to this manuscript.

Algae Materials: Applications Benefitting Health. https://doi.org/10.1016/B978-0-443-18816-9.00020-4

others have the ability to perform both photosynthesis and heterotrophy, a process called mixotrophy (Raven & Giordano, 2014). In the photoautrophy, their facility to fixate high amounts of CO_2 by consuming only basic nutrients (light, water, carbon) and fast growth have driven the attention to produce these organisms, in particular cyanobacteria and microalgae in controlled cultivation conditions (i.e., raceways, photobioreactors) (Assunção & Malcata, 2020); in the case of seaweed, they can be collected in situ for farming, produced in tanks in land or can be directly collected in open waters or intertidal areas; those directly collected from the nature, preferentially assuring the ecosystem equilibrium and the collection (Torres et al., 2014). Their widespread distribution (e.g., freshwater, marine environment, and even terrestrial habitats) diversity, and amazing adaptability to the environment allowed them to develop several strategies, paving the way for metabolite chemical diversity with a wide range of functional properties. Thus, algae are promising sources of novel bioactive-based natural compounds with the potential for preventing and treating diseases (Levasseur et al., 2020). Their biomass and extracts are rich in proteins, peptides, fibers, pigments, minerals, vitamins, extracellularpolysaccharides, antioxidants, polyunsaturated fatty acids (PUFAs), and phenolic compounds, which potentially possess antitumor, antiinflammatory, antiviral, antioxidant, antiaging, and immunomodulatory properties (Eppink et al., 2013; Lau et al., 2015; Levasseur et al., 2020; Pagels et al., 2019; Sathasivam et al., 2019; Vijayakumar & Menakha, 2015). Therefore, they have the potential to be used in a variety of industries, including food, nutraceuticals, pharmacology, cosmetics, animal feed, and even for the production of biomaterials (i.e., bioplastics) and energy for the production of biodiesel (Brennan & Owende, 2010; Pagels et al., 2019; Rizwan et al., 2018).

Most of the seaweed uses in industry are limited to the extraction of polysaccharides and phycocolloids, for example, *Eucheuma* spp., *Kappaphycus alvarezii, and Chondrus* sp. for carrageenans, the kelp *Laminaria japonica* and red alga *Gracilaria* spp. for agar (Kaliaperumal & Uthirasivan, 2001; Rahikainen, 2021) or *Sargassum* sp. and *Lessonia* spp. for alginates (Rahikainen, 2021). Other algae with commercial interest encompass *Undaria pinnatifida, Porphyra* sp., and *Sargassum fusiforme,* which are directly consumed as food or ingredients in human feed (Buschmann et al., 2017; Rahikainen, 2021). The aforementioned polysaccharides possess important rheological properties that can be used for thickening and emulsification within the food and cosmetics industry (Cruz et al., 2020; Levasseur et al., 2020; Rosales-Mendoza et al., 2020). These biomaterials have also biosurfactant abilities potentially useful for water and oil bioremediation (Zhang, Song, et al., 2021). In addition, some of these (e.g., fucoidan) also own important antioxidant, antitumor, anticoagulant, antiproliferative, antiviral, antiinflammatory, or immunostimulatory properties (De Jesus Raposo et al., 2013, 2015) (see Table 23.1). Seaweeds such as *Undaria pinnatifida* and *Laminaria japonica* also have an interest in fucoxanthin extraction (D'Orazio et al., 2012).

Microalgae (e.g., *Dunaliella salina, Chlorella vulgaris*) and cyanobacteria strains (e.g., *Arthrospira platensis*) have been extensively used for functional and nutritional components for the development of functional foods and aquaculture products (Levasseur et al., 2020) (Table 23.1). They are mainly used for other specific compounds in particular carbohydrates, lipids, vitamins, minerals, and pigments—those can be used as bulk or commodities for food additives, nutraceuticals, feed industry, and even for biofuel manufacturing (Borowitzka, 2013; Levasseur et al., 2020; Rosales-Mendoza et al., 2020). For instance, the microalga *D. salina* is used for the extraction of β-carotene, an orange pigment that possesses important antioxidant features that can be used for food, nutraceutical, or feed (Campo et al., 2007); another example is the prospection of zeaxanthin—a yellow/brown pigment that is used in the supplements for regular maintenance of normal and accurate visual function (Christaki et al., 2013)—from

Table 23.1 Main metabolites produced by each group of algae: cyanobacteria, microalgae and seaweed; respective potential bioactive properties, and industrial applications.

Alga sources	Metabolite compound	Bioactive potential	Application	Ref.
Cyanobacteria				
Arthrospira platensis *Arthrospira maxima* *Arthrospira fusiformis* *Oscillatoria tenuis* *Anabaena* sp. *Nostoc* sp.	Phycocyanin Allophycocyanin	Antioxidant Anti-inflammatory Anticancer Immunomodulation antiallergic Antidiabetes Antiobesity Hepatoprotective effect Antibacterial Antifungal Antiviral Photosensitizer	Food: blue colorant, antioxidant additive Cosmetics: colorant, antioxidant Diagnostic tool Medical therapeutics: cancer therapy	Assunção et al. (2022a), Pagels et al. (2019), Romay et al. (1998), Sheu et al. (2013), Shih et al, (2003)
A. platensis *Aphanizomenon flos-aquae*	PUFAs (GLA)	Antioxidant Anti-inflammatory Antiarteriosclerotic Neuroprotection	Pharmaceutical Nutraceutical: supplements, infant-enriched formulas Aquaculture	Sathasivam et al. (2019), Spolaore et al. (2006)
Microalgae				
Dunaliella salina *Dunaliella bardawil*	β-carotene	Provitamin A Antioxidant Hepatoprotective Antihypertensive Antiaging Antiallergic Immunomodulatory Eye protection Cardiovascular diseases prevention Photoprotection	Pharmaceutical Cosmetics: UV filter, antiaging, hydration agent, antipigmentation formulas Nutraceutical and food: food additive and colorant Feed: enrichment of animal feed	Campo et al. (2007), Jomova and Valko (2013), Neuman et al. (1999), Stahl and Sies, (2012), Vo et al. (2012), Yaakob et al. (2014)
Scenedesmus almeriensis *Nannochloropsis oculata* *Chlorella* sp.	Zeaxanthin	Antioxidant Chemoprotective Hepatoprotective Eye protection Antiaging Atherosclerosis prevention Antihypertensive	Pharmaceutical Cosmetics: antiaging formulation Food and nutraceutical Food and feed colorant	Bhalamurugan et al. (2018), Christaki et al. (2013)
S. almeriensis *Muriellopsis* sp.	Lutein	Antioxidant Chemoprotective Hepatoprotective	Food additive (enrichment of infant formulas)	Sathasivam and Ki (2018)

Continued

Table 23.1 Main metabolites produced by each group of algae: cyanobacteria, microalgae and seaweed; respective potential bioactive properties, and industrial applications.—cont'd

Alga sources	Metabolite compound	Bioactive potential	Application	Ref.
		Anti-inflammatory Eye and cardiovascular protection Antibacterial Chemotherapeutic agent (tumor sensitizer) Antiaging	and colorant Feed: enrichment of animal feed and colorant Medical therapeutics: tumor sensitizer	
Haematococcus pluvialis *Chlorella zofingiensis*	Astaxanthin	Antioxidant Antiaging Anti-inflammatory Antitumor Antihypertensive Antiobesity Antidiabetic Antibacterial Eye and cardiovascular protection Prevention of bone loss and muscle fatigue Immunomodulatory effect Neuroprotection Photoprotection	Pharmaceutical Food, nutraceutical for human nutrition Cosmetics: hydration formulas, skin texture improvement, reduction of wrinkles Feed: additive and supplement, food colorant	Campo et al. (2007), Guedes et al. (2011a), Sathasivam and Ki (2018)
Phaeodactylum tricornutum *Odontella aurita* *Isochrysis galbana*	Fucoxanthin	Antioxidant Antitumor Antimalarial Antiobesity Antidiabetic Antiangiogenic effect Neuroprotection	Pharmaceutical Food, nutraceutical for human nutrition	Neumann et al. (2019), Tan and Hou (2014)
Coelastrella striolata var. *multistriata* *Chlorella vulgaris* *C. zofingiensis*	Canthaxanthin	Antioxidant Antiinflammatory Chemoprotective Neuroprotection Prevention of eye-related and cardiovascular diseases Immunomodulatory effect Photoprotection	Pharmaceutical Food, nutraceutical for human nutrition Feed additive	Sathasivam & Ki, 2018

Table 23.1 Main metabolites produced by each group of algae: cyanobacteria, microalgae and seaweed; respective potential bioactive properties, and industrial applications.—cont'd

Alga sources	Metabolite compound	Bioactive potential	Application	Ref.
Porphyridium cruentum	Phycoerythrin	Antioxidant Antitumor Antidiabetic immunosuppressive and antihypertensive Antiaging Photosensitizer agent	Pharmaceutical Food: antioxidant and colorant Cosmetics: red colorant and antioxidant Research tool: fluorescent dyes for diagnostic assays and research applications (e.g., cell sorting in flow cytometry, immunoassays, and fluorescence microscopy)	Assunção; et al. (2022a), Bei et al. (2002), Dumay et al. (2014)
Phaeodactylum tricornutum *Odontella aurita* *Nannochloropsis* sp.	PUFAs (EPA)	Antioxidant Anti-inflammatory Antiarteriosclerotic Neuroprotection	Pharmaceutical Food: supplements, infant-enriched formulas Nutraceutical Aquaculture	Bhalamurugan et al. (2018), Borowitzka (2013), Suganya et al. (2016)
Crypthecodinium cohnii *Schizochytrium* sp. *I. galbana*	PUFAs (DHA)	Antioxidant Anti-inflammatory Antiarteriosclerotic Neuroprotection	Pharmaceutical Food: supplements, infant-enriched formulas Nutraceutical Aquaculture	Bhalamurugan et al. (2018), Borowitzka (2013), Suganya et al. (2016)
Botryococcus braunii	Hydrocarbons	n/a	Biodiesel (gasoline, kerosene, and diesel)	Ranga Rao et al. (2012)
Seaweed				
Sargassum sp. *Laminaria* spp. *Lessonia flavicans* *Macrocystis* spp. *Ascophyllum* spp. *Lessonia* spp.	Alginates	Antioxidant Chemoprotective Protection of gastrointestinal tract Bioadhesive Biosurfactant Gelling agent Thickener	Pharmaceutical: inhibits gastroesophageal reflux Food and cosmetic industry: stabilizer thickener, gelling agent	Achparaki et al. (2012), Ahmadi et al. (2015), Khanra et al. (2018), Wiraningtyan et al. (2020),

Continued

Table 23.1 Main metabolites produced by each group of algae: cyanobacteria, microalgae and seaweed; respective potential bioactive properties, and industrial applications.—cont'd

Alga sources	Metabolite compound	Bioactive potential	Application	Ref.
		Emulsifier Hydration agent	Textiles dye Bioplastics Bioremediation Therapeutics: drug delivery, encapsulation, dental impression wound welling	Zhang, Song, et al. (2021)
Laminaria japonica *Gracilaria* sp. *Gelidium* sp.	Agar	Emollient Thickener Stabilizer Bioadhesive Biosurfactant	Cosmetics: emollient agent Food: thickening agent, gelling properties Bioremediation Agriculture: carrier for nutrients and growth substance Feed: supplement	Zhang, Song, et al. (2021)
Eucheuma spp. *Kappaphycus* spp. *Chondrus* sp. *Mastocarpus stellatus*	Carrageenan	Gelling agent Thickener Emulsifier Bioadhesive Biosurfactant Antioxidant Antiviral Immunomodulator Antiproliferative Anti-inflammatory	Bioplastics Pharmaceutical Food: stabilizer, emulsifier thickener in dairy products and meat Bioremediation Cosmetics: thickener (e.g., shampoo, toothpaste) Therapeutics: wound coverage and drug delivery	De Jesus Raposo et al. (2015), Rahikainen (2021), Zhang, Song, et al. (2021)
Undaria pinnatifida *Fucus vesiculosus* *L. japonica*	Fucoidan (sulfated polysaccharides)	Antioxidant Antitumor Immunomodulator Anti-inflammatory Anticoagulant Anticholesterol Antithrombotic effect Antiviral Antimalarial Antiparasitic Antiaging	Pharmaceutical Nutraceutical: supplements Cosmetics: antiaging, skin whitening, skin cell regeneration; UV protection, reduction of wrinkles	Chen et al. (2009), Narayanaswamy et al. (2013), Rahikainen (2021), Wang et al. (2019)

Table 23.1 Main metabolites produced by each group of algae: cyanobacteria, microalgae and seaweed; respective potential bioactive properties, and industrial applications.—cont'd

Alga sources	Metabolite compound	Bioactive potential	Application	Ref.
		Probiotic Favors angiogenesis Benefit bone health		
Sargassum heterophyllum *Eisenia bicyclis* *L. japonica* *Undaria pinnatifida*	Fucoxanthin	Antioxidant Antidiabetic Antiobesity Antimalarial Anticancer Photoprotector	Pharmaceutical Nutraceutical: supplements Cosmetics: skin whitening agent, UV filter	Afolayan et al. (2008), D'Orazio et al. (2012) Rahikainen (2021)

N/A, *Not Applicable.*

Scenedesmus almeriensis and *Nannochloropsis oculata* (Bhalamurugan et al., 2018). On the other hand, *Botryococcus braunii* is rich in hydrocarbons and is being explored for the production of biodiesel (Ranga Rao et al., 2012). Microalgae have been also established to produce PUFAs; for instance, the dinoflagellate *Crypthecodinium cohnii* and a unicellular green microalgae *Schizochytrium* sp. are explored commercially for docosahexaenoic acid (DHA) and eicosapentaenoic acid (EPA) (Bhalamurugan et al., 2018; Borowitzka, 2013), but only the former is explored at commercial scale (Rizwan et al., 2018).

In the case of cyanobacteria, the main example is *Arthrospira platensis* (commercially known as *Spirulina*) is an excellent source of phycobiliproteins (PBPs), particularly phycocyanin—this is a bright blue pigment with bioactive properties commonly explored in food and cosmetics as a colorant and antioxidant but is also used in the diagnostic tool as a photosensitizer in photodynamic therapy for the treatment of cancer (Assunção et al., 2022a; Borowitzka, 2013). This cyanobacterium is also considered one of the main sources of γ-linolenic acid (GLA), a PUFA possessing a relevant role in human and animal metabolism maintenance (Sathasivam et al., 2019). More, the whole biomass of this cyanobacterium is also explored as powder or tablets as a functional ingredient owing to their high content in proteins, vitamins, and minerals—working as a superfood in human and animal health (Costa et al., 2019). Table 23.1 presents some of the main algal sources and their metabolites of interest, their respective bioactive potential, and the main applications in industry.

To face the expected demands for algal biomass and metabolites of interest is crucial to increase the commercial feasibility and sustainability of the process. Most algal bioprocessing is currently based on the production of whole dried biomass, to be included in the algae-based products, i.e., *Spirulina* or *Chlorella* as a powder for nutraceutical, dietary supplements, or feed (Jacob-Lopes et al., 2019). But there are other strategies in which algal-based businesses focus on a specific metabolite for one-specific application. For instance, the cost of a more refined process such as PBPs can reach a maximum of 12.300 € per kg when compared with the whole freeze-dried biomass used by the food industry, which can reach 15€ per kg (for *Chlorella*) (Jacob-Lopes et al., 2019). The market prices can vary according to application, quantity, and purity of the metabolite at stake (Assunção et al., 2022b). Yet the focus on target-specific applications in which metabolites can be worth several thousand does necessarily become the process more competitive, economically feasible, or sustainable. There are still several technical and economic challenges that hinder the generation of bioproducts at a commercial level, in part because of the high costs of cultivation, low yields of biomass production, and intensive downstream processing, the latter accounting for almost 70% of the total costs of production (Assunção et al., 2022b; Costa et al., 2019).

The idea to overcome these problems and capitalize on this industry toward more economical and sustainable feasibility is related to the application of a biorefinery approach. This concept encompasses the production of multiple products from the same biomass, with a strategy of multiproduct recovery associated with minimal waste generation and energy consumption (Assunção et al., 2022b). Thus, the algal biorefinery envisages an integrative and multifunctional process, which is applied to a diverse range of raw materials and generates a spectrum of different intermediates and products, allowing the full potential of biomass to be achieved (Assunção et al., 2022b; Markou & Nerantzis, 2013; Torres et al., 2019). The idea stems from petroleum refineries, which produce not only fuels but also numerous molecules for the chemical industry. More, the algal biorefinery also pictures a combination of processes and technologies to efficiently convert raw materials into various components (Torres et al., 2019)—in which the conversion is based on the fractionation of the biomass into individual constituents or sequential processes that do not affect the quality of substrates for subsequent steps, thereby increasing resource efficiency. Therefore, this approach should be viewed as an array of several multiunit operations for biomass valorization and product optimization from the upstream process (i.e., algal cultivation and biomass production) to the downstream processing (harvesting, extraction, and purification) (Mitra & Mishra, 2019). Besides, the integration of efficient equipment with minimal energy consumption is part of the concept of a greener, safer and sustainable algal-based process (Assunção et al., 2022b). The principle of circularity is adjacent to this approach, which considers waste reduction or elimination, product reuse and recyclability, and the supply of energy from renewable and sustainable sources (Torres et al., 2019). Its processes generate the least amount of waste for the environment, lowering the environmental impact and putting less stress on ecosystem services (Zhu, 2015).

Currently, the biorefinery approach within the algal industry is not yet implemented for any macro-, microalga, and cyanobacteria here described. Yet, several considerations should be addressed: (1) the selection of the type of alga and their biological features (i.e., morphology, type of metabolites produced); (2) selection of suitable (photo)bioreactor; (3) optimal cultivation conditions; (4) harvesting and biomass processing including disrupting, extraction, and eventually purification; and (5) biomass and metabolite specifications for targeted applications. Fundamental knowledge and deep R&D studies on the algal biological features (i.e., morphology, physiology) and their biotechnological potential, including screenings to preselect bioactive compounds that may exhibit a potential market value, are fundamental. Besides, every alga has specific cultivation conditions in terms of media, temperature, light, pH, and suitable (photo)bioreactor for growth. The biomass processing also requires attention, as the selection of cell disruption equipment(s) depends on the selected alga. For instance, having more filamentous or mucilaginous algae will have a different extraction proceeding from algae that do not possess such characteristics. The former will probably require more processing and require disruption equipment more efficient toward cell wall rupture, which can become more expensive (Assunção et al., 2022b). From an engineering point of view, this also requires a deep analysis regarding all stages of the production process, from the input, mass, energy balances, and the details of the equipment and technologies to be used (Acién et al., 2017). All this balance will dictate the most suitable type of biorefinery for each type of alga. Hence, this chapter suggests some approaches and strategies for a further economic and feasible biorefinery taking into account the specifications of each type of algae: macro-, microalgae, and cyanobacteria.

2. Biorefinery in cyanobacteria

Cyanobacteria are photosynthetic microorganisms with rapid growth and simple nutritional needs. They are very well adapted to the external environment as they possess several mechanisms of defense and flexibility in their metabolic routes that allow them to survive. This flexibility makes these organisms quite suitable to be a "microcell biofactory," quite suitable for genetic engineering toward the production wide range of value-added compounds from a biotechnological perspective. These include phycobiliproteins (e.g., phycocyanin, phycoerythrin), carotenoids (e.g., β-carotene, canthaxanthin), vitamins (e.g., C, D), cyanotoxins, polysaccharides, UV protectant compounds such as scytonemin and mycosporine-like amino acids (MAAs) (Demay et al., 2019; Li et al., 2021; Mandal et al., 2020; Pagels et al., 2019; Richa et al., 2018), which can be applied in wide range industries such as of food, feed, nutraceuticals, cosmetics, pharmaceuticals, or even bioplastics (e.g., polyhydroxybutyrate or PHB). Apart from this, cyanobacteria are also producers of several different bioactive molecules such as anticancer (e.g., dolastatin), antiviral (e.g., microviridin) antibacterial (e.g., bastadin), antiparasitic (e.g., calothrixin A, B), and antiinflammatory (e.g., malyngamide F) (Abed et al., 2009; Prasanna et al., 2010; Vijayakumar & Menakha, 2015). As a result of the high demand for certain cyanobacterial metabolites within the industry and their commercialization, mainly to be introduced in several formulas (e.g., phycocyanin used as a colorant in food, scytonemin as UV protector antioxidant in sunlight cosmetics), the concept of biorefinery in cyanobacterial-based process becomes a priority. This addresses an approach of multiple product recovery to save time and money, with a maximum valorization of biomass and ultimately approaching sustainable models for circular bio-economy. Thus, the conversion into high added-value products (phycobiliproteins, carotenoids, PUFAs, antioxidants, etc.), as well as the generation of various other coproducts (i.e., biodiesel, biooil, biochar), and co-electricity production and heat to sustain all steps of the process, is the ultimate goal of a cyanobacterial biorefinery (Prabha et al., 2022). Most of cyanobacterial biorefineries aim for a single-product recovery failing to meet the major purpose of valorizing biomass, which ends in a nonfeasible process and economic failure, owing to the elevated costs of production (i.e., photobioreactor maintenance, low biomass yields, etc.) and downstream processing (i.e., extraction and purification of the compound) (Prabha et al., 2022). Successive extraction procedures could be a strategy to implement based on high to low market value, envisaging a circular bioeconomy that could ultimately lead to a minimal-waste biorefinery. Till now, there is no information regarding a cyanobacterial biorefinery fulfilling or operating under these criteria. The available reports only present fundamental studies mainly at laboratory or small scale and mainly cover *Arthrospira platensis* (Costa et al., 2019; Esquivel-Hernández et al., 2021; Ferreira-Santos et al., 2021) as this is considered an efficient producer of multiple products (including phycobiliproteins) and is the only cyanobacterium commercially produced for food applications due to their unique nutritional composition (higher protein content, carbohydrates, lipids, and vitamins) and functional properties (Ferreira-Santos et al., 2021; Prabha et al., 2022). This gap allows unveiling other non—so-explored strains with equal or higher ability to produce multiproducts of interest. Some studies have also started to disclosure the potential of other species/strains such as *Synechocystis salina* (Assunção et al., 2021b; Meixner et al., 2018), *Chroococcidiopsis* sp. (Assunção et al., 2021a), *Cyanobium* sp. (Pagels et al., 2020), *Lyngbya* (Chandra et al., 2019), or *Anabaena* sp. (Ferreira et al., 2012) under a biorefinery context (Table 23.2). This diversity of cyanobacteria (i.e., morphology, physiology, metabolic routes) is expected to create a net of effective and different types of biorefineries with the recovery of multiproducts from different industrial spectra applications. Increased economic feasibility of a cyanobacterial biorefinery can be achieved by coupling a cyanobacterial farm with wastewater (i.e., industrial or municipal waste) and/or a flue gas unit, or even a unit that can capture the atmospheric CO_2. These microorganisms can grow under high concentrations of organic phosphorous and nitrogen nutrients concomitant with a high ability to fixate carbon dioxide, which could take advantage of a flue gas treatment unit (Singh et al., 2019). The option of using a wastewater unit for complete or partial replacement of synthetic growth culture medium could minimize the production costs and the usage of freshwater, as in some cases, total costs of cyanobacterial production are related to excessive consumption of nutrients and water accounting up to 80% (Costa et al., 2019).

Biomass can be processed and fractioned based on the chemical properties of compounds of interest (i.e., lipophilic vs. hydrophilic properties) to turn the process safer within the industry (e.g., food, feed). Extractions could be performed in successive order to obtain compounds with different chemical properties, by playing with different polarities of the solvents used, for instance, ethanol used for carotenoids and chlorophylls and a successive extraction with water or phosphate buffer for phycobiliproteins. Furthermore, the specificity of the solvent may allow a higher purification level of the compound(s) involved (Assunção et al., 2022b). An additional coproduct, with less value, is the processing of lipids (i.e., UFAs) into biodiesel by the process of transesterification. The spent biomass from value-added products and the processing of biodiesel (i.e., biomass oilcake) may undergo processes of thermal conversion as pyrolysis, liquefaction or gasification to be transformed into biooil and biochar (nutrient-rich material used as fertilizer), biogas (e.g., biomethane, syngas) or can be processed by fermentation for the production of bioethanol or hydrogen (Guedes et al., 2013). More, as the energy can represent around 15% of the costs in a cyanobacterial culture (Torzillo, 1997), in particular for stirring, pumping, and drying, the cogeneration of biofuel and bioenergy is a way to exploit the benefits of the production of heat and electricity that can satisfy the energy requirements of production operation and downstream processing units (e.g., harvesting and dewatering biomass, extraction). This is an approach to have a more sustainable biorefinery; for instance, Costa et al. (2019) explained that is possible to save 33% of the overall cost of production (both cultivation and downstream) for a biodiesel-based biorefinery, by integrating a unit of energy generation by anaerobic digestion (i.e., biomethane production). Additionally, saving costs in the process can also be attained by the solar panels or wind generator units that can contribute to clean energy entering the platform. More, from a perspective of bioeconomy and sustainability, a medium/nutrient recycling unit could also be added as an operating approach. A scheme for the multirecovery process and valorizing the biomass with electricity cogeneration is presented in Fig. 23.1.

As aforementioned, an integrated cyanobacterial biorefinery platform is for now economically unfeasible, because of the higher investment in the production of biomass, extraction, and purification procedures. Yet, in the literature, some approaches were reported to make this strategy economically more attractive, and be applied in a concept of the minimum waste model to promote a circular bioeconomy (Table 23.2).

Table 23.2 Multiproduct recovery for several species of cyanobacteria for biomass valorization under a biorefinery approach.

Species	Culture medium for cultivation	Cultivation conditions	Multirecovered products	Residual biomass coproduct(s)	Remarks	References
Anabaena sp. PCC 7120	$BG11_0$	**R**: 2L-flask; **T**: 25°C **LI**: 240 W, 384 W, 678 W **L/D**: 24h/0h	Hydrogen	Hydrogen	Fermentation by *Enterobacter aerogenes* bacteria	Ferreira et al. (2012)
Arthrospira platensis	Modified Jourdan	**R**: open raceway pond with greenhouses; **T**: n/a **LI**: n/a **L/D**: n/a	Value-added metabolite	Bioethanol Lactic acid	*Saccharomyces cerevisiae* LPB-287 and *Lactobacillus acidophilus* ATCC 43121	Esquivel-Hernández et al. (2021)
A. platensis	Zarrouk medium	**R**: 3000 L-open raceway tanks; **T**: 25 ± 10°C **LI**: n/a **L/D**: n/a	Proteins	Bioethanol	Fermentation by *Saccharomyces cerevisiae*	Peron-Schlosser et al. (2021)
A. platensis	Wastewater (secondarily treated wastewater)	**R**: 300 L-open raceway pond; **T**: 38 ± 3°C; **LI**; 418.12 ± 15 $\mu mol_{photons}\cdot m^{-2}.s^{-1}$ **L/D**: n/d	Biodiesel	Biomethane	Anaerobic digester slurry for biomethane production	Chavan and Mutnuri (2019)
A. platensis	n/d	n/d (powder from Azienda Agricola Prato della Voja was used)	Phycobiliproteins Phenolics	Chl *a* Carotenoids Lipids (PUFAs)	Sequential multistage procedure (physical/thermal (OH), enzymatic treatments, solvent extraction)	Ferreira-Santos et al. (2021)
Cyanobium sp.	BG11	**R**: 5-L glass flat bottom round flasks **T**: 25°C **LI**: 200 $\mu mol_{photons}\cdot m^{-2}.s^{-1}$ **L/D**: 16h:8h	Carotenoids Phycobiliproteins Phenolic compounds	n/a	Successive extractions with generally recognized as safe (GRAS) solvents (acetone, ethyl acetate, ethanol, and water)	Pagels et al. (2020)
Chroococcidiopsis sp.	Z8	**R**: 5-L glass flat bottom round flasks **T**: 25°C **LI**: 100 $\mu mol_{photons}\cdot m^{-2}.s^{-1}$ **L/D**: 16h:8h	Carotenoids Phycobiliproteins Phenolic compounds Polysaccharides	n/a	Successive extractions with GRAS solvents (methanol, acetone, ethanol, and phosphate buffer saline (PBS))	Assunção et al. (2021a)

Continued

Table 23.2 Multiproduct recovery for several species of cyanobacteria for biomass valorization under a biorefinery approach.—cont'd

Species	Culture medium for cultivation	Cultivation conditions	Multirecovered products	Residual biomass coproduct(s)	Remarks	References
Lyngbya sp.	BG11	**R**: 5-L bottle **T**: 25°C **LI**: 110 $\mu mol_{photons} \cdot m^{-2}.s^{-1}$ **L/D**: 16h:8h	MAAs	Lipids for biodiesel	High oxidative stress showed to be critical for an increase in MAAs and lipid production	Chandra et al. (2019)
Nostoc sp.	BG11	**R**: 5-L bottle; **T**: 25°C **LI**: 110 $\mu mol_{photons} \cdot m^{-2}.s^{-1}$ **L/D**: 16h: 8h	Polysaccharides, phytochemicals, and lipids	Lipids for biodiesel	-	Pekkoh et al. (2022)
Synechococystis PCC 6803	Municipal wastewater	**R**: 5-L bottle; **T**: 25 ± 1°C **LI**: 30 $\mu mol_{photons} \cdot m^{-2}.s^{-1}$ **L/D**: 10h:14h	Biodiesel	Bioethanol	*Saccharomyces cerevisiae* for bioethanol production	Ashokkumar et al. (2019)
Synechocystis salina	Z8	**R**: 5-L glass flat bot- tom round flasks **T**: 25°C **LI**: 100 $\mu mol_{photons} \cdot m^{-2}.s^{-1}$ **L/D**: 16h:8h	Carotenoids Phycobiliproteins Phenolic compounds Lipids	n/a	Successive extractions with GRAS solvents (hexane, acetone, ethanol, and PBS)	Assunção et al. (2021b)
S. salina	BG11	**R**: 200-L tubular glass photobioreactor **T**: n/d; **LI**: 9–11 W m^{-2} **L/D**: n/d	Chlorophylls Carotenoids Phycocyanin PHA	Proteins carbohydrates and lipids for biomethane and/or fertilizer	Pigments were extracted with solvent in a pretreatment Fermentation with *Cupriavidus necator* for PHA (polyhydroxyalkanoates) production	Meixner et al. (2018)

R, *type of reactor and volume;* T, *temperature;* LI, *light intensity;* L/D, *light/dark cycle;* N/A, *not applicable;* N/D, *not determined.*

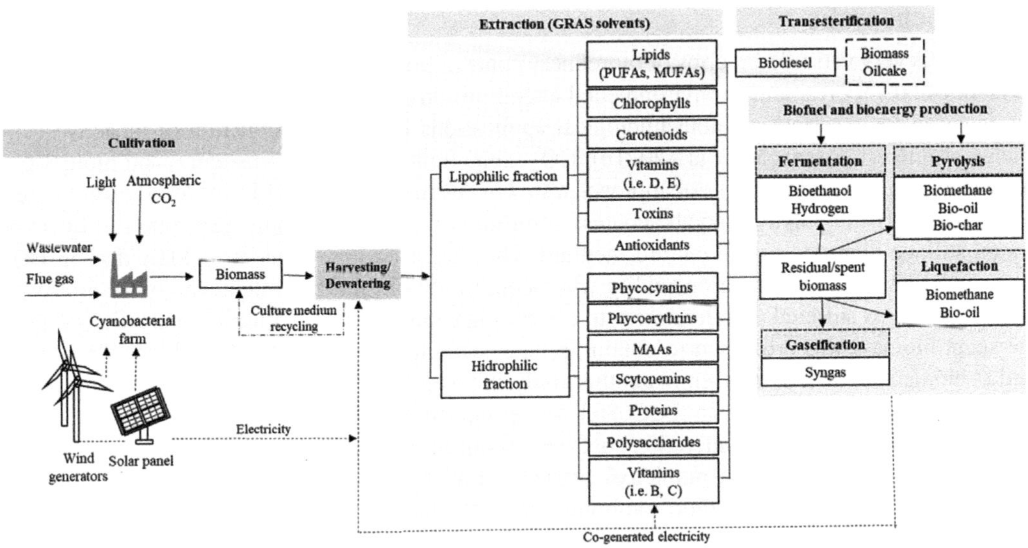

FIGURE 23.1

Overview of a Cyanobacterial Multirecovered Product Biorefinery with Cogenerated Bioenergy for the Maximum Valorization of Biomass.

2.1 Arthrospira

Regarding the *Arthrospira genus*, which encompasses *Arthrospira platensis*, one of the most important species of cyanobacteria produced commercially for food applications (Ferreira-Santos et al., 2021; Prabha et al., 2022) and which will probably assume a prominent position within the context of the biorefinery (Costa et al., 2019), since it is one of the most developed at the industrial level.

An approach was followed in a multiproduct biorefinery for *Arthrospira platensis* for ethanol and lactic acid production (Esquivel-Hernández et al., 2021). Biomass was firstly harvested to extract bioactive added-value compounds, and the residues were further valorized for fermentation with *Saccharomyces cerevisiae* to produce bioethanol and *Lactobacillus* to produce lactic acid. Economic analysis for these two products concerning fermentation scale, fermenter costs, production titer, fermentation time, and cyanobacterial biomass production was assessed and showed that fermenter scale, equipment cost, and product titer are critical variables. The study points out that by increasing product generation, costs tend to stabilize at a large scale with a high decrease in the unit product supporting this approach (Esquivel-Hernández et al., 2021). The production cost is estimated to be US$ 1.27 and US$ 0.39 per unit of product, for bioethanol and lactic acid, respectively.

On the other study, Chavan and Mutnuri (2019) used domestic wastewater to produce biomass of *Arthrospira platensis* and take advantage to extract lipids for biodiesel production and the residual biomass for biomethane production, in a sequential way. The study claims that *Arthrospira platensis* cultivated under wastewater streams can reduce not only the cost of wastewater treatment but also earn revenue from biodiesel and the production of biogas.

2.2 Synechocystis

This genus is very versatile in terms of biorefinery context. Some species such as *Synechocystis salina* are marine and very resilient to environmental alterations, in particular to salinity changes. Besides, it was reported for a variety of biotechnological applications including production of bioactive compounds of interest (Assunção et al., 2021b), CO_2 mitigation, wastewater treatment, and biofuel production (Gonçalves et al., 2016). An approach tested by Meixner et al. (2018) used *S. salina* to produce high-quality PHAs (polyhydroxyalkanoates) resorting to photoautotrophic fermentation by using diluted stillage digestate as the culture medium. They reported high content of PHA than in BG_{11} culture medium and with superior quality when compared with the PHA achieved by heterotrophic bacteria. Biomass suffered a pretreatment to extract pigments and then go for the extraction of PHA. The spent biomass was processed to obtain proteins, carbohydrates, and lipids and considered to be used as biomethane or biofertilizer. In another study, not with *S. salina*, but with a species belonging to the same genus, *Synechocystis* sp. PCC6803 was evaluated for sustainable bioenergy production. A reduction of up to 35% in the cultivation cost was obtained by growing this cyanobacterium in a large-scale open raceway pond with municipal wastewater with estimations to obtain 94.5 tons of dry biomass. $ha^{-1}.yr^{-1}$. A cheap method combining iron (III) chloride and a biopolymer recovered almost 98% of the biomass, and the transesterification used tungstated zirconia with high biodiesel yields of 91%. Lipid biomass residues were considered for bioethanol extraction to increase the feasibility of the process (Ashokkumar et al., 2019).

In another study, Beattie et al. (2021) assessed the economic and environmental impact of a cyanobacteria-based biorefinery (for *Synechocystis* sp. PCC 680) for the production of fuels and methyl laurate as oleochemical—including all aspects of cultivation, harvesting, extraction of methyl laurate, and fuel processing by hydrothermal liquefaction of the dewatered biomass. The study showed that increasing biomass and methyl laurate productivities can benefit the system meeting 0.80 dm^{-3} and reduce the global warming potential (GWP) of fuels. The study showed that a high environmental burden impact is related to off-gassing of delivering of CO_2 to the process, with high energy demand. The study points out that increasing the biocrude yields, on one hand, becomes the process more economically feasible, with lower prices of fuel but on the other hand increases GWP.

3. Biorefinery in microalgae

Microalgae have a similar market interest as cyanobacteria, but with a more complex structure (eukaryotic cells) and a larger group of already approved species for human consumption. They are found on a large scale worldwide to produce biomass or specific compounds. Species such as *Chlorella* spp., *Dunaliella* spp., and *Haematococcus pluvialis* have been mass-produced, owing to their nutritional value and pigment content (β-carotene and astaxanthin). Furthermore, due to their high lipidic content, species such as *Isochrysis* spp. and *Scenedesmus* spp. have been used in the feed industry (Guedes et al., 2011).

In terms of biorefineries, microalgal processes, in general, have focused on cultivating microalgae in wastewater for water treatment, along with biomass generation from the microalga–bacteria consortium, which can then be used to produce biofuels (Giraldo-Calderón et al., 2018). Microalgae are promising candidates for wastewater treatment and biofuel production because they grow quickly, produce no waste, and can be grown on nonagricultural land. As a result, combining wastewater treatment with microalgae cultivation could be a future green technology for biofuel and energy generation. The combination of wastewater treatment and microalgae cultivation for biofuel production results in environmental and economic benefits (Mehariya et al., 2021).

A great example of combining wastewater treatment and biofuel is reported by Hamalatha et al. (2019), which proposed a system (modeled in life cycle assessment (LCA) and technoeconomic analysis) using dairy wastewater integrated with bioethanol production. At the end of treatment, organic carbon removal was observed to be 90% with simultaneous removal of nutrients, with a biomolecular composition of microalgae (mixture of species) comprised of 38% carbohydrates, 15% proteins, and 22% lipids. The lipids are extracted, and the deoiled microalga is fermented using *Saccharomyces cerevisiae*, which results in the production of ethanol (third generation) (Hemalatha et al., 2019).

Due to socioeconomic concerns about environmental issues, the search for renewable energy is increasing, and microalgae-based processes and products play an important role in this. In their current state, microalgae-based processes face significant constraints. Because the technological routes are still in their infancy, large-scale production of algae biofuels is unlikely to be feasible. As a result, because microalgae have a diverse and rich chemical composition, they could be an effective feedstock for many high-value-added products (Deprá et al., 2018). Fig. 23.2 depicts an overview of the microalgae biorefinery.

Most studies focus on the approaches used in the recycling and reuse of the residual biomass, as up to 70% of the entire microalgae biomass can be obtained as residues after they serve their primary purpose. Economic analysis indicates that biofuel production from microalgae is mainly dependent on

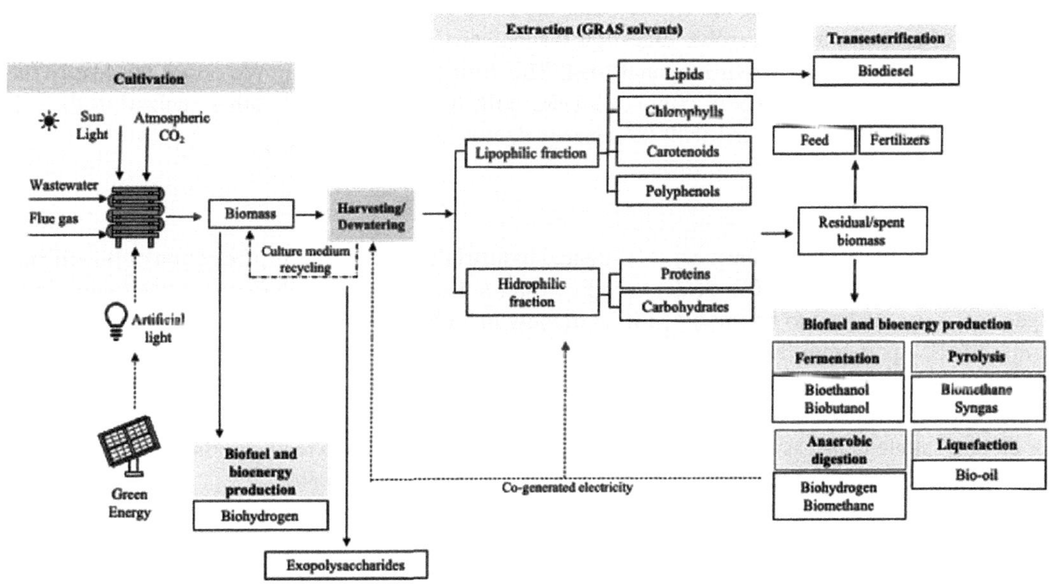

FIGURE 23.2

Overview of a Microalgal Multirecovered Product Biorefinery with Cogenerated Bioenergy for the Maximum Valorization of Biomass.

the production of valuable coproducts from the residual biomass in an algal biorefinery concept. This residual biomass can be carbohydrates, proteins, polyunsaturated, fatty acids, pigments, and other pharmaceutically active compounds, depending on the species in question (Sankaran et al., 2018). In this section, a summary of the main microalgae biorefinery will be discussed, considering specific species. Table 23.3 summarizes the studies that proposed biorefinery routes.

3.1 Chlorella

One of the most significant commercial microalgae is the coccoid green microalgae genus *Chlorella*. The dried biomass of *Chlorella* is processed into powder, capsules, or tablets for human nutrition (Rani et al., 2018). In terms of biorefinery propositions, *Chlorella* microalgae have been selected for a few LCA simulations and biorefinery propositions, mainly regarding its content of lipids, proteins, and carbohydrates.

Bussa et al. (2021) performed an LCA based on *C. vulgaris* biorefinery, with four different pathways being evaluated: two options to isolate soluble proteins and polysaccharides (membrane filtration and chemical separation), and two options to separate lipids from the insoluble fraction of the algal biomass (solvent extraction and supercritical extraction). The combination of supercritical extraction for lipids, and membrane filtration of polysaccharides and proteins led to the lowest environmental impact, mainly due to the high use of ammonium sulfate in the chemical separation and ethanol in the solvent extraction.

In a different strategy, Preat et al. (2020) evaluated the energy demand for different biorefinery scenarios for *C. vulgaris*, considering the viability of the obtention of lipids, proteins, methane, fertilizers, and dried residue. Results showed that the route of lipid extraction using hexane, acid precipitation of proteins, and anaerobic digestion for methane and fertilizers is ideal in terms of energy demands and consequently environmental load. The drying of residual biomass does not justify the emissions. Moreover, the authors compared bead mill, and high-pressure homogenization for cell disruption, the last being the one with less energy demand.

3.2 Dunaliella

Dunaliella spp. have been extensively investigated industrially due to an overproduction of β-carotene, a bioactive carotenoid (ca. 10% of dry weight). *Dunaliella* strains are eukaryotic unicellular microalgae that have evolved to be produced in extremely halophilic or hypersaline conditions (Goswami et al., 2022). Apart from β-carotene, this microalga is rich in lipids (especially triacylglycerols) and carbohydrates, indicating that the residual biomass after β-carotene extraction is ideal for the production of biofuels. *D. salina* biorefinery techniques have concentrated on the recovery of carotenoids for dietary supplements, as well as the utilization of residual biomass to recover other coproducts such as biogas, glycerol, and fertilizer.

Monte et al. (2019) combined solvent extraction with organic solvent nanofiltration to yield proteins, carotenoids, and glycerol. A saponification procedure is also suggested to separate carotenoids from glycerides. The suggested biorefinery leads to a recovery rate higher than 85%. In another study, Espada et al. (2020) simulated a *D. salina* biorefinery based on the production of β-carotene by supercritical extraction, solvent extraction, and saponification. The spent biomass was anaerobically digested to generate biogas, and the solid waste was proposed as fertilizer. Similarly, Thomassen et al.

Table 23.3 Multiproduct recovery for several species of microalgae for biomass valorization under biorefinery approach.

Species	Multirecovered products	Residual biomass coproduct(s)	Remarks	References
Chlorella vulgaris	Lipids	Polysaccharides and proteins	LCA	Bussa et al. (2021)
C. vulgaris	Lipids	Proteins, methane, and fertilizer	Energy demand evaluation	Préat et al. (2020)
Dunaliella salina	β Carotene	Glycerol	Solvent extraction combined with organic solvent nanofiltration	Monte et al. (2019)
D. salina	β-Carotene	Biogas and fertilizer	LCA	Espada et al. (2020)
D. salina	β-Carotene	Fertilizer	LCA	Thomassen et al. (2018)
Dunaliella tertiolecta	β-carotene	Biooil and biochar	Pyrolysis	Francavilla et al. (2015)
Haematococcus lacustris	Astaxanthin	Biochar	This biochar can be used in the iron-making industry	Ashokkumar et al. (2021)
Haematococcus pluvialis	Astaxanthin	Biomethane	n/a	Hosseini et al. (2020)
Isochrysis galbana	Lipids	Methane	Ethyl acetate for lipids extraction and anaerobic digestion of the spent biomass	Sánchez-Bayo et al. (2020)
I. galbana	Proteins, arabinose-, and glucose-rich polysaccharides	n/a	Use of ionic liquid (IL)-based aqueous biphasic systems (ABS)	Santos et al. (2018)
I. galbana	Carotenoids, lipids, antioxidants, carbohydrates, and proteins	n/a	Successive extraction	Gilbert-López et al. (2015)
Scenedesmus dimorphus	Biodiesel	Dihydroxyacetone	Economic evaluation (IRR of 19.8%)	Tejada Carbajal et al. (2020)
Scenedesmus obliquus	Treated wastewater	Polyphenols, flavonoids, biohydrogen, biooil, biochar, biogas, and fertilizer	A circular approach to the brewery industry	Ferreira et al. (2019)
Scenedesmus obliquus	Lipids, proteins, and carbohydrates fractions	n/a	Evaluation of different biorefinery routes	Ansari et al. (2017)

N/A, *Not Applicable.*

(2018) describe a biorefinery simulation in which β-carotene is extracted from *D. salina* biomass utilizing organic solvents and membrane filtering for further carotenoids separation, while biomass leftovers are employed as fertilizer.

Furthermore, Francavilla et al. (2015) proposed a biorefinery for *Dunaliella tertiolecta* using the residual biomass from the extraction of added-value compounds (β-carotene, phytosterols, and fatty acids) for fast pyrolysis targeting biooil and/or biochar as pyrolysis end products. At pyrolysis temperature of 600°C, biooil yielded 45%, while char reached 29%.

3.3 Haematococcus

Haematococcus pluvialis is a unicellular microalga that can synthesize astaxanthin up to 5% of its dry weight, making it the best and richest source of natural astaxanthin. Yet, astaxanthin generated from microalgae accounts for only a small portion of the market, which is estimated to be around 1% (Nishshanka et al., 2022).

In terms of biorefinery, a few studies have been performed regarding *Haematococcus* spp. Ashokkumar et al. (2021) proposed a biorefinery based on *Haematococcus lacustris* for the obtention of astaxanthin and biochar coproduction through pyrolysis of the residual biomass for metallic iron (Fe) production. As a result, using algal biochar in the iron-making industry would reduce energy use, cost-effectiveness, and CO_2 emissions.

In another approach, Hosseini et al. (2020) proposed the use of residual biomass of *H. pluvialis* for biomethane after astaxanthin extraction. The residual biomass was hydrolyzed leading to an efficiency of 97% of the total glucan, which was then anaerobically digested to produce biomethane.

3.4 Isochrysis

Isochrysis is a flagellated marine algae genus that has been reported to collect a high quantity of lipids for biofuel generation. *Isochrysis* strains are also rich in docosahexaenoic acid (DHA, 22:6n-3), an omega-3 polyunsaturated fatty acid, and fucoxanthin, a xanthophyll that acts as an important antioxidant in feed and food supplements (Sun et al., 2019). The studies regarding *Isochrysis* were usually considering experimental aspects of the process, to suggest the optimal route of extraction.

A biorefinery route combining lipid extract and residual biomass anaerobic digestion using *Isochrysis galbana* microalga has been proposed by Sánchez-Bayo et al. (2020). In this report, the authors compared different solvents for lipid extraction, where ethyl acetate and chloroform:methanol mixture yielded the best lipid extraction results. Moreover, the spent biomass from ethyl acetate yielded more methane after the digestion, being the suggested route for an energy-driven microalga biorefinery.

Santos et al. (2018) evaluated the use of ionic liquid (IL)–based aqueous biphasic systems (ABS) to separate proteins from arabinose- and glucose-rich polysaccharides from *I. galbana*. For this purpose, the nature of the inorganic salt, the IL structural features (anion and alkyl chain length), and the extraction point (using different inorganic salt concentrations) were the conditions optimized. The authors proposed a complete downstream process, in which the isolation of the target biomolecules (proteins, arabinose- and glucose-rich polysaccharides) and the reuse of the phase formers and main solvents applied were achieved.

Finally, Gilbert-Lopez et al. (2015) used four sequential steps in *I. galbana* using supercritical CO_2 extraction without/with ethanol, ethanol extraction, and water extraction. This biorefinery route targeted the obtention of carotenoids, lipids, antioxidant-rich extract, and carbohydrates + proteins, respectively. As a result, a good strategy for complete valorizing microalgae biomass was given by the authors, based on techniques that adhere to green chemistry principles.

3.5 Scenedesmus

Scenedesmus is a genus of green algae, in the class Chlorophyceae. *Scenedesmus* is widely produced for aquaculture purposes, mainly due to its high lipidic content. *Scenedesmus* biomass has been proposed for the obtention of a variety of biofuels including biohydrogen, biodiesel, and bioethanol. However, most studies of *Scenedesmus* are focused on a laboratory scale, and like all algal systems, face obstacles in large-scale production (Christenson & Sims, 2011).

In terms of biorefinery, Tejada Carbajal et al. (2020) have evaluated in terms of economic viability, five scenarios using *Scenedesmus dimorphus* for the production of fishmeal, algae oil, biodiesel, glycerol, and/or dihydroxyacetone with the bacteria *Gluconobacter oxydans*. The biorefinery with the best economic performance integrated reactive distillation for the production of biodiesel and dihydroxyacetone, achieving an internal rate of return (IRR) of 19.8%.

Moreover, Ferreira et al. (2019) evaluated the treatment of brewery effluent using the cultivation of *Scenedesmus obliquus* microalga for a circular biorefinery. The treatment proved efficient in the removal of nutrients, together with several compounds that could be extracted from the biomass, including polyphenols and flavonoids (subcritical water extraction), biohydrogen (dark fermentation), biooil, biochar and biogas (pyrolysis), and fertilizer (residual biomass). By using the residual biomass as fertilizer and the treated water for barley and wheat crops, the process becomes circular, which could offer economic and environmental benefits to the brewery industry.

Using a different approach, Ansari et al. (2017) evaluated six different sequential extractions of lipids, proteins, and carbohydrates from *Scenedesmus obliquus*, being the sequence proteins—lipids—carbohydrates the optimal for products recovery, which yielded a total of 28.14%, 22%, and 10.17%, respectively. The extracts were performed using chloroform:methanol (2:1, v/v) for lipids, cold ultrapure water for proteins, and H_2SO_4 (2%, v/v) for carbohydrates.

4. Biorefinery in seaweed

Seaweed, also known as macroalgae, has the potential to be a valuable feedstock for biorefinery because it produces diverse and versatile biomass that can be used for a variety of applications. They can be used directly for human consumption or processed into food additives and nutraceuticals, feeds, fertilizers, biofuels, cosmetics, and medicines, among other things. Global demand for seaweeds has increased in tandem with increased usage beyond previously traditional applications (Balina et al., 2017; Buschmann et al., 2017).

The seaweed biorefinery framework presents a conceptual model for high-added-value product production as well as fluid or gaseous biofuel production. This, in turn, lowers the cost of fuel production while maximizing the utilization of biomass (Balina et al., 2017).

Based on their pigment composition, seaweed can be classified into three groups: green (Chlorophyta), brown (Ochrophyta), and red (Rhodophyta) (van Hal et al., 2014). Depending on the type and species of seaweed, various fatty acids, oils, natural pigments, antioxidants, high-value biological components, and other substances can be extracted and potentially used in an industrial production system (Balina et al., 2017). Overall, the use of seaweed in biorefineries can be summarized as using the biomass to extract high-added-value compounds such as polysaccharides or pigments, while the residue can be used as fertilizer or to generate energy (Fig. 23.3). An interesting choice proposed by some authors is the recovery as a seaweed-flavored salt, for application in the food industry, using the washing water, or even the residual biomass (Baghel et al., 2020).

Furthermore, by using Integrated Multi-Trophic Aquaculture (IMTA) for cultivation, the concept of sustainability can be even more present, as most nutrients can come from other "discarded" water from fish farms (Czyrnek-Delêtre et al., 2017). Practical experiments and simulation studies on the use of seaweed as biorefinery feedstock are typically focused on one of the taxonomic groups, as the target compounds differ (Table 23.4), as will be discussed in the following.

4.1 Brown seaweeds

Brown seaweeds include many edible seaweeds. All brown algae contain alginate in their cell walls, which is extracted commercially and used as an industrial thickening agent in food and for other uses. Moreover, brown seaweeds are recognized sources of other compounds with a wide range of bioactive properties and applications, including the carotenoid fucoxanthin and some phlorotannins (Abraham et al., 2019; van Hal et al., 2014).

Alkaline conditions are used in traditional alginate extraction from brown seaweed to produce soluble sodium alginate as a single product. Abraham et al. (2019) proposed and optimized a seaweed biorefinery to extract alginate, fucoidan, and laminarin polysaccharides from the Australian seaweed *Durvillaea potatorum*. The addition and optimization of acid extraction as the first step enabled the production of three polysaccharide products: fucoidan and laminarin, acidic extractable alginate, and

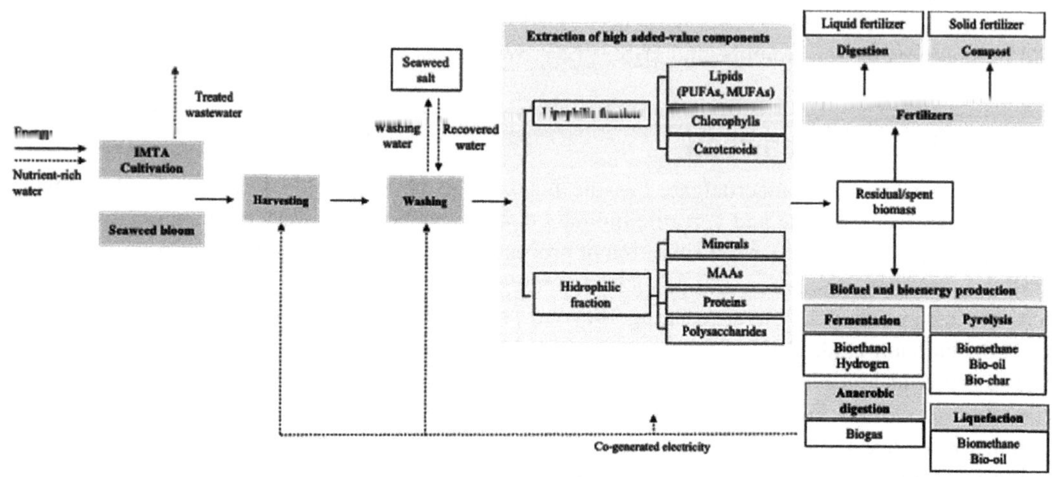

FIGURE 23.3

Overview of a Seaweed Multirecover Product Biorefinery With Cogenerated Bioenergy for the Maximum Valorization of Biomass.

Table 23.4 Multiproduct recovery for several species of seaweed for biomass valorization under biorefinery approach.

Species	Biomass origin	Multirecovered products	Residual biomass coproduct(s)	Remarks	References
Brown seaweed					
Durvillaea potatorum	Field sampling	Alginate	Fucoidan and laminarin	Addition of an acid extraction	Abraham et al. (2019)
Ecklonia maxima	n/a	Alginate	Fucoidan and laminarin	LCA	Zhang, Border, et al. (2021)
Fucus vesiculosus, Saccharina latissima, and *Laminaria digitata*	n/a	Laminarin	Fucoidan and feed supplement	LCA	Zhang and Thomsen (2021)
S. latissima	Offshore cultivated	Alginate	Lignocellulosic fraction	n/a	Nilsson et al. (2022)
S. latissima	n/a	Alginate	Biogas and fertilizer	LCA	Langlois et al. (2012)
Sargassum tenerrimum	Field sampling	Alginate	Protein concentrate, cellulose, and salt	Successive extraction with alkaline and acid solutions	Baghel et al. (2020)
Undaria pinnatifida	Field sampling	Phlorotannins	Antioxidant extracts	Successive extraction	Ferreira et al. (2021)
Green seaweed					
Ulva ohnoi and *Ulva tepida*	Field sampling	Seaweed salt	Fertilizer, feed, or biofuel	Recovery of water during the processing	Magnusson et al. (2016)
Ulva lactuca	Field sampling	Protein concentrate	Biofuels and organic solvents	Use of hydrolysis	Bikker et al. (2016)
U. ohnoi	Inland cultivation	Salts, starch, lipids, ulvan, proteins, and cellulose	n/a	n/a	Prabhu et al. (2020)
Red seaweed					
Gracilaria verrucosa	Field sampling	Agar	Bioethanol	Enzymatic hydrolysis and fermentation	Kumar et al. (2013)
Kappaphycus alvarezii	n/a	Carrageenan	Bioethanol	Fermentation	Ingle et al. (2018)

N/A, Not applicable.

alkaline extractable alginate. When compared with conventionally extracted alginate, the two alginates have significantly different biochemical properties, and the proposed biorefinery yields more total polysaccharides (43.57%, *w/w*) than the conventional processing for a single alginate product (38.97%, *w/w*).

In another study, Ferreira et al. (2021) proposed a biorefinery based on *Undaria pinnatifida* biomass for the production of bioactive compounds such as phlorotannins. A biorefinery was established to obtain the extracts using a route of water, ethanol:water (50:50, *v/v*), and ethanol sequentially. The antioxidant and antimicrobial capacities of the fractions were assessed. As the best extract in the evaluated assays, the ethanolic extract proved to be a promising candidate for further inclusion as an antioxidant and antibacterial agent, while the remaining fractions examined are potential sources for other biotechnological applications, such as antibacterial and as source of phlorotannin emphasizing the importance of a biorefinery and circular economy framework to add value to this seaweed and the industry.

Moreover, Baghel et al. (2020) evaluated the biorefinery route of *Sargassum tenerrimum* for the obtention of alginate, together with other valuable products, including protein concentrate, cellulose, and salt. The refinery route uses an alkaline extraction for the alginate extraction, and acid extraction for the protein concentrate and cellulose. The salt is recovered from liquid effluent through the process.

Nilsson et al. (2022) also proposed a biorefinery for *Saccharina latissima*, focused on the obtention of sodium alginate, together with recovered products such as lignocellulosic fraction, biogas, and fertilizer. The extraction is performed by acid pretreatment and alkaline extraction, and after the precipitation, alginate is recovered. The liquid fraction during the extraction is considered a lignocellulosic fraction, which can be used for biomaterials, or converted into biogas and fertilizer by anaerobic digestion.

In terms of modelized biorefineries, Langlois et al. (2012) proposed a similar biorefinery route for *S. latissima*, aiming for the obtention of sodium alginate. Then, the residual biomass goes through anaerobic digestion to produce biogas and liquid fertilizer, and the concept was evaluated under an LCA. The results validate the route suggested, highlighting the environmental impact caused by the electricity and energy used to heat the anaerobic digesters. The authors suggest that ecodesign can make algal biomethane competitive in several categories compared with natural gas: a decrease of 54.2% in greenhouse gas (GHG) emissions and 68.7% in fossil depletion. This is possible by designing the systems with a clean and efficient source of electricity (offshore wind farms) on-site and to heat the digesters.

Zhang, Song, et al. (2021) proposed a cascade valorization of *Ecklonia maxima*, an underutilized endemic brown macroalga in South Africa, for the production of high-value outputs such as alginate, laminarin, and fucoidan via a biorefinery approach guided by LCA. SWE (subcritical water extraction) and HWE (hot water extraction) were proposed and compared. When compared with regular alginate industry values, the results showed that the suggested systems have a lower water footprint but a higher carbon footprint, due to the use of energy. Moreover, the side-stream valorization scenario based on zero waste and closed-loop resource management principles demonstrated overall improvements, indicating a promising system optimization strategy and opportunity for business cobenefits.

Finally, Zhang and Thomsen (2021) intend to assess the economic viability and environmental sustainability of three conceptual macroalgal biorefinery systems based on the endemic brown macroalgae *Laminaria digitata*, *Fucus vesiculosus*, and *Saccharina latissima*. Laminarin, fucoidan, and a feed supplement are the main products recovered. The economic analysis determined that the major cost and revenue drivers are feedstock costs and laminarin and fucoidan sales. A sensitivity analysis of the economic viability of industrial-scale systems revealed that the break-even price for the *F. vesiculosus*, *S. latissima*, and *L. digitata* systems was US$36, 57, and 68 per kilo of feedstock, respectively. The net carbon footprint and net water footprint of industrial-scale systems, respectively, range from 3.8 to 11 kg CO_{2eq}/kg of feedstock and 0.1–0.2 $m^3{}_{water}$/kg of feedstock. The authors proposed that internal process and system optimizations, as well as greening the electricity mix, can help drive further progress toward environmental sustainability.

4.2 Green seaweeds

Green seaweeds have a distinct chemical composition and can contribute to marine biorefinery systems through another variety of potential products, particularly polysaccharides (sulfated polysaccharides, starch, and cellulose) and polyphenols (Zollmann et al., 2019). The most used and studied green seaweeds are belonging to the *Ulva* genera.

Magnusson et al. (2016) proposed a biorefinery for wastewater bioremediation as well as large-scale on-land *U. ohnoi* and *Ulva tepida* production. The authors proposed that during biomass cleaning, the water used was evaporated, trapped, and reused, and the accumulated salt be used in food processing (similar to the proposed in Fig. 23.3). The production of seaweed salt is a novel first step in a cascading biorefinery model for the utilization of macroalgal biomass, improving the quality of the processed biomass for the production of fertilizer, feed, or fuel due to a lower level of minerals in this biomass.

In another report, Bikker et al. (2016) proposed a biorefinery for *Ulva lactuca* containing 22.5% dry biomass of proteins. The biorefinery route was suggested as first, the sugars in the biomass are solubilized by hot water treatment followed by enzymatic hydrolysis and centrifugation resulting in a sugar-rich hydrolysate, containing glucose, rhamnose, and xylose, and a protein fraction. The sugar-rich hydrolysate can then be processed in anaerobic fermentation to produce acetone, butanol, ethanol, and propanediol.

In a different approach, Prabhu et al. (2020) developed a protocol for the coproduction of six different products from the green macroalga *U. ohnoi* using green extraction methods (i.e., salts, starch, lipids, ulvan, proteins, and cellulose). The based route uses a homogenization in water, which generates a liquid phase (salts + starch fractions) and solid residue. Then the biomass is extracted using ethanol (lipid fraction), followed by the ulvan extraction using oxalate salt, the protein extraction using the alkaline treatment, and cellulose extraction using water. In the end, 90% of the initial biomass is recovered into the six products.

4.3 Red seaweeds

The third group of seaweeds distinguishes itself due to the presence of phycobiliproteins, in special, phycoerythrin (pink/red pigment). Red algae such as *Porphyra* spp. and *Palmaria palmata* are a traditional part of European and Asian cuisines and are used to make other products such as agar, carrageenan, and food additives.

Regarding biorefineries of red algae, only a few examples were proposed to date. Kumar et al. (2013) proposed a biorefinery of *Gracilaria verrucosa* for the production of agar and bioethanol. The processed biomass resulted in the extraction of 30% of agar. The residual biomass was submitted to enzymatic hydrolysis and fermentation with *Saccharomyces cerevisiae* to produce ethanol with a yield reaching 0.43 g/g sugars.

Moreover, in a similar approach, Ingle et al. (2018) proposed a biorefinery of *Kappaphycus alvarezii* for the obtention of carrageenan and ethanol through conversion modeling and performance prediction. For that, the spent biomass after carrageenan extraction was fermented using *S. cerevisiae* and *Escherichia coli* for the production of ethanol. This specie of red seaweed is of particular interest, as it is currently one of the most widely cultivated macroalgae species.

Finally, in alternative to biofuels, Karunarathne and van Walsum (2022) proposed the conversion of the spent biomass of carrageenan (Algefiber) into carboxylic acids and derived compounds, using an open culture of microorganisms. Digestion produced carboxylic acids ranging in chain length from one to seven carbons. The thermal decomposition of carboxylate salts produced a mixture of ketones (application as industrial solvents, especially in industries of manufacturing plastics, explosives, coatings, adhesives, paints, and textiles).

5. Challenges and future perspectives

Most of the strengths of algal biorefineries relate to their intrinsic biotechnological value, with the obtention of compounds and products from natural sources, by assuring maximum valorization of algal biomass, with minimum waste by using biomass residuals for fertilizer application and the use of renewable energy (e.g., bioethanol, biodiesel) to feed the overall process. These platforms could provide an ecosystem service with a preponderant role in bioremediation (cultivation of alga in wastewater or municipal waters or use of IMTA) and decreasing carbon (and other contaminants) contents from flue gas streams from industries nearby. The reuse of water streams for nutrient recycling, or even the recuperation and reuse of solvents, in the process of extraction, can make this approach more feasible and environmentally friendly. In addition, energy demands could be suppressed by other means of renewable and green energy by resorting to wind generators or photovoltaic panels that could help in any step of the process, from the cultivation part (e.g., photobioreactors feeding, use of pumps, etc.) or in the downstream process (e.g., dewatering, extraction equipment used).

This is an industrial strategy quite appealing for the algae-based processing that could substantially benefit the green and circular economy. Despite the advantages, till now, algal biorefinery approaches can work in theory, but their implementation in practice is still with some upcoming challenges. The implementation of these units and the financial feasibility and stability still depend on the availability and efficiency of technologies (e.g., suitable photobioreactors, extraction, biomass conversion into energy), the complete optimized integration of these operation units as a whole, and still require huge investments and capitalizing funds to face the overall costs of the process in this type of platform. A SWOT analysis here presented (Table 23.5) describes the strengths, weaknesses, opportunities, and threats regarding the process of implementation of algal biorefineries.

An important aspect also relates to the quality and safety of algal biomass and final products, obtained through coupling the process of cultivation with wastewater and or CO_2 (i.e., flue gas), in particular for final applications encompassing functional foods, and cosmetics, or even pharmaceutical. Algal biomass can be subjected to oscillations in their chemical compositions; thus, biorefinery units should be prepared to deal with this issue and make a regular analysis to prevent the presence of any contaminants, in the final product, assuring consistent final quality, which reveals to be quite important.

Table 23.5 SWOT analysis for implementation of algal biorefineries.

SWOT analysis	
Strengths	**Weaknesses**
• Intrinsic biotechnological value • Production of unique bioactive compounds • Algal low requirements for growth (i.e., sunlight, CO_2, few minerals, use of wastewater or marine water; flue gas) • Ecosystem/ environmental services (CO_2 mitigation and bioremediation) • Maximum valorization of biomass production (fractionation of high- and low-value-added products) • Reduce waste production (recycling water-nutrient streams, reuse of solvents, biomass residuals for fertilizer application) • Cogeneration of energy (e.g., biodiesel, biooil) • Use of green and renewable energy (wind and photovoltaics) • Contribution to circular economy	• High initial investments and capital costs • Some technologies (e.g., photobioreactors, extraction equipment) are not yet fully developed at large scale • Efforts to assure biomass quality under different compositions of wastewater and/or flue gas streams (e.g., presence of contaminants) • Efforts to assure high-value-added compound/products quality and safety, if biomass is produced with wastewaters or flue gas streams, or from cyanobacteria sources in particular for food, cosmetic, and pharmaceutical application • High maintenance costs

Continued

Table 23.5 SWOT analysis for implementation of algal biorefineries.—cont'd

SWOT analysis	
Opportunities	**Threats**
• R&D research (i.e., use of promising alga species, genetic, molecular engineering) • Use of genetically modified organisms (GMOs) for enhanced production • Exploration of new bioactive compounds for development of new products and new markets • Optimization of algal cultivation conditions (photobioreactors, nutrients, light patterns, etc.) • Establishing an optimized multiextraction procedure (i.e., sequential) for several algal metabolites • Development of efficient biomass conversion technologies (e.g., energy) • Reinforcement and optimization of preexisting technologies (i.e., extraction) • Multidisciplinary approach (biology, chemistry, engineering, etc).	• Safety and legislation issues • Low acceptance from the public regarding bioproducts (e.g., GMO) • Gradual or phased implementation can take long years (e.g., economic and environmental studies) • Continuous injection of capital to reach an optimum investment return (long payback period) • High maintenance costs • Energetic demands and ecologic footprint can be a problem if integrated systems and units are not fully optimized

Regarding the opportunities, an implementation of a biorefinery unit could give some opportunities to start using novel algal species for commercial purposes, with new/better biotechnological potential. Besides, it could be an opportunity to develop genetic and molecular tools to improve biomass yields and enhance the production and recovery of certain metabolites, at the same time. This is an open window to explore new bioactive compounds with commercial interest, develop new products, and explore new markets. Furthermore, the use of genetically modified (micro)organisms (GMOs) can indeed speed up the production of several metabolites of interest to putatively high yields; yet, the costs involved to achieve an integrated platform of production with this can be extremely high. However, the use of such organisms can be threatened as they pose safety issues regarding their containment and downstream processing. On one hand, if GMOs are to selected to integrate an IMTA system or open ponds for cultivation, there is a risk of environmental contamination, with unknown consequences for the surrounding ecosystems (di Visconte et al., 2019). On the other hand, incorporation of bioactive metabolites from GMO strains into consumable products can be hard-road because the public in general usually does not understand the underlying technology, and this becomes even worse with the lack of associated legislation of the process (Assunção et al., 2022b).

The implementation of algal biorefinery platforms also can help in establishing optimizing cultivation, and in developing and improving integrated extraction technologies to fractionate the biomass into several compounds of interest. These platforms will have probably R&I strategy to optimize the production and develop extraction techniques to improve not only cultivation but also fractionation and new algal-based products.

The optimized integration of the different units of cultivation, downstream processing, retrofeed nutrient recycling, and energy production is a process that possibly has to be done phased-stages or gradually, which can take long years—probably with huge expenses for economic and environmental studies. This poses a risk for the investor in the medium- and long-term because there is a need for continuous injection of capital, with a later recovery from the initial and continuous investments. This type of platform will probably need a high maintenance provision, which has costs, and if inefficiencies occur at the level of the process (e.g., energy and spent CO_2), the ecologic footprint can be unbalanced and the process may not be sustainable as it should be.

Despite the challenges, the perspectives on algal biorefineries are starting to evolve positively. Many researchers and engineers have been putting a lot of effort into this topic, because of the great biotechnological potential of algae and a view of a more sustainable process, with minimum waste. The integration of knowledge within multiple disciplinary groups from biology and chemistry to engineering is essential as well as the synergetic effect within industry and academia (and vice versa) is seen as an opportunity to still push forward algal biorefinery to the next level and is expected to surpass the economic and technical constraints.

Acknowledgments

This work was financially partially supported by the following: national funds through FCT, within the scope of UIDB/04423/2020 and UIDP/04423/2020 granted to Interdisciplinary Centre of Marine and Environmental Research (CIIMAR) and UIDB/00511/2020 granted to Laboratory for Process Engineering, Environment, Biotechnology and Energy (LEPABE)—funded by national funds through FCT/MCTES (PIDDAC). A Ph.D. fellowship (PD/BD/137887/2018) for author J.A. and a Ph.D. fellowship (SFRH/BD/136767/2018) for author Fernando Pagels was granted by FCT, under the auspices of Programa Operacional Capital Humano (POCH), supported by the European Social Fund (ESF) and Portuguese funds (MECTES).

References

Abed, R. M. M., Dobretsov, S., & Sudesh, K. (2009). Applications of cyanobacteria in biotechnology. *Journal of Applied Microbiology, 106*, 1–12. https://doi.org/10.1111/j.1365-2672.2008.03918.x

Abraham, R. E., Su, P., Puri, M., Raston, C. L., & Zhang, W. (2019). Optimisation of biorefinery production of alginate, fucoidan and laminarin from brown seaweed *Durvillaea potatorum. Algal Research*, 38. https://doi.org/10.1016/j.algal.2018.101389

Achparaki, M., Thessalonikeos, E., Tsoukali, H., Mastrogianni, O., Zaggelidou, E., Chatzinikolaou, F., Vasilliades, N., Raikos, N., Isabirye, M., Raju, D. V. N., Kitutu, M., Yemeline, V., Deckers, J., & Poesen Additional, J. (2012). Algal alginate in biotechnology: Biosynthesis and applications. *Intech*, 13.

Acién, F. G., Molina, E., Fernández-Sevilla, J. M., Barbosa, M., Gouveia, L., Sepúlveda, C., Bazaes, J., & Arbib, Z. (2017). Economics of microalgae production. In *Microalgae-based biofuels and bioproducts: From feedstock cultivation to end-products* (pp. 485–503). Elsevier Inc. https://doi.org/10.1016/B978-0-08-101023-5.00020-0

Afolayan, A. F., Bolton, J. J., Lategan, C. A., Smith, P. J., & Beukes, D. R. (2008). Fucoxanthin, tetraprenylated toluquinone and toluhydroquinone metabolites from Sargassum heterophyllum inhibit the in vitro growth of the malaria parasite Plasmodium falciparum. *Zeitschrift fur Naturforschung - Section C Journal of Biosciences, 63*, 848–852. https://doi.org/10.1515/znc-2008-11-1211

Ahmadi, A., Zorofchian Moghadamtousi, S., Abubakar, S., & Zandi, K. (2015). Antiviral potential of algae polysaccharides isolated from marine sources: A review. *BioMed Research International*, 825203. https://doi.org/10.1155/2015/825203

Ansari, F. A., Shriwastav, A., Gupta, S. K., Rawat, I., & Bux, F. (2017). Exploration of microalgae biorefinery by optimizing sequential extraction of major metabolites from Scenedesmus obliquus. *Industrial & Engineering Chemistry Research*, 56. https://doi.org/10.1021/acs.iecr.6b04814

Ashokkumar, V., Chen, W. H., Kumar, G., Satjarak, A., Chanthapatchot, W., & Ngamcharussrivichai, C. (2021). A biorefinery approach for high value-added bioproduct (astaxanthin) from alga Haematococcus sp. and residue pyrolysis for biochar synthesis and metallic iron production from hematite (Fe2O3). *Fuel*, 304. https://doi.org/10.1016/j.fuel.2021.121150

Ashokkumar, V., Chen, W. H., Ngamcharussrivichai, C., Agila, E., & Ani, F. N. (2019). Potential of sustainable bioenergy production from Synechocystis sp. cultivated in wastewater at large scale – a low cost biorefinery approach. *Energy Conversion and Management, 186*, 188–199 https://doi.org/10.1016/j.enconman.2019.02.036

Assunção, J., Amaro, H. M., Lopes, G., Tavares, T., Malcata, F. X., & Guedes, A. C. (2021a). Exploration of marine genus Chroococcidiopsis sp.: A valuable source for antioxidant industry? *Journal of Applied Phycology, 33*, 2169–2187. https://doi.org/10.1007/s10811-021-02435-x

Assunção, J., Amaro, H. M., Lopes, G., Tavares, T., Malcata, F. X., & Guedes, A. C. (2021b). *Synechocystis salina*: Potential bioactivity and combined extraction of added-value metabolites. *Journal of Applied Phycology.* https://doi.org/10.1007/s10811-021-02558-1

Assunção, J., Amaro, H. M., Malcata, F. X., & Guedes, A. C. (2022a). Cyanobacterial pigments: Photosynthetic function and biotechnological purposes. In G. Lopes, M. Silva, & V. Vasconcelos (Eds.), *The pharmacological potential of cyanobacteria*. Elsevier Academic Press.

Assunção, J., Amaro, H. M., Malcata, F. X., & Guedes, A. C. (2022b). Extraction of microalgal bioactive compounds towards functional ingredients: A biorefinery approach, with prospects and challenges. *Algal Functional Foods and Nutraceuticals: Benefits, Opportunities and Challenges, 1*, 120–175.

Assunção, J., & Malcata, F. X. (2020). Enclosed "non-conventional" photobioreactors for microalga production: A review. *Algal Research, 52*, 102107. https://doi.org/10.1016/j.algal.2020.102107

Baghel, R. S., Suthar, P., Gajaria, T. K., Bhattacharya, S., Anil, A., & Reddy, C. R. K. (2020). Seaweed biorefinery: A sustainable process for valorising the biomass of brown seaweed. *Journal of Cleaner Production*, 263. https://doi.org/10.1016/j.jclepro.2020.121359

Balina, K., Romagnoli, F., & Blumberga, D. (2017). Seaweed biorefinery concept for sustainable use of marine resources. In *Energy procedia*. https://doi.org/10.1016/j.egypro.2017.09.067

Beattie, A., Vermaas, W., Darzins, A., Holland, S. C., Li, S., McGowen, J., Nielsen, D., & Quinn, J. C. (2021). A probabilistic economic and environmental impact assessment of a cyanobacteria-based biorefinery. *Algal Research, 59*, 102454. https://doi.org/10.1016/j.algal.2021.102454

Bei, H., Guang-Ce, W., Chen-Kui, Z., & Zhen-Gang, L. (2002). The experimental research of R-phycoerythrin subunits on cancer treatment: A new photosensitizer in PDT. *Cancer Biotherapy and Radiopharmaceuticals, 17*, 35–42. https://doi.org/10.1089/10849780252824055

Bhalamurugan, G. L., Valerie, O., & Mark, L. (2018). Valuable bioproducts obtained from microalgal biomass and their commercial applications: A review. *Environmental Engineering Research, 23*, 229–241. https://doi.org/10.4491/eer.2017.220

Bikker, P., van Krimpen, M. M., van Wikselaar, P., Houweling-Tan, B., Scaccia, N., van Hal, J. W., Huijgen, W. J. J., Cone, J. W., & López-Contreras, A. M. (2016). Biorefinery of the green seaweed Ulva lactuca to produce animal feed, chemicals and biofuels. *Journal of Applied Phycology*, 28. https://doi.org/10.1007/s10811-016-0842-3

Borowitzka, M. A. (2013). High-value products from microalgae-their development and commercialisation. *Journal of Applied Phycology, 25*, 743–756. https://doi.org/10.1007/s10811-013-9983-9

Brennan, L., & Owende, P. (2010). Biofuels from microalgae—a review of technologies for production, processing, and extractions of biofuels and co-products. *Renewable and Sustainable Energy Reviews, 14*, 557–577. https://doi.org/10.1016/j.rser.2009.10.009

Buschmann, A. H., Camus, C., Infante, J., Neori, A., Israel, Á., Hernández-González, M. C., Pereda, S. V., Gomez-Pinchetti, J. L., Golberg, A., Tadmor-Shalev, N., & Critchley, A. T. (2017). Seaweed production: Overview of the global state of exploitation, farming and emerging research activity. *European Journal of Phycology, 52*, 391–406. https://doi.org/10.1080/09670262.2017.1365175

Bussa, M., Zollfrank, C., & Röder, H. (2021). Life cycle assessment with parameterised inventory to derive target values for process parameters of microalgae biorefineries. *Algal Research*, 57. https://doi.org/10.1016/j.algal.2021.102352

Campo, J. A., García-gonzález, M., & Guerrero, M. G. (2007). Outdoor cultivation of microalgae for carotenoid production. *Current State and Perspectives*, 1163–1174. https://doi.org/10.1007/s00253-007-0844-9

Chandra, R., Das, P., Vishal, G., & Nagra, S. (2019). Factors affecting the induction of UV protectant and lipid productivity in Lyngbya for sequential biorefinery product recovery. *Bioresource Technology, 278*, 303–310. https://doi.org/10.1016/j.biortech.2019.01.084

Chavan, R., & Mutnuri, S. (2019). Tertiary treatment of domestic wastewater by Spirulina platensis integrated with microalgal biorefinery. *Biofuels, 10*, 33–44. https://doi.org/10.1080/17597269.2018.1461509

Chen, J. H., Lim, J. D., Sohn, E. H., Choi, Y. S., & Han, E. T. (2009). Growth-inhibitory effect of a fucoidan from brown seaweed Undaria pinnatifida on Plasmodium parasites. *Parasitology Research, 104*, 245–250. https://doi.org/10.1007/s00436-008-1182-2

Christaki, E., Bonos, E., Giannenasa, I., & Florou-Paneria, P. (2013). Functional properties of carotenoids originating from algae. *Journal of the Science of Food and Agriculture, 93*, 5–11. https://doi.org/10.1002/jsfa.5902

Christenson, L., & Sims, R. (2011). Production and harvesting of microalgae for wastewater treatment, biofuels, and bioproducts. *Biotechnology Advances*. https://doi.org/10.1016/j.biotechadv.2011.05.015

Costa, J. A. V., Freitas, B. C. B., Rosa, G. M., Moraes, L., Morais, M. G., & Mitchell, B. G. (2019). Operational and economic aspects of Spirulina-based biorefinery. *Bioresource Technology, 292*, 121946. https://doi.org/10.1016/j.biortech.2019.121946

Cruz, D., Vasconcelos, V., Pierre, G., Michaud, P., & Delattre, C. (2020). Exopolysaccharides from cyanobacteria: Strategies for bioprocess development. *Applied Sciences, 10*, 1–20.

Czyrnek-Delêtre, M. M., Rocca, S., Agostini, A., Giuntoli, J., & Murphy, J. D. (2017). Life cycle assessment of seaweed biomethane, generated from seaweed sourced from integrated multi-trophic aquaculture in temperate oceanic climates. *Applied Energy*, 196. https://doi.org/10.1016/j.apenergy.2017.03.129

De Jesus Raposo, M. F., De Morais, R. M. S. C., & De Morais, A. M. M. B. (2013). Bioactivity and applications of sulphated polysaccharides from marine microalgae. *Marine Drugs, 11*, 233–252. https://doi.org/10.3390/md11010233

De Jesus Raposo, M. F., De Morais, A. M. B., & De Morais, R. M. S. C. (2015). Marine polysaccharides from algae with potential biomedical applications. *Marine Drugs, 13*, 2967–3028. https://doi.org/10.3390/md13052967

Demay, J., Bernard, C., Reinhardt, A., & Marie, B. (2019). Natural products from cyanobacteria: Focus on beneficial activities. *Marine Drugs, 17*, 1–49. https://doi.org/10.3390/md17060320

Deprá, M. C., dos Santos, A. M., Severo, I. A., Santos, A. B., Zepka, L. Q., & Jacob-Lopes, E. (2018). Microalgal biorefineries for bioenergy production: Can we move from concept to industrial reality? *Bioenergy Research.* https://doi.org/10.1007/s12155-018-9934-z

D'Orazio, N., Gemello, E., Gammone, M. A., De Girolamo, M., Ficoneri, C., & Riccioni, G. (2012). Fucoxantin: A treasure from the sea. *Marine Drugs, 10*, 604–616. https://doi.org/10.3390/md10030604

Dumay, J., Morançais, M., Munier, M., Le Guillard, C., & Fleurence, J. (2014). Phycoerythrins: Valuable proteinic pigments in red seaweeds. In *Advances in botanical research* (pp. 321–343). Academic Press Inc. https://doi.org/10.1016/B978-0-12-408062-1.00011-1

Eppink, M. H. M., Barbosa, M. J., & Wijffels, R. H. (2013). Biorefinery of microalgae: Production of high value products, bulk chemicals and biofuels. *Symposium Biorefinery for Food, Fuel and Materials.*

Espada, J. J., Pérez-Antolín, D., Vicente, G., Bautista, L. F., Morales, V., & Rodríguez, R. (2020). Environmental and techno-economic evaluation of β-carotene production from Dunaliella salina. A biorefinery approach. *Biofuels, Bioproducts and Biorefining*, 14. https://doi.org/10.1002/bbb.2012

Esquivel-Hernández, D. A., Pennacchio, A., Torres-Acosta, M. A., Parra-Saldívar, R., de Souza Vandenberghe, L. P., & Faraco, V. (2021). Multi-product biorefinery from Arthrospira platensis biomass as feedstock for bioethanol and lactic acid production. *Scientific Reports, 11*, 1–15. https://doi.org/10.1038/s41598-021-97803-5

Ferreira-Santos, P., Miranda, S. M., Belo, I., Spigno, G., Teixeira, J. A., & Rocha, C. M. R. (2021). Sequential multi-stage extraction of biocompounds from Spirulina platensis: Combined effect of ohmic heating and enzymatic treatment. *Innovative Food Science & Emerging Technologies*, 71. https://doi.org/10.1016/j.ifset.2021.102707

Ferreira, C. A. M., Félix, R., Félix, C., Januário, A. P., Alves, N., Novais, S. C., Dias, J. R., & Lemos, M. F. L. (2021). A biorefinery approach to the biomass of the seaweed undaria pinnatifida (Harvey suringar, 1873): Obtaining phlorotannins-enriched extracts for wound healing. *Biomolecules*, 11. https://doi.org/10.3390/biom11030461

Ferreira, A. F., Marques, A. C., Batista, A. P., Marques, P. A. S. S., Gouveia, L., & Silva, C. M. (2012). Biological hydrogen production by Anabaena sp. – yield, energy and CO2 analysis including fermentative biomass recovery. *International Journal of Hydrogen Energy, 37*, 179–190. https://doi.org/10.1016/j.ijhydene.2011.09.056

Ferreira, A., Ribeiro, B., Ferreira, A. F., Tavares, M. L. A., Vladic, J., Vidović, S., Cvetkovic, D., Melkonyan, L., Avetisova, G., Goginyan, V., & Gouveia, L. (2019). Scenedesmus obliquus microalga-based biorefinery –

from brewery effluent to bioactive compounds, biofuels and biofertilizers — aiming at a circular bioeconomy. *Biofuels, Bioproducts and Biorefining*, 13. https://doi.org/10.1002/bbb.2032

Francavilla, M., Kamaterou, P., Intini, S., Monteleone, M., & Zabaniotou, A. (2015). Cascading microalgae biorefinery: Fast pyrolysis of Dunaliella tertiolecta lipid extracted-residue. *Algal Research*, 11. https://doi.org/10.1016/j.algal.2015.06.017

Gilbert-López, B., Mendiola, J. A., Fontecha, J., van den Broek, L. A. M., Sijtsma, L., Cifuentes, A., Herrero, M., & Ibáñez, E. (2015). Downstream processing of Isochrysis galbana: A step towards microalgal biorefinery. *Green Chemistry*, 17. https://doi.org/10.1039/c5gc01256b

Giraldo-Calderón, N. D., Romo-Buchelly, R. J., Arbeláez-Pérez, A. A., Echeverri-Hincapié, D., & Atehortúa-Garcés, L. (2018). Microalgae biorefineries: Applications and emerging technologies. *Dyna*, 85. https://doi.org/10.15446/dyna.v85n205.68780

Gonçalves, A. L., Pires, J. C. M., & Simões, M. (2016). Biotechnological potential of *Synechocystis salina* co-cultures with selected microalgae and cyanobacteria: Nutrients removal, biomass and lipid production. *Bioresource Technology, 200*, 279—286. https://doi.org/10.1016/j.biortech.2015.10.023

Goswami, R. K., Agrawal, K., & Verma, P. (2022). Microalgae Dunaliella as biofuel feedstock and β-carotene production: An influential step towards environmental sustainability. In *Energy conversion and management: X* (p. 13). https://doi.org/10.1016/j.ecmx.2021.100154

Guedes, A. C., Amaro, H. M., & Malcata, F. X. (2011). Microalgae as sources of high added-value compounds-a brief review of recent work. *Biotechnology Progress, 27*, 597—613. https://doi.org/10.1002/btpr.575

Guedes, A. C., Amaro, H. M., Sousa-Pinto, I., & Malcata, F. X. (2013). *Applications of spent biomass, biofuels from algae*. Elsevier B.V. https://doi.org/10.1016/B978-0-444-59558-4.00010-3

Guedes, A. C., & Malcata, F. X. (2012). Bioreactors for microalgae: A review of designs, features and applications. In Z. Liu, & P. G. Antolli (Eds.), *Bioreactors: Design, properties and applications* (pp. 1—52). New York: Nova Science Publisher.

van Hal, J. W., Huijgen, W. J. J., & López-Contreras, A. M. (2014). Opportunities and challenges for seaweed in the biobased economy. *Trends in Biotechnology*. https://doi.org/10.1016/j.tibtech.2014.02.007

Hemalatha, M., Sravan, J. S., Min, B., & Venkata Mohan, S. (2019). Microalgae-biorefinery with cascading resource recovery design associated to dairy wastewater treatment. *Bioresource Technology*, 284. https://doi.org/10.1016/j.biortech.2019.03.106

Hosseini, A., Jazini, M., Mahdieh, M., & Karimi, K. (2020). Efficient superantioxidant and biofuel production from microalga Haematococcus pluvialis via a biorefinery approach. *Bioresource Technology*, 306. https://doi.org/10.1016/j.biortech.2020.123100

Ingle, K., Vitkin, E., Robin, A., Yakhini, Z., Mishori, D., & Golberg, A. (2018). Macroalgae biorefinery from Kappaphycus alvarezii: Conversion modeling and performance prediction for India and Philippines as examples. *Bioenergy Research*, 11. https://doi.org/10.1007/s12155-017-9874-z

Jacob-Lopes, E., Maroneze, M. M., Deprá, M. C., Sartori, R. B., Dias, R. R., & Zepka, L. Q. (2019). Bioactive food compounds from microalgae: An innovative framework on industrial biorefineries. *Current Opinion in Food Science, 25*, 1—7. https://doi.org/10.1016/j.cofs.2018.12.003

Jomova, K., & Valko, M. (2013). Health protective effects of carotenoids and their interactions with other biological antioxidants. *European Journal of Medicinal Chemistry*. https://doi.org/10.1016/j.ejmech.2013.09.054

Kaliaperumal, N., & Uthirasivan, P. (2001). *Commercial scale production of agar from thered alga Gracilaria edulis (Gmelin) Silva. undefined.*

Karunarathne, S. A., & van Walsum, G. P. (2022). Integrating the carboxylate platform into a red seaweed biorefinery. *Applied Biochemistry and Biotechnology*, 194. https://doi.org/10.1007/s12010-021-03699-2

Khanra, S., Mondal, M., Halder, G., Tiwari, O. N., Gayen, K., & Bhowmick, T. K. (2018). Downstream processing of microalgae for pigments, protein and carbohydrate in industrial application: A review. *Food and Bioproducts Processing, 110*, 60—84. https://doi.org/10.1016/j.fbp.2018.02.002

Kumar, S., Gupta, R., Kumar, G., Sahoo, D., & Kuhad, R. C. (2013). Bioethanol production from Gracilaria verrucosa, a red alga, in a biorefinery approach. *Bioresource Technology*, 135. https://doi.org/10.1016/j.biortech.2012.10.120

Langlois, J., Sassi, J. F., Jard, G., Steyer, J. P., Delgenes, J. P., & Hélias, A. (2012). Life cycle assessment of biomethane from offshore-cultivated seaweed. *Biofuels, Bioproducts and Biorefining*, 6. https://doi.org/10.1002/bbb.1330

Lau, N., Matsui, M., & Abdullah, A. A. (2015). Cyanobacteria: Photoautotrophic microbial factories for the sustainable synthesis of industrial products. *BioMed Research International*, 1–9.

Levasseur, W., Perré, P., & Pozzobon, V. (2020). A review of high value-added molecules production by microalgae in light of the classification. *Biotechnology Advances*, 107545. https://doi.org/10.1016/j.biotechadv.2020.107545

Li, Z., Zhang, L., & Zhao, Z. (2021). Malyngamide F possesses anti-inflammatory and antinociceptive activity in rat models of inflammation. *Pain Research and Management, 18*, 4919391. https://doi.org/10.1155/2021/4919391

Magnusson, M., Carl, C., Mata, L., de Nys, R., & Paul, N. A. (2016). Seaweed salt from Ulva: A novel first step in a cascading biorefinery model. *Algal Research*, 16. https://doi.org/10.1016/j.algal.2016.03.018

Mandal, M. K., Chanu, N., & Chaurasia, N. (2020). Cyanobacterial pigments and their fluorescence characteristics: Applications in research and industry. In P. K. Singh, A. Kumar, V. K. Singh, & A. K. Shrivastava (Eds.), *Advances in cyanobacterial biology* (pp. 55–72). Academic Press Inc. https://doi.org/10.1016/B978-0-12-819311-2.00005-X

Markou, G., & Nerantzis, E. (2013). Microalgae for high-value compounds and biofuels production: A review with focus on cultivation under stress conditions. *Biotechnology Advances, 31*, 1532–1542. https://doi.org/10.1016/j.biotechadv.2013.07.011

Mehariya, S., Goswami, R. K., Verma, P., Lavecchia, R., & Zuorro, A. (2021). Integrated approach for wastewater treatment and biofuel production in microalgae biorefineries. *Energies*. https://doi.org/10.3390/en14082282

Meixner, K., Kovalcik, A., Sykacek, E., Gruber-Brunhumer, M., Zeilinger, W., Markl, K., Haas, C., Fritz, I., Mundigler, N., Stelzer, F., Neureiter, M., Fuchs, W., & Drosg, B. (2018). Cyanobacteria Biorefinery — production of poly(3-hydroxybutyrate) with Synechocystis salina and utilisation of residual biomass. *Journal of Biotechnology, 265*, 46–53. https://doi.org/10.1016/j.jbiotec.2017.10.020

Mitra, M., & Mishra, S. (2019). Multiproduct biorefinery from Arthrospira spp. towards zero waste: Current status and future trends. *Bioresource Technology, 291*, 121928. https://doi.org/10.1016/j.biortech.2019.121928

Monte, J., Sá, M., Parreira, C., Galante, J., Serra, A. R., Galinha, C. F., Costa, L., Pereira, V. J., Brazinha, C., & Crespo, J. G. (2019). Recycling of Dunaliella salina cultivation medium by integrated membrane filtration and advanced oxidation. *Algal Research*, 39. https://doi.org/10.1016/j.algal.2019.101460

Narayanaswamy, R., Jo, B. W., Choi, S. K., & Ismail, I. S. (2013). Fucoidan: Versatile cosmetic ingredient. An overview. *Journal of Applied Cosmetology, 31*, 131–138.

Neuman, I., Nahum, H., & Ben-Amotz, A. (1999). Prevention of exercise-induced asthma by a natural isomer mixture of β- carotene. *Annals of Allergy, Asthma, & Immunology, 82*, 549–553. https://doi.org/10.1016/S1081-1206(10)63165-1

Neumann, U., Derwenskus, F., Flister, V. F., Schmid-Staiger, U., Hirth, T., & Bischoff, S. C. (2019). Fucoxanthin, a carotenoid derived from Phaeodactylum tricornutum exerts antiproliferative and antioxidant activities in vitro. *Antioxidants, 8*, 1–11. https://doi.org/10.3390/antiox8060183

Nilsson, A. E., Bergman, K., Gomez Barrio, L. P., Cabral, E. M., & Tiwari, B. K. (2022). Life cycle assessment of a seaweed-based biorefinery concept for production of food, materials, and energy. *Algal Research, 65*, 102725. https://doi.org/10.1016/j.algal.2022.102725

Nishshanka, G. K. S. H., Liyanaarachchi, V. C., Nimarshana, P. H. V., Ariyadasa, T. U., & Chang, J. S. (2022). Haematococcus pluvialis: A potential feedstock for multiple-product biorefining. *Journal of Cleaner Production*, 344. https://doi.org/10.1016/j.jclepro.2022.131103

Pagels, F., Guedes, A. C., Amaro, H. M., Kijjoa, A., & Vasconcelos, V. (2019). Phycobiliproteins from cyanobacteria: Chemistry and biotechnological applications. *Biotechnology Advances, 37*, 1–22. https://doi.org/10.1016/J.BIOTECHADV.2019.02.010

Pagels, F., Salvaterra, D., Amaro, H. M., Lopes, G., Sousa-Pinto, I., Vasconcelos, V., & Guedes, A. C. (2020). Bioactive potential of Cyanobium sp. pigment-rich extracts. *Journal of Applied Phycology, 32*, 3031–3040. https://doi.org/10.1007/s10811-020-02213-1

Pekkoh, J., Lomakool, S., Chankham, J., Duangjan, K., Thurakit, T., Phinyo, K., Ruangrit, K., Tragoolpua, Y., Pumas, C., Pathom-aree, W., & Srinuanpan, S. (2022). Maximizing biomass productivity of cyanobacterium Nostoc sp. through high-throughput bioprocess optimization and application in multiproduct biorefinery towards a holistic zero waste. *Biomass Conversion and Biorefinery*. https://doi.org/10.1007/s13399-021-02285-0

Peron-Schlosser, B., Stobienia, M., Bispo, L., Colla, L. M., Baraldi, I. J., & Colla, E. (2021). Residual fractions from Arthrospira platensis protein extraction as feedstock for ethanol production. *Biocatalysis and Biotransformation, 39*, 378–388. https://doi.org/10.1080/10242422.2021.1931145

Prabha, S., Vijay, A. K., Paul, R. R., & George, B. (2022). Cyanobacterial biorefinery: Towards economic feasibility through the maximum valorization of biomass. *Science of the Total Environment, 814*, 152795. https://doi.org/10.1016/j.scitotenv.2021.152795

Prabhu, M. S., Israel, A., Palatnik, R. R., Zilberman, D., & Golberg, A. (2020). Integrated biorefinery process for sustainable fractionation of Ulva ohnoi (Chlorophyta): Process optimization and revenue analysis. *Journal of Applied Phycology*, 32. https://doi.org/10.1007/s10811-020-02044-0

Prasanna, R., Sood, A., Jaiswal, P., Nayak, S., Gupta, V., Chaudhary, V., Joshi, M., & Natarajan, C. (2010). Rediscovering cyanobacteria as valuable sources of bioactive compounds (Review). *Applied Biochemistry and Microbiology, 46*, 119–134. https://doi.org/10.1134/S0003683810020018

Préat, N., Taelman, S. E., de Meester, S., Allais, F., & Dewulf, J. (2020). Identification of microalgae biorefinery scenarios and development of mass and energy balance flowsheets. *Algal Research*, 45. https://doi.org/10.1016/j.algal.2019.101737

Puh, B. (2016). Consumers' purchase intentions towards natural cosmetics. *Ekonomski Vjesnik/Econviews: Review of Contemporary Business, Entrepreneurship and Economic Issues, 29*, 53–64.

Rahikainen, M. (2021). *Global production of macroalgae and uses as food, dietary supplements and food additives* (pp. 1–20).

Ranga Rao, A., Ravishankar, G. A., & Sarada, R. (2012). Cultivation of green alga Botryococcus braunii in raceway, circular ponds under outdoor conditions and its growth, hydrocarbon production. *Bioresource Technology, 123*, 528–533. https://doi.org/10.1016/j.biortech.2012.07.009

Rani, K., Sandal, N., & Sahoo, P. K. (2018). A comprehensive review on chlorella-its composition, health benefits, market and regulatory scenario. *The Pharma Innovation Journal, 7*, 584–589.

Raven, J. A., & Giordano, M. (2014). Algae. *Current Biology, 24*, 590–595. https://doi.org/10.1016/j.cub.2014.05.039

Richa, J. P., Sonker, A. S., Singh, V., & Sinha, R. P. (2018). Potential applications of natural bioactive cyanobacterial UV-protective compounds. In S. La Barre, & S. S. Bates (Eds.), *Blue biotechnology: Production and use of marine molecules* (pp. 683–707). Wiley-VCH Verlag GmGH & Co. KGaA. https://doi.org/10.1002/9783527801718.ch21

Rizwan, M., Mujtaba, G., Memon, S. A., Lee, K., & Rashid, N. (2018). Exploring the potential of microalgae for new biotechnology applications and beyond: A review. *Renewable and Sustainable Energy Reviews, 92*, 394–404. https://doi.org/10.1016/j.rser.2018.04.034

Romay, C., Armesto, J., Remirez, D., González, R., Ledon, N., & García, I. (1998). Antioxidant and anti-inflammatory properties of C-phycocyanin from blue-green algae. *Inflammation Research, 47*, 36–41. https://doi.org/10.1007/s000110050256

Rosales-Mendoza, S., García-Silva, I., González-Ortega, O., Sandoval-Vargas, J. M., Malla, A., & Vimolmangkang, S. (2020). The potential of algal biotechnology to produce antiviral compounds and biopharmaceuticals. *Molecules, 25*, 1–25. https://doi.org/10.3390/molecules25184049

Sánchez-Bayo, A., López-Chicharro, D., Morales, V., Espada, J. J., Puyol, D., Martínez, F., Astals, S., Vicente, G., Bautista, L. F., & Rodríguez, R. (2020). Biodiesel and biogas production from Isochrysis galbana using dry and wet lipid extraction: A biorefinery approach. *Renewable Energy*, 146. https://doi.org/10.1016/j.renene.2019.06.148

Sankaran, R., Show, P. L., Nagarajan, D., & Chang, J. S. (2018). Exploitation and biorefinery of microalgae. In *Waste biorefinery: Potential and perspectives*. https://doi.org/10.1016/B978-0-444-63992-9.00019-7

Santos, J. H. P. M., Trigo, J. P., Maricato, É., Nunes, C., Coimbra, M. A., & Ventura, S. P. M. (2018). Fractionation of Isochrysis galbana proteins, arabinans, and glucans using ionic-liquid-based aqueous biphasic systems. *ACS Sustainable Chemistry & Engineering*, 6. https://doi.org/10.1021/acssuschemeng.8b02597

Sathasivam, R., & Ki, J. S. (2018). A review of the biological activities of microalgal carotenoids and their potential use in healthcare and cosmetic industries. *Marine Drugs, 16*, 1–31. https://doi.org/10.3390/md16010026

Sathasivam, R., Radhakrishnan, R., Hashem, A., Fathi, E., & Allah, A. (2019). Microalgae metabolites: A rich source for food and medicine. *Saudi Journal of Biological Sciences, 26*, 709–722. https://doi.org/10.1016/j.sjbs.2017.11.003

Sheu, M., Hsieh, Y., Lai, C., Chang, C., & Wu, C. (2013). Antihyperlipidemic and antioxidant effects of C-phycocyanin in golden Syrian hamsters fed with a hypercholesterolemic diet animal preparation. *Journal of Traditional and Complementary Medicine, 3*, 41–47. https://doi.org/10.4103/2225-4110.106545

Shih, S. R., Tsai, K. N., Li, Y. S., Chueh, C. C., & Chan, E. C. (2003). Inhibition of enterovirus 71-induced apoptosis by allophycocyanin isolated from a blue-green alga Spirulina platensis. *Journal of Medical Virology, 70*, 119–125. https://doi.org/10.1002/jmv.10363

Singh, J. S., Kumar, A., & Singh, M. (2019). Cyanobacteria: A sustainable and commercial bio-resource in production of bio-fertilizer and bio-fuel from waste waters. *Environmental and Sustainability Indicators, 3–4*, 100008. https://doi.org/10.1016/j.indic.2019.100008

Spolaore, P., Joannis-cassan, C., Duran, E., Isambert, A., Génie, L., & Paris, E. C. (2006). Commercial applications of microalgae. *Journal of Bioscience and Bioengineering, 101*, 87–96. https://doi.org/10.1263/jbb.101.87

Stahl, W., & Sies, H. (2012). β-Carotene and other carotenoids in protection from sunlight. *American Journal of Clinical Nutrition*. https://doi.org/10.3945/ajcn.112.034819

Suganya, T., Varman, M., Masjuki, H. H., & Renganathan, S. (2016). Macroalgae and microalgae as a potential source for commercial applications along with biofuels production: A biorefinery approach. *Renewable and Sustainable Energy Reviews, 55*, 909–941. https://doi.org/10.1016/j.rser.2015.11.026

Sun, Z., Wang, X., & Liu, J. (2019). Screening of Isochrysis strains for simultaneous production of docosahexaenoic acid and fucoxanthin. *Algal Research*, 41. https://doi.org/10.1016/j.algal.2019.101545

Tan, C. P., & Hou, Y. H. (2014). First evidence for the anti-inflammatory activity of fucoxanthin in high-fat-diet-induced obesity in mice and the antioxidant functions in PC12 cells. *Inflammation, 37*, 443–450. https://doi.org/10.1007/s10753-013-9757-1

Tejada Carbajal, E. M., Martínez Hernández, E., Fernández Linares, L., Novelo Maldonado, E., & Limas Ballesteros, R. (2020). Techno-economic analysis of Scenedesmus dimorphus microalgae biorefinery scenarios for biodiesel production and glycerol valorization. *Bioresource Technology Reports*, 12. https://doi.org/10.1016/j.biteb.2020.100605

Thomassen, G., van Dael, M., & van Passel, S. (2018). The potential of microalgae biorefineries in Belgium and India: An environmental techno-economic assessment. *Bioresource Technology*, 267. https://doi.org/10.1016/j.biortech.2018.07.037

Torres, M. D., Kraan, S., & Domínguez, H. (2019). Seaweed biorefinery. *Reviews in Environmental Science and Biotechnology*. https://doi.org/10.1007/s11157-019-09496-y

Torres, F. A. E., Passalacqua, T. G., Velásquez, A. M. A., de Souza, R. A., Colepicolo, P., & Graminha, M. A. S. (2014). New drugs with antiprotozoal activity from marine algae: A review. *Revista Brasileira de Farmacognosia, 24*, 265–276. https://doi.org/10.1016/j.bjp.2014.07.001

Torzillo, G. (1997). Tubular bioreactors. In A. Vonshak (Ed.), *Spirulina platensis (Arthrospira) phisiology, cell-biology and Biotechnology* (pp. 101–115). Taylor & Francis.

Vijayakumar, S., & Menakha, M. (2015). Pharmaceutical applications of cyanobacteria—A review. *Journal of Acute Medicine, 5*, 15–23. https://doi.org/10.1016/j.jacme.2015.02.004

di Visconte, G. S., Spicer, A., Chuck, C. J., & Allen, M. J. (2019). The microalgae biorefinery: A perspective on the current status and future opportunities using genetic modification. *Applied Sciences*, 9. https://doi.org/10.3390/app9224793

Vo, T. S., Ngo, D. H., & Kim, S. K. (2012). Marine algae as a potential pharmaceutical source for anti-allergic therapeutics. *Process Biochemistry, 47*, 386–394. https://doi.org/10.1016/j.procbio.2011.12.014

Wang, H. M. D., Chen, C. C., Huynh, P., & Chang, J. S. (2015). Exploring the potential of using algae in cosmetics. *Bioresource Technology, 184*, 355–362. https://doi.org/10.1016/j.biortech.2014.12.001

Wang, Y., Xing, M., Cao, Q., Ji, A., Liang, H., & Song, S. (2019). Biological activities of fucoidan and the factors mediating its therapeutic effects: A review of recent studies. *Marine Drugs, 17*, 15–17. https://doi.org/10.3390/md17030183

Wiraningtyan, A., Ruslan, R., Ayu Mutmainnah, P., & Perkasa, M. (2020). Application of natural dyes and sodium alginate from Sargassum sp. Sea weed in coloring bima woven fabric. *Oriental Journal of Chemistry, 36*, 964–967. https://doi.org/10.13005/ojc/360524

Yaakob, Z., Ali, E., Zainal, A., Mohamad, M., & Takriff, M. S. (2014). An overview: Biomolecules from microalgae for animal feed and aquaculture. *Journal of Biological Research, 21*, 1–10. https://doi.org/10.1186/2241-5793-21-6

Zhang, X., Border, A., Goosen, N., & Thomsen, M. (2021). Environmental life cycle assessment of cascade valorisation strategies of South African macroalga Ecklonia maxima using green extraction technologies. *Algal Research*, 58. https://doi.org/10.1016/j.algal.2021.102348

Zhang, H., Song, T., Chen, X., & Li, P. (2021). An exploration of seaweed polysaccharides stimulating denitrifying bacteria for safer nitrate removal. *Molecules, 26*, 1–13. https://doi.org/10.3390/molecules26113390

Zhang, X., & Thomsen, M. (2021). Techno-economic and environmental assessment of novel biorefinery designs for sequential extraction of high-value biomolecules from brown macroalgae Laminaria digitata, Fucus vesiculosus, and Saccharina latissima. *Algal Research*, 60. https://doi.org/10.1016/j.algal.2021.102499

Zhu, L. (2015). Biorefinery as a promising approach to promote microalgae industry: An innovative framework. *Renewable and Sustainable Energy Reviews, 41*, 1376–1384. https://doi.org/10.1016/j.rser.2014.09.040

Zollmann, M., Robin, A., Prabhu, M., Polikovsky, M., Gillis, A., Greiserman, S., & Golberg, A. (2019). Green technology in green macroalgal biorefineries. *Phycologia, 58*. https://doi.org/10.1080/00318884.2019.1640516

CHAPTER

24 Algae materials: an alternative feed ingredients for the sustainable aquaculture

P. Santhanam[1], T. Muralisankar[2], K. Nanthini Devi[1], N. Krishnaveni[1], M. Divya[3], A. Gowthami[1], J. Roopavathy[4] and P. Perumal[1]

[1]*Marine Planktonology & Aquaculture Laboratory, Department of Marine Science, School of Marine Sciences, Bharathidasan University, Tiruchirappalli, Tamil Nadu, India;* [2]*Aquatic Ecology Laboratory, Department of Zoology, Bharathiar University, Coimbatore, Tamil Nadu, India;* [3]*TNJFU-Fisheries Business School, Tamil Nadu Dr. J. Jayalalithaa Fisheries University, DIVA Campus, Chennai, Tamil Nadu, India;* [4]*PG & Research Department of Zoology, Nirmala College for Women (Autonomous), Coimbatore, Tamil Nadu, India*

1. Preamble

Fish waste and small fish are used to produce fishmeal, which is necessary for aqua feed formulation because of its excellent digestion, deliciousness for fish and shrimp, balanced recipe, high concentrations of minerals, protein, and vital amino acids and fatty acids, which reduce waste and boost survival rates. Fishmeal has been a popular element for the creation of compound feeds because of its nutritive qualities and accessibility. One of the key ingredients in feed, fishmeal, has seen an upsurge in demand (Indexmundi, 2021). Currently, overfishing and regulated fishing methods are used to prepare fishmeal and fish oil from wild fish (Bakun & Broad, 2003). However, this specific element has raised significant worries on both the economy and the ecology. As a result, numerous sectors and researchers have already started exploring for appropriate and sustainable alternatives to fishmeal.

Due to their vast availability, low cost, and generally good nutritional composition, microalgae have likely attracted the most attention. Microalgae have significant levels of protein and amino acid compositions that are equivalent to those of other reference food proteins (Becker, 2007). Microalgae have gained interest as an animal food source because they are a rich source of vitamins, minerals, vital fatty acids, antioxidant, and pigments like carotenoids, in addition to their high protein content (Belay et al., 1996; James et al., 2006). It might have the potential to be employed in aquafeed as an antibacterial agent (Pradhan et al., 2012). Important biochemical components found in microalgal biomass include lipid, protein, carbohydrates, and a variety of pigments (Apandi et al., 2019). They make appropriate ingredients for the manufacturing of aquaculture feed since they are a significant source of food for shrimp in their natural habitat. Nearly all necessary nutrients, including polyunsaturated fatty acids (PUFA), amino acids, vitamins, and minerals are found in algae (Becker, 1994; Benemann, 1992; Knuckey et al., 2005). Microalgae are quickly gaining popularity as a feed additive due to their balanced nutritional makeup, with the potential to replace fishmeal and other traditional components in

Algae Materials: Applications Benefitting Health. https://doi.org/10.1016/B978-0-443-18816-9.00001-0

aquaculture. Due to their superior nutritional content and adaptability, algae such as *Spirulina* sp., *Chlorella* sp., *Scenedesmus* sp., *Dunaliella* sp., and *Nannochloropsis* sp. are frequently utilized in aqua feed formulation (Khatoon et al., 2009).

Microalgae have been utilized extensively as feed supplements for fishes, crustaceans and mollusks (Belay et al., 1996; Borowitzka, 1997). The demand for microalgae is anticipated to increase by 4.3% over the next seven years, reaching US$3.4 billion by 2020 (Globe Newswire, 2021). The microalgae business has so far mostly focused on food and goods, some important species such as *Spirulina* sp. and *Dunaliella salina* that are related to cosmetics. Additionally, fish hatcheries feed their larvae with live feed such as *Isochrysis* sp., *Pavlova* sp., and others, although these businesses are typically small scale. Microalgae, are currently being researched as a potential bulk feed component for adult fish, shellfish, and fish fingerlings (Hodar et al., 2020). The assumption that microalgae will be a promising alternative fish feed has a number of justifications. The net biomass productivity of microalgae is greater than that of any other terrestrial plant or animal (Rizwan et al., 2018). Microalgae do not require fertile ground to thrive, unlike plants that do; in addition, they can be grown in wastewater or even seawater (Li et al., 2019). Comparatively, the starch concentration of some species, such as *Tetraselmis subcordiformis*, *Chlamydomonas einhardtii*, and *Chlorella vulgaris* is higher (30–49%) than that of other microalgae (Dragone et al., 2011; Yao et al., 2012). Microalgae contain between 5% and 18% of a complex carbohydrate called fiber (Matos et al., 2016). Microalgal fiber, which differs from plant fiber in that it lacks lignin and contains little hemicellulose, suggests improved digestion (Niccolai et al., 2019). Additionally, certain microalgae create a lot of vitamins and immunostimulants in their cells and have a variety of pigments with antioxidant properties, which might benefit the health of aquatic species (Prabha et al., 2020; Zhou et al., 2019). Certain microalgae contain high concentrations of omega-3 fatty acids including docosahexaenoic acid (DHA) and eicosapentaenoic acid (EPA), which are good for human health in addition to fish health (Ryckebosch et al., 2012). Omega-3 fatty acids are naturally present in high proportions in a few types of marine fish that eat algae. Aqua feed may eventually contain microalgae as an antibacterial agent due to its high nutritional value and ability to effectively modulate fish immune systems (Pradhan et al., 2012; Watanuki et al., 2006). It has been demonstrated that the usage of microalgae improves shrimp's innate immunity and increases their resistance to pathogenic infection (Lin et al., 2010; Tayag et al., 2010).

2. Scope and objectives

The fastest-growing food-producing sector in recent decades has been the global cultivation of fish and shellfish, which has grown to be a significant industry in many nations. Aquaculture provides 50% of the fish used for human consumption, and the trend has been increasing steadily over the past few years (Prem & Tewari, 2020). To meet the global per capita fish consumption rates by 2050 without further depleting wild fisheries, aquaculture production must double from its current level (Waite et al., 2014). Currently, in the aquaculture sector, commercial species make up 70% of all fish (Tacon, 2020). Aquaculture uses prepared feed to produce high yields and quick growth to marketable size. Due to balanced nutrient composition, palatability, and excellent digestibility, fishmeal has traditionally been the main component of aquafeeds. Fishmeal is reported to contain more HUFAs, primarily docosahexaenoic acid (DHA) and eicosapentaenoic acid (EPA), minerals (Ca, P, and Mg), vitamins (A, B12, C, D3, choline, and inositol), as well as trace minerals (Zn, Cu, Se, Fe, etc.). It also has a high biological value because of its composition of essential amino acids, which is well-balanced (Higuera &

Cardenete, 1993). Fishmeal, which is the main and most expensive ingredient, often makes about 25–50% of commercial fish or shrimp feed formulas (Rodriguez et al., 2004). Fishmeal use in aquaculture diets is being reduced or eliminated due to rising fishmeal demand and overfishing of fish populations, which has prompted a quest for less expensive and sustainable protein components (Kiron et al., 2012; Salze et al., 2010). Because they are the main producers of the freshwater and saltwater ecosystems that create the food resources that shrimp are accustomed to ingest, microalgae are regarded as a great alternative protein source.

Microalgae have high protein concentrations and amino acid compositions that are comparable with other reference dietary proteins (Becker, 2007). Microalgae have gained interest as an animal food source in addition to their high protein content because they are a rich source of vitamins, minerals, essential fatty acids, amino acids, antioxidants and pigments (Belay et al., 1996; James et al., 2006). In addition to having a high nutritional value, microalgae have the potential to be used in aquafeed as an antibacterial agent and are an efficient immune modulator in fish (Pradhan et al., 2012; Watanuki et al., 2006). It has been demonstrated that the usage of microalgae improves aquatic animals' innate immunity and increases their resistance to pathogenic infection (Lin et al., 2010; Tayaget al., 2010). Microalgae are frequently added to prepared meals in the aquaculture sector to promote biological activity and improve the pigmentation of fish and shrimp (Ju et al., 2012).

3. Culture methods of microalgae

The different techniques and environmental factors can be used for cultivation of microalgae, which requires light as an energy source to convert the water (H_2O) and carbon dioxide (CO_2) into biomass through photosynthesis (Ozkurt, 2009). There are about 20–50% of the overall biomass is made up of photosynthetic products, which can take many different forms including cell components or storage resources (Chisti, 2007). In addition, nitrogen and phosphorus, which make up 10–20% of algae biomass, are essential nutrients for algae (Benemann & Oswald, 1996). The macronutrients such as Na, Mg, Ca & K, the micronutrients namely Mo, Mn, B, Co, Fe, Zn, and other trace elements are also necessary for the growth of microalgae. The different phases of algae growth are required for the successful culture conditions for example, lag, exponential, stationary, and death phase. The requirements for growth media can differ among microalgal species. However, practically all species share similar fundamental needs, which include basic nutrients, a source of carbon, either organic or inorganic, as well as requirements for nitrogen, phosphorus, and iron (Grobbelaar, 2004). Due to their effective photosynthetic processes and high nutritional content, microalgae are recognized as a potential source of food and energy. Various microalgae species with high protein levels can be used for the potential replacement for fishmeal protein or useful additives in aquafeeds (Ju & Dominy, 2012). Furthermore, microalgae are a rich source of vitamins, necessary amino acids, minerals, essential fatty acids, and carotenoid pigments (Takeuchi & Satoh, 2002), they may provide a great substitute ingredients for sustainable aquaculture feeds (Becker, 2007).

3.1 Collection of microalgae

The microalgae samples can be collected from the open sea, shrimp or fish farms. Early in the morning, samples of microalgae will be collected from different locations using a phytoplankton net (No. 25, bolting silk cloth with mesh size 48 μm). The samples can be filtered using a 158 μm zooplankton mesh filter to remove any remaining zooplanktons. Later, the microalgae filtrate will be added to the filtered seawater and brought to the laboratory. Agar plating and serial dilution techniques can be used in the lab to isolate the different types of microalgae (Robert, 2005).

3.2 Culture conditions

At prior to the culture of microalgae, the glasswares can be sterilized by autoclaving at 121°C and 15 lbs pressure. The different media such as Conway's, Zarrouk and BG11 can be used for the culture of algal strains (Huang & Liao, 1970; Zarrouk, 1966; Walne, 1974). In the autotrophic culture, algal cells in the stationary phase will be inoculated into 250 mL Erlenmeyer flasks containing sterilized seawater. The microalgae can be maintained at 23±1°C in the thermostatically controlled room with serial fluorescence light of around 2000 lux under 12/12 h light/dark photoperiod. The flasks are hand shaken twice a day (without aeration) to avoid the adherence of the cells to the walls and bottom of flasks. The media compositions of each medium are described in Tables 24.1–24.3.

3.3 Strain identification

For preliminary identification and culture confirmation, samples of the various microalgae culture will be morphologically inspected under a light microscope. Using a field guide, the algal cultures can be identified using standard references (Santhanam & Perumal, 2008; Subrahmanyan, 1946) and the algal strains will be categorized using the taxonomical keys.

3.4 Biomass harvesting

The microalgae culture can be harvested during stationary phase and rinse with distilled water being centrifuge at 3000 rpm for 10 mins. The biomass will be vacuum desiccated and dry at 60°C.

Table 24.1 Composition of Conway's medium.

Solution A		g/L
Potassium nitrate	–	100
Sodium orthophosphate	–	20
EDTA (Na)	–	45
Boric acid	–	33.4
Ferric chloride	–	1.3
Manganese chloride	–	0.36
Solution B		
Zinc chloride	–	4.2
Cobalt chloride	–	4.0
Copper sulfate	–	4.0
Ammonium molybdate	–	1.8
Solution C		
Vitamin B1 (thiamine)	–	20
Vitamin B12 (cyanocobalamine)	–	10

Table 24.2 Composition of BG11 medium.

BG11 medium		g/L
Sodium nitrate	–	1.500
Dipotassium hydrogen phosphate	–	0.0314
Magnesium sulfate	–	0.036
Calcium chloride dehydrate	–	0.0367
Sodium carbonate	–	0.020
Disodium magnesium EDTA	–	0.001
Citric acid	–	0.0056
Ferric ammonium citrate	–	0.006

Table 24.3 Composition of Zarrouk's medium.

Macronutrients		g/L
Sodium bicarbonate	–	16.8
Sodium nitrate	–	2.5
Sodium chloride	–	1.0
Potassium sulfate	–	1.0
Dipotassium phosphate	–	0.5
Magnesium sulfate	–	0.2
Ferrous sulfate	–	0.01
Calcium chloride	–	0.04
EDTA	–	0.08
Micronutrients		
Boric acid	–	2.86
Manganese chloride	–	1.81
Zinc sulfate	–	0.222
Sodium molybdate	–	0.0177
Copper sulfate pentahydrate	–	0.079

4. Microalgae used as a fishmeal replacement

The production of capture fisheries is constrained, which gives rise to aquaculture and has been expanding substantially globally. For the first time, especially in 2014, the aquaculture sector's contribution to the world's human food supply surpassed that of the capture fisheries (FAO, 2017). The need for aqua feed has grown in tandem with the expansion of aquaculture on a larger scale (Abasubong et al., 2018). Direct-fed species account for approximately 70% of all aquaculture production, or 35.7 million tons of product. These direct-fed species consumed an estimated 39.6 million tons of aquafeed in 2012, with 68% of them being dependent on commercially generated feed supplies (Tacon & Metian, 2015). In the past 10 years, the aquafeed market has experienced exponential expansion, and this trend is anticipated to continue due to overfishing, climate change, fishmeal shortages, and a rise in its price. Additionally, due to the water and gas pollution that occurs during the production of fishmeal, the excessive usage of fishmeal may result in a number of environmental issues (Sales, 2009). Therefore, a focus area for study in aquatic animal nutrition and feed is developing substitutes for fishmeal.

The fishmeal is the most excellent ingredient due to its compliance with the protein requirements of fish (Alam et al., 2005). An approach for sustainable aquaculture is the substitution of fishmeal in fish and shrimp feed with less expensive components of plant origin. It has been established over time that carefully crafted and attractive fishmeals are extremely important for fish culture (Omoregie, 2001). If the aquaculture sector is to continue to thrive, it is crucial to find substitute protein sources for fishmeal in fish feed (Francis et al., 2001). Aquaculture primarily uses extrusion cooking to create feeds that are nutritiously enough, have a proper mix, and can float (move slowly through the water column) before being consumed. Therefore, choosing materials when extruded, will process just the appropriate buoyancy and not dissolve nutrients into the water with high palatability for specific species is crucial in aquaculture. The secret to aquaculture success is the use of economical meals created based on the dietary needs of potential species (Tacon et al., 2008).

The protein source is the most expensive and crucial element of aquaculture feeding. Generally, fishmeal has been acknowledged as the main source of protein due to its abundance in vital elements. However, it has been challenging to meet the expanding production need of the aquafeed industry due to inconsistent output and a higher price for fishmeal. Therefore, it has become a top research goal to identify substitute sources of protein for fishmeal in diets. Over the past two decades, India has made fantastic advancements in the aquaculture feed industry. Currently, the aquaculture industry uses roughly 20 million tons of produced aquafeed, the majority of which is utilized in shrimp culture (Chandrapal, 2005). The cost-effective food production depends on a system that produces high-quality nourishment. The first factor in feed formulation is the quality of the components, which must be high-grade, healthful items. The use of microalgae in the creation of low-cost, highly nutritious diets for fish nutrition, particularly for the growth of oysters, tilapia, salmon, trout, etc., all of which are of major economic interest (Fleming et al., 1996; Horn, 1989).

The ability of microalgae to supply protein, lipids, vitamins, carotenoids, and energy in feed is very promising. In the late logarithmic growth phase, microalgae typically include 30–40% protein, 10–20% lipid, and 5–15% carbohydrate (Brown et al., 1997). Some commercially significant microalgal species have chemical compositions that are comparable with those of the available feed. Numerous previous publications provide thorough documentation of the chemical–nutrient composition of microalgae, including proximate analysis, amino acid and fatty acid compositions, vitamin, and mineral concentrations (Becker, 2004, 2007; Brown, 2002; Kent et al., 2015; Shields & Lupatsch, 2012). The food value of microalgae is influenced by the presence of protein and polyunsaturated fatty acids such as eicosapentaenoic acid (EPA), arachidonic acid (ARA), and docosahexaenoic acid (DHA) (Reitanet al., 1997). Microalgal protein might be a good substitute for fishmeal protein due to their excellent quality and amino acid profiles that are comparable with those of other reference protein sources (Becker, 2007). Studies on microalgae have also found significant amounts of essential amino acids (Brown et al., 1997). Recent research (Tibbetts et al., 2015) that assessed the biochemical characteristics, amino acid profiles, fatty acids, carotenoids, and elemental content of freshwater and marine microalgal biomass for animal feed applications showed the viability of employing microalgal protein in feed. There have been numerous studies on the use of biomass from various microalgae species, including *Tetraselmis suecica*, *Arthrospira* sp., *Chlorella* sp., *Scenedesmus* sp., and *Nanofrustulum* sp., as valuable supplemental protein sources or partial substitutes for fishmeal protein (Badwy et al., 2008).

The nutritional characteristics of some microalgae such as *Arthrospira* sp., *Dunaliella* sp. and *Chlorella* sp., are well known, and they have a long history of use as sources of animal feed. According to research on aquatic animals, the growth-enhancing impact is probably connected to the enhanced palatability and hence greater feed intake (Espe et al., 2006; Guillaume, 1997).

4.1 Source of feed ingredients

The components of feed formulation including groundnut oil cake, soyabean meal, rice bran, tapioca flour, vitamin mix, dry fish, egg albumin, and cod liver oil were bought from vendors; vitamins and minerals mix were acquired from pharmacies. For the feed preparation purpose, these ingredients were air-dried and kept in a desiccator in the laboratory.

4.2 Pellet feed preparation (Plate 24.1)

4.2.1 Grinding and mixing

The feed ingredients such as dry fish or shrimp, soy bean meal, groundnut oil cake, rice bran, tapioca flour, microalgae biomass, and other materials (Plate 24.2; Table 24.4) were finely processed using a micropulverizer before being sieved with a commercial sieve. The required level of each ingredient was measured to make optimal protein for the intrested fish or crustacean species.

PLATE 24.1

Feed formulation methods.

(a)

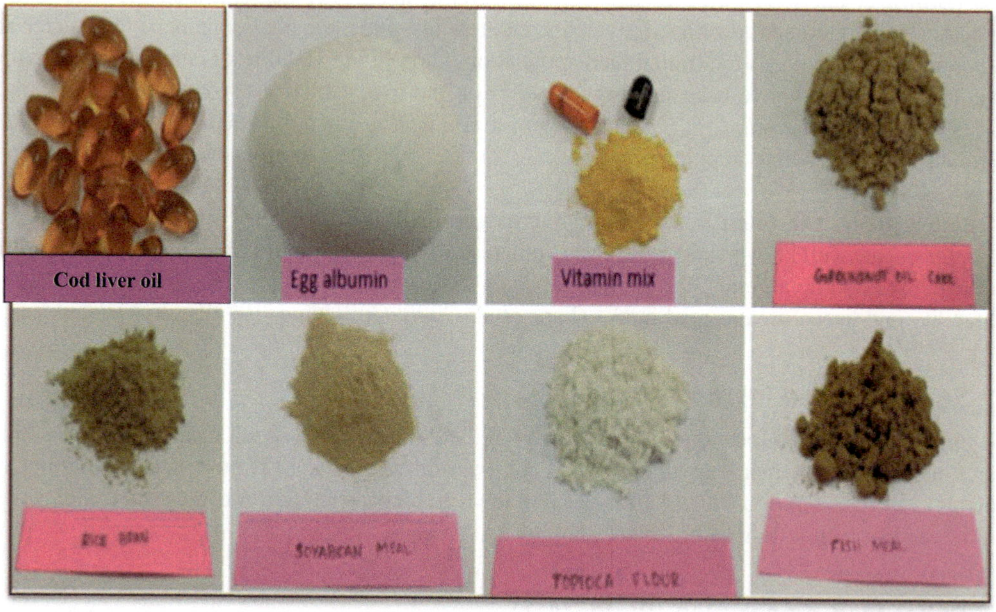

(b)

PLATE 24.2

Note: (A) Ingredients of control feed; (B) Ingredients of experimental feed.

Table 24.4 Composition of feed formulation. Replacement of fishmeal by using microalgae.

	g/100g				
Composition	**Control**	**25%**	**50%**	**75%**	**100%**
Dry fishmeal	40	30	20	10	0
Microalgae	0	10	20	30	40
Groundnut oil cake	10	10	10	10	10
Soybean meal	15	15	15	15	15
Rice bran	15	15	15	15	15
Tapioca	10	10	10	10	10
Cod liver oil	4	4	4	4	4
Egg albumin	5	5	5	5	5
Vitamins and minerals mix	1	1	1	1	1
Total	100	100	100	100	100

4.2.2 Steam cooking

The feed mixtures were combined and steam-cooked for 20 mins. The feed mixtures were further cooled, vitamins were added, and minerals and cod liver oil were mixed and then completely homogenized in a dough mixer. To make the finished combination into a paste, 10% distilled water and egg albumin were added to the feed mixture and finally homogenized for an additional 10 mins.

4.2.3 Pelletization and drying

The formulated feed mixtures were poured into the manual pelletizer, which was fastened with a disc of 3 mm in diameter, and the pellets were collected in an aluminium tray. The moist feed was loaded onto electrical trays at a temperature of between 75—80°C, where it was allowed to dry until the moisture content was below 10%.

4.3 Checking the quality of feed

The visual characteristics of the dry feed pellets such as homogeneity, color, smooth surface, and pieces fragrance were physically inspected. The proximate parameter of the pellet feed was examined. The pellets' water stability was also evaluated after being prepared for 24 hrs.

Vitamins and minerals mix (each capsule contains thiamine mononitrate 1P 10 mg; riboflavin 1P 10 mg; pyridoxine hydrochloride 1P 3 mg; vitamin B12 (as tablets 1:100) 1P 15 mcg; niacinamide 1P 100 mg; calcium pantothenate 1P 50 mg; folic acid 1P 1.5 mg; biotin USP 100 mcg; and ascorbic acid 1P 150 mg).

5. Benefits and future outcomes

Fishmeal is a good source of high-quality protein, provides a fair amount of energy, and is a good supply of key nutrients such as phosphorus, B vitamins, and essential fatty acids. Fishmeal costs have significantly increased over the past 10 years as a result of the constant expansion of aquaculture and the resulting rise in consumer demand (Duarte et al., 2009; FAO, 2009, p. 15). Although fishmeal is a scarce natural resource, it is a good source of protein and other vital components for aquaculture species. Due to its balanced composition of essential amino acids, fatty acids, vitamins, minerals, and palatability, fishmeal is regarded as an essential component in the diets of marine fish and shrimp (Suarez et al., 2009; Tacon et al., 2008, p. 564). In 2006, the aquaculture sector accounted for 68.2% of the world's fishmeal production.

Fishmeal is widely used in the nutrition of fish, shrimp, and other aquatic creatures. Dry fishmeal is a readily available but pricey form of an extremely high-protein animal by-product. Regardless of the culture system in which they are raised, the growth, health, and reproduction of fish, shrimp, and other aquatic creatures depend primarily on an adequate supply of nutrients through feed in terms of both quality and quantity (Kader et al., 2012). For aquatic animals, the utilization of alternative protein sources, particularly microalgae, is essential commercially and lowers the cost of feed production. But to get the greatest results, it is also crucial to implement tactics that maximize the quantity of alternative protein sources in the basic feed and encourage the growth of aquatic animals.

Microalgae are frequently added to prepared meals in the aquaculture sector to promote biological activity and improve the pigmentation of fish and shrimp (Ju et al., 2012). For the reason that microalgae are one of the richest sources of protein, vitamins, minerals, essential amino acids, fatty acids, and antioxidant pigments such as carotenoids, some authors have reported that pure microalgae meal could be partially substituted with fishmeal in the diets of fish, shrimp, and mollusks (Hanel et al., 2007; Ju et al., 2012). It is the richest source of vitamins B12 and beta-carotene (20 times more than carrots), contains 60–70% protein by weight, and is crammed with minerals and important fatty acids (James et al., 2006). These features make this microalga intriguing in terms of substituting fishmeal (Hanel et al., 2007). Experimental fish and shrimp feeds have used microalgae meal with some degree of success (Kiron et al., 2012). *Acipenser transmontanus*, a Siberian sturgeon, has been shown to benefit from the use of this microalga species as a partial substitute for fishmeal (Palmegiano et al., 2005). The giant freshwater prawn, *Macrobrachium rosenbergii* (Radhakrishnan et al., 2014), white shrimp, *Litopenaeus schmitti* (Ceballos et al., 2007), pacific white leg shrimp, *Penaeus vannamei* (Hanel et al., 2007; Macias Sancho et al., 2014), green tiger prawn, *Penaeus semisulcatus*, and an ornamental fish, *Puntius sophore* have all diets containing *Spirulina* sp. showed incredible promise for fostering greater growth (Bagre et al., 2012).

The nutrients that fish needs are the same as those other creatures need. These consist of vitamins, minerals, fats, carbs, and protein (Royes & Chapman, 2009). As a result of research showing that some vitamins interact with one another, a vitamin mixture is added to the feed formulation process. This is a crucial element to take into account when preparing the vitamin-rich meal. One of the most expensive elements in diets for cooked fish and shrimp is fishmeal. Fish nutritionists have experimented with substituting some or all of fishmeal with less expensive plant proteins. If the aquaculture business is to continue to thrive, finding substitute proteins for fishmeal in fish feed is futile. Due to the significant amount of carbohydrates contained in microalgae, replacing the diet's protein-sparing effect with carbohydrates-rich plant sources would be a realistic choice. Another strategy to improve food stability is the small-scale addition of a binder to the diet (Brinker et al., 2005). The experimental diets that included lipid-extracted algae (LEA) have a greater protein and carbohydrate content, according to the proximate composition of the formulated feeds. In aquatic animal feed, which frequently relies on formulated feeds to boost the growth, health, and quality of the farmed aquatic animal, microalgae are typically supplied as a partial source rather than a total replacement for artificial minerals or vitamins.

As a result of their greater fishmeal composition, control feed had a higher lipid content than LEA feed, which is likely because lipids were previously taken for the creation of biofuel. For various aquatic animal life phases, their high nutritional qualities can offer a high-quality nutritional package (Banerjee et al., 2010; Khatoon et al., 2009).

Feed stability in water is a crucial quality indicator, when creating aquaculture diets. A pellet feed's physical integrity is retained in water with little disintegration or nutrient loss. Shrimp will develop and survive more slowly if the pellets are unstable because they will quickly leak nutrients and contaminate the water in the culture environment. Water stability is influenced by a number of variables, including diet preparation and processing techniques, constituent kinds, binding agents, diet contents, and coating materials such as microcapsules (Belay et al., 1996; Teimouri et al., 2013). During the extrusion process, the semicrystalline structure of starch can be altered, which could encourage the interactions between the components and aid in stability. Additionally, microalgae may help with the capacity of components to bind together by adding functional qualities such as antioxidant value and vitamins. One of the main benefits of artificial feed over natural feed is said to be its high-water stability. The created diets with the lowest leaching % demonstrated how the chosen basic components and binding agents (tapioca flour, egg albumin, and cod liver oil) helped to achieve water stability that is sufficient to minimize water pollution and nutrient leaching. According to Radhakrishnan et al. (2014), diets containing tapioca flour, egg albumin, and fish oil demonstrated minimal leaching and good water stability. The biggest and most expensive part of aquaculture diets is protein. Up until now, fishmeal has been the main source of protein for aquafeed composition. However, fishmeal is expensive and in great demand in aquafeed formulations, nutritionists are looking for cheaper feed ingredients to substitute expensive fishmeal. According to the feeding experiment, using algal meal, which has a large amount of nutritional content and a high concentration of fatty acids and amino acids, can be a potential substitute for expensive fishmeal. Algal meal is also seen as being less expensive than fish meal or other types of animal meal used in the aquaculture sector. The addition of microalgae ingredients can drastically lower the cost of producing formulated feed, according to the feed developer.

References

Abasubong, K. P., Liu, W. B., Zhang, D. D., Yuan, X. Y., Xia, S. L., & Xu, C. (2018). Fishmeal replacement by rice protein concentrate with xylo oligosaccharides supplement benefits the growth performance, antioxidant capability and immune responses against Aeromonas hydrophila in blunt snout bream (Megalobramaamblycephala). *Fish & Shellfish Immunology, 78*, 177–186.

Alam, M. S., Teshima, S. I., Koshio, S., Ishikawa, M., Uyan, O., Hernandez, L. H. H., & Michael, F. R. (2005). Supplemental effects of coated methionine or lysine to soy protein isolate diet for juvenile kuruma shrimp, *Marsupenaeus japonicus. Aquaculture, 248*(1–4), 3–19.

Apandi, N. M., Mohamed, R. M. S. R., Gheethi, A., & Kassim, A. H. M. (2019). Microalgal biomass production through phycoremediation of fresh market wastewater and potential applications as aquaculture feeds. *Environmental Science & Pollution Research, 26*(4), 3226–3242.

Badwy, T. M., Ibrahim, E. M., & Zeinhom, M. M. (2008). Partial replacement of fishmeal with dried microalgae (Chlorella spp. and Scenedesmus spp.) in Nile tilapia (Oreochromis niloticus) diets. In H. Elghobashy, K. Fitzsimmons, & A. S. Diab (Eds.), *From the pharaohs to the future: Proceedings of the 8th international symposium on tilapia in aquaculture. Egypt, Cairo* (pp. 801–810).

Bagre, P., Parashar, A., Vyas, V., Bhargava, S., & Saroch, J. D. (2012). Comparative study of different percentage of natural carotenoid source Spirulina platensis on growth and survival of Puntius sophore. *Journal of Chemical, Biological and Physical Sciences, 2*, 2210–2216.
Bakun, A., & Broad, K. (2003). Environmental 'loopholes' and fish population dynamics: Comparative pattern recognition with focus on El Nino effects in the pacific. *Fisheries Oceanography, 12*(4-5), 458–473.
Banerjee, S., Khatoon, H., Shariff, M., & Yusoff, F. M. (2010). Enhancement of Penaeus monodon shrimp postlarvae growth and survival without water exchange using marine *Bacillus pumilus* and periphytic microalgae. *Fisheries Science, 76*(3), 481–487.
Becker, E. W. (1994). *Microalgae: Biotechnology and microbiology* (Vol 10). Cambridge University Press.
Becker, W. (2004). Microalgae in human and animal nutrition. In A. Richmond (Ed.), *Handbook of microalgal culture: Biotechnology and applied phycology* (pp. 312–351). Cambridge: Blackwell Science Ltd.
Becker, E. (2007). Micro-algae as a source of protein. *Biotechnology Advances, 25*, 207–210.
Belay, A., Kato, T., & Ota, Y. (1996). *Spirulina* (*Arthrospira*): Potential application as an animal feed supplement. *Journal of Applied Phycology, 8*(4), 303–311.
Benemann, J. R. (1992). Microalgae aquaculture feeds. *Journal of Applied Phycology, 4*(3), 233–245.
Benemann, J., & Oswald, W. (1996). *Final report to the US department of energy. Grant No. DEFG22-93PC93204*. USA: Pittsburgh Energy Technology Center.
Borowitzka, M. A. (1997). Microalgae for aquaculture: Opportunities and constraints. *Journal of Applied Phycology, 9*(5), 393–401.
Brinker, A., Koppe, W., & Rosch, R. (2005). Optimised effluent treatment by stabilised trout faeces. *Aquaculture, 249*(1–4), 125–144.
Brown, M. R. (2002). Nutritional value and use of microalgae in aquaculture. In , *3. Avances en NutricionAcuicola VI. Memorias del VI SimposiumInternacional de NutricikonAcuicola* (pp. 281–292).
Brown, M. R., Jeffrey, S. W., Volkman, J. K., & Dunstan, G. A. (1997). Nutritional properties of microalgae for mariculture. *Aquaculture, 151*(1–4), 315–331.
Ceballos, B., CiveraCerecedo, R., Villarreal, H., Galindo López, J., & Perez-Jar, L. (2007). Use of *Spirulina platensis* meal as feed attractant in diets for shrimp Litopenaeusschmitti. *Hidrobiologica, 17*(2), 113–117.
Chandrapal, P., & Kumar, P. (2005). A scalable multi-level distributed system-level diagnosis. In *International conference on distributed computing and internet technology* (pp. 192–202). Berlin, Heidelberg: Springer.
Chisti, Y. (2007). Biodiesel from microalgae. *Biotechnology Advances, 25*(3), 294–306.
Dragone, G., Fernandes, B. D., Abreu, A. P., Vicente, A. A., & Teixeira, J. A. (2011). Nutrient limitation as a strategy for increasing starch accumulation in microalgae. *Applied Energy, 88*(10), 3331–3335.
Duarte, C. M., Holmer, M., Olsen, Y., Soto, D., Marba, N., Guiu, J., Black, K., & Karakassis, I. (2009). Will the oceans help feed humanity? *BioScience, 59*(11), 967–976.
Espe, M., Lemme, A., Petri, A., & Mowafi, A. (2006). Can Atlantic salmon (*Salmo salar*) grow on diets devoid of fishmeal. *Aquaculture, 255*(1–4), 255–262.
FAO. (2009). *Food and agriculture organisation of the United Nations*.
FAO, Food and Agriculture Organization. (2017). *Fishery and aquaculture statistics. Rome, Italy.*
Fleming, A. E., Van Barneveld, R. J., & Hone, P. W. (1996). The development of artificial diets for abalone: A review and future directions. *Aquaculture, 140*(1–2), 5–53.
Francis, G., Makkar, H. P., & Becker, K. (2001). Antinutritional factors present in plant-derived alternate fish feed ingredients and their effects in fish. *Aquaculture, 199*(3–4), 197–227.
Globe Newswire. 2020. GW Pharmaceuticals Plc and its U.S. Subsidiary Greenwich Biosciences, Inc.
Grobbelaar, J.,U. (2004). Algal nutrition. In A. Richmond (Ed.), *Handbook of microalgal culture: Biotechnology and applied phycology* (pp. 97–115). Oxford: Blackwell.
Guillaume, D. M., Dacorogna, M. M., Dave, R. R., Muller, U. A., Olsen, R. B., & Pictet, O. V. (1997). From the bird's eye to the microscope: A survey of new stylized facts of the intra-daily foreign exchange markets. *Finance and Stochastics, 1*(2), 95–129.

Hanel, R., Broekmann, D., De Graaf, S., & Schnack, D. (2007). Partial replacement of fishmeal by lyophylized powder of the microalgae *Spirulina platensis* in Pacific white shrimp diets. *Marine Biology, 1*, 1−5.

Higuera, M., & Cardenete, G. (1993). La proteínaen la nutrición de lospeces. In F. Castello Orvay (Ed.), *Acuicultura marina: Fundamentosbiogicos y tecnologia de la produccion* (pp. 195−226). Barcelona: Universidad de Barcelona.

Hodar, A. R., Vasava, R. J., Mahavadiya, D. R., & Joshi, N. H. (2020). Fishmeal and fish oil replacement for aqua feed formulation by using alternative sources: A review. *Journal of Experimental Zoology India, 23*(1), 13−21.

Horn, D. (1989). Preparation and characterization of microdisperse bioavailable carotenoid hydrosols. *Makromolekulare Chemie, 166*(1), 139−153.

Huang, J. C., & Liao, C. S. (1970). Adsorption of pesticides by clay minerals. *Journal of the Sanitary Engineering Division, 96*(5), 1057−1078.

Indexmundi. (2021). *Fishmeal yearly price*. www.indexmundi.com.

James, R., Sampath, K., Thangarathinam, R., & Vasudhevan, I. (2006). Effect of dietary *Spirulina* level on growth, fertility, coloration and leucocyte count in red swordtail, Xiphophorus helleri. *Israeli Journal of Aquaculture-Bamidgeh, 58*, 97−104.

Ju, Z. Y., Deng, D. F., & Dominy, W. (2012). A defatted microalgae (*Haematococcuspluvialis*) meal as a protein ingredient to partially replace fishmeal in diets of Pacific white shrimp (*Litopenaeusvannamei*, Boone, 1931). *Aquaculture, 354*, 50−55.

Kader, M. A., & Koshio, S. (2012). Effect of composite mixture of seafood by-products and soybean proteins in replacement of fishmeal on the performance of red sea bream, *Pagrus major*. *Aquaculture, 368*, 95−102.

Kent, M., Welladsen, H. M., Mangott, A., & Li, Y. (2015). Nutritional evaluation of Australian microalgae as potential human health supplements. *PLoS One, 10*, e0118985.

Khatoon, H., Banerjee, S., Yusoff, F. M., & Shariff, M. (2009). Evaluation of indigenous marine periphytic *Amphora, Navicula* and *Cymbella* grown on substrate as feed supplement in *Penaeus monodon*postlarval hatchery system. *Aquaculture Nutrition, 15*(2), 186−193.

Kiron, V., Phromkunthong, W., Huntley, M., Archibald, I., & De Scheemaker, G. (2012). Marine microalgae from biorefinery as a potential feed protein source for Atlantic salmon, common carp and whiteleg shrimp. *Aquaculture Nutrition, 18*(5), 521−531.

Knuckey, R. M., Semmens, G. L., Mayer, R. J., & Rimmer, M. A. (2005). Development of an optimal microalgal diet for the culture of the calanoid copepod *Acartiasinjiensis*: Effect of algal species and feed concentration on copepod development. *Aquaculture, 249*(1−4), 339−351.

Li, K., Liu, Q., Fang, F., Luo, R., Lu, Q., Zhou, W., Huo, S., Cheng, P., Liu, J., Addy, M., & Chen, P. (2019). Microalgae-based wastewater treatment for nutrients recovery: A review. *Bioresource Technology, 291*, 121934.

Lin, P., Engling, G., & Yu, J. Z. (2010). Humic-like substances in fresh emissions of rice straw burning and in ambient aerosols in the Pearl River Delta Region, China. *Atmospheric Chemistry and Physics, 10*(14), 6487−6500.

Macias Sancho, J., Poersch, L. H., Bauer, W., Romano, L. A., Wasielesky, W., & Tesser, M. B. (2014). Fishmeal substitution with *Arthrospira* (*Spirulina platensis*) in a practical diet for *Litopenaeusvannamei*: Effects on growth and immunological parameters. *Aquaculture, 426*, 120−125.

Matos, A. P., Feller, R., Moecke, E. H. S., de Oliveira, J. V., Junior, A. F., Derner, R. B., & Sant Anna, E. S. (2016). Chemical characterization of six microalgae with potential utility for food application. *Journal of the American Oil Chemists Society, 93*(7), 963−972.

Niccolai, A., Zittelli, G. C., Rodolfi, L., Biondi, N., & Tredici, M. R. (2019). Microalgae of interest as food source: Biochemical composition and digestibility. *Algal Research, 42*, 101617.

Omoregie, E. (2001). Utilization and nutrient digestibility of mango seeds and palm kernel meal by juvenile *Labeo senegalensis* (Antheriniformes: Cyprinidae). *Aquaculture Research, 32*(9), 681−687.

Ozkurt, I. (2009). Qualifying of safflower and algae for energy. *Energy Education Science and Technology, 23*, 145–151.

Palmegiano, G. B., Agradi, E., Forneris, G., Gai, F., Gasco, L., Rigamonti, E., Sicuro, B., & Zoccarato, I. (2005). Spirulina as a nutrient source in diets for growing sturgeon (*Acipenser baeri*). *Aquaculture Research, 36*(2), 188–195.

Prabha, S. P., Nagappan, S., Rathna, R., Viveka, R., & Nakkeeran, E. (2020). Blue biotechnology: A vision for future marine biorefineries. *Environmental Progress and Sustainable*, 463–480.

Pradhan, C., Mohanty, M., & Rout, A. (2012). Phytochemical screening and comparative bioefficacy assessment of *Artocarpus altilis* leaf extracts for antimicrobial. *Frontiers in Life Science, 6*(3), 71–76.

Prem, R., & Tewari, V. K. (2020). Development of human-powered fish feeding machine for freshwater aquaculture farms of developing countries. *Aquacultural Engineering, 88*, 102028.

Radhakrishnan, S., Bhavan, P. S., Seenivasan, C., Shanthi, R., & Muralisankar, T. (2014). Replacement of fishmeal with *Spirulina platensis, Chlorella vulgaris* and *Azolla pinnata* on non-enzymatic and enzymatic antioxidant activities of *Macrobrachiumrosenbergii*. *Journal of Basic Applied Zoology, 67*(2), 253.

Reitan, K. I., Rainuzzo, J. R., oie, G., & Olsen, Y. (1997). A review of the nutritional effects of algae in marine fish larvae. *Aquaculture, 155*, 207–221.

Rizwan, M., Mujtaba, G., Memon, S. A., Lee, K., & Rashid, N. (2018). Exploring the potential of microalgae for new biotechnology applications and beyond: A review. *Renewable and Sustainable Energy Reviews, 92*, 394–404.

Robert, B. J. (2005). Serial dilution with a confirmation step. *Food Microbiology, 22*(6), 547–552.

Rodríguez-Iturbe, B., Quiroz, Y., Ferrebuz, A., Parra, G., & Vaziri, N. D. (2004). Evolution of renal interstitial inflammation and NF-κB activation in spontaneously hypertensive rats. *American journal of nephrology, 24*(6), 587–594.

Royes, J. A. B., & Chapman, F. A. (2009). *Preparing our own fish feeds*. Gainesville: Institute of Food and Agricultural Sciences, University of Florida.

Ryckebosch, E., Bruneel, C., Muylaert, K., & Foubert, I. (2012). Microalgae as an alternative source of omega-3 long chain polyunsaturated fatty acids. *Lipid Technology, 24*(6), 128–130.

Sales, J. (2009). The effect of fishmeal replacement by soyabean products on fish growth: A meta-analysis. *British Journal of Nutrition, 102*, 1709.

Salze, G., Mclean, E., Battle, P.,R., Schwarz, M. H., & Craig, S. R. (2010). Use of soy protein concentrate and novel ingredients in the total elimination of fishmeal and fish oil in diets for juvenile cobia, Rachycentron canadum. *Aquaculture, 298*, 294–299.

Santhanam, P., & Perumal, P. (2008). Marine plankton in Indian waters. In John Milton (Ed.), *Training manual on GIS and marine biodiversity* (pp. 1–12). Lyola College Pub.

Shields, R. J., & Lupatsch, I. (2012). Algae for aquaculture and animal feeds. *Animal Science, 21*, 23–37.

Suarez, L. E., Tapia-Salazar, M., Villarreal-Cavazos, D., Beltran-Rocha, J., Nieto-Lopez, M. G., Lemme, A., & Ricque-Marie, D. (2009). Apparent dry matter, energy, protein and amino acid digestibility of four soybean ingredients in white shrimp *Litopenaeusvannamei* juveniles. *Aquaculture, 292*(1–2), 87–94.

Subrahmanyan, R. (1946). A systematic account of the marine plankton diatoms of the Madras coast. *Proceedings of the Indian Academy of Sciences - Section B, 24*(4), 85–197.

Tacon, A. G. (2020). Trends in global aquaculture and aquafeed production: 2000–2017. *Reviews in Fisheries Science Aquaculture, 28*(1), 43–56.

Tacon, A. G., Hasan, M. R., & Metian, M. (2008). *Demand and supply of feed ingredients for farmed fish and crustaceans: Trends and prospects*. FAO Fisheries and Aquaculture Technical Paper.

Tacon, A. G. J., & Metian, M. (2015). Feed matters: Satisfying the feed demand of aquaculture. *Fisheries Science and Aquaculture, 23*, 1–10.

Takeuchi, T., Lu, J. U. N., Yoshizaki, G., & Satoh, S. (2002). Effect on the growth and body composition of juvenile tilapia Oreochromis niloticus. *Fisheries Science, 68*(1), 34–40.

Tayag, C. M., Lin, Y. C., Li, C. C., Liou, C. H., & Chen, J. C. (2010). Administration of the hot-water extract of *Spirulina platensis* enhanced the immune response of white shrimp *Litopenaeusvannamei* and its resistance against Vibrio alginolyticus. *Fish & Shellfish Immunology, 28*(5–6), 764–773.

Teimouri, M., Amirkolaie, A. K., & Yeganeh, S. (2013). The effects of *Spirulina platensis* meal as a feed supplement on growth performance and pigmentation of rainbow trout (Oncorhyn-chus mykiss). *Aquaculture, 396*, 14–19.

Tibbetts, S. M., Milley, J. E., & Lall, S. P. (2015). Chemical composition and nutritional properties of freshwater and marine microalgal biomass cultured in photobioreactors. *Journal of Applied Phycology, 27*, 1109–1119.

Waite, R., Beveridge, M., Brummett, R., Castine, S., Chaiyawannakarn, N., Kaushik, S., Mungkung, R., Nawapakpilai, S., & Phillips, M. (2014). *Improving productivity and environmental performance of aquaculture*. World Fish.

Walne, P. R. (1974). *Culture of bivalve molluscs, 50 years' experience at Conway*. Fishing News (Books) Ltd.

Watanuki, H., Ota, K., Tassakka, A. C. M. A., Kato, T., & Sakai, M. (2006). Immunostimulant effects of dietary *Spirulina platensis* on carp, *Cyprinus carpio*. *Aquaculture, 258*(1–4), 157–163.

Yao, C., Ai, J., Cao, X., Xue, S., & Zhang, W. (2012). Enhancing starch production of a marine green microalga *Tetraselmissubcordiformis* through nutrient limitation. *Bioresource Technology, 118*, 438–444.

Zarrouk, C. (1966). Contribution a L'etude D'une Cyanobacterie: Influence de Divers Facteurs Physiques et Chimiques sur la Croissance et la Photosynthese de Spirulina maxima (Setchell et Gardner) Geitler. https://scholar.google.com/scholar?hl=en.

Zhou, R., Zhou, R., Zhang, X., Fang, Z., Wang, X., Speight, R., Wang, H., Doherty, W., Cullen, P., & Ostrikov, K. (2019). High-performance plasma-enabled biorefinery of microalgae into value-added products. *ChemSusChem, 12*, 4976–4985.

Further reading

Han, X., Rosati, R., & Webb, J. (1996). Correlation of particle size distribution of solid waste to fish feed composition in an aquaculture recirculating system. In G. S. Libey (Ed.), *Successes and failures in commercial recirculating aquaculture, northeast regional agricultural engineering* (pp. 257–278). Roanoke, VA, USA: Service Virginia Polytechnic Institute and State University.

CHAPTER

Phycoremediation: algae-based bioremediation

25

Rachel Carmelita Mathias, Sharanya Kushalan, Himamshu Hegde, Neetha Nirupam Jathanna, Sowmya Sharada P and Smitha Hegde

Nitte (Deemed to be University), Nitte University Centre for Science Education and Research (NUCSER), Deralakatte, Mangaluru, Karnataka, India

1. Introduction

Freshwater scarcity has been apparent in recent decades due to the constant increase in demand. It poses an impending threat to humankind's long-term, sustainable growth, and development. According to the recent report of World Economic Forum, the annual risk assessment, water crises are the highest worldwide risk in terms of potential impact (Van Der Vegt et al., 2015). The driving causes behind the increased worldwide demand for water are the growing global population, better living standards, changing consumption patterns, and intensive irrigated agriculture (Brooks et al., 2002; Ercin & Hoekstra, 2014; Mekonnen & Hoekstra, 2016). Contamination and redistribution of toxic metals, metalloids, and radionuclides in the environment, as well as the entry of a diverse range of organic pollutants, need ever-higher standards of pollutant identification and treatment. Organic and inorganic pollutants have well-documented negative effects on ecosystems and human health, and much money gets invested on industrial treatment procedures to avoid or reduce discharges (Gadd, 2009). As a result, it is essential to emphasize the concept of reduce, recover, reuse, and recycle water. Water supplies near sectors such as textile, painting, glass, mining, and battery manufacture are the main source of contamination. Kidney disease, nervous system failure, and bone damage are all caused by the accumulation of dangerous heavy metals such as lead, iron, copper, zinc, chromium, and mercury. Heavy metals, which are highly mobile, nonbiodegradable, persistent, and hazardous at higher levels of exposure, must be removed from the point source to prevent biomagnification (Ashraf et al., 2017). Any metal (or metalloid) species might be called a "contaminant" if it appears in an unaccepted region or in a form or concentration that harms humans or the environment (Reena et al., 2011). Researchers are developing new technology to remove heavy metal ions from an aqueous solution. Several methods have been used to remove heavy metals from contaminated water. They include chemical precipitation (Mauchauffée & Meux, 2007; Rubio et al., 2002), ion exchange (Chen & Liu, 1996; Verma et al., 2008), adsorption (Cochrane et al., 2006), membrane filtration (Landaburu-Aguirre et al., 2010; Rahmanian et al., 2011), reverse osmosis (Mohsen-Nia et al., 2007; Yoon et al., 2009), solvent extraction (Lertlapwasin et al., 2010), and electrochemical treatment (Akbal & Camcidotless, 2011; Dharnaik & Ghosh, 2014). Many of these methods have considerable investment and ongoing

Algae Materials: Applications Benefitting Health. https://doi.org/10.1016/B978-0-443-18816-9.00005-8

expenditures. Furthermore, these techniques are further expensive and/or ineffective at very low concentrations of heavy metals, specifically below the concentration of 100 mg/L (Birungi & Chirwa, 2015; Jaafari & Yaghmaeian, 2019).

Many organic substances released into the environment are degraded by natural microbial populations, and this biodegradation potential is at the core of many existing and developing treatment methods. However, biodegradation products can be dangerous in some situations, and some organic molecules are particularly resistant to biodegradation. Biosorption is being suggested once again as a viable technique for removing these and other organic compounds from waste streams and effluents. Dyes, phenolic chemicals, and insecticides are among the substances that have been studied (Aksu, 2005). Dye-containing wastewaters are difficult to treat because the dyes are recalcitrant molecules (especially azo dyes), resistant to aerobic digestion, and durable to oxidizing agents, and may be present in small amounts. The most common methods for removing dyes may be cost-prohibitive and/or technically challenging. Many physicochemical procedures for treating dyes in wastewater have not been widely used because of their high prices; hence a mix of processes is frequently used to obtain the desired water quality. For dye-contaminated effluents, biosorption has been proposed as an effective decolorization approach (Crini & Badot, 2008). Selection of a suitable remediation option is critical for removal of the metal ions or organic pollutants from their point source and checks their mobility.

Bioremediation is a pollution control technique that provides biological processes to expedite the degradation or transformation of certain toxic substances into less harmful forms. This natural process is intended to effectively clean up the environment as an alternative to the conventional remediation approaches (El-Kassas & Mohamed, 2014). Bioremediation of heavy metals employing diverse microorganisms such as bacteria, algae, yeasts, and fungi has been widely used as an alternative to conventional treatments due to its environmental friendliness and cost efficiency at low heavy metals concentrations (Leong & Chang, 2020).

Algal biomass can remove heavy metals from the environment in two ways. On the one hand, live algae can be used. It has been discovered that heavy metal consumption occurs in two stages. The first is "physical adsorption," also known as "biosorption," which is a quick and reversible response on the cell surface. The second phase is the absorption of heavy metals into cells. It is referred to as chemisorption or bioaccumulation and is a more time-consuming process (Aksu & Kutsal, 1990). Biosorbents are beneficial when the biomass is inexpensive, simple to cultivate or harvest, readily available, and of similar quality. Metals and related elements, such as actinides, lanthanides, metalloids, and various radioisotopes of these compounds, have been the focus of most biosorption research. Particulates and colloids, as well as organometal(loid) and organic chemicals, such as dyes, have all been studied (Aksu, 2005). Overall, algae are the most commonly employed biosorbent compared with other biosorbents because of their wide availability, low processing costs, and excellent performance (Al-Homaidan et al., 2014).

1.1 Bioaccumulation and its mechanisms

Bioaccumulation defines the overall progress in accumulating an anthropogenic chemical uptake by organisms from their surroundings and is successfully assimilated and disseminated into the tissues of a living system. Bioaccumulation of heavy metals is reported in green algae *Cladophora and Spirulina* sp. (Doshi et al., 2008). The mechanism of heavy metal uptake is aided by phosphate and peptide moieties. Increased concern on bioaccumulation is because the accumulated anthropogenic chemical appears in the food chain in a quantitatively magnified ratio and is termed biomagnification (Kurade et al., 2019). However, the targeted use of algae in bioremediation is conducted in controlled conditions where the algae are treated appropriately, and care is taken to prevent the entry of the toxicant into the food chain. The uptake rate of xenobiotics has to surpass the rate of elimination of the same (Wang et al., 2019). Bioaccumulation and incorporation of a xenobiotic extensively depend on the route of accumulation through passive diffusion, special transport, and adsorption. The other factors known to influence the accumulation rate are xenobiotics' availability and the lipophilic nature of the contaminant (Blanco et al., 2019). Process of bioaccumulation of xenobiotics directly absorbed in case of algae, fish, and mollusks through routes such as passive diffusion or adsorption. In water as a medium, the cases of passive diffusion are prevalent. This process is independent of energy and functions against the concentration gradient without energy expenditure. The lipophilicity of the xenobiotic enhances the rate of bioaccumulation of xenobiotics. Yet another mechanism of intake of xenobiotics has facilitated transport with energy expenditure. Another most common surface phenomenon of uptake of xenobiotics in the aquatic medium is adsorption (Zhu et al., 2020). Direct adsorption involves the forces such as covalent and electrostatic forces. Adsorption, in particular, serves as the first step toward bioaccumulation. Species-specific bioaccumulation of heavy metals is reported in sea algae (Alamin et al., 2022). The difference in preference for bioaccumulation was observed in red, green, and brown algae. At the same time, a few algae such as *Padina crassa, Sargassum thunbergii, Caulerpa racemosa*, and *Asparagopsis taxiformis* showed comparable heavy metal accumulating behavior. Thus, selective use of algal candidates for detoxification and remediation is the potential green technique of bioremediation that is economical and eco-friendly.

1.2 Biosorption and its mechanisms

Biosorption has been used to remove potentially hazardous components via various techniques. Biosorption was formerly classified as either metabolism-dependent or metabolism-independent (Veglio' & Beolchini, 1997). Bioaccumulation was later coined for metabolism-dependent activities, while biosorption was coined for metabolic-independent processes (Chojnacka, 2010; Volesky, 2007). Metal biosorption involves various mechanisms that vary qualitatively and quantitatively depending on the species used, the source of the biomass, and the method used to prepare it (Volesky & Holan, 1995). Biosorption is a passive process that occurs at a faster rate than bioaccumulation. It entails several mechanisms, including adsorption, chelation/complexation, ion exchange, and surface precipitation (Fig. 25.1) (Abd El Hameed et al., 2015; Farooq et al., 2010).

Among them, ion exchange is considered the most important biosorption mechanism, as it occurs through diverse functional groups on the biomass surface (Vijayaraghavan & Raja, 2015). Adsorption mediated by van der Waals forces is known as physical adsorption. Copper, cadmium, zinc, uranium, and cobalt sorption by dead algal cells/biomasses is said to occur due to electrostatic interactions between metal ions in solutions and microbial cell walls (Kuyucak & Volesky, 1988). Copper biosorption by *Chlorella vulgaris* occurs with the interactions of electrostatic forces (Aksu & Kutsal, 1991). Algae have a high biosorption ability due to the nature of their cell wall, which is made up of chitin, polysaccharides, proteins, and lipids, all of which include important functional groups that aid in biosorption (Davis et al., 2003). The key mechanism for removing metal from a solution is complex formation. Complexes are formed between the active groups of the cell surface and metal ions during this removal process (Treen-Sears et al., 1984). Copper biosorption by *C. vulgaris* and *Zoogloea*

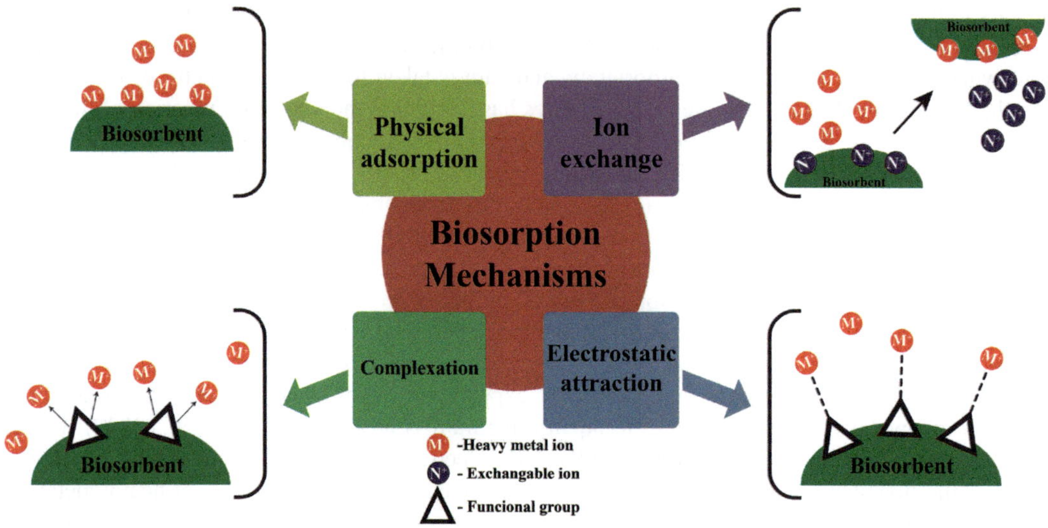

FIGURE 25.1

A schematic representation of the mechanisms involved in the biosorption of potentially hazardous substances.

ramigera is claimed to occur via adsorption and the creation of coordination interactions between metals and amino and carboxyl groups of cell wall polysaccharides (Viraraghavan, 1995). *Pseudomonas syringae* accumulates calcium, magnesium, cadmium, zinc, copper, and mercury by a complexation mechanism. The poisonous metal chelates with the organic molecules to generate metalloorganic compounds. Microorganisms produce an organic acid that functions as a chelating agent (e.g., citric, oxalic, gluonic, fumaric, lactic, and malic acids) (Han et al., 2006). The organic acid also aids in the solubilization and leaching of metal complexes from their surfaces. Metals in microbial polysaccharides and other polymers can be biosorbed or complexed by carboxyl groups (Wang & Chen, 2009). The thermodynamic adsorption investigations of Acid Blue 161 (AB-161) on microalgal biomass by Da Fontoura et al. (2017) stated that the primary mechanism is an electrostatic interaction with a spontaneous endothermal process. For marine algae, the active molecular entities involved are believed to be carboxyl and sulfate groups (Davis et al., 2003). The carboxyl groups of alginic acid and the carboxyl and sulfate groups of fucoidan in marine algae are involved in ion exchange. Biosorbents cells contain polysaccharides that can be exchanged with bivalent metal ions (Treen-Sears et al., 1984). The biosorption mechanisms are same for both heavy metals and organic pollutants, but unlike heavy metals, organic pollutants can be degraded. Once the pollutants get sorbed onto the biosorbent, there is a high possibility that they will react with functional group present on the biosorbent surface and change their chemical structure. Many scientists have seen that oxygen present in the biosorbent plays an important role in neutralizing organic pollutants. Increasing the oxygen containing functional groups by chemical modification helps to degrade the organic pollutants efficiently. Furthermore, hydrolysis of lignin present in algal cells is also an example of biosorbent modification that will improve biosorption. These modifications are done by chemicals such as $ZnCl_2$, H_3PO_4, polymerization, and so on (Bouzikri et al., 2020; Coelho et al., 2020; Lebron et al., 2019; Mokhtar et al., 2017).

The biosorption process is commonly determined by the biomass employed to remove potentially harmful components (Chojnacka, 2010). For instance, the cell walls of bacteria (peptidoglycan), fungi (chitin), and algae (alginate, sulfonated polysaccharides) differ in composition; thus, variations in the presence of functional groups on the cell wall surface are responsible for the differences in mechanisms (Farooq et al., 2010).

Ion exchange was reported by Verma et al. (2008) as a key process for the biosorption of potentially hazardous components by macrophyte dry biomass. Aquatic macrophytes were thought to act as natural ion exchangers. Because monovalent ions (H^+, Na^+/K^+) have a weak attachment to biomass compared with divalent ions, they have been reported to be engaged in the ion exchange process (Verma et al., 2008). The chemical groups acetamido, amino, amido, sulfhydryl, sulfate, and carboxyl can attract and sequester metals in biomass (Gardea-Torresdey et al., 1990; Schiewer & Volesky, 2014, pp. 329–362; Volesky & Holan, 1995). Metal sequestration during biosorption is mediated by several mechanisms, the most important of which are ionic interactions and the formation of complexes between metal cations and ligands found in the cell wall biopolymers' structure as precipitation on the cell wall matrix (Crist et al., 1994). Ahmad et al. (2018) recently reported that sulfate, carboxyl, and hydroxyl groups are involved in microalgae biosorption. The facilitation of metal ion sorption by pores existing on the surface was also demonstrated by changes in morphological structure seen by scanning electron microscopy (SEM) and/or energy-dispersive X-ray (EDX) after biosorption (Ahmad et al., 2018).

2. Factors affecting biosorption

It is observed that the sorption of heavy metal ions onto the algal biomass's surface depends on many factors such as pH, the temperature of the medium, contact time between biosorbent and the metal solution, initial metal concentration, biosorbent dosage, and presence of other heavy metals (Fig. 25.2). The increase and decrease in these parameters have a remarkable effect on adsorption. The effects of each of these factors are explained in brief in the following.

3. Effect of pH

Sorption uptake of heavy metals is deficient at low pH, and it increases with an increase in pH up to a specific pH value. The phenomenon is observed because of the competition of H^+ ions with positively charged heavy metal ions for binding sites. As pH increases, the concentration of H+ ions will decrease; thus, there is less competition for binding sites. After a particular pH, heavy metals' adsorption starts to decrease again due to the precipitation of insoluble metal hydroxides (i.e., there will be a change in speciation of heavy metals as the pH of the system changes) (Bulgariu & Bulgariu, 2012; Gupta & Rastogi, 2008; Romera et al., 2007; Tavana et al., 2020). The optimum pH value for heavy metal intake varies with each heavy metal and also depends on the biomass taken (Romera et al., 2007; Dönmez et al., 1999).

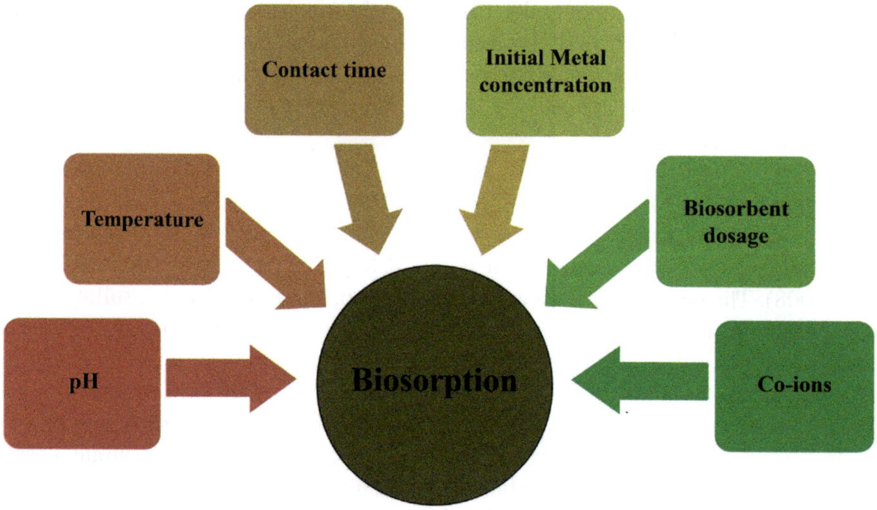

FIGURE 25.2

Factors affecting the biosorption efficiency.

4. Effect of temperature

Heavy metals' adsorption normally decreases with increased temperature, indicating that the reaction is exothermic (Aksu, 2001; Bulgariu & Bulgariu, 2012).

5. Effect of contact time

More than 90% of adsorption occurs within a few minutes of contact time, and then equilibrium reaches slowly, indicating that there is more than one step in heavy metal uptake: initial stage where rapid adsorption takes place on the cell wall surfaces of algal biomass, followed by slow intake of heavy metals by cells (Bulgariu & Bulgariu, 2012; Vilar et al., 2006).

6. Effect of initial metal concentration

The initial metal concentration provides a driving force that will result in overcoming the mass transfer resistance between biosorbents (solid phase) and metal solution (liquid phase) (Vilar et al., 2006). An increase in initial metal concentration increases the loading capacity of biomasses because of the increased probability of collision of metal ions with the active sites (Dönmez et al., 1999). But at the same time, increase in the initial metal concentration decreases the removal percentage. As metal concentration increases, the number of saturated active sites increases, resulting in the unavailability of binding sites for metal ions at higher concentrations (Tavana et al., 2020).

7. Effect of biosorbent dosage

An increase in biosorbent dosage increases heavy metal removal to a threshold level, after which the adsorption decreases. With an increase in biosorbent dosage, the active site also increases; hence, there will be an increase in adsorption capacity, but as biosorbent dosage crosses a specific limit, the active sites are blocked due to aggregation of biosorbent particles. If the active sites are blocked, there will be fewer sites available for metal ion adsorption, and hence, there will be a decrease in adsorption capacity (Gupta & Rastogi, 2008; Dönmez et al., 1999).

8. Effect of competing ions

The presence of other heavy metal ions also affects the adsorption capacity of biosorbents in the same way as H^+ ions, i.e., they compete for the active sites on the biosorbent surfaces. In a system containing more than one type of metal ions, the metal ion with a relatively high initial concentration has more chance of coming in contact with the active sites and hence an increased chance of adsorption compared with metal ions with relatively lower initial concentration (Aksu & Dönmez, 2006).

9. Microalgal bioremediation

Contamination diffusion can be caused by diverse activities of agriculture and livestock polluting groundwater and surface water in rural areas. Microalgae are appropriate for pollutant elimination due to their biosorption ability. The carbohydrate of microalgal cell wall aids in the biosorption of toxic impurities. Biosorption in microalgae is used for the removal of pesticide especially atrazine, carbofuran, dimethoate, and simazine by living *C. vulgaris* (García-Galán et al., 2020). During bioremediation of contaminants, microalgae are known to release oxygen as metabolite during effluent treatment, and this is consumed by aerobic bacteria to promote degradation of the residual organic loads. It is also observed that the high concentration of suspended chlorides, nitrates, manganese, sodium, lead, copper, and chromium can be utilized to promote their growth and transform them into value-added secondary metabolite. Green microalga, *C. vulgaris,* has demonstrated the ability to utilize the contaminant and produces an appreciable amount of biodiesel (Fazal et al., 2021). Apart from metals, nitrate contributes to the most common contaminant available in ground water. Petrovič & Simonič (2015) studies state that the application of nitrate-contaminated groundwater-based media with a naturally supplemented minerals and satisfactory addition of phosphate can facilitate the growth of microalgae in multiapproach applications. The application of the commonly used microalgae with different properties including *C. vulgaris*, *Chlamydomonas reinhardtii*, and *Ettlia* sp. in water and wastewater treatment can increase the likelihood of an effective nitrate bioremediation. *Ettlia* sp. exhibited the maximum nutrient assimilation and growth rate between the three microalgae during bioremediation (Rezvani et al., 2018). Microalgal biofilm is considered as the novel and efficient type of phycoremediation method that can be used to effectively treat water from the river. The successful application of microalgal biofilms has structural stability comprising of rough surface (Ugya et al., 2021). The rigid morphology of algal biofilms has its applicability extended for anaerobically digested food waste centrate (Sutherland & Bramucci, 2022). Yet another microalga, *Scenedesmus obliquus*, has proved itself to be a self-sustainable, economical and environmental friendly candidate for treating poultry wastewater. *S. obliquus* has achieved progressive results when grown in highly polluted agroindustrial effluent originating from the poultry industry. Microalgae were observed to be a potential bioremediator in case of both ammonium and phosphate (Oliveira et al., 2019). *Galdieria* is a polyextremophilic red microalga belonging to the class Cyanidiophyceae. *Galdieria* thrives in geothermal volcanic soils with temperatures of 50°C and extraordinary sulfuric concentrations. Galdieria possesses extraordinary metabolic versatility. *Chlorella pyrenoidosa* was used to estimate the removal efficiency of artificial tobacco wastewater. Tobacco wastewater is too difficult to purify, posing a significant environmental problem due to the harmful and toxic components. Acidic pH 5.0 was optimal for *C. pyrenoidosa* to grow and remediate toxic pollutants (Hao et al., 2022). The use of microalgae in phycoremediation is mentioned in Table 25.1.

Table 25.1 Bioremediation ability of microalgae under varying conditions.

Name of the algae	Pollutant	Optimum pH	Contact time	Biosorption capacity	Reference
C. vulgaris	Nitrate phosphate	7	–	31.26 mg/g 11.07 mg/g 9.12 mg/g	Fazal et al. (2021)
Chlorella vulgaris *Chlamydomonas reinhardtii*	Nitrate	6.7 ± 0.3	3 days	0.21 g/L/d 16.6 g/L/d 3.06 mg/L/d	Fazal et al. (2021)
Scenedesmus obliquus	Poultry wastewater	6	24 h	11.07 mg/g 9.12 mg/L	Oliveira et al. (2019)
Chlorella sp., *Scenedesmus* sp., *and Muriellopsis* sp.	Chloride	7.86	55 days	0.5 g/L	Ramírez et al. (2017)
C. vulgaris	Crude petroleum oil	6.8	5 days	222 mg/L	Kuttiyathil et al. (2021)
Scenedesmus vacuolatus, Chlorella, Skeletonema costatum, Microcystis flos-aquae	Pharmaceuticals and personal care products	4.0	5 days	18 mg/L 14 mg/L 5 mg/L 0.05 mg/L	Miazek and Brozek-Pluska (2019)
Scenedesmus obliquus, Chlorella sorokiniana	Nitrogen, phosphorus, and other trace elements	8.6 7.5	24 h	88 mg/L 54.5 mg/L	Al-Jabri et al. (2021)
Oedogonium	Al, Cu, Fe, Mn, and Zn	6.8	16 days	46.8 ± 9.45 gL^{-1}, 253.66 ± 40.62 gL^{-1}, and 355.5 ± 50.69 gL^{-1}	Aslam et al. (2019)
Chlorella sp., *Rhodomonas* sp., and *Thalassiosira* sp.	Phenolic compounds	8 4 9	24 h	2000 mg/dL 150 mg/dL 50 mg/dL	Surkatti and Al-Zuhair (2018)
Chlorella sp. and *Nannochloropsis* sp.	Leachate	5–7	12 days	5% 10% 15%	Shaari et al. (2021)

10. Macroalgal bioremediation

Due to anthropogenic activity, heavy metals, organic pollutants, and dyes are released into water bodies. Their transport and bioavailability have a massive effect on the functioning of ecosystems (Ekanayake et al., 2021; Gomes & Asaeda, 2013; Karaś et al., 2021). One of the main promising techniques for removing these pollutants from the environment is via phycoremediation, i.e., by using macro- or microalgae. It is a promising agent for the remediation of a wide range of contaminants, including hazardous metal ions, herbicides, insecticides, radionuclides, oil spills, and petroleum chemicals. The use of macroalgae as an effective phytoremediation agent has increased as it has greater potential to detoxify pollutants due to its size, as it has a bigger surface area and may give a wider biosorption area for hazardous chemicals (Shackira et al., 2021, pp. 273–293). Extensive literature is available on the bioremediation of heavy metals, dyes, or organic pollutants using macroalgae.

A comparative study conducted by Salem et al. (2021) on two different macroalgae *Colpomenia sinuosa* and *Ulva fasciata* reports that *C. sinuosa* has the highest biosorption capacity of cobalt from the aqueous solution (62.5 mg/g) when compared with *U. fasciata* (40.0 mg/g). The contact time was 60min, and the optimum pH for biosorption was 7.0 and 6.0 for *C. sinuosa* and *U. fasciata,* respectively (Salem et al., 2021). Similarly, a comparative study was conducted by Bonanno et al. (2020) on seagrass *Cymodocea nodosa* and green algae *Ulva lactuca* sequestering trace elements such as cadmium and chromium, copper, nickel, lead and zinc in seawater and sediments of the benthic region. *C. nodosa* has higher bioaccumulating capacity with constant temperature, salinity, and pH (Bonanno et al., 2020). The combination of *Tribonema vulgare*, *Tribonema microchloron*, and *Trichoderma viride* has higher chromium biosorption capacity, i.e., around 16,230 mg/kg from the aquatic reservoir by the reduction mechanism (Ociński et al., 2021). *Ulva ohnoi* is reported to have phytoremediation potential as it can uptake cadmium in seawater and brackish water due to its high nutrient absorption rate and high growth rates (Bastos et al., 2019). The biochar synthesized from *U. lactuca* has the potential ability to absorb the organic pollutant Remazol dyes from the wastewater at acidic pH (Gokulan et al., 2019). Similarly, the biochar synthesized from *Ulva reticulata* can absorb arsenic from contaminated waters (Senthilkumar et al., 2020). *Chara vulgaris* can remediate the textile effluents from water bodies by decreasing the BOD, COD, pH, EC, and TDS of the 10%–50% concentrated textile effluent within 120 h of contact time. Furthermore, it has increased the dissolved oxygen level of the water (Mahajan et al., 2019). *C. vulgaris* can also remediate the methyl red dye by checking the effect of different parameters such as contact time, pH, temperature, and concentration. The results suggest that the maximum absorption was achieved at 48 hours contact time of macroalgae with dye and the favorable pH was 5.0. The mechanism involved is via the formation of electrostatic and hydrogen bonds formed among functional algae (Mahajan & Kaushal, 2020). Brown algae's *Cystoseira barbata* and *Cystoseira crinita* can potentially remove chromium (Cr III and IV) from the wastewater by reducing at pH of 4.5 (Yalçın & Özyürek, 2018). The living *U. lactuca* has a high biosorption capacity for trace elements arsenic, cadmium, lead, copper, chromium, mercury, manganese, and nickel from the wastewater (Henriques et al., 2019). The calcification of *Chara braunii* algae, when immobilized with manganese, has enhanced the remediation of arsenic from the water system by 51.7% (Amirnia et al., 2019). A comparative study using *Chaetomorpha linum* and *Chaetomorpha prolifera* demonstrates that the former has high potential in remediation of aquaculture waste water via a two-step system by removing ammonium, nitrate, and phosphate step by step in 48 hours (Aquilino et al., 2020). The release of endocrine-disrupting chemicals (EDCs) such as 17a-ethinylestradiol (EE2) and bisphenol-A (BPA) to water bodies has caused harm to aquatic organisms. However, the red seaweed *Gracilaria* species has the potential to remediate these toxicants by the process of bromination (Astrahan et al., 2021). Similarly, intertidal macroalgae could remediate the phenolic EDCs such as BPA and nonylphenol (NP) from the coastal water systems (Zhang et al., 2021). *Gracilaria* can remove methylene blue and nickel from wastewater at pH 8.0 and has the highest percentage removal efficiency of nickel and methylene blue 98.90% and 97.31%, respectively (El-Naggar & Rabei, 2020). Correspondingly, the use of *Sargassum latifolium* could remediate methylene blue from the industrial wastewater effluent by the mechanism of physisorption (Mansour et al., 2022).

The bioproducts of *Fucus spiralis* (polyphenols, fucoidan, and alginate), brown algae, are reported to have a biosorbent ability by taking the lead from the contaminated water by ion-exchange mechanism (Filote et al., 2019). A study by Libatique et al. (2020) reports that *Sarcodia suiae,* red macroalgae, can absorb total arsenic and inorganic arsenic accumulation from the wastewater, stating that temperature affects the absorption. In contrast, absorption was independent of light (Libatique et al., 2020). The cultivated *Gracilaria* showed a good biosorbent activity as it could remediate the heavy metals from the coastal marine sediment, improving the seawater quality (Luo et al., 2020). Freshwater macroalgae also can bioremediate the heavy metals or any other kind of pollutants. In his study, Roberts et al. (2018) focused on *Oedogonium* algae in remediation of the aluminum present in wastewater. A comparative study among green, red, and brown macroalgae reports that green algae have a higher biosorption capacity of iron, nickel, and lead (Sun et al., 2019). These studies collectively indicate that macroalgae have excellent potential in bioaccumulating heavy metals, organic pollutants, etc. The mechanism behind the biosorption varies among algae, and the biosorption is affected by varying factors such as contact time, pH, and temperature. It can be used as an alternative material in adsorbing harmful pollutants without affecting the aquatic ecosystem. The use of macroalgae in phycoremediation is described in Table 25.2.

Table 25.2 Bioremediation capacity of macroalgae used at optimal conditions.

Name of the algae	Pollutant	Optimum pH	Contact time	Biosorption capacity	Reference
Colpomenia sinuosa *Ulva fasciata*	Cobalt	7 6	60 min	62.5 mg/g 40.0 mg/g	Salem et al. (2021)
Cymodocea nodosa *Ulva lactuca*	Cadmium Chromium Copper NickelLead Zinc	7.90–8.25	–	2.12 mg/kg 7.70 mg/kg 19.5 mg/kg 22.5 mg/kg 6.10 mg/kg 13.9 mg/kg	Bonanno et al. (2020)
	Cadmium Chromium Copper NickelLead Zinc	7.90–8.25	–	0.35 mg/kg 2.80 mg/kg 8.10 mg/kg 10.1 mg/kg 4.50 mg/kg 44.3 mg/kg	

Table 25.2 Bioremediation capacity of macroalgae used at optimal conditions.—cont'd

Name of the algae	Pollutant	Optimum pH	Contact time	Biosorption capacity	Reference
Macroalgal community *Tribonema vulgare*, *Tribonema microchloron*, and *Trichoderma viride*	Chromium	8.0	24 hours	16230 mg/kg	Ociński et al. (2021)
Ulva ohnoi	Cadmium	8.7	8 days	4.199 ± 0.835 μg/g	Bastos et al. (2019)
Ulva lactuca	Remazol dyes RBO3R RBBR, RBV5R, and RBB	2.0	60 min	0.301, 0.292, 0.265, and 0.224 mmol/g	Gokulan et al. (2019)
Ulva reticulata	Arsenic	4.0	13 hours	8.12 mg/g	Senthilkumar et al. (2020)
Chara vulgaris	Textile effluent	7.27	120 hours	-	Mahajan et al. (2019)
Chara vulgaris	Methyl red dye	5.0	48 hours	20 mg/L	Mahajan and Kaushal (2020)
Cystoseira barbata, *Cystoseira crinita*	Chromium Cr III Cr IV	4.5	24 hours	100.0% 80.3% 100.0% 76.5%	Yalçın and Özyürek (2018)
Ulva lactuca	Trace elements Arsenic Cadmium Lead Copper Chromium Mercury Manganese Nickel	7.8	72 hours	1−10 g/L	Henriques et al. (2019)
Chara braunii	Arsenic	8.0	24 hours	51.7%	Amirnia et al. (2019)
Chaetomorpha linum *Chaetomorpha prolifera*	Ammonium Nitrate Phosphate	–	48 hours	198 μg/g 28 μg/g 37 μg/g 48 μg/g 14 μg/g 20 μg/g	Aquilino et al. (2020)
Gracilaria species	17a-ethinylestradiol Bisphenol-A	–	48 hours	–	Astrahan et al. (2021)
Gracilaria	Nickel methylene blue	8.0	180 min	98.90% 97.31%	El-Naggar and Rabei (2020)

Continued

Table 25.2 Bioremediation capacity of macroalgae used at optimal conditions.—cont'd

Name of the algae	Pollutant	Optimum pH	Contact time	Biosorption capacity	Reference
Sargassum latifolium	Methylene blue	6.0	15 min	94.88%	Mansour et al. (2022)
Fucus spiralis	Lead	4.5	6 hours	132 mg/g	Filote et al. (2019)
Sarcodia suiae	Arsenic	–	7 days	250–500 μg/L	Libatique et al. (2020)
Oedogonium	Aluminum	<2	8 weeks	467 mg/kg	Roberts et al. (2018)

11. Conclusion

Rapid urbanization, progressive agricultural practices, and anthropogenic developmental activities have poised the mammoth challenge of treating municipal, industrial, and agricultural waste that contains organic matter, heavy metals, and other toxic chemicals. Untreated waste can further the damages caused to an already fragile public health and natural ecosystems. Use of micro- and macroalgae in bioremediation of polluted sites is referred to as phycoremediation. It is an emerging area of research that has several advantages alongside few disadvantages. Limited knowledge in its application in bioremediation, mechanisms, information on species specificity, and long-term toxicity studies are few of its disadvantages. The ease of generation of biomass, fast growth, and upscaling cultures in controlled conditions are its advantages. Most importantly, the innate natural ability of the algae to accumulate and degrade toxicants is suited for its application in bioremediation. It is more acceptable as a green option with no intervention of transgenic or gene editing technology, thereby making it an affordable, acceptable, and sustainable technology for treating contaminated wastewater.

References

Abd El Hameed, A. H., Eweda, W. E., Abou-Taleb, K. A., & Mira, H. I. (2015). Biosorption of uranium and heavy metals using some local fungi isolated from phosphatic fertilizers. *Annals of Agricultural Science, 60*(2), 345–351. https://doi.org/10.1016/j.aoas.2015.10.003

Ahmad, A., Bhat, A. H., & Buang, A. (2018). Biosorption of transition metals by freely suspended and Ca-alginate immobilised with Chlorella vulgaris: Kinetic and equilibrium modeling. *Journal of Cleaner Production, 171*, 1361–1375. https://doi.org/10.1016/J.JCLEPRO.2017.09.252

Akbal, F., & Camcidotless, S. (2011). Copper, chromium and nickel removal from metal plating wastewater by electrocoagulation. *Desalination, 269*(1–3), 214–222. https://doi.org/10.1016/j.desal.2010.11.001

Aksu, Z. (2001). Equilibrium and kinetic modelling of cadmium(II) biosorption by C. vulgaris in a batch system: Effect of temperature. *Separation and Purification Technology, 21*(3), 285–294. https://doi.org/10.1016/S1383-5866(00)00212-4

Aksu, Z. (2005). Application of biosorption for the removal of organic pollutants: A review. *Process Biochemistry, 40*(3–4), 997–1026. https://doi.org/10.1016/J.PROCBIO.2004.04.008

Aksu, Z., & Dönmez, G. (2006). Binary biosorption of cadmium(II) and nickel(II) onto dried Chlorella vulgaris: Co-Ion effect on mono-component isotherm parameters. *Process Biochemistry, 41*(4), 860–868. https://doi.org/10.1016/J.PROCBIO.2005.10.025

Aksu, Z., & Kutsal, T. (1990). A comparative study for biosorption characteristics of heavy metal ions with C. vulgaris. *Environmental Technology, 11*(10), 979–987. https://doi.org/10.1080/09593339009384950

Aksu, Z., & Kutsal, T. (1991). A bioseparation process for removing lead(II) ions from waste water by using C. vulgaris. *Journal of Chemical Technology & Biotechnology, 52*(1), 109–118. https://doi.org/10.1002/jctb.280520108

Al-Homaidan, A. A., Al-Houri, H. J., Al-Hazzani, A. A., Elgaaly, G., & Moubayed, N. M. (2014). Biosorption of copper ions from aqueous solutions by Spirulina platensis biomass. *Arabian Journal of Chemistry, 7*(1), 57–62. https://doi.org/10.1016/j.arabjc.2013.05.022

Al-Jabri, H., Das, P., Khan, S., Thaher, M., & AbdulQuadir, M. (2021). Treatment of wastewaters by microalgae and the potential applications of the produced biomass—a review. *Water (Switzerland), 13*(1), 27. https://doi.org/10.3390/w13010027

Alamin, M., Tsuji, S., Hata, A., Hara-Yamamura, H., & Honda, R. (2022). Selection of surrogate viruses for process control in detection of SARS-CoV-2 in wastewater. *Science of the Total Environment, 823*, 153737.

Amirnia, S., Asaeda, T., Takeuchi, C., & Kaneko, Y. (2019). Manganese-mediated immobilization of arsenic by calcifying macro-algae, Chara braunii. *Science of the Total Environment, 646*, 661–669. https://doi.org/10.1016/J.SCITOTENV.2018.07.275

Aquilino, F., Paradiso, A., Trani, R., Longo, C., Pierri, C., Corriero, G., & de Pinto, M. C. (2020). Chaetomorpha linum in the bioremediation of aquaculture wastewater: Optimization of nutrient removal efficiency at the laboratory scale. *Aquaculture, 523*, 735133. https://doi.org/10.1016/J.AQUACULTURE.2020.735133

Ashraf, A., Bibi, I., Niazi, N. K., Ok, Y. S., Murtaza, G., Shahid, M., Kunhikrishnan, A., Li, D., & Mahmood, T. (2017). Chromium (VI) sorption efficiency of acid-activated banana peel over organo-montmorillonite in aqueous solutions. *International Journal of Phytoremediation, 19*(7), 605–613. https://doi.org/10.1080/15226514.2016.1256372

Aslam, A., Thomas-Hall, S. R., Mughal, T., Zaman, Q. U., Ehsan, N., Javied, S., & Schenk, P. M. (2019). Heavy metal bioremediation of coal-fired flue gas using microalgae under different CO2 concentrations. *Journal of Environmental Management, 241*, 243–250. https://doi.org/10.1016/J.JENVMAN.2019.03.118

Astrahan, P., Korzen, L., Khanin, M., Sharoni, Y., & Israel, Á. (2021). Seaweeds fast EDC bioremediation: Supporting evidence of EE2 and BPA degradation by the red seaweed Gracilaria sp., and a proposed model for the remedy of marine-borne phenol pollutants. *Environmental Pollution, 278*, 116853. https://doi.org/10.1016/J.ENVPOL.2021.116853

Bastos, E., Schneider, M., de Quadros, D. P. C., Welz, B., Batista, M. B., Horta, P. A., Rörig, L. R., & Barufi, J. B. (2019). Phytoremediation potential of *Ulva ohnoi* (Chlorophyta): Influence of temperature and salinity on the uptake efficiency and toxicity of cadmium. *Ecotoxicology and Environmental Safety, 174*, 334–343. https://doi.org/10.1016/J.ECOENV.2019.01.130

Birungi, Z. S., & Chirwa, E. M. N. (2015). The adsorption potential and recovery of thallium using green micro-algae from eutrophic water sources. *Journal of Hazardous Materials, 299*, 67–77. https://doi.org/10.1016/j.jhazmat.2015.06.011

Blanco, J., Martín, H., Mariño, C., & Rossignoli, A. E. (2019). Simple diffusion as the mechanism of okadaic acid uptake by the mussel digestive gland. *Toxins, 11*(7), 395. https://doi.org/10.3390/toxins11070395

Bonanno, G., Veneziano, V., Raccuia, S. A., & Orlando-Bonaca, M. (2020). Seagrass *Cymodocea nodosa* and seaweed *Ulva lactuca* as tools for trace element biomonitoring. A comparative study. *Marine Pollution Bulletin, 161*, 111743. https://doi.org/10.1016/J.MARPOLBUL.2020.111743

Bouzikri, S., Ouasfi, N., Benzidia, N., Salhi, A., Bakkas, S., & Khamliche, L. (2020). Marine alga "*Bifurcaria bifurcata*": Biosorption of reactive blue 19 and methylene blue from aqueous solutions. *Environmental Science and Pollution Research, 27*(27), 33636–33648. https://doi.org/10.1007/S11356-020-07846-W

Brooks, S. S., Palmer, M. A., Cardinale, B. J., Swan, C. M., & Ribblett, S. (2002). Assessing stream ecosystem rehabilitation: Limitations of community structure data. *Restoration Ecology, 10*(1), 156–168. https://doi.org/10.1046/j.1526-100X.2002.10117.x

Bulgariu, D., & Bulgariu, L. (2012). Equilibrium and kinetics studies of heavy metal ions biosorption on green algae waste biomass. *Bioresource Technology, 103*(1), 489–493. https://doi.org/10.1016/J.BIORTECH.2011.10.016

Chen, H., & Liu, W. (1996). Cellulose-based photocatalytic paper with Ag2O nanoparticles loaded on graphite fibers. *Journal of Bioresources and Bioproducts, 1*(4), 192–198. https://doi.org/10.21967/jbb.v1i4.63

Chojnacka, K. (2010). Biosorption and bioaccumulation - the prospects for practical applications. *Environment International, 36*(3), 299–307. https://doi.org/10.1016/j.envint.2009.12.001

Cochrane, E. L., Lu, S., Gibb, S. W., & Villaescusa, I. (2006). A comparison of low-cost biosorbents and commercial sorbents for the removal of copper from aqueous media. *Journal of Hazardous Materials, 137*(1), 198–206. https://doi.org/10.1016/j.jhazmat.2006.01.054

Coelho, C. M., de Andrade, J. R., da Silva, M. G. C., & Vieira, M. G. A. (2020). Removal of propranolol hydrochloride by batch biosorption using remaining biomass of alginate extraction from *Sargassum filipendula* algae. *Environmental Science and Pollution Research, 27*(14), 16599–16611. https://doi.org/10.1007/S11356-020-08109-4

Crini, G., & Badot, P. M. (2008). Application of chitosan, a natural aminopolysaccharide, for dye removal from aqueous solutions by adsorption processes using batch studies: A review of recent literature. *Progress in Polymer Science, 33*(4), 399–447. https://doi.org/10.1016/J.PROGPOLYMSCI.2007.11.001

Crist, R. H., Martin, J. R., Carr, D., Watson, J. R., Clarke, H. J., & Carr, D. (1994). Interaction of metals and protons with algae. 4. Ion exchange vs adsorption models and a reassessment of scatchard plots; ion-exchange rates and equilibria compared with calcium alginate. *Environmental Science and Technology, 28*(11), 1859–1866. https://doi.org/10.1021/es00060a016

Da Fontoura, J. T., Rolim, G. S., Mella, B., Farenzena, M., & Gutterres, M. (2017). Defatted microalgal biomass as biosorbent for the removal of Acid Blue 161 dye from tannery effluent. *Journal of Environmental Chemical Engineering, 5*(5), 5076–5084. https://doi.org/10.1016/J.JECE.2017.09.051

Davis, T. A., Volesky, B., & Mucci, A. (2003). A review of the biochemistry of heavy metal biosorption by brown algae. *Water Research*, 37, 4311–4330.

Dharnaik, A. S., & Ghosh, P. K. (2014). Hexavalent chromium [Cr(VI)] removal by the electrochemical ion-exchange process. *Environmental Technology, 35*(18), 2272–2279. https://doi.org/10.1080/09593330.2014.902108

Dönmez, G.Ç., Aksu, Z., Öztürk, A., & Kutsal, T. (1999). A comparative study on heavy metal biosorption characteristics of some algae. *Process Biochemistry, 34*(9), 885–892. https://doi.org/10.1016/S0032-9592(99)00005-9

Doshi, H., Seth, C., Ray, A., & Kothari, I. L. (2008). Bioaccumulation of heavy metals by green algae. *Current Microbiology, 56*(3), 246–255. https://doi.org/10.1007/s00284-007-9070-z

Ekanayake, M. S., Udayanga, D., Wijesekara, I., & Manage, P. (2021). Phytoremediation of synthetic textile dyes: Biosorption and enzymatic degradation involved in efficient dye decolorization by *Eichhornia crassipes* (mart.) solms and *Pistia stratiotes* L. *Environmental Science and Pollution Research International, 28*(16), 20476–20486. https://doi.org/10.1007/S11356-020-11699-8

El-Kassas, H. Y., & Mohamed, L. A. (2014). Bioremediation of the textile waste effluent by Chlorella vulgaris. *Egyptian Journal of Aquatic Research, 40*(3), 301–308. https://doi.org/10.1016/j.ejar.2014.08.003

El-Naggar, N. E. A., & Rabei, N. H. (2020). Bioprocessing optimization for efficient simultaneous removal of methylene blue and nickel by Gracilaria seaweed biomass. *Scientific Reports, 10*(1), 1–21. https://doi.org/10.1038/s41598-020-74389-y

Ercin, A. E., & Hoekstra, A. Y. (2014). Water footprint scenarios for 2050: A global analysis. *Environment International, 64*, 71–82. https://doi.org/10.1016/j.envint.2013.11.019

Farooq, U., Kozinski, J. A., Khan, M. A., & Athar, M. (2010). Biosorption of heavy metal ions using wheat based biosorbents—a review of the recent literature. *Bioresource Technology, 101*(14), 5043–5053. https://doi.org/10.1016/j.biortech.2010.02.030

Fazal, T., Rehman, M. S. U., Javed, F., Akhtar, M., Mushtaq, A., Hafeez, A., Din, A. A., Iqbal, J., Rashid, N., & Rehman, F. (2021). Integrating bioremediation of textile wastewater with biodiesel production using microalgae (*Chlorella vulgaris*). *Chemosphere, 281*, 130758. https://doi.org/10.1016/J.CHEMOSPHERE.2021.130758

Filote, C., Volf, I., Santos, S. C., & Botelho, C. M. (2019). Bioadsorptive removal of Pb(II) from aqueous solution by the biorefinery waste of *Fucus spiralis*. *Science of the Total Environment, 648*, 1201–1209. https://doi.org/10.1016/J.SCITOTENV.2018.08.210

Gadd, G. M. (2009). Biosorption: Critical review of scientific rationale, environmental importance and significance for pollution treatment. *Journal of Chemical Technology & Biotechnology, 84*(1), 13–28. https://doi.org/10.1002/JCTB.1999

García-Galán, M. J., Monllor-Alcaraz, L. S., Postigo, C., Uggetti, E., de Alda, M. L., Díez-Montero, R., & García, J. (2020). Microalgae-based bioremediation of water contaminated by pesticides in peri-urban agricultural areas. *Environmental Pollution, 265*, 114579. https://doi.org/10.1016/J.ENVPOL.2020.114579

Gardea-Torresdey, J. L., Becker-Hapak, M. K., Hosea, J. M., & Darnall, D. W. (1990). Effect of chemical modification of algal carboxyl groups on metal ion binding. *Environmental Science and Technology, 24*(9), 1372–1378. https://doi.org/10.1021/ES00079A011/ASSET/ES00079A011.FP.PNG_V03

Gokulan, R., Prabhu, G. G., & Jegan, J. (2019). Remediation of complex remazol effluent using biochar derived from green seaweed biomass. *International Journal of Phytoremediation, 21*(12), 1179–1189. https://doi.org/10.1080/15226514.2019.1612845

Gomes, P. I. A., & Asaeda, T. (2013). Phytoremediation of heavy metals by calcifying macro-algae (nitella pseudoflabellata): Implications of redox insensitive end products. *Chemosphere, 92*(10), 1328–1334. https://doi.org/10.1016/J.CHEMOSPHERE.2013.05.043

Gupta, V. K., & Rastogi, A. (2008). Sorption and desorption studies of chromium(VI) from nonviable cyanobacterium Nostoc muscorum biomass. *Journal of Hazardous Materials, 154*(1–3), 347–354. https://doi.org/10.1016/J.JHAZMAT.2007.10.032

Han, R., Li, H., Li, Y., Zhang, J., Xiao, H., & Shi, J. (2006). Biosorption of copper and lead ions by waste beer yeast. *Journal of Hazardous Materials, 137*(3), 1569–1576. https://doi.org/10.1016/J.JHAZMAT.2006.04.045

Hao, T. B., Balamurugan, S., Zhang, Z. H., Liu, S. F., Wang, X., Li, D. W., Yang, W. D., & Li, H. Y. (2022) Effective bioremediation of tobacco wastewater by microalgae at acidic pH for synergistic biomass and lipid accumulation. *Journal of Hazardous Materials, 426*, 127820. https://doi.org/10.1016/J.JHAZMAT.2021.127820

Henriques, B., Teixeira, A., Figueira, P., Reis, A. T., Almeida, J., Vale, C., & Pereira, E. (2019). Simultaneous removal of trace elements from contaminated waters by living *Ulva lactuca*. *Science of the Total Environment, 652*, 880–888. https://doi.org/10.1016/J.SCITOTENV.2018.10.282

Jaafari, J., & Yaghmaeian, K. (2019). Optimization of heavy metal biosorption onto freshwater algae (Chlorella coloniales) using response surface methodology (RSM). *Chemosphere, 217*, 447–455. https://doi.org/10.1016/j.chemosphere.2018.10.205

Karaś, M. A., Wdowiak-Wróbel, S., & Sokołowski, W. (2021). Selection of endophytic strains for enhanced bacteria-a ssisted phytoremediation of organic pollutants posing a public health hazard. *International Journal of Molecular Sciences, 22*(17). https://doi.org/10.3390/IJMS22179557

Kurade, M. B., Xiong, J. Q., Govindwar, S. P., Roh, H. S., Saratale, G. D., Jeon, B. H., & Lim, H. (2019). Uptake and biodegradation of emerging contaminant sulfamethoxazole from aqueous phase using Ipomoea aquatica. *Chemosphere, 225*, 696–704. https://doi.org/10.1016/j.chemosphere.2019.03.086

Kuttiyathil, M. S., Mohamed, M. M., & Al-Zuhair, S. (2021). Using microalgae for remediation of crude petroleum oil–water emulsions. *Biotechnology Progress, 37*(2), e3098. https://doi.org/10.1002/BTPR.3098

Kuyucak, N., & Volesky, B. (1988). Biosorbents for recovery of metals from industrial solutions. *Biotechnology Letters, 10*(2), 137–142. https://doi.org/10.1007/BF01024641

Landaburu-Aguirre, J., Pongrácz, E., Perämäki, P., & Keiski, R. L. (2010). Micellar-enhanced ultrafiltration for the removal of cadmium and zinc: Use of response surface methodology to improve understanding of process performance and optimisation. *Journal of Hazardous Materials, 180*(1–3), 524–534. https://doi.org/10.1016/j.jhazmat.2010.04.066

Lebron, Y. A. R., Moreira, V. R., & Santos, L. V. S. (2019). Studies on dye biosorption enhancement by chemically modified Fucus vesiculosus, Spirulina maxima and Chlorella pyrenoidosa algae. *Journal of Cleaner Production, 240*, 118197. https://doi.org/10.1016/j.jclepro.2019.118197

Leong, Y. K., & Chang, J. S. (2020). Bioremediation of heavy metals using microalgae: Recent advances and mechanisms. *Bioresource Technology, 303*, 122886. https://doi.org/10.1016/j.biortech.2020.122886

Lertlapwasin, R., Bhawawet, N., Imyim, A., & Fuangswasdi, S. (2010). Ionic liquid extraction of heavy metal ions by 2-aminothiophenol in 1-butyl-3-methylimidazolium hexafluorophosphate and their association constants. *Separation and Purification Technology, 72*(1), 70–76. https://doi.org/10.1016/j.seppur.2010.01.004

Libatique, M. J. H., Lee, M. C., Yeh, H. Y., & Jhang, F. J. (2020). Total and inorganic arsenic biosorption by *Sarcodia suiae* (Rhodophyta), as affected by controlled environmental conditions. *Chemosphere, 248*, 126084. https://doi.org/10.1016/J.CHEMOSPHERE.2020.126084

Luo, H., Wang, Q., Liu, Z., Wang, S., Long, A., & Yang, Y. (2020). Potential bioremediation effects of seaweed *Gracilaria lemaneiformis* on heavy metals in coastal sediment from a typical mariculture zone. *Chemosphere, 245*, 125636. https://doi.org/10.1016/J.CHEMOSPHERE.2019.125636

Mahajan, P., Kaushal, J., Upmanyu, A., & Bhatti, J. (2019). Assessment of phytoremediation potential of *Chara vulgaris* to treat toxic pollutants of textile effluent. *Journal of Toxicology*, https://doi.org/10.1155/2019/8351272.

Mahajan, P., & Kaushal, J. (2020). Phytoremediation of azo dye methyl red by macroalgae Chara vulgaris L.: Kinetic and equilibrium studies. *Environmental Science and Pollution Research, 27*(21), 26406–26418. https://doi.org/10.1007/S11356-020-08977-W

Mansour, A. T., Alprol, A. E., Abualnaja, K. M., El-Beltagi, H. S., Ramadan, K. M., & Ashour, M. (2022). Dried Brown seaweed's phytoremediation potential for methylene blue dye removal from aquatic environments. *Polymers, 14*(7), 1375. https://doi.org/10.3390/POLYM14071375

Mauchauffée, S., & Meux, E. (2007). Use of sodium decanoate for selective precipitation of metals contained in industrial wastewater. *Chemosphere, 69*(5), 763–768. https://doi.org/10.1016/j.chemosphere.2007.05.006

Mekonnen, M. M., & Hoekstra, A. Y. (2016). Sustainability: Four billion people facing severe water scarcity. *Science Advances, 2*. https://doi.org/10.1126/sciadv.1500323

Miazek, K., & Brozek-Pluska, B. (2019). Effect of phrs and pcps on microalgal growth, metabolism and microalgae-based bioremediation processes: A review. *International Journal of Molecular Sciences, 20*(10), 2492. https://doi.org/10.3390/ijms20102492

Mohsen-Nia, M., Montazeri, P., & Modarress, H. (2007). Removal of Cu2+ and Ni2+ from wastewater with a chelating agent and reverse osmosis processes. *Desalination, 217*(1–3), 276–281. https://doi.org/10.1016/j.desal.2006.01.043

Mokhtar, N., Aziz, E. A., Aris, A., Ishak, W. F. W., & Ali, N. S. M. (2017). Biosorption of azo-dye using marine macro-alga of *Eucheuma Spinosum*. *Journal of Environmental Chemical Engineering, 5*(6), 5721–5731. https://doi.org/10.1016/J.JECE.2017.10.043

Ociński, D., Augustynowicz, J., Wołowski, K., Mazur, P., Sitek, E., & Raczyk, J. (2021). Natural community of macroalgae from chromium-contaminated site for effective remediation of Cr(VI)-containing leachates. *Science of the Total Environment, 786*, 147501. https://doi.org/10.1016/J.SCITOTENV.2021.147501

Oliveira, A. C., Barata, A., Batista, A. P., & Gouveia, L. (2019). *Scenedesmus obliquus* in poultry wastewater bioremediation. *Environmental Technology, 40*(28), 3735–3744. https://doi.org/10.1080/09593330.2018.1488003

Petrovič, A., & Simonič, M. (2015). The effect of carbon source on nitrate and ammonium removal from drinking water by immobilised *Chlorella sorokiniana*. *International Journal of Environmental Science and Technology, 12*(10), 3175–3188. https://doi.org/10.1007/s13762-014-0747-0

Rahmanian, B., Pakizeh, M., Esfandyari, M., Heshmatnezhad, F., & Maskooki, A. (2011). Fuzzy modeling and simulation for lead removal using micellar-enhanced ultrafiltration (MEUF). *Journal of Hazardous Materials, 192*(2), 585–592. https://doi.org/10.1016/j.jhazmat.2011.05.051

Ramírez, M. E., Vélez, Y. H., Rendón, L., & Alzate, E. (2017). Potential of microalgae in the bioremediation of water with chloride content. *Brazilian Journal of Biology, 78*(3), 472–476. https://doi.org/10.1590/1519-6984.169372

Reena, S., Neetu, G., Anurag, M., & Rajiv, G. (2011). Heavy metals and living systems: An overview. *Indian Journal of Pharmacology, 43*(3), 246–253. https://doi.org/10.4103/0253-7613.81505

Rezvani, F., Sarrafzadeh, M. H., Seo, S. H., & Oh, H. M. (2018). Optimal strategies for bioremediation of nitrate-contaminated groundwater and microalgae biomass production. *Environmental Science and Pollution Research, 25*(27), 27471–27482. https://doi.org/10.1007/S11356-018-2777-5

Roberts, D. A., Shiels, L., Tickle, J., De Nys, R., & Paul, N. A. (2018). Bioremediation of aluminium from the waste water of a conventional water treatment plant using the freshwater macroalga *Oedogonium*. *Water, 10*(5), 626. https://doi.org/10.3390/W10050626

Romera, E., Gonzalez, F., & Ballester, A. (2007). Comparative study of biosorption of heavy metals using different types of algae. *Bioresource Technology, 98*(17), 3344–3353. https://doi.org/10.1016/j.biortech.2006.09.026

Rubio, J., Souza, M. L., & Smith, R. W. (2002). Overview of flotation as a wastewater treatment technique. *Minerals Engineering, 15*(3), 139–155. https://doi.org/10.1016/S0892-6875(01)00216-3

Salem, D. M. S. A., Moawad, M. N., & El-Sayed, A. A. M. (2021). Comparative study for bioremediation of cobalt contaminated aqueous solutions by two types of marine macroalgae. *The Egyptian Journal of Aquatic Research, 47*(1), 13–19. https://doi.org/10.1016/J.EJAR.2021.02.002

Schiewer, S., & Volesky, B. (2014). Biosorption processes for heavy metal removal. *Environmental microbe-metal interactions*. https://doi.org/10.1128/9781555818098.CH14

Senthilkumar, R., Reddy Prasad, D. M., Govindarajan, L., Saravanakumar, K., & Naveen Prasad, B. S. (2020). Synthesis of green marine algal-based biochar for remediation of arsenic(V) from contaminated waters in batch and column mode of operation. *International Journal of Phytoremediation, 22*(3), 279–286. https://doi.org/10.1080/15226514.2019.1658710

Shaari, A. L., Sa, S. N. C., Surif, M., Zolkarnain, N., & Ghazali, R. (2021). Growth of marine microalgae in landfill leachate and their ability as pollutants removal. *Tropical Life Sciences Research, 32*(2), 133–146. https://doi.org/10.21315/TLSR2021.32.2.9

Shackira, A. M., Jazeel, K., & Puthur, J. T. (2021). Phycoremediation and phytoremediation: Promising tools of green remediation. *Sustainable environmental clean-up: Green remediation*. https://doi.org/10.1016/B978-0-12-823828-8.00013-X

Sun, X., Liu, Z., Jiang, Q., & Yang, Y. (2019). Concentrations of various elements in seaweed and seawater from Shen'ao Bay, Nan'ao Island, Guangdong coast, China: Environmental monitoring and the bioremediation potential of the seaweed. *Science of the Total Environment, 659*, 632–639.

Surkatti, R., & Al-Zuhair, S. (2018). Microalgae cultivation for phenolic compounds removal. *Environmental Science and Pollution Research, 25*(34), 33936–33956. https://doi.org/10.1007/S11356-018-3450-8

Sutherland, D. L., & Bramucci, A. (2022). Dissolved organic phosphorus bioremediation from food-waste centrate using microalgae. *Journal of Environmental Management, 313*, 115018. https://doi.org/10.1016/J.JENVMAN.2022.115018

Tavana, M., Pahlavanzadeh, H., & Zarei, M. J. (2020). The novel usage of dead biomass of green algae of Schizomeris leibleinii for biosorption of copper(II) from aqueous solutions: Equilibrium, kinetics and thermodynamics. *Journal of Environmental Chemical Engineering, 8*(5), 104272. https://doi.org/10.1016/J.JECE.2020.104272

Treen-Sears, M. E., Volesky, B., & Neufeld, R. J. (1984). Ion exchange/complexation of the uranyl ion by Rhizopus biosorbent. *Biotechnology and Bioengineering, 26*(11), 1323–1329. https://doi.org/10.1002/bit.260261109

Ugya, A. Y., Ajibade, F. O., & Hua, X. (2021). The efficiency of microalgae biofilm in the phycoremediation of water from River Kaduna. *Journal of Environmental Management, 295*, 113109. https://doi.org/10.1016/J.JENVMAN.2021.113109

Van Der Vegt, G. S., Essens, P., Wahlström, M., & George, G. (2015). Managing risk and resilience. *Academy of Management Journal, 58*(4), 971–980. https://doi.org/10.5465/amj.2015.4004

Veglio', F., & Beolchini, F. (1997). Removal of metals by biosorption: A review. *Hydrometallurgy, 44*(3), 301–316. https://doi.org/10.1016/s0304-386x(96)00059-x

Verma, V. K., Tewari, S., & Rai, J. P. N. (2008). Ion exchange during heavy metal bio-sorption from aqueous solution by dried biomass of macrophytes. *Bioresource Technology, 99*(6), 1932–1938. https://doi.org/10.1016/j.biortech.2007.03.042

Vijayaraghavan, K., & Raja, F. D. (2015). Interaction of vermiculite with Pb(II), C+d(II), Cu(II) and Ni(II) ions in single and quaternary mixtures. *Clean - Soil, Air, Water, 43*(8), 1174–1180. https://doi.org/10.1002/clen.201400423

Vilar, V. J. P., Botelho, C. M. S., & Boaventura, R. A. R. (2006). Equilibrium and kinetic modelling of Cd(II) biosorption by algae Gelidium and agar extraction algal waste. *Water Research, 40*(2), 291–302. https://doi.org/10.1016/J.WATRES.2005.11.008

Viraraghavan, A. (1995). Fungal biosorption-an alternative treatment option for heavy metal bearing waste water. *Bioresource Technology, 53*, 195–206. https://doi.org/10.1016/0960-8524(95)00072-M

Volesky, B. (2007). Biosorption and me. *Water Research, 41*(18), 4017–4029. https://doi.org/10.1016/j.watres.2007.05.062

Volesky, B., & Holan, Z. R. (1995). Biosorption of heavy metals. *Biotechnology Progress, 11*(3), 235–250. https://doi.org/10.1021/bp00033a001

Wang, J., & Chen, C. (2009). Biosorbents for heavy metals removal and their future. *Biotechnology Advances, 27*(2), 195–226. https://doi.org/10.1016/J.BIOTECHADV.2008.11.002

Wang, S., Li, H., & You, J. (2019). Enantioselective degradation and bioaccumulation of sediment-associated fipronil in Lumbriculus variegatus: Toxicokinetic analysis. *Science of the Total Environment* (Vol 672,, 335–341. https://doi.org/10.1016/j.scitotenv.2019.03.490

Yalçın, S., & Özyürek, M. (2018). Biosorption potential of two brown seaweeds in the removal of chromium. *Water Science and Technology, 78*(12), 2564–2576. https://doi.org/10.2166/WST.2019.007

Yoon, J., Amy, G., Chung, J., Sohn, J., & Yoon, Y. (2009). Removal of toxic ions (chromate, arsenate, and perchlorate) using reverse osmosis, nanofiltration, and ultrafiltration membranes. *Chemosphere, 77*(2), 228–235. https://doi.org/10.1016/j.chemosphere.2009.07.028

Zhang, C., Lu, J., & Wu, J. (2021). Enhanced removal of phenolic endocrine disrupting chemicals from coastal waters by intertidal macroalgae. *Journal of Hazardous Materials, 411*, 125105. https://doi.org/10.1016/J.JHAZMAT.2021.125105

Zhu, M., Wang, Z., Chen, J., Xie, H., Zhao, H., & Yuan, X. (2020). Bioaccumulation, Biotransformation, and multicompartmental toxicokinetic model of antibiotics in sea cucumber (*Apostichopus japonicus*). *Environmental Science and Technology, 54*(20), 13175–13185. https://doi.org/10.1021/acs.est.0c04421

Index

Note: Page numbers followed by "*f*" indicate figures and "*t*" indicate tables.

B

C

D

V

W

Z

9780443188169